Veterinary Pathology

Seventh Edition

Part I: General Pathology
Part II: Special Pathology
Part III: Infectious Disease

Ganti A Sastry
GMVC, BVSc, MSc, MS, PhD (Kansas)
Retired Principal and Professor of Pathology
College of Veterinary Science, Tirupati 517 502

Revised and Enlarged by

P Rama Rao
MVSc, FRVCS (Sweden), PhD
Principal, Andhra Pradesh Agricultural University
College of Veterinary Science, Tirupati 517 502

CBSPD

CBS Publishers & Distributors Pvt Ltd

New Delhi • Bengaluru • Chennai • Kochi • Kolkata • Lucknow • Mumbai
Hyderabad • Jharkhand • Nagpur • Patna • Pune • Uttarakhand

Veterinary Pathology

Seventh Edition

ISBN: 978-81-239-0738-3

Copyright © Authors & Publisher

Seventh Edition: 2001
Reprint: 2002, 2003, 2004, 2005, 2006, 2007, 2008, 2009, 2010, 2011, 2012, 2013
2014, 2015, 2016, 2017, 2018, 2019, 2020, 2021, 2022, 2023, 2024, **2025**
First Edition: 1968
Second Edition: 1969
Third Edition: 1971
Fourth Edition: 1975
Fifth Edition: 1979
Sixth Edition: 1983

Published by **Satish Kumar Jain** and produced by **Varun Jain** for
CBS Publishers & Distributors Pvt Ltd
4819/XI Prahlad Street, 24 Ansari Road, Daryaganj, New Delhi 110 002, India
Ph: 011-23289259, 23266861   Website: www.cbspd.com
                               e-mail: delhi@cbspd.com

Corporate Office: 204 FIE, Industrial Area, Patparganj, Delhi 110 092, India
Ph: 011-4934 4934   Fax: 011-4934 4935   e-mail: publishing@cbspd.com:
                                             publicity@cbspd.com

Branches

- **Bengaluru:** Seema House 2975, 17th Cross, KR Road, Banasankari 2nd Stage, Bengaluru 560 070, Karnataka, India
  Ph: +91-80-26771678/79   Fax: +91-80-26771680   e-mail: bangalore@cbspd.com
- **Chennai:** 7, Subbaraya Street, Shenoy Nagar, Chennai 600 030, Tamil Nadu, India
  Ph: +91-44-26680620, 26681266   Fax: +91-44-42032115   e-mail: chennai@cbspd.com
- **Kochi:** 42/1325, 1326, Power House Road, Opp KSEB, Power House, Ernakulam Kochi 682 018, Kerala, India
  Ph: +91-484-4059061-65,67   Fax: +91 484-4059065   e-mail: kochi@cbspd.com
- **Kolkata:** 147, Hind Ceramics Compound, 1st Floor, Nilgunj Road, Belghoria, Kolkata-700056, West Bengal, India
  Ph: +033-25633055, 033-25633056   e-mail: kolkata@cbspd.com
- **Lucknow:** Basement, Khushnuma Complex, 7 Meerabai Marg (Behind Jawahar Bhawan), Lucknow-226001, UP India
  Ph: +91-522-4000032   e-mail: tiwari.lucknow@cbspd.com
- **Mumbai:** PWD Shed, Gala no 25/26, Ramchandra Bhatt Marg, Next to JJ Hospital Gate no. 2, Opp. Union Bank of India Noorbaug, Mumbai-400009, Maharashtra, India
  Ph: 022-66661880/89   e-mail: mumbai@cbspd.com

Representatives

- Hyderabad   0-9885175004
- Patna   0-9334159340
- Jharkhand   0-9811541605
- Pune   0-9664372571
- Nagpur   0-8692091830
- Uttarakhand 0-9716462459

Printed at Chaman Enterprises, Daryaganj, Delhi, India

Dedicated to my late father

G.Y. Somayajulu Garu

CONTRIBUTORS

Part I

Dr. D. Anjaneya Prasad, BVSC, MVSC, PhD (Kansas) Professor of Animal Sciences, College of Veterinary Science, Tirupati (A.P.)

Dr. P. Babu Rao, MSC, PhD, Lecturer in Physics, S.V. College of Arts and Science, Tirupati (AP.)

Dr. B Sambamurthy, GMVC, BVSC, MS, (Kansas) Retired Associate Professor of Microbiology, College of Veterinary Science, Tirupati (A.P.)

Dr. K. Sathyanarayana Rao, GMVC, BVSC, MS, (Kansas) Principal & Professor of Physiology, College of Veterinary Science, Tirupati (A.P.)

Dr. A. Sreeramamurthy, BVSC, MSC (Vet), PhD, Head, Department of Animal Genetics and Breeding, College of Veterinary Science, Tirupati.

Dr. D. Sunderasiva Reo, M D. Retired Principal and Professor of Pathology, Rangaraya Medical College. Kakinada, A P. *Now* Pathologist, Amala Cancer Hospital and Research Institute, Puzhakkal, Kerala.

Dr. J.L. Vegad. BVSC, ASSOC, IVRI; PhD. (New Zealand). University Professor and Head, Department of Pathology, College of Veterinary Science & AH. Jabalpur, M.P.

Part II

Dr. M.S. Kwatra, MVSC, PhD. Senior Veterinary Investigation Officer, College of Veterinary Science, Ludhiana. (Mulberry Heart Disease and Manchester wasting Disease of cattle).

Dr. B.B. Mallick, D.S.C. (Sarborne); FCAI., Assoc IVRI, BSC, GVSC (Hons) Head, Division of Virology, IVRI, Mukteswar. (Immunological Diseases).

Dr. A. Rajan, BVSC, MVSC, PhD. Professor of Pathology, College of Veterinary Science, (PO) Mannuthy, Trichur, Kerala. (Revision of "Endocrine Glands").

Dr. P. Rama Rao, BVSC, MVSC, FRVCS (Sweden) PhD. Head, Department of Pathology, College of Veterinary Science, Tirupathi. (Revision of "The Reproductive System").

Dr. J.L. Vegad. BVSC, Assoc. IVRI, PhD (New Zealand). University Professor and Head, Department of Pathology, College of Veterinary Science, & AH, Jabalpur (The Nervous System).

Part III

The autor is greatly indebted to

Dr. T.N. Ghosh, PhD, Professor of Protozoology, Calcutta School of Tropical Medicine. Calcutta, for his note on "Toxoplasmosis",

Dr. D.S. Kalra PhD., Professor & Head Department of Pathology, Veterinary College, Hissar, for the Chapter on "Diseases caused by Fungi",

Dr. S. Krishnaswamy MVSC, DVPH, & FH (Den), PhD, Professor and Head Department of Microbiology, College of Veterinary Science, Tirupathi, for reviewing chapters on 'Diseases caused by Bacteria, Viruses and Rickettsia.".

Dr. M. S. Kwatra, MVSC, PhD, Senior Veterinary Investigation Officer, College of Veterinary Science, Ludhiana, for contributing on 'Degnalla Disease',

Dr S. J. Seshadri, MVSC, Professor & Head. Department of Pathology Bangalore Veterinary College for his contribution on Bacterial and Viral Diseases of Poultry,

Dr. S. R. G. Muralidharam, BVSC, MVSC, PhD, Assistant Professor of Parasitology, College of Veterinary Science, Tirupathi, for reviewing chapters on Diseases caused by "Helminths and Insects and their larvae.",

Dr. A. Venkataratnam, GMVC, BVSC, DAP, MSC (L'pool) Associate Professor of Parasitology, College of Veterinary Science, Tirupathi, for reviewing chapter on "Diseases caused by Protozoa.",

PREFACE FOR THE SEVENTH EDITION

In an effort to update the informations on poultry diseases according to VCI syllabus and needs of poultry industry. Dr. P. Rama Rao, Dr. Gopal Yadgirkar and Dr. S. Rafeeque Ahmed former staff members of the Department of Pathology in the faculty of Veterinary Science, Acharya NG Ranga Agricultural University have revised the book.

The information is compiled from the following books beside incorporating the experiences and observations of above authors.

1. Diseases of Poultry, edt. B.W. Calnek and associates, Affiliated East West Press, Iowa State University Press, 9th edition, 1994.
2. Poultry Diseases, FTW Jordan,
   English Language Book Society,
   Bailliere Tindall, 3rd edition, 1990.

DR. P. RAMA RAO

PREFACE

PREFACE FOR THE SEVENTH EDITION

Continued demand for my book by the student body had necessitated the production of this Sixth edition. It is no little satisfaction to the author to find that his book is felt needed and is found useful by most of the Veterinary students of the country.

In this edition some parts are revised and many facts and conditions have been added to bring the contents of the book uptodate. As such the number of pages had to be increased. To make the information more authentic. I have made a departure in this edition, by requesting Professors of other colleges as well as some of our own college, to subscribe chapters or conditions in which they have specialised. By this means the students get benefitted by the experience and expertise of these Professors. I am grateful for the following persons for contributing on the subjects noted against their names to this part :

Dr. D. Anjaneya Prasad	— Bovine Ketosis and revision of chapter on Disturbances in Mineral Metabolism.
Dr. P. Babu Rao	— Ionizing radiation.
Dr. B. Sambamurthy	— Viruses and Tumorigenesis.
Dr. K Sathyanarayana Rao	— Electron Microscopic structure and function of normal cell.
Dr. A. Sreerama Murthy	— Hereditary defects and disease resistance.
Dr. D. Sunderasiva Rao	— Revision of chapter on Neoplasms.
Dr. J.L. Vegad	—Chapters on Inflammation and Healing.

Acknowledgement of contribution of others will be made in the subsequent parts.

I am grateful to Dr. P. Rama Rao, Head and the staff of the Department of Pathology of this college, for their help in the production of this edition.

I am indebted to the following Professors of Veterinary Pathology for prescribing my book in their colleges and rendering me advice and help whenever soght for : Dr. S. Domodaran and Dr. P. Ramakrishnan of Madras ; Dr. M. Krishnan Nair and Dr. A. Rajan of Kerala ; Dr. S.J. Seshadri of Bangalore ; Dr. S.S. Bhagwat of Nagpur ; Dr. B.C. Nayak of Bhubaneswar ; Dr. S.N. Nandi of Calcutta ; Dr. S. Kwatra, formerly of Assam ; Dr. J.L. Vegad of Jabalpur ; Dr. U.R.K. Rao. of Mhow ; Dr. P.K. Ramachandran and Dr. N.S. Parihar of IVRI, Izatnagar; Dr. P.N. Mehrotra and Dr. P.L. Arya of Bikaner; Dr. S.M. Ajinkya of Bombay ; Dr. D.N. Kalra of Hissar ; Dr. Balwant Singh of Ludhiana ; Dr. B. Deshpande of Parbhani ; Dr. J.M. Dwivedi of Mathura ; Dr. D.D. Heranjal of Anand and Dr. M. Mahender of Hyderabad.

It is the fond hope of the author that this book will continue to be useful to the Veterinary Students and that it will help lay a sound foundation in the better and rational understanding of disease processes.

GANTI A. SASTR

CONTENTS

PART I

General Pathology

INTRODUCTION

Pathology literally means "discourse of disease" (Pathos-disease; logos-discourse (Greek) and so may be defined as a study of disease. It is "that branch of medicine which treats of the essential nature of disease especially of the structural and functional changes in tissues and organs of the body which cause or are caused by disease" (Dorland). In pathology, the causes, nature and evolution of the disease are studied as well as changes in the anatomy, physiology and chemistry that result from it.

To understand pathology, one should know what disease is and so one should have a knowledge of what health is. Health is defined as "a normal condition of the body and mind, i e., with all the parts functioning normally". That means to say the body exists in complete harmony with its surroundings or environment. This mechanism by which the body is kept in equilibrium is very delicate and is known as **Homeostasis**, adapting it to the constant changes in its internal and external environment. Pathology may, therefore, be defined as "the inadequate adaptation" to such changes. To be a good pathologist, therefore, one should have a good knowledge of the body in health. That means to say, he should have a thorough knowledge of anatomy, histology, physiology and biochemistry.

Disease : (Dis-negative, ease) is any departure from healthy state. That means, it is not in harmony with its environment It is opposite of health and so the body and mind are not in a normal condition.

The following terms are used in the study of Pathology.

General Pathology deals with fundamental processes that are common to more than one tissue or organ.

Special Pathology is the study of diseases peculiar to certain systems or organs. Depending on the system, this may be subdivided into genito-urinary pathology, surgical pathology, gynaecological and obstetrical pathology etc.

Clinical Pathology is that branch of pathology used in the diagnosis of diseases in the hospital, at the patient's bedside.

Nutritional Pathology is the study of disease processes resulting from deficiency or excess of essential food stuffs.

Pathological Physiology is the study of changes in physiology consequent on the disease processes.

Chemical Pathology deals with alterations in biochemical processes in diseases

Experimental Pathology means the study of disease artificially produced in animals.

Comparative Pathology is the study of diseases of animals and comparing them to those occurring in man.

Oncology is the study of tumors.

The study of pathology of diseases comprises of the following sub-heads:

Etiology is the study of causation of disease. A knowledge of the causative factors of diseases is essential for their prevention. These may originate from within the body (intrinsic) or may enter from outside (extrinsic). Depending on their nature, the etiological agents may be classed as physical, chemical, nutritional and biological (bacterial, viral, fungal or animal).

Predisposing causes are those that make the animal susceptible to the disease and are also known as **remote, distant** or **preparatory causes.** These make the animal react easily to the exciting causes which actually incite the disease processes. The **exciting** causes are also known as **proximate, determining, immediate** or **direct** causes. The science of etiology includes the study of diseases in relation to geographical location, season, breed, age, sex and the environmental factors.

Pathogenesis means the mechanism by which the causes produce diseases. As an example, the action of ultra-violet light on white skin may be cited. In the absence of the filtering pigment the ultra-violet rays of the sun penetrate deeply and injure the endothelium of the capillaries of the dermis, so that their permeability is increased, exudation occurs and so red swelling results. (Sun-burn)

Symptoms or Signs are the outward manifestations of the patient suffering from diseases while alive. For example, in pneumonia, the symptoms seen are fever, laboured breathing, fast pulse, expectoration, abnormal sounds on auscultation etc.

Lesions : These are the alterations in structure, detectable macroscopically by the naked eye or microscopically. In small pox, the vesicles on the skin are lesions. In tuberculosis the nodules in lungs and other organs are lesions.

Diagnosis is the art of determination of the nature of disease—its causes, lesions, symptoms etc.

Incubation period is the time that elapses between the action of a cause and manifestation of disease. For example, in Rinderpest, it may take 4 to 9 days for the animals to manifest symptoms after they are infected with the virus and so the incubation period of Rinderpest is 4 to 9 days. This varies from a few hours as in Infectious Canine Hepatitis to some years as in Johne's Disease in cattle and Leprosy in man.

The **course** of the disease is the duration of time through which the series of changes characteristic of a disease pass through to their ultimate end.

The beginning of the manifestation of symptoms is called the onset of disease. This may be sudden or gradual.

A disease may terminate in **recovery or death** or may prolong for a considerable length of time as in chronic diseases.

Prognosis of a disease is the estimate by a clinician of probable severity and outcome of disease.

Morbidity is the percentage of exposed animals that get affected. For example, in a herd of 100 animals exposed to infection of Rinderpest virus, 50 may actually suffer. Here the morbidity is therefore 50%.

Mortality rate of a disease is the percentage of deaths among animals affected by that disease.

Morbidity may be high and mortality low in some diseases (Foot and Mouth Disease, Vaccinia) while in others morbidity may be low but mortality high (Mucosal Disease).

INTRODUCTION

Autopsy (seeing with ones own eyes). The pathologist cuts open a carcass to see the lesions in diseases This is also known as necropsy, (seeing a corpse). **Biopsy** means examination of tissues received from living animals.

Purpose of study of Pathology :

In short, the study of Pathology involves the study of the etiology, pathogenesis and effects of disease. It tries to explain the modes of action of harmful agents in the environment on the body and changes that are caused or might ensue. Knowing these methods of disease production, prevention can be rationally thought of and implemented. The main purpose of pathology is to explain rationally the causes of symptoms seen in disease and to predict what changes might supervene if treatment is not given. It, therefore, helps a clinician not only in his correct diagnosis of the condition but also in the prognosis and regulation of his treatment. So, to this extent, pathology is helpful in the practice of Veterinary Medicine and Surgery.

CHAPTER 2

ETIOLOGY

Predisposing causes	Local defence mechanisms
Exciting causes	Resistance to infection
Portals of infection	Establishment of infection.

We have already noticed that the causes of diseases may be broadly divided into predisposing and exciting causes. Now we shall study in greater detail these factors.

Predisposing causes :

These may also be termed as internal or abnormal constitutional factors and may be divided into :–

I. Genetic or inherited causes : These are transmitted to the offspring through the germ plasm and may be,

(a) Lethal characters which are inherited and invariably cause death of the animal, either in-utero or after birth. These factors may be Mendelian-dominant or recessive. eg. atresia coli in horses, parrot beak in cattle.

(b) Sub-lethal factors which are inherited and interfere with the function of the body but do not cause death : eg. imperforate anus, infertility in cattle, scrotal hernia in pigs, deafness in white cats

(c) Defects that are inherited errors in structure or function :– cryptorchidism, webbed digits

II. Non-genetic or non-inherited defects : are abnormalities not transmitted via the germ plasm.

A. Anomalies : An anomaly is developmental defect affecting an organ . part of the body. These may be

1. **Disturbances in development**
   (a) Arrest of development
   i) **Agenesia or aplasia**—is complete absence
   ii) **Hypoplasia**—is reduction in size.
   iii) **Atresia**—is closure of the lumen of a hollow organ or duct
   iv) **Fissure** on the median line
   v) **Fusion of paired organs**—horse-shoe kidneys,
   (b) Excessive development : (i) Congenital hypertrophy
   (ii) Increase in number——polydactyla.
2. **Persistance of fetal structures**—persistant urachus.
3. **Displacements during development**
4. **Fusion of sexual characters :** (i) Hermaphrodite
   (ii) Pseudohermaphrodite (iii) Freemartin.

B. Monster is an animal in which is present extensive abnormal development. These do not survive if born alive.

Besides the above, the following intrinsic causes are recognised :

Genus : Certain diseases are peculiar to certain genera of animals while others are not affected. For example, man is immune to Rinderpest that affects cattle and cattle are immune to cholera of man :

Breed : Certain breeds, either due to selective breeding or inherent characteristics, are more susceptible to some diseases than others. For example the dairy breed of cattle as a whole, are more susceptible to diseases than the beef breed. The heavier built breeds of dogs—the German shepherd, Great Danes-are more susceptible to bone diseases. Bulldog breeds suffer more often of brain tumors.

Age : Different diseases affect different age groups. Due to contact in childhood and so having acquired immunity, older people are immune to certain diseases that easily affect children. Strangles affects the younger horses. So also canine distemper is a disease of pups. On the other hand, tumors. in general, are found more frequently in older age groups.

In the younger animals, deficiencies play a major role If the minerals and vitamins that are required for a growing animal are deficient in the diet, deficiency diseases are manifested quickly in the young animals.

Sex : Each sex is susceptible to diseases of its organs of reproduction. Apart from these, sexes differ in their susceptibility to diseases. Males are more susceptible to night blindness (man) and nephritis (dog) (but cows are more susceptible to nephritis). **Females** are more susceptible to goiter and liver diseases (human).

Color: Color appears to play a part in the causation of disease by virtue of the presence or absence of melanin pigment. Grey horses suffer from malignant melanoma more frequently than other colored horses. Animals with white skin are more susceptible to photodynamic diseases.

Exciting causes of diseases:

1. **Physical causes:**
   - (a) **Those that injure by radiation:**
     - i) Thermal—burns
     - ii) Light and ultraviolet rays—sunburns or photosensitisation.
     - iii) X-rays
     - iv) Gamma rays
     - v) Particulate radiation
   - (b)**Cold :** freezing injures tissues, especially blood vessels, causing necrosis—frost bite
   - (c) **Electricity:** (i) Electrical burns (ii) Electrocution.
   - (d) **Decreased atmospheric pressure** (Brisket Disease)
   - (e) **Increased atmospheric pressure:** (Caisson Disease)
   - (f) **Mechanical injuries:** by various means—automobile accidents, gunshots. cutting by wires, razors etc. giving rise to the following conditions:

Perforation: This is a wound caused by a bullet or a nail. This type of wound in usually infected by a anaerobes (*Clostridium tetani* or *welchii*) producing severe intoxication.

Laceration: This is a wound in which there is tearing of tissues, eg. wire cuts on legs of horses, automobile accidents.

Concussion: This is a violent jar or shock caused by an injury and is usually applied to injuries of the head. There may or may not be loss of consciousness.

Sprain : This is an injury of joint in which there may be stretching or rupture of ligaments, muscles or tendons   In this the anatomical relationship of the structures is maintained.

Luxation or dislocation · In this injury of the joint, the anatomical relationships of various tissues are not maintained and the ligaments may be torn.

Fracture· This is breakage of a bone, tooth, claw, hoof, cartilage or horn.

2. **Chemical causes** :

(a) **If in excess certain chemicals cause poisoning** :

i) **Inorganic** acids, alkalies, heavy metals and their salts (As, P, Hg etc.)

ii) **Organic** : Insecticides, rodenticides, fungicides, drugs (Chloroform etc.) alkaloids (plant poisons), Vitamin D

(b) **Deficiency of certain chemicals cause poisoning** :

i) **Inorganic** : Deficiency of calcium, phosphorus, sodium, iron, copper, zinc, manganese, iodine, fluorine, magnesium.

ii) **Organic**: Deficiency of vitamins, fats, carbohydrates and proteins.

3. **Biological causes (animate)**

(a)	Bacteria	(f)	Arthropodes
(b)	Viruses	(g)	Fungi
(c)	Helminths	(h)	Rickettsiae
(d)	Trematodes	(i)	Protozoa
(e)	Cestodes		

These animate causes may produce disease by

  i)   mechanical irritation or obstruction

  ii) production of chemical toxins

  iii) allergic or immune mechanism or

  iv) by causing hormone imbalance.

It is the bacteria, viruses and protozoa of the animate causes that are responsible for most of the diseases. We shall now consider the routes (or portals) of infection, the resistance and defence mechanisms of the host and finally the method of establishment of infection by these agents.

Portals of infection :

The skin : Skin is the largest tissue exposed and it harbours normally many kinds of bacteria as 'resident flora'.   Some other bacteria are found for shorter duration as 'transients'. But, the skin is able to prevent the entry of these bacteria by processes we will detail under defence mechanism.

Being such a large exposed area, skin is vulnerable to trauma, facilitating entry of the organisms such as *Staphylococci, Streptococci, Pseudomonas pyocyanea, Leptospira icterohemorrhagia. Corynebacterium pyogenes* and *ovis, Cryptococcus Clostridium welchii, Clostridium chauvoei and Clostridium tetani*. Such organisms might have been residents or 'transients', or might have originated from surgeons and attendants while dressing the wound

Infection through intact skin can occur in parasitic infections. The infective larvae of strongyl worms can penetrate the skin.

Bites of ectoparasites—flies, mosquitoes, ticks mites and bugs—may be source of infection. Babesiosis, trypanosomiasis and other protozoan diseases are examples of this type of infection.

ETIOLOGY

The alimentary tract.

     The Mouth : As in the skin, in the mouth also are found certain resident flora which may gain entry into the tissues and blood through wounds in the buccal mucosa—injury by awns, sharp bits, glass pieces, blades, pieces of wire as well as in the condition called black-tongue. Actinomycosis and Actinobacillosis are diseases that are caused by entry of the causative bacteria through such wounds.

     The intestine : Microorganisms cause disease in two ways. It may be by proliferation in the gut and producing powerful toxins which are absorbed through the wall into the blood; for example—enterotoxemia in sheep. Or, the bacilli may enter the wall of the gut and set up inflammation locally or may be blood borne or lymyh borne and finally reach various tissues. How is the passage of the organisms into the bowel wall brought about is not clear in all cases.

     Endamoeba and Balantidium coli are able to move about and so affect the epithelial cells causing parasitism. So also, the merozoites of the coccidia are able to infect the epithelial cells by their motility.

     Tubercular organisms as well as the organism of Johne's Disease are probably conveyed into mucous and submucous tissues by migrating macrophages.

     Wounds caused by biting nematodes may be portals of entry of various organisms from the intestines.

     The conjunctiva : *Brucella* and *Cryptococcus* can infect the intact conjunctiva and cause disease.

     The respiratory tract : Inspite of the fact that the mucous and the ciliary movements of the respiratory passages help in ridding these tissues of infective agents, still, respiratory tract is an important route through which infection occurs. For example, infection of tuberculosis, glanders and various viral diseases occurs through this route.

     Genito-urinary tract : Dourine in horses, brucellosis in pigs, and trichomoniasis in cattle are transmitted by coitus. Venereal tumor in dogs is an example of a neoplasm transmitted by copulation.

     Injuries of the uterus and vagina during parturition are sources of infection to these organs.

     Mammary gland : Infection can occur through the teat canal, causing mastitis.

Mode of transmission :

1. By direct contact :

     a. Genital tract-Dourine, brucellosis, trichomoniasis and venereal tumors are transmitted by copulation.

     b. Skin –bites of rabid animals disseminate rabies. Anthrax in man by handling infected wool ('wool-sorter's disease".) and pox diseases are also transmitted through skin.

     c. Infection of wounds–by tetanus-contaminated earth in horses.

     d. By droplet infection : When animals cough or sneeze, the infective bacteria are sprayed into the air in which they become suspended as minute droplets. These droplets may be inhaled by animals nearby and so get infected. Examples are contagious bovine pleuro-pneumonia, tuberculosis and glanders.

2.   By congenital transmission :

Infection may be conveyed to the offspring in-utero in two ways :

(i) **through the ovum** : The best example of this is Bacillary White Diarrhoea in fowls caused by *Salmonella pullorum*. The egg may be infected either while still in the ovary, or when it is passing through the oviduct or by the sperm of the cock.

(ii) **Through the placenta** —from the blood of the mother to that of the fetus. Johne's Disease has been found to be an example. Tuberculosis is another.

3.   Diseases transmitted indirectly :

(a)   **by food and water** :   Food and water may be contaminated by the excreta and saliva of infected animals and so be sources of infection.   Examples are the enteric fevers, Rinderpest, Hemorrhagic septicemia, Tuberculosis, Johne'. Disease, liver fluke infection, Ranikhet Disease.

(b)   **By milk** : Tuberculosis, when udder is affected ; virus of Foot and Mouth Disease may be voided through milk, ingestion of which will cause disease.

(c)   **By Fomites** : Inanimate objects like feed troughs and attendent's towels etc.   may be contaminated by the excretions and secretions and hence be sources of infection—eg. Foot and Mouth Disease, Rinderpest, Mucosal Disease·

4.   Diseases transmitted by arthropod vectors :

(a)   **Acting as mechanical carriers:** Anthrax and Surra may be transmitted mechanically by biting flies.

(b)   **Multiplication of the parasite in the vector :**   Rickettsia multiply in the lice and ticks and then are transmitted to the hosts by bites of these vectors

(c)   **Transmission of the parasite from the insect to its next generation through its egg :** Babesiosis.   The protozoan passes through the ova of the tick to the larvae of next generation, which are sources of infection.

(d)   **A portion of the life cycle of parasite is undergone in the vector :** Trypanosomiasis. The sexual life-cycle takes place in the flies. So also the sexual life cycle of piroplasms takes place in the ticks.

5.   By air :

We have already studied spread of disease by droplet infection. Another method how air may be infected is by the feces, urine, and sputum that may be voided on the ground getting mixed with dust which when dry may be blown by wind and so be a source of infection.   Pox diseases are disseminated this way.

With so many portals of entry, varied methods of infection and surrounded by teeming pathogens, how are animals able to protect themselves and withstand infection? The answers to this question are : (1) The local defence mechanisms and (2) the resitance to infection. We will now consider these :

1.   The local defence mechanisms :

(a)   **Skin** :   Though a large area of skin is continuously exposed to bacteria, infection does not occur because of autogenous disinfection of the skin that is brought about in the following ways :

(i) **The superficial squamous epithelium** is lost by normal wear and tear and with it the bacterial population is got rid of.

(ii) **Dessication of the skin** : occurs due to its exposed situation and this is not a congenial environment for the bacteria to thrive.

(iii) **The pH of the skin** : the reaction of the skin is acidic and this prevents the establishment and growth of bacteria.

(iv) **Secretion of fatty acids**– the long chain unsaturated fatty acids present on the skin rapidly kill certain bacteria-*Streptococcus pyogenes, Corynebacterium diphtheriae* etc.

(v) **Bactericidal agents** : substances produced by some bacteria may inhibit the growth of others. The resident bacteria may 'resent a foreign intruder, quite as much as the occupants of a crowded railway carriage resent an extra passenger', and so may "crowd out" others. Hence it is rather difficult for the establishment of a new species of bacterium on the skin.

(b) **The Mouth** : The resistance to infection in the mouth is due to

i) **Saliva,** which destroys some microorganisms and inhibits growth of many others and which also washes away the bacteria into the pharynx and thence into the stomach

ii) **The buccal mucosa**, which is lined by thick squamous stratified epithelium, is resistant to infection.

(c) **The Stomach** : In the stomach most of the bacteria that enter are destroyed. The sterilising property of the gastric juice is due to the hydrochloric acid (which is responsible to the low pH in the stomach) but not to the enzymes. This property of the stomach is known as "gastric germicidal barrier", which however, may be broken down in states of hypochlorhydria, in which gastric secretion is diminished This condition of decreased gastric secretion occurs in pyrexia and other minor infections, when microorganisms enter the bowel and cause enteric disorders

(d) **Intestines** : In milk-fed young, the fermentation of lactose by *Lactobacillus bifidans* produces acid, lowering the pH of the bowel and so renders the medium inimical to pathogenic bacteria.

(e) **The conjunctival sac** : The conjunctiva is exposed to dust and bacteria. It gets rid of these by (i) the mechanical washing down by tears into the inner canthus and from there into the nostril.

(ii) **The action of lysozyme** : This is a protein formed in the tears and has the property of digesting the polysaccharides of the cell walls of some bacteria and so brings about their lysis.

(f) **The Respiratory tract** : The respiratory tract is able to get rid of the foreign matter by the following ways : (i) The mucous secreted washes out the foreign material; (ii) the foreign material gets adherent and trapped on to the moist mucous surfaces; (iii) the cilia remove these by their motion; (iv) the mucous may contain antiviral and antibacterial antibodies, destroying the pathogens; (v) reflex actions like coughing and sneezing help in the removal of bacteria and other irritating materials from the respiratory tract.

(g) **The genito urinary tract** : The urethra is kept sterile by the mechanical cleansing action of the urine.

The vagina in adults is kept free of pathogenic organisms by the presence of a lactobacillus (Doderlein's bacillus) which, by converting glycogen into lactic acid lowers the pH of vagina and so renders it uncongenial for other bacteria to thrive.

2. Resistance to infection :

The following processes are responsible for rendering an animal resistant to infection ;

a) Phagocytosis that occurs naturally in normal animals.

b) Phagocytosis that occurs as immunity develops in the animals and so forms part of immune body reaction.

c) Tissues may contain antibiotic substances that restrict bacterial growth.

The severity and extent of bacterial infection is determined and modified by the following factors.

(a) **Age :** Usually young animals are more susceptible because their mesenchymal tissues have not become mature enough for the production of phagocytes and immune bodies. Susceptibility is greater after the antibodies acquired from the dam are used up.

Older animals, on the other hand, acquire immunity by exposure to the pathogens in small doses. Very old animals are also more susceptible, since in these the metabolic processes slow down and so the production of phagocytes and immune bodies is retarded and their movement is also sluggish. Senile atrophy of the mesenchymal tissue as well as cardiac weakness that may develop in old age render them more prone to respiratory affections.

(b) **Nutrition :** Good nutrition of an animal favours resistance to infection while poor nutrition renders it more susceptible.

If the animal is not well-fed and if the food is deficient in proteins and vitamins, the body tissues atrophy. Atrophied mesenchymal tissues cannot produce phagocytes and antibodies. Moreover, hemopoietic tissues are also atrophied. Hence is the need for good nourishing diet.

(c) **Altered metabolism :** In diabetes, the sugar concentration of tissues is high and so microorganisms are able to grow well giving rise to suppurative infections. Similarly, in renal disease, loss of protein may result in hypoproteinemia and hypoglobulinemia. So the protective antibodies are lost rendering the animal more susceptible to disease.

(d) **Phagocytes :** Phagocytes (neutrophils) have been termed the *first line of defence* in the body. These engulf the bacteria and arrest their growth. Any conditions that may arrest their formation in the bone marrow (damage of hemopoietic system by radiation or chemicals) producing leucopenia, is attendant with danger of speedy widespread infection. So it is clear that leucocytosis or increase in the number of leucocytes is a favourable sign while leukopenia (decrease) is an unfavourable one.

Immunity in an animal is concerned with the location and speedy arrest of the growth of microorganisms by phagocytosis of pathogen. If immunity is absent, phagocytosis does not occur, the organisms grow quickly, produce toxins damaging the cells, liberate fibrinolytic substances and cause edema They may grow and spread by contiguity to the lymph vessels and blood. Hence spread of infection is prevented by localisation of infection and this is brought about by phagocytosis and antibiodies. The speedy arrival of these to the site of infection is facilitated by increased permeability of the vascular endothelium and formation of the exudate.

Fever in resistance :

Elevation of temperature of the body is probably beneficial since it adversely affects bacterial infection in the following ways : (a) Bacteria cannot grow at high temperature (ii) phagocyte production, phagocytosis and antibody production are enhanced at elevated temperature.

But it should be borne in mind that too high a temperature is harmful since at that temperature antibody production is stopped and antibodies that are already formed may be destroyed.

Long continued fever is harmful to the body since there is great break down and loss of proteins leading to loss in antibody producing tissues. Fever also causes degeneration of heart, liver and kidneys. Production of plasma proteins and hemoglobin may be interfered with.

A question may now be raised as to how with various local defence mechanisms and the resistance of the body by various means described, infection can occur at all and the organisms are able to grow establishing infection.

This is brought about by (a) interference with phagocytosis, (b) neutralisation of various anti-bacterial substances that are formed in the body and (c) facilitating the speedy spread of infection by obstructing the localisation mechanism.

a) Obstruction of phagocytosis: This is brought about by the bacterial antigens and haptens which are found on the surface of bacteria. These make it difficult for the phagocytes to engulf the bacteria

b) Neutralisation of antibodies: The antigens found on virulent bacteria neutralise the antibody-complement mixture preventing thereby the bactericidal effect of the antibodies.

c) Mechanism of obstructing localisation of infection :

i Staphylocoagulase: A coagulase produced by staphylococci clots blood and so indirectly hinders phagocytosis and so localisation is obstructed.

ii) Leucocidins: Some organisms like the pyogenic cocci produce toxins that kill leucocytes thus preventing phagocytosis and so obstructing localisation.

iii) Hyaluronidase: The tissues are kept together by a cement substance which is hyaluronic acid polysaccharide. Some bacteria like *Streptococci*, *Staphylococci* and *Clostridium welchii*, produce a toxin (hyaluronidase) which is able to break down the cement substance by depolymerisation of hyaluronic acid. This permits the easy spread of the organisms.

iv) Kinases : *Streptococci* and *Staphylococci* produce streptokinase and staphylokinase respectively that are able to digest the fibrin formed in inflammation, destroying the protective barrier and so help in spread of infection.

v) Capsules : Certain organisms are able to form capsules which protect them from phagocytic activity of the the leucocytes. Though engulfed by the leucocytes the organisms resist digestion because of the capsules and are transported to other regions where they may set up fresh centers of infection.

vi) Proteolytic enzymes The anaerobic bacteria produce proteolytic enzymes, which digest the muscle and collagen liberating certain non–specific toxins that cause toxemia and also help in spread of infection.

vii) Nucleases : Some pathogenic bacteria. (*eg: Streptococcus pyogenes*) produce desoxyribonucleases which destroy leucocytes and so help in obstructing localisation of infection.

The above factors, therefore, help in inhibiting phagocytosis and prevent the localisation of infection and thus help in the establishment of infection.

Iatrogenic disease : This is a condition produced by the physician by over or needless medication. At the present time the number of drugs put into the market are legion and their actions are not entirely known. It should be realised by the student that each animal reacts differently to drugs. This condition was known as "idiosyncrasy". This is now explained by the absence of alteration in certain enzyme systems which metabolise the drugs, resulting in harmful products in the body, causing hemolysis etc. This altered state may even be inherited.

One can never caution too much on 'overmedication'. The danger of the pathogenic organisms becoming drugfast must always be borne in mind. This is especially the case while dealing with coccidiosis in poultry. The danger of treatment by antibiotics, either in too high adose or in too protracted a treatment should be brought home to the student since anemia may be caused. The caution appears to be needed now since indiscriminate use of these drugs is an everyday experience. The danger of the patient becoming allergic to such medication must also be borne in mind.

ELECTRON MICROSCOPIC STRUCTURE OF A NORMAL CELL

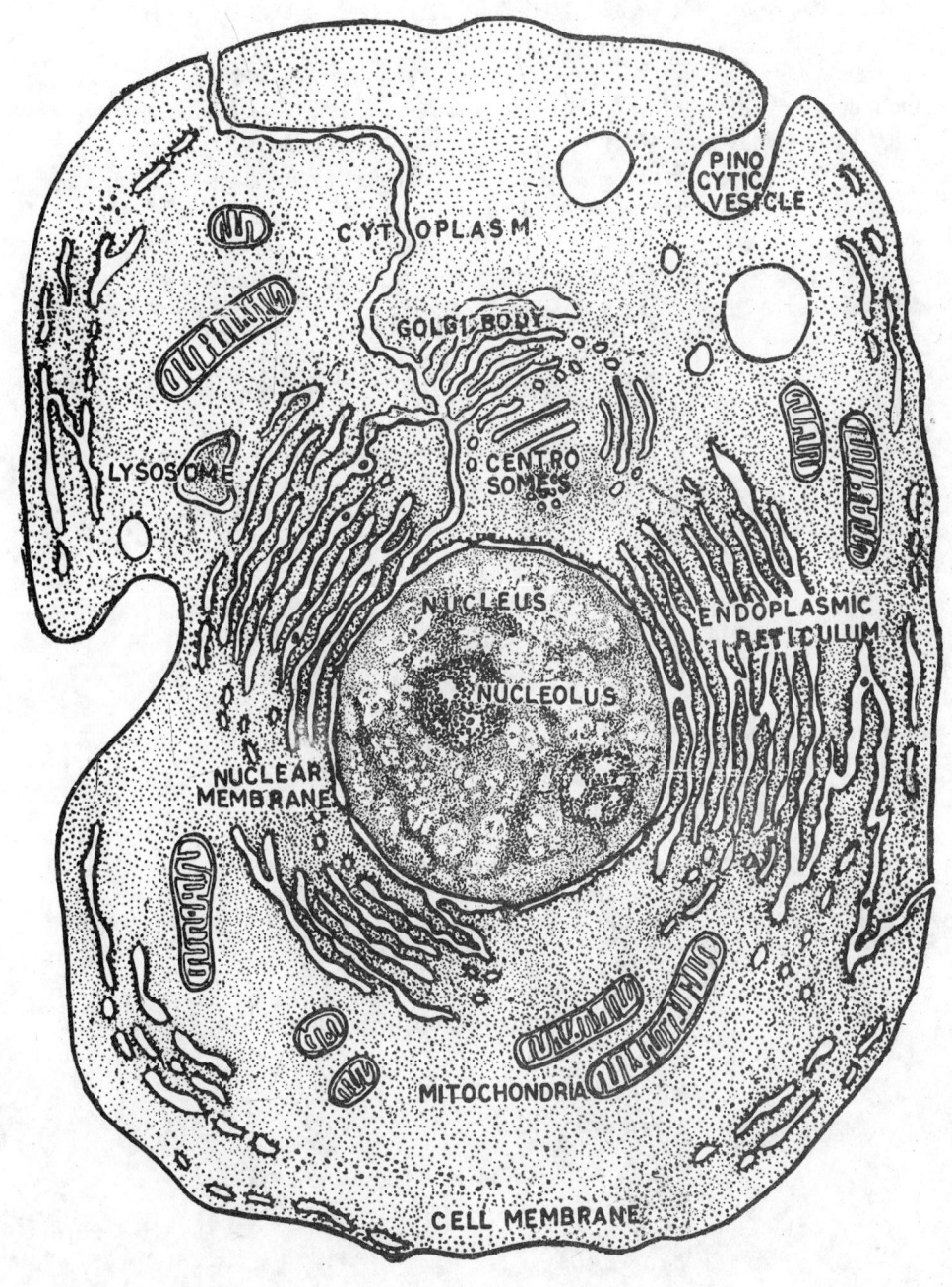

CHAPTER 3

THE CELL IN HEALTH AND DISEASE

Electron microscopic structure and function of normal cell	Mucinous degeneration
	Mucinoid degeneration
Unit membrane	Amyloid infiltration
Endoplasmic reticulum	Fibrinoid infiltration
Golgi apparatus	Gout
Lysosomes	Glycogen infiltration
Mitochondria	Fatty change
Centrioles	Ketosis in cattle
Kinetosomes	Pregnancy toxemia in ewes
Nucleus	Necrosis
Nucleolus	‹ angrene
Cloudy swelling	Postmortem changes
Hyaline degeneration	

Virchow had laid the foundation of modern pathology and he had enunciated that the disease of an organism is due to the pathological changes in the constituent cells and for this we owe him endless homage. He is, therefore known as "Father of Cellular Pathology". Due to advances in science and technology, the study of disease processes has now been extended to the molecular constituents and biochemical processes of the cells and so the modern concept of diseases is extension of Virchow's theory, namely, that disease is due to changes in the molecular structure and biochemical function of a cell. Hence it is appropriate to study the electron microscopic structure and function of a cell before we proceed to study the disease processes that may result due to changes in them.

Electron microscopic structure and function of a normal cell

(Contributed by Dr. K. Satyanarayana Rao)

The cell is a structural unit of a tissue. Although it has been possible to identify many structural components of the cell under light microscope, with the advent of electron microscope many subcellular particles have been brought into light.

Under the light microscope cell appears to be composed of an external membrane and a jelly-like cytoplasm with a central nucleus. Many of the organellae are enveloped by the cytoplasm and each organelle is enveloped in a membrane. The ground substance or the cytoplasmic matrix is a viscous fluid.

The unit membrane (The cell membrane): Every cell is bounded by a membrane called as plasma membrane hitherto. Under the electron microscope this membrane appears as two dark bands on either side of a light band. The same membrane is continuous and also forms a boundary between the subcellular particles like the endoplasmic reticulum, mitochondria, microsomes and other organellae and extend into the interior of the cell upto the perinuclear zone. All cells inclusive of plant cells, bacteria and protozoa also possess such a membrane. In the plant cells and bacteria this membrane is seen under the polysaccharide cell wall. So this is a biological constant common to all kinds of cells. In view of the ubiquitous nature of this membrane in the cell it is known as a **Unit membrane** (I. D. Robertson).

The thickness of this unit membrane varies from 70 to 100A°(Angstrom units). The inner dark band and the outer dark band are 20A° units in thickness and the central light zone is 35 A° units in thickness on an average. The membrane consists of a double layer of lipid molecules covered by a layer of protein on either side. The lipid molecules are oriented at right angles to the plane of the membrane with their non-polar ends pointing towards one another and the polar groups pointing towards the surface of membrane and get adsorbed on the protein chain. This endows the membrane with certain degree of elasticity, mechanical resistance and surface tension. Fat soluble hydrocarbons dissolve in the lipids of the membrane and so enter the cell

The membrane also presents a number of pores formed by invagination of the membrane and are continued into the endoplasmic reticulum. The diameter of the pores is about 7A° units and they are lined by a protein possessing positive charges. These pores lead a tortuous pattern inside the cell leading to a depth of about 300A° units.

The asymmetry of the unit membrane is attributed to the difference in macromolecular material(lipoprotein) lining the external and cytoplasmic surfaces of a lipid micille.

Accessory structures of cell membrane : The microvilli and vesicles are found on the surface by projections and invaginations on the unit membrane. The microvilli (so characteristic of intestines) increase the cell surface and give a brush border appearance (kidney tubule cells).

Interdigitary membranes increase the contact surface between adjacent cells. Inpocketing of basement membrane allows for more effective transport of substances. Jelly layer or mucous coat is peculiar to mucosal surfaces. This does not take part in active transport. Cilia or flagella arise from kinetosomes.

Various reactions take place at the cell surface : chemical dyes become attached there; drugs react first with surface constituents; virus particles get adsorbed on the cell membrane before entering into the interior; poisons alter the permeability and so some ions enter and get attached to the mitochondria while some come out causing sickness.

Functions of cell membrane :

A. **Transport :** 1. Transport of metabolites into the cell and outside the cell by active transport as well as by passive transport like osmosis, diffusion, filtration and solvent drag.

2. The lipoprotein structure of the unit membrane helps the transport of lipid soluble substancess in the cell.

3. Passive transport is also brought about by phagocytosis and pinocytosis (In pinocytosis the fluid is taken as an invagination and then cut off in the inside)

4. Active transport is energy dependent and carrier mediated. Energy for active transport across the membrane is obtained through break down of ATP to AMP.

5. Certain enzymes present at the cell surface known as permeases specific for each substrate are induced by the presence of the substrate and may operate in the membrane transport without energy.

B. Conduction, contraction and movements of the cell.

C. Regulation of cell volume by adjusting the water content of the cell

D. Secretion : Membrane may itself perform active secretion as in the case of HCl secretion by the gastric glands.

E. Antibody formation : Antigens react at the surface of the cell-wall with the formation of antibodies which later enter the blood stream.

CELL ORGANELLES

These are subcellular particles operating in a discrete manner performing a series of functions which are necessary for the normal maintenance of the cell activity. The structure of each is characteristic of its function.

Endoplasmic reticulum : This is a series of internal membranes which extend between the cell membrane and the nuclear membrane and is particularly well developed in granular tissue cells where protein synthesis occurs. A number of sacs or cysterne and vesicles are formed which are interconnected with each other and are lined by a single layer of unit membrane material. The endoplasmic reticulum probably constitutes an intracellular circulatory system. The membrane consists of lipids, proteins and RNA. The membrane may appear rough or smooth and when smooth has the same structure as plasma membrane; roughness is due to the lining *ribosomes* or *Palade granules*. The endoplasmic reticulum connects the outer membrane, the nucleus and Golgi apparatus and so it helps to keep the interior of the cell in contact with the nucleus. The endoplasmic reticulum is not evident in the kidney cells.

The fluid cytoplasm outside the membrane of the endoplasmic reticulum is called **Hyloplasm,** Basophilia of the cytoplasm of a cell is due to the ribosomes containing the RNA.

Function of the endoplasmic reticulum : This system conducts the excitation potential generated in the outer membrane to the interior of the cell. but also responds to hypertonic and hypotonic concentration of salts and metabolites. Some portions of the membrane are capable of electron transport, ribosome synthesis and protein synthesis. Being an intracellular circulatory system, it brings the organelles in contact with each other.

Mycrosomes : These are small dense particles lining the endoplasmic reticulum. They are concentrations of enzymes concerned with cholesterol and fatty acid synthesis.

Ribosomes : These are microsomes concerned in protein synthesis and contain RNA. They also line the cytoplasmic side of the endoplasmic reticular membrane or they may be free in the cytoplasm. Individual ribosome particles are not able to synthesise protein, but when they are aggregated to form *polysomes or polyribosomes,* they are able to function together and synthesise protein. Magnesium ions are necessary for binding the ribosomes into polysomes. The membrane associated with the ribosomes do transport the protein formed into particles to that portion of the cell requiring the material. The ribosomes form a pattern or template for protein synthessis. The amino acids are first arranged in proper order and then united by enzymes on the wall of the endoplasmic reticulum to form protein.

Golgi Apparatus is a system of flattened sacs with vesicles and vacuoles of different sizes which are bounded by unit membrane. These are involved in the storage of secretory granules like zymogens until they are released. Golgi apparatus probably functions as a packaging area, to wrap the secreted protein to prevent its coming into contact with the cytoplasm. Then the wrapped protein is transported to the cell surface for release. Probably large carbohydrate mole-cules are built and fat stored here.

LYSOSOMES : These are sac-like structures containing a variety of acid hydrolases and surrounded by a membrane similar to a single unit membrne. These together with some closely related vacuolated structure devoid of hydrolases form an intracellular digestive system. Their size is about 0.5 u. The granules of polymorphs are lysosomes. Lysosomes are involved in a wide variety of animal cells in storage, processing and digestion of extracellular materials. The exogenous material is frequently formed in membrane-lined vacuoles devoid of acid hydro-lase activity known as *phagosomes* Lysosomes are important in the digestive processes of phagocytic cells The granules discharge their enzymes into the phagocytic vacuoles. Any foreign protein first accumulates in the cytoplasmic granules, which stain negatively for acid phosphatase i.e., in the phagosomes. These then merge with the acid phosphatase containing lysosomes forming phago-lysosomes. Extracellular material enters into the phagosomes by vesiculation of the plasma membrane and phagocytosis Lysozymes are the enzymes present in the lysosomes.

Functions dependent on heterophage or heterolysis

a) Heterotropic nutrition by intracellular digestion and under special ecological condition by extracellular digestion is a key function of lysosomes in unicellular organisms and metazoan cells.

b) Defence against bacteria and other microorganisms like viruses or toxic micromolecules is a protective function aided by special cytocidal agents like phagocytins. The extracellular release of lysosomal enzymes has been suspected in spermatozoa, osteoclasts and in tumor cells. Invasive enzymes like lysozymes, hyaluronidase, collagenase and the permeability increasing factor (a protease) are localised in lysosomes of certain cells.

c) Invasion by lysis obstructing structures leading to the disappearance of certain intracellular organelles or structures are important in cellular differen-tiation or metamorphosis as exemplified in erythrocyte maturation, skin keratini-zation, follicular atresia, uterine cycles, involution of the uterus and mammary gland etc.

Functions dependant on autophagy. The nutrition under unfavou-rable conditions of food supply is by resorbing proteins by autophagy. Intracell-ular scavenging as a part of cell rejuvination of long lived cells and self cleara-nce of dead cells are examples of these

Lysosomes in pathology : Lysosomes produce cellular alterations either by failing in their adequate lytic function or by performing them in an injurious fashion.

In infection and intoxication there is a congestive enlargement of lyso-somes with an abnormality of enzymatic equipment and by overloading of diges-tive material or undigested material. Injurious lytic activity is noticed due to hydrolytic formation of injurious entities in viral infections or intoxications. Injurious self digestion of cells also occurs by autophagy or autolysis following lysosomal rupture Damage of extracellular structures is seen due to enzyme leakage or excessive enzyme extrusion. Glycogen storage disease, metachromatic leucodystrophy, gargoylism etc. are inborn lysosomal diseases.

Stabilizers and labilizers of lysosome membranes. Cholesterol, cortisone and cortisol stabilize the lysosomal membranes and hence are important in inflammatory conditions. Free in tissues, lysosomal substances may invoke inflammation. The stabilizers mentioned above are anti-inflammatory and so quieten the inflammation. Hypervitaminosis A and avitaminosis E may weaken the lysosomal membrane bringing about extrusion of enzymes and autolysis of cells. Chlorpromazine at high concentration has similar effect. Lysosomal membranes are made more permeable by anoxia, ischemia and necrosis.

MITOCHONDRIA. A liver cell may contain about 2500 mitochondria. These are rod or cucumber shaped and consist of double layered membranes and appear as sac within sac containing fluid matrix between the outer and inner sacs. Their size varies from 1.5 to 3 u in length and 0 5 to 1 u in diameter. Their shape and size vary from cell to cell. The inner membrane is folded to project inside as shelves or crystae of varying sizes. In some the crystae appear as small nubs and in some extend from side to side and lie parallel to each other. The crystae may flatten out to form streaks which are fenestrated into pores. The intracellular distribution of mitochondria is also constant. They are often polarized and are often localised in the direction in which metabolites enter the cell and are associated with active part of the cell.

Mitochondria are not rigid and their plasticity allows movements within the cell and facilitates variations in shape. A characteristic protein appears to be present in the membrane and this helps in the movement. The number and types of crystae may also vary. The mitochondrial membrane consists of protein-lipid-protein layers. The lipid layer is chiefly a phospholipid like lecithin, cephalin inositol, phosphatid or cardiolipin. It is suggested that the mitochondrial membrane in contrast to the unit membrane has seven layers, with two triple layer arrangement separated by a lip d layer.

The crystae offer large surface area and house respiratory assemblies like electron transport particles (ETP) and produce ATP. In highly active cells like the liver cells, the number of crystae are larger, so that the inner membrane has four times the surface area of the outer membrane. Distributed on either surface of the membrane are a number of mitochondrial particles known as elementary particles on the sites of ATP synthesis.

Enzymes of Kreb's and fatty acid cycles are localised as discrete particles on the outer membrane.

Functions. The mitochondria are described as the 'power house' of the cell. They are concerned with the ATP synthesis and electron transport systems They are concerned in the Kreb's cycle, fatty acid oxidation, ketone body oxidation and formation etc., as they contain many enzymes concerning these metabolic reactions.

The mitochondrial membrane transports many substances through it for the organellae to function. Because of the contractile protein in the membrane the membrane can change its dimensions. The membrane also acts as an osmometer allowing the mitochondria to swell or shrink.

Centrioles. Two bodies lie at right angles to each other near the nucleus. Each is a hollow cylinder with nine pairs of fine tubules or filaments. Centrioles are involved in the organisation of spindles in cell division.

Kinetosomes or basal bodies. These are seen in ciliate or flagellate cells. The cilia or flagella extend from kinetosomes. They have a fibrillar structure and resemble centrosomes. This fibrillar organisation converts chemical energy into mechanical work. This fibrillar organisation is also seen in muscle cells.

NUCLEUS.

Nuclear membrane : The nucleus is covered by a nuclear membrane. This membrane consists of two unit membranes, separated by a distance of 150 A° units. The space in between the two nuclear membranes is known as *perinuclear space*. The membrane has a large number of pores in it, which provide continuity between the cytoplasm and the intranuclear material. Large molecules may move in either direction through these pores It is also said that the endoplasmic reticulum is probably formed by the nuclear membrane.

The **nucleus** contains the chromatin filaments in which the entire DNA of the cell is localised. When the cell is at rest the chromatin material is diffusely distributed in the nucleus. Thus the nucleus carries the genetic material of the cell. Before cell division starts, the chromatin material coils up tightly to form the chromosomes, which are always fixed in number for each species

There are a number of DNA molecules on the genes, (situated on the chromosomes) having a specific sequence in a coded form for the thousands of protein molecules synthesized by the cell.

The **nucleolus** appears as a spherical body inside the nucleus and is visible under the light microscope. Under the Electron microscope it appears to contain tightly packed granules similar to the ribosomes of the cytoplasm and these are rich in RNA. So the nucleolus is concerned with the synthesis of messenger RNA which carries the genetic code from the DNA into the cytoplasm for protein synthesis. The nucleoli and chromatin material are together in a protein matrix—the cell sap. Under the direction of DNA the RNA of the nucleolus synthesizes RNA, which diffuses through the pores into the cytoplasm, and this messenger RNA forms the basis on which the ribosomes synthesize proteins.

Functions of the nucleus : 1. It is concerned in cell division.

2. It contains the genetic material.

3 It controls the cell metabolism by providing the genetic code for the synthesis of proteins like the enzymes needed for metabolic activity of cell and its growth.

Structure of the DNA molecules : The DNA molecule (Deoxyribonucleic acid) has a double helical structure Ribose or deoxyribose and phosphate form the back bone of the helix with the purine or pyramidine bases—Adenine, Thymine, Cytosine and Guanine forming the side groups. These four letters (A T, C&G) spell out the genetic message The exact sequence of these bases determines the structure of the particular portein molecule. In the DNA molecule the bases Adenine and Thymine, and Guanine and Cytosine are bonded through hydrogen bridges. The genetic code in the DNA is a triplet code for each amino acid and is called *codon*-the structural gene. The *operon* gene or *operation* gene controls the activity of the other genes and coordinates the functions of the adjacent structural genes. There is another gene—the *Regulatory repressor gene*, which functions in the control of the operator gene-operon-by binding itself to it

The DNA molecule of the codon is first transcribed into a similar molecule—the messenger RNA—in the nucleolus. The RNA is liberated into the cytoplasm and then taken to the ribosomes where the code is translated into protein synthesis with the help of Ribosomal RNA and transfer RNA.

The sequence of the nucleotides in the nucleic acid and the arrangement of the amino acids in the protein is maintained by the strict directions of the genetical code. If there is misplacement of even one of these components, disease will result. For example, there are 574 amino acids in one molecule of hemoglobin. Even if one of these amino acids is wrongly placed, sickle cell anemia results.

Cellular proteins are mostly synthesised in the rough endoplasmic reticulum and on the free cytoplasmic polysomes. Some of the proteins so synthesized remain near the site of synthesis (sedentery protein) while a great majority move away to other organelles--the migratory protein. A portion of the migratory protein known as non exportable protein is distributed to the various cell organelles. The other part of the migratory protein includes the secretory products and is referred to as exportable proteins which are stored in the Golgi apparatus as secretory granules. Some protein synthesis can also take place in mitochondria.

Phagocytosis (eating) : When a foreign body enters, the unit menbrane forms an invagination by throwing out pseudopodia and form pockets that draw the foreign material into the interior of the cell. The membrane then envelops the material in a vacuole or vesicle which is pinched off and then this vacuole floats free in the cytoplasm.

Pinocytosis means drinking of liquids.

DEGENERATIVE PROCESSES

If the injury is of sufficient intensity, the cells undergo degeneration, which is a retrograde change. All degenerations do not end in the death of cell (necrosis). Such non-fatal degenerations are reversible and if the injury is removed the affected cells return to their normal state.

The degenerative changes that the cells undergo are many and varied. The type of degenerative change that the cells may undergo depends upon the following factors.

1. Cells themselves— both kind of cells and their number. Some cells are more susceptible to injury than others. Eg. the hepatic cell is susceptible to even mild irritants while the more sturdy fibroblast is resistant.

2. Quaity of injurious agent : Some irritants are more toxic than others.

3. Quantity of the injurious agent : An irritant may produce mild degeneration if in small doses, but may cause necrosis in larger quantities.

4. Length of time : If the duration of the action of an injurious agent is short, the degeneration is mild. But if acting over a longer period, the damage may be severe.

In degenerations, any or all of the following aspects of cellular activity may be altered, temporarily or permanantly :

1. nutrition, growth and maintenance,   2. reproduction and
3. specialised function.

Function and structure of the cells are so interdependent that change in one (however slight) produces a corresponding change in the other. So from altered function in an organ, we may, with fair amount of accuracy, predict the alteration in structure such an organ may have. Similarly, by microscopical examination of altered tissues, we may conjecture the change in function that might have been present in life.

In injury the intra cellular systems that are mostly affected are (1) aerobic respiration which involves oxidative phosphorylation for the production of ATP, (2) maintenance of the integrity of the cellular membrane, which is responsible for the maintenance of ionic and osmotic homeostasis, (3) synthesis of enzymes and proteins and (4) preservation of the genetic apparatus of the cell.

Injury at any one point of the cell leads to a chain reaction resulting in the disruption of the normal physiological activity of the cell. For example, if aerobic respiration is affected, impairment of synthesis of ATP results. In the absence of ATP synthesis sufficient energy is not available for the proper functioning of the "sodium pump", by which intracellular ion content is maintained. In the absence of aerobic respiration, anerobic glycolysis occurs, when the pH is lowered due to excessive production of lactic acid. It is well known that for the normal action of enzymatic and biochemical reactions and functions in a cell optimum pH is required and if this is altered, then the normal biochemical reactions stop This results in the stoppage of protein synthesis and the integrity of the cell membrane is lost and so the function of the cell is affected. The membrane becomes more permeable so that all the intracellular enzymes leak out of the cell into the blood. Estimation of such enzymes in the blood like the GOT, GPT, LDH etc. gives a clue as to the damage of the cells.

DISTURBANCES OF PROTEIN METABOLISM

Parenchymatous degeneration, Cloudy swelling, albuminous degeneration. This is the earliest morphologic evidence of cellular degeneration seen. This is mild and easily reversible. But with the persistance of the irritant this degeneration may progress even to necrosis. In cloudy swelling there is an alteration in the physical state of the protein and is seen in the highly specialised cells such as cord cell of the liver, tubular epithelial cells of the kidneys and cardiac muscle.

Causes : The two most important causes are :

I. **Hypoxia :** This may be encountered in

1. Respiratory diseases in which air cannot reach alveoli (pneumonia.)
2. Deficient atmospheric oxygen as on mountains.
3. Conditions where there is deficient hemoglobin-anemia, hemorrhages or carbon monoxyhemoglobin
4. Cardiac disease—blood flow retarded, chronic venous congestion develops.
5. Local disturbances affecting circulation.

II. Toxins : Bacterial and viral, plant alkaloids, Chemicals-P, As, Cu, bile salts. Hence, cloudy swelling is seen in most of the acute infectious diseases. Fever is cited to be one of the causes and starvation is yet another

Macroscopically, the affected organ is swollen, has a pale or par-boiled appearance. On section, the cut surfaces bulge and cannot be brought back into exact opposition.

Microscipically, the cells are swollen and the cytoplasm has a "ground-glass" appearance. The cytoplasm, which must normally be clear, is granular. In the kidneys, the lumen of the tubules is narrowed due to the swelling of the cells, which in the proximal convoluted tubules, lose their brush border. The following is one of the explanations for swelling of the cells. The mitochondria, which are the store house of cellular enzymes, become swollen and decreased in density. The cytoplasmic lipids and proteins are absorbed on these mitochondria. The enzymatic activity responsible for energy production is decreased. Hence there is accumulation of intracellular sodium due to a failure of the sodium pump (as energy is decreased). There is passage of sodium ions into cell since it is not extruded when energy yielding metabolism is disrupted Accumulation of sodium ions in the cells raises the osmotic pressure and so water passes into the cells from the intercellular fluid, resulting in swelling of the cells. Increased permeability of the plasma membrane may also be a factor in this condition.

Sequelae : The causes of necrosis are the same as those of cloudy swelling but at a higher concentration and acting for a longer time. So, if the causes for cloudy swelling continue, the degenerative process may end up in necrosis. No symptoms or harmful effects are noticed in cloudy swelling, which being the mildest of degenerations is readily reversible. After death, the cytoplasm of the parenchymatous organs assume a granular condition due to alterations in the enzymes which are protein in nature. It is difficult to differentiate this post mortem change from ante-mortem cloudy swelling. So autopsy must be performed immediately after death and the tissues fixed by suitable preservatives.

HYDROPIC DEGENERATION

This is closely related to cloudy swelling. The cytoplasm contains one or two large vacuoles or few small ones. The causes are similar to those of cloudy swelling. Because of a more severe irritant, the permeability of the cell membrane is increased accompanied by loss of potassium ions and passage of sodium ions into the cell. The mitochondria are altered and the sodium pump is damaged with the passage into and accumulation of sodium in the cell.

Hydropic degneration is well exemplified by the blisters on the skin (burn) or mucous membranes (due to chemical irritants) and the vesicles that form in Pox diseases and Foot and Mouth Disease. Hydropic change is seen in the renal tubule cells of animals that are given glucose or sucrose intravenously and in the tubules of animals suffering from potassium deficiency nephrosis (hypokalemia due to diarrhoea) and in starvation. Liver cells manifest hydropic degeneration after administration of ether or chloroform and in poisoning by carbon tetrachloride.

Sequelae : Though reversible, because of the severity of the irritants necrosis may supervene.

HYALINE DEGENERATION

This is the name applied to a condition of cells and connective tissue which become converted into a homogeneous, glassy material, (Hyolos = Glass). This is a name given to the physical condition of the protein that has resulted due to irritants. The actual chemical structure is not known. This material takes pink stain with eosin. The involved tissue dies before hyaline change takes place.

Three types of hyaline are described :

1. **The connective tissue hyaline :** The fibrous collagen fibres in the walls of arteries are affected in arteriosclerosis Similar change of connective tissue is seen in chronic lymphadenitis and in fibroids. Glomeruli in chronic nephritis undergo hyaline degeneration. Scar tissue is yet another example.

2. **Epithelial hyaline :** Stratum corneum of the skin is the best example of epithelial hyaline. But this is physiological. This becomes pathological when excessive keratinisation occurs (hyperkeratosis) as seen in Vitamin A deficiency. Pathological epithelial hyaline is seen in the structures called *Corpora amylaceae* which stain deeply with iodine like starch. These are concentric layers of hyalinised epithelial cells seen in the prostate glands, old infarcts of lungs, the brain and udder of animals. In the udder the desquamated alveolar epithelial cells are compressed and kneaded together tightly which later urdergo hyaline degeneration. The presence of these indicate an earlier inflammation in the udder.

In the liver and kidney, small hyaline droplets may form intracellularly as a result of more severe irritants than those causing cloudy swelling. In diabetes the cells of islets of Langerhans may be converted into hyaline.

3. **Muscle hyaline :** The muscle fibres become converted into pale or homogeneous glassy material. This is seen in Zenker's degeneration of the *rectus abdominis* muscle in human typhoid, in the affected muscles in "equine azoturia, or in the "white-muscle disease" in calves and "Stiff lamb disease" in lambs. The muscles resemble "fish-flesh" and lose their fibrillar staining. While deprivation of oxygen and trauma may produce hyaline degeneration of muscle, this condition is also met with in a variety of febrile diseases, in anaphylaxis and around tumors.

MUCINOUS DEGENERATION

Normally, mucin or mucous, which is a viscid glycoprotein (a combination of protein and mucopolysaccharides of high molecular weight) is produced by the columnar or cuboidal epithelial cells of the mucous membranes and their glands. This is probably produced by the mitochondria and modified by the Golgi apparatus.

The epithelial cells are said to undergo mucinous degeneration when they produce and secrete excessive quantities of mucin while manifesting degenerative changes.

Causes : Any mild irritant, mechanical, thermal, chemical (disinfectants etc.,) and viral (canine distemper and virus diarrhoea of cattle) may cause this degeneration.

In neoplasms of columnar epithelium, mucinous degeneration may sometimes be found—eg. Cancer of the stomach, large intestines and mammary gland Catarrhal inflammation of the mucous membranes of the respiratory, digestive and reproductive systems is characterised by excessive production of mucous.

Macroscopically, the mucous is a slimy, glassy fluid, precipitated by acetic acid. In acute rhinitis (common cold in man) there is profuse watery mucous secretion. Secondary infection by pus-forming organisms turns the mucous to a turbid material.

Microscopically, mucous is evident in the cytoplasm of the cells as small droplets or, when these coalesce. as. a large droplet, compressing and displacing the nucleus to a side. With increasing accumulation of mucin the cell assumes a goblet form (goblet cell), which may ultimately rupture. The ruptured cell dies and is desquamated. The mucin stains faintly blue with hematoxylin (basic dye). It takes a purplish red colour when stained by Periodic Acid Schiff reagent (PAS) and a blue colour when stained by Alcian Blue.

Note : The cystadenoma of the ovary contains a viscid material called Pseudomucin which differs from normal mucin, in

(1) its staining pink with eosin (acid dye) but not by hematoxlin and (2) its not being precipitated by acetic acid.

Sequelae : As soon as the causative factor is removed, overproduction of the mucin stops and lost epithelium is repaired by regeneration of the surviving cells so that no permanent damage occurs.

MUCOID DEGENERATION OR MYXOMATOUS DEGENERATION

Connective tissue cells also, especially of the fetus, produce mucin-like glycoprotein, which is not normally present in the adult tissue. Sometimes, this substance is found in the adult when tissues are said to undergo mucoid degeneration.

Mucoid degeneration may be met with in the following conditions :

1. **Neoplasms :** Mucoid degeneration is found in myxomas and myxosarcomas (connective tissue tumors). The cells are stellate (star-like) with long branching processes in the meshes of which is found the faintly blue mucin.

2. **Malnutrition :** In malnutrition as in chronic debilitating diseases, mucoid degeneration may be seen in the bone marrow, cartilage and adipose tissue. The fat in the coronary groove and omentum are frequently affected.

3. **Myxedema :** In man, in deficiency of thyroid secretion, the connective tissue of the skin and other places (especially of the face) undergoes mucoid degeneration and so the parts appear swollen and gelatinous.

AMYLOID INFILTRATION

Amyloid is a homogeneous, translucent pink-staining material found in the ground substance, between the cells (unlike hyaline, glycogen and fat which collect within the cells). Because the affected organs appear waxy (especially on a cut section) this disorder is also known as **waxy degeneration.**

Chemically, amyloid is a protein, pobably globulin conjugated with sulfated polysaccharide. Amyloid infiltration is not common among animals. Dog and horse are more often affected naturally than other animals.

Under the electron microscope amyloid is found to be made up of mostly non-branching fibres and a few that appear twisted.

Cause : The exact causes are not known. But it is found that the condition is the result of an antigen-antibody reaction. This disorder is met with in almost all horses used for anti-sera production. It is thought, that when there is hyperglobulinemia, the antibody is deposited around the blood vessels where it comes into contact with the antigen and amyloidosis results. Amyloidosis occurs when there is proliferation of protein producing plasma cells. Amyloid is identical with alpha globlun, is carried by blood from its site of production and synthesis and leaks through vascular linings and assumes a fibril form from a soluble state at the place of deposit.

In man prior to effective sulphonamide and antibiotic therapy, numerous cases of amyloidosis used to be encountered in such chronic suppurative conditions like osteomyelitis or chronic diseases like tuberculosis, syphilis or long continued sepsis Now it is found in rheumatoid arthritis, pyelonephritis, Hodgkin's disease and plasma cell myelomatosis.

Experimentally, amyloidosis can be produced by
(1) repeated injection into animals bacteria or their toxins.
(2) injection of neutrose (sodium caseinate) into mice.
(3) feeding large quantities of egg albumin or cheese to mice.

Classification : In human pathology, amyloidosis is divided into

(1) Primary, in which no predisposing causes (as in the secondary) could be found.

(2) Secondary, when amyloidosis is found in association with chronic inflammatory conditions especially with prolonged suppuration such as tuberculosis, syphilis, osteomyelitis or in conditions like multiple myeloma, disseminated lupus erythematosis, Hodgkin's disease, renal-cell carcinoma etc.,

(3) Familial This condination has been found to be hereditory in some families :- Familial amyloid nephropathy, polyneuropathy, cardiac amyloidosis etc.,

Such a classification is not possible in animals in which amyloidosis is not associated with chronic debilitating disease (except in the fowl, in which amyloidosis, especially of the spleen. is met with in tuberculosis).

The changes that occur in the tissues in amyloidosis are attributable to three factors :

1. The amyloid which is deposited around the blood vessels, compresses them and causes stenosis, resulting in ischemia and consequent hypoxia of the affected organs.

2. The increasing deposition of the amyloid produces pressure atrophy of the sorrounding tissue cells. This is especially the case in liver.

3. The deposition of the amyloid on the walls of capillaries makes them thicker and so rendering it impossible for the transfer of nutrients, gases and metabolic waste products between the capillary blood and the tissue cells.

Organs affected : The liver, kidney and spleen are more frequently affected. Other organs that may also be involved are :- adrenals, pancreas, lymph-nodes and the intestines. Since the disease is progressive, all organs may be affected ultimately.

Spleen :

Macroscopically. amyloid spleen is enlarged. paler than normal and firm. The edges are sharp.

Microscopically the amyloid may be distributed in two ways.

1. In the common variety it is deposited around the central arteries of the Malpighian corpuscles. Gradually, lymphoid tissue is replaced by amyloid and this gives an appearance of grains of boiled sago scattered over the organ. So this form is known as 'Sago-spleen".

2. In the rare variety, amyloid is laid in the connective tissue of the sinuses and the reticulum of the pulp Here the infiltration is more diffuse and so the organ is much enlarged. It is called "bacon spleen".

Liver ; Macroscopically, the liver is greatly enlarged and the edges rounded. It is friable and so rupture is common and is an usual cause of death. On section, the cut surface is waxy and smooth.

Microscopically, the amyloid is deposited between the endothelium of the sinusoids and the hepatic cells and so the cord cells suffer from hypoxia and lack of nutrition, resulting in fatty degeneration.

Kidney :　Macroscopically, the secondary changes that occur due to deposition of amyloid complicates the picture. The kidney may be enlarged or be normal or be even contracted in size. Fatty changes that occur in the tubular epithelium give the kidney a spotty appearance instead of a waxy look.

Microscopically, the amyloid is deposited in the connective tissue of the glomerular capillaries (that is, between the endothelium and the epithelium) in the walls of the cappillaries in the interstitial tissue found between the straight tubules and under the basement membrane of the collecting tubules.

The glomerular capillaries that are stenosed gradually become obtsructed and obliterated, though a few remain patent. Blood flow, therefore, is very much retarded and reduced. Since the efferent artery from the glomerulus is the nutritive vessel for the tubules, ischemia of these vessels results in lack of nutrition and oxygen supply to the tubules and so their epithelium undergoes fatty degeneration, atrophy and even necrosis giving rise to the "nephrotic-syndrome", characterised by edema, albuminuria and hypercholesterolemia. By the deposition of amyloid in the intertubular vessels also greater injury to the tubules is caused. The atrophied and necrosed tubules are replaced by fibrous tissue and this is the cause of shrinkage in size of the kidneys giving rise to hypertension and such animals may ultimately die of renal failure.

Pancreas : The amyloid is deposited between the capillaries and cells of islets of Langerhans. These cells gradually undergo nutritional and pressure atrophy, resulting in diabetes.

Perviscular amyloid infiltration has been described in the conjunctiva, in the submucosa of respiratory tract, in the dermis and subcutaneous tissues.

The amyloid takes a slightly pink color with hematoxylin and is homogeneous and structureless.

Staining properties :

Iodine gives a mahogany brown color to the amyloid. If this is followed by the application of a weak solution(1%) of sulphuric acid, a black or blue color is obtained. Hence the material was once thought to be starch, amylum, which gives a blue color with iodine and so it received the name of amyloid (starch-like).

Methly violet gives a metachromatic stain—the amyloid stains rose-red while the surrounding tissue is stained blue.

Congo red : Amyloid, which is stained red by Congo red has a great affinity for the dye and so when injected, this dye is rapidly absorbed by it. Congo red can be used clnically to assess the extent of amyloidosis by injecting a known quantity of the dye as a 1% solution intravenously and one hour later measuring the amount of dye still present in the blood. In amyloidosis almost all the dye will disappear from the blood (as it is removed by the amyloid) while normally only 15 to 30 per cent of the dye is lost. When sections are stained with Congo red, amyloid takes a deep red color.

At autopsy, a weak solution of iodine is useful. When this is poured on a cut surface of spleen and kidney the affected malpighian corpuscles and glomeruli respectively stand out as brownish spots.

Amyloidoma : This is an isolated deposition of amyloid material in a neoplastic form in the heart, tongue or the subcutaneous tissue.

Sequelae : Amyloidosis is a progressive process and amyloid once deposited stays for ever· If etiological factors are removed further desposition may be stopped. No clinical symptoms have been noticed since amyloidosis is only a post-mortem finding in most of the cases. It is not usually fatal, unless the affected organ is ruptured, as happens, sometimes, in hepatic amyloidosis f horses.

FIBRINOID DEGENERATION

In certain diseases of man, collectively known as Collagen Diseases (Polyarteritis nodosa, rheumatoid arthritis, systemic lupus erythematosis, sclerodermal dermatomyositis, glomerulonephritis and rheumatic fever) the collagen of the connective tissue in the wall of vessels, especially of the smaller arterioles, becomes swollen, hyalinised and necrosed. On this collagen is deposited a material known as 'fibrinoid' which, as per the present consensus of opinion, is derived from or is a modified form of plasma fibrin. Such connective tissue is said to have undergone fibrinoid degeneration and appears as a structureless, homogeneous material.

It is thought that some (if not all) of the collagen diseases mentioned above arise as a result of an antigen-antibody reaction (allergic reaction). But the precise antigen involved is not evident. In this reaction the mucopolysaccharide ground substance is depolymerised, thereby rendering the vessel wal, more permeable. The altered ground substance may itself act as an irritant causing inflammation with abundant protein-rich exudate collecting at the site, between the separated collagen fibrils.

Though fibrinoid degeneration is met with in necrotic areas in the muscles in Foot and Mouth Disease, in the mesenteric arteries in polyarteritis nodosa and in the 'diamond' lesions of swine erysipelas, no disease comparable to the human collagen diseases occurs in animals.

GOUT

Gout is a condition in which crystals of uric acid or urates of sodium and calcium are deposited in the tissues.

This condition is met with in man, apes and birds. In man, gout is associated with drinking large amounts of beer and red wine. This malady is universally found in people belonging to all walks of life. "It would almost seem, as if, in the past at least, genius and gout went hand in hand and that excess of uric acid affected on the brain as well as on the joints" (Boyd).

In man and animals, urea is the final degradation product of protein metabolism. Purines are converted into uric acid, which in animals is oxidised to harmless allantoin in the liver by uricase. But in man uric acid is excreted as it is, since the liver is deficient in this enzyme. On the other hand, in birds, uric acid is the final degradation product of protein metabolism. Nature has provided this mechanism wisely to suit a particular purpose as uric acid is not very soluble. In birds developing in the egg with hard shell, excretion of large amounts of fluid cannot be thought of. The embryo has to conserve all the fluid it has for development. So, the mechanism of uric acid excretion is innovated Uric acid is not very soluble and so most of the water from its solution is reabsorbed in the cloaca and hence a solid mass of uric acid accumulates in the allantois. If on the other hand, urea were to be excreted, all water could not have been reabsorbed due to the high solubility of urea and so the embryo would have died in the egg of uremia or dessication. This **uricotelic** mechanism is retained in the birds. In mammals because of the presence of effective fetal membranes disposal of urea is not a problem, and so the ammonia is effectively converted into urea. Only very small quantities of uric acid are formed from purines.

Even this small amount of uric acid is dangerous because, most of it is reabsorbed from the glomerular filtrate by the proximal convoluted tubules. In animals this presents no problem since it is converted into allantoin, a harmless substance. But in man and apes it is of importance, since the enzyme, uricase, is not present and so uric acid accumulates in the blood. In the Dalmatian dog uric acid is excreted in large quantities because the tubules are incapable of reabsorbing it (uricase is present in this breed also).

Why should there be the deposition of the uric acid and urates in tissues in man? There appears to be a genetical predisposition for this disease which along with errors in feeding habits (eating large quantities of protein) precipitates the disease. Probably uric acid is produced in larger quantities by the conversion of ammonia into uric acid.

In gout, there is hyperuricemia, which may be due either to (1) over production of uric acid, (2) to impaired destruction of uric acid because of deficient enzyme systems or (3) to its decreased excretion, either due to impaired filtration or altered excretion in the tubules as may occur in drug therapy like aspirin, mercuric salts etc. In man, the serum uric acid content is very high (normal range being 2.4 mg. to 6.4 mg. per 100 ml. of serum).

In birds, probably a sudden change in rations, especially in heavy laying hens, with a high protein diet may precipitate the onset of this disease. Vitamin A deficiency has been cited as a cause, in which the urethral ducts get occluded by the desquamated metaplastic squamous stratified epithelium preventing the excretion of uric acid and so facilitating its accumulation in the blood.

Gout appears in two forms in birds : the articular and visceral.

Macroscopically, in the articular form, the joints are much swollen, with deposition of masses of chalk-like material (tophi) in and around the joints. Usually the wing and leg joints are affected. Since this is a painful condition the affected birds cannot move about and so die of starvation.

In the visceral form chalk-like crystals are deposited on the serous membranes, or various organs and in tissues around the kindeys.

Microscopically, in the tissues of the joint, (the capsule or articular surface) a foreign body reaction is seen, inflammatory changes with infiltration of macrophages and giant cells are observed around crystals of uric acid or urates lymphocytes, plasma cells along with young fibroblasts.

In the kindeys degenerative changes of the lower part of the tubules are seen. Blood and pigment casts may be noticed in some of the tubules. The needle shaped (acicular) crystals may be found in the kidneys.

In man two forms of gout are seen : (a) **Primary** which is familial and hereditary. In this condition hyperuricemia is produced due to metabolic defects in purine metabolism (like absence of hypoxanthine-guanine phosoribosyl trans-ferase), (b) **Secondary** which occurs due to over production of uric acid with hyperuricemia, as in polycythemia, hemolytic anemias, leukemia, and rapidly growing tumors, in which diseases there is large scale breakdown of cells and greater nucleoprotein metabolism.

DISTURBANCES IN CARBOHYDRATE METABOLISM

Glycogen and glucose are two forms in which carbohydrates are found in the body. These are formed from the dietary starches, sugars and cellulose. In times of need, glucose can be formed from proteins and fats (gluconeogenesis).

Disturbances in the metabolism of carbohydrate are not common in animals.

GLYCOGEN INFILTRATION

Normally glycogen is present in the hepatic cells, muscle fibers and squamous epithelium of cervix uteri. In well nourished animals, there may be abundant glycogen infiltration in these tissues.

Pathologically, glycogen occurs in the epithelial cells of liver and loops of Henle in the kidney, in the neutrophils in inflamed and necrotic tissue, in some rapidly growing tumors and some times in the cardiac muscle. In diabetes the epithelial cells of the Henle's loops reabsorb the glycogen from the glomerular filtrate.

In man glycogen storage diseases (Van Gierke's disease, Pompe's disease) is a hereditary condition that occurs due a genetic defect resulting in deficiency of enzymes required for glycogen metabolism (like glucose-6-phosphatase, alpha-glucosidase, amylo 1, 6 glucosidase etc.,)

Microscopically, the cells show vacuoles, where glycogen was once present, as it is removed during preparation of the sections. This is stainable by special stains (with Best's carmine glycogen is stained a bright pink).

Among animals, presence of glycogen is not of much significance.

DISTURBANCES OF FAT METABOLISM

Body lipids serve the following functions :
1. They are sources of energy.
2. They form vital constituents of all tissue cells.
3. They insulate and so protect vital organs, acting as shock absorber or cushions.
4. They are stored energy.

The fats that are ingested are digested in the intestines and hydrolysed into fatty acids and glycerol. In the wall of the bowel, there is reformation of the neutral fats, three fourths of which enter the lacteals and reach the blood via thes thoracic duct. A quarter of the absorbed fat enters the capillaries and through the portal vein reaches the liver. So ultimately all the fat passes through the liver. That part of the fat that is not immediately required for the tissues, is stored in the fat-depots—the adipose or white fibrous connective tissue.

The fat depots of the body are :
1. Subcutis
2. Perirenal region
3. Pericardial subserosa
4. Orbital fossa
5. Bone marrow
6. Brisket of cattle
7. Jowl of Swine
8. Omentum
9. Mesentary

From these depots fat is transported to the liver, where, with the help of Choline (lipotropic factor) it is converted into a phospholipid in which form alone lipids can be metabolised by the tissue cells. If choline is absent, the neutral fat cannot be phosphorylated and so it stays in the liver, producing fatty infiltration. The liver cells store some of the fat temporarily. So there is a regular turnover of the fat in the adipose tissue, liver and body tissues. The fats of animals are derived from the feed they take and these animal fats are characteristic for each animal. It is the liver that is concerned in the resynthesis of fat that is peculiar to each animal.

Obesity is a condition in which there is exessive accumulation of fat in the fat depots. A certain amount of 'fatty' condition is desirable in food and show animals. But excessive fat is not relished.

Causes of obesity are : 1. Excessive intake of food when there may be relative decrease in lipotropic factors leading to decreased transport of fat out of the liver. In this condition there may be increased pinocytosis of chylomicrons from the intestines.

2. Lack of exercise (especially if coupled with the first). This is seen particularly in house dogs, stabled and breeding animals and laying hens.

3. Endocrine disorders in which there is decreased rate of metabolism-hypothyroidism, hypopituitarism, deficiency of testicular and ovarian hormones (castration). Obesity is frequently found in tumors of the Pineal gland and hypothalamus; in Frohlich's syndrome; in Cushing's syndrome, adreno-genital syndrome and in granulosa-cell tumors.

There are many disadvantages of obesity, these are :

1. **Interference with mechanical function :** The fat that accumulates between the cardiac fibers seperates them so that they are not able to contract as efficiently as they can when they are aligned with each other closely. So extra work load is imposed upon the heart. Besides, the extra blood supply needed by the adipose tissue throws burden on the heart still more.

2. Fat accumulated in the omentum and peritoneum decreases the capacity of the abdominal cavity thereby interfering with the action of diaphragm and lungs resulting in dyspnoea.

3. Increased weight prevents the normal movement of the body as in old house dogs.

4. Skeletal muscles become flabby. The above factors render the animal to be easily fatigued, with a rapid pulse and polypnoea.

5. In beef cattle, obesity predisposes the cows to prolapse of the vagina.

6. Fatal hemorrhage in fat hens may occur due to rupture of liver.

In man, obesity predisposes him to diabetes mellitus, arteriosclerosis, hypertension, hypertensive cardiovascular renal disease and arteriosclerotic heart disease.

FATTY CHANGES—FATTY INFILTRATION AND DEGENERATION

Normally fat is present in the protoplasm of cells in a colloidal state in combination with proteins, carbohydrates, inorganic salts and water. This cannot be stained, while the neutral fats of adipose tissue and that brought to the liver for phosphorylation can be stained.

Fat becomes visible in hepatic cells under two conditions :

1. When the mitochondria are damaged due to irritants, the enzymes concerned in the metabolism of fats are disrupted and so the fat that should be in a colloidal state in the protoplasm, accumulates and becomes visible. (Formerly this was called Fat Phanerosis = Unmasking of fat). This condition is known as **fatty degeneration**

2. When there is some systemic metabolic derangement by which large quantities of fat are transported from the fat depots to the healthy cells of the liver, there is over-loading in these cells of fat, which the liver is unable to speedily and effectively deal with and so fat accumulates in the liver cells. So this is called **fatty infiltration**. It is true that this loading of cells with fat may cause degenerative changes and so fatty degeneration may result later.

Sometimes it is difficult to assess, by an examination of a liver section if it is fatty degeneration or infiltration. In such cases, a non–commital term "Fatty change" is used.

Fatty degeneration : This is a manifestation of cell-sickness and is found in the parenchymatous cells of the liver, kidney and heart due to some injury. As in cloudy swelling so also in fatty degeneration, the causes are (1) hypoxia and (2) poisons.

Hypoxia is seen in conditions described under cloudy swelling. In chronic venous congestion due to congestive heart failure the liver cells near the central vein become loaded with fat and degenerate finally.

Poisons : Hepatotoxic agents damage or denature the enzymes and so cause fatty degeneration. These may be, chloroform, phosphorus, carbontetrachloride, essential oils, drugs, pregnancy toxaemias, bacterial toxins, viral diseases (infectious canine hepatitis); plant alkaloids, heavy metals and their salts.

Besides the above. fatty degeneration may be seen in diabetes mellitus, acetonemia and starvation or long continued dietary deficiency of protein.

Macroscopically, the liver is enlarged, soft, friable, and paler. In shape it becomes spherical. The borders become rounded. Being soft it is cut easily and the cut surface bulges. Droplets of fat, can be seen on the blade of the knife.

Similarly, the kidney is enlarged, pale and soft. Heart is flabby. Presence of fat on the ventricular endocardium gives it a speckled appearance and so is called 'thrush-breast heart".

Microscopically, the hepatic cells contain numerous vacuoles. These are the spaces occupied by globules of fat, which get dissolved by fat solvents during the processing of tissue. It requires special technique and dyes to stain the fat. The cytoplasm of the cells is granular. The nucleus is usually not displaced and may show pyknosis (condensation) or karyorrhexis (fragmentation).

In the kidney, the tubular epithelium is mostly affected, especially that of the proximal convoluted tubules and the ascending limb of Henle. Small fat droplets are seen in these cells.

In the heart, numerous fat droplets are found within the muscle fibers, arranged in longitudinal rows.

Sequelae : Fatty degeneration is a reversible process and so if the cause is removed, the cells return to normal. But if the cause is severe, necrosis will occur. The necrosed cells are removed by autolysis and phagocytosis. Regeneration occurs by the proliferation of the surviving cells since hepatic cells have retained their capacity to regenerate. So, no permanent damage is caused. But if the cause persists for a long time, complete regeneration is not possible and the hepatic cells are replaced by fibrous tissue.

The heart which becomes weak in fatty degeneration undergoes dilatation eventually leading to chronic venous congestion and cardiac failure.

Fatty Infiltration may occur when too much fat is loaded in to the hepatic cells and when there is deficiency of choline. In man fatty infiltration is seen in heavy beer drinkers in whom there is a chronic deficiency of dietary protein including choline and methionine. The liver is the only organ that suffers.

Macroscopically, there is great enlargement of the liver, which is soft, yellow and friable.

Microscopically, the picture is very characteristic. The liver cells show one or two large vacuoles (indicating the place occupied by fat) which displace the nucleus by thrusting it to a side. (In fatty degeneration. on the other hand, the fat globules are smaller and numerous and the nucleus is not displaced).

In long continued cases, fatty degeneration may coexist

Sequelae : If the cause is not removed (as in heavy drinkers) fatty infiltration progresses to cirrhosis of the liver.

Note : It is normal to find lipid vacuoles in the adrenal cortex of most animals and in the convoluted tubules of the kidney of cats

It is now considered that fatty degeneration and fatty infiltration are different manifestations of the same condition, viz. accumulation of fat in damaged and sick cells. The size and number of droplets have no significance. They may be single or multiple; they may be small or large in the cells. This derangement of fat metabolism is now designated **"fatty change"**

In Obesity, fat may accumulate in the connective tissue cells and convert them into fat cells. This is known as **stromal fatty infiltration** and is mostly seen in the heart, pancreas and voluntary muscles in muscular dystrophy. In the heart fat accumulates under the epicardium and reaches the endocardium extending between muscle bundles and so the action of the heart is interfered with in extreme cases. The pancreas may appear to be envoloped in a mass of fat and its glands are widely seperated by fat.

Fat Stains : During the preparation of sections, fat solvents like xylol remove the fat. So it is necessary to use the freezing microtome technique by which process passage of sections through fat solvents prior to staining is avoided.

The usual fat stains used are :

1. **Osmic acid :** This stains fats black. Osmic acid being a fixative, tissues get fixed by it. After fixation paraffin blocks can be made, cut and stained in the usual manner.

2. Sudan III and IV ⎫
   Scarlach R and   ⎬ Stain fats Red.
   Oil Red O   ⎭

Lipotropic factors : Mention has already been made that for proper utilisation of fats, they should be converted into phospholipids in the liver. For this they must be conjugated with choline when one fatty acid is removed from triglyceride and phosphoric acid is linked to a nitrogenous base. So choline which aids in the transport of fat is called a lipotropic factor. In the absence of choline neutral fats accumulate in the liver and fatty infiltration results. Methionine which is the precursor of choline is also called a lipotropic factor since it is sufficient to provide methionine in the diet for the formation of choline in the body.

In pregnancy, especially with multiple fetuses, choline is sidetracked to the fetus and the mother suffers from deficiency of lipotropic factors and so fatty liver results. Similarly new-born pigs and calves deprived of colostrum develop fatty liver since colostrum is a rich source of choline. In starved animals, fatty liver may be encountered due to lack of dietary choline or methionine.

BOVINE KETOSIS
(Revised by Dr. D. Anjaneya Prasad)

Bovine ketosis also known as acetonemia, is a metabolic disorder due to impaired carbohydrate and fat metabolism. It is generally considered to be the most serious malady affecting dairy cattle, after milk fever. This condition is characterised by Ketonemia, Ketonuria, Ketolactia, hypoglycemia, and low hepatic glycogen content. Primary Ketosis, develops independently of, or prior to some concurrent disease and always occurs during the first ten days to eight weeks after calving. Secondary ketosis develops as a consequence of appetite depression or anorexia along with some other primary disease such as abomasal disorders, reticulitis, reticulo-peritonitis, mastitis, metritis, retained placenta etc. Uncomplicated and complicated ketosis are used as synonyms for primary and secondary ketosis respectively. Nutritional ketosis is due to undernutrition. Alimentary ketosis is induced by certain silages rich in butyric acid. Thus primary ketosis may be subclinical or clinical, spontaneous or nutritional and uncomplicated or complicated; but secondary ketosis must necessarily be complicated.

To understand the etiology and pathogenesis of ketosis a brief review of carbohydrate and fat metabolism is necessary. The end products of carbohydrate digestion in the rumen are acetic (60-70%), propionic (15-20%) and butyric acids (10-14%). On a normal diet of hay and concentrates acetic and butyric acids are ketogenic while propionic acid is glycogenic as it is converted into glycogen in the liver. The quantity of acetic and butyric acids is about four fold the concentration of propionic acid. This proportion is sufficient to maintain health. Propionic acid is metabolized to glucose through methyl malonyl CoA, succinyl CoA, and oxaloacetate. Oxaloacetic acid is an intermediate in the tricarboxylic acid cycle. If glucose supply to tissues is curtailed due to failure of its production from propionic acid for some reason or other, then glucose will be formed by gluconeogenesis (from fats, amino acids and glycerol) in which process large quantities of oxaloacetate will be utilised.

The ketogenic acids (acetic and butyric acids) are metabolised through acetyl CoA in the TCA cycle and electron transport system to CO_2, water and energy. If sufficient amount of oxaloacetate is not available the acetyl CoA accumulates and two molecules of acetyl CoA condense to form acetoacetyl CoA and then by deacylation free acetoacetate is formed. Acetoacetate is reduced enzymatically to hydroxy butyrate. Acetone is the decarboxylation product of acetoacetate. Thus ketosis develops.

Fatty acids can be converted into glucose if sufficient coenzyme A is present. In its absence or deficiency, fatty acids are converted into ketones.

Adrenaline is intimately concerned with glucose metabolism, especially in times of stress. Among others adrenaline has the following functions in respect of glucose metabolism : (1) The blood sugar level is increased by the conversion of liver glycogen to glucose by the action of adrenaline.

(2) The muscle glycogen is converted into lactic acid by adrenaline. And lactic acid is, in its turn converted by the liver into glycogen.

(3) Adrenaline stimulates the anterior pituitary, at times of stress (parturition, onset of heavy lactation) to liberate Adreno-corticotropic hormone, which acts on the adrenal cortex, thereby increasing its cortisone production. Cortisone stimulates the formation of glucose from fats and proteins (gluconeogenesis).

For the synthesis of adrenaline, Vitamin B_{12} is necessary. And Vitamin B_{12} is synthesized in the rumen by bacteria provided these are supplied with sufficient carbohydrates, proteins and minerals (especially cobalt, calcium and phosphorus).

The normal blood glucose level in the cow is 40 to 50 mg. per 100 ml. (In man and dog it is 100 to 120 mg.) So this low blood glucose level predisposes the animal for hypoglycemia.

With a knowledge of the above factors, we will now consider pathogenesis of ketosis :

1. **Lack of nutritive food** : Over 1 Kg of glucose per day may be needed for lactose synthesis in high producing dairy cows. The lactational drain of glucose and the high rate of gluconeogenesis are conditions predisposing to ketosis. Balance of antiketogenic and ketogenic nutrients is very important. In the first two months of lactation, cows producing 20 to 40 litres milk/day utilise 1 to 2 Kg of body fat/day and upto 0 4 Kg/day of body protein. Fat mobilization provides substrate for hepatic ketogenesis and protein provides substrates for ketogenesis as well as gluconeogenesis. Of the rumen volatile fatty acids propionate is a precursor of oxaloacetate and is antiketogenic, while butyrate is a precursor of acetyl CoA and thus is ketogenic. A reduction in propionate production and an increase in butyrate favours development of ketosis. High butyric acid silage is a sufficient predisposing factor for ketosis. If the body is not well stocked with glycogen (in liver and muscle) at parturition and when milk is secreted in large quantities after parturition, hypoglycemia develops, ketogenic fatty acids cannot enter TCA cycle resulting in formation of ketone bodies.

2. **Deficiency of Coenzyme A** : If this occurs then fatty acids cannot be converted into glucose but are converted into ketones.

3. **Deficiency of cobalt** : Vitamin B_{12} cannot be synthesized and so epinephrine is not formed. So carbohydrate metabolism is affected, gluconeogenesis does not occur and hypoglycemia may result.

4. **Adrenocortical insufficiency** : It is postulated that the adrenal cortex may become exhausted due to over activity in conditions of stress prevailing during pregnancy and parturition. In the absence of cortisone, gluconeogenesis does not occur when it is greatly needed during the onset of lactation to supply increased glucose that is necessary for lactose formation.

5. **Hypothyroidism** : This can be a contributory factor but has not been confirmed.

6. **Lack of exercise** : Usually, the ketones are oxidised by muscles. But in animals that are kept indoors with very little exercise, the ketones that may form accumulate and cause the condition.

7. Hepatic insufficiency : A healthy liver is necessary for the conversion of lactic acid to glycogen and for the storage of glycogen But if it is diseased, hypoglycemia may develop.

8. Loss of appetite : In such post–parturient conditions like metritis and severe conditions like traumatic reticulitis and abomasal displacement, the animal may develop anorexia and so may rot consume feed, hypoglycemia and secondary ketosis developing thereby In starvation propionic acid concentration of the rumen is lowered resulting in ketosis.

9. Type of food : Ensilage with its preformed butyric acid is more ketogenic than h y. Similarly high protein feeds lead to the formation of greater quantities of butyric acid in the rumen.

10. Fluorosis : In fluorosis, due to dental lesions, the animals may not eat their food. Due to lameness caused by pain in limbs they may be confined to their stalls and so do not get any exercise.

The above factors, therefore, may cause ketosis.

Symptoms : Cows with acetonemia usually exhibit ketonuria, hypolactia and anorexia. The odour of a cow with acetonemia resembles that of acetone or chloroform. The faeces are firm and covered with mucus, yet abnormally fluid faeces are not uncommon. A vulvar discharge is common. Udder is often swollen, with apparent distension of the superficial veins. Loss of condition, dehydration and coma, culminating in death are also common. Clinically hypoglycemia (25 mg% of glucose as against 50 mg% in normal animals), ketonemia (ketotic animals have as high as 50 mg% as against less than 10 mg% in normal animals , and Ketonuria (urine may contain 500 mg% as against a normal level of 0 to 15 mg%) and decreased blood calcium levels (9 mg%). The free fatty acid level in ketotic animals will be as high as 50 mg% as compared to less than 10 mg% in normal cows Serum glutamic oxaloacetic transminase activity is often increased, plasma levels of free 17–OH corticosteroids are elevated and the protein bound iodine level is decreased.

Eosinopenia, lymphocytosis and neutropenia are the other changes noticed. The reasons for the toxic symptoms are :

1. Decrease in the alkali reserve. Aceto-acetic acid and beta-hydroxy butyric acid are acids and so by combining with bicarbonate, deplete the alkali reserve resulting in acidosis.

2. Aceto-acetic acid is toxic and is responsible for coma.

3. Production of isopropyl alcohol In the rumen aceto-acetic acid may be broken down to isopropyl alcohol, which is toxic and causes nervous symptoms.

Lesions : Fatty degeneration of liver and kidney is noticed The liver may undergo even necrosis if the disease is severe and prolonged. In one study 14 cows with ketosis were sacrificed to examine the endocrine glands, and other organs. The adrenals were enlarged and flabby, with marked fatty infiltration, with partial degeneration of the cortex, but the medulla was not affected Regressive changes of the anterior lobe of the pituitary gland, involution of the thymo–lymphatic system. acute involution of the pancreas, gastro-intestinal inflammation, ulcers, nephrosis and fatty changes in the liver were observed.

Treatment : Recovery in ketosis is usually the rule if treatment is instituted promtly. No single treatment cures all affected cows. However 100% success has been claimed by some for one therapy or another.

1. About 500 ml of a 50% glucose solution given by intravenous drip or calcium borogluconate intravenously to correct hypoglycemia.

2. Peroral administration of glucose precursors viz., sodium propionate, lactates, glycerol or propylene glycol has been suggested. About 125 to 250 grams of propylene glycol or sodium propionate is recommended twice a day orally.

3. **Hormones :** One gram of cortisone given i/m or i/v and 0.5 units of insulin/kg body weight given i/m. About 200 to 600 units of ACTH is also used. Five mg of flumethasone given as one injection superseded dexamethasone TMA (15 mg)in mobilising glucogenic amino acids. Thyroxine given intramuscularly may help.

4. **Lipotropic factors :** Claims of efficacy for choline, cysteamine, L-methionine and hydroxy anologue of methionine have also been made.

5. **Vitamins :** Some workers have claimed beneficial effects with B_{12}, thiamine hydrochloride and nicotinic acid.

Prevention : By balancing feed intake with milk production, so that blood glucose level is maintained and by minimising the ketogenic materials the the cow is forced to handle, ketosis can be prevented. Suggested methods in achieving this end are 1) cows should not be excessively fat at calving time 2) Increased level of concentrates to be provided from the later part of the dry period 3) High butyric acid silage to be avoided (4) sufficient energy, protein minerals and vitamins commensurate with the cow's needs have to be provided. 5) Propylene glycol or sodium propionate could be fed at recommended levels to susceptible cows.

Pregnancy toxemia in Ewes : This condition is also a variety of ketosis occurring in late pregnancy of ewes. Those having 2 to 3 lambs are more prone to suffer. Faulty dentition, old age, bulky innutritious food, lack of exercise, starvation, exposure to cold and inclement weather and heavy hemonchous infestation are other factors in the causation of this condition.

Pathogenesis : In this condition also, the essential features are hypoglycemia and depletion of hepatic glycogen. If the food is deficient or lacking in sufficient carbohydrate liver glycogen cannot be formed. Normally, as in the cow, the glucose level of the blood is low (40-50 mg %). In animals having more than one fetus, the glycogen of the mother is diverted to the fetuses, which store glycogen for their use, thereby depriving the dam of its hepatic store. The place of glycogen in the liver is taken up by the fat from depot fat. Again, the maternal choline (a lipotropic factor) is side tracked to the fetus and so the dam suffers from choline deficiency. Choline is necessary for the formation of phospholipids. In its absence liver gets loaded with fat. The normal fat content of liver may be 3% to 4% while in this condition it may be as much as 35% by weight.

This excess fat is oxidised to ketones, which would normally have been used up by the muscles. But if the animals are not exercised, the ketones accumulate, resulting in ketosis. Ketosis causes acidosis, ketonemia and ketonuria.

Symptoms : There is depression, sleepiness and coma. Mortality may be heavy.

Lesions : The liver is highly fatty—being yellow in color and ver friable. Advanced pregnancy with multiple fetuses is observed. Subepicardial petechiae are noticed.

Clinical : Urine reveals ketone bodies by Rothera's test. (Ketonuria)· Plasma cortisol levels increased probably due to the failure of liver to metabolise cortisol.

Control measures must be directed towards provision of nutritious food and sufficient exercise during pregnancy.

NECROSIS AND SOMATIC DEATH

Local death of cells or tissue in a living body is **necrosis.** The death of the body as a whole is called **Somatic death.** The changes that take place gradually in the cells while they are dying are known as **necrobiosis.** This is also called physiological death. This term is not used now.

Necrosis of cells is usually preceded by various degenerative changes studied earlier, viz , cloudy swelling, fatty change and hyaline degeneration. It is not essential, however, that all cells should pass through these degenerative changes before becoming necrosed, for a severe irritant can cause necrosis directly.

Causes : 1. Poisons.

a) Chemical Poisons : These may be drugs, strong acids and alkalies, insecticides, fungicides and other toxic chemicals like lead arsenate (an insecticide) phenol, mercuric perchloride, etc These cause death of cells either by coagulation of cellular constituents or by poisoning the enzyme systems.

b) Poisons of Pathogenic microorganisms : The toxins produced by bacteria, viruses, fungi, rickettsia, protozoa and metazoan parasites may cause necrosis, for example the exotoxin produced by the bacillus producing diphtheria inhibitis oxidative processes and so protein synthesis in the cell stops resulting in the death of the cell.

The viruses and rickettsia being intra-cellular parasites, deprive the cells of their essential nutrients and so the metabolism of the cells is affected resulting in their death. The viruses may reproduce themselves so rapidly that the cells die. Or the viral RNA and DNA may combine with the cellular RNA and DNA so that the cell dies. The toxins may cause thrombosis of the vessels leading to ischemia and subsequent necrosis. Or the toxins may be directly lethal to the cells. Besides, they may particularly be toxic to tissues which become sensitive (allergic) to the foriegn protein of the toxins.

c) Plant poisons : Alkaloids produce necrosis. For example, plants of Senecio species in large amounts are hepatotoxic and produce necrosis of liver cells. Mushrooms contain a toxic glycoside, phallin, which causes renal tubular necrosis

d) Toxins produced within the body : Degeneration of cells liberates certain toxins which are injurious. For example, in severe burns, necrosis of eiver cells occurs. Uremia and icterus are toxemias of endogenous origin.

e) **Animal poisons** : Necrosis of skin occurs when cantharidin from cantharides beetles is applied. It is common-day experience to see local necrosis in bee stings.

2. **Loss of blood supply** : If oxygen supply and nutrition to tissues are cut off by blocking blood supply, necrosis results. Even if blood supply is restored subsequently, some cells may die. The time necessary for the deprivation of blood supply to cause death of the cells varies with cells. For example, while the tubular-epithelial cells of the kidney die due to lack of blood the more sturdy connective tissue cells survive. The following conditions may cause necrosis.

a) **Passive hyperemia** : In persistent passive hyperemia with sluggish flow and deficient oxygen and nutrition, when the oxygen and nutrition available are used up, necrosis results, eg. volvulus, torsion of intestines and strangulated hernia.

b) **Ischemia** : Decreased blood supply to a part is produced by
   i) Thrombosis and embolism;
   ii) Compression of the artery by tumor, ligature, tourniquet, abscesses, cysts :
   iii) Volvulus, intussusception of intentines and
   iv) Ergot poison which causes contraction of the smooth muscle of arteries.

3 **Mechanical injuries** : Produce necrosis by crushing and cutting off of blood supply.

4. **Physical Agents** : Excessive *heat* may denature and coagulate the protoplasm while excessive *cold* kills tissue cells by stopping their enzymatic activity and changing their colloidal constitution. Ice crystals that may form tear the tissues on thawing. Frost bite causes thrombosis of blood vessels resulting in necrosis of the affected parts. *Electric current, x-rays, ultra violet rays* and *radiation* may also coagulate the cytoplasm or cause ionisation and so produce death of cells.

Macroscopically, it is easy to recognise a necrotic area since it stands out distinctly by its change in color. The tissue is usually paler in color and has a 'cooked appearance'. The necrotic area is usually surrounded by a red zone of hyperemia. It loses its tensile strength and so is soft and friable. It is easily torn or broken.

If pus forming organisms are present, abscess may be present. But if putrefactive bacteria invade the area foul smell is obtained and the area turns green or black. (Gangrene).

Microscopically, the changes that occur are nuclear and cytoplasmic.

Nuclear changes : These are more striking.

(a) **Pyknosis** : The nucleus becomes smaller, rounded and condensed. Becoming homogeneous, internal structure is lacking and nucleoli are not visible. It stains dark due to liberation of nucleic acids by hydrolysis of nucleoprotein. Nucleic acid stains deeply with basic stains.

(b) **Karyorrhexis** : Here fragmentation of the nucleus occurs. This may be preceded by *karyoschisis* wherein cracks appear and at these places the nucleus may break. The fragments of the nucleus may be scattered in the cytoplasm. This is found in caseous necrosis

(c) Karyolysis : This is dissolution of the nucleus. The nuclear material may disappear leaving the nuclear membrane, which may also disappear later on.

(d Chromatolysis : is the disappearance of nucleolus, the chromosomes and other stainable material of the nucleus.

Changes in the cytoplasm : The cytoplasm is swollen, becomes homogeneous and stains more pink. The acidophilia of the cells is attributed to the decreased basophilia that results as a consequence of disorganisation and lysis of ribosomes by the action of the lysosomal enzymes of the cells themselves (autolysis) as well as to the action of the polymorphs that emigrate (heterolysis) to the necrotic area. The normal cytoplasmic structure is lost. The normal sharp contour of the cells cannot be seen and the cell outlines disappear. Muscle cells lose their characteristic striations, swell and become hyalinised (Zenker's degeneation).

Types of necrosis :

1. Coagulation necrosis : In this type though the cells have become necrosed and their cellular details cannot be distinguished yet the architectural detail of the tissue is preserved (eg. glomeruli of kidneys).

This type is seen in infarcts of spleen and kidney (due to deprivation of blood); in the muscles in "white-muscle disease", in the kidneys of animals poisoned by mercury, in necrotic livers of cattle infected by *Spherophorus necrophorus* and in places (skin, mucosa) where phenol is applied.

Macroscopically, the necrotic tissue is dry, homogeneous and white or grey. It may be a little depressed than the healthy area.

Microscopically, the cytoplasm has no structure. There is coagulation of the protein of the cytoplasm by the intracellular enzymes. Probably lymph may be coagulated, the mechanism being similar to clotting of blood. Since autolytic enzymes are destroyed, the necrotic material is liquefied only very slowly. Around the necrotic area, in the healthy tissue, an inflammatory reaction may be seen.

Sequelae : Gradually, the infarcted area may be liquefied by the proteolytic enzymes of leucocytes and absorbed. Sometimes calcification may occur

2. Caseation necrosis : This is seen in tuburculosis in animals and caseous lymphad nitis in sheep in which the toxins act for a long time.

Macroscopically, the tissue is dry, Cheese-like and granular. The color is white. grey or yellow. Calcium salts may be deposited in the caseous materia due to its high fat content.

Microscopically, all details of structure are lost. In the hematoxylin and eosin s ained sections, caseous material takes a faintly blue color. Since caseous material is not chemotactic, leucocytes are not present and so liquefaction and absorption do not occur. Calcification is identified by the presence of blue staining granular material.

Sequelae : The caseous material may stimulate the fibroblasts and so a fibrous capsule may be formed around the lesion, which persists for a long time.

3   Colliquative or Liquefaction necrosis : Liquefaction of a tissue may occur in *two* ways (1) by the action of intracellular autolytic enzymes and (2) by the proteolytic enzymes of the leucocytes. The first method occurs in the central nervous system, which is high in lipid and water content, but low in coagulable albumin. It may occur in (a) areas of infarction and injury by trauma.

(b)   cyanide poisoning, hypoxic states and carbon monoxide poisoing,

(c)   crazy chick disease—Vitamin E deficiency in fowl.

(d)   thaiamine deficiency in the cat and

(c)   mouldy corn poisoning in horses.

The second method of liquefaction (by proteolytic enzymes) occurs in absceses and in places where positively chemotatic agents (turpentine) are injected. (Neutrophils accumulate and these liberate the proteolytic enzymes). Larvae of *Oesirus ovis* may also produce liquefactive necrosis of the brain in sheep by liberating proteolytic enzymes.

Macroscopically there may be liquefied necrotic material which may be watery or tenacious, white or yellow. In long standing cases, a fibrous capsule may be present. Sometimes the fluid may be absorbed leaving a cavity. The lung lesions in tuburculosis may break down, be liquefied and absorbed. leaving cavities (Vomicae in man and dog).

Microscopically, a cyst-like space is visible since the liquefied material is usually lost during processing of the tissue. In the abscesses, the pyogenic membrane is visible.

4.   Fat necrosis : Necrosis of fat in adipose tissue may be of two types : (1) traumatic fat necrosis and (2) enzymatic fat necrosis.

Traumatic fat necrosis : This occurs usually in the subcutaneous adipose tissue that is exposed to external trauma—fighting, biting etc (breasts in ladies) and sites of insulin injection in diabetes.

Mechanical injury during parturition may traumatise the fat around the vagina (perivaginal fat) in cows.

The traumatic injury ruptures the fat cells, which while dying (Traumatic necrosis) release their fat content, which incites a foriegn-body reaction in which macrophages engulf the fat. Foreign body giant cells may be seen. This inflammation lasts for a long time and is seen as a nodule.

Macroscopically, the necrotic fat appears as a white, opaque firm mass. Traumatic necrotic perivaginal fat can be felt as masses while examining per rectum.

Enzymatic fat necrosis : This is due to the action of pancreatic enzymes on the surrounding adipose tissue when they escape out. This can be brought about by injury to the pancreas. Other causes for escape of the enzynes are regurgitation of the bile up the pancreatic duct, infection or infarction of the pancreas, and obstruction of the pancreatic duct by neoplasms.

The lipase hydrolyses the fats, liberating glycerol and fatty acids. Fatty acids combine with calcium, potassium and sodium to form soaps, which are visible as white chalky material. Spilling of pancreatic juice into the peritioneal cavity causes necrosis of adipose tissue of the omentum and mesentary with the deposition of white chalky masses. Transport of the lipase by lymphatics to distant places like the mediastinum may cause fat necrosis there.

Macroscopically, the necrotic areas are characterised by the presence of white chalk-like masses in the pancreatic, omental and mesenteric fat. Around the masses an inflammatory reaction is often seen with macrophages, giant cells and a fibrous capsule.

Microscopically, lime salts may be deposited on the masses. The fibrous tissue nearby may be converted into bone (metaplasia).

Note : After death, lipase may be liberated from the pancreas and cause fat necrosis. But as this takes place after death, when circulation has stopped, there is no leucocytic infiltration around the lesion. Ante-mortem fat necrosis on the other hand is accompanied by an inflammatory reaction around it.

Autolysis : It is pertinent to discuss this subject here. Cells contain some enzymes, which after death, produce certain changes in the cytoplasm and nucleus. The enzymes which are acid hydrolases leak out of the lysosomes which become either more permeable or may even rupture due to anoxia that occurs on cessation of circulation. That autolysis is caused by intracellular enzymes can be proved by a simple experiment. Take two pieces of fresh liver. Boil one piece for a few minutes. Insert both the pieces into the abdominal cavity of an animal. The boiled piece, in which the enzymes are destroyed remains intact for months and the nuclei can be well stained. But on the other hand the unboiled piece undergoes changes similar to those of necrosis. These enzymes convert the highly complex organic structures to simple inorganic compounds like water, hydrogensulphide, carbondioxide, nitrites etc. The changes produced in the cells (cytoplasm and nucleus) are similar to those produced in necrosis but more exaggerated. Nuclei disappear and there is loss of cellular details. Autolysis is most rapid in those organs in which metabolism is very active—liver, kidneys, adrenals and bone marrow. Autolysis of skin and brain occurs very slowly. For a pathologist it is important to distinguish between autolysis and necrosis. The following features are helpful.

1. In autolysis, the whole section will be a uniform dead tissue. But in necrosis, one can see, side by side, both living and dead tissues.

2. In autolysis there is no inflammatory reaction while usually, around a necrotic area a zone of inflammatory reaction is seen.

GANGRENE

Necrosis of tissues with putrefaction by saprophytic bacteria is called gangrene.

Causes : Gangrene is seen most often in the lungs, intestines, mammary glands, large muscles (of the shoulder and thigh) and extremities.

1. **Lungs :** (i) Gangrene occurs most commonly due to faulty drenching of medicines. The irritant may pass directly to the lung and cause necrosis which is cotaminated by air-borne saprophytes or the drench itself may contain the bacteria and gangrene results.

ii) Food may pass into the trachea in paralysis and infectious disease of the throat and produce gangrene.

iii) In the horse, careless passsage of the stomach-tube into the trachea allows the introduction of medicines into lungs with resultant gangrene.

iv) In severe infections of the lung, bacteria may cause necrosis which is subsequently infected by air-borne saprophytes.

2. Intestines : (i) Gangrene of the intestines occurs in the malpositions—volvulus and intussusception. The resulting acute passive hyperemia causes necrosis which is invaded by the saprophytes already present in the intestines.

ii) In the horse, in infection by *Strongylus vulgaris* which produces aneurysm and thrombosis of the anterior mesenteric artery, infarction of the bowel occurs which is invaded by saprophytes producing gangrene.

3 Extremities : Low temperatures (freezing, frost-bite) cause thrombosis of the vessels and so necrosis of tips of ears, snout, feet, tail and wattle ensues. Infection by putrefactive organisms results in gangrene.

Ergot, which causes contraction of the smooth muscle, produces constriction of the arterioles of the limbs, thereby causing ischemia and necrosis which with infection become gangrenous subsequently. Ergot poisoning occurs when grasses on which it grows are eaten by animals.

Senile gangrene occurs in man in old age due to arteriosclerosis, in which narrowing of arteries causes ischemia. Diabetic gangrene occurs due to narrowing of arteries. Bacteria grow well in the tissues due to their high sugar content. Thromboangitis obliterans in man is again another cause for gangrene of the extremities since blood supply is arrested due to the occlusion of the arterioles.

Acids which coagulate the fluids cause dry gangrene (see below), while alkalies which liquefy the tissues cause moist gangrene.

4. Mammary gland : In mastitis, caused by staphylococcus, necrosis occurs either due to toxins or due to thrombosis of the mammary vessels (infection spreading to the vessel walls). Subsequent infection by saprophytes results in gangrene.

In tissues that are poor in moisture and from those in which evaporation of moisture is possible (eg. extremities and surfaces of body) the gangrenous tissue becomes dry and so is called *dry gangrene* in contrast to the very moist condition of the gangrene of internal organs (lungs and intestines), which is called *moist gangrene*

Dry gangrene : Macroscopically, dry gangrene is seen in the extremities. When necrosis occurs, there is no longer circulation of blood and the necrotic tissues which are exposed become dehydrated due to evaporation of moisture. For the growth of microorganisms two conditions are essential : (a) moisture and (b) optimum temperature. Since circulation has stopped and since evaporation occurs, the necrosed area is dry and cold and these conditions are not conducive to the growth of organisms. So saprophytes that invade the area grow very slowly Hence spread of the gangrenous area is very slow.

The part is cold to the touch, the skin has a leathery feel and the tissues are shrivelled or mummified. The red blood cells are lysed liberating hemoglobin which combines with hydrogen sulphide produced by the putrefactive bacteria forming the black iron sulphide which imparts a greenish or blackish color to the gangrenous area. The tissues emit a foul odour.

The gangrene extends slowly in the direction of the body upto a point where sufficient blood circulation is present to keep the part alive. At this site, the dead tissue is demarcated from the living by a line of inflammatory granulation tissue. The neutrophils and macrophages in this inflammatory zone produce lysis of the necrotic tissue which is therefore slowly seperated and natural amputation occurs.

Microscopically, a structureless necrotic area, stained pink is seen with numerous bacteria. A few gas bubbles are evident by clear spaces.

Moist gangrene : This variety is seen in internal organs where conditions are conducive for the rapid growth of organisms, viz. abundant moisture (which does not get evaporated) and optimum temperature (the part is kept warm by the surrounding organs). Under these conditions the bacteria multiply fast and spread rapidly to the adjacent tissues. The dead tissues are broken down and liquefied by the organisms with the production of protein degradation products—indol and skatol, to which the foul odour is due.

Macrocopically, the necrotic tissue is greenish or black in color due to the formation of iron sulphide from hemoglobin released from hemolysed erythrocytes. Foul smell is present. Since there is rapid spread, no demarcation is visible between the dead and living parts. In the case of the bowel, the putrifactive process may cause rupture of the bowel wall with infection spreading to the peritoneum and other viscera.

Sequelae : In the moist gangrene toxins produced by the disintegration of tissues as well as by the organisms are absorbed in large quantities resulting in fatal toxemia. The condition in which there is circulation in and intoxication of the body by the products liberated by the growth of saprophytes in the necrotic tissue is called **sapremia.** In the case of intestinal gangrene death may occur due to septicemia, peritonitis or shock.

Gas gangrene : Many anaerobic spore-forming bacteria of the Clostridium group can cause necrosis of tissues and then grow in them. So they not only cause necrosis but cause putrefaction of the tissues as well. Therefore they are both pathogenic and saprophytic organisms. The peculiar features of these Clostridia are : (1) that they are normally found in the soil, in the intestines and probably also in tissues, especially muscles and (2) that they produce large amounts of gas.

Gas gangrene used to be a very common complication of war wounds in man. The organism responsible is *Clostridium welchii.*

In animals, the organisms responsible for gas gangrene are *Clostridium chauvoei* (causing *black quarter*), *Clostridium septicum* (causing *malignant edema*) and *Clostridium novyi* (causing *black disease*). These may enter wounds caused by docking, shearing, castration, ear notching or through the needle in injections. Trauma may cause necrosis of the tissues in which the anaerobes grow, multiply and through their powerful exotoxins kill the adjacent tissues, which are invaded by these bacteria and thus the organisms spread throughout the body. The organisms produce gas which accumulating in the tissues gives a characteristic crackling noise when pressed. The reason for the thigh and shoulder muscles to be more often affected is that these are more prone to trauma.

Macroscopically, the affected muscles are black in color, emit a foul odour and show evidence of gas. On section a serosanguineous, foul smelling fluid is found to exude.

Microscopically, the muscle cells appear ruptured. The sarcolemma is separated from the fibre, which dies as it is deprived of blood supply. The necrotic tissues are edematous, with non-specific inflammatory reaction. The edema fluid as well as the necrotic tissues show numerous rod-shaped, gram-positive spore-containing organisms

Results of necrosis Necrosis of different tissues due to various causes have different terminations. These may be summarised as under.

1   Liquefaction and removal : Liquefaction of the necrotic tissue is brought about by the proteolytic enzymes liberated from the neutrophils when they are destroyed. This fluid material is subsequently removed by blood and lymph. The complete digestion and removal is possible if the necrotic area is small But if larger this is not possible and the following processes take place.

2.   Liquefaction and formation of cyst : This occurs sometimes if the liquefied material is formed too fast to be removed and so accumulates forming a cyst, around which a fibrous capsule is formed.

3.   Liquefaction and abscess formation : When necrosis is due to pyogenic bacteria, neutrophils are attracted to the area. These with their enzymes liberated after their death produce liquefaction of the dead material to form pus, which may be discharged on the surface when the superficial epithelium sloughs after becoming necrotic This is how the abscesses and boils open up. To relieve pain and to provide drainage the surgeon may lance the abscess.

4.   Encapsulation without liquefaction : In tissues which do not have liquefying enzymes, the necrotic tissue acting as a foriegn body, incites an inflammatory reaction whereby there is infiltration of leucocytes and the fibrous tissue proliferates forming a capsule. Thus the encapsulated necrotic material may persist for years.

5.   Sloughing or desquamation : The necrotic tissue, if present on the skin or intestines, may be cast off (sloughed off). Similarly necrotic epithelium of ducts and tubules may be sloughed into their lumens.

6.   Organisation or replacement by scar tissue : The necrotic tissue may be invaded by fibroblasts, capillaries and leucocytes. The leucocytes liquefy and remove the necrotic tissue while new vascular fibrous tissue takes its place. This is known as **organisation.**

7.   Calcification : Calcium salts may be deposited on the necrotic tissues since conditions for such a deposition are obtaining in the necrotic material, viz., dead tissue, deficient circulation and decreased carbon dioxide tension in the area.

8.   Gangrene : If necrotic areas are exposed, saprophytic bacteria may invade and cause gangrene. The outcome of gangrene will be fatal if internal organs are affected.

9.   Death of the individual if extensive necrosis occurs in vital organs such as liver, kidney or heart

10. **Regeneration** : Depending on their nature, the cells, lost or necrosed, can be, replaced by the proliferation of the adjacent surviving cells. The ability to regenerate differs with different tissue cells. Neurones have lost their ability to proliferate and so when once dead they are lost for ever. On the other hand, the fibroblast and the hepatic cell have retained their regenerative capacities.

SOMATIC DEATH

When respiration and cardiac action have stopped, the animal is said to have undergone somatic death. After death, the cells undergo certain changes (postmortem changes) which a pathologist must have knowledge of to distinguish them from lesions found in disease. By a careful study of the postmortem changes one can determine the probable time of death and this is of great importance in medicolegal cases.

Cooling of the body *(algor mortis)* commences at or before the stoppage of blood flow. The rate of cooling depends on the following factors; (i) external atmospheric temperature; (ii) air currents: (iii) the thickness of hair coat or wool; (iv) adiposity of the animal. (v) amount of fermentable ingesta in the digestive tract. Larger animals cool slowly; so also in sheep, with thick wool cooling occurs slowly. Limbs and other extremities cool more rapidly than the trunk.

The rate at which postmortem changes take place depends on the rate of cooling and other factors detailed below.

1. **Surrounding atmospheric temperature** : Since the postmortem changes are brought about by enzymatic and bacterial activity, high temperature that accelerates this activity will naturally bring on the postmortem changes soon. So in summer, the carcases putrefy quickly. Cold on the other hand retards the enzymatic and bacterial activity. Freezing and deep freezing may stop that activity completely. Hence is the perfect state of preservation of carcases under polar ice-caps for considerable length of time.

2. **State of the body at the time of death** : Higher the temperature at death, sooner do postmortem changes commence.

3. **State of muscular activity of animal prior to death** : In animals that have been very active prior to death postmortem changes commence quicker. This is found in animals that die in chase. Similarly, animals that are killed or die of strychnine poisoning and in animals that die of tetanus, postmortem changes appear early. The reasons are (i) higher body temperature and (ii) greater production of lactic acid in muscular contractions and exercise.

4. **Size of animal** : Since body cools slower and so heat is retained longer in larger animals, post mortem changes appear quicker in them.

5. **External coverings** : Since thick hair or wool retard heat dissipation, postmortem changes are seen sooner in thick haired or coated animals.

6. **Fatness of animals** : Fat is a poor conductor of heat and so heat loss in fat carcases is slow, with resultant speedier onset of postmortem changes.

7. **Infection of animal** : Widespread bacterial infection, especially septicemic in character, at the time of death brings on postmortem changes earlier.

The following are the changes noticed after death.

1. Rigor mortis : This is the contraction of muscles after death so that the joints become stiff and the body is rigid. Rigor mortis develops first in those muscles that are very active, eg. heart, palpebral muscle, muscles of the head and neck. Gradually other muscles of the fore limbs, the trunk and the hind limbs, are affected in that order. It passes off also in this order, starting first in the head. Usually, rigor mortis appears in 1 to 8 hours after death and may disappear from 20 to 30 hours. The following factors hasten the onset of rigor mortis. (a) high atmospheric temperature. (b) active exercise—hunting, fighting, racing or strugging; (c) strychnine poisoning or (d) tetanus.

Causes of rigor mortis : The exact mechanism is not known though many theories have been propounded. After death, there is a great turnover of high energy phosphate bonds in the muscles. Adenosine triphosphate which breaks down is resynthesized by the energy derived from glycolysis. So long as ATP is present rigors do not occur. With the exhaustion of glycogen, all of ATP is degraded and rigors occur, since in the absence of ATP relaxation of muscles cannot occur. For the relaxation of muscle to occur, a considerable quantity of ATP must be absorbed to the muscle proteins. Hence onset of rigor is delayed in well-fed animals with large quantities of stored muscle glycogen. But in starved animals rigor naturally commences earlier. Subsequently when there is no longer any energy necessary for keeping up the chemical activity in the muscle fibers, rigors pass off.

Onset of rigor mortis is slow in cold weather and in emaciated and cachectic animals. In the latter it is due to the complete exhaustion of chemical systems producing energy.

2 Postmortem staining : Hemolysis occurs after death and the liberated hemoglobin stains the endothelium of large vessels pink. Hemolysis is hastened by infection with hemolytic microorganisms, in which case there may also be ante-mortem staining of the vessels.

When putrefaction occurs, hydrogen sulphide is liberated. This is especially the case in intestines where the hydrogen sulphide combines with hemoglobin to form black iron sulphide staining the intestine black—*Pseudomelanosis coli*. All tissues may thus be stained black or green ultimately.

3. Postmortem softening : After death the tissues are softened by the action of autolytic enzymes of the cells and the proteolytic ferments of the saprophytes and infecting bacteria. Therefore softening occurs earlier in animals that have died of septicemic diseases, since in such animals the putrefactive bacteria invade the body even before death. The tissues are subsequently liquefied.

During this process, the wall of the stomach or intestine may become so thinned that rupture may occur. But this can be distinguished from ante-mortem rupture in the absence of inflammatory reaction on the borders of the postmortem rupture.

Postmortem softening and liquefaction of the pancreas may release lipase which acting on adipose tissue may cause fat necrosis This is distinguishable from the antemortem fat necrosis in that the postmortem variety does not have leucocytic infiltration.

Mucosa of the intestines peels off easily due to postmortem autolysis.

4. **Postmortem clotting of blood :** After death the endothelial cells of the intima of blood vessels become degenerated due to lack of oxygen and thromboplastin is liberated. Similarly disintegration of leucocytes also liberates thromboplastin which clots the blood.

In slow death when there is opportunity for stagnation and subsequent clotting, heavier red cells are found at the bottom, the lighter white cells and platelets above it and on top is the serum. So the clot containing red cells is red while other parts of it are white. The white clot is called the *'chicken fat'* clot while the red *'current jelly'* clot.

Postmortem clot is soft and elastic and does not get attached to the endothelium which is smooth. But the antemortem clot (thrombus) is friable, inelastic and attached to the endothelium, which becomes roughened when the clot is peeled off.

In anthrax no clot forms because fibrinolysin produced by the bacteria liquefies the fibrin. In sweet clover poisoning clotting does not occur since prothrombin activity is inhibited.

5. **Hypostatic congestion :** In the carcase due to gravity, blood accumulates in vessels that are found on the side nearest to the ground (that is on the lower side). This is called hypostatic congestion.

6. **Postmortem bloat :** Gas may accumulate in the rumen and intestines due to fermentation of food after death. This may be distinguished from the antemortem bloat by the absence of congestion in other abdominal viscera in the postmortem variety.

7. **Postmortem imbibition of bile** occurs in tissues around gall bladder due to their absorbing the bile.

8. **Postmortem displacements :** The intestines may be displaced after death due to rolling of the carcase or by the accumulation of gas. Those displacements can be differentiated from the antemortem displacement (like torsion or volvulus) in that in the latter acute passive congestion is noticed but none in the former.

In equine practice, stud fee is payable only on the birth of a live foal. So, the veterinarian may be required to certify as to whether a foal was born alive or dead. The two criteria to be looked for are (1) does the lung float in water? If it floats, the foal was born alive since presence of air renders the lung buoyant. Air can be present in the lung only if the animal had breathed and breathing can occur only if the foal was born alive.

(2) Did it suckle? Presence of milk or curds in the stomach is valid evidence that the foal was alive at birth and had suckled.

CHAPTER 4

INFLAMMATION

By Dr. J. L. Vegad

Causes of inflammation	Classification of inflammation
Cardinal signs of inflammation	Various acute exudative inflammations
Vascular phenomena	Terminology of inflammation
Chemical mediators	Chronic inflammation
Cellular phenomena	Allergic inflammation
Inflammatory exudate	Viral inflammation
Cells of exudate	Rickettsial inflammation
Function of exudate	Granulomatous inflammation
Functions of the fibrin	Fate of inflammation

Inflammation may be defined as the reaction of the living tissue to injury or as the reactive process which begins following a sublethal injury to tissue and ends with complete healing. Healing, which is the end result of inflammation, is a part of the dynamic process and not a distinct entity in itself.

Inflammation is fundamental to the survival of the organism. It cannot be over emphasized that without it there could be neither protection against the effects of noxious external stimuli nor repair of damaged tissue. Inflammation is, therefore, on the whole, a beneficial process. However, recently it has been realized that inflammation may, at times, wander away from its beneficial path and may become considerably more harmful to the body than the noxious stimulus which initiated the reaction. Allergic and rheumatic diseases fall into this category.

Inflammation can be provoked by any noxious stimulus, called an irritant. The reaction is complex response and involves a series of events rather than a single event.

The **purpose of inflammation** is to destroy and remove the irritant, and to repair the damaged tissue.

The **causes of inflammation** are :

1. **Pathogenic organisms :** These include bacteria, viruses fungi, rickettsia, protozoa and parasitic metazoa.

2. **Chemical poisons :** These are of endless variety, and include acids, alkalies and other poisons.

3. **Mechanical and thermal injuries :** These include burns by heat, electricity, light or other radiant energy such as X-rays, Excessive cold and injuries caused by trauma are also included.

4 **Immune reactions :** Inflammation is associated with antigen-antibody interactions, which occur under various circumstances. These include allergic diseases. delayed hypersensitivity, the Arthus reaction, serum sickness and certain of the autoimmune diseases

The cardinal signs of inflammation were first described by Celsus in the first century A. D. He described four, a fifth was added by the Greek physician, Galen (A. D. 130–200). They are the clinical signs which characterize inflammation, and are :

1. **Redness (Rubor) :** This is due to a great increase of blood in the inflammed area as the result of hyperaemia.

2. **Swelling (Tumor) :** This is due mainly to hyperemia. There is more blood in the inflammed tissues. Blood has volume and therefore the part becomes larger than normal. Secondly, humoral and cellular substances exude from the inflammed blood vessels into the surrounding tissues (called exudate), and contribute to swelling.

3. **Heat (Calor) :** At the site of inflammation there is increased heat. This also results from the increased blood flow through the area, which carries warmth to the periphery from the higher interior temperature of the body. In addition, since in inflammation the rate of metabolism is increased, there is greater production of heat.

4. **Pain (Dolor :** The inflammed area is painful. Pain occurs as the result of increased pressure upon and injury to nerve endings. Histamine, 5-Hydroxytryptamine and kinins (bradykinin) are released following injury, and produce pain by stimulating the pain fibres directly. Pain is also caused by stretching of tissues when exudate accumulates in the area of inflammation. The accumulated potassium (which escapes out of the cells) and the increased osmotic pressure may also cause pain.

5. **Loss of function (Functio-laesa) :** This is due partly to mechanical swelling and partly to destruction of tissues. For example, in lesions of kidney, with damage to renal parenchyma, anuria may develop.

TISSUE ALTERATIONS IN THE INFLAMMATORY PROCESS These can be placed into two groups—the circulatory changes and the cellular events.

CIRCULATORY CHANGES IN INFLAMMATION

The circulatory or vascular changes were first studied by Cohnheim in 1877. He spread the mesentery of a curarized frog across the stage of a microscope and observed the flow of blood through the vessels, following the application of a drop of dilute acetic acid. Cohnheim's original description is still fully applicapable today. Based on his observation the circulatory changes are :

1. **CHANGES IN THE BLOOD VESSELS :**

(a) **Momentary constriction :** Immediately upon application of the irritant to the tissue, the blood vessels are constricted. This is considered to be due to a kind of stimulant action of the irritant before its full effect is felt. The constriction of vessels is very short-lived, and is therefore of not much consequence.

(b) **Dilatation :** The momentary constriction of the vessels is quickly followed by their dilatation, which occurs in arterioles and venules, and not capillaries. Dilatation is caused by the vasodilatory nerve impulses and the local vasodilatory action of substances, such as histamine, which are formed at the site of inflammation. These substances are known as **chemical mediators of** inflammation. Vasodilatation leads to hyperaemia or increased blood flow. A number of capillaries previously dormant or collapsed, now fill up and may even get engorged.

(c) **Increased vascular permeability** : Along with the vasodilatation, there is also increased vascular permeability. The endothelial wall of capillaries and venules forms a semipermeable barrier that allows free movement of small molecules but normally restricts the passage of plasma proteins. A cardinal feature of the inflammation is the striking increase in the permeability of these vessels to plasma proteins. This is referred to as increased vascular permeability. Because of this alteration in the wall of the vessel, plasma constituents, leucocytes and erythrocytes, pass through the wall of the vessel into the surrounding tissue. Normally the endothelial cells of capillaries are fused by tight intercellular junctions In inflammation these are loosened to permit out flow of fluid and protein. These cells contain myofibrils and myosin which make the cells to contract thereby providing intercellular passages. The exudative stage is thus instituted. The chemical mediators of inflammation (to be discussed shortly) are the principal factors in increasing permeability. Increase in blood pressure and stretching of vessel walls also contribute to increased permeability.

2. CHANGES IN THE RATE OF FLOW :

(a) **Acceleration of the blood flow** : The vasodilatation results in increased blood flow, but is soon followed by,

(b) **Retardation of the blood flow** : This change is essential for emigration of the leucocytes. Retardation is accomplished in four ways : (i) **by increasing the capillary bed** in the area. When a large number of hitherto collapsed capillaries are suddenly opened and when blood has to flow through them, the rate of blood flow naturally diminishes. (ii) by the **swelling of the endothelial cell**s lining the capillaries. As the endothelial cells swell and increase in their diameter, they project into the lumen, offer resistance to the flow of blood, and so the rate is diminished (iii) **Haemoconcentration** which occurs following the passage of plasma out of the blood vessels increase the viscosity of the blood, and this leads to further retardation of the flow. (iv) **Margination of the leucocytes** : Following a marked reduction in the flow, the leucocytes adhere to the endothelial wall, narrow the lumen and their presence on the vessel increases the roughness of the endothelial surface. This increases the peripheral resistance of the vessel wall and thereby retards the rate of flow.

(c) **Stasis** : When the above factors markedly reduce the flow, blood barely moves through the vessel, and stasis is produced. This situation is ideal for the escape of molecular and cellular elements essential for the formation of inflammatory exudate.

CHANGES IN THE BLOOD STREAM :

The main change consists of a **redistribution** of the cellular elements of the blood stream Normally in the blood stream of a vessel two distinct zones can be seen. In the centre are found the cellular elements (erythrocytes and leucocytes). This part is called the **axial stream**. The cellular elements are held in the centre by the centripetal force of the flowing blood. External to the axial stream is the **plasmatic stream, a clear zone** consisting mainly of plasma, which is in contact with the wall of the vessel. As the blood flow slows down, the centripetal force of the blood stream is overcome by the certrifugal force, and the leucocytes tend to drop out of the axial stream. They then come into the plasmatic zone, drag themselves along the vessel wall and finally adhere to the endothelium. This adherence of the leucocytes to the wall is called 'pavementing' or **margination of the leucocytes**. This change is essential for their emigration. Pavementation is prevented or inhibited by adrenal steroids and so these are anti-inflammatory.

4. **Exudation of plasma :** Following increased vascular permeability, the fluid part of the blood exudates into the inflamed area. The accumulated plasma outside the vessel is known as an **inflammatory exudate**. As we shall see later, its formation is most beneficial to the body.

5. **Emigration of the leucocytes :** The process of leucocytes moving outside the vessels is known as emigration of leucocytes. Leucocytes, by means of their power of amoeboid movement, migrate through the narrow spaces between the endothelial cells of the vessel wall and proceed towards the irritant. The force which attracts them into the inflamed tissues is called **chemotaxis,** a chemical attraction. Leucocytes are needed in the area of inflammation to phagocytose the irritant (especially bacteria).

It is undisputably established by experiments using Boydin Multipore Chamber technique that some chemotactic factors by the influence of which the leucocytes are attracted towards the irritant are present. It is also proved that some of these factors are effective on only granulocytes, some only on mononuclears while some others on both. The attractants are some components of the complement and bacterial toxins. Another factor from sensitized lymphocytes is chemotactic for macrophages and monocytes but not for granulocytes. So it is that the macrophages are a feature in hypersensitivity reaction.

The lysosomal enzymes of neutrophils activate a component of complement and the resultant factor is chemotactic to macrophages. So adrenal steroids which stabilise the lysosomal membrane prevent the release of such activating enzymes and so are anti-inflammatory.

It has now been established that increased vascular permeability is not the cause of leucocyte emigration and that both of them are seperate phenomena. The mechanism of leucocyte emigration is still not completely understood.

6. **Diapedesis of the erythrocytes :** During stasis, distribution of erythrocytes in the peripheral part of the blood stream favours the escape of an occasional red cell through the permeable capillary wall along with emigrating leucocytes. This is called **diapedesis**. Erythrocytes have no power of movement, and therefore diapedesis is a passive phenomenon. The red cells may also enter the tissues by **rhexis** (through a break of the capillary wall). If their number becomes very great, the condition is known as **haemorrahagic inflammation**.

Chemical mediators of inflammation : These are the factors that are responsible for increasing the vascular permeability. They also elicit certain changes of the inflammatory process.

a) **Histamine** is widely distributed in tissues, in the granules of mast cells and basophils and platelets. Following injury, histamine is released from the granules probably by enzymatic action, and this produces vasodilatation and increased vascular permeability and induction of pain but has no effect either on the function or migration of leucocytes. It initiates the early vascular response only, and is soon superseded by other mediators which then sustain or induce the delayed vascular reaction.

b) **5-Hydroxytryptamine (5-HT; Serotonin)** : Its roll is less certain in inflammation than that of histamine. It is present in the mast cells of rat and mouse only, and does not seem to play any significant role in inflammation in mammals.

c) **Kinins** : The kinins are a group of straight-chain polypeptides, which resemble bradykinin in structure and pharmacological activity. Of the group, **bradykinin** and **kallidin** are currently considered to be the most likely mediators which sustain the vascular reactions after the initial histamine response. Kinins are produced following injury from normal serum precursors by the action of several enzymes in a sequence of events remarkably similar to the coagulation mechanism.

d) **Globulin permeability factor (GPF)** : The normal plasma contains a substance (globulin), which on activation following injury increases the permeability of capaillary endothelium. It now seems certain that GPF participates in inflammation by acting as an enzyme in kinin formation.

e) **Prostaglandins** :   These are fatty acid derivatives, and are biosynthesized from unsaturated essential fatty acids. Unlike histamine and 5-HT, they are not stored in the body but are synthesized immediately prior to release by enzymatic action. They increase vascular permeability, induce emigration of leucocytes and produce pain.

Some prostaglandins have opposite functions. All these prostaglandins are situated on the cell surface, especially on leucocytes and so are widely distributed by the mobility of these cells. The prostaglandins, therefore, may initiate, maintain or suppress inflammation according to the necessity arising.

f) **Complement** : The complement system consists of a group of serum proteins that interact sequentially, following injury, to produce a variety of factors that participate in inflammatory process. They are capable of releasing mastcell bound histamine and chemotactic factors from the neutrophils, and also increase vascular permeability

g) **Lysosomal enzymes** : Neutrophilic granules, under the electron microscope. are seen as sac-like structures, containing a large number of hydrolytic, oxydative and proteolytic enzymes. These sac-like bodies are called lysosomes, and their enzymes are known as lysosomal enzymes. Recently it has been shown that lysosomal enzymes participate actively in the inflammatory process, mainly through kinin formation and activation of complement.

h) **Direct action of the irritant** on the capillaries may increase their permeability.

PHAGOCYTOSIS

Amoeba obtains its nutrition by taking into its own body foreign food particles. This process of taking food into the body by cells is called **phagocytosis (Phagen=I eat)**. This process of phagocytosis also involves the movement of the cells to the food particles by the cytoplasm flowing to one side—a process known as "movement by pseudopodia". The cytoplasm encircles the foreign particle and takes it into itself where it is digested by enzymes. It is clear that prior to ingestion of the particle, the leucocyte should spread itself around it. Serum protein that is deposited on the foreign particle, helps in the spreading of the leucocytes. Similarly, opsonins by coating the bacteria, facilitate the neutrophils to spread over them.

The ability of phagocytosis of bacteria by leucocytes is restricted. Neutrophils can ingest streptococci or gonococci but they cannot assimilate tubercular organisms. The macrophages (the sessile histiocytes and wandering monocytes) take up only tubercular organisms but not streptococci or gonococci.

Other cells in the body that serve as phagocytes are the capillary endothelial cells, especially of sinusoids of liver, spleen and lymph glands, squamous epithelium of the serous surfaces and the septal cells of the lung alveoli.

THE INFLAMMATORY EXUDATE : The inflammatory exudate largely consists of (i) the blood plasma and (ii) the cellular elements.

(i) **The plasma :** The plasma flows out of the capillaries due to (a) increased permeability of the capillary wall, the causes for which have been detailed above.

(b) Increased osmotic pressure of tissue fluid. The formation of the tissue fluid depends on the balance between the hydrostatic pressure in and outside the vessel wall. While fluid passes out of the capillary at its arterial end, lymph passes into it at its venous end. It is the osmotic pressure contributed by the plasma proteins and tissue fluid that maintains this balance. If the osmotic pressure of the tissue fluid is raised, more of fluid will pass out of the vessel. This is what happens in acute inflammation. Because of irritation, the tissue cells are damaged and the large and complex molecules of the protoplasm are broken down to smaller molecules, mostly acid in reaction. These smaller molecules raise the osmotic pressure of the interstitial fluid, which draws out the plasma from the capillaries.

ii) **The cellular elements :** These are derived from the blood and from local tissues. The following cells are involved in the cellular changes in inflammation.

1. **Neutrophils :** These are also known as polymorphonuclear leucocytes. They are actively amoeboid and phagocytic. The neutrophilic granules are supposed to be lysosomes and contain a host of 'digestive' enzymes with which neutrophils are able to destroy the bacteria. The function of neutrophils is engulfing and digestion of degenerating, necrotic or foreign material. This phenomenon is known as **phagocytosis**. On this account the cells are also called **microphages** of Metchnikoff, since as phagocytic cells they are smaller in size. They may ingest any foreign particulate matter such as bacteria, cellular debris,

pigments and even carbon. To phagocytize bacteria, certain serum proteins called **opsonins** are required. Most bacteria are killed following phagocytosis, mainly by the action of the enzymes, lysozyme and phagocytin, present in the neutrophilic granules. If the material is 'undigestible' like carbon particle, it is released upon the death of the neutrophil. Neutrophilic granules also release endogenous pyrogen (fever producing substance) which produces fever.

The neutrophils are highly chemotactic and are usually the first to arrive at the site of inflammation. They are therefore called the **first line of cellular defence.** Neutrophils cannot multiply and are drawn from the bone marrow where they are in great store. They are the characteristic feature in acute inflammation. An increase in the number of leucocytes in the blood is known as **leucocytosis,** and decrease **leucopenia. Schilling index** helps in determining the immature forms of neutrophils in the blood and increase in their number is known as 'shift to the left'. Neutrophils constitute the 'pus cells' in purulent exudate.

The production of neutrophils is under the control of a granulopoietin, known as **colony stimulating factor (CSF),** which is produced by the macrophages of the bone marrow. The synthesis of CSF is stimulated by bacterial products. CSF is necessary for the stimulation of mitosis of stem cells.

Release of neutrophils from the bone marrow is promoted by a factor in the plasma known as **leucocytosis inducing factor (LIF),** the concentration of which is increased by bacterial products.

Even after the neutrophils are destroyed they are useful since they liberate the proteolytic enzymes which are useful in the digestion and liquefaction of the dead material and thus help in its removal, paving way for regeneration and repair. After the inflammation subsides, when no longer required, the neutrophils pass back into circulation.

2. Eosinophils : In the blood of domestic animals eosinophils are normally between 1–7 per cent. These are very short lived and are found more in tissue fluid than in blood and are also found in the epithelial lining of the intestines, the respiratory tract and skin. They are motile and appear early in an area of inflammation by their amoeboid movement. However, they are mainly present in various allergic inflammations, important examples being asthma and hay fever in human beings. Eosinophils are also very numerous in the vicinity of animal parasites in parasitic infections. This is particularly prominent in *Trichinella spiralis* infection and in intestinal worms. The finding of an increased number of eosinophils in the blood (eosinophilia) is an indication that parasitism or an allergy may be present.

The functions of these cells still remain obscure. Though the lysosomal granules of the eosinophils contain many of the catabolic enzymes of neutrophils, they lack lysozyme and phagocytin. They are phagocytic but this property is not so important as the phagocytic property of the neutrophils. Probably this property is directed toward engulfing and destroying antigen–antibody precipitates. It is presumed that eosinophils also play a role as an antagonist to the inflammatory response. Recent evidence has shown that eosinophil extracts are antagonistic to histamine, 5-hydroxytryptamine and bradykinin, all of which are mediators of of the inflammatory response. However, the exact role that eosinophil granules play in inflammation is still not clear. They are rich in the enzyme peroxidase, but its role remains unknown.

Charcot-Leydon crystals are the crystalline proteins derived from the nucleus of eosinophils in areas where these are broken down and seen in the sputum of man in asthma. Eosinophils disappear from the blood when ACTH or cortisone are administered.

3. Basophils : Basophils are few in the blood making up to only 0.5 to 1.0 per cent of leucocytes. They are not phagocytic. In the blood stream this cell is known as the basophil, and in the tissue as the **mast cell**. While the former arises from the bone marrow, the latter is a connective tissue cell found throughout the connective tissues in every organ, and are most numerous around the small blood vessels and in serous membranes. Both basophil and mast cell contain a large number of deeply basophilic and metachromatic granules. These granules, besides heparin, also contain histamine and 5-hydroxytryptamine, and possibly also bradykinin. It is these compounds which relate the function of the basophil and mast cell to the inflammatory process. Following trauma, mast cells degranulate and release their vasoactive amines. 5-hydroxytryptamine is, however, confined to rat and mouse mast cells only. By preventing clotting of blood and exudate, heparin facilitates absoption and phagocytosis.

Tumor of mast cells, mastocytoma, is seen in dogs.

4. Lymphocytes : Lymphocytes comprise 40 to 60 percent of total blood leucocytes. They are not phagocytic and possess only limited powers of amoeboid movement. At best they succeed in getting just outside the blood vessel, constituting perivascular lymphocytic infiltration or 'perivascular cuffing'. They usually appear late in inflammation.

Lymphocytes are produced by the lymphoid tissue of the body. It is now established that there are two populations of lymphocytes, one dependent upon the thymus, for development, called T-lymphocytes, and the other independent of the thymus, called B-lymphocytes. In birds, development of B-lymphocytes is dependent upon the bursa of Fabricius, but no such single organ or tissue has been identified in mammals. Most of the lymphocytes present in blood are T-lymphocytes and long lived, while B-lymphocytes are largely restricted to lymphoid tissue and short lived. B-lymphocytes further differentiate into plasm cells which synthesize immunoglobulins. In contrast the T lymphocytes are responsible for cell-mediated immunological reactions; they do not form immuno-globulins. Thus, on the whole, the role of lymphocytes in inflammation is that they participate in the immune response. They are mainly found in chronic inflammation like tuberculosis and in certain viral infections.

5. Plasma cells : Plasma cells are not found in the blood, but are present only in tissues. They are an important constituent of many inflammatory reactions. Plasma cells possess more cytoplasm than the lymphocytes and are therefore larger. The nucleus is eccentrically placed in the cell and is spherical.

The arrangement of chromatin granules along the nuclear membrane imparts it a clock-face or cart–wheel appearance. They are not phagocytic and possess only limited amoeboid movement.

Plasma cells do not undergo mitosis; they originate from B-lymphocytes. The function of the plama cell is the production and transport of antibodies (immunoglobulins). Plasm cells are numerous in some types of chronic inflammation like Johne's disease, actinomycosis and actinobacillosis.

In the neoplastic disease called multiple myeloma of man, plasma cells are found in large numbers and give rise to Bence-Jones protein in the urine. This protein is precipitated at 48°C but dissolves at 100°C. The lymphocytes and the plasma cells constitute the small round cells of chronic inflammation.

6. Macrophage : It is a large mononuclear cell containing a spherical nucleus. This cell is also known as large mononuclear cell, histiocyte, resting wandering cell, reticuloendothelial cell and **macrophage** of Metchnikoff, in contrast to the smaller **microphage** (neutrophil). In the circulating blood, the macrophages are known as the **monocytes**, and comprise 1 to 5 per cent of the total leucocytes. These monocytes when they come out of the blood stream constitute the macrophages in inflammation. The other source of their derivation is through active cell division locally, from the pre-existing macrophages. Macrophages appear late in inflammation.

Macrophages have tremendous phagocytic powers since they are very rich in lysosomal enzymes. They are the main phagocytic cells of the body. Macrophages are the cells which eventually complete the destruction of the irritant through their powerful enzymes, and remove the necrotic tissue from the area. On this account they are called the **second line of cellular defence**, and are at times also referred to as scavenger cells. Macrophages are involved in both acute and chronic inflammation. They are, however, particularly abundant in certain chronic inflammatory diseases like tuberculosis and Johne's disease. They constitute the epithelioid cells in tuberculosis; heart failure cells found in lungs (with ingested pigment) in a failing heart and compound granular corpuscles seen in an area of softening of brain. Microglia are the macrophages of the brain.

The macrophages are tough having a longer life. It is due to the property of resynthesis of the enzymes and other cellular structures (not possessed by granulocytes) that the toughness of these cells is partly due.

Macrophages synthesize lysozyme, certain components of complement, endogenous pyrogen, transferrin, granulopoietin and interferon.

The endothelial cells of the reticulo-endothelial system (Kuppffer's cells of liver, sinusoidal cells of spleen, lymph nodes, adrenal cortex and bone marrow), endothelial cells lining the capillaries and septal cells of the lung also function as phagocytic cells like macrophages.

In an inflammation, neutrophils, predominate first (since they are brought from the blood) while macrophages replace them soon. The shorter life of the neutrophils is also a reason for the greater number of the macrophages seen in the later part of inflammation. The delay in the appearance of macrophages is because it takes a little time for the histiocytes to proliferate and supply them.

7. **Giant Cells :** When macrophages fuse together to form a large phagocytic cell, it is called a giant cell. It contains multiple nuclei and an abundance of cytoplasm. There are two basic types of giant cells : **(i) foreign-body giant cell** and (ii) **tumour giant cell.** The first type is produced by the fusion of macrophages and is evoked in response to foreign material in the tissues. It contains multiple nuclei, 50, 100 or even more. They may be arranged around the periphery of the cell, clustered at one or both poles of the cell or scatterred throughout the cytoplasm. A very good example of the foreign-body giant cell is Langhan's giant cell of tuberculosis. They are also seen in Johne's disease, actinomycosis and blastomycosis.

The second type, tumour giant cells. are found only in tumours. The cells are large and possess multiple nuclei, but the number is relatively few (2, 4, 8 or 16), Tumour giant cells are neoplastic cells and result from nuclear division being not followed by cytoplasmic division. As a result the cytoplasm contains several nuclei.

A third type, known as the miscellaneous group : The **Reed-Stenberg cells** of Hodgkin's disease which are mesodermal cells with twin nuclei which are identical—the mirror image of the other.

(ii) Touton giant cells : These are large cells containing lipid material and found in xanthomas. The cells are characteristic in having a ring of nuclei at the periphery.

THE EXUDATE

As a result of circulatory and cellular changes described above, humoral and cellular substances accumulate in an area of inflammation. This is known as the **inflammatory exudate,** and the process of its formation exudation. It is composed of five major constituents : (1) the irritant (2) injured tissue cells (3) leucocytes (also macrophages and plasma cells) (4) plasma constituents (water, protein, fibrin and antibodies), and (5) erythrocytes.

Functions of the exudate : Exudate is actually beneficial to the body. Its functions are as follows :

1. It dilutes the irritant. The irritant is thus changed from severe to mild, and thus less damage is caused to the tissues. Moreover, irritants like bacteria are dispersed and so better phagocytized.

2. The exudate also mechanically carries the irritant away, especially to the exterior, in the case of cutaneous or mucous surfaces. This helps in getting rid of the irritant from the body.

3. It brings phagocytes (neutrophils, macrophages) to the area to destroy the irritant.

4. Exudate also brings fibrin to the area. Fibrin is formed from the fibrinogen that leaves the vessels following increased permeability. Fibrin performs several useful functions. a) It entraps the irritant and thus retards its spread This facilitates their phagocytosis : b) It forms a kind of layer around the cells, thus protecting them from the damaging effect of the irritant. (c) It seals the lumens of the lymphatics effectively. This keeps irritants like bacteria from entering into them, thus stopping their spread to the regional lymph nodes.

(d) Fibrin helps in healing and repair. It does so by forming a kind of a scaffold for fibroblasts and angioblasts to work. (e) Fibrin also has a stimulating effect on the proliferation of fibroblasts. This further helps in healing. (f) for the movement of leucocytes, fibrin is required and acts as a scaffold. Otherwise they cannot move in a fluid medium.

5. The exudate brings antibodies to the area of inflammation, which are most effective against bacteria and viruses.

6. Finally, the exudate brings to the area of inflammation increased amount of nutrients and oxygen that are so much needed by the tissues in their enhanced activity in defending the body as well as for regeneration and repair. Exudate also drains the area of products of tissue destruction and altered metabolism which are acidic in nature. The optimum H-ion concentration for phagocytosis of the irritant is thus maintained. Exudate also provides a suitable medium for the working of phagocytes, enzymes, and antibodies under optimum conditions.

In the early stages of inflammation, fibrin may be useful as the adhesions that are formed by it limit the area of inflammation. But later, when the fibrin is not completely removed, but becomes organised, the parts involved are constricted by the contraction of the granulation tissue and so the function of the parts is affected; eg. constrictive pericarditis.

The proteolytic ferments of neutrophils liquefy the fibrin and so it is that pus does not clot since it contains large number of neutrophils, which liquefy the clot.

Spread of inflammation : The ground substance of the tissues is made of a polymerised mucopolysaccharide—hyaluronic acid. The higher the polymerisation the more viscid it is and so more effective is its use as a barrier againt invasion by microorganisms. Some bacteria which are highly invasive produce a hyaluronidase which depolymerises the hyaluronic acid and so it becomes less viscid thereby enabling the organisms to spread. The protective action of the hyaluronic acid is under the control of pituitary and adrenal since cortisone increases the permeability of the barrier.

The fibrin barrier that is formed in inflammation may be removed by the action of fibrinolysin (kinases) produced by certain bacteria thereby facilitating their spread.

Similarly hemolysins and leucocydins of certain bacteria weaken the system and kill the defensive leucocytes Toxins may enter the blood stream, reach vital organs like the heart and bone marrow and damage them with grave results.

Classification of inflammation : Inflammation is variously classified. It can be mild or severe depending upon the nature of the irritant. According to its duration inflammation may be peracute, acute, subacute or chronic in its course. The duration is governed by the fact that whether the irritant is severe or of low intensity. Acute inflammation is of a comparatively short duration and is characterized by marked vascular changes, whereas the chronic inflammation is of a relatively long duration and is accompanied by marked proliferation of connective tissue, blood vessels and epithelium. The vascular changes are less prominent and there is scanty exudation.

The acute inflammation, which is always accompanied by exudate formation, is further classified as follows, based upon the principal constituent of the exudate.

1. Catarrhal or Mucous inflammation : This is present when the principal constituent of the exudate is mucous. Mucous comes from the blood. Catarrhal inflammation occurs only in those areas where cells capable of producing mucous are present. It is therefore limited to mucous membranes. It is produced by irritants which are mild in nature such as mild irritating chemicals (formalin, phenol, detergent), irritating foods in the digestive tract, inhaled dust, cold air, bacterial and viral infections of low virulence of respiratory tract. "Catarrhal inflammation" is also applied to inflammation of epithelium lining ducts and tubules (eg. uriniferous tubules).

In this condition there is proliferation of the epithelium which is desquamated into the exudate, which consists of these cells, polymorphs and mucous which is a clear, transparent glistening slimy material, containing water and mucin (a compound containing nucleoprotein). Microscopically this stains blue with hematoxylin. The secretion becomes **mucopurulent** following infection by pyogenic oragnisms.

In result, if the cause is removed, recovery occurs quickly. But if it persists, it may progress to a chronic stage when the epithelium of the mucosa is denuded and its wall becomes thickened due to fibrosis.

2. Serous inflammation : This is present when the principal constituent of the exudate is plasma or lymph, a clear watery fluid. It is caused by moderately severe irritants Various chemical irritants when applied to skin cause blisters. Traumatic injuries of rubbing nature and second degree burn also form blisters on the skin. Certain viruses (foot and mouth disease, vesicular stomatitis) also produce blisters in skin or mucous membrane. Serous inflammation is common in the serous membranes (pericardium, pleura and peritoneum) and in joint spaces. It is often the first stage in many inflammatory processes. In some cases of inflammation of serous membranes, the exudate may contain enough fibrin so as to give a frosted glass or shaggy appearance to the smooth surface.

Microscopically, in H & E. stained sections, the exudate appears as a homogeneous to finely granular material which stains pink with eosin; the intensity of eosin-staining depending upon the amount of protein present. Other alterations characteristic of inflammation are present.

Serous inflammation being mild, the outcome is favourable. The fluid is promptly absorbed if the cause is overcome. If it persists, since the number of neutrophils, which could remove the exudate by autolysis is far too few, it usually becomes organised and so adhesions may occur,

3. Fibrinous inflammation : This is present when the principal constituent of the exudate is fibrin. This type of inflammation is caused by a more violent type of injury. The consequent marked increase in vascular permeability enables fibrinogen to escape into the surrounding tissue. It is observed in various viral diseases such as infectious feline enteritis and malignant catarrhal fever and when mucous membranes are invaded by *Corynebacterium diphtheriae*, various Salmonella or *Sphaerophorus necrophorus*.

The organ is firmer and tenser than normal. This is well seen when fibrin accumulates in the alveoli of lung in pneumonia. Lung then acquires both consistency and appearance of liver (hepatization). On epithelial surface (mucous, serous or cutaneous) fibrin is seen as stringy, yellowish, netlike material. In these places it entraps desquamated epithelium and foreign material. In tubular organ it forms cast of the organ.

Masses of fibrin on epithelial surface may either form a **pseudo-membrane** or a **croupous membrane**, when it is easily peeled away or a **diphtheritic membrane** when it is quite firmly attached to the underlying tissue. The latter is the result of epithelium having undergone coagulative necrosis. If the necrotic, denuded epithelial cells are also included in the mass, it is a **"true membrane"**. Otherwise, without the epithelial cells, it is called a **"false membrane"**. The best examples of diphtheritic membrane formation are pharynx in calf diphtheria and intestinal tract in swine fever. Fibrin stains dirty pink with eosin. In result, in this type of inflammation the tissue destruction is so much that the animal may not survive. Fibrin on epithelial surfaces is desquamated. In body cavity it is removed by phagocytes. In peritoneal cavity it may result in **adhesions**, in which case fibrin gets organized and connective tissue is deposited in the area. Adhesions then interfere with the intestinal motility and thus create further complications.

4. **Suppurative or purulent inflammation :** This is characterised by the presence of pus, which is the result of softening and liquefaction of a tissue. It consists of the exudate, liquefied tissue, polymorphs and organisms.

For suppuration to occur the following three factors are essential and if any one of these is absent suppuration will not occur: (a) necrosis, (b) presence of sufficient number of polymorphs and (c) digestion of necrotic material by proteolytic enzymes. The presence of polymorphs alone does not constitute suppuration. Hence any irritant causing positive chemotaxis and necrosis will produce suppuration.

The chief causes of suppurative inflammation are :

(1) Pyogenic organisms : eg. Staphylococci, Streptococci, and members of the coli group.

(2) Specific organisms like *C. pyogenes, P. mallei, Actinomyces bovis* etc.

(3) Chemicals — $ZnCl_2$, $HgCl_2$, croton oil etc.

The proteolytic enzymes are produced mostly by the leucocytes and to a lesser extent by the infecting bacteria and the necrosed tissue cells themselves. Serum contains an antienzyme which tends to inhibit the action of the prortease of the leucocytes. The rabbit serum is particularly rich in the antienzyme and poor in leucocytes and so it is not common to see suppurative conditions in this animal in infections by ordinary pyogenic organisms. The antienzyme is lipoid in character, rich in unsaturated fatty acids. Caseous material of tuberculosis is also rich in unsaturated fatty acids and so acts as an antienzyme. Hence suppuration is not encountered in tuberculosis.

Pus is composed of necrotic neutrophils, necrotic tissue cells, more or less liquefied, and minor amounts of other constituents of inflammatory exudate including serum. Pus which is alkaline, is usually of creamy colour. It may, however, be white, yellow, green, red, black or blue depending upon the species of animal and the aetiological agent present. Streptococci and staphylococci produce white or yellow pus. Corynebacteria, particularly in cattle, produce greenish pus; blue–green colour comes from the pigment forming pyocyaneus bacillus; and black colour from disintegrating hoof material (iron sulphide). Pus is red when there is haemorrhage. Consistency may vary from thin and watery to creamy, thick, viscid or granular depending upon the species of animal, amount of necrotic material and dehydration of the exudate. Pus is liquid because of the proteolytic enzymes of the neutrophil. Canine pus is thin and watery due to extremely proteolytic neutrophilic enzymes. Bovine pus is rather viscid. Avian pus has a dry, caseous consistency due to the presence of antitriptic enzymes. The pus serum, **liquor puris**, does not coagulate because the fibrin of the exudate gets digested by the proteolytic enzymes of the leucocytes.

Cellulitis (Phlegmon) is a diffuse spreading suppurative inflammation of connective tissue caused by sptreptococci and has red raised margins.

Abscess : Collection of pus locally within a closed cavity in an organ, or tissue is called abscess. When pyogenic organisms enter an organ, an acute inflammation results with death of cells in the centre. This dead material is liquified by the proteolytic ferments mainly from the polymorphs, resulting in pus in a cavity. The neighbouring tissue partly damaged and partly living constitutes the wall of this cavity and at this wall active warfare is going on to limit the spread of infection. So, dead tissue and dead inflammatory cells are continually shed into the cavity from this zone thereby increasing the quantity of pus and thus the abscess becomes enlarged. The limiting zone is, therefore, called **the pyogenic membrane.** The abscess becomes bigger and bigger till it reaches the surface, (skin or mucous membrane) points and opens, discharging the pus. The discontinuity of the skin or mucous membrane that results by the opening of an abscess on to the surface is called an **ulcer,** and the base lies in the subcutaneous tissue. Ulcer may also form due to the direct action of the irritant on the skin or mucosa.

Sinus : Is the name of the track in the tissues communicating with an epithelial surface discharging pus from an abscess.

Fistula is the track that connects two epithelial surfaces, skin and mucous membrane, for the discharge of pus from an abscess.

Boil (or Furuncle) is a small suppurative inflammation in the skin which involves a hair follicle or a sebaceous gland, casued by *Staphylococcus aureus.*

Pustule is a circumscribed cavity in the epidermis with pus.

Microscopically, the principal cell of the exudate is the neutrophil in various stages of disintegration. In significance, the presence of pus indicates the presence of bacteria in the area of inflammation. If an abscess persists for a long time, bacteria are destroyed and it would yield no organisms on culture (sterile abscess). Persistence of confined pus can lead to toxaemia.

5. **Haemorrhagic inflammation :** This is present when the principal constituent of the exudate is the erythrocyte. It is caused by a violent type of irritant which causes serious damage to blood vessels. Bacterial, viral and protozoal diseases (black quarter, anthrax, haemorrhagic septicemia, infectious laryngotracheitis and coccidiosis) will casue such severe injury that haemorrhage occurs. The result of haemorrhagic inflammation is extremely unfavourable. Due to severe haemorrhage, the animal may die from anaemia as in coccidiosis in the chicken.

6. **Gangrenous inflammation :** Consequent on venous stasis and thrombosis in some acute inflammations, the tissues may become necrosed and invaded by saprophytic organisms resulting in characteristic discoloration of the area known as **gangrene**. This is a characteristic appearance in infections by the **Clostridia** eg. Black quarter.

The above six forms constitute the acute exudative inflammation. Under natural conditions they may occur as a combination of two or more types and then we may describe the exudates by using such adjectives as mucopurulent, serofibrinous, fibrinopurulent or serosanguineous. Also, a reaction may change from a mild to a more severe form and vice-versa.

In addition to the above, when the lymphocyte is the predominant cell in the exudate the inflammation is called as **lymphocytic inflammation,** and if the cause is an antigen–antibody reaction it is known as an **allergic inflammation**.

The terminology of inflammation : Inflammation of an organ or tissue is designated by adding suffix ' itis ' after its name in Latin or Greek. Using the following table, the term acute diffuse serofibrinous peritonitis indicates an acute inflammation of the entire peritoneum characterized by serofibrinous exudate. The term chronic diffuse suppurative interstitial hepatitis signifies chronic inflammation of the interstitial tissue of the liver characterized by pus formation.

To describe combination of exudate, the least important constituent of the exudate is placed first. For example, the term serofibrinous would indicate that fibrin was the predominant constituent. The term mucopurulent would mean more pus in the exudate than mucous.

TABLE

Time	Extent	Exudate	Position in organ	Anatomy	Suffix
Acute	Focal	Serous	Parenchymatous	Nephr-	itis
Chronic	Diffuse	Fibrinous	Interstitial	Hepat-	itis
		Catarrhal		Rhin-	itis
		Suppurative		Periton-	itis
		Haemorrahagic		Enter-	itis
		Gangrenous		Mast-	itis

CHRONIC INFLAMMATION

Inflammation is said to be chronic when the irritant persists for a long period of time and the body responds by producing excessive amounts of connective tissue and epithelium in the area. The tissue alters itself in some permanent manner to the presence of the irritant. Skin, mucous membrane of the gastrointestinal tract and lymph nodes are more frequently involved.

A chronic inflammation may follow an acute inflammation. When the body is unable to remove and destroy the irritant, it continues to persist in the area, interferes with healing and causes constant irritation. This leads to proliferation of excessive amounts of tissue in the area. Even prolonged persistence of necrotic tissue in the area, by acting as a foreign body, may induce chronic inflammation. Further, if the irritant is of low intensity, it fails to stimulate body defences to cause its destruction and removal and chronic inflammation occurs, examples being tuberculosis, Johne's disease, actinomycosis and actinobacillosis.

In general, the irritants causing chronic inflammation are :

1. **Bacteria**—causing acute septicaemic diseases but have, later on, become attenuated and localised; eg. *Pasteurella aviseptica* (liver) and *Erysipelothrix rhusiopathiae* (heart valves—vegetation, and in joints).

2. **Phytotoxins**—of certain plants belonging to the genus Senecio and crotalaria producing lesions in the liver of horses.

3. **Foreign bodies**—eg. sharp objects as in Traumatic reticulitis and pericarditis in cattle; dust in pneumoconiosis; encysted larvae in trichinosis. Inert material that is difficult to be phagocytosed (splinters, thorns, suture material, dead parasites and necrotic tissue) persists in the tissue and induces a proliferative type of reaction.

4. **Constant and repeated mechanical irritation (trauma)** to a wound by preventing or delaying healing, induces chronic inflammation—saddle and collar galls; kennel granuloma in the dog.

When incised the fibrous granulation tissue is white, tough and hard in the case of mature variety; white, smoothly dense and perhaps watery(oedematous) if it is newly formed; yellowish, soft and easily cut when composed of the reticulo-endothelial variety. Due to the granulomatous character of the newly formed fibrous tissue, this inflammation is also at times referred to as 'Granulomatous inflammation' : examples being tuberculosis, actinobacillosis, and actinomycosis. The precise nature of 'granulation tissue' will be explained more fully in the next chapter.

In chronic inflammation cardinal signs of inflammation, though present, are not easily distinguishable. There is scanty exudation; the main character of the chronic inflammation being marked proliferation of connective tissue and epithelium. Microscopically also, circulatory and cellular alterations characteristic of acute inflammation are much less prominent. Fibrous granulation tissue consists of newly formed fibrous tissue and numerous young capillaries. Macrophages, giant cells, lymphocytes and plasma cells predominate in the cellular exudate. Neutrophils are present if bacteria exist in the area. The area of inflammation may be surrounded by a zone of proliferating white fibrous tissue and capillaries.   The epithelial structures in the area show hyperplasia, hypertrophy or even metaplasia.   In conclusion, the criterian used to decide whether chronic inflammation is present or not is the proliferation and deposition of excessive amounts of connective tissue, reticuloendothelial tissue and epithelium.

In significance, presence of chronic inflammation indicates that healing has been unduly prolonged.   The deposition of the tissue in the area is a permanent change and persists throughout life.   Hyperplastic and hypertrophic epithelial structures, however, do tend to diminish in amount.   The proliferation of connective tissue and epithelium may lead to considerable distortion and disfigurement of the part. As white fibrous connective tissue shrinks and contracts the part may be unable to function, motility of the organ is prevented and it may become impossible to flex a joint.   However, the greatest danger in chronic inflammation may lie in the fact that the hyperplastic tissues may progress to metaplasia and neoplasia

ALLERGIC   INFLAMMATION

When an animal or person, previously sensitised to a foreign protein—animal, vegetable or bacterial in origin—is injected with the same protein local inflammatory response is noticed which is known as allergic inflammation and the animal is said to be in a state of allergy.   This is a result of antigen-antibody union.

Essentially this process is a necrotising inflammation, necrosis occurring due to the union of antigen and antibody within the cells.   This phenomenon is put to practical use in the diagnosis of T. B., J. D., glanders etc. by the allergic tests.

In an allergic test the following reactions take place :

1.   At the site of injection of the antigen into the dermis of a sensitised animal edema containing neutrophils occurs, especially perivascularly and perineurally.

2.   Number of neutrophils gradually increases upto 48 hours, thrombi may form in the veins while in the arteries the endothelial cells proliferate. Degeneration and necrosis of the vascular endothelium and smooth muscle occurs.

3.   After the 2nd day, the neutrophils gradually decrease in number while macrophages and eosinophils begin to increase.   A few giant cells may be formed.

4.   By the 72nd hour the allergic reaction has attained its maximum intensity, viz. the presence of a hot, painful, diffuse, swelling.

5.   After this stage, the swelling slowly subsides.

VIRAL   INFLAMMATION

Viruses are obligatory parasites of living cells and so cannot thrive outside these cells.   Once inside the cells, they are protected from antibodies. Immunity, therefore depends on the availability of circulating antibodies that can neutralise the viruses while still in circulation or in tissues, before their entry into the susceptible cells.

Some viral infections are characterised by the formation of "inclusion bodies", which are supposed to be colonies or aggregates of viruses, when they are basophylic or they may be just completed replicated viruses when they will be acidophylic. These inclusion bodies may be :-

1. **Intracytoplasmic** : Guarnieri bodies of vaccinia; Bollinger bodies in fowl pox ; Negri bodies of rabies.
2. **Intranuclear** : Infectious canine hepatitis.
3. **Both intracytoplasmic and intranuclear** : Small pox and dog distemper.

The cells of the body may be broadly classified as :

i) **The renewing cell population (Labile cells)** : These cells proliferate throughout life to replace those lost. The best examples are the epithelium of the skin and the intestinal mucous membrane. So in healing, complete regeneration of this type of cells is possible.

ii) **The expanding cell population (Stable cells)** : These cells normally lie quiescent but have retained their capacity for proliferation by mitosis so that they can rapidly multiply when need arises, and readily repair their losses. Examples of this variety are the connective tissue cells (fibroblasts) and, at least in part, the liver, the pancreas, the kidney and many of the endocrine glands repair ther losses by cellular proliferation.

iii) **The static cell population (permanent cells)**: These cells are highly differentiated and have lost their capacity to proliferate during such differentiation. They are so highly specialized that they have sacrificed proliferation for the sake of function. The examples of this category are the nerve and muscle cells. When lost. the cells of the brain or the smooth and cardiac muscle cells can never be replaced; once destroyed they are lost for ever.

Reactions of cells to invasion by virus : The tissue reaction in viral infection differs among the three types of cells mentioned above. The patterns of reaction by these cells respectively are (a) hyperplasia only (Shope papilloma) (b) hyperplasia followed by necrosis (fowl pox and vaccinia) and (c) necrosis alone (Foot and Mouth Disease, Rabies).

These patterns are well exemplified by viruses affecting the epithelium of skin.

1. **Only hyperplasia** : The best example is Shope papilloma that occurs naturally in the cotton-tail rabbits, In this condition infection by the virus of the epithelium of the skin results in a papillomatous out-growth of the epithelial cells. No necrosis is found.

2. **Hyperplasia followed by necrosis** : In fowl pox, there are warty growths caused by proliferation of the cells, which initially show vacuolisation of the cytoplasm. Later by keratinisation and by formation of inclusion bodies, the structure of the cells is disrupted and necrosis occurs late in infection. But, by and large. the predominant change is proliferation.

3. **Proliferation and necrosis** : In vaccinia there is initial, brief period of epithelial proliferation at once followed by necrosis of affected cells.

4. **Necrosis only** : In foot and mouth disease the cellular reaction consists entirely of degeneration and necrosis of the affected epithelial cells which become greatly swollen with pyknotic nuclei. There is no proliferation of cells at all. Viruses under this category are called **cytocidal**.

The above lesions of the cells are primary. The inflammatory changes that may be encountered are to be considered as secondary response to the primary degenerative lesion caused by the direct action of virus on the cells. It is safe to say that wherever such mesenchymal :. ˙tion is encountered, the infectious viral agent can be found in high concentration. For example, in viral infections of the central nervous system, places where "cuffing" of vessels occurs, are rich in the virus. At these places nerve cell damage is encountered. The inflammatory cells that infiltrate a tissue affected by virus are the lymphocytes, macrophages and plasma cells.

One noteworthy feature of viral infection is the absence of neutrophils (which are a feature of bacterial infections). Suppuration in viral diseases is absent unless there is secondary infection by pyogenic organisms.

It should be understood that the above reactions are common to all viral infections and diagnosis cannot be arrived at by histopathological examination alone. Confirmation of diagnosis is made by serological tests—complement fixation and H. I. Test, etc.

RICKETTSIAL INFLAMMATION

Rickettsia are organisms midway between bacteria and viruses. They are obligatory parasites of the cells and are transmitted from one animal to another through an arthropod vector.

Rickettsia gaining entry into the body reach the capillaries of various organs and then entering the endothelial cells cause their swelling and proliferation. This leads to thrombosis of the vessel. Perivascular infiltration by mononuclear cells is present. The vascular lesion is responsible for necrosis and certain amount of proliferation in the tissues involved. This histological picture is common to all the diseases caused by different rickettsia. So a diagnosis on histopathological examination alone is not possible. Diagnosis is confirmed by serological tests.

GRANULOMATOUS INFLAMMATION

This is a variety of chronic inflammation, circumscribed and consisting of inflammatory cells but without true exudate and vascular changes. The cells noticed are mostly of Reticulo-endothelial origin. The histiocytes in the lesion have large amount of cytoplasm and resemble epithelial cells and so are called epithelioid cells. These cells together with fibroblasts and lymphocytes gather around the irritant. These lesions at first are very small and resemble granules, and later fuse to form nodules.

The change of the histiocyte into an epithelioid cell is attributed to the lipid content of the cytoplasm. These epithelioid cells may fuse together to form Langhan's type of giant cells.

Depending on the causal agent, the granulomata may be infectious or foreign body type.

The causes of infectious granulomata may be bacteria (M. tuberculosis. P. mallei, Br. abortus), fungi (Coccidioidimycosis) or viruses—Lymphogranuloma Venereum (in man).

The causes of foreign body granulomata may be any foreign bodies notably, sutures, talcum powder, silica etc. The microscopical picture is very characteristic. In the centre may be found the histiocytes and epithelioid cells surrounded by proliferating fibroblasts. Around these are found lymphocytes and plasma cells and the whole lesion is enclosed by fibrous tissue. Giant cells may be found within the central zone or at the periphery. In tuberculosis calcification may occur in the centre of the lesion. In the foreign body type, especially in silicosis, the entire lesion may be converted into dense scar tissue.

FATE OF INFLAMMATION

The purpose of inflammation, as we have already seen, is to rid the body of the irritant and to restore the tissues to their normal condition.This is achieved by the vascular and cellular responses. If the irritant is mild and not too abundant, then the vascular and cellular responses are adequate to sufficiently deal with it, destroy and remove it before it causes too much damage. The dead tissue is removed partly by phagocytosis by the macrophages and partly by the lymphatics after it is liquefied by the proteolytic enzymes. The inflammatory cells that have infiltrated into the area wander away, the blood vessels return to normal and thus the whole tissue regains its normal state. The process is called resolution.

How exactly is the part restored to as near a normal functional state as possible constitutes the fascinating subject of healing and is dealt with at some length in the next chapter.

CHAPTER 5

HEALING

By Dr. J. L. Vegad.

As we have seen, the primary objective of the inflammatory process is to destroy and remove the irritant from an area. After the injurious agent has been overcome, repair of the damaged tissues takes place. Healing is the process whereby the body restores the injured part to as near its previous normal condition as possible.

Healing, therefore, is closely associated with inflammation. In fact, the two processes are inseperable, and occur simultaneously. It is often impossible to determine where inflammation stops and repair starts.

Healing is a fundamental process in nature, and occurs both in animal and plant kingdoms. The lower an animal in evolution the greater are the powers of repair. For example, when an earthworm is severed, a new part grows in its place. Similarly, the tail of a salamander (a lizard-like amphibian) or the limb of an amphibian will grow when amputated. Also, in young goats a remarkable regeneration of the rumen and reticulum takes place after their complete removal.

Tissue or organ involved also influences the process of healing. If the tissue or organ is very highly specialised, it has less ability to regenerate. For example, in brain and spinal cord the lost nerve cells cannot be replaced. Similarly, the age of an animal also influences healing. The younger the animal the more rapid and complete is healing.

Before healing can take place, the products of inflammation such as exudates and dead cells have to be removed from the area. This is accomplished by liquefaction of the dead tissue. This, in turn, is attained by the autolytic enzymes of the dead tissue itself (**autolysis**), and also by enzymes derived from inflammatory leucocytes (**heterolysis**). The liquefied material (fluid) is then readily absorbed into lymph and blood and paves the way for healing. Healing is then accomplished in two ways. In **repair by regeneration,** the lost cells and tissues are replaced by others of the same kind, while in **repair by substitution,** healing is accomplished by proliferation of fibrous tissue, that is, highly specialized cells are replaced by less specialized connective tissue cells.

CAUSES OF HEALING

Regardless of how the healing is achieved to repair the damage, the cells of the area must proliferate. But then what is the stimulus for this proliferation? Or, why does the healing occur after all? Unfortunately, answers to these questions are still largely unknown. Probably, there appears to be mechanical and chemical basis for the proliferation of cells during healing.

1. **Ribbert's tissue tension theory** suggests that normally the cells are at an equilibrium due to their exerting a sort of restraining pressure on each other. When this pressure is abolished in wounds due to destruction of tissue, the cells begin to proliferate and continue to do so until the previous restraining pressure is restored. More recently, Abercrombie has put forward the concept of 'contact inhibition', which indicates that short feedback loops link the genes which trigger DNA synthesis to the cell membrane. That is, when cells are in contact, the signal to these genes is to stop their expression. When cells become separated or lost, these genes are activated and proliferation starts.

2. On the chemical front, **trephone theory of Carrel** proposes that degenerating cells in the area produce a growth-stimulating substance called trephone, and this acts as a stimulus for proliferation of cells. Later Hammet suggested that the trephone or wound hormone described by Carrel is the **sulphy-dryl (SH) group.** That the degenerating cells themselves release the growth-stimulating substance is borne out by the fact that in an aseptic wound, if all cell debris is removed, repair does not occur quickly. If, on the other hand, a few bacteria are added to the wound, cells are destroyed and repair begins soon. More recently, Bullough has postulated the existence of mitotic inhibitors called **chalones.** The concentration of such inhibitors is believed to decrease during wound healing, and this triggers off the healing process.

REPAIR BY REGENERATION

This type of repair is governed by the fact whether the cells possess the capacity to proliferate or not. If the cells have lost this capacity, they cannot repair themselves by regeneration. The power of regeneration differs widely with different cells. We have already seen that the cells of the body can be grouped into three classes with respect to their ability to proliferate, as follows : (Page 65)

1. **The renewing cell population : (Labile cells)**
2. **The expanding cell population : (Stable cells)**
3. **The Static cell population : (Permanent cells)**

Thus, considering the ability of the body cells to proliferate under the three different groups narrated above, it is clear that in those tissues where the cells have retained their powers to proliferate, healing occurs by regeneration, but in tissues where the cells have lost their powers of proliferation, the place of such cells is filled by less specialized connective tissue cells.

REPAIR BY SUBSTITUTION

As we have seen, in this type of repair, the destroyed tissue is replaced by new tissue which consists mainly of proliferating young connective tissue cells **(fibroblasts)** and young blood-vascular endothelial cells **(angioblasts)**, the latter forming the young capillaries. The basic process is the same whether it is healing of a closed wound, an open wound, an abscess, an ulcer, a blood clot, a thrombus or an infarct.

HEALING OF A WOUND

The wound caused by the surgeon's sterile scalpel and aseptically sutured is the best example of a clean closed wound. When there is no added infection and the wound is not open, wound healing takes place without interference. This is known as **healing by first intention** or **by primary union.** If, on the other hand, as a result of infection or because large segments of tissues are lost, the wound remains open, repair is complicated and takes place by what is called **healing by second intention** or **by granulation tissue formation.**

Healing of wounds by first intention or by primary union :

This occurs in a closed wound where there is little loss of tissue and very slight bleeding. Infection is absent due to aseptic precautions. Inflammatory reaction is mild and repair begins in about twelve hours by proliferation of fibroblasts and angioblasts. The fibroblasts bridge the gap between the two cut surfaces. In the beginning the fibroblasts are large, oval or fusiform (tapering at both ends or spindle-shaped) cells consisting mainly of nuclei. Soon they start forming collagen fibres and become gradually swollen. The newly formed collagen fibres separate them from one another, and soon surround them by wavy bundles. In the later stages, nuclei of fibroblasts become less prominent and bundles of fibres more prominent. The fibres then shrink and the resulting tissue is called **scar** or **cicatrix.**

When the fibroblasts are proliferating to bridge the gap between the cut surfaces, the angioblasts form buds by the side of or at the end of capillaries. Blood is pushed through them and angioblasts continue to proliferate to form new capillaries among the fibroblasts. These new capillaries connect with each other and form a network. This new tissue is, therefore, very vascular in the beginning and this vaecularity imparts it a red colour. This vascularity is required since the newly growing cells require increased nutrition. As the tissue matures, its nutritive requirements are less and it becomes avascular. It then looks white in colour.

When the healing by fibroblasts and angioblasts is completed, the surface epithelium regenerates from the margins of the wound and covers it by about the fourth day. The scar formed is white and puckered (wrinkled), and contains neither hair follicles nor sweat glands.

Healing by second intention or by granulation tissue formation :

This occurs in an open wound where there is excessive loss of tissue. In such cases the wounds are infected, many blood vessels are torn and the wounds contain areas of necrosis and inflammation. Following injury inflammatory reaction and haemorrhage occur. Blood clots and neutrophils gather into the area in large numbers to destroy the irritant (usually bacteria). By the second day pus may be visible in the wound. In 48 to 72 hours macrophages and lymphocytes appear in the area and gradually outnumber the neutrophils. Macrophages actively involve themselves in liquefying and removing the necrosed tissue and cellular debris.

Underneath the mass of blood covering the wound, are tiny red granules or buds giving the surface a granular appearance. It is on this account that this tissue is called **granulation tissue.** The red granules are made of proliferating capillaries. Along with the newly formed capillaries there is proliferation of fibroblasts. The gap is then completely filled in with a layer of fibroblasts and proliferating capillaries. As already mentioned under inflammation, fibrin helps in healing by forming a kind of a framework for fibroblasts and angioblasts to work. Fibrin also has a stimulating effect on the proliferation of fibroblasts, although at present little is known of the factors stimulating fibroblast proliferation in the healing wound. Many substances—especially products of cell necrosis—are claimed to be effective in stimulating fibroblast proliferation. The newly formed tissue does not contain nerves and is therefore without sensation.

In the granulation tissue fibroblasts and capillaries are arranged in a definite order. In the wound, the capillaries grow at right angles to its base and project towards its surface. Fibroblasts, in the deeper part of the wound, are arranged at right angles to the capillaries and parallel to the surface. Here they exert their pull laterally so as to bring the tissues closely together to avoid any gap during healing. Towards the surface of the wound, fibroblasts are arranged parallel to the proliferating capillaries. In this position they exert their tension towards the wound surface so as to cling on to the tissues and prevent them from growing beyond the surface. This definite arrangement of fibroblasts to capillaries helps in differentiating excessive granulation tissue from fibrosarcoma, which completely lacks any such orderly arrangement.

After the wound has been covered by granulation tissue, the surface epithelium from the margins regenerates and grows over it. Sweat glands, sebaceous glands, hair and hair follicles and melanoblasts do not regenerate. The skin is therefore not pigmented and is dry because it lacks sweat and sebaceous glands.

The area gets devascularized after the completion of healing, and a relatively avascular scar is left. Following the shrinkage of the collagen, the area may get considerably distorted and disfigured. The scar is therefore white and puckered.

If the irritant, movement or trauma prevent the healing, granulation tissue continues getting produced and may be present in abnormally large amounts. This is called excessive granulation tissue or proud flesh.

Sometimes, for reasons unknown, the connective tissue under the scar continues to grow even after the epithelium covers it. This mass of proliferating connective tissue is known as a keloid. Keloids are not true tumours, though they may recur after excision. They are particularly common among Negroes, and may have a genetic or familial predisposition. Among animals, keloids have been reported in horses.

Systemic metabolic response to injury :

Injury especially severe injury (abdominal operations or fractures) is accompanied by certain systemic effects in the body and these are mostly mediated by hormones.

1. There is great breakdown of muscle protein brought about by adrenal corticosteroids and hence there is loss of nitrogen.

2. Through the release of adrenaline and nor-adrenaline, consequent on fright, carbohydrate from the liver is mobilised.

3. Since new tissue is built up in the healing of a wound, protein subs-trates are, therefore, diverted from other parts of the body to the site of the wounds.

4. Increased aldosterone production in trauma results in greater sodium resorption by the renal tubules and this leads to decreased urinary sodium excretion.

5. Increased production of antidiuretic hormone in trauma causes reab-sorption of water by the renal tubules and so there is water retention in the body.

FACTORS AFFECTING WOUND HEALING

These may either be general or local.

General factors :

1. **Age :** With advancing age, the rate of healing may be considerably impaired, an important factor being the inadequate blood supply in the old age due to generalised vascular disease (arteriosclerosis).

2. **Nutrition :** Nutrition plays a very important role in healing, espe-cially proteins. Low protein levels in the diet have an adverse effect on healing. Of special importance are the two sulphur-containing amino-acids, methionine and cystine. In the absence of these amino-acids connective tissue of weak tensile strength is formed. Methionine also increases the rate of utilisation of the protein and its sulphur radicle may also be used for the formation of chondroitin sulphate which imparts firmness to the ground substance. In protein deficiency, few fibroblasts form and synthesis of collagen in inhibited.

Probably **Zinc** is necessary in wound healing as a cofactor in the enzyme processes involved in wound healing.

3   **Vitamins :**

Vitamin C (ascorbic acid) : Its deficiency is an important cause of poor and delayed wound healing. In the early stages of wound healing acid mucopolysaccharides (hyaluronic acid) are produced. In the absence of vitamin C, this substance is lacking, but it reappears on the injection of the vitamin. Under vitamin C deficiency fibroblasts produce little collagen, and what is produced is of poor quality. When the supply of this vitamin is inadequate, the healing of wound is delayed and they tend to break open again. Therefore an adequate supply of vitamin C is necessary for good healing.

In vitamin C deficiency, there may be capillary rupture and hemorrhage, which interfere with healing process.

Vitamin C may be necessary for the formation of specific messenger or transfer RNA necessary for the formation of the polypeptides.

In the healing of fractures, **vitamin D** in adequate quantities is essential.

Thyamine by increasing the rate of utilisation of available protein helps in the healing process.

4. **Hormones :** Of the various hormones that influence wound healing, cortisone is the most important. Cortisone interferes with the process of repair. It probably does so by inducing chemical changes in the mucopolysaccharides of the ground substance of connective tissue. Besides, during its presence less

collagen and fewer blood vessels are formed, and there is dearth of fibroblasts. Glucocorticoids produce the following changes in the inflammatory reaction :

    a) Decrease in the permeability of the blood vessels by inhibiting hyaluronidase, the spreading factor.

    b) Impeding margination and migration of leucocytes.

    c) Giant cell formation retarded.

    d) Impairment of phagocytic powers of polymorphs and macrophages.

    e) Almost complete lack of fibroblastic and vascular proliferation.

    f) Depletion of eosinophils.

    g) Lysosomes contain digestive enzymes. In an inflammation when the cells are damaged and ruptured, these enzymes are released. Corticosteroids prevent the rupture of lysosomes by strengthening the membranes.

    h) Corticoseteroids deplete the lymphocytes in the inflammed area by injuring them in the lymphopoietic tissues, decreasing their number in the circulation and inhibiting their migration.

    (i) Cortisone and hydrocortisone depress protein and polysaccharide synthesis. Cortisone hinders the action of histidine decarboxylase thus interfering with the formation of histamine locally and so it is anti–inflammatory since histamine is necessary for the inflammatory process.

    Thyroxine is necessary for rapid healing. In myxedema repair is slow and incomplete.

    5. **Drugs** : Penicillin interferes with repair, possibly by preventing collagen cross linking. This linking is of particular significance in wound healing because this molecular reaction is responsible for building a network of collagen fibres that give the unique, resilient tensile strength to the connective tissue. Penicillin is probably converted to penicillamine, which then chelates copper and other metals indispensable for the cross-link reaction.

    Local factors :

    1. **Ischaemia** : Ischaemia is a local anaemia, a cutting-off of the arterial blood supply to a part. Under such conditions normal healing cannot occur

    2. **Local irritants** : These, such as bacterial infection, presence of necrotic debris, pus or foreign bodies in a wound markedly interfere with healing.

    Healing of some special tissues :

    Epithelium can, in general, regenerate with considerable ease. Like fibrous tissue, the epithelium of the skin, alimentary, respiratory and urogenital tracts, has retained its regenerative properties. When the epithelium is lost, repair occurs by proliferation of epithelium from the margin of the wounds. Secretory epithelium of glands, however, is commonly not replaced, examples being mammary gland, gastric glands, seminiferous tubules etc In the liver, epithelial cells regenerate to a limited extent. In the kidney, no new nephrons can be formed. The cells can undergo only hypertrophy.

    Masothelium of the serous surfaces is quickly regenerated.

    Connective tissue : As we have already seen, fibrous tissue proliferates rapidly, replacing its own kind and others which are not able to regenerate.

When young it is more cellular and rich in young capillaries. With age, it becomes denser and less vascular. Old scars may be very dense and even hyaline with very poor blood supply.

Cartilage and bone : Due to avascularity, repair in cartilage is very slow and imperfect. It is usually replaced by fibrous tissue. Repair is very good and complete in the bone, osteoblasts playing the key role.

Tendons and ligaments regenerate slowly due to avascularity but healing is quite perfect.

Elastic tissue is also replaced rather slowly but completely.

Blood vessels are easily replaced by newly formed capillaries. However, the muscular coat (characteristic of arteries and veins) is seldom added.

Muscle : Lost muscles are reunited by fibrous tissue. Smooth and cardiac muscles never regenerate. A certain amount of regeneration sometimes occurs in voluntary striated muscle. Buds of muscle cells develop into the tissues of the fibrous union. These buds have many nuclei and little sacroplasm. If sarcolemma remains intact, then the sacroplasm can be replaced.

Nerve cell : Cannot be replaced; once destroyed is lost for ever.

Neuroglia proliferate readily.

Nerves : If the nerve cell is intact, repair of the peripheral nerves can occur. When a peripheral nerve is cut, the distal portion first dies but is slowly regenerated by new growth from the proximal end; and, although rarely, union can occur between proximal and the original distal end. If, on the other hand, a nerve cell body dies, then the whole neuron dies and is not replaced. The peripheral nerve then undergoes a series of retrogressive changes known as 'Wallerian degeneration". The axis cylinder becomes fibrillated and disintegrates. The myelin sheath breaks up into droplets and is removed by the macrophages and the sheath cells, the latter being converted into phagocytes.

FEVER

Fever, the abnormal elevation of central body temperature, is one of the most common and well-known manifestations of disease.

In generalized inflammatory diseases, there is an increased metabolic rate, and as a result the body temperature rises. Body temperature is normally maintained by a complex series of feed-back reactions that control the dissipation and production of body heat. Heat is normally lost by radiation, conduction, evaporation of sweat, pulmonary ventilation, and by the loss of heat when secretory (milk) or excretory (urine) substances are eliminated from the body. In very severe inflammatory reactions these processes of heat dissipation are not able to keep pace with the increased metabolic rate and fever occurs. Or, it may so happen that the heat-regulating centre in the hypothalamus is injured by the irritants and so heat regulation cannot be properly controlled, again resulting in fever.

Thus, fever is a syndrome in which there is, besides rise of body temperature (pyrexia), a disturbance in metabolism, and various functional disturbances such as increased pulse rate, anorexia, nausea, vomiting, constipation, increased thirst, scanty urine and dehydration.

Hyper~~semia~~ is merely an increase of the body temperature due to heat storage without systemic disturbances and hence differs from fever or pyrexia.

CAUSES OF FEVER

Fever has long been attributed to products of tissue destruction. One common denominator of nearly all diseases associated with fever is tissue injury. Therefore the causes of fever are varied and include a number of agents that can inflict tissue injury. Among the most common causes of fever are bacteria. Most of the known pyrogenic(fever-producing) agents are derived from microbes or their products. In fact, all infections whether bacterial, viral, protozoal, fungal or rickettsial may cause fever. And, so the fever may be produced by :

1. **Endotoxins of Gram-negative bacteria** : These are also called bacterial pyrogens. They form part of the cell wall of Gram-negative bacteria. They are present in all Gram-negative organisms (both rough and smooth forms), but Gram positive bacteria and other microorganisms lack them. The endotoxins are macromolecular complexes and are chemically lipopolysaccharide (LPS).

2. **Gram-positive bacteria** : A fundamental difference between Gram-negative and Gram-positive bacteria is that Gram-positive bacteria do not seem to possess the pyrogenic lipopolysaccharide endotoxins present in cell walls of Gram-negative bacteria. Therefore, the fever-inducing activities of Gram-negative and Gram-positive bacteria are markedly different; the former being more potent fever-producers. Fever resulting from Gram-positive bacteria appears to be due to multiple causes.

3. **Viruses** : The pyrogenic factor is closely associated with the viral particle. However, little is known at present about the specific factors in viruses that cause fever.

4. **Protozoa**, examples being trypanosomes, piroplasma and anaplasma.

5. **Fungi and rickettsiae.**

6. **Hypersensitivity** : The mechanism of fever has been studied in several types of delayed (cellular) hypersensitivity. The essential mechanism appears to be the formation of antigen-antibody complexes which are the initial stimulus for pyrogen release. The reaction of antigen with antibody results in release of a serum pyrogen from the damaged tissues of the host, which then induces the fever.

7. **Mechanical injuries**, such as severe crushing, extensive surgical operations.

8. **Vascular disorders** producing infarcts, such as, myocardial infarction.

9. **Neoplasms.** Most of the malignant tumours may cause fever due to their undergoing degenerative and necrotic changes.

PATHOGENESIS OF FEVER

Recently new facts have come to light regarding the production of fever. It is now clear that the various factors enumerated under the causes of fever do not themselves produce the fever directly, but that they induce the fever indirectly by releasing a pyrogen from the tissues of the host. This so-called 'endogenous pyrogen' EP) can be detected in the circulation, and this, in turn, is directly responsible for producing the fever by acting on the heat-regulating centre in the hypothalamus.

Neutrophils are the most important pyrogen-producing cells. Monocytes and macrophages also play a very important role in the production of endogenous pyrogen . Other cells include Kupffer cells in liver and fixed tissue macrophages (in lung, spleen, lymph node, liver, skeletal muscle, kidney). In contrast with the neutrophil and monocyte, the lymphocyte has not been demonstrated to release endogenous pyrogen. However, lymphocytes do play a role indirectly in fever production in states of delayed hypersensitivity. In these conditions, antigen stimulates sensitized T-lymphocytes to release a 'lymphokine' which, in turn, activates neutrophils and monocytes to release endogenous pyrogen.

Eosinophils, basophils and mast cells also do not release endogenous pyrogen.It is interesting to note that cells which release endogenous pyrogen have proved so far to be those that are capable of phagocytosirs. However, the meaning of this association,if any, is not as yet clear. It is not well understood at present as to how actually do the above mentioned cells become activated to produce and release the endogenous pyrogen. Probably phagocytosis may be acting stimulus in the process of pyrogen release. But then, the mechanism of pyrogen release still remains unknown.

Thus, to conclude, fever most often arises principally because of reduced dissipation of heat through functional changes controlled by the heat-regulating centre in the hyphothalamus, and that this centre is influenced by a common endogenous pyrogen released mainly from neutrophils (also monocytes, macrophages and certain other cells) by blood-borne toxins from pathogenic micro-organisms, or from protein decomposition following tissue injury.

CHANGES IN FEVER

Circulation : Increased pulse rate, lowered blood pressure.

Respiration : Increased respiration. This may be brought about by the action of CO_2 on the respiratory centre. Due to increased metabolism in fever more CO_2 is produced.

The functions of various organs and glands are diminished. This results in diminished formation of saliva, bile, pancreatic juice and urine. Symptoms of vitamin deficiency may be noticed due to anorexia. Excessive sweating may produce acute fluid loss resulting in circulatory and renal failure. Parenchymatous degeneration of liver, kidney and heart may be noticed. Basal metabolic rate is increased at the rate of 10 per cent for each degree of Centigrade rise and so body proteins are consumed at a fast rate, resulting in emaciation noticed in protracted febrile conditions.

The Course of Fever : There are three stages in fever :

1. **The Cold stage or period of rising temperature :** This is also the initial stage or the stage of shivering. During this period the temperature rises but the person feels cold and rigors occur due to contraction of cutaneous blood vessels.

2. **The Hot stage (Fastigium) :** is the period of sustained hight temperature, the fastigium or flush. Here the temperature reaches maximum. The person feels hot as the cutaneous vessels are dilated.

3. **The Sweating phase :** is the period of falling temperature or defervescence. The temperature begins to fall suddenly by **crisis** or **slowly by lysis** and the patient sweats profusely

FUNCTIONS OF FEVER

These are beneficial in nature and include :

1. Increased phagocytosis; moderate rise in temperature increases the activity of neutrophils.

2. Increased production of neutrophils.

3. Distribution of the leucocytes is accelerated due to increased velocity of blood.

4. Formation of antibodies more quickly and in larger quantities.

5. Bacteria cannot thrive at high temperature and so to a certain extent fever is bacteriostatic.

6. Antigen–antibody reactions occur more rapidly.

CHAPTER 6

DISTURBANCES OF GROWTH

APLASIA AND AGENESIS : Agenesis means "without beginning". Aplasia means "without formation" (Plasin = to form). That means to say, the organ in question had a beginning but due to some reason it had failed to develop.

In both the conditions, the tissue or organ, is absent. In the place of the aplastic organs, a fatty or fibrous mass may be present. Usually aplasia of one of paired organs occurs, eg. adrenal, kidney. No clinical effect is felt since the other organ carries on the work of the two.

Aplasia or agenesis of a vital organ is incompatible with life. The fetus dies immediately after birth if it is born alive.

HYPOPLASIA : In this condition, the organs fail to develop to their full normal size though there was a beginning. This is a congenital defect The affected organ, which is usually of the paired organs is reduced in size.

Causes of Aplasia and Hypoplasia are :

1. Heredity. Aplasia or hypoplasia of an organ is inherited. These may be sex linked also (cerebellar hypoplasia, taillessness, hypoplasia of ovary, abrachia).

2. Injury or stres to the embryo may cause these defects. Injury may be caused by the following : (a) mechanical injury; (b) ionizing radiation; (c) ischemia: (d) chemical poisons; (e) diseases of the dam infecting the fetus—bacterial, viral or spirochetal infections and (f) malnutrition.

3. Abnormal location. In cryptorchidism the testes that are in the abdominal cavity are smaller and hypoplastic.

ATROPHY is decrease in the size of a tissue after it has attained its full growth It differs from hypoplasia in that the latter has never attained the normal full size at all, but was under-developed from the beginning.

The reduction in size of the tissue may be due to (1) decrease in number of constituent cells or (2) it may be due to decrease in size of individual cells. The first variety is called *numerical atrophy* while the second *quantitative atrophy*. Atrophy is usually a slow process.

Types : Different types are recognised depending on the causation.

1. **Physiological atrophy :** In involution of various organs, physiological atrophy is normally seen. For example, atrophy of thymus during growth from infant to an adult; atrophy of the ovary and breast after menopause.

2. **Senile atrophy**: This may be included in the physiological types because when animals grow older atrophy of lymphoid tissue, breast, bones and ovary occurs. This may partly be due to arteriosclerosis which is an aging process.

3. **Starvation atrophy :** In starvation fat is used up to yield energy. After all the fat is used, the muscular and glandular tissues are utilised to supply energy. The central nervous system and the bones are the last to be affected.

4. **Malnutrition atrophy :** There may be derangement in the assimilation of food as in chronic diseases or the food may not be utilised.

5. **Atrophy due to lack of adequate blood supply :** In this may be included conditions like anemia and chronic venous congestion and lesions producing ischemia that ultimately cause atrophy due to lack of oxygen and blood supply. In CVC of the liver, the cells around the central vein are atrophied because of lack of oxygen and nutrition. So also narrowing of blood vessels in arteriosclerosis causes ischemia and atrophy. In amyloidosis ischemia occurs locally with atrophy of affected parenchyma of liver and kidney.

6 **Disuse atrophy :** Activity and use of a part acts as a stimulus for the growth of an organ. As a corollory. the opposite is also true, viz, if a part is not used, it becomes wasted. This is seen in muscles of limbs immobilised in the treatment of fractures.

7. **Neurotrophic atrophy :** If a trophic nerve is injured, the corresponding muscle atrophies : eg. in injury to *Suprascapular* nerve the *Supraspinatus* muscles atrophy. Similarly atrophy of the *Crico-arytenoideus* muscle occurs due to injury of the recurrent laryngeal nerve (resulting in roaring). Injury to a nerve causes its degeneration and so impulses to muscles cannot pass to them for contraction and so they become atrophied doing no work. The mechanism, therefore, is similar to disuse atrophy, eg. atrophy of limb muscles in poliomyelitis.

8. **Pressure atrophy :** Prolonged pressure produces atrophy. Examples : tissue pressed upon by growing neoplasms, cysts, aneurysms etc. produce atrophy. Hydatid cyst of liver or lung causes pressure atrophy. Actually the pressure may cause local ischemia, which may produce atrophy due to dificient nutrition and oxygen.

When ducts of excretory or secretory organs are occluded accumulating material causes pressure atrophy of the organ.

9. **Exhaustion atrophy :** Prolonged over work may cause exhaustion atrophy. Examples : atrophy of thyroid in exophthalmic goitre. Here the thyroid is continuously stimulated and after a time, it is no longer able to cope up, exhaustion sets in and it gets atrophied.

Similarly, exhaustion atrophy of pancreas can occur in continued hyperglycemia due to gigantism resulting in diabetes.

10. **Feed-Back atrophy :** May occur in endocrine glands if their normal hormones are supplied continuously either by a tumor or in therapy. For example, atrophy of adrenal cortex occurs if a tumor of cortisone producing tissue is present.

11. **Endocrine atrophy :** Hormones are necessary for the normal growth and health of certain cells. These are called trophic hormones. In their absence, the concerned cells are atrophied. For example, in hypopituitarism atrophy of the thyroid, testes ovary, mammary gland and adrenal cortex occurs.

Macroscopically, the affected atrophied organ is smaller in size, and may be flabby. Capsules when present are shrivelled. Some organs may show relatively greater amount of fibrous tissue and so may appear firmer. The tissues are decreased in size because the protein catabolism in the cells is faster than protein synthesis.

Microscopically, the cells may be smaller than normal, and may be fewer in number. Sometimes some may disappear altogether. Since the cytoplasm is decreased, the nuclei appear to be relatively larger in size and the number of the nuclei also appear to be more. Some cells (heart and liver) contain a yellow, granular lipochrome—brown atrophy or pigment atrophy. In some organs, fibrous tissue is more conspicuous as in spleen where trabeculae appear more prominent. As the parenchyma shrinks, fat accumulates in the stromal connective tissue such as in pancreas, striated muscle and breast. This is sometimes known as fatty atrophy.

Serous atrophy of fat is seen in debilitating and cachectic diseases. The fat depots reveal this change, especially the cardiac, renal, pancreatic and omental fat. The fat assumes a gelatinous appearance and in its place a thin watery fluid is seen.

HYPERTROPHY : Hypertrophy is the enlargement of a tissue or organ due to increased size of the individual functional cells or fibers without disruption of normal architecture.

Hypertrophy is always a result of increased function demanded of a tissue. It is never due to an irritant and inflammation. Therefore, the increase in size of the cells is due to higher rate of metabolism which is reflected by increased work. Hypertrophy occurs in cells that have lost the capacity to multiply by mitosis. There is an increase in the amount of protein and RNA in the cell and also increase in organellae.

Three varieties are described.

1. **Physiological hypertrophy** : This variety occurs in normal health. The best example is the increase in size of uterus during pregnancy. Similarly, the biceps of the black-smith is enlarged. The muscles of draft horses become large.

2. **Compensatory hypertrophy** : In this condition if some tissue of an organ is lost the remaining tissue becomes enlarged to compensate for tissue lost. This is exemplified by the hypertrophy of one of paired organs when the other is atrophied or damaged. In the absence of or damage to a kidney, the other becomes hypertrophied since it has to carry on the work of both. Here no new elements are formed. The existing glomeruli and tubules become increased in size.

3. **Adaptive hypertroyhy** : This is adaptation of an organ to certain pathological conditions. For example, when there is stenosis of the left auriculoventricular opening the left auricle hypertrophies since the muscles have to contract more forcibly to overcome the obstruction and so the muscle cells do more work and become hypertrophied due to increased metabolism. Similarly in aortic valvular stenosis the left ventricle hypertrophies: hypertrophy of the stomach may occur in pyloric obstruction: hypertrophy of ileum may occur in a defective ileo-cecal valve which prevents free passage of ingesta: in asthma and pulmonary adenomatosis the musculature of lung hypertrophies; in partial obstruction of urethra, hypertrophy of bladder occurs.

Pseudohypertrophy : In this condition, increase in the size of the cells involved does not occur but the whole organ appears larger in size due to the increase in some other tissue. For example, in pseudohypertrophy of the muscle, (called *pseudohypertrophic muscular dystrophy* in man) the volume of the muscular tissue is increased not because of increase in size of the muscle cells but because of the presence of adipose tissue or granulation tissue. Actually the muscular tissue is atrophied and decreased in amount.

HYPERPLASIA : Hyperplasia is an increase of a tissue or organ due to increase in the number of cells. There may be disruption of architecture.

We have already noticed that in an adult animal, only certain cells have retained their powers of proliferation while others have lost this property. For example, the hepatic cell and the fibroblast can proliferate while the nerve cell and the muscle cell do not. So hyperplasia of the latter cannot occur but only hypertrophy. In many instances, hypertrophy and hyperplasia may coexist together (liver).

Causes . 1. Chronic irritation : The irritant may be

i) **Mechanical :** Chronic mechanical irritation on skin stimulates the proliferation of connective tissue and the epithelial cells, eg. kennel granuloma of the elbows in dogs; calluses of hands of working man. ii) *Toxic chemicals :* Toxic substances in *Senecio* plants cause hyperplasia of hepatic connective tissue. iii) *Bacterial infections :* During infections, irritants may be produced causing hyperplasia eg. lymphoid hyperplasia in chronic infections.

2. Endocrine imbalance : eg. hyperplasia of the prostate which is cured by castration or injection of estrogens. Hyperplasia of the endometrium and mammary gland is produced by injection of estrogens.

3. Excessive Endocrine Secretion : i) *Gigantism* is found in tumors of the eosinophile cells of the anterior pituitary. In this condition excess of somatotrophic hormone is secreted resulting in excessive growth of the skeleton and macrosplanchnia. ii) Hyperplasia of the penis and clitoris occurs in tumors of the adrenal cortex since androgens are secreted in excess.

4. Deficiencies : Deficiency of iodine produces hyperplasia of the thyroid.

Deficiency of vitamin A results in hyperplasia and hyperkeratinisation of the epithelium of skin (This same effect is produced by poisoning with chlorinated naphthalenes which destroy vitamin A in the body). In nutritional roup of chicken, caused by vitamin A deficiency, hyperplasia of the esophageal epithelium is seen.

5. Viruses : Viruses cause proliferation of cells and so hyperplasia results. Examples are : Hyperplasia of the epithelium by pox viruses and wart viruses.

Hyperplasia is functional and so is found in the bone marrow in **myeloid hyperplasia** which occurs at times of need. In the adult, in the long bones, the marrow contains fat But in anemic states, this is converted to red marrow which is erythropoietic (myeloid hyperplasia).

Hypertrophy and hyperplasia, it should be understood, are purposive and result as a response to increased demands. For example, when a portion of the liver is lost, there is hyperplasia and hypertrophy of the surviving cells. Again these conditions are controlled processes since they are beneficial to the body. But the cell proliferation that occurs in haphazard and autonomous manner in tumors, is not beneficial to the body and serves no purpose.

METAPLASIA

Even in the adult, there are certain undifferentiated cells which have potentialities to develop into diverse tissues. When need arises, these are stimulated and differentiate into an entirely new tissue in place of the existing one.

Metaplasia is the transformation of one type of tissue into another. But this transformation is limited. One type of ectodermal cells can produce only another type of ectodermal cells; for example columnar epithelium can be transformed into squamous stratified. One type of mesodermal cell can be transformed into another. Fibrous tissue may be converted into bone. But an ectodermal cell cannot be transformed into a mesodermal cell.

Metaplasia is seen in only two types of tissue, viz. epithelium and connective tissue.

NOTE :— Metaplasia is different from anaplasia. In the latter, there is absence of differentiation of the neoplastic cells and so they resemble embryonic type of cells. This change is noticed in fast growing malignant tumors.

Epithelial metaplasia : This arises due to (i) Chronic irritation and (ii) Vitamin A deficiency.

Chronic irritation : Metaplasia that arises due to chronic irritation is essentially a protective mechanism since the new type of tissue is usually more resistant. For example, in prolapse of the uterus, the columnar epithelium is changed into squamous startified form, which can withstand infection and irritation better. In gall stones, the columnar epithelium of the gall bladder is converted into squamous stratified.

" In avitaminosis A, " the simple columnar or ciliated epithelium is converted into squamous stratified. This may not be beneficial to the animal but on the other hand may be harmful. Because, if the ciliated epithelium of the bronchi is replaced by squamous stratified epithelium the foreign bodies and secretions are not removed and this facilitates infection, resulting in pneumonia and gangrene.

Connective tissue metaplasia : Connective tissue may be converted to cartilage or bone. This occurs in the course of healing as in the wounds of abdominal wall. In old age ossification of the laryngeal and tracheal cartilages is common. In old cattle thickened pulmonary alveolar septa may be converted into spicules of bone. Metaplasia is very commonly seen in the mixed tumors of the mammary glands of the bitch. The stroma may be converted into cartilage and bone and so it is common to find a fibro-chondro-osteo-adenocarcinoma.

Sequelae : Metaplastic epithelium may become neoplastic as is seen in metaplastic bronchial epithelium of man, from which bronchial carcinoma may arise.

Mesothelial metaplasia : When irritated with the accumulated fluid. the flat cells lining the peritoneum, pleura and pericardium become cuboidal or columnar.

Endometriosis in women is an example of the metaplasia of peritoneal mesothelium.

DYSPLASIA : This is a name given to the alteration in the size, shape and orientation of adult cells. Usually epithelial cells are affected though sometimes mesenchymal cells may also be involved.

Cause is usually chronic irritation and inflammation. This is seen in the epithelium of the skin, cervix and esophagus.

Microscopically, the normal epidermis contains, from below upwards the basal-cell layer, the prickle-cell layer and the cornified-cell layer. Each layer is also composed of definite number of rows of cells and the cells are uniform in size and shape. But in dysplasia the number of rows of cells in each layer is increased and the size and shape of cells are also altered. Instead of a single layer of basal cells, there may be many layers. Mitotic figures are many. The cells of prickle-cell layer are changed in size and shape.

In the glandular epithelium the polarity of the cells is lost and so their normal parallel arrangement is disrupted.

Sequelae : If the cause is removed, the cells return to normal. It should be remembered that dysplasia is a stage seen early in the development of cancer—precancerous stage—and so should always be viewed with concern.

CHAPTER 7

CIRCULATORY DISTURBANCES

Hyperemia
   Active hyperemia
   Passive hype emia
     Acute local passive hyperemia
     Acute general passive hyperemia
     Chronic local passive hyperemia
     Chronic general passive hyperemia
Hypostatic congestion.

Ischemia
Hemorrhage
Thrombosis
Embolism
Infarction
Edema
Shock

Hyperemia or congestion is the condition in which there is increased amount of blood in the blood vessels of the body.

Physiological hyperemia is found normally in actively functioning organs:eg. mammary gland during lactation; intestinal mucosa during digestion.

Active hyperemia · This is a condition in which there is increased flow of blood in the arterial system due to dilatation of arterioles and capillaries. This condition is an acute phenomenon.

Causes : 1. Stimulation of vasodilator centre. Blushing, in which there is reddening of the face, neck and ears is of this category and is psychic.

2 **Paralysis of vasoconstrictor nerves:** eg paralysis of sympathetic nerves causes hyperemia.

3. **Reflex nervous mechanism** ; as in application of heat; injury by trauma or chemicals that occurs in inflammation, through the mediation of metabolites.

Active hyperemia is the first stage in acute inflammations and is the cause of redness, heat and swelling seen in that condition.

Macroscopically, the part is red and warm.

Microscopically, the capillaries are dilated and contain red cells, which are not normally seen in histological sections. Also capillaries appear to be numerous, since many of them, which are closed normally, are now engorged and come into view.

Passive hyperemia: in this condition there is decreased out flow of blood from an area or organ and so blood accumulates in the venous side of blood vessels.

Passive hyperemia may be acute or chronic.

Acute passive hyperemia : This may be local or chronic.

Acute local passive hyperemia occurs due to sudden blocking or pressure on veins. Examples of this condition are found in strangulated hernia, volvulus and intussusception. In these conditions, the thin walled veins are compressed while the thick walled arteries are still patent. So blood flows into the area but is not drained from it. So hyperemia results. Because of the increased tissue tension, arterial flow also is diminished subsequently. Hence the part suffers from lack of nutrition and oxygen, necrosis resulting. Invasion by saprophytic bacteria results in gangrene. Thrombi of veins and pressure by ligatures, tourniquets or rubber bands may occlude the veins with resultant local hyperemia.

Acute general passive hyperemia : The most important cause is acute heart failure in which there is diminished blood pressure. Due to insufficient force of cardiac contraction blood stagnates in the great veins and the venous system. This condition is seen just before death.

Similarly acute general passive hyperemia may be met with in asphyxia and shock.

Macroscopically, the part involved is swollen and bluish-red in color (cyanotic) due to the presence of large amount of venous blood. The large veins are engorged and dilated. On section dark blood oozes from cut surfaces.

Microscopically, the veins and capillaries are filled with blood. If the liver and spleen are affected, the sinusoids are engorged with blood.

Chronic local passive congestion : The causes are:

1. Obstruction of a vein from within: This is usually brought about by thrombosis.

2. Pressure of a vein from the outside: The following produce pressure on veins.

(a) expanding tumor. (b) enlarging lymph node or abscess (c) tight bandages on legs; **(d)** contracting scar tissue may compress local veins. This is very well seen in the cirrhosis of liver where the contracting fibrous tissue compresses the sinusoids, obstructing flow of blood and so hyperemia of the portal vein occurs.

(e) hydrostatic force as in varicose veins of the legs in man. Here due to the great dilatation of the veins, the valves are no longer functional and so the blood stagnates due to gravity, especially in the leg veins.

Macroscopically, the organs involved are enlarged in the early stages. Later, when fibrosis results, the part becomes smaller. The affected organ and the blood are bluish—cyanotic—due to the presence of unoxygenated blood.

Since the obstruction develops gradually, collateral circulation is established and so the congestion will be moderate. Due to hypoxia, the capillary endothelium is damaged and so edema and hemorrhages in the tissue are seen. Specialised parenchymatous cells suffer because of hypoxia and so they may undergo atrophy. Fibrous tissue is stimulated by hypoxia and reduced nutrition and so fibrosis of the part occurs which is responsible for the smallness and firmness of the organs.

Microscopically, engorgement of the venules and capillaries is seen. Edematous fluid separates the cells and fibers of the tissue. The cells may be atrophied and increase in the fibrous tissue is apparent.

Sequelae : The changes and damage are, more or less, permanent. Function of the tissue or organ is impaired.

Chronic general passive hyperemia: This condition is produced if there are some lesions hindering flow of blood in the heart and lungs, two organs through which large quantity of blood passes.

The lesions of the heart are: 1. **Stenosis** of valvular openings. Usually stenosis (or narrowing) of the auriculo-ventricular openings is more common. Vegetative endocarditis is the commonest cause of stenosis of the auriculo-ventricular openings. Because of the stenosis (narrowing) of the A–V openings during diastole of the ventricles all the blood in the auricles does not empty into the ventricles during auricular systole and so some of the blood is retained in the auricles Hence, during auricular diastole, all the blood contained in the vena cavae and the pulmonary veins does not empty into the auricles (since it is not completely empty) and so blood is retained in these veins and hence stagnation of blood in them results.

2. **Valvular insufficiency :** The valves do not close properly. Usually mitral and aortic valves are affected. Because mitral valves do not close properly, blood escapes into the left auricle during ventricular systole and so blood accumulates in the auricle and during auricular diastole, the pulmonary veins are not able to empty themselves completely and so blood stagnates in them Hence blood flow through alveolar capillaries is hindered. This leads to improper emptying of blood from the right ventricle during systole. Eventually, blood stagnates in the right auricle and the great veins.

In the case of tricuspid incompetency (which is common in animals) during systole of right ventricle, blood escapes into the right auricle and so auricle is not empty during its diastole.   Hence during the auricular diastole blood in vena cavae does not empty completely into the right auricle and so there is stagnation of blood in them.

3. **Chronic myocardial failure**: Degeneration or necrosis of the myocardium may occur due to arteriosclerosis of the coronary arteries, toxins as in diphtheria; deficiency of thiamine (beri beri in man), deficiency of vitamin E (white-muscle disease). hypertension and congenital anomalies. The heart is too weak to over-come the elasticity of the arterioles and force blood through them So, blood pressure falls. In an effort to maintain the blood pressure the arterioles contract reflexly thereby forcing all the blood into the veins, which dilate and accommodate all the blood. Hence venous congestion occurs, Among the anomalies that may be responsible are the persistent foramen ovale, defects in the interventricular septum and malposition of the aorta.

4. **Constrictive pericarditis:** In this condition, the expansion of the heart is hindered and flow of blood into the heart through vena cavae is blocked and so stagnation of blood is seen in the veins. This condition is met with in traumatic pericarditis, where there are adhesions between the pericardium and epicardium.

The pulmonary lesions that cause obstruction of flow of blood are

(i) **Emphysema**: In the horses in the condition called **broken wind** (chronic alveolar emphysema) the alveoli are dilated and ruptured thereby compressing the capillaries or even destroying some of them.   So the capillary bed is decreased and therefore blood stagnates in the right ventricle and pulmonary artery as all the blood cannot be pumped into the lungs from the heart. Eventually blood stagnates in the right auricle and great veins.

(ii) **In chronic interstitial pneumonia:** as well as in **pneumconioses**, the contraction of the fibrous tissue constricts and obliterates the capillaries and so blood stagnates in the heart and veins.

Macroscopically, in general, the affected tissues are cyanotic (bluish) due to the presence of venous blood. The veins stand out, being engorged with blood. Edema of the limbs and ventral portion of the abdomen is common.

Serous cavities contain increased amount of fluid. Three organs are mostly affected: the lungs, liver and spleen.

Lungs : The lungs are heavier, appear solid looking and leathery. They are dark brown in color. On section, blood oozes out. Only very little crepitation is present.

Microscopically, the capillaries and venules are engorged and tortuous. The alveolar septa are thickened due to increased fibrous tissue. Some capillaries rupture and hemorrhages into the alveoli occurs. The erythrocytes are hemolysed liberatirg hemoglobin, which is split up by enzymes into hemosiderin (iron containing part) and hematoidin (the iron-free part). The hemosiderin is engulfed by the macrophages and these are called the **heart failure cells.** They are so named because their presence in the sputum indicates cardiac damage and dysfunction. Such a heart will fail sometime in the future.

By Perl's stain (Prussian blue stain) the iron can be stained within the macrophages. These macrophages carry and deposit the pigment in the lymphatics and throughout the lung parenchyma, where a mild chronic irritation is set up by the pigment and so fibrosis occurs. This is the cause for toughness or hardness **(induration)** of the lung.Since the pigment also gives a brownish color the condition caused by chronic venous congestion of the lung is called **"brown induration"**. Certain amount of edema is present in the submucosa of the bronchi. In mitral stenosis, pulmonary arterioles develop a media to withstand the increased blood pressure.

Liver: Macroscopically, the liver is enlarged in size and increased in weight. The color is dark brown. The edges are rounded. On section, large amount of blood exudes and the cut surface has a peculiar mottled appearance, resembling cut section of a nutmeg. This is because, the central third of the lobules are darker in color due to engorgement of sinusoids of blood while the peripheral area, in which the sinusoids are not engorged, is lighter in color. Besides, the cells of the periphery have increased lipid content (fatty degeneration due to hypoxia) which also contributes to the paleness of the periphery.

Microscopically, there is atrophy and sometimes even necrosis of the centrilobular cells. This may be due to hypoxia, which renders the hepatic cells more susceptible to the action of toxic substances absorbed by the hepatic cells brought to the liver. The cells at the periphery may be loaded with fat. In more chronic cases there will be fibrous thickening of the walls of the central vein, collapse of the lobules due to necrosis of the cells and then the fibrous tissue may extend into the parenchyma giving rise to **'Cardiac cirrhosis'**

Spleen : Chronic venous congestion of the spleen is seen usually in chronic venous congestion of the portal vein as a result of hepatic cirrhosis. This is a complication of vegetative endocarditis in swine and traumatic pericarditis in cattle.

Macroscopically, the spleen is enlarged, three to four times its normal size (splenomegaly). It is deep reddish-purple in color, and is hard and indurated-**cyanotic induration**. The edges are rounded and on section large amounts of blood ooze out.

Microscopically, the splenic sinusoids are engorged and their walls are thickened with fibrous tissue.   There is deposition of large quantities of hemosiderin pigment in reticulum cells.   Trabeculae are thickened.

Kidneys :   **Causes** :   Apart from cardiac and pulmonary lesions that give rise to chronic general venous congestion, pressure on the renal veins by tumors of adrenals or abscesses may cause venous congestion of kidneys.

Macroscopically, the kidneys are enlarged and dark purple in color. The cortico-medullary junction zone is dark-red.

Microscopically, the glomeruli are engorged and the tubular epithelium shows degenerative changes due to anoxia. The fibrous tissue content of the glomeruli may be increased. The basement membrane of the glomeruli may also be thickened.

In general, the changes that take place in chronic venous congestion may be summarised as follows:

Decreased circulation gives rise to

   a) hypoxia
   b) decreased nutrition   } These cause atrophy and necrosis
   c) accumulation of catabolites   } of the cells followed by fibrosis.
   d) increase in vascular permeability   }
       (as the endothelium is damaged)   } This gives rise to hemorrhages.
   e) Sodium retention : Damage of kid-
       neys, liver and adrenals, affects sodium
       excretion. Renal tubules reabsorb   This results in edema
       the sodium. Along with sodium water   } along with (d) above
       is reabsorbed and this increases the
       blood volume and so raised pressure
       in the venous blood results.

Hypostatic congestion: Due to gravity, blood may accumulate in the organs and tissues of the lower side of a recumbent animal.   This is called hypostatic congestion.

Hypostatic congestion is seen in animals suffering from diseases in which cardiac damage occurs giving rise to slow or sluggish movement of blood.   This is also common in animals made to lie down for a long time in a very inconvenient posture from which they are not able to get up   In horses that are made to stand for long periods hypostatic congestion of the limbs is frequent.   Just before death, due to loss of vascular tonus, hypostatic congestion may be seen.

Of all organs hypostatic congestion is seen commonly in the lungs because here the thin walled capillaries have no support and so can dilate accomodating the blood.

Congestion that occurs on the lower parts of an animal after death due to gravity is a form of hypostatic congestion.

Ischemia is the local deficiency of arterial blood in an organ.   This is generally due to blocking of some part of arterial system.   Obstruction of a blood vessel may be due to :

   1) blockage of the lumen by a thrombus or embolus;
   2) alteration of the wall of the artery: this may be due to (a) atheromatous plaques or (b) to increased tonus of the musculature (ergot poisoning) and

3) pressure from the outside—by a tumor, cyst, abscess. neoplasm or tourniquet etc. If ischemia is continued for some time, necrosis or atrophy of the part rusults. This depends on the rate of closure and the amount of anastamoses developed. If the closure is rapid and anastamoses poor necrosis of the affected part occurs. Ischemic necrosis called **infarction** is common in heart, spleen, kidney and brain. But if closure is slow and anastamoses poor, atrophy will result. The atrophied parenchymatous tissue is replaced by fibrous tissue.

Macroscopically, the ischemic tissue appears pale and is cold to the touch.

HEMORRHAGE

Hemorrhage is escape of blood from an artery, vein or capillary to the outside, (external hemorrhage) or into a body cavity or into tissues : (internal hemorrhage). Depending upon the site hemorrhage is designated as:

epistaxis—bleeding from nose

hematemesis—blood in vomit

hemoptysis—blood in sputum

metrorrhagia—bleeding from uterus

enterorrhagia—bleeding from intestine

melena—blood in stools

hematuria—blood in urine

hemothorax—blood in the thoracic cavity

hemopericard—blood in the pericardium

hematocele—bleeding into tunica vaginalis

hemosalpinx—bleeding in oviducts

hematoma—a tumor-like accumulation of blood in the tissue

Apoplexy is hemorrhage into the brain with loss of consciousness. Depending on the extent, hemorrhage may be classified as:

petechiae - these are small and pin-point hemorrhages.

ecchymoses—these are extensive hemorrhages in the tissue or on body surface.

extravasation—these are extensive hemorrhages in the tissues.

Depending on the mode of formation, capillary hemorrhages may be formed by **rhexis**—that is by the break in the wall of the capillary or by

diapedesis—in this there is no actual break in the capillary wall. But the red cells pass either through the endothelial cells, or between them when the permeability of the endothelium is increased or along with leucocytes.

Causes : Hemorrhage may be due to either conditions affecting blood vessels or conditions affecting the blood itself.

Conditions affecting the blood vessels 1. Trauma: mechanical injuries including laceration, incisions, contusions and ruptures disrupt the vessel wall thereby enabling escape of blood. Clumps of bacteria as in Swine erysipelas, anthrax and hemorrhagic septicemia may block the capillaries and so hemorrhage may result.

2. **Necrosis** of the vessel wall—by an ulcer of gastric mucosa or by a spreading neoplasm may result in hemorrhage.

3. **Diseases of the vessel walls: aneurysm :** the wall in this condition becomes weakened, bulges and may rupture. This occurs in the anterior mesenteric artery in the equines due to *Strongylus vulgaris* infection.

Atheroma of the aorta and larger arteries is a weak spot from where hemorrhage may occur. There is degeneration, necrosis and weakening of the wall of the blood at this place.

4   **Toxic injury to the capillary endothelium :** The capillary endothelium is easily damaged by various toxins which may be:

a) Bacterial; common in such conditions like anthrax, hemorrhagic septicemia, black quarter.

b) Viral—hog cholera. The viruses injure the endothelial cells.

c) Chemicals: arsenic, phosphorus, chloroform, cyanide.

d) Enterotoxemias as in sheep and calves due to *Clostridium welchii* type D

e) Asphyxia—lack of oxygen injures the endothelium.

5. **Increased blood pressure** In exercise and excitement, as in race horses, there may be bleeding due to rupture of blood vessels in which the blood pressure has enormously increased

6 **Hypoxia and lack of nutrition:** This is a condition of importance in capillary hemorrhages. This occurs in passive congestion. The capillary endothelium is damaged and so hemorrhage occurs.

Conditions affecting blood constituents:

i) **Hemophilia:** This is a hereditary and sex linked disease in which clotting is very much delayed. Probably antihemophilic globulin is deficient.

ii) **Thrombocytopenic purpura:** There is decrease in the number of platelets. This occurs as a post—infection toxemia as in strangles, pneumonia etc.

Thrombocytopenia may also be due to injury to bone marrow by

(a) irradiation; (b) chemicals—benzol etc. and   (c) replacement by leukemic cells.

In hypersplenism thrombocytopenia occurs since the thrombocytes may be destroyed by the over-activity of the spleen.

iii) **Nutrition:**

a) **Deficiency of Vitamin K.** Vitamin K is required for the formation of prothrombin. In its absence prothrombin cannot be formed and so clotting does not take place.

In long continued sulfonamide therapy hemorrhages may occur because sulfonamide does not permit the intestinal flora that synthesize vitamin K to flourish resulting in vitamin K deficiency.

b) **Deficiency of Vitamin C:** Vitamin C is required for the ground substance of the wall of the capillaries. In vitamin C deficiency, the capillaries become more fragile and so hemorrhages occur.

iv) **Heparinoid state:** In anaphylactic shock and after severe exposure to ionising radiation, an excess of heparin is found that inhibits clotting of blood and so hemorrhages may occur. Heparin prevents agglutination of platelets and so platelet thrombi are not formed. It neutralises the action of thrombin preventing, thereby, the formation of soft fibrin clot.

(v) **Plant toxins**—Bracken fern and sweet clover prevent the formation of prothrombin.

Arrest of hemorrhage: Various mechanisms come into play in the arrest of hemorrhage.

1. Vascular contraction: This is the first stage in the arrest of hemorrhage from small vessels. This is very effective in the early stages. Because when the blood flows, clot cannot form as thromboplastin is washed away.

2 Platelet agglutination; Platelets agglutinate and occlude the bleeding area. This is very important in small punctate hemorrhages. This is the **white clot.**

3. Clot Formation : When the blood flow slows down, clot forms and occludes the rupture in the vessel wall thereby stopping bleeding. This is the **red clot.**

4. Tissue pressure: As blood passes into the tissue, the perivascular pressure increases and this opposes the intravascular blood pressure and so bleeding is arrested.

5. Decreased blood pressure: This is a very important factor in the control of massive hemorrhage. When large qnatities of blood are lost, blood flow is slowed as blood pressure is decreased and so no more bleeding occurs.

The clot; The clot that forms is of two types: temporary and permanent. The temporary clot may be again of 2 types: the red and white clots.

The red clot occurs due to the formation of fibrin in the meshes of which erythyrocytes are trapped. Due to trauma that is responsible or hemorrhage thrombokinase is liberated from the injured vessel wall and the surrounding tissues. This thrombokinase puts into motion the formation of clot from fibrinogen. It takes about 4 minutes for the blood to coagulate.

The white clot is formed by the conglutination of the platelets, which stick together at the roughened area of the endothelium where rupture has occurred and forms a plug arresting the bleeding. Subsequently thrombokinase liberated from the platelets may coagulate the blood that stagnates at the area. This clot is like a "nail head". The head of the nail that plugs the rupture is made up of the mass of agglutinated platelets and the tail is made up of the coagulated blood.

The permanent clot is formed by the organisation of the temporary clot. Around this temporary clot is formed an inflammatory exudate as a result of injury. From the surrounding area, new capillaries and fibroblasts grow into the clot. The macrophages that arrive remove the temporary clot by phagocytosis. So vascular fibrous tissue is formed. Since there is no longer any need for the macrophages these wander off. Similarly as blood vessels are not required any more, they disappear. The fibroblasts lay down collagen and the ruptured area is firmly sewn and repaired.

If the wound is infected, this type of complete healing cannot take place since an abscess occurs. The temporary clot is softened and so secondary hemorrhage may occur from this place

Changes in the extravasated blood: What happens to the extravasated blood depends on its size and location. Blood that escapes onto the surface of the body or into the lower portions of intestines or vagina or urethra is lost. Blood found in the lumens of stomach or intestines may be digested

When hemorrhage is very small, i.e , petechial in nature, the fluid portion is absorbed by lymphatics and blood vessels and the red cells and fibrin are removed by phagocytosis.   The leucocytes wander away.   But in large hemorrhages, the red cells cannot be completely removed and so they become hemolysed, the liberated hemoglobin staining the tissue.   By the enzymes of the tissues and phagocytes, hemoglobin is split into heme and globin.   Globin is a soluble histone and so is removed.   Heme is split into two parts, an iron containing hemosiderin which is insoluble and is present as golden yellow granules   Some of these may be engulfed by macrophages. in which they can be stained by Prussian blue stain.   Hematoidin (iron free) is soluble and so diffuses into the surrounding tissues through the fluid, staining them yellow or green.   The fibrin is removed by phagocytosis. Most of the leucocytes wander off.

Sequelae of hemorrhage: The sequelae depend upon the rate and quantity of blood lost. If the blood loss is slow, the blood volume will be made up from the tissue fluids.   But if the rate is fast and if $\frac{1}{3}$ to $\frac{1}{4}$ of blood in the body is lost, the animal may go into 'shock' and die.   Symptoms of hemorrhagic shock are decreased blood pressure, increased rate in pulse and respiration, pallor, faintness and cold skin.   The brain suffers from hypoxia resulting in fainting

A sudden hemorrhage into the pericardium causes such pressure that the heart is prevented from dilatation during diastole and so it does not get filled up.   So no blood is pumped out resulting in arrest of circulation and death. This condition is known as **cardiac tamponade.**

Hemorrhages into trachea or larger bronchi may cause asphyxia— drowning by blood.

THROMBOSIS

Intravital, intravascular clotting of blood is called **thrombosis. That means** to say that the clotting of blood occurs within blood vessels while the animal is alive.

Causes: 1. Injury to the endothelium :

a) Trauma: Any injury that damages the endothelium of the blood vessel can produce thrombosis. eg. lacerations, contusions, ruptures etc.   Frequent puncture of veins for intravascular injections is also a cause.

b) Toxins: Bacterial toxins injure the endocardium and produce thrombosis. Streptococci and erysipelothrix injure the endocardium and produce the vegetations which are thrombi.

c) Degenerative disease of vessel wall, especially atherosclerosis and damage to the intima.

d) Viruses: Some viruses, especially Hog Cholera virus, injure the endothelium causing thrombosis (in spleen).

e)Parasites: Parasites that pierce through the vessel walls and those that inhabit the vessel walls cause injury to the intima. For example *Strongylus vulgaris* is pierces through and inhabits in the wall of the anterior mesenteric artery in the horse and causes thrombosis.

f) Tumors: Invading tumors injure the endothelium of the blood vessels.

The endothelium involved in the circumstances mentioned above is damaged and destroyed leaving bare the cement substance on which platelets and fibrin become attached, so as to seal it off as a protective mechanism. When the endothelium is injured, ATP is broken down to ADP and the released ADP causes platelet agglutination.

2. **Alteration in the blood flow:** In the normal or fast blood flow thrombokinase which triggers the clotting mechanism, is washed out. Similarly the blood constituents are removed from the site in a rapid stream. So for clotting to occur, the rate of blood flow must be slowed down. Besides, when the blood flow slows down, the cells from the axial stream fall out to the periphery and platelets being the outermost in the axial stream are the first to fall out and stick to the endothelium by virtue of their adhesive capacity. So thrombosis is more common in veins in which blood flow is slower. Blood flow is slowed in the following conditions.

a) Chronic venous congestion, b) In the aged and debilitated animals, c) Aneurysm, d) Varicose veins: In animals thrombosis is common in the vascular sinuses in the submucosa of the nasal passages in cattle and horses; in the scrotal plexuses of the horses and in the large veins of the broad ligaments of bovine uterus.

e) It is most common in persons suffering from congestive heart failure and in those confined to bed. It is in the leg veins that thrombi are seen in such persons since they are superficial and easily damaged by compression.

f) In shallow breathing due to pulmonary disease, abdominal distension or shock, venous flow is retarded.

3. **Alteration in the constituents of blood:**

a) **Increase in the number of thrombocytes:** After parturition and severe surgical operations, there is a steep rise in thrombocyte count and hence the reason for thrombosis encountered in these conditions.

b) **Increase in the adhesiveness of the platelets :** This occurs also after parturition and operations.

c) **Decrease in the heparin:** Heparin is an anticoagulant. If it is decreared, coagulation may occur. It is very likely that in certain diseased states heparin may be decreased in amount.

d) **Increased amount of Plasma fibrinogen and prothrombin.** In trauma the fibrinogen and prothrombin content of the plasma is increased, which enhances the chances for thrombosis.

e) **Increased viscosity of blood:** Increased viscosity as occurs in hemoconcentration (due to dehydration that occurs in severe injury or following major surgery) and polycythemia, predisposes to clotting of blood.

f) **Sludging of blood:** Normally red blood cells repel each other. But in trauma, the red cells get agglutinated intravascularly and this is known as sludging of blood.

g) **Increased fragility of erythrocytes** Ruptured erythrocytes give rise to cellular debris which favours agglutination of blood elements and so helps in clot formation. Any increase in the fragility of red cells, therefore, a predisposing factor in thrombosis. This occurs in hemolytic anemias.

h) In a man, if cortisone is administered continuously (as therapy for rheumatoid arthritis), coronary thrombosis is apt to occur, probably by increasing the blood lipid content.   Lipaemia causes intravascular platelet aggregation leading to thrombus formation.

Mechanism of thrombus formation: When blood flows through the vessels, normally no thrombus forms, because the endothelium is smooth.   When it is injured it becomes roughened   At these places, the platelets get attached due to their inherent adhesive property. We have already studied that this is the way the minute hemorrhages are arrested by the formation of a plug of thrombocytes   Now, when the thrombocytes stick to the endothelium and form a small plug, more and more platelets that fall out get adhered to it, at right angles to the flow of blood, in ridges or lamelle called the lin s of zahn.   The small plaque or plug that is first found on the wall is called a **mural thrombus.** The presence of this mass of platelets obstructs the flow of blood and gives rise to swirls, that cause slowing of the blood stream.   This mass of platelets forms the **white or pale thrombus**

After agglutination, the platelets lose their morphological individuality and are formed into a homogeneous mass in which the platelets cannot be identified.   During this process some of the platelets may disintegrate and liberate a thromboplastin which converts fibrinogen into fibrin.

For the clotting mechanism to be completed, it takes a minimum of four minutes. But if the blood flow is rapid, thrombokinase and fibrinogen will be washed away and no clotting occurs.   But in sluggish blood flow, fibrinogen is converted into fibrin which in its meshes entangles the erythrocytes which give red color to the thrombus—**the red thrombus**.

The thrombus increases in size in the direction of blood flow.   This still further slows the flow of blood thereby facilitating the thrombus to become larger. So, thrombus may ultimately become so enlarged that it may completely occlude the lumen (**occlusive thrombus**).

Varieties of thrombi : 1   Cardiac thrombi: These may be found in the (a) wall—**mural thrombi** or (b) Valves—**valvular thrombi** Valvular thrombi are common in the pig in Swine Erysipelas (endocarditis). Mural thrombi are common in cattle in Black Quarter caused by *Clostridium chauvoei.*

Ball thrombi are those that are found in the auricles.   These are unattached and are too large to pass through the auriculo-ventricular openings and cause intermittent valvular obstruction.

2. Arterial thrombi : Most common example is the thrombus found in the anterior mesenteric artery in equines, caused by *Strongylus vulgaris.* A thrombomatous lesion of aorta caused by *Spirocerca lupi* in the dog or *Onchocerca armillata* in cattle is a frequent site of thrombi in these animals.

3. Venous thrombi: These are common in the leg veins of men, who are suddenly bedridden. The leg veins collapse and get pressed against the hard surface of the bed by the calf when the intimal endothelium gets damaged, releasing thromboplastin and subsequently thrombus forms. Venous thrombi may also be seen in chronic venous congestion.

Venous thrombi are not so common in animals. The scrotal plexuses of the horse, the vascular sinuses of the nose in the cow and the veins of the bovine uterine broad ligaments are the places where venous thrombi may be seen.

4   Capillary thrombi : These are seen in inflammations.

5. Lateral thrombi : These are found attached to a side of the wall of blood vessels.

6. Occlusive thrombi : These occlude a small artery or vein.

7. Propagating thrombi : These arise by progressive extension, proximally, of an occlusive thrombus.

8. Saddle thrombi : A thrombus that is spreading into both the branches at the bifurcation of an artery is called a saddle thrombus.

9. Canalised thrombi : Occlusive thrombi develop a passage through which partial blood supply is maintained. Such passages are lined by endothelium and such thrombi are called canalised thrombi.

10. Septic thrombi : A thrombus containing bacteria is called a septic thrombus. If thrombosis is the result of infection. the thrombus starts as a septic thrombus. But, a bland thrombus may be subsequently infected to become a septic thrombus.

11. Aseptic thrombi : These are opposite of the above and are bland.

12. Pale or white thrombi : As explained earlier, these are composed of aggregates of thrombocytes. These are seen in the heart, aorta and femoral arteries.

13. Red thrombi : These consist of all the constituents of the blood and are seen in the veins more commonly.

14. Mixed thrombi: These are a mixture of white and red thrombi,the white occurring during fast flow of blood while the red during its sluggish flow.

15. Laminated thrombi : These are a variety of mixed thrombi. But here the white and red thrombi are formed as alternate layers. When the flow of blood is rapid, as in exercise, white thrombus is formed, while at rest, the red thrombus forms. Again at exercise, a white thrombus forms over the earlier red one and so on and on.

Fate of thrombi : The fate of a thrombus depends on whether it is septic or aseptic. **Aseptic thrombi : 1. Contraction :** The fibrin of the thrombus may contract due to the action of thromasthenin from platelets and so the thrombus may shrink, leaving a space for blood flow.

2. Absorption : Due to the activity of the leucocytls, the thrombus may be softened and absorbed. A danger in this process is that small pieces may be broken off and form **emboli**, which may get lodged in other organs and produce infection.

Organisation : The endothelium of the vessel wall proliferates and spreads over thrombus. The endothelial cells may even line any crevices that may be present Subsequently blood capillaries (from *vasa vasorum*) and fibro blasts from the wall of the vessel invade the thrombus. The fibroblasts elabo_ rate collagen and elastin. Along with this vascular fibrous tissue macrophages also invade the thrombus for its liquefaction and removal. Finally the new connective tissue, which on maturation contracts, provides enough space to enable blood flow just sufficient to keep the structures beyond alive.

4. Canalisation : During the organisation of the thrombus, we have observed that the endothelial cells proliferate and line the thrombus and the crevices. These cells may even invade the thrombus and anastamose with the capillaries of the connective tissue. During maturation and reorganisation of the organised thrombus, arterioles and venules are formed coaxial to the blood vessel. Some of these may run the entire length of the thrombus when it is said to be **recanalised** and so blood circulation is restored through these new blood vessels of the thrombus.

5. Calcification : The thrombus, especially of the veins, may be calcified. When found in the vein, it is known as a **phlebolith.** (Stone of vein).

Septic thrombi : If the thrombi are infected with bacteria, inflammatory reaction appears with exudate, which liquefies the thrombus, septic emboli result, giving rise to centers of infection wherever these emboli lodge.

If the infection is by pyogenic bacteria abscesses form. These organisms give rise to pyemia and metastatic abscesses elsewhere.

Effects of thrombi ; The effects of thrombi depend upon : 1. whether they are mural or occlusive; 2. The size of the vessel affected; 3. Whether an artery or vein is affected; 4. The site of occurrence (i. e., organ affected); 5. amount of collateral circulation present or may develop; 6. The rate of formation of the thrombus; 7. Whether it is disintegrated and form emboli and 8. Whether it is bland or septic.

The following effects are noticed : 1. No serious effects are seen especially in thrombosis of small veins, where sufficient collateral circulation is present. Even with thrombosis of larger vessels, no harmful effects are seen if the animal is kept at rest.

2. Passive congestion and edema occur when there is obstruction to venous return. This is seen in limbs mostly.

3 Infarction or ischemic necrosis occurs in thrombosis of the arteries—eg. infarction of kidney or myocardium.

4. Gangrene of the intestines or limbs : When main arteries of intestines or limbs are affected, necrosis occurs, which when invaded by saprophytic organisms results in gangrene.

5. Interference with cardiac function mechanically :

a) Ball thrombus in the auricles prevents filling of the atria and it may also cause obstruction to flow of blood from auricles to ventricles.

b) Mural thrombi in the ventricles may interfere with filling of the ventricles.

The above conditions finally result in chronic venous congestion.

6. Colic : In horses, thrombosis of the anterior mesenteric artery is accompanied by colic.

7. Lameness : Partial obstruction of iliac arteries causes lameness in animals and this is apparent when the animal is exercised.

8. Septicemia or Pyemia will result if the thrombus is infected. Metastatic abscesses may occur in various organs where septic emboli may lodge.

9 Sudden death: This occurs in thrombosis of coronary arteries, (more common in man but rare in animals).

Note: It should be understood that thrombosis is a vital dynamic phenomenon aimed at strengthening a weak or ruptured area of a blood vessel, primarily for arresting hemorrhage. Untoward or harmful effects that may arise are only secondary, unfortunate accidents

Post-mortem clot and thrombus : We have already studied that after death due to injury of the endothelium by hypoxic conditions, thromboplastin is liberated which causes clotting of blood. So, it is important to be able to distinguish between an ante-mortem clot (thrombus) and a post-mortem clot. The following are the distinguishing features.

Character	Thrombus	Post-mortem clot
Size	fills the vessel	is smaller than vessel
Consistency	friable, dry and crumbles when pressed	rubbery elastic and moist.
Surface	rough	smooth and glistening
Attachment	firmly attached to the endothelium	weakly attached to the wall
Structure	laminated-lines of Zahn seen	homogeneous
Color	red, white or mixed	red or "chicken fat"
Endothelium	endothelium is roughened.	smooth—not damaged.

EMBOLISM

The mechanism by which foreign material is transported through the circulatory system is known as **Embolism** and the foreign material transported is an **embolus** (emboli, plural). Emboli may be solid, liquid or gaseous.

Emboli are almost always lodged in the artery or capillary, since their lumina get progessively narrowed as the blood flows away from the heart and so afford suitable sites for the lodgement of emboli. Emboli of veins are rare since the width of the veins gradually increases as the blood flows towards the heart and hence the foreign materials do not get trapped but freely move into the heart.

Emboli may also arise in the lymph vessels. These are trapped in the sinuses of lymph nodes.

Sources and varieties of emboli:

1. Thrombotic emboli : Thrombi are the commonest sources. Due to force of blood stream, pieces of thrombi may be broken loose and swept away. **Septic thrombi** may disintegrate and form emboli. Emboli arising from the heart are disseminated throughout the body. Emboli originating from venous thrombi get lodged in pulmonary vessels. Septic emboli produce foci of infection wherever they may lodge.

Thrombotic emboli contain thrombokinase and so wherever they lodge, they produce thrombosis and so may cause complete occlusion if the original embolus has not already done so.

2. Fat emboli Fat of the animal may form emboli. Fat which is fluid in character and is immiscible with water remains separate and forms emboli. The following are the several ways by which fat emboli may arise.

i) Fat emboli are common in crushing injuries of the bones which are greatly moved and manipulated. The fat in the marrow cavity gets dislodged and finds its way into the vascular sinuses and from thence to the blood stream forming emboli. These are usually lodged in the lungs. Fat embolism is common in small animals involved in automobile accidents

Fat emdolism is more common in older people because in them the bones are brittle and so break easily and the bone marrow is almost totally fatty.

ii) In obese animals emboli may arise from trauma of subcutaneous fat.

iii) In fatty livers due to deficiency of lipotropic factors (choline, methionine or cystine as in chronic alcoholics) or due to poisoning by phosphorus or carbon tetrachloride, any injury to liver (trauma or biopsy) may produce fat emboli as the damaged liver cells rupture and liaberate the fat to form emboli.

iv) Extensive burns involving subcutaneous fat is yet another cause for fat emboli.

3 Air embolism: Formerly it was thought that a little air injected into the veins while giving intravenous injections might cause death. But it is now known that as much as 100 ml. of air can be injected intravenously without much consequence. It is only when air much in excess of this volume finds its way into the blood circulation that damage occurs.

The following are the various methods how air emboli may arise.

(i) In incision of the large neck veins during surgery or in suicide, air may be sucked into the veins during inspiration.

(ii) In criminal abortion involving pumping of air into the uterus, air may enter the large uterine veins and cause embolism

In these cases, the air that enters into the circulation reaches the heart where during the cardiac contractions, blood is churned to a thick foam which prevents the filling of the right auricle. This causes acute heart failure, death supervening within a few minutes.

(iii) **Caisson disease:** The solubility of gases in a fluid is proportiona to the pressure to which the fluid is exposed. The pressure under water is very great. So, in divers, at great depths, the air that enters the blood from lungs is dissolved in greater amounts than on the land. When the diver surfaces rapidly, the pressure in lungs is suddenly decreased and the dissolved gases are liberated quickly. Oxygen is utilised by the tissues. But the nitrogen which bubbles ou in large quantities forms emboli, occluding and stopping the flow of blood in the capillaries with dire results. To prevent this calamity, the divers are brought to the surface by stages slowly so that the nitrogen is liberated in small quantities slowly when it does not do any damage.   Caisson is a bell shaped contrivance in which the divers are brought to the surface slowly, in stages.

(iv) When people ascend suddenly by air planes to high altitudes, the pressure diminishes and so dissolved gases of blood are released suddenly with effects similar to those described under caisson disease.

In these cases, nitrogen that is suddenly released stops the circulation in the capillaries and necrosis results. This may occur in the brain and spinal cord where small infarcts may be produced causing cerebral symptoms. Finally coma results. The suddenly released nitrogen may tear the fat containing cells so that fat emboli form in venules and capillaries. This is especially the case with persons having fatty livers. Gas bubbles liberated in the nervous tissue cause its disintegration.

In man, caisson disease is also known as **bends** as the patient assumes a bent posture due to cramps produced in this condition.

4. Bacterial emboli: Bacteria may be detached mechanically in sites where they are present as heavy infection and transported to distant organs.

These bacteria which are finally arrested in capillaries, proliferate and produce fresh foci of infection, known as metastatic foci of infection.

Emboli arising from septic thrombi are also sources of bacterial emboli. Usually cardiac vegetations in vegetative endocarditis are the sources of such septic emboli producing foci of infection in the liver, kidneys, spleen or lungs.

5. Tumor emboli: A growing tumor may invade a vein and clumps of tumor cells may then be transported to other areas where new growths occur. This is a method in the spread of tumors

6. Parasitic emboli. *Dirofilaria immitis* may form an embolus in the pulmonary artery of dogs. Schistosome species form emboli in the mesenteric, portal and nasal blood vessels. The larvae of the round worms during their migration may form emboli. eg. ascarides, strongyles. Similarly masses of trypanosomes may form emboli. This is of clinical importance in the treatment of surra by tartar emetic. If the drug is given at a rapid rate, a large number of the protozoa may be killed suddenly, which may form emboli in the coronary vessels with fatal termination. Becoming emboli, nematode larvae may reach the brain, where *cerebrospinal nematodiasis* or *Kumri* is produced. Filaria are found as emboli in lymphatics.

7. Lymphatic emboli are those that are transported by lymph vessels Mostly tumor cells are the chief lymphatic emboli. These emboli are trapped in the nearest lymph node

8 Amniotic embolism: During difficult parturition in women, with forceful uterine contractions, when there is a tear in the membrane, amniotic fluid, meconium and desquamated epithelial cells of the fetus may be forced into the venous system forming emboli which get lodged in the lungs.

9. Paradoxical emboli: These are emboli that pass directly into the left auricle from the right auricle through a patent foramen ovale. So an embolus originating in a vein is found lodged in systemic vessels rather than in pulmonary vessels.

Effects of embolism The effects of embolism depend on the nature and size of the embolus as well as on the site of its lodgement. The results may be 1. Formation of new foci of infection, if the embolus is septic. 2. Formation of metastatic tumours, if the embolus is neoplastic 3. Focal ischemia is produced by impaction with medium or large sized emboli. 4. Infarction or focal necrosis occurs if an artery is occluded and sufficient collateral circulation is not present.

INFARCTION

An infarct is an area of coagulative necrosis that results due to the sudden blocking of an artery which has no collateral circulation. Such an artery is called an "end artery". But, strictly speaking, these arteries have a few collaterals which are not, however, sufficient to keep the tissue alive if the artery is obstructed.

The causes for arrest of blood supply are :

1   Thrombus and embolus: These are the most common causes for arrest of blood supply.

2. Pressure on the vessel wall causing ischemia: a) Ligatures or tourniquets; b) Tumors; c) Abscesses or cysts: d) Volvulus and intussusception of the intestines; e) Decubitus—bed sores in bedridden persons in whom the continued pressure produces ischemia and necrosis of the part involved.

3. Contraction of the vessel wall: The walls of the vessels may be contracted so much that ischemia results. For example, in ergot-poisoning, the smooth muscle of the media contracts with narrowing of the wall and so ischemia results. This condition is seen in animals in the extremities (legs, ears, tail, wattle) where collateral circulation is limi ed

4. Hypotension; Severe hypotension that may occur in shock or anesthesia may cause ischemia of the brain resulting in necrosis (infarction).

Pathogenesis: Due to sudden blockade of an end artery, say by a thrombus, blood flow stops and so there is stagnation of blood. The tissues that have been hitherto supplied by this artery, suffer from hypoxia since the available oxygen in the stagnant blood is gradually used up. Hypoxia of the capillaries and the venules results in their becoming atonic and so they dilate. They are filled with blood from the new anastamosing vessels present (These are not sufficient to keep the part alive). As a result of anoxia and of irritation by the chemical metabolites that accumulate there the cells die. The specialised parenchymatous cells die first and later cells of other tissues. Necrosis starts at the centre within 24 hours and extends to the periphery. By 72nd hour the whole area has undergone coagulative necrosis and so the architecture of the tissue can be discerned For example, in an infarct of kidney, glomeruli and tubules can be discerned though the whole area takes on a pink stain, showing that it is necrotic. The blood vessels are also damaged due to anoxia and so their endothelium is damaged giving rise to edema and hemorrhage by diapedesis· So, the area appears stuffed with blood and it is for this stuffing that the process gets its name of infarction (infarcire means to stuff). In organs in which capillaries are less supported and in which there is more interstitital space (as in lungs) this stuffing with blood is greater than in the more solid organs like the liver The part that is stuffed with blood is now red in color and so is called a red infarct. Subsequently the red cells undergo hemolysis and the hemoglobin diffuses into the surrounding tissues. So now the infarcted area becomes pae and is called a pale infarct.

The dead tissue acts as a foreign body and so incites an inflammatory reaction Hence, around the necrotic area, in the healthy tissues is found a zone of hyperemla. Macrophages migrate to this area. From this zone angioblasts grow into the necrotic area and form capillaries. So also fibroblasts enter and form fibro-vascul.r tissue. The necrotic tissue is liquefied and removed by the macrophages and then the granulation tissue occupies the whole area. When this occupation is complete the capillaries disappear. Macrophages wander away and the fibroblasts lay down collagen, which contracts. Hence a healed infarct is depressed in level than the surrounding healthy tissue.

Macroscopically, the infarcts are red or pale in color and are cone shaped—the apex of the cone being at the point of obstruction of the vessel and the base towards periphery.

Infarcts are most commonly seen in kidneys, spleen, intenstines and lungs. They may be found less frequently in the liver in which they are red.

Infarcts of the kidney are most common in the cows and pigs: These are yellow or pale and wedge-shaped and are found in the cortex. The apex of the wedge is at the arcuate arteries while the base is at the capsular end of the cortex. The capsule having separate blood supply of its own, does not become necrosed.

Usually, renal infarcts are not very harmful and one notices the healed depressed areas at postmortem. The sources of emboli are usually cardiac vegetations, in cattle due to *C pyogenes* or *Streptococci* and in pigs due to *Erysipelothrix rhusiopathiae.* In cows, in which it occurs most often, infarcts occur due to emboli arising from uterine vein after parturition.

Infarcts of the spleen : These may be pale or because of plentiful blood content may sometimes be red. Splenic infarcts are found at the borders more frequently. In the dog it is sometimes seen as a band encircling the whole organ These infarcts result from cardiac thrombi.

Infarcts of the heart ; In animals, infarcts of the heart are not as common as in man. The most common cause in man is arteriosclerosis, which is not encountered among animals. The infarct may be red or pale depending on its age. It may involve the whole thickness of the wall. Animals may die, even before necrosis occurs if infarction is extensive. Healed infarcts are recognised by depressed scars. One of the sequelae of cardiac infarction is *myomalacia cardis.*

Infarcts of the liver : When there is some disturbance of the blood supply, arterial and portal, as from pressure by tumors or in extensive thrombosis of the hepatic vessels that occurs in infection by *Clostridium hemolyticum* in bovines, infarcts of the liver may be encountered. These are usually red in color due to the double blood supply.

Infarcts of the lungs : In spite of the double blood supply by the pulmonary and bronchial arteries, infarction of the lungs is common. These are cone shaped and red in color. Pulmonary infarcts may arise when emboli lodge in pulmonary vessels. Emboli may originate from the uterine veins and the posterior vena cava in cows, and from the mesenteric veins in horses. Thrombosis of the posterior vena cava arises due to spread of inflammation from hepatic abscesses to this vessel. In the pigs, hog cholera virus may injure the endothelium of the pulmonary vessels resulting in thrombosis and infarction.

Hypostatic congestion, pneumonia and chronic venous congestion may so alter the pulmonary circulation, that thrombosis and embolism of pulmonary vessels may occur. In cattle, sheep and pigs *pasteurella* infection is a frequent cause of pulmonary infarction. Pulmonary infarction is a complication in bronchopneumonia. Here the infarcted area undergoes suppuration and liquefaction necrosis, resulting in an abscess.

Infarcts of the intestines These are common in the horse due to thrombosis of the anterior mesenteric artery by the migrating *Strongyle* larvae. The infarct may involve the whole circumference of the bowel and is usually red.

Infarction also occurs in volvulus, intussusception and strangulation. In there situations, the veins are compressed while the arterial blood flow is unaffected leading to acute local venous congestion and ultimately to necrosis and gangrene. The wall of the affected portion of the intestine is red and thickened due to infiltration by blood and plasma. The lumen of the bowel contains blood and exudate. As explained earlier, infarction of intestines is fatal due to toxemia, shock and peritonitis.

Infarcts of the brain; These are common in man and result from arteriosclerosis. Among animals, cerebral infarction is seen in dogs that meet with automobile accidents. Infarcted area undergoes colliquative necrosis (softening). The myelin is engulfed by the microglia which thus constitute the ''**Compound granular corpuscles**'. The softened area may either be organised or cyst may be formed by the neuroglia which proliferate and enclose a cavity containing a yellow fluid the ''**apoplectic cyst**''. Nervous disorders may be seen in animals with this condition.

Sequelae of infarcts 1. The tissue may be organised and a scar may form.

2, Gangrene This is common in infarcts of the intestines or extermities due to invasion of the necrotic tissues by saprophytes

3. Death: If infarction occurs in brain or heart or intestines, death occurs due to shock. Death may also be due to toxemia or septicemia that results when bacteria invade and multiply in an infarct.

EDEMA

Edema is an abnormal accumulation of fluid in the intercelluar spaces and body cavities.

In the body, water is found within the cells (intracellularly), between the cells (inter-cellular spaces) and in the blood plasma Nearly 2/3 of this water is intracellular, about one fourth is interstitial while the water of blood-plasma makes up to only 1/15 of the total quantity.

The extracellular or interstitial fluid serves the following functions: (i) It forms the environment of the cells being in contact with them. (ii) It forms a medium for transfer of metabolites from the cells and supply of nutrients to them (iii) It functions as a buffering mechanism and maintains the normal pH.

The intracellular water content is almost constant. It is the interstitial fluid content that may vary

Formation of tissue fluid: As per Starling's hypothesis, the fluid exchange between the blood and tissues takes place in the capillaries The capillary endothelium is semipermeable. That means to say, it is permeable to water and crystalloids but not for colloids. So while water and salts freely pass out into the tissues, plasma proteins are held back. The following forces play in the formation of the tissue fluid.

1. **Hydrostatic pressure of blood in the capillary**: The capillary has two ends—the arterial and the venous. At the arterial end, the hydrostatic pressure of the blood in animals is about 45 mm. Hg., while at the venous end 15mm. Hg

2. **The extracellular fluid out side the capillary** has a pressure of about *one* or *two* mm. Hg So, the fluid in the capillary is at a higher pressure and because of this filtration occurs and fluid moves out into the tissue spaces.

3. **Colloidal osmotic pressure of the blood plasma.** The plasma proteins are water-binding and are said to have osmotic effect. By this osmotic pressure, water can be drawn into the capillary The osmotic pressure exerted by the plasma proteins is equivalent to 30 mm. Hg.

4. **The tissue fluids** also have certain amount of proteins possessing osmotic pressure. This is equivalent to about two to three mm. Hg. Due to increased os pr. of blood colloids, water tends to flow into the capillaries. Therefore the osmotic pressure of the colloids opposes the filtration effect of the hydrostatic pressure.

The plasma colloids are : albumin (Mol. wt. 40,000). globulin (Mol. wt. 170,000), fibrinogen (Mol. wt. 500,000), prothrombin etc Of these, albumin having a smaller molecular size possesses higher osmotic activity, nearly four times that of globulin.

Now let us consider what happens at the **arterial end** of a capillary Here the hydrostatic pressure of blood is 45 mm. Hg. The hydrostatic pressure of tissue fluid=1 to 2 mm. Hg

Therefore filtration pressure = 45 minus 2 = 43 mm. Hg.
Colloidal osmotic pressure of the proteins = 30 mm. Hg.
Colloidal osmotic pressure of tissue proteins = 2 to 3 mm. Hg.
Therefore, absorption effect = 30 minus 3 = 27 mm. Hg.

Now this absorption effect is lower than the filtration pressure by 16mm. Hg. So water passes out into the tissues at the arterial end of the capillary.

At the venous end : As the blood flows, the hydrostatic pressure falls to about 15 mm Hg.

Here the osmotic effect is = 30 minus 3 = 27 mm. Hg. So the effective absorption pressure = 27 minus 15 = 12 mm. Hg. and so fluid is sucked into the capillaries at the venous end.

Normally, there is no accumulation of interstitial fluid. That which is filtered is drained away by lymph vessels. In certain conditions there is accumulation of this fluid giving rise to edema.

Edema is also known as **dropsy or hydrops**, Generalised subcutaneous edema is known as **anasarca**. Edema of certain regions is known by different names :—

Hydrosalpinx - edema of oviduct.

Hydroperitoneum or Ascites —accumulation of fluid in the peritoneal cavity.

Hydropericardium—fluid in pericardium.

Hydrocele—fluid in the tunica vaginalis.

Hydrocephalus—accumulation of fluid in the ventricles of the brain.

Hydrothorax - fluid in thoracic cavity.

When fluid accumulates in the loose tissue, it can be dispersed by pressure and so at the point where pressure is exerted, a depression occurs. This is known as *pitting on pressure* When pressure is removed, the pit disappears as the fluid returns to its original site.

Causes of edema: 1. **Increased permeability of the capillary endothelium**: The endothelium is normally permeable to water and crystalloids. It is not permeable for colloids. But when the endothelium is injured by chemicals or anoxia, then it becomes permeable to plasma colloids, which pass out into the interstitial space. It is suggested that the cement substance of the endothelium is so altered that it allows the passage of proteins. Dilatation of the capillaries consequent on the relaxation of the Rouget cells is one of the important factors in the passage of protein into the edema fluid.

When the plasma proteins leak out, the osmotic pressure of the plasma is lowered while that of the tissue fluid is raised. So, this raised osmotic pressure exerts a great sucking pull on the fluid from the capillaries into the tissue. The drain of plasma proteins into tissues is called "albuminuria into the tissues". This mechanism is similar to albuminuria that occurs in acute nephritis in which albumin escapes into the urine from the glomerular capillaries, which are injured by toxins etc,

2. **Decrease of the colloid osmotic pressure** : This is by far the most common cause of edema in domestic animals. Loss of plasma proteins may occur in several ways :

i) **Continuous loss of blood** : a) In stomach worm infection and ancylostomiasis, blood is sucked by the worms Each worm is estimated to suck about .8 ml. of blood in 24 hours. The blood that passes through the gut of the worm serves to supply it necessary oxygen. So, when infection is heavy, with millions of worms in the stomach, the loss of blood is enormous. The worms have a habit of changing their position often and so inflict numerous punctures on the gastric mucosa From these minute wounds bleeding occurs since the worms are known to secrete an anticoagulant that prevents clotting of blood. So, the net result of these is that there is continuous drain of blood and with it plasma proteins are lost over months and years, finally resulting in hypoproteinemia.

b) Bleeding can also occur from chronic gastric ulcers in pigs and dogs.

ii) **Renal lesions** : Albuminuria occus in acute glomerulonephritis, when the glomerular capillaries are damaged by the toxins. But this condition is not of importance in animals, which do not usually suffer from glomerulonephritis. When albumin is lost in the urine hypoproteinemia occurs.

iii) **Hepatic disease** : Liver manufactures plasma proteins from the amino acids. If liver is damaged as in liver fluke infestation and cirrhosis this synthesis **cannot** occur and so hypoproteinemia occurs.

iv) **Insufficient intake** : In stravation. famine, malnutrition and malabsorption, sufficient amount of proteins are not either ingested or absorbed and so hypoproteinemia results.

We have studied that it is the osmotic effect of colloids that opposes the filtration effect of hydrostatic pressure and so prevents the free flow of water out of the capillary. We have also noted that it is the albumin fraction that possesses the highest osmotic effect, about 4 times that of globulin. So, the over all osmotic sucking effect of albumin fraction is atleast nine times than that of plasma globulin since albumin in blood is nearly three times the quantity of globulin (albumin:globulin ratio is 3:1). If, therefore, albmuin is lost then the osmotic pressure of blood plasma falls, the sucking effect on water is diminished or lowered and so water flows out of capillaries in larger quantities, resulting in edema. It is estimated that if plasma protein level falls below 5 percent edema will result.

3. Increased hydrostatic pressure of blood: Normal passage of fluid from the capillary to the tissues occurs because the higher hydrostatic pressure of the blood over comes the osmotic effects of the plasma proteins. It, therefore, stands to reason that if this hydrostatic pressure is raised as occurs in general or local venous congestion, edema will result. In cardiac failure, therefore, there is raised venous pressure and edema results. The stagnation of blood may also dilate the capillaries, making them more permeable, which yet may be another cause for edema formation.

4. Obstruction of lymph vessels: Normally, the tissue fluid is drained by lymph vessels. But if these are blocked, edema develops. Drainage by lymph vessels may be obstructed by

i) pressure from *outside* by tumors, abscesses, cysts, tourniquets or tight fitting harness. (ii) Obstruction of lumen *from within* by a) tumor cells; b) thrombus; (c) parasties—*Filaria* or *Wuchereria bancrofti* in man and *Demodex canis* in dogs.

iii) Inflammatory conditions as in farcy and ulcerative lymphangitis, in which the lumen is blocked by the exudate, debris etc.

The accumulating fluid may further press on the lymphatics and so may accentuate the condition still further. Edema that develops due to obstruction of lymphatics is local.

Macroscopically, the edematous part is swollen, is increased in weight and is smooth to the touch. If not accompanied by inflammation, the part is cold. The area pits on pressure and is painless. On section, clear fluid flows out from the cut surface. The tissue has a water-logged appearance. In edema of lungs frothy fluid is present in the trachea and bronchial tree.

Microscopically, the intercellular spaces are widened. Since the fluid contains proteins it stains pink and is granular.

Sequelae: If the cause is removed edema disappears. But if the edema persists, the fluid acts as a mild irritant and so fibrosis supervenes.

Different types of edema: 1. Inflammatory edema: We have studied that at the site of inflammation, exudate forms which is the cause of the swelling of the area *(tumor)*. The permeability of the capillary endothelium is increased by the action of the toxins and so fluid, rich in protein, passes out. Due to pressure of this fluid on local nerve endings, pain *(dolor)* is produced.

The exudate that forms in inflammation differs from the tissue fluid (transudate) in many respects. The following are the differences.

		Transudate.	Exudate
1	Color	No color, clear	Turbid and white, may be red if mixed with blood
2	Odor	Not present	Odor may be present
3	Reaction	Alkaline	Acid
4	Specific gravity	Lower than 1015	Higher than 1018
5.	Protein content	Low—less than 3%	More than 4%
6	Coagulation	Does not coagulate, fibrin content low	Coagulates both within the body and without, due to high fibrin content
7.	Enzyme content	Low	High
8.	Cell count	None or a few lymphocytes or mesothelial cells present	High. Neutrophils and erythrocytes present.
9.	Bacterial content	Sterile	Bacteria usually present

Since inflammatory exudate contains fibrin net work, the fluid is not displaceable and so it does not pit on pressure.

2. **Cardiac edema** In congestive cardiac failure, due to damage of mitral and aortic valves (in man due to rheumatic fever) chronic general venous congestion develops. Due to insufficient renal circulation in this condition (consequent on decreased cardiac force, which is responsible for the hydrostatic pressure of the blood in the vessels) oliguria with diminished chloride excretion occurs. This results in sodium-ion retention, which in the tissues raises the osmotic pressure of the tissue fluid, aggravating edema.

The causes for the formation of edema are;   i) Increased hydrostatic pressure of blood in the capillaries due to venous congestion; ii) Hyperpermeability of the endothelium, which becomes damaged due to the hypoxic condition that prevails due to a) stagnation and non oxygenation of blood. b) edema of lungs that also occur in congestive heart failure, prevents the entry of air into the alveoli and so oxygenation does not occur; iii) Retention of sodium ions by the dysfunction of kidneys.

In cardiac edema fluid accumulates in the dependent parts—in the legs and under the abdomen. This is because of increased venous pressure that develops due to gravity. If the patient is recumbent, edema of sacral and abdominal (ventral) regions is noted.

In animals, cardiac edema may be met with in traumatic pericarditis of bovines and chronic vesicular emphysema of horses. In these two conditions chronic venous congestion finally develops (as explained earlier under chronic general venous congestion) and transudate accumulates in the serous cavities—ascites and hydrothorax developing.

3. **Renal edema:** Edema is seen in acute, subacute and chronic nephritis. The genesis of edema in these conditions is different.

Acute glomerulonephritis. This type is not common among animals. In man edema first appears on the eyelids and face, because in these places the

tisssue is loose.   Usually edema is seen in the morning in the face but when the
patient gets up and assumes an errect posture edema of the face may disappear.
The following factors are thought to be responsible for the formation of edema.

a) Decreased colloidal osmotic pressure of blood :   Due to damage
of the glomerular capillaries by the toxins, albuminuria occurs,   resulting in
hypoproteinemia and edema.

b) Increased osmotic pressure of extracellular fluid :   In acute nephritis there is oliguria and anuria resulting in sodium-ion retention, which raises
the osmotic pressure of the tissue fluid and so helps in the formation of edema.

c) Increased capillary permeability :   The same toxins that injure the
glomerular capillaries, also injure the endothelium of other capillaries making
it hyperpermeable and so resulting in the passage of greater quantities of fluid
outside the capillaries.

d) Rise in capillary blood pressure :   The irritants that damage the
kidney also damage the heart causing cardiac failure, the outcome of which is
venous congestion and so raised capillary blood pressure.

ii) Subacute Nephritis and Nephrosis :   The factors that are responsible
for edema in these conditions are :

a) Decreased colloidal osmotic pressure of the blood :   In subacute
nephritis and nephrosis, due to damage of the glomerular capillaries. large
quantities of albumin are lost in the urine and so the albulin : globulin ratio of
normal plasma which should be 3 : 1 is reversed, resulting in decreased colloidal
osmotic pressure of blood.   Hence edema results.

b) Salt retention :   Hypoalbuminemia is accompanied by decreased
plasma volume which stimulates the adrenal cortex to secrete more of aldosterone.
This mineralocorticoid helps in the reabsorption of sodium chloride and so this
salt is retained in the tissue fluid, raising its osmotic pressure and causing the
formation of edema.

c)   There is probably increased production of antidiuretic hormone (ADH).

iii) Chronic glomerulonephritits :   Chronic glomerulonephritis
causes hypertension.   Prolonged hypertension throws a great strain on the heart,
which fails.   Cardiac failure results in chronic venous congestion and so there is
raised venous pressure.   Edema occurs due to raised capillary blood pressure.

4 Hunger or famine edema or war edema :   When people starve or are
fed with diets deficient in protein, edema develops.   This is due to hypoproteinemia and consequent fall in the colloidal osmotic pressure of the blood.   In
famine and war, availability of proteins is limited and so hypoproteinemia
develops.

5 Edema of the lungs .   The following peculiar features of the lungs
are responsible for the development of edema there :   i)   The texture of the lung
parenchyma is loose and so edema can develop easily;   ii)   The capillary surface
from which fluid can escape out is far greater in lung than in any other organ;
iii)   When the lung alveoli are inflated the alveolar endothelium becomes very
thin and is not capable of supporting the capillary endothelium against the
intracapillary pressure.   Hence the fluid easily flows out and enters the alveoli.

Edema of the lungs is a serious problem since the alveoli and bronchi get filled by the fluid which displaces air and so the lungs become useless functionally. Dyspnoea, cyanosis and death may follow.

Causes : a) Cardiac failure—hypertension, valvular disease, pericarditis (raising pulmonary capillary blood pressure).

b) Renal lesions leading to hypertension and increased capillary blood pressure.

c) Pressure on pulmonary veins by neoplasms.

d) Injury to brain and intracranial hemorrhage.

e) Rapid removal of effusions from pleural or peritoneal cavities.

f) Poisons damaging the capillaries—the poisons may be inhaled (war gases) or may reach lungs via blood stream (ANTU—alphanaphthyl thiourea—a rat poison, if ingested accidentally by dogs causes fatal edema of lungs).

g) Infections—as in pneumonias caused by various organisms and in Mulberry Heart Disease.

The edema fluid being rich in protein is a good medium for bacteria to thrive and so the animal easily succumbs to infection

6. Cachectic edema - Edema develops in the later stages of many wasting diseases and anemias. This may be attributable to cardiac weakness that develops as well as to malnutrition which causes damage to the capillaries.

7. Myxedema : This is a curious form of edema that occurs in chronic thyroid deficiency. The tissue metabolism is some how altered so that there is accumulation of protein and mucopolysaccharides in the tissue fluids, raising the osmotic pressure of the fluid locally and so large quantities of water are drawn into the site.

8. Parasitic edema : This is the most common form seen in animals suffering from infestations by stomach worms, liver-flukes and immature amphistomes. The stomach worms, *Trichostrongylus*, *Hemonchus* and *Osteriagia SP*. suck blood and cause hypoproteinemia.

The liver-flukes damage the liver. During the migratory life of the cercaria, hemorrhage and necrosis of liver occur. When the adult flukes inhabit the bile ducts, they cause irritation of the lining mucosa of the ducts by their spines and cause chronic irritation resulting in cirrhosis. The parasites are thought to liberate some kind of toxin which, acting on hemopoietic tissues, produces anemia. The damaged liver parenchyma is not able to synthesize the plasma proteins and so ultimately hypoproteinemia and edema result. The edematous fluid collects in the lower jaw, especially in sheep. The accumulation of fluid gives the area a peculiar appearance, as though a bottle is inserted under the skin and so the condition is known as *"Bottle Jaw"*. At this place the tissue is loose and so accumulation of fluid is facilitated.

9. Angioneurotic edema: In man sudden local edema develops on the lips, glottis and throat, which is believed to be due to the increased capillary permeability that results as a reaction of hypersensitivity to an unidentified allergen. Neurogenic action on the capillaries is also postulated. Edema occurring in snake bite is considered to be of this variety.

A similar condition is met with in cattle and horses. The allergen may be either endogenous or exogenous. The exogenous allergin may be a plant protein since edema is noticed after the animals graze on pasture when it is in flower· Fishmeal also is suspected to be a cause.

The histamine released causes dilatation of blood vessels and damages the capillary endothelium and so edema results. Edema is seen on the head, udder and perineum Some times only the conjunctiva and eye lids may be affected. Lachrimation is profuse. The affected parts are not painful.

Recovery is spontaneous.

10. Brisket disease: When cattle are moved to hill stations, 9000 ft. above sea level, they develop edema which is prominent subcutaneously over the abdomen, brisket, neck and jowl. This is known as *brisket disease*. A few fatalities may be seen· A small number of sheep and horses may also suffer.

At high altitudes, the partial pressure of oxygen is low and so oxygen supply to the tissues is diminished. When the animal is in a state of partial hypoxia, polycythemia develops, which increases the viscosity of blood. This rise in viscosity throws greater work on heart since it has to pump more forcibly.

Hypoxia causes polypnoea, which condition again taxes the hear further by making it work faster, and with greater force This, therefore, causes hypertrophy of the myocardium.

When the animals are in hills, they have to go up and down, entailing greater strain on the heart. Now, the heart also receives blood deficient in oxygen. So, the cardiac muscle working hard in hypoxic conditions becomes degenerated and hence the hypertrophied heart slowly dilates *(decompensation)*. When the ventricle dilates, the valvular ring is drawn downwards and so the cusps do not close perfectly and thus **valvular incompetency** arises This condition gives rise to chronic venous congestion.

Hence edema develops in high altitudes for the following reasons:

1. Hypoxia injures the capillary endothelium, making it more permeable.
2. Chronic venous congestion that develops causes a) increased capillary blood pressure, as well as b) hypoxia.

Because of gravity, the fluid accumulates at the region of brisket and hence the name of "Brisket Disease" to the condition.

SHOCK

Shock may be defined as "a common grave medical emergency charaeterzied basically by a reduction in the effective circulating blood volume and in the blood pressure" (Robbins) or as a "disparity between the volume of blood and the volume capacity of the vascular system" (Anderson).

Shock may be caused by severe hemorrhage, traumatic injury, severe burns, poisons and in man psychic stimuli.

Shock is divided into a) primary and b) secondary forms.

Primary shock : This appears immediately after injury, especially when it is extensive. This is of nervous origin, stimuli causing wide spread capillary paralysis.   Rough handling of animals and undue manipulation of intestines in abdominal operations may bring on this condition.

In man psychic states—fear, excitement, apprehension—may also result in shock.   Among animals restraint of wild and timid animals may cause a comparable condition.

Primary shock is similar to syncope or fainting.

The psychic neurogenic impulses cause vasodilatation of the splanchnic vessels with resultant lowering of blood pressure, cerebral ischemia and loss of consciousness.   The affected patient has a pallid face,   slowed breathing and slow, feeble pulse.   This is a transient condition and the patient recovers with rest.   No morphological changes are noticed in the organs.

Secondary Shock: This form is more serious, some times terminating fatally.   The essential feature is that there is disproportion between the volume of blood and the volume of the blood-vascular space.   Sufficient blood is not present to fill the blood vessels and maintain the blood pressure.   As such lesser amount of blood is available for the heart to pump out.   So circulation is inadequate.   This results in fall in peripheral blood pressure. To remedy this there is reflex sympathetic vasoconstriction to help maintain blood supply to heart and brain.   **This is the** cause of pallor of skin in man.   To conserve fluid ATP, aldosterone,renin- angiotensin system mechanisms are stimulated.But vasoconstriction causes renal ischemia and damage leading to death finally, as described later.. To correct this defect, heart's action is accelerated but the pulse is weak since the force of the heart is proportional to the quantity of blood that enters it. Finally death occurs.

Causes: 1. Reduction in blood volume: This may occur in two ways;   (a) loss of blood from injuries and (b) loss of fluid into injured tissues.   Examples of blood loss are hemorrhages within or without the body.   Total quantity of blood in the vascular system is therefore diminished.   Loss of erythrocytes and plasma proteins also add to the loss of space-occupying materials from the vascular system.

Loss of fluid occurs in the following situations:

i) severe burns $\left.\begin{array}{l}\\ \\\end{array}\right\}$ Plasma leaks out of capillaries
ii) crushing injuries into the tissues in large quantities.

iii) Persistent vomiting $\left.\begin{array}{l}\\ \\\end{array}\right\}$ Fluid is lost from the body.
and diarrhoea

iv) Sodium deficiency $\left.\begin{array}{l}\\ \\\end{array}\right\}$ This occurs in pyloric stenosis and consequently diminished fluid intake and dehydration occur.

v) Addison's disease $\left.\begin{array}{l}\\ \\\end{array}\right\}$ Dehydration occurs since there is excessive loss
and diabetic coma of fluids and electrolytes.

vi) Severe water deprivation.

vii) Intense edema that occurs by poisoning by war gases (phosgene, mustard gas and lewisite) and poisons (ANTU) reduces the blood volume.

2 Capillary bed dilatation: Normally, only some of the capillaries are patent while others are closed.   But if all the capillaries are dilated, blood will be pooled into them and so diminished blood is available for the heart to pump.

Capillary dilatation may occur when the tonus of the walls of the capillaries is lost thereby making them relax. This dilatation may be brought about by the following conditions

i) Neurogenic stimuli: In states of anxiety, fear and intense pain, vascular tone may be lost due to the action of neurogenic stimuli. Such stimuli also arise with severe wounds when along with bleeding, the disparity between the blood volume and blood vascular space is enhanced

ii) Bacterial toxins and toxic metabolic products: It is postulated that some bacterial toxins absorbed from the gut may accentuate the vasomotor collapse resulting in vasodilatation. (as in burns and crushing injuries).

Trauma of tissues is supposed to liberate histamine-like substannce which not only dilates the capillaries but makes them more permeable and so fluid passes out into tissues resulting in decreased blood volume.

iii) Anoxia : In shock anoxia occurs which causes capillary dilatation.

3. Acute circulatory failure: If the heart suddenly fails as may occur in infarction of the myocardium, paroxysmal tachycardia, cardiac tamponade and massive pulmonary embolism, circulation cannot be maintained, blood pressure falls and blood volume diminished as cardiac output is low and thus shock results,

Predisposing factors Cold, exhaustion, depression and general anesthesia predispose the animals to the development of shock.

Cold, exhaustion and depression are factors to be taken cognizance of while transporting animals for long distances and in surgical operations on animals which struggle hard during restraint

Pathogenesis of shock : Though various theories have been propounded and explanations offered, yet the exact mechanism how the process is brough about is still obscure. The following are some of the theories.

Ischemia : In extensive hemorrhage, burns etc. in which there is reduced blood volume, the blood circulation is not adequate There is reflex contraction of some arterioles so as to help maintain blood pressure. This causes ischemia with resultant degeneration, necrosis and loss of function of tissues and organs like liver, kidney and heart.

2. Toxins: (septic shock) Normally, the muscles and liver harbour certain bacteria, In state of shock, tissues are invaded by greater number of such bacteria from the intestines. Besides, in this condition the virulence and toxicity of these organisms is increased with greater production of toxins, which enhance and maintain the vasomotor collapse. In the shock-state, the detoxicating mechanism of the body is damaged and so the condition is aggravated and a state of **irreversible** shock ensues.

3. Vasotropic principles: Whenever there is anoxia. (ischemia) the kidney (cortex) produces a vaso excitor material (VEM) and the liver, skeletal muscle and spleen produce a vaso depressor material (VDM), under anerobic conditions VDM is probably ferritin. Healthy kidneys and liver are capable of destroying VEM and VDM respectively.

In the initial stages of shock with decreased blood volume, VEM is produced to effect vasoconstriction as a compensatory mechanism and also the capacity of the normal kidney to destory it is lost.

As the shock-like state continues with mounting tissue hypoxia, VDM is produced causing capillary dilatation and pooling of blood in the capillaries. Since liver damage occurs in continued circulatory deficiency, it is not able to destroy VDM, which accumulates, causing further dilatation of the capillaries and the patient enters into a state of **irreversible** shock. Due to fall in blood pressure, the tissues become ischemic and so hypoxia results. This causes decreased aerobic respiration of the cells and so anerobic glycolysis occurs, resulting in loss of energy for protein synthesis since ATP is not formed. Accumulating pyruvic acid and lactic acid due to anerobic glycolysis results in acidosis which causes release of lysosomal enzymes which injure the cells. Such changes contribute to the condition of irreversible shock.

Symptoms: The animal is lethargic and recumbent. It is weak and pallid with a rapid, weak pulse The extremities are cold. Skin is cold and clammy. The animal has an anxious expression. Breathing is shallow. Blood pressure is low and the animal becomes progressivelly duler and weaker and ultimately dies of circulatory failure

Macroscopically, the lesions depend upon the cause. If due to hemorrhage the tissues appear pale. If due to increased permeability of the capillaries with passage of fluid into the tissues, the tissues appear water—logged.

Passive congestion of liver, kidneys, lungs and intestines is seen. Serous membranes often contain petechiae. Fatty degeneration and early necrosis may be observed in the liver, kidneys and heart. Pulmonary edema and congestion are most characteristic gross changes seen. The spleen is small in size.

The kidneys show changes of lower nephron nephrosis, with slight enlargement, pale cortex and reddish-blue pyramids.

The adrenal cortex, in the early phases of adaptive response, is brilliantly yellow. Later, in the exhaustion phase, the cortex in diminished is size and loses its yellow colour.

Microscopically, the venules and capillaries are engorged. Lungs may show fat embolism, which is especially common in traumatic shock. In the liver fatty degeneration is noticed in the central zone of the lobule in the early stages. With the persistence of the state of shock the whole of the lobule may show this degenerative change. Cardiac muscle may also show fatty degeneration.

In the kidney are seen amorphous and granular pigment casts in the lower portions of the nephron viz, distal convoluted tubules and collecting tubules. These casts consist of hemoglobin. The epithelium of the tubules containing these casts may reveal necrosis and degeneration. The cortical cells of the adrenal lose their foamy appearance due to depletion of cholesterol.

Effects: If the blood volume and blood pressure are restored promptly by injecting whole blood or synthetic plasma substitutes, recovery results.

supportive treatment includes keeping the animal warm and administering antihistamine drugs and adrenaline

But death occurs if the animal passes on to irreversible shock. Here due to continued ischemia vital organs are damaged. Causes of death may be :

1. **Renal insufficiency:** In this case, oliguria and anuria develop due to the following reasons: (a) The pigment casts may block the tubules so that anuria and uremia develop.

(b) Renal parenchyma may be compressed by inflammatory edema to such an extent that renal function is suppressed.

(c) **Ischemia:** When there is vascular collapse, there is renal vasoconstriction and the blood is shunted directly from the cortex to medulla and so cortex and pheripheral glomeruli become ischemic and hence renal function is impaired. The low blood pressure that is present in shock is another contributary cause for anuria.

(d) Damaged tubular epithelium reabsorbs the glomerular filtrate and hence anuria results.

2. **Cardiac failure:** Fatty degeneration of the myocardium may cause cardiac failure and death.

3. **Cerebral ischemia:** Due to decreased blood pressure, sufficient blood is not supplied to the brain which suffers from anoxia. This results in neuronal degeneration and cerebral edema followed by encephalomalacia, resulting in death.

4. **Pulmonary infection:** In pulmonary edema, the edematous fluid is a good medium for bacteria, which thrive well in it and so may cause fatal respiratory affection when there is superimposed infection.

CHAPTER 8

PHYSICAL AND CHEMICAL INJURIES

Heat	Chemical injuries
Heat stroke	Corrosives—acids and alkalies
Burns	Mercuric chloride
Excessive cold	Lead
Light	Arsenic
Photosensitization	Phosphorus
Electricity	Carbon monoxide
Ionizing radiation	Barbiturates
Dosage	Death from anesthetics
Biological effects of ionizing	Hydrocyanic acid
radiation on macromolecules	Nitrates and nitrites
Total Body radiation	Chlorinated hydrocarbons
Atomic Bomb injury	Organic phosphates

HEAT

Due to metabolic reactions that take place in the body, heat is produced. But man and domestic animals are homothermic and so the body temperature must be maintained within certain limits. Life may be endangered if the temperature goes beyond these limits. Excess of heat produced is got rid of by radiation, conduction, convection and vaporisation from skin and respiratory tract. Regulation of body heat is under the control of the heat-regulating centre located in the hypothalamus. This centre is assisted in heat regulation by the vasomotor centre, the respiratory centre, the sweat centre as well as the pilomotor machanism.

When the environmental temperature equals that of body temperature very little loss by radiation, convection and conduction can occur. So heat is lost in such a circumstance by sweating only. If the external temperature is higher than the body temperature. increased sweating occurs. In animals panting with increased respiration serves the same purpose. Since in animals sweating does not take pluce as well as in man (except in the horse) loss of heat takes place by panting, polypnoea and evaporation from the tongue. In the dog, blood supply to the tongue is increased and heat is lost by evaporation of saliva.

Heat stroke; When the humidity of the environment is high, heat loss from the body is interfered with, espacially if the environmental temperature is high and the animal is worked in confined or enclosed areas. In such a situation the animals suffer from *heat stroke.* If this occurs when exposed directly to sun it is knowa as *sun stroke.* In these conditions, the heat regulating centre is over-taxed and so is not able to cope up with the situation and so temperature rises.

The following symptoms may be noticed Dullness, depression, palpi-tation ofthe heart, rapid and weak pulse, dyspnoea, congestion of mucous mem-branes and elevation of temperature. Later death follows trembling and con-vulsions of the animal.

Lesions: These are indefinite. The following may be noticed—pete-chiae on mucous membranes and skin, congestion of blood vessels; dilatation of the right side of the heart; cerebral and pulmonary edema; cloudy swelling of the liver, heart and kidneys; hyperemia or hemorrhages in the brain. The vascular changes are characteristic of circulatory failure as a result of shock. Focal necrosis and fragmentation of myocardial fibers are common. Sometimes lower nephron nephrosis is seen. Plasma potassium level is increased. Clotting of blood is slow and incomplete. Hemorrhages in heat stroke may be due to fibrinolysis, hypofibrinogenemia, hypoprothrombinemia and increased capillary fragility.

In animals that die, rigor mortis sets in rapidly to a marked degree and postmortem decomposition occurs quickly.

BURNS

The tissue changes that occur on excessive absorption of heat by skin are known as burns. The extent and depth of injury to burns depends on the intensity of heat and the period for which it is applied. The nature of heat modifies the character of lesions. Dry heat causes dessication and charring while moist heat causes boiling or cooking of tissues resulting in opaque coagulation.

Being the first tissue exposed to the full intensity of heat, the epidermis is the most severely affected. Tissues lying deeper are injured less.

Based upon the depth of tissue changes, burns of the body are classified into four degrees.

The first degree burn : Here the epidermis alone is affected. There is **erythema** of the skin due to hyperemia of the cutaneous vessels. No other damage is noticed and the cells show no changes. A mild acute inflammatory exudate may be present in the dermis. After a few days, desquamation of surface epithelium may occur.

The second degree burn : In this type along with erythematous changes coagulative necrosis of the epidermal cells takes place. **Bullae or vesicles** form due to infiltration of blood serum into epidermis or below it separating it from the dermis. The vesicle contains a ˈpink-staining granular debris, fibrin and neutrophils. The cytoplasm of the [epithelial cells is coagulated and the nuclei are pyknotic. Healing is prompt and complete by the regeneration of epithelium unless there is secondary infection.

The third degree burn : The heat penetrates into the dermis causing destruction of not only the epidermis but also the dermis and its components. The epidermis is dessicated and charred and so a black layer is seen on the skin. In the dermis, the connective tissue which normally has a characteristic fibrillar appearance, is transformed into a swollen, amorphous, acidophilic granular mass. The blood vessels and the adnexa (hair follicles, sweat and sebacious glands) are destroyed. Acute inflammatory reaction evidenced by edematous fluid and neutrophilic infiltration is present.

After a few days, the necrotic tissue sloughs and healing occurs by granulation tissue. Permanent scarring occurs. The destroyed skin appendages do not regenerate and so are lost for ever.

The fourth degree burn : The changes are similar to those described in the third degree burn but extend to the sub-cutaneous facia and deeper tissues.

So, muscles, bone and even the central nervous system may be destroyed. Repair is by scar formation preceded by sloughing of the necrotic tissue.

Burns are now classified as "partial" and "full thickness" burns. In the latter, due to destruction of dermal appendages, closure of wound by epithelium does not occur and so grafting is necessary.

Effects of burns : If one third to one fourth of body surface is affected even in first or second degree burns death may result within 24 hours. The symptoms seen are : difficulty in respiration, fall in blood pressure and body temperature—symptoms seen in cases of traumatic shock. In milder cases edema of lungs and nephritis may cause death after several days.

Pathology of burns : When tissues are injured, histamine or histamine-like substance is liberated, which on absorption into the circulation causes dilatation of splanchnic capillaries which thus become engorged with blood. The permeability of the endothelium of the capillaries is increased resulting in escape of fluid into the tissues. If large blebs are ruptured, fluid may be lost from the exposed raw surface by evaporation. The result of these phenomena is hemoconcentration and decreased blood volume, which result in shock. There is severe pulmonary edema due to hypoprotenemia consequent on loss of fluid from burn wound. Renal function fails because of changes in circulation that may occur in ischemic shock and that may develop consequent on intravascular agglutination of red blood cells that are damaged by heat. Anoxia, deficient nutrition and accumulating metabolites cause degeneration and necrosis of various organs, death resulting ultimately. The burned wound may also be infected by bacteria and so suppuration, toxemia and septicemia may result. These conditons may also be responsible death for of the animal.

The following changes may be noticed in the organs : Hyperemia of the viscera and central nervous system; atrophy and loss of lipid in adrenal cortex; damage to the epithelium of the proximal convoluted tubules of the kidney; hemoglobinuria or lower nephron nephrosis; focal hepatic necrosis (the cells show intranuclear inclusions similar to the Councilman bodies seen in Yellow Fever of man) hyperemia and appearance of numerous, shallow ulcers in the stomach and deuodenum (Curling's ulcers in man).

EXCESSIVE COLD

Animals are able to withstand cold provided they are fed with sufficient nutritious diet for production of heat. Heat production is enhanced by exercise and heat dissipation is reduced by thick hair coat.

But if animals are exposed to very low temperatures for a considerable ttme, especially those that do not receive nutritious diet, their body temperuture may go down and the muscles become stiff rendering the animal to move with difficulty, which immobility still further lowers the temperature. Under these conditions, the blood from skin is driven to the interior of the body and so the heart and lungs are engorged and oveburdened. Rate of metabolism slows down and the cardiac and respiratory centres are depressed. If the temperature of the body falls below 70°F the heart stops,

Frost bite : This is the condition seen locally on parts exposed to cold, viz, nose, ears, scrotum, snout, tail, extremities, comb and wattles. Various degrees of frost bite are recognised.

1 Mild: First due to contraction of blood vessels, the parts appear white. Paralytic dilatation that follows causes engorgement of blood vessels, thereby rendering the parts red and swollen During thawing great pain is felt.

If the loss of heat is prevented at this stage, return to normal condition occurs. But if exposed for longer duration, the following stage is reached.

2. Moderately severe: When the tissues are exposed to temperatures below 0°C for longer periods than the first variety, injury of the vessel wall occurs and inflammation of the tissues ensues. Epidermis shows redness together with certain amount of necrosis and blister formation, followed by desquamation.

3. Severe: When the temperature falls far lower than freezing point, circulation of blood and lymph stops and the part undergoes necrosis with gangrene supervening.

Mechanism of damage: The following are the causes of damage by freezing : 1. The cells may die or be damaged by freezing because their vital metabolic processes may be suppressed thereby slowing or blocking their metabolism. 2. The water of the cells may become crystalized. These crystals may tear the cells. 3. By withdrawal of water (by crystalization) the electrolytes in the cells become highly concentrated and so become toxic. 4. Vascular damage occurs and so plasma and serum escape out causing agglutination of red cells into a thick jelly-like material (sludging of blood) that occludes the vessels resulting in necrosis.

LIGHT

Photosensitization :

Dermatitis may be produced by the action of sunlight on certain photodynamic substances that may be present in the skin. Naturally therefore, this occurs in parts of the body that are exposed to the sunlight and which are not protected either by pigment, wool or blankets. As such, in sheep the following areas are affected: face, muzzle: ears and back (if it is uncovered by wool); in cattle : the teats, udder, vulva, perineum and other unpigmented parts. That natural skin-pigment is protective is well exemplified in photosensitization in a pie-bald horse, in which the black parts are healthy while the gray or white parts are affected.

Lesions : The severity of lesions depend upon (1) the length of exposure to sunlight and (2) the concentration of the photodynamic substance in the skin. It does not depend upon the kind of the sensitising agent.

Pathogenesis : Four factors are necessary; (1) Susceptible animal (i.e. with no protective hair or wool or pigment) (2) Oxygen—furnished by hemoglobin (3) Sunlight (red-orange spectrum), (4) Photodynamic substance (of plant origin or endogenous, metabolic origin).

Due to the Photodynamic effect, the permeability of the lysosomes of endothelial cells and the adjacent connective tissue cells is increased, and so one or more chemical mediators are released and these are responsible for the lesions.

Radiant energy after penetrating into the small cutaneous blood vessels catalyses the oxidation of the photodynamic substances, producing some toxic material, which affects the vascular endothelium.   Consequently, congestion, edema and even thrombosis results.   Edematous fluid may even ooze out of the skin.   Necrosis of the skin occurs which may be sloughed,   ulcer being formed. Scab develops from the drying exudate.   Infection of the wound by secondary organisms may also occur, gangrene supervening.

In sheep, the affected edematous ears droop. Dyspnoea may occur due to swelling of muzzle.   There may be swelling and closure of eyelids and lachrymation may be present. Due to edema, the face appears swollen and so the condition is known as "big head". Affected animals seek shady places and cows plunge into water to dip their udders. Mastitis is a frequent sequel   when teats are severely affected with open raw sores.

If the affection is mild, healing occurs without leaving any scar.   But if widespread necrosis has occurred, healing takes place by granulation tissue, leaving a scar.

Urine may be brown in color.

Sources of photodynamic substances: Depending upon the sources of photodynamic substances photosensitization is drscribed under three types.

1. Primary photosensitivity: In this category, the photosensitivity is due to exogenous materials, which are absorbed directly by the intestinal mucosa and reach the skin   On exposure to sunlight these substances in the skin produce the lesions. Examples of such substances are: hypericin from plants of genus hypericum and photodynamic principle from buckwheat (*Fagopyrum esculentum*)

Phenothiazine is converted into sulfoxide in the intestines from where it is conveyed in large quantities to the liver which is not able to convert all of it into harmless compounds. So the sulfoxide enters the circulation and reaches the skin where it produces photosensitization. It is also excreted through the tears and so the eye shows keratitis and ulceration.   The urine is red in colour from the presence of the dye.

2. Photosensitivity due to adnormal endogenous pigments: The uroporphyrins are the photodynamic substances involved.   These are probably basic pigments in the formation of hemoglobin. Though their origin and genesis are not clear, whether due to lack or derangement of certain enzymes, these are found in excess in the body and in the skin   In cattle, *congenital osteohemochromatosis* occurs in which the pigment is present in the bone, skin and teeth. The color of the teeth turns to pink on exposure to light—"pink tooth".   Such animals if protected from sunlight, do not appear to suffer from any discomfort and do not manifest any symptoms.

3.   Hepatogenous photosensitivity: In this variety, photosensitization is conditional to hepatic dysfunction. Normally chlorophyll is converted into phylloerythrin (the sensitising pigment) by ptotozoa in the alimentary tract. Some of this is absorbed and excreted into the bile.   If by any chance there is obstruction to the excretion of bile (as in obstructive jaundice or hepatic necrosis) phylloerythrin accumulates in the blood and produces photosensitivity. Agents that produce hepatic necrosis (hepatotoxins) may be bacterial, chemical or of plant origin.   Phytotoxins are most common.   The following are some among a great many hepatotoxic plants that produce photosensitization:

Calatrops (*Tribulus terrestris*) in South Africa produces icterus acompanied by photosensitization.

Fungus, *Pithomyces chartarum*, growing on rye grass produces "Facial eczema" in sheep in Newzealand.

The hepatotoxin directly acts on the liver cells and the epithelium of bile ducts, causing them to swell and so the passage of bile is prevented leading to stasis and jaundice.   Microscopically the cells are swollen and manifest fatty degeneration.   Inflammation of the bile ducts, cholangitis and pericholangitis are present.

Atomospheric pressure: Injuries due to high and low atmospheric pressure have been studied earlier under "Caisson Disease" and "Brisket Disease" respectively.

ELECTRICITY

Electricity, artificially produced, or naturally occurring in lightning, may cause injury and death, if strong enough.   Sudden death (electrocution) may occur if animals come into contact with high-tension wires,   The cardiac and respiratory functions are suddenly arrested, causing cardiac fibrillation, stoppage of cardiac contraction (the pace maker and conduction system are affected) and respiratory failure.   The local effects of electric injury are similar to those of burns, since much of the damage is caused by production of heat. The changes are noticed at the points of entry and exit of the current.   Feet are usually the points of exit.

The nature and severity of the damage produced by electricity depend upon the following;

1 **Type of electric current:** An alternating current is more dangerous than direct current since the former is produced at high tension.

2. **Path of the current through the body:** Current that passes through the left side of the body or through the brain is more dangerous than the one of equal magnitude that passes through the extremities.

3. **Amonnt of current**: Higher the amperage greater is the damage.

4. **Duration of the flow of current:** The amount of heat generated incresses directly with time.

The burns are first dry, followed in thirtysix hours by hyperemia and edema.   Epidermis may show small cavities, believed to be caused by sudden generation of steam.   Ulcers that form by sloughing of necrotic tissues heal slowly.

Lesions seen internally are:— petechiae on the serous membranes, edema of lungs and engorgement of the right ventricle, rupture of some blood vessels and thrombosis of others.

Necrosis produced directly by electricity may be due to severe and sudden shift of ions intracellularly.   The selective permeability of the cellular membranes is seriously affected as the electric current severely alters their polatily resulting in death of cells.

In *lightning injury*, characteristic *lightning figures* namely, red lines on the skin resembling trees are seen.   There is usually unconsciousness.   Due to medullary paralysis apnvea occurs.   There is violent contraction of the heart. Death is due to myocardial anoxia and occurs several minutes after the lightning stroke.   Extensive hemorrhages, laceration of meninges and disorganisation of brain may be noticed,

IONIZING RADIATION

Radiation is of two types· (1) One type is that propagated by wave motion the "*electromagnetic radiation*" (eg., radio waves. infrared rays, visible radiation, ultraviolet radiation, X-and Gamma radiation) and (2) the second arises from the movement of particles such as the Alpha and Beta particles, neutrons and deuterons—the "*particulate radiation*".

The Atom: The atom consists of a positively charged central core. the "*nucleus*" surrounded by one or more negatively charged planetary "*electrons*" (which orbit around the nucleus) Almust all the mass of the atom resides in the nucleus, which is composed of two different types of stable parti— cles, of almost equal mass, the "*protons*", which are positively charged and the "*neutrons*", which are electrically neutral. The mass of the electron is $\frac{1}{1836}$th of that of the proton   Although its   charge is   opposite in sign it is numerically equal to that of proton. "*Atomic number*" (z) refers to the number of planetary electrons in the electrically neutral atom and equals the number of protons in the nucleus. Therefore the higher the atomic number of an element the greater the number of electrons it has.   The "*atomic weight*" is the measure of total number of protons and neutrons.

Alpha particle: This is a helium nucleus, i. e. a close combination of two neutrons and two protons.   Alpha particles though having high ionizing properties have low penetrating power and so have very little biological effect.

Beta particles or rays: These are electrons or positrons (which have mass of electrons and a charge equal to but opposite to that of an electron). emitted by a radio-active nucleus.   Beta particles do not exist within the nucleus but are created at the instance of emission. The emission of an electron involves the change of a neutron into a proton within the nucleus, while the emission of a proton involves the change of a proton into a neutron.   Beta particles have deeper penetrating power, than the alpha particles and have ionizing property. Normally beta rays are of little danger except when they are emitted by radio- active isotopes.   Upon contact lesions may be noticed on the skin.

Gamma rays: These are short wave electromagnetic radiation emitted from the nuclei of radio isotopes.   These have greater penetrating power but are poorly ionizing.   These may penetrate into deeper parts of the body and produce total body effect.

X-rays or Roentgen rays are highly penetrating electromagnetic radia- tion.   Their energy is lower than that of gamma rays.

Neutrons are nuclear particles and have very low range and deep penet- ration power.   They cause indirect ionization.

Isotopes : Atoms of an element which have the same number of protons (z) in their nuclei but a different number of neutrons (n) are colled isotopes of that element. The mass number M is defined as $M = Z$ plus n. When a particular isotope is being considered the following notation is used ; to the chemical symbol of the element, the mass number of the isotope is added as a superscript. The atomic number of the element is added as a subscript. eg, $_1^1H$; $_6^{12}C$; $_{79}^{197}Au$ are respcetively the most abundant isotopes of Hydrogen, Carbon and Gold.

An isotope results by addition of extra neutrons and so the atomic weight is also altered. Artificially isotopes can be made by forcing extra neutrons into the nucleus of the atoms (directly or indirectly).

Radioactivity : This refers to the property of spontaneous "disintegration" possessed by certain unstable types of atomic nuclei. This disintegration is accompained by the emission of either alpha or beta particles and gamma rays. An element becomes radioactive if its nucleus contains an excess of neutrons or protons thereby necessitating redistribution of the particles, when energy is emitted in the form of radiation.

The isotope (unstable) of an element which is radio active is termed "radio isotope".

Ionizing radiation : This refers to radiation (electromagnetic or corpuscular) which is capable of causing "ionization" (i. e. forming ions), either directly or indirectly, of the molecular structure of the cells. Electrons and alpha particles are considerably more effective in this respect than neutrons or gamma rays. Irrespective of the source of radiation or its particulate or electro-magnetic nature the biological activity of all types of radiant energy is the same namely ionization and thus producing the tissue changes

If an atom happens to be in the path of ionizing radiation, an electron is ejected from the atom which thereby becomes electrically unbalanced and so positively charged ion results. The detached electron gets itself attached to another atom thereby disturbing its electrical neutrality, a negatively charged ion resulting. Hence "ion pairs" are created. Ionization, therefore, is loss or gain of electrons. If such ionization were to take place in a molecule there will be disruption of molecular aggregate manifested by chemical change. The biological effects of radiation on cells could be attributed to the chemical changes in the molecules of cells, especially chemical changes of water, which forms the bulk of the celluar protoplasm. The responses following the exposure of human body or any vertibrate to ionizing radiation may be divided into two types : (1) somatic or body effects which occur in the individual and (2) genetic effects which are transmitted to future generations. The somatic responses in animals include such phenomena as loss of hair, skin disorders, dysfunction of the systems manufacturing blood cells, complete destruction of certain tissues and induction of malignant growths. Genetic effects are due to alterations in the chromosomes.

Dosage : A practical unit of radiation dose for monitoring purposes is the "rad". If 1 gm, of biological matter receives a radiation dose of 1 rad, then 100 ergs of energy will te dissipated (i.e. energy lost by ionizing radiation).

The rad is a modificaion of the older unit called the roentgen(r) used to specify the dosage of X-rays. 1 r liberates 83.4 ergs per gm. The rad unit is indipendent of nature of radiation.

Rem : The rem is that unit of Dose Equivalent (DE), a quantity which expresses the biological damage to an exposed person on a common scale of all ionizing radiation. Dose Equivalent is obtained by multiplying the above dose (in rads) by quantity termed quality factor (QF) which is dependent on the linear energy transfer of the type of radiation.

DE in rems $=$ Dose in rads X QF. The quality factor for X, Gamma and Beta radiation is 1, for Thermal neutrons is 3; for fast neutrons 10; for protons 10; for Alpha radiation 10 and for Heavy recoil nuclei 20.
Danger of isotopes ingested.

	Isotope
Very highly toxic	Sr^{90}
Highly toxic	Ca^{45}, Sr^{82}, Ba^{140}, I^{181}
Moderately toxic	Na^{22}, Na^{24}, P^{82}, S^{95}, Cl^{86}, K^{42}, Mn^{82}
	Mn^{54}, Mn^{56}, Fe^{55}, Co^{60}, Zn^{65}, Br^{92}
	Rb^{86}, Mo^{99}, Cs^{187}, Ba^{197}
Slightly toxic	H^3, C^{14}.

The hazard involved when radioisotopes are ingested or inhaled will depend on a number of factors, such as (i) the half-life and energy of the iso-tope; (ii) the biological half-life, i.e. the time taken for the elimination of half of the ingested material from the body; (iii) the accumulation of isotopes in critical organs and (iv) the formation of toxic by-products as a result of (a) splitting of molecules by radiation or (b) reaction of free radicals.

The highly toxic elements such as Se^{89}, Ca^{45} and Sr^{90} accumulate in bones and produce damage to the blood forming cells. I^{181} accumulates in the thyroid gland. The moderately toxic elements do not accumulate to such high degree in critical organs and have a relatively short biological half life. Tritium (H^3) and C^{14} are usually only slightly toxic because of their rapid biological turnover. However, these become very toxic under conditions of slow turnover (eg. in nucleic acids or as $BaC^{14}O_3$ dust lodged in lungs.). High doses of radiation can produce cancer with latent period of up to 20 years. The most serious risk appears to be lukemia which normally appears within a few years after irradiation.

A common misconception is that radiation below the permissible dose does no harm. The fact is that any radiation does biological damage. It is certain that natural radiation (due to cosmic rays etc.) contributes to the natural mutation rate and that any increase in radiation however small, will increase this rate. The natural mutation rate in humans is so high that it results in serious abnormalities in about 3% of all births. Thus nuclear warfare is to be feared for accompanying increase in world wide radiation level and so to resulting increase in abnormal births

Half life of a radioactive substance. This is the length of time (measured in in seconbs, hours, days or years) required for the activity of a radio-isotope to decay to half its original value, i.e. for half of the atoms to disintegrate.

Half-lives vary from isotope, to isotope, some being less than a millionth of a second and some more than a billion years. The half-lives of some of the radio-isotopes are given below :

Iodine[193] is 6 seconds
Phsophorus[82] is 14 days
Strontium[90] is 28 years
Radium[238] is 1500 years
Carbon[14] is 5700 years
Uranium is billion years

Mode of action of Radiation on tissue cells :

Direct biological effect : Two theories are propounded to explain the direct injuries caused to living cells. It should be understood that these two theories do not oppose each other but both may be useful to explain the different types of biological response,

The target theory : Many particulate structures are found in the cells. The cytoplasm contains the mitochondria with many enzymes on their surface; the centrosomes containing the spindle of mitosis and the nucleus containing the chromosomes with numerous genes on them. When an ionizing ray is directed on such a cell, the electrons may hit any of these particulate structures, which are the targets affected. If the target were to be an enzyme no biological effects may be caused since other mitochondria may have similar enzymes to carry on the biochemical reaction. But if the target were to be a gene, which is not usually duplicated, a changed or destroyed gene results in **mutation.**

Poison theory : Water forms a large part of every cell and so when cells are exposed to ionizing radiation, there is disruption of the water molecules. Ionization of water results in the formation of free H and OH radicals which combining with free oxygen form HO_2, HO_3 and $H_2 O_2$ which are highly oxidising agents. These toxic products cause widespread changes (in enzymes and metabolites) which are supposed to be responsible for the acute symptoms seen and known as "acute radiation poisoning,"

Indirect biological effects : Here the damage does not occur directly to the cell. But changes occur in the connective tissue and blood vessels, the integrity of which is very essential for the normal health of the tissues.

The 'indirect' effects of radiation are very pronounced on the skin. The connective tissue and blood vessels in the corium provide support and nutrition to the epidermis.

The epidermal cells are vulnerable to ionizing radiation in small or larger doses, the division of the cells being stopped by the former and partial or complete destruction of the cells by the latter. But if penetrating radiation is administered in doses too insufficient to cause injury to the epithelial cells, the connective tissue and blood vessels may be injured. With repeated doses of such radiation, inflammation progressing to scarring of dermis results. With changes in the corium the overlying epidermis shows degenerative changes and necrosis. So such changes in the epidermis are caused by "indirect" action.

Lesions of Nuclei : The nucleus is most vulnerable for ionizing radiation during the prophase of mitosis. The chromosomes and chromatids are the targets. They become fragmented and later may unite. It is also possible that combination of different pieces of chromosomes may occur. So a cell with new characteristics may be evolved. If the sex cells are thus affected mutations result since different combinations of genes occur.

A small dose of radiation may stop mitosis, followed by increased division. Synthesis of mRNA and the production of such vital enzymes like DNA polymerase and thymidinekinase is prevented and so DNA synthesis is inhibited resulting in stoppage or retardation of mitosis. But a larger dose causes complete stoppage of mitosis and cell death. Repeated and continuous radiation results in neoplasia.

Lesions of cytoplasm : Probably the cell membrane is injured, resulting in alteration of its permeability. Cytoplasmic vacuoles may appear. Mitochondria and golgi apparatus may become damaged.

Biological effects of ionizing radiation on macromolecules.

Chemically both proteins and nucleic acids belong to a general class of compounds known as polymers (or high polymers), A polymer is made up of a repeating type unit (the monomer), which is duplicated again and again. The repeating units making up proteins are the amino acids. The monomers of the nucleic acids are the nucleotides.

In proteins, long chain of amino acids occur in characteristic proportions and in specific sequences, linked together by the peptide linkage CONH (because a molecule of water is eliminated for each peptide bond formed) it is customary to say that the amino acids condense to form the proteins. Besides, the nucleic acids are also condensation type polymers, a molecule of water being eliminated for each nucleotide joined to the chain. Side chains protruding from the main chain can form cross linkages between neighbouring chains. In solution they form finely dispersed colloid systems. The variety of existing proteins is very great but all proteins have one reaction in common-viz., the property of denaturation. This is usually an irreversible change which occurs when proteins combine with certain chemicals or when heat or radiation is applied.

The denaturation manifests itself as coagulation with subsequent insolubility. It has been found that ionizing radiations lower the resistance of proteins to thermal denaturation. After irradiation, protein solutions contain different denatured protein derivatives and show a marked decrease in energy content signifying deep-seated structural changes. It has been established that the oxidative attack by OH radicals is directed toward the peptide linkage. This results in the formation of high M.W. carbonyl (C=O) compounds and of keto acids and the release of ammonia. The radiation sensitive SH groups of cysteine (an amino acid occurring in the egg albumin) can be oxidised to a disulphide (S-S) which can form a cross linkage between neighbouring chains. Oxidation can occur even further, beyond S-S stage: without denaturation or marked instability of the protein. It has further been observed that after hydrolysis of irradiated serum albumin, several amino acids are partly destroyed. Examination of irradiated protein solutions by spectrophotometry also reveals changes. Bovine serum albumin, serum globulin and egg albumin show an increase in optical density which is apparently due to the action of radiation on their tyrosine component; if however, the proteins contain more tryptophan than tyrosine, a decrease in density is observed.

Enzymes are proteins which differ from other proteins in their ability to act as catalysts. Enzymes speed up specific chemical reactions. They either have special active groups in their make up enabling them to combine with their specific substrates on which they act or they are more or less firmly linked to a non-protein partner or prosthetic group which in cooperation with the protein part, functions as a highly specific catalytic system.

As far as their protein nature is concerned, enzymes will undergo the same general changes as described for other proteins with consequent loss of activity. There are some enzymes, the S-H enzymes, in which that group is essential for enzymic activity. The S-H group is particularly sensitive to radiation and may undergo changes before deeper seated modification of the protein has taken place. If the inactivation of the S-H enzyme has not gone far, it can be restored to its original activity by the addition of the tripeptide glutathione which contains S-H.

An example of an enzyme containing prosthetic group is D-amino acid oxidase, which specifically oxidises D-amino acids only. This enzyme can be split into flavin adenine dinucleotide, a non-protein and a specific protein. Neither part, on its own, has enzyme activity. Each part can be chemically changed by radiation and when rejoined shows a lower activity than after irradiation of the complete enzyme.

Enzymes are known to have various sensitivities to radiations, at least in dilute solutions. The data on these are somewhat suspect because of the marked effect of impurities. However, one of the most sensitive enzymes seems to be carboxypeptidase; ribonuclease is ten times more resistant than this while catalse is hundered times more resistant. Some enzymes are inactivated even when in the dry, crystalline state (the results support the target theory). All enzymes studied are inactivated in aqeous solutions by ionizing radiations—his can mean direct target action or attack by radiation produced radicals or probably both.

In the solid state, large doses of radiation (of the order of 20×10^6 rads of x-rays) may be needed to bring about a substantial change in the enzymes. Doses as high as 10,000 rads are required to produce changes in the phosphorylating ability of mitochondria as revealed by *in vitro* studies of irradiating mitochondria. In those animals (rats) that had received doses of 700 rads the mitochondria isolated from the spleen or thymus showed less capacity for generating ATP compared to those isolated from normal animals.

The nucleic acid DNA has been studied intensely for irradiation effects. The molecule is huge (MW is 5 millions of that obtained from leucocytes). The standard methods of determining the MW were used. One by measurement of the viscosity of DNA solutions and measurement of the speed with which the molecules settle out in a high-speed centrifuge, showed that the MW falls during irradiation, as though the big molecules were being split into pieces. The other method however, by light-scattering techniques gave a constant MW during irradiation. The implication is that the molecule is broken up but the pieces do not completely uncoil. With such loosened structure, easier degradation by heat results (as was also found for several enzymes).

It is an accepted fact that ionizing radiation exerts deleterious effects upon living tissues in proportion to dosage although not necessarily in direct proportion The rationale of radiation therapy rests upon this premise. International safety standards are periodically reviewed and revised. Although all radiation absorbed is believed to be harmful to living cells in some degree, it is fair to assume, in the absence of concrete evidence to the contrary, that a threshold level for the dosage can be found below which such deleterious effects may be unrecognisable. This is the basis for the maximum permissible dose concept.

Total dose of radiation : No radiation worker may be permitted to accumulate a total dose to his whole body, gonads or red bone marrow which exceeds the value given by the formula $D = 5 (N-18)$, where D is the dose in rems and N is the age of the radiation worker in years and fraction there of. (Persons under 18 years of age therefore should not be occupationally exposed to ionizing radiations.) Based on these criteria, it is advisable that the average yearly bose to be received by a worker should not exceed 5 REM and the average weekly dose should remain below 0.1 REM For one quarter year it should be less than 3 REM. The permissible exposure for other tissues are: for skin, bone and thyroid — 15 REMs for a quarter or 30 REMs for a year. For hands and forearms, feet and ankles— 40 REMs for a quarter and 75 REMs for a year.

In nature, the following quantity of radiation is absorbed by the body:

Natural background : Dose to gonads ⎫ (milli rads per year)
                         and soft tissues ⎭

External radiation :
Cosmic rays.	28
Local Gamma rays.	47
Radon in air	8

Internal radiation :
K^{40}	19
C^{14}	1
Radon & disintegration products	2

Man-made contributions :
Medical radiology.	100
Shoefitting fluoroscope machines.	1
Luminous and watch dials	1
Occupational exposure.	2
Television sets.	1
Fall-out from weapon tests.	2
Total :	204

It is now apparent that the natural background level has been materially increased by man-made radiation, particularly medical exposure; however, the contribution from weapon fall-out is almost negligible to date, despite its notoriety. It is also evident that the average cumulative dose received by the germ cells during the first 30 years of life is in the neighbourhood of 3 to 8 rads.

There are three major categories of radiation damage: i) direct local destruction or atrophy from heavy exposure; ii) premature ageing and increased incidence of leukemia or other malignant tumors from radiation to which the entire body is exposed and iii) the production of undesirable mutations in reproductive cells as the result of cumulative absorption of radiation by gonads.

The first form of radiation damage is inexcusable except in those instances where very large therapeutic doses applied locally are needed to control lethal lesions. It is to be remembered that the second form of damage relates to total body radiation. Barring occasional massive exposures from nuclear accidents or atomic warfare, the hazards of radiation result from small doses accumulated over a long period of time. As already mentioned the effects of such doses fall into two categories, genetic effects and delayed somatic effects.

Of the various possible types of injury, genetic effects have been considered more important than somatic effects since they are essentially irreversible and threaten future generations, whereas somatic effects occur only in the exposed subjects. The Federal Radiation Council has, therefore, recommended that the maximum permissible gonadal radiation dose for the general population not exceed 5 REMs during the first 30 years of life, this dose being below that estimated to double the natural mutation rate (10—100 REMs)

Concerning the hazard of delayed somatic radiation injury such as life-shortening and cancer, there is no conclusive evidence as yet that doses only slightly above the natural background are damaging. Hence until more is known about the quantitative relation between these effects and dose, the risks of low-level exposures cannot conceivably account for more than small fraction of the spontaneous incidence of cancer and degenerative changes associated with senescence.

Anatomic distribution of isotopes : In the distribution of injury, the area damaged depends on the localisation of the radio-active element in the organism and the energy of the emitted radiation. In general, elements that are concentrated in the bone tend to be eliminated very slowly and are therefore particularly hazardous.

Distribution and excretion of some radio-active elements :

Radio-active element.	Principal organ of deposition	Half life in days	
		Physical	Biological
Ba^{140}	Bone	12.8	200
Ca^{45}	Bone	152	18,000
C^{14}	Bone	2.09×10^6	180
Cs^{137}	Muscle	12,000	17
I^{131}	Thyroid	8	180
$Plutonium^{239}$	Bone	8.8×10^6	43,000
$Polonium^{210}$	Spleen	183.3	57
$Radium^{226}$	Bone	5.9×10^5	16,000
$Strontium^{90}$	Bone	9,100	3,900
$Uranium^{238}$	Bone	5.9×10^7	300
$Zenon^{133}$	Body	5.3	0.1

Radiosensitivity of tissues : All cells are susceptible to ionizing radiation but the degree of susceptibility varies with different cells. There appears to be a direct relationship between radiosensitivity and metabolic activity. So in conditions where metabolic activity is increased as in fever, acute inflammation, and hyperthyroidism cells become more susceptible to radiation. Tissues that are actively proliferating are also more susceptible to ionizing radiation. So cells that are able to reproduce are highly radiosensitive while those that are well differentiated are radioresistant. This is stated as follows in the Law of Bergonie and Tribondeau, "Immature cells and cells in an active state of division are more sensitive to irradiation than are those that have acquired adult morphological and physiological characteristics". So such cells which multiply throughout life due to their short span of life are very radiosensitive.

The body tissues may, therefore, be broadly divided under the following categories.

1. **Radiosensitive: less than 250r kills or seriously injures many cells.**
    Lymphoid tissue
    Hemopoietic tissue
    Gastro-intestinal epithelium
    Germ cells (of ovary and testicle)

2. **Radioresponsive : 2500 to 5000r kills or seriously injures many cells.**
    Epithelium of skin and skin appendages
    Blood vascular endothelium
    Salivary glands
    Cornea, conjunctiva and lens
    Growing bone and cartilage
    Collagen and elastic tissue (not fibroblasts)

3. **Radioresistant : More than 5000r kills or seriously injures many cells.**
    Kidneys
    Liver
    Thyroid
    Pancreas
    Pituitary, Adrenal
    Parathyroids
    Mature bone and cartilage: muscle
    Brain and nervous tissue,

Cancer cells : Cancer cells also like the tissues from which they originate may be radiosensitive, radioresponsive or radioresistant. The highly immature and rapidly proliferating cells are most radiosensitive. But there are exceptions to this rule Malignant melanoma and osteogenic sarcoma may be teeming with mitoses but they are very radioresistant. Lymphosarcomas are very radiosensitive But even in very radiosensitive tumors, after radiation, a few cells may be radioresistant and so do not die. These by proliferation bring forth newer generations of highly radioresistant cells and so the patient may ultimately die inspite of radiation therapy.

Total body radiation : In animals exposed to total body radiation, there is lymphocytopenia followed by neutropenia and thrombocytopenia Anemia develops slowly. *Acute radiation syndrome*: The acute effects of total body radiation are due to depletion of cells: cells of the G. I. tract, cells of hemopoietic system etc. Two factors decide the onset of syndrome (1) the dose and (2) individual susceptibility.

There is distinct relationship between dose and the effects seen. In the low dose range (100 r.) death is due to damage to the hemopoietic system and depression of hemo and leukopoiesis. Death occurs in 2 months. Symptoms seen are leukopenia, purpura, hemorrhage and infection. In the middle dose range (500 r.) death is due to denudation of the gastrointestinal epithelium and its inflammation. Symptoms seen are diarrhoea, fever and disturbances in electrolyte balance. Death occurs in two weeks. In the high dose range, (2000 r.) death is the result of failure of central nervous system. Symptoms seen are convulsions, tremors, ataxia and lethargy. Death occurs in two days.

The following are the lesions found in different tissues in total body radiation.

Lymphoid tissues: such as lymph nodes, spleen, thymus and tonsils are highly radiosensitive. *Microscopically*, the lymphocytes show pyknosis, karyorrhexis and karyolysis of nuclei. The cytoplasm appears granular, swollen or even coagulated. Ultimately all lymphoid tissue disappears leaving only reticuloendothelial cells which are either not affected or only slightly. So the germinal centres have a "washed out" appearance. A few remaining lymphocytes may be enlarged in size with many nuclei. The sinuses are filled with macrophages that have engulfed erythrocytes. Hemorrhages and retained pigment may also be noticed. Peripheral blood will show progressive fall in lymphocyte count due to failure of formation. The life span of lymphocytes is short being only a few hours. So severe injury to lymphoid tissue is followed immediately by significant drop of lymphocytes in the peripheral blood.

Blood and marrow: Hemopoietic cells are even more vulnerable to radiation than lymphoid. Granulocytes and thrombocytes of the bone marrow may be killed or they may be so damaged that they cannot mature. The marrow may become acellular and in a week or two only reticular tissue containing serum may be left. Secondary infection by bacteria or fungi may occur and so colonies of these may be found in the marrow. Regeneration from cells that have escaped radiation injury is possible. Sometimes hyperplasia of bone marrow may occur.

The life of granulocytes and platelets, being some days, is longer than that of lymphocytes. So the peripheral blood picture shows granulocytopenia and thrombocytopenia after about a week while lymphocytopenia occurs within a few hours after exposure. Due to initial stimulation, transient leucocytosis may be present at first.

Due to thrombocytopenia, hemorrhages are common. Sometimes, especially due to repeated exposures, radiation may cause hyperplasia of the bone marrow. In certain cases, this hyperplasia becomes so uncontrolled that frank neoplasia occurs viz. leukemia.

Digestive system : The epithelium of the gastrointestinal tract is highly radiosensitive. But certain protection is afforded to it by virtue of its deep location inside the body.

The epithelial cells undergo degeneration. Vacuolization of the cytoplasm and pyknosis, karyorrhexis and karyolysis of nuclei are noticed. The mucosal cells later become shrunken and necrotic. Sloughing of the necrotic tissue results in ulcer formation. There may be submucosal and subserosal edema with leucocytic infiltration around the ulcerated area. Bacterial infection is very frequent. The blood vessels are dilated and contain fibrin thrombi. Healing may be followed by scarring resulting in thickening and stenosis of the bowel.

Nausea and vomition result usually within a few hours after exposure to doses above 200 rads. After several times this much dose sloughing of the intestinal lining may lead to ulceration, intractable diarrhoea, dehydration and invasion of the blood stream by bacteria that normally inhabit the lumen of the bowel. This sequence of events is usually fatal and constitutes one of the major causes of death after massive irradiation of the whole body.

Germ cells The germ cells of the testes and ovaries are highly susceptible to radiation. The spermatogonia and spermatocytes show degenerative and necrotic changes (while spermatozoa may be unaffected) So also the ova are markedly degenerated and necrotic. The result of these changes is that sterility may supervene. Changes in the chromosomes may cause mutations to occur.

Blood vessels: The vascular endothelium is very radiosensitive. Transitory dilatation of blood vessels causing erythema of the skin is one of the earliest known reactions to ionizing radiation. It occurs after only a few hundred rads and may be accompanied by increased permeability of blood capillaries. With larger amounts of radiation the endothelial cells may become swollen and necrotic. Sloughing of the dead cells occurs. There may be permanent dilatation of the blood vessels, as well as scarring. The intima may be nfiltrated by leucocytes, exudate and fibrin. The collagen and elastic fibers of the vascular wall become degenerated. Usually thrombosis occurs and later organisation of this thrombus leads to occlusion of the vessels resulting in metabolic disturbances and atrophy of the tissues involved.

Bone : The growing bone of young animals is more vulnerable than mature adult bone. Only a few hundred rads applied to bone and tooth forming cells in infancy or early childhood cause disturbances of dentition and skeletal growth. In contrast mature bones and teeth are relatively radioresistant. Large amounts of radiation, however, such as may accumulate from locally deposited radio-isotopes or from the treatment of cancer produce demineralisation and necrosis of bone that can lead to fractures, loosening of teeth, bone cancer and other complications.

Lungs: Although relatively radio-resistant, the lungs may be injured by intensive irradiation. The result is a chronic pneumonitis, when the lungs become pale and dry. The alveolar walls are thickened with hyaline membranes, and there may be extensive scarring of the lung parenchyma and blood vessels, giving the organ a rubbery consistency. A complication of the locally deposited radio-isotopes is cancer of the lung, which has been noted in miners of radio-active ore and in experimental animals.

Nervous system : Although only relatively large amounts of radiation will kill nerve cells in the adult, the developing nervous system is highly radio-sensitive. Even in the adult, transitory functional disturbances may be elicited by relatively low doses and after intensive exposure of the brain (100-10,000 rads, depending upon the species) incapacitating neurological effects may lead to death within minutes or hours.

Eye : The part of the eye most easily injured by radiation is the lens, opacity of which (cataract formation) has been observed after exposure to 200 rads of X-rays. Smaller doses of neutrons are estimated to have caused cataracts; several examples of such induction have been noted among nuclear physicists. The cornea, conjunctiva and the retina withstand much more radiation. The retina however, is highly radiosensitive early in its embryonic development. Minute amounts of ionizing radiation are visible through radio chemical reactions in the retina.

Endocrine glands : The endocrine glands have traditionally been regarded as radioresistant because of their ability to withstand relatively large amounts of radiation without developing morphological lesions. There is growing evidence, however, that rather small doses may elicit :changes in endo-crine function. Apart from radiation injury itself, the endocrine system's ada-ptational response to acute effects of radiation resembles its response to other types of stress.

Urinary system : The kidney and lower urinary tract are relatively radioresistant. Depending on the species, however, doses in excess of 500-2000 rads may cause gradually progressive scarring and atrophy of the kidney, which can lead to fatal loss of renal function.

Atomic bomb injury : Injury by atomic bomb explosion may be caused in three ways :—

1. **Mechanical blast injury :** by a wave of compression of air and also by falling structures.

2. **Burn injuries :** These may be a) Flash burns due thermal radiation from the bomb and b) flame from incendiary action on combustible materials.

3. **Radiation effects :** Total body radiation may occur due to the various forms of radiant energy that are liberated during the bomb burst. The effects of radiation may be two fold. One is by direct radiation on the body through the skin—*the external radiation.* The second is by ingestion of materials contaminated by radioactive substance. It is called *internal radiation.*

The external radiation may be caused primarily by the radio-activity released during the explosion, by neutrons and gamma rays. The fast moving neutrons that are liberated may enter nearby inert, stable matter and can induce radioactivity into it. This is *induced* or *secondary radioactivity.*

During explosion, some of the original radioactive material may remain. This and the induced radioactive materials may be transformed as a radioactive cloud by the power of the explosion and this cloud may be transported to great distances and then settle down as particulate matter or may be washed down by rain. This is called **radioactive fall-out** and is a serious hazard to life. The fall-out may cause external radiation or internal radiation

The lesions of the external radiation are seen on the skin. The effects of the radioisotopes on the skin depend upon :

1) The type of radiation (eg. alpha or beta or gamma rays) and their energy. 2) The length of time the radioisotopes have been in contact with the skin. 3) Half-life of the radioisotopes and 4) Possible chemical injury the isotopes may also cause.

The first change noticed on the skin is a slight erythema, which occurs within 6 to 8 hours. These changes become progressively intense and in 2 to 3 weeks come to resemble a severe sunburn. *Microscopically*, there is marked vasodilatation with dermal and subcutaneous edema. Infiltration by inflammatory cells may be observed. The collagen of the dermis assumes an amorphous, acidophilic-mass-like appearance, losing its fibrillary character. Vesicles on the epidermis and later ulcers may develop which are very resistant to healing. The epidermis becomes thinned and the zone of malpighian cells is narrowed due to loss in cell division.

The skin appendages show degenerative changes and may disappear. Hair may be lost due to atrophy of dermal papillae. This is known as *epilation*. Scars form and the regenerative epithelium is very thin, becoming susceptible to injury and infection. Parakeratosis with brown scale formation may be noticed as a lesion of **chronic radiation dermatitis**. In some areas of the skin there may be abnormal melanin production while in others, there may be failure of melanin production and so these areas appear white.

Internal radiation : The radioisotopes that may be ingested from fallout cause considerable damage since they are in contact with the tissues for a long time causing lesions. The isotopes may be deposited in critical tissues, i.e. radioactive iodine may reach the thyroid while "bone-seekers" (strontium, uranium, radium, plutonium) are deposited in the bone. Calcium may be eliminated in the feces and milk. In the bones, the hemopoietic tissue may be destroyed resulting in agranulocytosis, leucopenia, aplastic anemia and sometimes ending in fatal leukemia.

CHEMICAL INJURIES

In man poisoning occurs under suicidal, homicidal, or accidental conditions. So, the number and variety of poisons are many. Fortunately, among animals, poisoning occurs mostly accidentally when animals ingest poisonous substances. Criminal poisoning does occur sometimes and is of chemico-legal importance.

Mode of action : Poisons may act at the point of entry (war gases in lungs) or they may be transported to certain seats of predilection, where they cause damage, without affecting the portal of entry (benzene though absorbed through the lungs, affects only the hepatic cells and bone marrow). In some cases a combination of both processes may occur. These poisonous effects may either be acute, subacute or chronic.

Some poisons are protoplasmic poisons and so affect all tissues while others have selective toxicity. For example, heavy metals affect all organs. But methyl alchol affects the cells of retina and optic nerve.

Poisons are able to exert their toxic property by being able to alter the chemical and physical characters of protoplasm. The physiological actions of the poisons also differ. For example, morphine and barbiturates depress specifically the cells of the central nervous system, strychnine on the other hand causes hyperactivity of these cells; carbon monoxide produces a state of anoxia by preventing hemoglobin from taking up oxygen; cyanide inhibits tissue oxidation; some poisons may remove vital chemicals, for example oxalates and fluorides remove calcium causing hypocalcemia.

Corrosives : Acids and alkalies : These destroy the protoplasm at the point of contact. The acids are : sulphuric, nitric, hydrochloric, oxalic, acetic and carbonic. The alkalies are : caustic soda (NaOH), caustic potash (KOH), calcium oxide, barium chloride, mercuric chloride, zinc sulphate. These produce inflammation, necrosis and later sloughing of the necrotic tissue and ulcers, depending upon the concentration and length of exposure to the corrosives. Acids produce coagulation necrosis while alkalies liquefaction variety. The lesion produced by carbolic acid is white and opaque. Carbolic acid being a good fixative, the tissue is well preserved without autolytic changes supervening.

Mercuric chloride : This produces coagulative necrosis in virtue of the protein precipitating action of mercuric salts. Mercury inactivates enzymes, specially cytochrome oxidases. Besides it may damage cellular membranes by combining with sulphydryl and phosphoryl groups. It is the mercuric ion that is toxic. Pure mercury is not poisonous. The mercuric ion is absorbed from the skin and mucous membranes of stomach and mouth and temporarily stored in the liver. It is excreted through the large intestines and kidneys.

In the colon hemorrhagic colitis is found. In the stomach necrosis of the gastric mucosa and ulcer formation occur.

In the kidneys, the epithelium of convoluted tubules becomes necrotic and suppression of urine results. In the epithelial cells of proximal convoluted tubules there is disruption of endoplasmic reticulum, breakage of plasma membrane, and swelling of mitochondria. Eosinophilic droplets are formed probably the result of reabsorbed proteins from the glomerular filtrate. Anuria may develop, as the lumen is filled with granular casts. Infiltration by granulocytes occurs around the necrotic tubules. Calcium salts may be deposited on the necrotic epithelial cells and the surrounding tubular basement membrane within 5 to 7 days. The reason for this calcification is obscure.

Lead : Animals may sometimes suffer from lead poisoning.
Sources of poison :
   A: **By ingestion.** i) Licking lead paint off poles, fences, barns, etc.
        ii) Accidental swallowing of lead paint or putty.

iii) Contamination of fodder or water from oil fields.

iv) Contamination of fodder from sprays of insecticides containing lead salts.

B. **By inhalation.** i) Fumes from burning storage batteries.

ii) Exposure to fumes from lead smelters.

Ingestion is the more comon method of poisoning.

Lead is a protoplasmic poison. It is poorly excreted and so cumulative poisoning may occur (but this does not seem to be seen in cattle). Lead poisoning may be acute or chronic. These differ only in their rate of development. Otherwise their pathology is similar. The important lesions are found in the gastrointestinal, hemopoietic and nervous systems.

The symptoms seen when the gastro-intestinal system is affected are colic and abdominal pain. There is catarrhal enteritis, ruminal atony followed by constipation and diarrhoea. A 'blue line' is seen at the junction of the teeth and gums due to formation of lead sulphide. This forms by the interaction of hydrogen sulphide (that forms due to the decaying of food materials) and circulating lead.

When the hemopoietic system is affected the erythrocytes become more fragile and so hemolysis occurs resulting in anemia, which causes increased hemopoietic activity. Basophilic stippling is seen. The lead 'poisons' certain of the enzymes required in hemoglobin synthesis and so anemia may result this way also.

The renal lesions though not so severe as in the hemopoietic system, consist of damage to the proximal convoluted tubules. So glycosuria, phosphaturia and aminoaciduria result. Lead salts inhibit a dehydrogenase that is essential in pyruvic acid metabolism and electrolyte transport and so mitochondrial respiration is damaged.

In chronic plumbism one can see intranuclear acid fast inclusions in the proximal convoluted tubular epithelium and in the hepatic cells. These inclusions are believed to be a compound of protein and lead that is reabsorbed.

The changes noticed when the nervous system is affected are: diffuse edema of the grey and white matter, increased cerebro-spinal fluid and degeneration of neurons throughout. The symptoms seen are: intermittent convulsions, opisthotonus, frenzy, mania, blindness, fear, coma, pushing the head against solid objects and circling movements.

There is demyelination of the motor nerves of the limbs and so paralysis and knuckling occur. Roaring may occur in horses in which the laryngeal nerves are affected. Hepatic and renal tubular epithelial cells contain acid-fast inclusions.

In chronic lead poisoning, there may be osteoporosis and lameness. Calcium is replaced by lead in the bones.

Lead can be detected in the urine, feces, blood and milk.

Arsenic: This is of importance because it is used in criminal poisoning. Arsenic is a protoplasmic poison and combines with suphydril groups and so invactivates SH—cotaining enzymes and hence inhibits cellular resparation. The poisoning can be acute if in a large doses or it may be chronic if ingested in small quantites, because it accumulates in the body. Poisoning may occur accidentally by ingesting rat poisons or paints.

Being an active irritant, arsenic produces severe hemorrhagic gastroenteritis. Due to necrotising lesions of the walls of capillaries, petechiae are noticed on the serous membranes and skin. Thrombosis may occur in vessels of brain causing infarcts of the brain substance. Fatty degeneration of the liver, kidney and myocardium may be seen. Death in acute poisoning may be due to the depressant action of the chemical on the central nervous system.

In chronic poisoning, the lesions are mainly on the skin, gastrointestinal tract and the nervous system. In the stomach and intestines, congestion, edema and small ulcers may arise. Myelin degeneration and destruction of the axis cylinders are the lesions seen in the nervous system. Pigmentation and severve keratinisation of the skin are found.

Phosphorus: Poisoning with phosphorus can occur in dogs and cats by accidental injestion of rat poison.

Phosphorus is a violent poison and causes extreme fatty degeneration of heart, kidney and liver, as well as of the axis cylinders of the nerve fibres in the central nervous system and peripheral nerves. Acute necrosis of the liver, jaundice and hemorrhages are also seen in acute poisoning.

Carbon monoxide: Carbon monoxide is an odorless, non-irritating gas. Combining with hemoglobin it forms a stable compound since it has 200 to 300 times greater affinity for hemoglobin than oxygen and so asphyxia occurs as no hemoglobin is available to form oxyhemoglobin. The formation of carbon-monoxyhemoglobin is also very rapid and hence the speed at which death may occur. Carbonmonoxyhemoglobin is cherry-red. Hyperemia of all tissues occurs. Due to hypoxia petechiae, edema and hyperemia of the brain are common. If death is delayed there is degeneration and necrosis of the lenticular nuclei due to thrombosis of cerebral vessels. Anoxia may also cause fatty changes in the heart, liver and kidney. But these lesions may not be seen if death occurs quickly.

In *chronic poisoning* by carbon monoxide, the lesions described above are seen being the result of low grade anoxia. If death occurs it is due to failure of cardiac, hepatic or renal function. But more often recovery is the rule.

Barbiturates Poisoning by any overdose of barbiturates may occur in dogs and cats. Liver is the organ that detoxifies barbiturates. So if barbiturates are given to animals with damaged livers, toxic effects and death may occur even in therapeutic doses. Barbiturates are powerful depressants of the central nervous system The respiratory centre may be depressed and death may occur due to respiratory failure Shock-like symptoms may be noticed due to vasodilatation. Anoxia caused thus produces renal lesions.

Death from anesthetics: An overdose of anesthetics may cause death by producing respiratory failure or cardiac fibrillation. Aspiration of gastric contents, excessive mucous secretions or blood may cause asphyxia and a sudden rise or fall in blood pressure may also contribute to death during anesthesia.

Hydrocyanic acid poisoning: The usual cause of cyanide poisoning in animals is by ingestion of cyanogenetic plants like young shoots of Sorghum.

The poison causes histotoxic anoxia by preventing intracellular oxidation processes being toxic to cytochrome oxidase and so tissue asphyxia results. But the blood does not lack oxygen and so it is bright red. Cerebral anoxia occurs manifested by muscle tremors, convulsions and dyspnoea,

Death may be instantaneous but in some cases symptoms may be manifested after some minutes or an hour. The animal falls down with convulsions, dyspnoea, anxiety, restlessness, frothing at the mouth, dilatation of the pupils involuntary defecation and micturition and finally opisthotonus. Mucosa is bright red in color The characterisitic almond odor is not conspicuous.

Lesions are not very characteristic. Congestion and petechiae may be found in patches in the abomasum and small intestines. Sub-epicardial and sub-endocardial hemorrhages are constantly seen. Blood clots slowly.

Nitrate and nitrite poisoning: Poisoning may occur in animals by ingesting food and water containing nitrates. Poisoning may also occur by animals eating or licking fertilisers. But the usual source of nitrate poisoning in animals is ingestion of plants containing high percentage of nitrates, chiefly potassium nitrate. Some soils are very rich in this chemical and so the plants grown there have high nitrate content.

Nitrates are irritants to the kidney and urinary tract and large amounts of nitrates cause hemolytic anemia. Gastro-enteritis may also be caused by the ingestion of nitrates.

The nitrates derived from the plants are reduced to nitrites probably with the aid of bacteria in the rumen. Nitrites produce methemoglobin. Nitrites are also vaso dilators and so lower blood pressure

Symptoms: Cyanosis, dyspnoea, very rapid pulse and respiration are most important symptoms. There may be extreme weakness and recumbency, frequent micturition followed by terminal convulsions.

Lesions: The characteristic appearance is the dark brownish colour of the blood due to methemoglobin. The mucous membranes are cyanotic. Gastro enteritis may also be seen. Patechiae are seen on the heart muscle and trachea.

Chlorinated hydrocarbons: The most important member of this group of insecticides is DDT (Dichloro-Diphenyl-Trichloroethane). The others are Benzene hexachloride (and Lindane which is its pure gamma isomer). Aldrin, Endrin, Chlordane, Dieldrin and Toxaphene. Poisoning may occur when these compounds are applied on the skin.

Being soluble in fat these poisons are concentrated in fat depots. Dangerous quantities may be excreted in the milk which may therefore be a source of poisoning in the young animals.

Symptoms include spasmodic twitching and quivering of various groups of muscles which may be seen within minutes of poisoning. • Apprehensivenes and frenzy followed by convulsions are the terminal symptoms seen. Some animals may be showing symptoms of "Blind staggers". Death occurs due to respiratory failure.

Lesions: Though nervous symptoms are appreciable, lesions in the Central Nervous System are obscure. Petechiae and ecchymoses are found in the heart and other places. Pulmonary congestion and edema are invariably noticed. In the animals that have survived a day or two, acute, toxic hepatitis and acute tubular nephritis may be seen. In animals that have ingested the poison enteritis is encountered.

Carbon tetrachloride: Since this drug is used widely as an anthelmintic, poisoning may occur sometimes. By whichever route it may enter the body by ingestion, inhalation or by injection the seats of predilection of the toxic drug are the liver and kidney. CCl_4 may directly injure the cells in higher concentrations. But it is now believed that it is the metabolic products of the drug (like free Cl or CCl_3) that are toxic by peroxidation of lipids and shifting double bonds of polyunsaturated fatty acids of the cell membranes. In the renal tubular cells protein synthesis is impaired due to dissociation of polysomes and shedding of ribosomes from the rough endoplasmic reticulum. Dissolution of mRNA results. Stoppage of protein synthesis results in accumulation of lipids in the cells since lipoproteins are not synthesized. Cell death in CCl_4 poisoning may be due to injury to cell membranes (of endoplasmic reticulum and mitochondria) and depletion of ATP reserves. Hence irreversible injury to the cellular respiration occurs (See page 20 for changes in anoxic injury that results thereby)

Microscopically: In the liver fatty change and coagulative necrosis will be seen in the hepatic cells. Entire lobules may be lost.

In the kidney intense necrosis of renal tubules, mostly in the distal convoluted tubules (some times in the proximal tubules also) may be observed. There is sloughing of cells with granular casts, red cell casts and casts of hemoglobin in the lumen of these tubules. If damage is not too severe, regeneration of the liver and kidney tubules is possible, restoring normal function.

Organic phosphates: In this group of insecticides are the Parathion and Malathion. These poisons inhibit the action of cholinesterase and so acetylcholine acts continuously. Therefore there are symptoms of excessive stimulation of parasympathetic nerves exhibited by increased peristalsis, salivation, bronchial constriction, increased secretion of bronchiolar mucous glands, constriction of the pupils and sweating (the muscarin effect.) Hyperstimulation of cholinergic-nerves of the Sympathetic nervous system is demonstrated by twitching of muscles, tetany and weakness followed by paralysis (the nicotinic effect).

Symptoms include the above together with drowsiness, convulsions and coma. Cause of death is asphyxia.

Lesions : No pathognomonic lesions are observed. Hemorrhages may be found on the heart, gastro-intestinal tract and the lungs. Depletion of lymphocytes in the thymus and spleen may be noticed.

CHAPTER 9

DISTURBANCES OF MINERAL METABOLISM
(Revised by Dr. D. Anjaneya Prasad)

Potassium	Iron
Sodium and Sodium chloride	Copper
Calcium	Cobalt
Pathological calcification	Zinc
Milk fever	Iodine
Phosphorus	Fluorine
Magnesium	Selenium
Manganese	Molybdenum
Sulphur	

Various elements are considered essential for the body and the following are their functions :

a) They regulate the osmotic pressure of body fluids. For example, sodium, potassium and chlorine maintain the electrolyte equilibrium of the extra and intracellular fluid.

b) They form building materials and so constitute structural components of tissues. For example, calcium and phosphorus are important constituents of bones and teeth.

c) They are of great importance in enzyme-catalysed reactions. They act as important cofactors. For example, magnesium and manganese act as cofactors in carbohydrate metabolism.

d) Some minerals are an integral part of some enzymes. For example, zinc enters into the composition of carbonic anhydrase; cobalt is a part of vitamin B_{12}, copper is a part of tyrosinase, molybdenum that of xanthine oxidase and iron that of catalase and peroxidase.

e) They help in oxygen transport: for example, iron in hemoglobin.

f) They may form an integral part of hormones; for example, zinc is found in insulin, iodine in thyroxine.

g) They help maintain the acid-base equilibrium. For example, sodium and potassium help maintain the pH of the body fluids

h) As constituents of the proteins and lipids that make up the muscles, connective tissues, skin, hair etc. For example sulphur in proteins and phosphours in phospholipids.

Deficiencies of the elements, therefore, result in the retardation or even cessation of the above functions and disease develops. Though minerals are essential for the body, too excess an amount upsets the delicately balanced biochemical ractions and pathological changes may thus result.

POTASSIUM

Potassium is found intracellularly. Its functions are:

1. Relaxation of cardiac muscle,
2. Taking part in carbohydrate and protein metabolisms.
3. Playing a part in the secretion of acid by the cells of gastric mucosa, for, in potassium deficiency, acidity is decreased.
4. Taking part in certain metabolic aspects of muscle and nerve.
5. Is the chief basic ion of the cell.

Potassium is mostly excreted by the kidneys. Aldosterone which controls reabsorption of sodium also controls the excretion of potassium. Usually potassium deficiency is not met with in animals.

Experimental hypopotassemia has been produced in laboratory animals, dogs and calves. In the heart, the lesions comprise of loss of striation, necrosis and infiltration by reticulo-endothelial cells followed by scarring. Since the bulk of potassium is in the muscle, deficiency is manifested by loss of muscular tone, flabbiness, weakness and paralysis.

Hypopotassemia may arise in medication with desoxycorticosterone under the action of which large quantities of potassium are excreted by the kidneys with resultant cardiac lesions.

Hypopotassemia is not of much importance.

SODIUM AND CHLORINE (Salt)

Sodium is present in the extracellur fluid and in blood plasma and forms bulk of the cation there. It is present as chloride and so depletion or excess of sodium occurs with depletion or excess of sodium chloride.

Functions : 1. Sodium chloride helps in the maintenance of the osmotic pressure of blood, interstitial tissue and cells.

2. Salt in plasma decreases the viscosity of blood.

3. The chlorine part of hydrochloric acid in gastric juice is derived from the sodium chloride ingested.

4. Maintenance of acid-base balance in the extracellular fluid is by sodium chloride.

Salt deficiency : Usually with balanced diets salt deficiency does not occur. But some times in herbivores, deficiency arises if ration is not supplemented by salt since plants are usually poor in sodium content.

Deficiency may also occur in the following conditions :

 a) Excessive vomiting and diarrhoea.

 b) Excessive sweating (especially in horses).

 c) In sows, excessive amounts are lost through milk if litters are large.

 d) Loss in urine of excessive amounts in Addison's disease. Here, in the absence of aldosterone, sodium chloride cannot be reabsorbed by the tubular epithelium and so is lost in urine.

Heat exhaustion : In hot weather when working horses sweat much, salt is lost from the body and so the osmotic pressure of the extracellular fluid is decreased. This upsets the normal osmotic equilibrium. So there is hindrance to the exchange of oxygen, metabolites and nutrients between the cells and the extracellular fluids. Neurones show lipoidal degeneration, which is manifested by nervous symptoms This syndrome is known as *heat exhaustion.*

Dehydration : When there is deficiency of salt, the osmotic pressure of extracellular fluid is diminished. To correct this, kidney removes much water from the blood and so the blood volume is diminished and hemoconcentration develops. The concentration of plasma proteins increases (due to loss of water) resulting in increased colloidal osmotic pressure. Therefore, water from the interstitial tissues is drawn into the capillaries, depleting the extracellular fluid.

Again since the osmotic pressure of extra cellular fluid is diminished, water passes into the cells to make it isotonic with the extracellular fluid. So extracellular fluid diminishes. This condition constitutes dehydration.

Chronic salt deprivation results in poor growth, unthriftiness, a rough coat and poor reproductive capacity. But it would require many months of salt deprivation to cause symptoms.

Salt poisoning : Fowls and pigs are particularly susceptible for salt poisoning Fowls may have access to brine vats used for pickling meat or fish and may poison themselves by drinking the brine. In self-feeding hoppers. salt may settle at the bottom by ingestion of which chicks may be poisoned. Such chicks show ascites and nephritis.

Swine also may be poisoned by drinking brine used for salting meats or by eating a ration containing excessive amount of salted fish meal. Affected animals show gastroenteritis and eosinophilic meningoencephalitis. In peracute cases, death occurs in a few hours. In acute cases, animals die after one or two days. **Symptoms** seen are anorexia, rapid pulse, rapid respiration, severe convulsions and sometimes paralysis In chronic cases somnolence, drooping of ears incoordination of movements, circus movements and sometimes blindness, deafness and inability to grunt occur. **Macroscopically,** there may be hyperemia, edema of leptomeninges and brain, hyperemia of gastrointestinal mucosa, conjunctiva, liver and lungs and myocardial degeneration. **Microscopically,** lesions of "eosinophilic" meningoencephalitis are noticed consisting of hyperemia of the leptomeninges and the cortex, edema and perivascular infiltration of eosinophils and some lymphocytes. The endothelial cells of the capillaries and small veins swell and proliferate. Pyramidal cells of the cortex are degenerated. Areas of softening may be seen on the floor of sulci and in the cerebellun and spinal cord—polioencephalomalacia.

The sodium ions readily penetrate the blood-brain barrier and produce edema which damages ganglion cells and causes degeneration. Probably some toxic substances are produced by the presence of salt and so for detoxication eosinophils infiltrate the area.

It should be stressed here that though excessive salt may be ingested by animals, so long as sufficient and unlimited quantity of water is available. poisoning does not occur It is only when water is in short supply and kidneys are damaged that salt poisoning may occur if ingested in excessive quantities.

Salt poisoning may also occur if large quantities of physiological saline are administered to sick animals.

CALCIUM

Calcium is an important element and the vegetation obtains it from the soil. While the leaves and stems of plants are rich in calcium, the grains are poor in this mineral. It is a critical nutrient in the ration of dairy cattle.

Calcium is excreted from the body by the intestines and kidneys. Calcium metabolism is under the control of Vitamin D, parathyroid hormone and calcitonin.

Vitamin D appears to control the absorption of calcium by the intestines by increasing the absorption of the element by the cells of intestinal mucosa. How this is brought about is not clear,

Parathyroid hormone and thyro calcitonin control the calcium content of blood. Parathyroid hormone, in a manner not known, stimulates calcium absorption. The normal serum calcium level is 9 to 11 mg. per 100 ml of blood. If this is decreased, parathyroid is stimulated, and so more of parthyroid hormone is secreted. The action of parathyroid hormone depends on two fractions of the hormone, one acting on the bone and another on the kidney. The net result of both is to raise the blood calcium level so as to keep the solubility product of calcium and phosphorus to remain constant. Normally, the calcium and phosphate ions of the blood are in a state of equilibrium with those of calcium phosphate of the bone.These are subject to laws of ionic dissociation.That means to say if calcium level rises, phosphorus level must diminish and vice versa. If parathyroid hormone is injected the excretion of phosphorus is enhanced, (since by the action of one fraction, the renal threshold for phosphorus is lowered) and so hypophosphatemia results. Therefore there is compensatory increase in blood calcium which is derived from the bone. Under the action of the second fraction of parathyroid hormone, calcium is withdrawn from the bone (by the activity of osteoclasts) Calcitonin is the hypocalcemic hormone, produced by the ultimobranchial cells of the thyroid. It lowers the calcium to the normal level when hypercalcemia occurs.

Functions of calcium 1. Calcium with phosphorus forms the important mineral in the structure of bone and teeth and so is essential for ossification.

2. Calcium is required for normal clotting of blood.

3 It is essential for normal contraction of heart and for neuro-muscular excitability.

4. It helps in the maintenance of normal osmotic pressure and pH of blood.

5. It is required for milk production, egg laying and reproduction.

6. Calcium enters into the structures of intercellular substance.

7. It is essential for maintaining cell membrane permeability.

Excess of Calcium Calcium in the food must be in the proportion of 2:1 with phosphorus On the other hand if excess of calcium is given in the feed and that too with liberal quantities of vitamin D, too much of calcium is absorbed and so hypercalcemia occurs. This condition results in calcification of many tissues other than bone. Normally, calcium is present in the blood in a supersaturated condition. So if blood calcium level goes above 12 mg, it cannot exist in solution and so is deposited

Pathological calcification, Metastatic calcification : (General calcification) We have presently seen that increased absorption of calcium due to increased feeding of this mineral together with excessive amounts of vitamin D results in hypercalcemia (blood calcium level above 12 mg%). Hypercalcemia may also occur in tumors of parathyroid glands, in which condition, more of parathyroid hormone is secreted which, as we have seen earlier, causes hyperphosphaturia, hypophosphatemia and ultimately hypercalcemia by withdrawing calcium from bones. A third factor responsible for hypercalcemia is the presence of tumors in bone, especially those that replace bone. For example, in myeloid tumors, demineralisation of the bone occurs in the vicinity of tumors where there is abundant metabolic activity.

The fourth cause for hypercalcemia is renal disease in which there is retention of phorphorus The resulting hyperphosphatemia depresses calcium and the hrpocalcemia thus produced stimulates the parathyroid which undergoes hyperplasia and hypertrophy, liberating excessive amounts of parathyroid hormone which causes hypercalcemia.

The tissues affected are. i) **Kidney**: the tubular epithelial cells are calcified and casts are formed which plug the tubules.

ii) **Lungs**: the alveoli of lungs are the places where calcification occurs.

iii) **Stomach**: the mucous membrane is calcified. Calcification of kidneys, lungs and stomach is explained by the fact that while excreting acid by these organs, (Kidney— hippuric acid and uric acid; lungs —carbondioxide; stomach— HCl) local alkalinity is increased. And increased alkalinity lowers the solubility product of calcium and phosphate and hence these minerals are deposited.

Formation of kidney stones is a frequent feature noticed in hypercalcemia.

Dystrophic calcification (local calcification) Focal deposition of calcium salts may occur in tissues which are either undergoing degeneration or are necrotic Here, the calcium level of the blood is normal.

Why and how the necrotic tissues get calcified is not clearly explained though many theories are put forward. One of these is that of Klotz. Usually necrosis is preceded by fatty degeneration. In the necrotic area, the neutral fats are hydrolysed into fatty acids and glycerol. Glycerol is removed and calcium combines with the fatty acids to form calcium soaps. Subsequently, the fatty acids are replaced by phosphoric and carbonic acids to form the insoluble calcium phosphate and carbonates.

A second theory explains that in necrosis there is also autolysis and so there is break-down of nucleoproteins and phospholipids, thereby elevating the local concentration of phosphate ions. Again, in necrosis, phosphatase is liberated from the injured cells which in turn (as in normal bone) liberates phosphate ions. So, the net result is that the solubility product of the calcium and phosphate ions exceeds the normal concentration locally and so calcium phosphate is deposited.

Dystrophic calcification is met with in:- caseous tuberculosis; lesions in the walls of atheromatous blood vessels; abscesses; degenerating tumors; dead parasites (Trichinella); old thrombi; actinomycotic lesions; old areas of scarring; necrotic ganglion cells; necrotic renal tubular epithelium (in poisoning by mercuric perchloride), lithopedion; phleboliths.

Macroscopically, some lesions may only be microscopic and so not visible to the eye. Others appear as white or grayish masses in the tissue. When cut, gritty sound is heard and the characteristic gritty feeling is felt.

Microscpically, calcium salts that are deposited as granules or spheres take a blue stain with hematoxylin.

Significance Calcification by itself is not harmful. But mechanical interference of the local tissue may occur. Excess of calcium interferes with the action of zinc and may cause parakeratosis in swine.

Deficiency of Calcium: For normal functioning of the body, calcium must be present in the blood in a concentration of II mg. per 100ml. If this is reduced to 8 mg.% tetany and incoordination result and below 3 mg. % life cannot exist.

Calcium deficiency (hypocalcemia) may arise under following conditions:

A. Deficient intake: This may occur if animals are not fed green shoots and leaves but are maintained only on concentrates. Experimentally, hypocalcemia may be produced in laboratory animals by feeding calcium deficient diets.

B. Disturbances in absorption:

    i) Intestinal dysfunction

        a) Diarrhoea which removes calcium before it is absorbed.

        b) Malabsorption syndrome: in sprue, coeliac disease and steatorrhoea absorption is prevented for some reasons not yet clear.

    ii) Vitamin D deficiency: Calcium is not absorbed in vitamin D deficiency, and this may be brought about by

        a) deficiency in intake of the vitamin,

        b) deficiency of bile in hepatic diseases—bile is necessary for absorption of vitamin D.

    iii) Formation of insoluble complexes.

        a) Calcium may combine with oxalates and form insoluble compounds that cannot be absorbed.

        b) If excess of phosphorus is fed. insoluble calcium phosphate may be formed.

C. Excessive excretion by kidneys: In certain renal diseases like idiopathic hypercalcuria and renal tubular acidosis there is increased excretion of calcium.

D. Parathyroid deficiency: In parathyroid hypofunction, the calcium level of the blood is decreased. This hypofunction may be due to

        a) non-development or aplasia

        b) removal--accidentally while extirpating thyroids in treating goitre

Milk Fever—Parturient paresis – post parturient hypocalcemia;

Immediately after parturition. cows and to a lesser extent ewes, sows, goats and bitches, suffer from suddenly developing hypocalcemia. The syndrome noticed includes tetany, recumbency, coma and even death. This condition is known as *"Milk Fever"*.

Milk fever is especially common in heavy milking cows, particularly those that are not fed with sufficient amounts of calcium and phosphorus during pregnancy.

Pathogenesis: After parturition, when a large quantity of milk is secreted, large amounts of calcium are excreted through it. The source of milk-calcium is blood. Blood calcium is normally recouped by withdrawal from bones For this withdrawal parathyroid hormone is necessary. In animals suffering from milk fever, blood calcium is not recouped because of low parathyroid activity and so hypocalcemia develops.

Calcium intake during the dry period influences this problem. High calcium intake during the dry period (over 100–125 g. per cow per day) tends to increase the problem, while a low calcium diet (8 g. calcium daily per 450 Kg body weight) fed 14 days prepartum prevented the problem. Similarly, feeding a low calcium diet (33—44 g/day) prepartum and a calcium rich diet (143–197 g/day) postpartum prevented milk fever and excessively low plasma calcium. Although there has been some work indicating that the Ca:P ratio in the prepartum ration may be critical, other evidence suggests that calcium level may be more important.

So, immediately after parturition when there is a drain of calcium in milk, the parathyroid is not active enough to maintain calcium homeostasis and so hypocalcemia occurs. It is thought that excess of calcitonin may be responsible for hypocalcemia.

Symptoms: Excitement, tetany with hypersensitiveness and muscle tremors of the head and limbs are noticed first. Animal refuses to move. Anorexia, grinding of teeth and stiffness of limbs are also noticed. The eyes are dry and staring. These symptoms are followed by recumbency and drowsiness, progressing to coma. If prompt treatment is not taken up, animals may die. Blood calcium (total and ionized) and pohsphorus are lower than normal.

Lesions: No gross or microscopic lesions have been reported.

Treatment: Intravenous injection of calcium borogluconate (400 to 800 ml. of 25% solution for cattle) is universally adopted and the results are spectacular. Even while the drug is being administered, the animal stands up, takes interest of the surroundings and begins to eat.

Prevention: The condition can be prevented by feeding, during late pregnancy, a diet high in phosphorus and low in calcium so that the parathyroid may be stimulated and conditioned for its increased activity that will be required at parturition.

Soft-shelled eggs: Soft-shelled eggs may sometimes be laid by hens which are not fed with adequate quantities of calcium. Normally the calcium that is used in the formation of the egg-shell is drawn from the bones, which become smaller with thinner walls. The bones recoup their calcium content from feed. But if feed is deficient in calcium, the bird has no calcium to draw upon to form the egg—shell which consequently is soft and so soft-shelled egg results. This condition may also develop if a ration with high phosphorus content is fed. Calcium is withdrawn from bones to balance hyperphosphatemia that develops and so sufficient calcium is not avilable for the egg-shells.

Rickets: Calcium deficiency may be one of the causes of rickets, which is described in detail under musculo-skeletal system.

Phosphorus.

The normal inoragnic phosphorus content of blood is 4 to 6 mg. per 100 ml. Phosphorus is found along with calcium in the body—in bones and teeth mostly. Phosphorus is required for the synthesis of proteins and enzymes of the body and plays an important role in the intermediary metabolism of carbohydrates and creatine in reactions that occur in muscle contractions. It is also utilised in the formation of phospholipids of milk.

Excess of phosphorus: We have alredy noted that calcium and phosphorus must be fed in a definite ratio of 2:1. But if the phosphorus content is raised so as to make the ratio 2:2 or 2:4. then, hyperphosphatemia will result. To balance this increased phosphorus, calcium is withdrawn from the bones, which thus become porous and brittle--osteoporosis. This condition develops in animals fed with concentrates mainly. Horses that are maintained by feeding only bran develop a condition called osteitis fibrosa, also called *'big head'*, *'miller's disease'* or *'bran disease'*. This condition is described fully under musculo-skeletal system.

Animals fed with excess of phosphorus suffer from lameness. The bones becoming brittle are prone to fractures.

Hyperphosphatemia may also arise secondarily in diseases of kidney. In severe nephritis, phosphorus is not excreted, retention of which gives rise to hyperphosphatemia. To balance the excess of phosphorus, calcium is withdrawn from bones resulting in rickets (renal rickets) or osteomalacia.

In the grains, one half to two thirds of the phosphorus is in the phytate form. Cotton seed meal and some other plant protein supplements contain an even higher proportion of phytate phosphorus. Utilisation of phytic acid phosphorus is influenced by the level of vitamin D, calcium, and alimentary tract pH, the calcium: phosphorus ratio and by other factors.

Phosphorus deficiency: This is also known as *aphosphorosis* and is an endemic condition in certain parts of the world. The soil is deficient in phosphorus and so the plants grown there lack this element. Aphosphorosis is reflected by a) poor quality of flesh b) slowing down of rate of growth c) decreasd lactation d) irregular reproduction e) pica and f) absence of estrum,

Phosphorus deficiency is one of the causes of rickets, ostesmalacia and anemia due to post-parturient hemoglobinuria.

In South Africa, Sir Arnold Theiler had found that the disease known as "*Lamsrekte*" was due to aphosphorosis which develops due to soil being deficient in phosphorus. The animals suffer from lameness and so the name—lame sickness (lam=lame, siekte=sickness) was given to it. Animals suffering from aphosphorosis develop "pica" or "allotriophagy" or depraved appetite. They chew bones of dead animals on the fields. Some of these bones have the toxins of *Clostridium botulinum*, which thrives in the bones. Animals ingesting this toxin die of botulism.

Excessive amounts of iron may combine with phosphorus forming insoluble phosphate of iron, resulting in phosphorus deficiency, manifested by lameness and frequent fractures.

MAGNESIUM

Magnesium is widely distributed in the tissues and is present intracellularly. It is a cofactor for many enzymes especially in the glycolytic cycle and is a component of inorganic pyrophosphate.

Magnesium is of importance in the growth and development of bone as well as in muscle contraction. Magnesium and calcium have to be in definite proportion, viz., 1:3.5 for proper growth of bone.

Excess of magnesium: This may rarely happen. But if excessive amounts are fed, bone formation is interfered with since magnesium antagonises the action of calcium. The bones are soft and their deformities occur.

Deficiency of magnesium: Hypomagnesemia The normal magne-- sium content of blood is 2 to 3 mg. per 100 ml. If it is lower, tetanic symptoms develop.

Animals fed on lush grasses or young wheat shoots develop incoordi-- nation, recumbency and drowsiness. This is therefore known as '*grass tetani*, or '*wheat poisoning*' or '*grass staggers*'. Animals in pregnancy or at parturition are more susceptible. The cause of this condition is not known. Hypomagnese-- mia develops though the pasture and fodder have normal magnesium content. There is probably some factor in the diet which reduces the absorption or inter-- nal metabolism of magnesium.

A second factor may be the production of excessive quantities of ammonia in the rumen from protein--rich diets. The ammonia prevents the absorption of magnesium.

A third factor in the production of hypomagnesemia is the ingestion of large quantities of potassium (with wheat shoots) which depresses serum levels of magnesium

A fourth factor may be hyperactivity of thyroid. In hyperthyroidism, which occurs in cold weather, hypomagnesemia may occur. Hypomagnesemia may develop in calves raised wholly on milk since it is deficient in magnesium. These animals show excitement, tetany, convulsions and finally death due to respiratory failure. The magnesium-deficient pigs exhibit the "Stepping syndrome" which causes it to keep stepping or lifting its hind leg almost continu-- ously while standing. To date it is thought that magnesium supplementation to swine diets is not needed.

Lesions : Extravasations of blood may be seen in the subcutaneous tissues and under pericardium, peritoneum, pleura and mucosa of the bowel. Thrombosis of the venules of the heart may also be noticed microscopically. Agonal emphysema may also occur.

In alcholics there is excessive loss of magnesium in urine, vomition and diarrhoea, resulting in magnesium deficiency. Delerium tremens may be due to hypomagnesemia resulting thus.

MANGANESE

This is present in the soil and all forage plants contain sufficient amounts. In the body all tissues contain manganese, which is particularly abundant in the liver.

Manganese acts as a cofactor for many enzymes, especially those that take part in Krebs' cycle.

Deficiency of manganese : Perosis or Slipped tendon : This is of importance only in the growing fowl. In manganese deficiency, the normal growth of the bone does not occur (Manganese is necessary for the synthesis of cartilage matrix) The long bones are shortened. The epiphyseal cartilage fails to ossify (normally the ossification must be complete by 12 weeks of age). The groove of the tibial condyles is shallow and the gastrocnemius tendon slips out of this groove—"slipped tendon". Animals cannot walk and so die of starvation.

In pigs manganese deficiency causes decreased growth and feed efficiency, reproductive problems, lameness, shortening of legs, thickening of tarsal and carpal regions and bending of front legs. There is obesity. Enchondral ossification in the distal epiphyses of radius is stopped. Manganese is associated with fertility in ewes and young cattle.

SULPHUR

Deficiency of sulphur does not occur. Excess of sulphur may sometimes occur, when birds are fed large quantities of it as an anticoccidial drug. Sulphur replaces phosphorus and ' sulphur rickets" may supervene.

For efficient utilization of urea. a nitrogen: sulphur ratio of 10:1 has been suggested. The pig has little ability to use inorganic forms of sulphur.

IRON

Iron is universally found in the soil. It is found in the body in the following : hemoglobin, myoglobin, enzymes, cytochrome, catalase, peroxidase. It is stored as ferritin and hemosiderin.

Iron is absorbed from the intestines. The following factors influence its uptake :—

1) *Valence :* Ferrous iron is absorbed more readily than ferric.

2) pH . Acid medium is required since in such a medium the ferric form is converted into the ferrous form. Hence an alkaline medium decreases absorption.

3) *Phosphorus and phytic acid* - These form insoluble complexes with iron. which therefore cannot be absorbed.

4) *Presence of copper :* Copper mediates absorption of iron.

5) *Amount of iron present :* The higher the concentration of iron the greater is its absorption.

6) *Amount of iron present in the body :* The amount of iron absorbed is conditioned by the amount present in the body. Iron is a 'one-way' traffic element. Almost none of it is excreted. So only when some amount of it is used up that further quantities are absorbed. The cells of the intestinal mucosa have selective powers of absorption of iron. This is known as *mucosal block.*

Excess of iron : This may not usually occur. If, however, excessive amounts are present, lameness due to bone disease occurs, as iron forms insoluble phoshate and phosphorus deficiency results.

Deficiency of iron : In animals deficiency may not usually arise except in baby pigs that are raised on cement floors without access to soil. Sow's milk is deficient in iron and the piglets usually obtain their iron requirements by swallowing some soil (containing iron) while rooting. But piglets raised in houses with cement flooring cannot obtain their iron requirements unless specially supplemented. Such piglets suffer from anemia—"piglet anemia" which is a microcytic, hypochromic type of anemia and do not grow well, are stumpy and suffer from enteritis and pneumonia since anemic animals lose their vitality and are easily infected. At postmortem the blood is thin, watery and pale. Mild edema is found. The heart is dilated and liver is enlarged and fatty.

Iron deficiency seldom occurs in dairy cattle under natural conditions except in young animals fed only milk or where there is severe loss of blood because of parasitic infestations or disease.

Iron deficiency may also occur when there is impairment of absorption as in long standing diarrhoea or by lack of hydrochloric acid (achlorhydria). Iron deficiency may occur indirectly when animals are infested with blood sucking parasites — ancylostomes. stomach worms and ticks.

Iron in proper proportions appears to be necessary for the proper utilisation of folic acid and so in iron deficiency anemia, red cell aberrations characteristic of folic acid deficiency are seen.

It is found that the phagocytic activity of neutrophils is decreased in iron deficiency anemia due to the decreased activity of iron, containing enzyme-myeloperoxidase, in the neutrophils.

COPPER

Copper is present in most of the soils in adequate quantities. In some parts of the world some soils may be deficient.

Copper is present in the body in most of the tissues. It is present in the central nervous system in large quantities. The functions of copper are :-

1- It controls the absorption of iron from the intestines
2. It controls the utilisation of iron in hemoglobin synthesis.
3. It is a component of cytochrome oxidase.
4. It activates ascorbic oxidase, tyrosinase, succinic oxidase.
5. It is a component of mitochondria.
6. It regulates the rate of phospholipid synthesis.

Excess of copper is poisonous and the subject of copper poisoning is described fully under "Diseases of hemopoietic system".

Deficiency of copper: In most regions of the world, the soil has adequate amounts of copper. But in some places, the soil is so poor that deficiency occurs in herbivorous animals. The lesions are found in the hemopoietic tissues, hair and central nervous system. The lesions seen in copper deficiency may be explained by decrease in tissue respiration due to lack of cytochrome oxidase which is not formed in sufficient quantities in this deficiency.

Depression of osteoblastic activity in copper deficiency may produce osteoporosis. Osteoid is not deposited on the cartilageous spicules.

Copper being necessary for the formation of hemoglobin, copper deficiency causes anemia, which is microcytic and hypochromic in type. Animals so affected are unthrifty, do not gain weight and are anemic. Treatment is by way of providing copper and iron in "salt-licks".

In sheep, a condition known as "steely wool" is supposed to be caused by copper deficiency in which the wool loses its crimp and becomes hair-like. It loses its color—*achromotrichia*. Greying of wool is attributed to deficient melanin production, because copper-containing tyrosinase is required for the normal formation of melanin. Molybdenum has an antagonistic action against copper. Molybdenum accelerates, the excretion of copper from the liver and thus if a ration of forage having large amounts of molybdenum is ingested, copper deficiency may arise. Such a condition develops in sheep. If sheep are grazed on land with molybdenum-rich grasses, lambs are born of them manifesting copper deficiency nervous symptoms, known as Sway back. Since copper is necessary for the development of nervous system copper deficiency produces lesions in it. The condition is described fully in the chapter "Diseases of Nervous System"

Deficiency in copper may cause fatal syncope in cattle (falling disease), The affected animals die suddenly with or without symptoms of diarrhoea. emaciation and hypochromic microcytic anemia. Heart shows fibrosis in places where myofibrils have atrophied. Probably the cytochrome system of the myocardium is destroyed in copper deficiency. Degeneration of internal elastic laminae ef great vessels and heart leads to their rupture.

A swelling of the ends of the long bones, especially above pasterns is a chracteristic symptom in cattle. The bones become fragile, which often result in multiple fractures of ribs, femur or humerus. Cows in a copper - depleted condition may fail to conceive, may have difficuly at calving (retained placenta) or may give birth to calves with congenital rickets.

COBALT

Sufficient amounts of cobalt are present in the soils of most parts of the world.

Cobalt is a constituent of vitamin B_{12}, which is necessary for hemopoiesis. So cobalt deficiency results in anemia.

Cobalt deficiency: This is found only in the ruminant. Cobalt is necessary for the maintenance of ruminal flora which synthesize vitamin B_{12}. In cobalt deficiency, the ruminal microorganisms cannot thrive and so vitamin B_{12} cannot be synthesized and so anemia develops. Symptoms seen in cattle and sheep are, loss of appetite, progressive emaciation, rough coat, pallid mucous membranes and death. In cobalt deficiency, the ruminant is unable to metabolise propionic acid and so the animal suffers from anorexia and inanition. The hemoglobin content of the blood falls from a normal of 12 grams per 100ml to as low as 5 grams. Growth of wool is retarded and the fibers become weak Fatty degeneration of the liver and heavy deposition of hemosiderin in the spleen and liver are noticed at postmortem This condition is enzootic in certain parts of the world and is known as *"hill sickness"* in New Zealand and *"enzootic marasmus"* in Australia.

The intestinal microoganims of the pig are capable of producing some vitamin B $_{12}$ in the lower part of the intestinal tract when cobalt is present. How much if it is absorbed is not known. But it will be available in the faeces, and it is known that pigs will consume some faeces Cobalt at 400—600 ppm. in the diet was toxic to pigs resulting in anorexia, growth depression incoordination and extreme muscular tremors.

CHAPTER 9

ZINC

Zinc is of importance since it is a constituent of enzymes, (carbonic anhydrase, uricase, kidney phosphatase, alcohol dehydrogenase, carboxypeptidase, glutamic dehydrogenase and a hormone – insulin.

Excessive calcium and phosphorous may interfere with the action of zinc, producing zinc deficiency.

Deficiency of zinc: Zinc deficiency produces a condition called "*Parakeratosis*" which is particularly noticed in pigs, due to addition of excessive amounts of calcium and phosphorus to the ration. When the calcium content of the feed is high, the digestibility of fat in the ration is reduced and so deficiency of the essential fatty acids results. And fatty acid deficiency has been found to cause skin lesions. The affected animals are unthrifty and do not grow well. Diarrhoea and vomition may be present.

Microscopically the lesions are present on the skin of the face, legs perineal region, inguinal region, shoulders and ears.

To start with the skin is covered over by papules and pustules, which after becoming dessicated form into crusts These crusts may crack and infection may occur through them. Oral lesions consist of ulcers which prevent eating.

Macroscopically the stratified squamous epithelium is thickened in the skin and esophagus. In the skin all layers of the epidermis, except, the stratum lucidum are thickened. Hairs become sparse and walls of arterioles are damaged.

In cattle deficiency of zinc may cause parakeratosis as in pigs and alopecia. The lesions are seen on the vulva, anus, muzzle, ears, back, flank, neck, knees and backs of legs.

In sheep deficiency of zinc causes loss of wool and thick wrinkled skin.

Zinc is used for speedier healing of wounds. In man zinc deficiency causes poor gonadal maturation, stunted growth and disappearance of axial, facial and pubic hair.

In birds zinc is necessary for egg production, for the formation and maintenance of epithelial tissues and for the development of the skeleton. Deficiency of zinc in the young birds causes abnormality of the skeleton leading to leg weakness and ataxia. The long bones become shorter, thicker and. crooked The joints become enlarged and rigid. Due to hyperkeratosis, feather development is impaired and on the legs and feet necrotic dermatitis may be seen. In the embryo skeletal development may be interfered with and so entire limbs may be absent.

Excess of zinc produces anemia by interfering with iron-porphyrin-forming function of copper.

IODINE

Iodine enters into the molecule of thyroxine, a hormone secreted by the thyroid.

Excess of Iodine: Usually iodine is not ingested in large quantities. But if given over a considerable period of time as a therapeutic agent in the form of iodide, poisoning by iodine, known as "**iodism**" may occur. The symptoms of iodism are:– conjunctivitis and excessive lachymation, hyperkeratosis, rhinitis and mild cough. Discontinuance of iodide therapy abates these symptoms.

Deficiency of iodine: Soils and waters of some areas in the world (Great Lake regions in the U S. Swiss Alps, Himalayan region etc) are deficient in iodine. Animals and men inhabiting such areas suffer from iodine deficiency, the condition also being known as *goitre*. Iodine deficiency can also occur in high intake of calcium and when feed and water are polluted with bacteria. Goitre and hypothyroidism are described fully in the chapter "The Endocrine System".

Recent research has shown that cows receiving corn silage as sole forage and soya bean meal gave birth to calves with enlarged thyroid glands. Goitrogenic substances were present in the soya bean meal. Lactation and gestation increase the demand for iodine. Adding iodine to feed increases the iodine content of milk.

FLUORINE

Fluorine is present in the soil and water. Some soils have greater amounts than are safe for the body.

Excess of fluorine; Fluorine poisoning is a chronic condition and is met with in areas where fluorine content of the soil is high. In India there are areas where ' fluorosis" (which is the condition of fluorine poisoning) is met with in animals and men. Animals are poisoned by drinking water and eating forage containing large amounts of fluorine. Pastures may be contaminated by top dressing with phosphatic limestone containing fluorine; by smoke, vapour or dust from the following industrial plants: aluminium, copper, glass and enamel; iron and steel, superphosphate, and also from dust in volcanic eruptions. Fluorosis may also occur in animals fed rock phosphate containing large amounts of fluorine. Rock phosphate is fed to correct phosphorus deficiency. Fluorine is a cumulative poison. Fluorine in excess of 2 ppm in water is toxic to animals.

The lesions in fluorosis are found in the teeth and bones.

In teeth, lesions are found in those that are developing at the time of poisoning. So deciduous as well as permanent teeth that have grown prior to poisoning, are free from lesions. The lateral incissors of cattle show most pronounced changes. The affected teeth show opaque. chalk-like areas, some show mottling, having linear pigmented streaks and pits. Teeth exhibiting these changes suffer from hypoplasia of enamel and are so soft that they may wear down to the gums. Presence of multiple caries is a feature of this poisoning.

The bones, especially the metacarpal and metatarsal bones, the sternum, mandible and the phalanges, become shorter, thicker and broader. The cortex becomes thickened, encroaching on the medullary cavity, which therefore becomes narrower. The bones may sometimes become porous and brittle, and so prone to fractures. Periosteal hyperostosis or exostoses form especially around brittle joints and at the places of attachment of ligaments and tendons, giving rise to pain causing lameness and stiffness in animals. This is due to excessive mobilisation of calcium and phosphorus (giving rise to hypercalcemia) to compensate for the increased excretion of these elements in urine in conjunction with fluorine. 'Shifting lameness' is a characteristic symptom of fluorosis. Fluorine probably interferes with proper mineralisation of bone matrix by poisoning the alkaline phosphatase and so causes osteodystrophy described above. Degenerative changes in the heart, liver, bone marrow, kidneys, adrenals and the central nervous system may also be seen In some cases of chronic poisoning, anemia may be seen due to the suppression of hemopoietic activity of the bone marrow.

Acute fluorine poisoning can occur if an overdose of sodium fluoride (which is used as a vermicide in pigs and poultry) is administered. The symptoms are those of gastro-intestinal irritation due to the formation of hydrofluoric acid viz., vomition. Nervous symptoms—tetany and hyperasthesia—may occur by the formation of a physiologically inactive calcium fluoride in the blood plasma. (calcium must be in an ionized form to be active physiologically). Convulsions followed by death may occur.

Fluorine appears to be necessary for the development of hardness in bones and teeth, especially in growing animals. Fluoridation (i. e. addition of traces of sodium fluoride to drinking water) is practiced in some countries to correct possible deficiency.

SELENIUM

Selenium is a trace element found in the soil and plants and is required for the biological function of the cells. Some of its functions are similar to those of Vitamin E but it cannot entirely replace this vitamin. Also Vitamin E cannot replace entirely selenium since the vitamin cannot have the growth promoting effect in selenium deficient animals.

Selenium enhances the activity and improves the transport and retention of vitamin E. Probably it is necessary for the transport of the vitamin across the cells.

The alkaline soils and excess of sulphate fertilisers deplete the plants of selenium since this element is displaced. So animals fed on such plants suffer from selenium deficiency. Rain may wash the element from the soils.

Selenium deficiency : Selenium deficiency occurs if the feeds of animals have less than 0.05 p.p.m. of this element. Besides producing muscular dystrophy (white muscle disease), selenium deficiency also causes atrophy of the exocrine glandular tissue of the pancreas, resulting in loss of zoning of the cells. Acini shrink and become uneven in size. There may be ultimately fibrous tissue replacement. Hence in this deficiency, there is decreased production of the enzymes. So due to lack of the lipase, absorption of fats and fat soluble vitamins decreases and so the requirement of vitamin E is higher in selenium deficiency. In such animals the tocopherol level of the blood is lower.

Non-specific conditions like neonatal mortality and unthriftiness in weaner calves, lambs and goats; chronic diarrhoea in calves, infertility due to resorption in ewes and dietetic hepatosis in swine can be much improved and alleviated by selenium supplements. These are known as 'selenium responsive, diseases, because they are not known or proved to be caused by selenium deficiency alone. Abortion in sheep and a predisposition to retention of placenta in cows are supposed to arise in selenium deficiency.

In animals selenium improves growth and in fowls it improves the hat-chability of eggs.

Selenium poisoning . In certain soils the selenium content is high and so the plants that grow there contain a high percentage of selenium, so that animals that eat such plants suffer from selenium toxicity. Plants of the genus **Astragulus** and **Oxytropis** contain 500 to 15000 p p.m. of selenium.

Selenium interferes with the action of the enzymes that are made up of sulphur containing amino acids, because selenium is antagonistic to sulphur. Hence the biological processes are affected resulting in necrosis

Acute poisoning with selenium is noticed in animals that consume a large quantity of seleniferous weeds. Gastro-intestinal symptoms, besides symtoms arising from respiratory and myocardial failure are seen within a few hours or within a day or two. Polydipsia and polyurea are noticed In cattle due to discomfort in the abdominal area the animals seek relief by continuous walking. They may press their heads on hard objects. As vision is affected, the animals walk into obstacles. Paralysis and weakness follow Due to respiratory failure dyspnoea, cyanosis and death supervene. This is commonly known as "blind staggers".

Microscopically, hemorrhagic enteritis and proctitis, venous congestion of the lungs and abdominal viscera, acute toxic hepatitis, tubular nephritis and hemorrhages on the epicardium and endocardium and other organs are noticed. The mucous membranes of the bladder and the folds of the omasum are inflammed, since probably, the element is excreted through these places.

Chronic selenium poisoning may occur if the element is ingested in small amounts for a long period, when it becomes cumulative and causes chronic poisoning. This condition is known by the laymen as **'alkali disease'.** In these animals alopecia and rough coat are seen. The long hairs are lost——in the mane and tail in horses. in the tail of cattle and body hair of swine. On the hoof. grooves are formed parallel to the coronary band. These grooves may be so deep, and cracked sometimes, that the hoof may get detached from the sensitive laminae and slough off. The toe grows very long. The changes· of the skin and hoof are attributed to the modification in the structure of the keratin consequent on replacement of sulphur by selenium. Lameness may also occur due to affection of the joints

Microscopically, there is chronic venous congestion of the lungs and other viscera due to myocardial insufficiency. Besides focal necrosis, there is sero fibrinous infiltration of the myocardium. Later, lymphocytic infiltration also occurs. Edema of the pleura, peritoneum, pericardium and the brain occurs due to the cardiac lesions. Petechiae are seen on the epicardium and endocardium. Liver shows congestion and degenerative changes and finally cirrhosis results. Hyaline casts are found in the renal tubules, the epithelium of which shows swelling and hyaline changes. The articular cartilages are eroded causing lameness

Malformations are encountered in fetuses. In fowls poisoned by selenium hatchability is decreased and the chicks that may be hatched are weak.

Selenium reduces the sulphur and protein content of sheep s liver. In all poisoned animals, a moderate anemia is seen and the hemoglobin level falls to 7 gms. percent.

CHAPTER 9

MOLYBDENUM

Molybdenum is a constituent of xanthine oxidase and is an essential element. Levels below 3 mg/Kg of forages are normal. Levels above 20 mg Kg in forages are associated with occurrence of Teart pasture and toxicity. Copper supplementation was beneficial. If copper is low in diet a smaller amount of molybdenum is poisonous, but as copper increases, so does tolerance to molybdenum. A normal level of inorganic sulfate provides protection against high intakes of molybdenum by reducing molybdenum retention in the body. Excess sulfur may enhance molybdenum toxicity and modify copper utilization. Symptoms of chronic molybdenum poisoning (molybdenosis) are scouring, unthriftiness, rough-hair coat, loss of hair colour, dehydration, arching of the back, emaciation and in extreme cases death.

CHAPTER 10

DEFICIENCY DISEASES

Vitamin A	Niacin
Hypervitaminosis A	Pantothenic acid
Vitamin D	Pyridoxine
Hypervitaminosis D	Biotin
Vitamin E	Choline
Vitamin K	Folic acid
Vitamin C	Vitamin B_{12}
Thiamine	Proteins
Riboflavin	Fats

The term *vitamine* was first coined in 1912 by Casmir Funk to denote certain substances that are of vital importance to the body. He thought that these substances were amines. But later it was found that they had no amine structure and so 'e' was omitted and the term *vitamin* was retained.

The vitamins are needed in minute quantities. They do not yield energy but are required for the biochemical reaction taking place in the metabolism of energy yielding substances: carbohydrates, lipids and proteins. They do not enter into the structure of cells. Some vitamins form part of molecules of some enzymes while others function as co-enzymes. So it is apparent that deficiency of vitamins interferes with biochemical activities of cells and so manifested by disease.

Vitamins are available in nature and the source of a majority of them is plants. Now most of them are synthesized.

Vitamin deficiency can occur in two ways : *1) By reduced intake :* This is not of much consequence nowadays, since synthetic vitamins are added to feeds to make up for possible deficiencies. This is called *primary* or *simple* deficiency. Alcoholics frequently suffer from vitamin deficiency (as well as other deficiencies) since they do not eat sufficient food (and so vitamins). More often this is due to their inability to purchase both drink and food. They prefer to subsist on drink rather than eat adequate and balanced food.

2) Secondary deficiency in which sufficient vitamin is ingested but due to various factors mentioned below they are not available to the body.

A) Deficient availability to the body : This may occur in long continued vomiting, painful lesions in the mouth that prevent prehension, anorexia and esophageal obstruction.

B) Malabsorption: This may occur in the conditions making up *malabsorption syndrome.* Under these may be listed.

i) **Steatorrhoea** (fat in stools). For normal digestion of fats, bile is necessary for emulcification and lipase for splitting the fats into fatty acids and glycerol. So digestion of fats may be interfered with when :

a) there is obstruction of bile ducts by worms, tumors or gall stones,

b) there is obstruction of pancreatic duct by tumor or stones.

c) there is chronic pancreatitis. Here lipase is not produced.

d) there is sprue—this is important in man. In this condition there is atrophy and thinning of the intestinal wall. The villi are short and blunt; the absorptive surface is greatly reduced.

ii) Chronic enteritis: Here the absorption is interfered with due to fibrous thickening of the instestinal mucosa.

iii) Neoplasms of the small bowel : Lymphocytoma is most common in which absorption of food is interfered with:

C) Increased demands: Increased demands for vitamins arise:

a) when the rate of growth is very fast as in infancy and puberty.

b) when there is greater utilisation and so need in the body as at pregnancy and lactation.

c) when the metabolic rate in the body is increased as in fever and hyperthyroidism.

D) Reduced storage: This is of particular importance to vitamin A, vitamin D and some members of vitamin 'B' Group which are stored in the liver. In diseases of liver in which parenchyma is damaged diffusely, as in cirrhosis, these vitamins cannot be stored and so deficiency may develop.

Vitamins are broadly divided in to 2 groups, viz. the fat soluble and water soluble. Vitamins A, D, E and K are fat soluble while vitamins C and B group are water soluble.

Vitamin A (Retinol or vitamin A$_1$ and Dehydroretinol or vitamin A$_2$)

Sources : Mostly green plants and fish liver oils. Carotene and cryptoxanthine from plants can be converted into vitamin by all species of animals except probably cat. Intestines and the liver are the places where this conversion takes place.

Storage : Vitamin A is stored in the liver in the Kupffer's cells in an esterified form, as **Retinol.** Metabolised to retinol, vitamin A takes part in the formation of Rhodopsin in the Rod cells and lodges in cone cells of Retina. Converted to retinoic acid it helps growth.

It is transported in the blood by a high density lipoproptein. It may also be found in adipose tissues.

Deficiency of vitamin A : Dificiency of vitamin A may occur in the following ways.

1. Insufficient supply in the ration: This may not be of much importance in adults which obtain sufficient amount through green feeds. But in drought periods deficiency may be encountered in cattle and sheep. Poultry and swine may suffer if they are not supplied with vitamin supplement or green feed.

2. Chronic diseases of intestines : Here vitamin A cannot be formed from carotene nor can it be absored.

3. Diseases of liver: When liver is diseased vitamin A is not stored. In the absence of bile, vitamin A cannot be absorbed.

4. Interference in the conversion of carotene: Chlorinated naphthalenes prevent the conversion of carotene into vitamin A and so animals poisoned by chlorinated naphthalene (contained in lubricating oils) suffer from vitamin A deficiency (X–disease or hyperkeratosis).

5. Excess of intake of inorganic phosphorus Excess of ingestion of phosphate diets diminishes vitamin A storage.

Pathology: Vitamin A is essential for the normal maintenance of epithelial tissues, for bone growth in the young and for the regeneration of visual purple that is necessary for night vision.

Vitamin A by maintaining a high state of health and by maintaining the epithelial surfaces' protects ahe body from infections and hence it is known. as "anti-infection vitamin" though no specific anti-infection property is noticed.

Night blindness : (Nyctalopia—night blindness) Rods of the retina are the receptors that are concerned with dim-light vision. These rods contain the visual purple—rhodopsin. On exposure to light, this rhodopsin is split into retenine (which is vitamin A aldehyde) and opsin (which is a protein). During this splitting, nervous impulses are initiated that travel by optic nerve to higher centers. In the dark, the reaction is reversed to make visual purple. During this reaction, some amount of retenine is lost. The loss is made good by drawing on vitamin A of blood. In deficiency, blood vitamin A is reduced and so visual purple is not formed in sufficient quantity as sufficient vitamin A is not available and so the animal cannot see in dim light since impulses are not initiated and hence the condition is called night blindness.

Epithelial tissues : Vitamin A is necessary for the maintenance of the specialised epithelial surfaces of the body, viz. the mucous membranes of the gastrointestinal, the respiratory and urogenital tracts, and the epithelium of the eye and the skin. In the absence of vitamin A, the epithelial cells atrophy, disappear and their place is taken up by keratinised, stratified squamous epithelium due to proliferation of the basal cells.

In the respiratory tract, the ciliated epithelium is replaced by the stratified squamous type. Foreign bodies are not removed in the absence of ciliated epithelium and secretions are not present. The foreign bodies produce irritation leading to infection and pneumonia.

In the urinary tract the transitional epithelium of the renal pelvis and bladder undergo metaplasia into keratinised stratified squamous epithelium. The keratinised debris forms a nidus for the formation of **urinary calculi**'.

Reproductive system: In the female though conception occurs abortion takes place due to placental degeneration. At the junction of the maternal and fetal membranes, necrosis occurs with superimposed infection leading to separation of the fetal from maternal placenta. So the nutrition of the fetus is affected and hence it dies. A dead fetus is a foreign body and so is expelled — abortion. Sometimes a weak or dead calf may be born. Retention of placenta is common.

In the bull, the epithelium of the seminiferous tubules undergoes degeneration and so there is reduction in the number of normal spermatozoa. Sterility may thus ensue though lidido is mintained. These changes are reversible (unlike those of vitamin E deficiency) since with vitamin A feeding, rapid return to normal occurs.

In the eye a condition called **xerophthalmia** occurs. The corneal and scleral membranes become karatinised and so the goblet mucous cells there disappear. Along with these changes, the lining of the lachrymal ducts is replaced by the keratinised stratified squamous epithelium. The keratinised debris occludes the duct and so the tears do not bathe the eye. Hence ths mucosal surface of the eye becomes dry (xeros=dry) and rough. Because the cornea is thickened vision is impaired. The corneal surface may ulcerate (because of friction), become opaque and then be infected. Inflammatory changes set in with infilatration by leucocytes. Softening of the cornea occurs (keratomalacia) and perforation may result.

Skin: changes in the skin lead to rough, dry coat and a shaggy appearance There may be hyperkeratinisation. Bran-like scales (pityriasis) are present on the skin. In the horse, hooves may mainfest vertical cracks.

Hyperkeratosis: This condition is seen in animals poisoned by chlo-rinated naphthalenes, especially those that are of high heat resistance. These chlorinated naphthalenes are used in lubricating oils, which are used in farm machinery Lubricants of the machinery contaminate the feed pellets while these are made. Similarly machinery used for mixing feeds may be a source of con-tamination. Chlorinated naphthalenes are excreted in the milk and so calves drinking such milk are poisoned. Chlorinated naphthalenes antagonise vitamin A. Probably they prevent the conversion of carotene into vitamin A, destroy vitamin A already formed and prevent its assinilation.

Symptoms: The first symptom noticed is profuse lachrymation. This is followed by anorexia and emaciation The skin of the withers, shoulders and back is usually affected showing a dry, inelastic, rough, often hairless, wrinkled skin. The skin in some places may show fissures through which secondary infection can occur. The mucosa of the lips and tongue may show white patches, which later may become ulcers Infection of these places by *Spherophorus necrophorus* as well as by virus of papular stomatitis may occur.

Microscopically, the change is seen in the epithelium wherever it is located. There is hyperkeratosis of the stratified squamous epithelium and meta-plasia of the ciliated and columnar epithelium to startified squamous variety. In the skin there is acanthosis and the stratum corneum is very much thicke-ned. In glands and ducts, hyperplasia and metaplasia of the columnar epithelium gives rise to obstruction of the ducts by the desquamated cells and so cysts form containing mucous and debris. This is found in the glands of digestive tract and in salivary glands leading to digestive troubles. Hyperplastic epithelium of the lips and tongue gives rise to plaques which may be 1 cm. thick and 2 cm. in diameter. The epithelium of the mucosa of bile duct and gall bladder may show hyperplasia and cyst formation. Some show papillary projection into the lumen. Smaller bile ducts proliferate with encircling fibrosis. The renal tubules show epithelial hyperplasia, leading to their dilatation. Fibrosis to certain extent, is present along with tubular changes. In the eye, due to obstruction of the lachrymal duct, tears cannot flow out and so keratitis and ulceration of the cornea result. Tubules and glandular organs of the male genitalia show squamous metaplasia and cornification. Similar squamous metaplasia is found in the endometrium and its glands.

Defects in bone growth: It is appropriate at this juncture to describe briefly the formation of bone so that the role of vitamins in the process may be better understood.

Normal development of bone: The bones of the body develop in two ways (1) The *intramembranous* bone formation by which the flat bones develop In this process the fibroblastic cells of the 'cambium layer of the periosteum differentiate into osteoblasts and osteocytes gradually. This method of bone formation is seen in the flat bones of the skull and pelvis. The shaft of the diaphysis widens by this process.

(2) The *endochondral* bone formation by which the bones increase in length.   Growth of the bone occurs at the epiphyseal cartilage, which is situated between the diaphysis and the epiphysis.   The cartilage cells proliferate, mature and finally become ossified at the diaphyseal end and so the bone grows.   The various steps in the formation of bone are as follows:—

The cartilage cells near the epiphysis divide and are dark staining. These are arranged in rows towards the diaphysis.   Those cells nearer the dia— ph sis are larger with a clear cytoplasm. These are the cells that are becoming degenerated. Some of these cells are calcified. Subsequently, capillaries invade and corrode the   calcified cartilagenous matter leaving longitudinal trabeculae. Accompanying blood vessels osteoblasts arrive and deposit osteoid on these trabeculae. Subsequently calcium salts are deposited on this osteoid—calcification. With Hematoxylin and Eosin method of staining, osteoid takes a pink color while calcified tissue (bone) a blue color.   More and more osteoid is deposited concentrically around the capillaries burying the osteoblasts, forming the Haver-- sian canals.   These osteoblasts, which produce phosphatase are kept alive by the canaliculi which connect the lacunae containing the osteoblasts with the capillaries, thus providing the cells with tissue fluid which transports necessary oxygen, food, calcium and phosphorus.

In health, there is a continuous remodelling and rebuilding of the bone and so it is said to be in a state of flux.

In young animals bone growth is retarded in vitamin A deficiency due to interference in the growth of epiphyseal bone.   The main defect appears to be that capillaries do not invade the cartilage and osteoblasts do not appear. In the absence of these two processes, growth of bone cannot occur. In calves, this is seen in the skull which does not grow. On the other hand, the brain continues to grow and so the skull becomes too small for the growing brain, parts of which, therefore, herniate into large foramina. Such a condition is mani- fested by nervous symptoms.   These include paralysis of skeletal muscles due to damage of peripheral nerve roots, blindness due to the pressure on the optic nerve by the narrow optic—nerve canal and encephalopathy.

Because the brain grows in a smaller skull, the intacranial pressure of the cerebro—spinal fluid increases.   This causes convulsions and paralysis.

Due to decreased production of acid mucopolysaccharides in young birds, derangement in bone growth and cartilage development may occur leading to skeletal abnormality.   Ataxia and incoordination is produced due to com— pression of the nervous system by the deformed vertebral canal and cranium.

Hydrocephalus   may result due to   increased production of cerebrosipinal fluid.

Papillidema occurs due to constriction of the optic-nerve passage, resulting in degeneration of the retina and complete blindness.

Nutritional roup: Among adult fowls, deficiency of vitamin A causes a condition known as *Nutritional roup* characterised by oculonasal discharge. conjunctivitis, sticking of the eye lids, presence of cheesy material in the eye and nasal sinuses and swelling of the face. The main feature is the keratinisation of the epithelium lining the ducts of the mucous glands opening into esophagus and pharnyx, resulting in the occlusion of these ducts and so stagnation and inspissation of the secretions of the glands occur.

Infection of the upper nasal passages occurs and inflammation results.

Other non-specific effects of vit. A deficiency in birds are fall in egg production and hatchability, poor feather development and retarded growth.

Congenital malformation: Besides blindness, dermoids in the eyes of calves, absence of eyes (anophthalmos) or smaller eyes (microphthalmos) may occur in piglets. Anasarca, palatoschisis (cleft palate; hare-lip) and malformed limbs are other congenital defects noticed.

Hypervitaminosis A: Large quantities of vitamin A in calves produces rickets-like deformities and osteomalacia. In large quantities, vitamin A disrupts the lysosomal membranes and so the enzymes, proteases, liberated destroy the protein mucopolysaccharide complexes of the matrix of the cartilage causing chondromalacia. In man hypervitaminosis A is manifested by persistent chronic headache and distorted vision which may be due to increased intracranial pressure consequent on either greater production or decreased absorption of cerebrospinal fluid.

Pain in bones may be felt due to calcification of tendinous, subperiosteal and pericapsular tissues. Pruritus, soreness of the corners of the mouth and coarseness of hair may be present. Exostoses in various places occur.

Among animals naturally occurring hypervitaminosis A has been reported in cats that were maintained by prolonged feed'ng on bovine livers. Deforming spondylosis was noticed. Animals showed lameness, ankylosis of cervical region, changes in posture and hyperasthesia of skin.

Vitamin D (Anti—rachitic factor)

Six to eight forms of vitamin D, all having basically the same sterol structure exist. Of these two are important: D_2 or calciferol formed in the body by the action of ultraviolet light on ergosterol and D_3 also formed in the body by the action of ultraviolet light on 7–dehydrocholesterol. D^3 is particularly useful in poultry nutrition.

Physiology: Vitamin D is fat soluble and so for its absorption bile salts are necessary. This vitamin is mostly stored in the liver.

Machanism of action of Vitaman D in : Bone and in intestinal mucosa

Vitamin D2 or D3 is first hydroxylated to 25 - hydroxy cholecalcifero' in liver and then to 1, 25 - hydroxy cholecalciferol in the kidney. This dihydroxy' lated compound has a hormone like action in the intestinal mucosa and bone.

The 1-25 dihydroxy. cholecalciferol is carried in association with a binding protein in blood plasma to the target tissues where it is transported to the nucleus. In the intesti- nal mucosa this stimulates the synthesis of a carrier protein which helps absorption of calcium and in bones this helps in calcium resorption in association with parathyroid hormone.

Sources: 1. Fish liver oil: 2. Sun-cured hay, 3. Exposure of body to sun.

Functions: 1. Vitamin D is concerned with the absorption of calcium and phosphorus from the intestines.

2. It raises the serum calcium and phosphorus levels. This is brought about as follows. By increasing the absorption of calcium, the serum level of the mineral is elevated. Elevation of serum calcium suppresses the parathyroid activity and this leads to increased renal tubular reabsorption of phosphorus and so the serum phosphorous level is elevated.

3 Retention of calcium and phosphorus in the body.

4. Mineralisation or deposition of calcium and phosphorus in the bone. Deficiency of vitamin D.

Causes: 1. Lack of Irradiation of skin: (a) This may occur in winter months in cold countries where the animals are kept indoors. (b) In places where there is a smoke-screen (due to industrial plants) ultraviolet rays cannot reach the earth. (c) Heavy coats and black skin of animals, prevent the penetration of ultraviolet rays to reach the body.

2. Diseases of the liver and intestines: The same circumstances described under deficiency of vitamin A that prevent its absorption, also prevent the absorption of vitamin D. Carotene, if fed in large quantities as lush green feed has anti-vitamin D potency. Vitamin A in large quantities markedly retards growth, Vitamin D deficiency causes rickets in the young and osteomalacia in the older animals. These diseases are described in the chapter on the Musculo-skeletal system. Besides lesions in the bones, vitamin D deficiency causes reduction in productivity, poor weight gain and poor reproduction efficiency

Massive doses (20, 000 000 IU of vitamin D per day) starting 5 days before expected calving date and continuing through the first day, have been helpful in controlling milk fever.

In the fowl, hypovitaminosis D is characterised by laying of thin she-lled eggs, leg weakness and reduced hatchability. The beak becomes very pliable and the keel bone bends.

Hypervitaminosis D: Large quantities of vitamin D are toxic producing death in some cases. Hypercalcemia and hypercalcuria occur. Metastatic calcification with calcification of the media of arteries and arterioles and renal calculi may result. Osteoporosis may finally develop in this condition due to intense osteoclastic activity. The resorbed bone is replaced by a poorly calci-fied osteoid beneath the periosteum and in the marrow cavity. Death is due to renal failure.

Vitamin E (Anti-sterility factor)

Vitamin E is fat soluble and under this term are included the alpha beta, gamma and delta tocopherols. These tocopherols are so named because, they are understood to maintain pregnancy (tokos=child-birth; pheros=carry). Of these, alpha tocopherol is most active and is found in germs of grains and in green plants. Now all are synthesised.

Physiology; Vitamin E is an antioxidant and controls oxidation in muscles, fats and liver. In its absence the oxidation is increased to nearly 400 times and necrosis occurs. Because of its antioxidant properties, vitamin E prevents the destruction of vitamin A in the gastrointestinal tract and so is called a " *Vitamin-A sparer* ".

Probably the main effect of tocopherol is to protect the vital systems of cells from oxidation by peroxidase. It also prevents oxidation of essential fatty acids in membranes and sulphide and selenide in non-heme ion enzyme system that takes part in mitochiondrial respiration and microsomal oxidation.

Excess of fats (unsaturated fatty acids) in cod liver oil destory alpha tocopherol and hence produce symptoms of vitamin E deficiency.So necrosis of heart and other muscles occurs and these subsequently undergo dystrophic calcification This is of importance in poultry nutrition, because when excess of cod-liver oil by an over zealous feeder is added it will produce the above condition. Traces of selenium are found to be able to replace vitamin E in preventing excessive oxidation. In some metabolic transformations, Selenium may act as a substitute for vitamin E since both participate in hydrogen transfer-electron transport. So selenium may be called a "Vitamin E sparer" since it enhances transport and retention of Vit E.

Selenium and Vitamin E have a metabolic inter-relationship. Selenium is a component of Glutahione peroxidase. Both the selenium and vitamin E prevent oxidative damage to the cell.

Vitamin E also protects the integrity of cell membranes and organell; membranes in association with Vitamin A

Vitamin E Deficiency: Vitamin E deficiency may occur in the following conditions: 1 When animals are fed only on concentrates (which are milled, removing the germ thereby) without green fodder.

2. Calves fed only on milk which is deficient in vitamin E.

3. Rations containing high percentage of unsaturated fatty acids which destory vitamin E or feeding with fish, meat and bonemeal containing sodium bisulphite.

Among animals, calves and lambs are affected by vitamin E deficiency. Muscular dystrophy known as '*white-muscle-disease*' in calves and ' *stiff-lamb disease*' in lambs is the condition encountered. This is described in the chapter "Musculo skeletal system."

Azoturia or equine myoglobinuria in equines, is suggested to be brough-about by vitamin E deficiency. This condition is also dealt with under the Must culo-skeletal system.

Vitamin E deficiency in experimental rats has been found to cause profound changes in the reproductive system. In the male the spetmatozoa are

destroyed and the entire seminiferous epithelium degenerates. This is not corrected by subsequent vitamin E restitution or therapy.

In the female, the fetus in early pegnancy dies and is absorbed. (In
vitamin A deficiency the fetus is not affected).

"Yellow fat disease or "Steatitis" or Hepatitis Dietetica. This condition is met with in minks and pigs fed with large quantities of fish oil or fish-
meal over a long time. The body fat is brownish in color.   This disorder develops
due to enhanced oxidation of the unsaturated fatty acids in the absence of alpha
tokopherol. *Microscopically*, a peculiar, amorphous substance, **ceroid**, is found
in the interstices of adipose cells.   This ceroid is also found in the hepatic cells,
Kupffer's cells and in macrophages. These macrophages develop into giant cells.

In pigs this condition is noticed in rapidly growing animals. The symptoms shown are : dyspnoea, diarrhoea, vomition, melena, paralysis and trembling. Icterus is seen in relapsing cases.

Affected animals are usually in good condition.   There may be massive
generalised edema, massive hepatic necrosis and ulceration of the gastric cardia.
Hemorrhagic diathesis may occur, manifested by extensive subendocardial hemorrhage and hemorrhage into the joints.   Serous cavities contain a small quantity
of protein rich fluid.   There may be pulmonary edema when heart is affected.
Bone marrow may show abnormal red cells.   Death may occur suddenly.
Poultry :   Vitamin E deficiency produces the following conditions :

a) **Crazy-chick disease (Encephalomalacia)** : Birds upto 8 weeks are
affected. They appear sleepy. Some are highly excitable and exihibit twisting of
head and neck, ataxia and convulsive movements. The major symptoms of this
deficiency disease are sudden prostration, with legs outstretched and toes flexed
and retraction and often lateral twisting of the head.

Microscopically, hemorrahagic areas of softening are found over the
cerebellum, medulla, mid— brain and cerebrum.

b) **Exudative diathesis :** This is found in chicks of 2 to 8 weeks of age.
The chicks lose condition suddenly. Accumulation of fluids is seen under the
wings, abdomen and other places.   Hemorrhages are found in the heart, liver,
brain, subcutaneous fat and tissues. Affected chicks suffer from microcytic anemia.

c) **Nutritional muscular dystrophy :** This condition is seen in birds
over 4 months of age and occurs in those fed excessive quantities of cod-liver oil.
Symptoms are vague with loss of condition and weight. Mortality rate rises. On
autopsy white necrotic areas are found in streaks and patches, over the heart,
breast and skeletal muscles. The lesion is primarily found in the mitochondrial
and lysosomal membranes.

Vitamin K

This is another fat-soluble vitamin   and exists in 2 forms: Vitamin K_1
(phylloquinone)and Vitamin K_2 (menaquinone). A synthetic form (Menadione) is
available which is water   souble.   Vitamin K_1 is found in all green
plants and in some fruits. Vitamin K_2 is synthesised in the intestines by
bacteria. For the absorption of vitamin K bile salts are necessary.

Functions : Vitamin K is required for the formation of prothrombin and factors VII, IX and X by liver. So, in vitamin K deficiency, blood does not coagulate and bleeding occurs.

Deficiency of vitamin K : In normal conditions, animals may not suffer from deficiency of the vitamin. Deficiency may, however, occur in the following conditions :

A) Failure of absorption from intestine :

i) Since bile salts are necessary for vitamin K absorption, deficiency of this vitamin (conditioned) may occur in obstructive jaundice.

ii) In malabsorption conditions like sprue and celiac diseae, when fat is not absorbed, this deficiency may be met with.

iii) **Intestinal disorders :** Diarrhoea, vomiting, ulcerative colitis. Here vitamin K is not absorbed,

In the above conditions the synthetic vitamin (which is water soluble and so does not require bile salts) may by given orally or the natural vitamin parenterally.

iv) **In diseases of liver:** In cirrhosis bile is not formed and so vitamin K is not absorbed. In this condition therapy by synthetic or natural vitamin will not be of any use since the liver is not able to function and so is not able to synthesize prothrombin.

B) Lack of bacterial synthesis : If the bacterial flora of the intestines is diminished by use of antibiotics or bacteriostatics, vitamin K deficiency results, This situation arises in poultry which are given sulphaquinoxaline as a coccidiostatic and in pigs fed on sulpha drugs. In pigs multiple hemorrhages are found throughout the body.

C) In the new born: In the new born pigs especially, the milk is deficient in vitamin K and the intestinal flora are absent in the first few days of life. Also in these baby pigs, the liver has not yet begun producing bile. These factors tend to produce bleeding in piglets from the umbilical cord or from wounds. Vitamin K deficiency as such is not of much importance in cattle. In sweet clover poisoning vitamin K medication is indicated for making up destroyed prothrombin and thus help clotting of blood.

In the poultry, large subcutaneous hemorrhages and hemorrahages in the abdominal cavity are found—*hemorrahagic syndrome.* Anemia may therefore develop. Mature birds apparently are not subjected to acute vitamin K deficiency, which suggests that they may synthesise the vitamin. However, it has been shown that birds fed a diet low in vitamin K produce eggs that are low in Vitamin K. When these eggs are incubated, chicks with very low vitamin K are hatched. As a consequence the chicks may bleed to death from an iujury similar to that caused by wing banding.

Vitamin C (Ascorbic acid, Antiscorbutic factor)

Vitamin C is water soluble and is found in all green plants and fruits. Citrus fruits are particularly rich in this vitamin.

Vitamin C is synthesized in the intestines of all animals except in the guinea-pig and primates. Hence it is unusual to find vitamin C deficiency in animals other than guinea-pigs which may suffer if not fed with sufficien amount of greens.

Vitamin C is easily absorbed from the intestines. All tissues may contain the vitamin but adrenals contain the largest quantities.

Crystalline vitamin C has been synthesized. Heating and cooking destroy this vitamin.

Functions : Vitamin C is a strong reducing agent. It is required for the normal metabolism of tyrosine, phenlyalanine and dehydroxyphenylalanine (DOPA). It is also necessary for the formation of hydroxyproline and hydroxylysine., which are the constituents of collagen. It is postulated that vitamin C may be required for the synthesis of the specific mRNA that is required for incorporation of hydroxyproline into the collagen molecule. So it is evident that deficiency of vitamin C results in imperfect formation of collagen which is necessary for proper wound healing.

In some way vitamin C is necessary for the activity and secretion of adrenal steroid hormones. It plays an important role in the formation of the derivatives of mesenchymal tissues—osteoid, collagen, dentine and intercellular cement substances. In some way it influences the mucopolysaccharides of the ground substance in keeping it healthy.

Deficiency of vitamin C in man produces scurvy. In this condition the mesenchymal tissue derivatives (Osteoid etc.) are not formed and so the following derangements may arise.

1. **Wound healing**: Wounds do not heal properly since collagen is not laid though firoblastic proliferation occurs. So the scars that may form do not have sufficient tensile strength and so they may break open often. Hemorrhages into the wound are also seen, due to lack of intercellular cement substance between the endothelial cells of the newly formed capillaries.

2. **Bone formation**: Vitamin C plays an important role in endochondra ossification. In the absence of vitamin C, osteoblastic activity does not occur and so osteoid is not laid though invasion of blood capillaries into the cartilage and deposition of minerals are normal. So in the absence of osteoid formation' bone prowth does not occur. Since the calcified cartilage is not absorbed, it persists and extends into marrow shaft. This is similar to the change seen in rickets. The bones are soft and become deformed on bearing weight. Massive hemorrhages occur in the epiphyses, particularly subperiosteally.

We have now learnt that vitamins A, D and C are required for bone growth. But their roles are different. We may summarise them as follows.

Vitamin A is required for the proliferation and invasion of capillaries into the cartilage and also for the appearance of osteoblasts. So in vitamin A deficiency osteoid formation and calcification are normal.

Vitamin D is required for the calcification of the osteoid. So in vitamin D deficiency, penetration of capillaries and osteoid formation are normal but calcification does not occur.

Vitamin C is required for the normal formation of the osteoid. In vitamin C deficiency, therefore, penetration of the capillaries and calcification are normal but osteoid formation does not occur.

3. **Hemorrhagic diathesis** . Due to deficiency of intercellular substance, the capillaries become weakened and fragile and so bleeding occurs subperiosteally subcutaneously and into the muscles and joints of legs. Bleeding from the gums is also present.

4. **Battery sickness** : In Germany, among fowls kept in batteries, within 3 to 4 months, paralysis and atrophy of striated muscles were considered to be due to vitamin C deficiency. Treatment consisted of intramuscular injection of 100mg. of vitamin C

Vitamin B complex

This includes several components which differ in their chemical structure as well as in their physiological functions. Majority of them function as coenzymes in the Krebs tricarboxylic acid cycle of carbohydrate metabolism. and known as "energy releasing vitamins". since by their participation they are required in the synthesis of high energy bonds in ATP. Most of these components are synthesized in the rumen and so ruminants seldom suffer from deficiency of these vitamins except in the case of very young animals in which the rumen has not yet attained its full functional capacity.

All the vitamins are water soluble and are found in yeast.

Thiamine (Antineuritic or anti-beri-beri factor), Vitamin B-1.

Thiamine is present in the whole grains and meats. Milled and polished grains are deficient. Thiamine is a component of Thiamine Pyrophosphate (TPP) which is a component of the Pyruvate dehydrogenase complex responsible for pyruvic acid breakdown. This vitamin therefore, plays an important role in the intermediary metabolism of glucose. Since nervous tissue derives most of its energy from glucose, metabolism of nervous tissue is greatly affcted by thiamine deficiency. It is necessary for the synthesis of acetylcholine and so nerve fibers may be affected in its deficiency.

Thiamine is necessary for the proper utilisation of lactic acid produced during cardohydrate metabolism. In its deficiency, lactic acid is not converted into glycogen and so it accumulates in the body.

Deficiency : Deficiency of thiamine is of importance only in poultry and swine.

In poultry. deficiency may arise if fed on milled products, without greens. The symptoms are extreme loss of appetite, polyneuritis and death. Paralysis of the legs occurs, first starting at toes. The affected bird sits on the flexed legs and bends the head back-wards, having a "star-gazing" attitude. Degeneration of the peripheral nerves is noticed. Atrophy of the heart muscles with auricular dilatation may be observed. Anemia, lymphocytopenia and hyperglycemia are the other changes noticed.

In **pigs** sudden death may be encountered with cardiac dilatation and areas of myocardial necrosis. In animals that die after sometime, inappetance, emaciation and leg weakness together with bradycardia and slower respiratory rate are found. Scarring of myocardium may be seen in some with repeated episodes of vitamin deficiency. The electrocardiogram is abnormal. Edema of the lungs is also seen.

Beri-Beri : This is a disease found in people of Asian countries who eat polished rice. While milling rice the aleuron layer of the grain containing most of the vitamin is removed with bran and so is lost. The symptoms are peripheral neuritis, edema and myocardial weakness.

Chastek paralysis : In a fur farm at Chastek (Minnesota) foxes developed paralytic symptoms when fed on some species of raw fish. The fish contained an enzyme, **thiaminase,** which destroyed thiamine in the animals causing acute thiamine deficiency. The affected animals died after manifesting anorexia emaciation, diarrohea, weakness and paralysis. Certain neurons in the ventricles of the brain undergo degeneration and necrosis. No changes are found in the ganglion cells and white matter

A similar disease is also found in cats that are fed raw fish. Focal degeneration and hemorrhages are found in the gray matter of the brain **Microscopically,** edema, gliosis and gitter cells are found in the lesions.

Bracken fern *(Preridium aquillinum)* and horse-tail *(Eguisetum arbense)* contain thiaminase. So when horses eat these plants, thiamine deficiency results manifested by incoordination of movemeut. and bradycardia· Treatment by parenteral administration of thiamine abates the symptoms.

Riboflavin or Vitamin B₂

This is found in green grass, meat and milk. Synthesis occurs in rumen.

Functions Riboflavin is found in flavoproteins: Warburg's cytochrome C, xanthine oxidase and other enzymes. It aids in the transfer of oxygen from the plasma to the tissue fluid and in transport of hydrogen.

Deficiency: Normally deficiency does not occur in animals. It is mostly a problem in poultry and pigs that are fed exclusively on grain and grain products. It is generally supplemented in poultry and swine diets.

Poultry· Diarrhoea and "Curled toe" paralysis occur in chicks. They walk on their hocks since the toes are curled inwards. Degeneration of myelin sheaths, proliferation of the cells of neurilemma with swelling and fragmentation of the axis cylinders are the changes noticed in the peripheral nerves. The affected nerves are swollen 4 or 5 times their normal size and are yellow in colour.

In the adult birds, decreased egg and antibody production result. There is decreased hatchability as well as embryonic mortality.

Swine: In pigs the following symptoms are seen: retarded growth, diarrhoea, rough, thin hair coat with ulcers of the skin in various places. The hooves show ridges and thickening probably causing pain and so crippling the animal. The ocular lesions include conjunctivitis, edema of the eyelids, vascularisation of the cornea and cataracts. A normocytic, normochromic anemia may be seen Still-births way be frequent.

In man chelitis with cracks at the corners of the mouth is a characteristic symptom. Glossitis, conjunctivitis, keratitis and dermatitis are other symptoms noticed.

Niacin: Nicotinic acid: (anti—pellagra factor) B₃

Niacin is a water soluble vitamin, found in alfalfa and animal byproducts. It not present in corn, oats and milk,

Function: Niacin enters into the structure of coenzymes NAD and NADP. These function as dehydrogenases in cellular respiration. So, it is of importance in cell respiration as well as in carbohydrate metabolism.

Deficiency: Normally deficiency should not occur if green alfalfa is fed to animals. But in pigs and dogs fed on corn only deficiency may arise since corn is deficient in niacin. Tryptophan, if available, can be converted into niacin (cats cannot convert tryptophan into niacin). But the protein of corn, *zein*, is deficient in tryptophan and so animals fed on corn alone develop deficiency symptoms.

Pigs: In young pigs nutritional enteritis develops manifested by chronic diarrhoea. The animals lose appetite, become emaciated and dehydrated and so their growth rate is poor. Severe anemia may also develop as well as scaby dermatitis and alopecia.

The lesions consist at first of severe patchy colitis with necrosis of the mucosa. This may subsequently be invaded by *Spherophorus necrophorus, Balan--tidium coli* or *Solmonella cholerae suis.* Later the necrotic areas may extend to form diffuse necrotic enteritis.

Dogs: In dogs deficiency of niacin causes "black tongue" or "Canine pellagra". Formerly this was called "*Stuttgart disease*". Probably this is only an experimental condition and no clinical entity may ever occur.

The symptoms seen are: Cyanosis of the tongue which may have ulcers at the borders, salivation, anorexia, diarrhoea and vomition leading to dehydration and emaciation and nervous symptoms. Secondary bacterial infection may bring on fever.

Macroscopically, hemorrhages may be noticed in the intestines and stomach. Ulcers may be present in the large intestines.

Microscopically, peripheral nerves may reveal degeneration of the myelin sheath while in the bowel is found severe mucoid degeneration and subsequent necrosis of the wall.

In birds vitamin B_3 deficiency may cause reduced egg production and hatchability, poor growth and feather development, diarrhoea and leg deformities.

Man : In man the deficiency is called "Pellagra" which means rough skin (pelle—skin, agra—rough). It is manifested by dermatitis, diarrhoea, and dementia. Dementia means mental deterioration due to degeneration of ganglion cells of the brain accompanied by degeneration of spinal cord.

Niacin deficiency may cause extreme fatty degeneration of the liver culminating in cirrhosis.

Pantothenic acid

This vitamin is found in alfalfa hay and cereals. Corn is deficient.

Function: Pantothenic acid forms an important part of co-enzyme A, that plays a vital role in intermediary metabolism. In ducks, this vitamin is essential in the synthesis of amino levulinic acid which is a precursor of heme.

Deficiency: Ducks may suffer from anemia since heme may not be synthesized in the absence of pantothenic acid.

Deficiency can occur in pigs that are maintained only on corn, which is deficient in this vitamin. The symptoms shown are: anorexia, retarded weight gain, dermatitis with collection around the eyes of a dark brown exudate, alopecia in patches, diarrhoea and a peculiar jerky *goose stepping* gait.

Microscopically, chromatolysis of the dorsal root ganglion cells is found in early stages. This is followed by demyelination with degeneration of axis cylinders of brachial and sciatic nerves as well as the dorsal root fibers. Intestinal mucosa reveals necrosis of its cells. Hyperplasia of the intestinal lymphoid tissue is observed. A normocytic anemia is also present.

Poultry: In chicks the growth rate as well as development of feathers are retarded. Dermatitis and broken feathers are seen. Dermatit s which is necrotic, affects the angles of the mouth, borders of the eye lids, the vent, the balls of feet and joints. A scab-like material is found at these places. Proventriculitis and stomatitis may be present. Myelin degeneration may be found in the spinal cord and peripheral nerves giving rise to incoordination of movement. In hens, egg production and hatchability are reduced. Antibody formation may be affected.

Pyridoxine (Vitamin B$_6$)

Pyridoxine is present in most of the cereal grains and so deficiency is seldom noticed.

Function: Pyridoxine is in necessary for the proper metabolism of amino acids, especially tryptophan. It is necessary for the formation of enzymes required for deamination, decarboxylation and transamination. It is also required for melanin formation, transmethylation of methionine, in biosynthesis of porphyrin, in the metabolism of unsaturated fatty acids and cholesterol and in antibody production.

Deficiency : Deficiency of this vitamin causes changes in the dermal, erythropoietic and nervous tissues.

In pigs, a severe microcytic anemia may be caused, showing anisocytosis and reticulocytosis. There is extensive deposition of hemosiderin, intra and extracellularly, in the spleen, liver and bone marrow. Hyperplasia of the bone marrow is common. Due to lesions in the nervous tissue, the movement is jerky and incoordinated with swaying of the posterior limbs. Convulsions may be seen. Microscopically demyelination and degeneration of the axis cylinders of the peripheral nerves as well as necrosis of the dorsal root ganglia may be observed.

Cutaneous hyperkeratosis and acanthosis are the lesions seen in the skin. Fatty degeneration of the liver is common. Antibody production may be curtailed.

Biotin

This is a sulphur containing vitamin present in green plants, liver-meal and yeast.

Function : Biotin has an important role in carbondioxide fixation.
Avidin from raw white of the egg is an antivitamin for Biotin.

Biotin is synthesized by microbial flora and so bactericides and antibiotics prevent the synthesis and so deficiency may arise with their use.

Deficiency :   Natural deficiency does not occur. Experimentally, deficiency may cause paralysis of hind quarters in calves; stomatitis, alopecia, dermatitis and cracking of soles and top surface of the hooves in pigs; dermatitis and
poor hatchability in poultry.   In fowls, *perosis* is caused as in manganese and
choline deficiencies.

Choline

Yeast, fish-meal are rich in choline.

Choline can be synthesized in the body from the sulphur containing
amino acid methionine.

Function : The main function of choline is in the liver, participating
in fat metabolism.

Neutral fats brought from the depots to the liver have to be phosphorylated and converted into phospholipids before they can be transported and
utilised by the tissues. Choline is involved in this phospholipid formation. In
the absence of choline, therefore, the phospholipid turnover is reduced and
neutral fats brought to the liver from the depots cannot be transported and so
accumulate in the hepatic cells, giving rise to massive fatty infiltration which
culminates in cirrhosis.

A second role of choline is its influence in the oxidation of fatty acids in
the liver. If choline is deficient, fatty acids accumulate in the liver cells—fatty
infiltration.

Choline deficiency : Normally choline deficiency does not occur. Defic
ency of choline may be one of the factors for extreme fatty livers found in
"**pregnancy toxemia**" of ewes. Here choline is diverted to the fetuses and so
the ewe suffers from choline deficiency and so fatty degeneration of liver results.

Fatty degeneration of liver may be found in puppies with inadequate
choline. In hogs also, sometimes, choline deficiency may cause ataxia, fatty
degeneration of liver and mortality. Clinically, hepatic damage can be assessed
by liver function tests : retention of bromsulphalein, increase in prothrombin
time and elevated serum alkaline phosphatase level.

In poultry *perosis* is caused by choline deficiency.

Folic acid

Folic acid is present in abundance in yeast and greens. All animals
except the pig, are able to synthesize the vitamin in their intestines.

Chemically, folic acid is pteroylglutamic acid. It is more active in a
reduced state as *folinic acid* and this reduction is brought about by ascorbic acid.

Function : Folic acid is concerned in the synthesis of methyl groups,
serine, purines and thymine.   Hence it is intimately concerned in Nucleic acid
synthesis and metabolism(DNA and RNA). In poultry, along with lysine, folic
acid is required for pigmentation of feathers.

Deficiency : Folic acid deficiency causes macrocytic anemia (Sprue-
pernicious anemia) and agranulocytosis. In turkey poultry, deficiency of folic acid
causes straight-neck paralysis, macrocytic anemia, agranulocytosis, increased

In man folic acid deficiency may be induced in therapy by contraceptives, antimalarial and anticonvulsant drugs.

Vitamin B_{12} (Cyanocobalamine ; Anti-Pernicious Anemia factor)

Vitamin B_{12} is a cobalt-containing vitamin found in meat and meat products and in milk.

For the absorption of vitamin B_{12} (extrinsic factor) in the intestines, an intrinsic factor (hemopoietin) secreted by the gastric mucosa is essential. The intrinsic factor-B_{12} complex formed is conveyed to the distal parts of the ileum where it is split at the plasma membrane of the epithelial cells and the liberated B_{12} is absorbed into the cells from where it is transported into the blood, which carries it into the liver via portal system and stored there. From the liver it is transported by blood, where it is bound to a protein, to various sites of DNA and RNA synthesis. B_{12} absorption in the intestines occurs only if the pH is 6.5 and calcium and magnesium ions are available. So in animals with deficient intrinsic factor (as in gastrectomy) deficiency of vitamin B_{12} may arise.

Vitamin B_{12} is also called animal protein factor (APF) since in the early days before its composition was known, this name was given when animal proteins were fed to correct the deficiency symptoms.

Function : Vitamin B_{12} is essential for biosynthesis of labile methyl groups and thymidine. So it is intimately connected with nucleic acid metabolism and production of RNA and DNA. It is also concerned in carbohydrate, fat and protein metabolisms It is essential for mitosis and maturation of cells. It is intimately linked with folic acid metabolism.

Deficiency : In animals natural deficiency does not occur since it is synthesized in the rumen and intestines, provided cobalt is available in food. In deficiency of cobalt vitamin B_{12} deficiency may arise. Deficiency symptoms in calves. pigs, and poultry are vague ; anorexia, slow rate of gain in weight and muscular weakness. Embryonic mortality may be high.

In man, macrocytic anemia (Pernicious anemia) results due to vitamin B_{12} deficiency.

Proteins

Proteins are the most important of the three principal food stuffs : proteins, fats and carbohydrates. Their functions are :

1. Forming one of the ingredients of the internal frame work of cells and he interstitial substances like collagen, osteoid and dentine,

2. From the amino acids of the proteins the following are synthesized: hormones, enzymes, plasma proteins, hemoglobin and other biologically active substances of the body

3. The following which help in the defence of the body are made from the proteins:

a) epidermis, b) lysozyme, c) mucous in mucous membranes. d) leucocytes, e) antibodies.

Usually deficiency of proteins should not occur if balanced rations are fed. But sometimes deficiency may occur under the following conditions:

1. Inadequacy of the protein fed. This is true of pail–fed calves in which the protein of the milk is not adequate to meet the needs of a growing animal.

Pregnant sows must be fed high value protein. Otherwise piglets that are born suffer from aplastic anemia.

2. Feeding rations that do not have required essential amino acids. Corn is deficient in tryptophan and so pigs fed exclusively on corn suffer from trypto--phan deficiency and also niacin deficiency since naicin is formed from tryptophan.

3. Animals in full production—cows in lactation and laying hens, must be fed adequate quntities.

4. In times of famine and war, cheap proteins only may be available and that too in insufficient quantities.

5. Anorexia may develop when animals are sick and so they may not consume sufficient feed. Similarly in diseases of teeth, tongue and esophagus, ingestion of feed is curtailed.

6. Even if sufficient quantities are ingested, proteins may be lost from the bowel when (a) there is increased motility as in diarrhoea and (b) when the wall of the intestines is so altered, as in Johne's disease, that the food is not absorbed.

7. In diseases of the liver, plasma and tissue proteins are not synthesized.

8. Infection with gastrointestinal parasites, which suck blood, results in loss of proteins along with the blood.

9. In man when kidney is damaged, large quantities of albumin are lost in the urine resulting in hypoproteinemia.

Effects of protein deficiency. 1. Slowing of growth. The proliferation of cartilage cells is decreased. 2. The osteoblastic activity is interfered with and so osteoporosis may result. 3. Thymus and lymphoid tissues become atrophied. 4· Formation of hemoglobin is decreased resulting in anemia. 5. Plasma proteins are decreased and edema results. 6. Hypoferremia and hypocupremia may result as normally iron and copper exist bound to protein in the blood. 7. Collagen may not be formed and so healing of wounds is not normal. 8. Fractures do not heal properly. 9. In children, Kwashiorkor is caused by protein deficiency, characterised by edema. dermatitis, retardation of growth, wasting of muscle with weight loss, cold hands and feet, changes in the hair, diarrhoea, enlarge—ment of liver, apathy and psychic changes.

Hypochromic protein deficiency anemia in young pigs is due to deficiency of iron, copper, vitamin B complex and animal proteins.

Macroscopically, there is pallor of skin, mucous membranes, muscles and organs. The coat of the animal is coarse. growth is retarded and emaciation is present. Edema of the eye lids and subcutaneous tissue over the neck is frequent. Organs appear to be devoid of blood. Heart is dilated. Hydropericardium and centrilobular hepatic necrosis are seen. Animals frequently suffer from secondary bronchopeumonia, rhinitis,enteritis, arthritis and inflam—mation of serous membranes.

Blood: Oligocythemia (Erythrocytes may be one million per c.mm.) anisocytosis, poikilocytosis and presence of annulocytes, reticulocytes, moderate shift to the left and lymphocytosis. Hemoglobin may be less than two gms/per 100 ml. of blood. Serum gamma globulin is decreased.

In piglets, physiological anemia occurs upto 10 days after birth. After his period, anemia is corrected and by 21st day the condition is normal.

FATS

Functions: 1. These are sources of energy, 2. They improve the taste of food, 3. Many natural fats contain fat-soluble vitamins and so are sources of these vitamins. 4. Some fatty acids cannot be synthesized in the body and so are essential for the body.

Essential fatty acids: These are unsaturated fatty acids and their physiological function is obscure. They are related to ceroid that is deposited in various cells of body.

Linoleic, linoleinic and arachidonic acids are the fatty acids having double bonds, that are essential. The last two can be synthesized from the first in the body, if it is supplied but cannot be formed in the body otherwise. The sources for these fatty acids are the vegetable oils.

Experimentally, deficiency of these fatty acids produce the following changes: 1 Scaling of skin—hyperkeratosis and pityriasis. 2. Alopecia 3 Loss of libido with atrophy of the testes. Tubular epithelium is degenerated and giant cells form. 4. Uterine mucosa is atrophied. Embryos are absorbed. or stay longer. Hemorrhagic placentitis is seen. 5. Fat content of the liver is elevated. 6. Retarded growth occurs 7. Bloody urine is encountered Epithelium of the renal tubules are degenerated and calcified with hyaline casts in the lumen.

DISORDERS OF PIGMENT METABOLISM

Endogenous pigmentation

    Melanin

    Hemoglobin   Exogenous pigmentation

      Hemosiderosis   Pneumoconiosis

    Bilirubin   Anthracosis

    Jaundice   Silicosis

    Hematoidin   Plumbism

    Lipochromes   Argyra

                                 Tatooing.

Various coloring agents are found in the body. Some like melinin and bilirubin are formed within the body and therefore are called *endogenous* while others like coal dust and lead enter into the body from outside and hence are termed *exogenous*.

Endogenous pigmentation

MELANIN

Melanin is an iron and sulphur containing brownish, granular pigment derived from amino acid Tyrosine. Melanin protects the body from the actinic rays of the sun and so skin cancers in Negroes are rare

Normal location : Normally, melanin is present in the skin, choroid of the eye, pia-arachnoid, the adrenal cortex and in substantia nigra in man. Sometimes it may be found in the lungs, liver, kidney, heart, brain and mammary gland in young animals without causing any lesions.

In the skin, among the cells of stratum germinativum of the epidermis, are found certain peculiar branching cells that are thought to be derived and then migrated from the neural crest. These possess an enzyme, *tyrosinase*, which can convert tyrosine (hydroxy phenylalanine) into melanin. These cells are called *melanoblasts*. Tyrosine, an essential amino acid, is present in the body and is the precursor of melanin, adrenaline and thyroxine. Melanin produced is dispersed to other cells. Macrophages may transport and deposit melanin to other parts of the body and are called *melanophores* (carriers of melanin). These cells do not have tyrosinase and so cannot convert tyrosine into melanin. These two different types of cells can be identified by incubating frozen tissues with dihydroxyphenylalanine (DOPA)when melanoblasts convert DOPA into melanin and so black spots are found wherever they are formed. Such cells that can convert DOPA to melanin are termed DOPA *positive*. Melanophores on the other hand, are DOPA *negative* since they cannot produce melanin having no tyrosinase. The melanin granules stain black with Fontana's silver impregnation method.

Copper ions are needed for activating tyrosinase. So copper deficient diets cause depigmentation Sulphur-containing compounds like thiouracil when administered, combine with copper and so make it unavailable for the activity of tyrosinase and so depigmentation occurs.

Sulphydryl groups may also combine with copper and render it unavailable for the activation of tyrosinase. Ultraviolet rays of the sun probably decrease the concentration of sulphydryl groups with consequent activation of melanin-producing mechanism and so melanin granules are found in the mela-noblasts and their branching processes. This is the mechanism of sun tanning in white skinned persons.

Melanocyte stimulating Hormone : The posterior pituitary, pars intermedia, produces a hormone which controls the production and dispersion of melanin—the Melanocyte Stimulating Hormone or M. S. H. This is of practical importance in lower animals like frog and chameleon which change the color of their skins. This hormone is probably inhibited by the adrenal hormones. Melatonin of Pineal gland antagonises the MSH.

Pathological conditions involve either too much or too little production of melanin. The following conditions are concerned with over production.

1. **Tumors** : Tumors comprising of melanin-forming cells are called melanomas. These may be benign or malignant.

In the malignant variety melanin is produced in such great quantities that it may be excreted in urine, which assumes a coffee color and is termed *melanuria*. The total amount of melanin in a darkest negro may not exceed 1 gm. But in a malignant melanoma as much as 300 gms. of the pigment may be present. Among animals the white and grey horses are frequently affected. In fact if these animals are allowed to grow to an old age, they invariably die of melanoma. Masses of the tumor can be seen under the tail, under the scrotum and in internal organs. This is described fully in chapter on Neoplasms.

2. **Addison's disease** : When there is bilateral destruction of the adrenals due to tuberculosis, atrophy or neoplasms, the skin of man assumes a deep tan color. The genesis of this color is explained in two ways : (a) When the adrenals are destroyed, the inhibitory mechanism of the M.S.H. by adrenal hormones is lost and so under the activity of M. S. H. greater quantity of melanin is produced in the skin

(b) Adrenal medulla normally utilises large amounts of tyrosine for synthesis of adrenaline. If adrenals are destroyed this tyrosine is diverted to the skin where it is converted into melanin and hence the deep coloration of the skin.

Melanosis coli : In man, when there is stasis of food as in constipation or obstruction, the intestinal mucosa takes on a black color which is visible through the wall. The mucosal epithelium is free It is the macrophages containing the pigment being found in the stroma of the mucous membrane that are responsible for the color. Melanin is either synthesized in the gut or is ingested with food. (*Pseudomelanosis coli* has already been studied under postmortem changes, vide page 46).

4. **Acanthosis nigricans** : In dogs suffering from a sertoli-cell tumor in which there is estrogen production, raised, rough, black patches of skin are found in the axilla, groin, under the belly and ventral thoracic region. There is hyper-keratosis with increased pigmentation. the pigment being found in the cells of the basal layer of the epidermis. In man it is supposed to be indicative of internal malignant neoplasm. Estrogenic hormones appear to have a role in the control of pigmentation.

Melanosis of the cornea: Blindness in certain breeds of dogs, viz. Boxers, Boston terriers and Pekingese, may be due to the presence of melanin in the superficial stroma of the cornea. The affection is bilateral and symmetrical
Conditions with little production of melanin:

1. **Albinism** This is a congenital condition in which there is complete absence of the pigment. Even the iris and choroid are deficient and so appear red, due to blood circulation being visible. Retina may have a little melanin. This condition is seen in, man, horses, some breeds of dogs (Bulldogs, Collies), cats, mice, rats and rabbits. In dogs deafness is associated with albinism There is probaly absence of tyrosinase since the epidermal cells are all DOPA negative.

2. **Leucoderma:** In this condition there is local loss of pigment and is seen in scars after healing of wounds. It may also be present as a congenital defect in certain breeds of dogs. Such areas become inflammed if exposed to sunlight.

HEMOGLOBIN AND ITS DERIVATIVES

Erythrocytes contain the pigment, hemoglobin, consisting of a protein—globin and a crystalline part, heme. Old erythrocytes are destroyed by the cells of the reticuloendothelial system. Since spleen possesses **much** of the R. E. tissue destruction of greater number of erythrocytes takes place there and hence spleen is known as the ": Graveyard of erythrocytes"

Globin is utilsed by the body. Heme is further broken down into (a) an iron-containing part—hemosiderin and (b) an iron-free, pigmentary part—hematoidin.

Hemoglobin has a small molecule, which can pass through the glomerular filter Normally, the hemoglobin that is liberated by the destruction of red cells is metabolised by the R. E. cells and so none is avilable for excretion by the kidney. But excessive destruction or hemolysis of red cells may occur in the following conditions: in infection by *Babesia* or *Clostridium hemolitycum* or *Leptospira pomona* and in poisoning by chlorate in cattle. In such cases, hemo globin may be excreted in the urine, giving rise to *hemoglobinuria*.

Hemosiderin containing iron is retained in the body and is utilised again in the synthesis of hemoglobin by tissues.

Hemosiderin is found usually as golden-yellow or golden-brown crystals within the macrophages in spleen. This can be stained by the prussian-blue method (Perls stain)

In the alveoli of lungs, hemosiderin-laden macrophages are known as 'heart-failure cells' since these are found in chronic venous congestion of lungs arising from cardiac lesions—mitral stenosis or incompetency and aortic valvular disease Since constant pressure on the walls of capillaries causes diffuse increase in connective tissue of the alveolar walls, the lung assumes a hard consistency, Coupled with the presence of the brown pigment. this condition is known as "brown induration"

Hemosiderosis: We have already noticed that hemosiderin is normally present in the R. E. cells of spleen. They may also be found in the R. E cells of liver and bone marrow to a lesser extent.

Excessive amounts of hemosiderin crystals may accumulate whenever there is increased hemolysis and in protein deficiency, in chronic anemia and in intravenous iron therepy. This condition is known as **hemosiderosis**

Hemosiderosis occurs in chronic venous congestion and the hemosiderin crystals are found in larger amounts in the R. E. cells of the spleen, liver and lung (we have already considered chronic venous congestion of lungs).

In local hemorrhage, the erythrocytes are hemolysed and the hemosiderin crystals are deposited locally within the histiocytes of the area. Hemosiderin by itself is harmless and is removed from the tissues gradually.

Transfusion hemosiderosis: When whole-blood transfusion is given repeatedly, hemosiderosis develops after about 100th transfusion. Here the exogenous iron brought into the system directly is stored by the R. E. cells first. When these become saturated, iron spills over to the parenchymatous cells of the liver and kidney and fibrosis of these organs results since the hemosiderin crystals act as mild irritants.

Hemochromatosis: This is a rare condition in man in whom large quantities of a pigment indistinguishable from hemosiderin are loaded in the parenchymatous cells of the liver, kidney, spleen and cells of other tissues. The accumulation of the pigment is followed by cirrhosis of the liver, diabetes mellitus, enhanced pigmentation of the skin and testicular atrophy resulting in gynecomastia and impotence Because both diabetes and deep coloring of the skin by melanin are present, the condition is also known as "bronzed diabetes" There seems to be an inborn error of metabolism so that large quantities of iron are absorbed by the intestines.

Normally, iron absorption is conditioned by the needs of the body. Even though excessive amouuts of iron are ingested, the inestinal mucosal epithelium does not absorb all of it. This is known as "mucosal block" And iron is also known as a one-way traffic element. That means very little of it is excreted. Further amounts are absorbed only when it is used up. Hence, excessive quantities absorbed in certain conditions are stored in the parenchymatous cells of organs (not in the R E. cells as in hemosiderosis). In this condition the source of iron pigment is not excessive hemolysis but enhanced absorption by duodenum. In the liver, Laennec type of cirrhosis develops and in the pancreas the deposits of iron destory the acinar cells and islets of Langerhans, leading to fibrosis and diabetes. Testicles atrophy. Prussian-blue reaction is given by deposited iron in the kidneys (proximal tubular cells), adrenals, spleen, heart voluntary muscles and thyroid. Only males are affected.

Among animals a condition similar to hemochromatosis occurs in equine infectious anemia in which large quantities of hemosiderin are deposited in the liver and kidneys.

Copper and cobalt are necessary for the utilisation of iron. So in deficiency of these elements iron accumulates in tissues.

In goats, in places where there is cobalt deficiency, large amounts of hemosiderin deposits are found in the epithelial cells of the proximnl convoluted tubules of the kidneys, which therefore take brown or black color. Kupffer's cells of the liver and the R. E. cells of the spleen and lymph nodes also contain this pigment. Necrosis and atrophy of the testes are noticed. No clinical symptoms are observed.

Bilirubin : We have already studied that heme of hemoglobin is split into hemosiderin and hematoidin by the R. E. cells, especially those of spleen and bone marrow.

Hematoidin is converted into bilirubin that is excreted in bile by the liver. The bilirubin atter it is formed is conjugated with a protein and so the molecule of this compound becomes bigger—too big to pass through the glomerular filter. This compound of bilirubin proteinate is also called **hemobilirubin**.

This hemobilirubin is brought to the liver where it is excreted by the hepatic cells into bile canaliculi as **cholebilirubin**. During this transformation, the protein is removed from the bilirubin, which is then conjugated with glucuronic acid with the aid of glucuronyl transferase and sodium salts. By this process, the bilirubin molecule is rendered smaller and more soluble in water so that it freely passes through the glomerular filter.

The hemobilirubin that passes through bile into the intestines is converted by the bacteria into **urobilinogen** which is further oxidised to **stercobilin** (urobilin) that stains the feces. Some of the *urobilinogen* is absorbed by the intestines and is transported to the liver again by the portal blood, from which it is re-excreted into the bile.

Some of the *urobilinogen* may be absorbed into the lymphatics from which it may enter the general circulation and hence may be excreted by the glomeruli into the urine, since the molecules of the *urobilinogen* are small enough to pass the glomerular filter. *Urobilinogen* imparts the yellow color to the urine and the presence of small quantities of this pigment in urine is normal.

Jaundice or Icterus : Jaundice or icterus is a condition in which the visible mucous membranes are stained yellow due to the presence of excess of hemobilirubin in the blood serum.

Excessive brilirubin may be present in the blood either because too much of it is produced (due to destruction of large number of erythrocytes) or it is not completely excreted from the blood by the liver. The latter may occur when the liver is damaged and so is not able to excrete even the normal amounts of bilirubin or when there is some obstruction to the outflow of bile.

The first variety occurs because there is too much hemolysis and is called **Hemolytic Jaundice**. Since the disorder is not due to any defects in the liver and since blood plasma contains more of bilirubin due to causes that occur before the blood passes through the liver, it is also known as **Pre-hepaticJaundice**.

The second variety, in which excess of bilirubin is found in the blood (even though normal quantities of it are produced) because of liver damage when the liver cells are not able to do their function properly in excreting the bilirubin is also known as **Intrahepatic Jaundice**. The damage to liver is usually due to some kind of toxin. So jaundice arising out of the action of toxins on liver is also known as **Toxic Jaundice**

The third variety of jaundice in which there is excessive bilirubin in blood due to obstruction in the biliary tract is known as **Obstructive Jaundice**. In this variety, the bilirubin of blood has already passed through hepatic cells. The defect is in bile ducts. So this variety is also known as **Post-hepatic Jaundice.**

Chemical test for bilirubin : The presence of bilirubin can be tested by van den Bergh test. In this test, a mixture of solutions of sulphanilic acid and sodium nitrite called the reagent is added to a sample of blood serum. A pink color develops if bilirubin is present.

Cholebilirubin gives immediate color reaction. It is therefore called **immediate test.** Since the color is obtained without adding any other reagent, the test is also called **direct.**

On the other hand, hemobilirubin, which is a conjugation of bilirubin and protein, does not give the immediate or direct test. If alcohol is added to the mixture of serum and reagent, color develops at once. Since the color develops after the addition of alcohol, the test is said to be **indirect.** The color may also develop, even if alcohol is not added, after the mixture of serum and reagent is kept for some time (ten minutes). So it is called **delayed reaction.**

In toxic jaundice, the serum contains both hemobilirubin and cholebilirubin. So there is appearance of color as soon as the reagent is added to the serum. And this color deepens on standing for sometime. So the test is said to be **biphasic.**

van den Bergh test, therefore, is useful clinically, to determine the type of jaundice the animal may be suffering from.

Hemolytic Jaundice : The causes of excessive hemolysis are :

1. Protozoa ;— Piroplasms, Anaplasma, Hemobartonella, Aegyptianella, Trypanosomes, Eperythrozoon.
2. Viruses :— Equine infectious anemia.
3. Bacteria :— *Clostridium hemolyticum,* hemolytic *streptococci, Leptospira*
4. Plant poisons :— Ricin, saponin.
5. Chemicals :— Sodium or potassium chlorate, nitrobenzene, pyrogallic acid, copper, chronic selenium poisoning in sheep, phenothiazine poisoning in horses,
6. Animal poisons :— Snake venom.
7. Incompatible blood transfusion.
8. Post parturient hemoglobinuria.
9. **Icterus neonatorum :** This is seen in foals born of some pedigree horses and in mules and is analagous to **erythroblastosis fetalis** of infants born of a Rh positive father and Rh negative mother. The erythrocytes of the horse contain certain unidentified antigens, which are inherited by the fetus and so the antigens develop in it. The antigens being water soluble diffuse through the placenta and enter the dam. If the dam happens not to possess the same.type of blood group iso-immunisation against the stallion's (and foal's) erythrocytes develops in the dam and so hemolysins circulate in the dam's blood. These antibodies cannot pass through the six layered **epitheliochoreal** placenta of the mare. But they are excreted in the colostrum. So the new born foal drinks this colostrum, hemolysins are absorbed in the intestines and produce hemolysis (iso-erythrolysis) and icterus results. The condition is usually fatal, the duration of illness being about 5 days.

A similar condition has been noticed in calves, puppies and piglets also in all of which the placenta has four or more layers. **Syndesmochorial** in cattle with 5 layers: **endotheliochorial** in dogs with 4 layers; **epitheliochorial** in pigs with 6 layers. It is only if the placenta is three layered that the antibodies can pass through the placenta to the fetus direct as occurs in man with **hemochorial** type of placenta. Neonatal immunolytic anemia and icterus have been reported in calves born of cows that were inoculated with an anaplasmosis vaccine of bovine origin.

Pathogenesis: Since there is greater hemolysis, greater amount of hemobilirubin is produced. Large quantities of it are brought to the liver. The liver excretes, therefore, a large amount of cholebilirubin into bile, which is therefore, intensely yellow or green. The gall-bladder may be full.

But the liver cannot deal with all the hemobilirubin brought to it. So some of it is left over in the blood and it is this excess amount of hemobilirubin that stains the tissues and mucous membranes yellow (Jaundice). Since the molecule of hemobilirubin is large and since it cannot pass through the glomerular filter, urine *does not* contain any hemobilirubin.

Since a large quantity of cholebilirubin is produced, large quantities of urobilinogen and stercobilin are formed. Feces are stained more intensely

A part of the urobilinogen is reabsorbed by the intestines. A fraction of it enters the portal circulation and reaches the liver, which is already performing its work at its maximum capacity and so is not able to carry any more load for excretion of the urobilinogen, which, therefore, finds its way into the general circulation and finally to the kidney, which excretes it in the urine. A part of urobilinogen that is absorbed by the intestines enters the lymphatics and ultimately reaches the kidney via general circulation and so is excreted in the urine. Urine, therefore, has a deep yellow color since it contains abnormal amounts of urobilinogen.

Clinically, the following observations may be made: (a) Serum gives an indirect van den Bergh test. (b) Urine is deeply yellow and contains urobilinogen. (c) Feces are intensely yellow. (d) Liver function tests are negative, indicating, thereby, that liver cells are not damaged. (e) Sometimes, when excessive hemoglobin is released and is not completely metabolised by the R. E. cells some of it may be excreted in the urine, which is, therefore coffee colored. (f) If the condition is due to protozoa, the parasites can be seen in the erythrocytes. (g) Blood picture shows changes of advanced anemia and the packed-cell-volume of blood is very low.

Toxic jaundice: This type is also called intrahepatic jaundice and arises when the liver cells are damaged by various toxins and so cannot perform its normal function. Here, formation of hemobilirubin is normal.

Causes: The toxins that may injure the hepatic cells and produce cloudy swelling, fatty degeneration and necrosis are varied and may be:—

(a) **Bacteria and viruses of infectious diseases:** (i) Leptorpirosis (ii) Salmonella. (iii) Virus of infectious canine hepatitis.

(b) **Plant toxins:** Toxins of family Senecio, Crotalaria and Astragalus ,

(c) **Cbemicals:** (i) phosphorus, (ii) Chronic copper poisoning, (iii) pitchclay pigeons, (iv) Chloroform.

Pathogenesis: Since there is degeneration and necrosis of hepatic cells the normal functions of the liver are affected. It is not able to convert all the normally formed hemobilirubin brought to it into cholebilirubin and so some of it is left in the blood.

The degenerated hepatic cells swell and so compress and block the bile capillaries. So much so, the cholebilirubin that is formed does not find its way into the biliary system and intestines but stays on in liver from where it is reabsorbed into the blood. In a liver with degenerated cells, the hepatic cords are not intact but become disorganised. The hepatic cords support the bile capillaries, and so when the cords are disrupted, the integrity of the bile capillaries is lost and so they rupture, spilling the bile into sinusoids. Hence the cholebilirubin that is formed enters the blood circulation and is excreted into urine since the cholebilirubin molecule is small.

In toxic jaundice, therefore, blood serum contains both hemobilirubin and cholebilirubin, which are responsible for the staining of the tissues.

Clinical findings:

(a) Serum containing both types of bilirubin gives a biphasic van den Bergh test. (b) Urine contains cholebilirubin. (c) Liver function tests reveal injury to the hepatic cells. (d) Blood picture does not reveal any changes.

Obstructive jaundice: This variety arises due to obstruction to the normal flow of bile. Obstruction may be due to:

1. **Blocking of bile duct from within:** (a) *Ascaris lumbricoides* in swine (b) *Thysanosoma actinioides* (fringed tape worm) in the sheep. (c) *Fasciola gigantica* in cattle. (d) Gall stones (not of importance in animals).

2. **Pressure on bile ducts from without:** (a) Tumors, (b) abscesses, (c) grannlomas, (d) fibrous tissue as in biliary cirrhosis. (e) enlarged and inflammed pancreas and portal lymph nodes.

3. **Inflammatory processes involving biliary system :** In cholangitis and cholecystitis, caused by fascioliasis and infection by *Dicrcoelium dendriticum*, pressure causes closure of the biliary tracts.

4. **Closure of the orifice of bile duct in the duodenum:** In duodenitis, the swelling of the mucosa of the duodenum closes the orifice of the bile duct.

Pathogenesis There is normal formation of hemobilirubin which is converted into cholebilirubin. But because there is obstruction there is stasis of bile, which on increasing pressure becomes absorbed into the blood and stains the tissues. Being a small molecule, cholebilirubin freely passes through the glomerular filter and so is found in the urine.

As no bile enters into the intestines, urobilinogen is not formed and feces are not stained and hence are clay colored. Urine does not contain any urobilinogen. In the absence of bile, fats are not emulsified and digested. So they are voided in the feces, which are therefore greasy. Fats may also be decomposed by intestinal bacteria. emitting foul odour. Feces in such cases have a very offensive smell.

Clinical findings:

(a) Serum gives a direct van den Bergh test, (b) Urine contains cholebilirubin. (c) Urine does not have urobilinogen. (d) Feces have a clay color, are greasy and emit a foul odour. (e) Liver function tests in the early stages are negative. But in later stages, due to pressure of accumulating bile, there may be atrophy and degenerative changes of the hepatic cells when liver function tests may reveal hepatocellular damage.

The following table depicts the differences various types of jaundice may show when tested by laboratory methods.

	Hemolytic (Pre-hepatic)	Toxic (Intra-hepatic)	Obstructive (Post-hepatic)
Serum - van den Bergh test	Indirect	Biphasic	Direct
Urine bilirubin	Not present	Present	Present
Urine urobilinogen	Slightly present	Present	Not Present
Feces	Intense yellow No smell	Normal	Clay colored Greasy Foul smell
Liver function tests	Negative	Positive	Negative (in early stages)
Blood prothrombin time	Normal	Prolonged	Prolonged
Total serum cholesterol	Normal	Decreased	Increased

In the absence of bile or in its deficiency, vitamin K is not absorbed. Vitamin K is required for the synthesis of prothrombin. So in obstructive jaundice, prothrombin is deficient and hence the clotting is delayed as revealed by prolonged prothrombin time.

Jaundice is also classified as *retention* and *regurgitation* jaundice. Hemolytic jaundice comes under the variety of retention jaundice since some quantity of hemobilirubin is retained in the blood without being dealt with by the liver. In regurgitation jaundice the cholebilirubin is converted into bilirubin or reenters the blood. This is therefore found in toxic and obstructive types of jaundice.

Hematoidin : In places where hemorrhages occur, hematoidin may be liberated and it is to this pigment that the color of a bruise or contusion is due. Hematoidin is first converted into biliverdin and so the bruise is green. Subsequently the color changes to yellow when biliverdin is converted into bilirubin. It is extra-cellular (while hemosiderin is intracellular)

Lipochromes: In old age and in wasting cachectic diseases, certain brownish granular pigment accumulates in the heart muscle, skeletal muscle, liver cells, nerve cells, in the cells of adrenal cortex, cells of Leydig, seminal vesicles and corpus luteum. Since this pigment is stained by Sudan it is believed to be derived from some fat pigment. Its exact nature is not known. It is believed to be degraded remnant of lysosomes. It is also known as 'wear and tear' pigment since it is noticed in old animals.

In the heart, the pigment is located at both the poles of muscle cells. This pigment imparts a brown color to the whole heart, which in old age undergoes senile atrophy, Such a condition of the heart is known as "brown atrophy" (Note. This should not be confused with "brown induration" of the lung which occurs in chronic venous congestion of the lung). Sometimes in old age similar pigment may be found in the parenchymatous cells of a shrunken liver, when it is known as "brown atrophy" of the liver.

Ceroid: In vitamin E deficient animals, a peculiar wax-like acidfast material accumulates in the uterine muscle fibers, ovary, interstitial cells of testes, lymph nodes, spleen, fat, macrophages of the liver, bone marrow, lung, kidney, as well as, skeletal and cardiac muscles. The exact nature of this pigment is not known. Similar material is found in choline deficient liver cells.

EXOGENOUS PIGMENTATION

Pneumoconiosis:

Dusts of various kinds—coal dust, iron dust, stone dust, asbestos dust—may be inhaled by animals. These dusts besides imparting a color, cause fibrosis of the lungs. This condition is called pneumoconiosis.

The condition caused by coal dust is known as *anthracosis*;that of stone dust—*silicosis*, that of iron dust—*siderosis:* that of asbestos dust—*abestosis*. that of fine stone dust or cement—*chalicosis:-* that of cotton dust or lint, *byssinosis:*.

Anthracosis: Anthracosis is a condition in which there is an accumulation of carbon particles in the lungs. Usually it is coal dust that accumulates.

Anthracosis is seen in horses, cattle, dogs and cats that live in industrial areas where soot and coal dust pollute the air. It is common in "pit ponies" used to haul in coal mines.

Macroscopically,the lungs show black streaks or spots, especially at the lower borders. The lungs have a "peppered" appearance. The medulla of the lymph nodes is darker in color. The carbon particles are only mild irritants and so large scale fibrosis is seldom seen.

Microscopically, the pigment is seen as black granules either between cells or within phagocytes. In the lungs, it is present in the alveolar walls and connective tissue septa. In the lymph nodes, the pigment is deposited in the medulla (in all animals other than the pigs, in which it is found in the cortex) between lymphoid cells. Macrophages engulfing the coal particles may transport them to other locations.

Results: Usually no serious damage occurs. But in heavy deposition, especially when combined with silicon dioxide, fibrosis of the lung takes place as is seen in miners.

Silicosis: In this condition, stone dust is inhaled and is more common in man than animals. Persons working in iron, gold and diamond mines, stone quarries, glazing and enamel industries are frequently affected.

Silica is insoluble and is a very powerful irritant, causing extensive fibrosis which predisposes the lung to tuberculosis.

Particles of silicates enter the alveoli where they are engulfed by macrophages, which transport them to the intrapulmonary lymphatics and then to the lymph nodes. The silicates are converted into silicic acid, which is highly toxic to the tissues. It is now postulated that silica may combine with body protein and form as an antigen against which antibody is formed and the result of antigen-antibody reaction is manifested by the tissue changes noticed. The macrophages that ingest the silica particles are converted into epithelioid cells with large cytoplasm filled with lipid material. Some of these cells form Langhans type of giant cells by the division of the nucleus. Due to irritation fibrosis occurs, forming granulomatous nodules. In time, much of the pulmonary tissue is replaced by fibrous tissue, which hindering blood circulation results in hypertophy of the right ventricle (*cor pulmonale*) and ultimately ending in chronic venous congestion. But the patient is likely to die far earlier of infection by tuberculosis.

Macroscopically, the pleura is thickened. Throughout the lung parenchyma are seen numerous circular nodules of varying sizes. Focal emphysema and bronchiectasis may be noticed.

Microscopically, the "silicotic nodule" has a characteristic appearance viz. around central particles of silica are formed concentric laminae of hyaline collagen. Pulmonary arteries are thickened (changes due to hypertension). Focal areas of emphysema and bronchiectasis are seen. Fetalisation of alveolar epithelium may be noticed.

Plumbism

Chronic poisoning by lead is known as plumbism. Lead may enter the body by ingestion of paints containing lead or ingestion of water and fodder contaminated by lead fumes from lead mines. Another source is water that passes through lead pipes from which lead may be dissolved in water and poison the body.

Lead combines with hydrogen sulphide wherever this is produced and forms black lead sulphide. This is very spectacularly noticed in the gums at their junction with teeth Here hydrogen sulphide is produced by putrefactive bacteria from food particles and this interacts with lead deposited there and gives a blue color, the "blue line". Similarly, the intestinal mucosa may be gray coloured. Feces may be gray due to the presence of lead sulphide.

Argyria

This is a condition that is noticed in long continued therapy with silver salts, which are deposited as finely granular albuminate. The skin and conjunctiva become gray or gray-blue. Internal organs may show similar pigmentation.

The pigment is not intracellular. It is deposited in the cement substance. In the skin it is found in the dermis just under the epidermis, while in the kidneys it is deposited in the lamina densa of glomeruli and outside the tubular epithelium. The black granular deposits are found in the Kupffer's cells and the ground substance of the arterioles and venules of the liver.

No harm appears to be caused except the disfigurement, for the face of man is gray or ash colored and this is a permanent blemish.

Tatooing

In tatooing, the pigment particles are transported by the macrophages to the connective tissue in the corium and deposited there This pigment is not found in the epithelial cells. The phagocytes may also transport and deposit the pigment in the regional lymph nodes.

CHAPTER 12

GENETICS AND DISEASE

Cytogenetics
    Autosomes
    Sex chromosomes
    Dominant
    Recessive
    Homozygous
    Hetrerozygous
    Diploid
    Haploid
    Genotype
    Karyotype
    Phenotype
    Mitosis
    Meiosis
Abnormalities of chromosomes
    Idiogram
    Heteroploidy
    Fuploidy
    Trisomy
    Monosomy
Methods of altering number
        of chromosomes
    Nondisjunction
    Mosaicism
    Simple loss of chromosomes
    Translocation
    Crossing over
    Deletion
    Inversion
    Sex-chromatin

Chemical structure of chromosomes
    DNA
    Protein synthesis

Causes of genetic injury
    Maternal age
    Radiation
    Viruses
    Chemicals
Anomalies of sex chromosomes
    Klinefelter's syndrome
    Turner's syndrome
    Adrenogenital syndrome
Autosomal anomalies
    Down's syndrome
    Malignant lymphoma
    Spontaneous human abortion
Disturbances characterised by metabolic
                   block
    Glycogen storage disease
    Galactosemia
    Phenylketonuria
    Alkaptonuria
    Amaurotic idiocy
    Porphyria
    Excess of phylloerythrin
    Purine metabolism
    Yellow fat in sheep
Metabolic diseases producing new end
                   products
    Sickle cell anemia
    Neoplasms
Lethal genes
Resistance to diseases
Hereditary defects and disease resistance
Heritable diseases of animals (lists)

    While describing the etiology of disease in the second chapter mention was made that heredity is one of the causes. In this chapter the genetical principles involved in the manifestation of heritable diseases are considered.

    Cytogenetiis. It is now well established that heritable characters ar governed by genes located in chromosomes, whose number for a particular spe es of animal is constant. These chromosomes are usually paired and the tota imber of chromosomes in different species of animals are :

185

Man	46	Dog	78	Fowl	77–78
Cattle	60	Cat	38	Turkey	81–82
Horse	64	Sheep	54	Duck	79–80
Mule	63	Goat	60	Rabbit	44
Donkey	62	Pig	38 or 40	Guinea-pig	64
				Mouse	40

In each species one set of two chromosomes, called *sex-chromosomes* differ from the rest. In the females these two are identical and are known as X chromosomes and so the female is said to have XX chromosomes. In the male the two sex chromosomes differ. One is identical with X and the other is different and is designated as Y chromosome and so the male has XY chromosomes. The chromosomes other than sex-chromosomes are known as *autosomes* and these contain most of the genetic material. The genes are located on the chromosomes. The potency of the genes may be of different grades Powerful genes are known as *dominant*, while the weaker ones are known as *recessive*. The chromosomes which are paired are usually identical and so are *homologous*. If the genes situated at the same locus in the homologous pair are both dominant or recessive, the condition is known as *homozygous*. But if one is dominant and the other is recessive, the condition is known as *heterozygous*.

If the gene is dominant, then its characteristics will be manifested in the animal irrespective whether the arrangement is homozygous or heterozygous. If it is recessive the characteristic will be manifested only in a homozygous arrangement, since the weak gene requires strengthening by another. That means it requires a double dose. In animals where the gene is recessive and the arrangement is heterozygous, the characteristic is not manifested and the animal is simply a *carrier* of that characteristic which will be manifested by its progeny if a homozygous situation for that gene arises.

The somatic cells have paired chromosomes. In a cat there are 19 pairs. They are known as *diploid*. The sex cells have only half the number of somatic cells, viz. only 19 single chromosomes and they are known as *haploid*. A *genotype* constitutes the animal's full set of genes. *Karyotype* means the animal's complete chromosomal picture. A *phenotype* represents the complete individual (physical, biochemical and physiological) that is controlled by the genes.

Mitosis and Meiosis. Mitosis is the process of cell division in which each of the two daughter cells receives identical amounts and kinds of DNA. Since the students have already studied this subject, mention will be made only of the *five* stages in mitosis: (i) *Interphase*, in which there is build up of DNA in the cell nucleus to twice its normal amount (so that the daughter cells can inherit equal amounts of it, as much as their parent cell had); (ii) *prophase*, in which the chromosomes are seen as double strands of dense chromatin joined at the centromere (each of these strands are known as *chromatids*) and the two centrioles migrate to either pole of cell (iii) *Metaphase*, in which the chromosomes align themselves at the equatorial plane and the spindle apparatus is formed, attaching the chromatids at the centromere to the centriole; (iv) *Anaphase*, in which the centromere splits longitudinally and the chromatids thus released migrate to the pole of the cell and become independent chromosomes, the spindle then disappears and two nuclear membranes are apparent and (v) *Telophase*, in which two daughter nuclei with separate nuclear membranes and cell membranes are formed and in which the chromosomes become dispersed throughout the nucleus.

Meiosis is the process of cell division encountered in the spermatozoa and ova in which only one half of the original chromosomes are obtained. For example, in the cat these gametes have only 19 single chromosomes. There are two separate divisons, Meiosis I and Meiosis II and each has the 5 phases similar to those described under mitosis. So each gamete gives rise to 4 haploid daughter cells.

Abnormalities of chromosomes: The chromosomes in a species of animal can be identified by the process of autoradiography and their pattern is known as *idiogram*. Metaphase is the most suitable period to take this autoradiography. There may be certain abnormalities in the chromosomes, which may be responsible for the death of the gamete or zygote, whih is known as "self cleansing mechanism". But some chromosomal alterations may persist and bring about changes in the phenotype of the individual. Such abnormal chromosomes are cause of disease in man and animals. The abnormality may be in the number or shape of the chromosomes. These abnormalities in the chromosomes may be familial (hereditary) or they may develop when the fetus is *in utero*. The latter are not familial. In man hemophilia and color blindness are diseases caused due to alteration in chromosomes of the former variety (familial) being transmitted through the gamete. Mongolian idiocy is an example of the latter, being found in a child born of perfectly normal healthy parents.

If the number of the chromosomes in the gamete or zygote is different from the normal, the condition is known as *heteroploidy*. If there are exact multiples of haploid sets the condition is said to be *euploidy* and the cells are *euploid*. Animals having this condition do not survive but die within the first few days of embryonic life. Alterations in the number of chromosomes that are not exact multiples of haploid number are known as *aneuploid heteroploidy*. *Trisomy* is a condition in which instead of 2 chromosomes, there are three present in a particular set of chromosomes. In man trisomy of group 21 chromosomes is responsible for Mongolian idiocy. *Monosomy* is a condition in which there is only one of a set of chromosomes. This condition may also cause disease. Trisomy and monosomy may affect the autosomes or sex-chromosomes.

Methods of altering the number of chromosomes:

Non-disjunction is the process in which there is no separation of the chormatid pair in mitosis or of the chromosomal pair during meiosis. In this situation one daughter cell may have trisomy while the second monosomy. Usually only one chromosome is affected in non-disjunction and so the resulting daughter cell will have 40 or 36 chromosomes in a cat. Therefore, if a gamete is involved then all the cells of the animals have abnormal number of chromosomes. In case this abnormality were to occur during the first division of the zygote, then the daughter cells will have (in the cat) 40 and 36 chromosomes and this gives rise to *mosaicism*. Sometimes non-disjunction can occur in somatic cells giving rise to another type of mosaicism. Such cells having abnormal number of chromosomes do not survive usually.

Simple loss of a chromosome: sometimes during metaphase one of the chromatids may fail to migrate to a pole and hence it is lost to that daughter cell. The other daughter cell has the normal number.

Translocation: occurs when a chromosome fractures and a fragment of it is attached to a different chromosome of a different pair. Sometimes the two chromosomes may fracture simultaneously and fragments of these may be exchanged. This condition is known as *reciprocal translocation. Crossing over* is a special form of normal reciprocal translocation in which the corresponding segments of homologous chromosomes are exchanged In translocation the number of chromosomes is not affected but the karyotype or chromosomal idiogram will be different. In such cases should the zygote survive, phenotype shows the abnormal genetic effects.

Deletion A chromatid may be fractured and a segment may not reunite and so a daughter cell will have an incomplete chromatid. Since the lost part contains some genes, the individual may be diseased.

Inversion: In this condition a chromosome fractures in 2 places giving rise to 3 segments. The middle part somersaults and then the three parts reunite, In such a condition the sequence of genes is altered.

Sex chromatin and Barr body: In the female majority of normal cells during intermitotic period have in their nuclei, attached to the innerside of the nuclear membrane, a lump of chromatin called the sex-chromatin. It is deeply stained and is 1 u by 1 u. This sex chromatin or Barr body is seen in all of the body cells, but more especially in the epithelial cells of the buccal mucosa and vaginal mucosa. Normally in females 85% of the nuclei in these cells have the sex-chromatin. Probably only 10% of the males may have it or not at all. The polymorphonuclear leucocytes of females have a small drum-stick shaped body attached to one of the lobes of their nuclei. Males do not have this body. This is also known as sex - chromatin.

Chemical structure of chromosomes. The chromosomes are composed of DNA, the structure of which we have already learnt in chapter three. It is this DNA that carries the genetic code, controlling the synthesis of proteins. The double helix molecule is composed of a large number of nucleotides, each containing the purines, *adenine* and *guanine* and pyramidines, *thymine* and *cytosine.* These nitrogenous bases are arranged in a definite sequence. Transposition of even one base in a nucleotide results in mutation. The genes are segments of DNA molecule and these are responsible for the synthesis of enzymes necessary for protein synthesis. So it is inferable that lack or loss or modification of a gene may result in the loss of that enzyme or enzymes and so physical aberration and disease may result. All the diseases are explained by loss or alteration of enzymes involved in vital biochemical reactions. Man and guinea-pig do not synthesise Vitamin C while all other animals are able to, probably due to loss of specific gene responsible for the synthesis of the specific enzymes required for vit. C synthesis.

Protein synthesis: DNA is responsible to determine the type, time and amount of protein to be synthesized. RNA fabricates the proteins. 3 types of RNA are present, all of which are synthesized in the nucleus. (1) Messenger RNA (mRNA) or template RNA (containing 1500 nucleotides) which transfer the genetic code from the DNA to the ribosomes,where protein is actually synthesized. (2) The transfer RNA (containing 70 nucleotides) which being a soluble fraction is known as sRNA. This is responsible for the transfer of amino acids. There is a specific sRNA for each of the 20 amino acids. (3) The ribosomal RNA which is located on the ribosomes and has 1650 to 3300 nucleotides and is stable

When protein is to be formed, the mRNA carrying genetic code from the DNA is formed from the DNA of the nucleus. It passes into the cytoplasm and is attached to a ribosomal particle. The amino acids activated by ATP are attached to their specific sRNA which carries them to the ribosome particle and attaches them to a corresponding location in the ribosome through the coupling in mRNA. So the amino acid is placed at the appropriate location. In this way various amino acids are brought together and linked by peptide bonds and polypeptides result and thus the particular enzyme or protein is synthesized

It is postulated that three nucleotides in the DNA molecule are needed to code for each amino acid (the triplet code) and a gene contains about 3000 nucleotide pairs. Any alteration in the structure of DNA naturally alters the genetical code and the mRNAs. When a virus containing DNA enters the cell nucleus, it merges with the cell DNA and so the genetical code may be altered.

Causes of genetic injury

Maternal age: A predisposing factor in genetical injury is the maternal age. It is found that in mongolism, congenital anomalies and many gonadal dysgeneses occur in the new born of aged mothers. mothers over 45 years of age. The ova which are present in the female from birth (it should be remembered that unlike spermatozoa, new ova are not formed in the female. There is only maturation of the ova that are already present at birth) are subjected to changes of aging and other adverse effects encountered during the life of the mother. So probably the genetic material is altered.

Radiation: Ionising radiation causes mutations and breaks in the chromosomes and so alters the genetical code, causing abnormality in the off-spring.

Viruses: We have considered above that viruses may alter the DNA of the cell and so genetical make up is altered.

Chemical: 5-bromouracil causes chromosomal abnormalities. Nitrogen mustard causes chromosomal breakage and hence such substances may cause genetical aberration.

Anomalies in sex chromosomes

Klinefelter's syndrome: The individuals are phenotypically males with eunuchoid features, gynecomastia, small testes. low urinary excretion of 17-ketosteroids but greater excretion of gonadotrophins, sterility, scanty sexual hair, lean and slender body and subnormal intellect. Many have sex-chromatin and they have a total chromosome count of 47 with a XXY karyotype. Probably the condition arises due to non-disjunction or crossing over.

Turner's syndrome: This is found in females, showing genital underdevelopment, webbing of the neck, amenorrhoea and sterility. Patients are short and subnormal in intelligence. No sex-chromatin is present. Somatic cells n-

45 chromosomes. Ovaries are absent. The karyotype is XO i.e. one X chromosome is missing.

Adrenogenital syndrome: This is female hermaphroditism. Here gonads are normal. There is hypertrophy of the clitoris. Karyotype is XX and there is sex chromatin. The cause is due to a mutant recessive gene which interferes with normal steroid metabolism in the adrenal cortex of the fetus and postnatally. There is hyperplasia of the adrenal cortex and excess of androgen production which causes the masculinising characters.

Autosomal anomalies : Though autosomes are more in number than sex chromosomes and theoretically abnormality in the autosomes must be much more than the abnormality of sex chromosomes, in actuality, this is not so since autosomal anomalies cause death of the zygote.

Down's syndrome or mongolism: The new born child has mongoloid or slanted eyes, retardation of growth and imbecility. These characteristics are due to malformation of the brain. There is trisomy of the acrocentric chromosomes No. 21 and so there are 47 chromosomes in the somatic cells. Usually children born of mothers who are aged 45 or above suffer from this disorder.

Malignant lymphoma : In cattle suffering from malignant lymphoma there were 3 large metacentric chromosomes and so somatic cells had 61 chromosomes instead of normal 60. In the cells of circulating blood, in a case of lymphosarcoma chromosomal abnormalities were noticed. In dogs suffering from lymphosarcoma, the somatic cells had one extra chromosome.

Spontaneous human abortions : Chromosomal defects cause death of the fetus which is therefore aborted. The defects noticed are : translocation on chromosome A; trisomy of group E: anencephaly, XO sex chromosomes; trisomy and XXY sex chromosomes etc.

Diseases characterised by metabolic block

Glycogen storage disease or von Gierke's disease : This is a hereditary disease, in which there is abnormal accumulation of glycogen in the liver and kidneys. The disorder is due to lack of glucose-6-phosphatase in the liver and kidneys. This enzyme is necessary for the dephosphorylation of glucose-6 phosphate a step necessary for the breakdown of glycogen to form glucose. Hence glycogen accumulates in the liver and kidney.

Galactosemia : This is a hereditary condition, transmitted as an autosomal recessive, due to lack of galactose-1-phosphate uridyl transferase which is necessary for the conversion of galactose-1-phosphate to glucose-1-phosphate. In children galactose accumulates, in the absence of this enzyme, in the liver causing cirrhosis and in the eye causing cataracts. Patients are mentally retarded probably due to damage of the central nervous system consequent on hypoglycemia that develops.

Phenylketonuria : This hereditary disease is due to the lack of the enzyme phenylalanine hydroxylase necessary for the conversion of phenylalanine into tyrosine. Hence phenylalanine accumulates in the blood and spinal cord. Some of it may be excreted in the urine as such but a large part is converted into phenyl-pyruvic acid. As these compounds damage the brain, patients manifest mental deficiency. The defect is transmitted through a single recessive autosomal trait.

Alcaptonuria. This is due to lack of homogentisic acid oxidase. Homogentisic acid is a normal degradation product in the metabolism of tyrosine and phenylalanine. But when the enzyme is absent, this product appears in the urine, which on exposure becomes black in color. In many patients polymerisation of the homogentisic acid results in the formation of an yellow pigment (o.hronosis) which accumulates in the cartilage, tendons and other collagen-rich tissues, imparting to them a brown color. This defect is due to an autosomal dominant factor with incomplete penetrance.

Another block in the metabolism of tyrosine causes albinism, which is also hereditary.

Amaurotic idiocy or Tay-Sach's disease. This is caused by an abnormality in the metabolism of fats and so sphingomyelin accumulates in the body tissues and the disease is characterised by idiocy, muscular weakness and blindness. There is degeneration of the cells of the brain, which is small and hard. Patients die before 3 years of age. The condition is due to a simple recessive character.

Porphyria. In this condition there is accumulation of porphyrin in the blood, deposition of the pigment in the bones, teeth and other tissues, and it is excreted in the urine. The condition is due to a simple recessive factor. There is defective metabolism of protein and the condition is seen in cattle and swine. The animals become photosensitive and skin exposed to the sun becomes blistered and necrotic with deep ulcers resulting. The teeth have a pink color (pink tooth).

Excess of phylloerythrin. We have already studied this condition under "hepatogenous photosensitisation" (Vide page 118). It transpires that in some sheep even without hepatic damage or obstruction to bile excretion, the phylloerythrin may not be excreted by the liver and this is due a hereditary defect in hepatic function due to a simple recessive mutation.

Purine metabolism. In the Dalmatian dog, major part of the uric acid is not converted into allantoin though possessing the needed enzyme, uricase. This is due to a simple recessive autosomal factor. There are no systemic changes and the dog is quite healthy. The spots on the body have no association with this genetic defect.

Yellow fat in sheep Some sheep form yellow fat in the body instead of the normal white fat and this is due to the influence of recessive mutation. There is probably some defect in the metabolism of carotinoid pigments. Though the health of the animals is not affected, people do not like to purchase the mutton because of abnormal color.

Metabolic diseases producing new end products.

In this category there is no deficiency of the enzyme concerned but there is some alteration in it so that end products are different. *Sickle-cell anemia* arises due to change in the composition of hemoglobin in which at one position (no. 6 in the alpha chain) *valine* is substituted for *glutamic acid* This is evidently due to alteration in the genetic code. The concerned factor is a recessive one.

Neoplasms. Some neoplasms are familial and are due to alteration and abnormalities in the karyotypes. In man the intestinal polyps are due to an autosomal non-sex-linked dominant.

Similarly pituitary neoplasms in the Boston terriers are hereditary.

Lethal genes. Some defective genes cause death of the zygotes. The lethal effect may also be manifested any time after fertilisation—in the embryo, fetus, newborn or adult animal. Usually these genes are recessive.

There are some other genes which though not producing death, cause undesirable characteristics in the animals and are called sub-lethal genes. Imperforate anus in the calf is an example of this variety.

An example of lethal gene is hemophilia in dog and man. This is due to a sex-linked gene carried in the X chromosome. The female can be homozygous or heterozygous for this gene while the male is hemizygous. So if the X chromosome of the male were to contain this gene he will be affected with hemophilia. In such persons anti-hemophilic globulin (AHG) is not produced. The female possessing normal homozygous or heterozygous chromosomes produces normal amounts of AHG and so do not suffer.

Resistance to disease. In nature during the process of natural selection, only animals that are resistant to diseases (bacterial, viral, rickettsial, protozoal, parastic and metabolic) survive and this resistance is inherited

In livestock industry profit depends on the production of the animals—rapid growth (in meat animals and table birds), production of large amount of milk, laying of large and more number of eggs, yielding good and large quantity of wool and ability to do sustained work (in the case of work bullocks for tilling and draft purposes). This could be attained only if the feed consumed is utilised to the maximum. For the utilisation of the feed, enzymes must be present in sufficient quantity for the various biochemical reactions in metabolism We have noted above that the synthesis of enzymes is genetically controlled. Hence profitable and economical utilisation of food stuffs with resultant greater production is heritable and so by selection animals having propensities for greater production can be developed. Actually it was in this way that the the present day breeds of animals have been developed from scrub animals of the wild. For example of better utilisation of feed, mention may be made of White Leghorn which among poultry, is the most efficient in this respect.

HEREDITARY DEFECTS AND DISEASE RESISTANCE
(By Dr. A. Sreerama Murthy)

Animals during their life-time are subjected to various stress factors namely climatic, nutritional, microbiologic and so on. The production in farm animals depends upon the extent to which these stress factors are overcome, which ability often has a genetic background, usually complicated by environmental interaction

The stress factors bring about deviations from the normal pattern of development with the result that the individual may either succumb during embryonic life itself or with lesser deviations result in less serious defects. This variation ranging from directly lethal malformation to defects reduce the fitness of the animal. The genetic basis of such anatomical malformations and metabolic disorders is often rather simple and it is possible to determine the inheritance of many of these defects.

It is often difficult to classify sharply the numerous defects occurring in all species and breeds of farm animals. The defects are classified in general as hereditary acquired or environmentally induced. It is possible, sometimes, that the same defect can be produced phenotypically by a mutant gene or a chromosomal aberration or through environmental agencies like certain chemicals acting during the pregnant period of the mother. Depending upon the type of manifestation the defects can be classified as morphological and physiological (biochemical). There is no fundamental difference in the hereditary basis of these defects. Many morphological malformations are observable in the new-born animal and therefore called congenital defects as differentiated from others which manifest themselves at later stages of life. Eg. genital malformations or functional sterility which can be recognised at the age of puberty.

Congenital defects can be induced experimentally. Such artificial induction of congenital malformations can be brought about by various agencies like :—

i) Administration of X-rays to pregnant mothers
ii) Injection of drugs like insulin, cortisone or sulphanilamide, as well as pituitary hormones, oestrogens and various antimetabolites to pregnant mother.
iii) Deficiencies in certain vitamins and aminoacids—in rats and pigs Vit. A deficiency has been shown to produce eye anamolies, cleft palate, hare-lip and leg defects.

Teratogenic agents are those substances which, when administered to the pregnant female give rise to malformations of the young.

It is also interesting to note that the whole array of congenital malformations in the same species either produced as a result of homozygosity for recessive genes or due to action of teratogenic agents are more or less similar. More or less perfect phenocopies can be produced by these experimentally induced influences on pregnant mother or the fertilized egg. The sensitivity of the foetuses to various teratogenic agents vary at different stages of gestation, being greatest in early stages of pregnancy immediately after implantation in the uterus. Also the same teratogenic agent may have a varying degree of effects at different stages of embryonic development. The teratogenic agents exert their influence by blocking gene action that control foetal development.

Experimental induction of congenital malformations by teratogenic agents does not mean that these defects are non-genetic if they appear spontaneously from time to time in all species of farm animals. Evidence from breeding experiments in rodents in fact indicate that the spontaneous occurrence of these induced defects is usually due to homozygosity for recessive genes. All spontaneous cases of congenital malformations, especially if the several malformations occur in the same individual, should be carefully scrutinised before they are considered as hereditary.

Genes with unfavourable effect are classified into lethal, semilethal and subvital. When a gene in an effective dose causes the death of the carrier before the attainment of puberty is said to be lethal. When the percentage of survivors is greater than 'O' but less than 50 the gene is said to be semilethal and when more than 50 but less than 100 per cent of the carriers survive the gene is called subvital. Depending upon the external environment and the genotype there may be overlapping in the above classification.

Genes mutate sometimes and their effect is always unfavourable to the individuals in the environment in which they are adapted. Very little is known about the frequency of gene mutations in farm animals, or about the frequency of genes with lethal or semilethal effects, the relative frequency vary from species to species.

The reasons for homologous defects in different species can be speculated, at least in part, that differentiation of one and the same organ in two closely related species is under the control of genes with a common origin and that one or some of these have a high frequency of mutation in both species. It is also probable that similar defects can result from mutations in completely independent gene systems

Intangible Causes:

In all species of farm animals a number of defects are the result of "accidents" in development (intangible causes) They appear with low frequency. Examples are

i) Symmetrical and asymetrical duplications: conjoined twins, duplication of hind or foreleg, etc.,

ii) acrania (absent or incomplete skull)

iii ectocardia (exposure of the heart)

iv) exencephaly (exposure of the brain)

v) situs inversus viscerum (lateral transposition of visceral organs)

These intangible defects may also be due to chromosomal abberration or dominant mutations which eliminate themselves as soon as their effect has been manifested.

The analysis of the genetic basis of congenital malformations is more complicated than has been generally appreciated during the past few decades. What is hereditary in one case may be a phenocopy in another. Furthermore a hereditary disposition for malformations may have an observable effect only under certain environmental conditions.

Disease Resistance

As far as resistance to disease is concerned, this is mainly a quantitative trait influenced by both heredity and environment. The importance of inheritance is evident from the fact that some species of farm animals have complete resistance to certain diseases whereas others are easily infected. An example of this is foot and mouth disease which attacks cattle and pigs but not horses.

The animals are subjected to multitude of stresses such as nutritional deficiencies, climatic sensitivity and pathogenic microorganisms. It is well known that to withstand these stresses is easier for some animals and breeds than for others.

1) Resistance to Nutritional Deficiencies:

a) The Leghorns are less sensitive to manganese deficiency (perosis) than heavier breeds like Rhode Island Red. There is a genetic tolerance to manganese deficiency.

b) Deficiency of thiamine (Vit. B) resulting in polyneuritis. The Leghorn is less demanding in its vitamin requirements than the heavy American breeds.

c) Deficiency of riboflavin lowers hatchability, growth rate and causes high mortality in chicks. Resistance to this deficiency is known to be genetically determined.

d) Deficiency of Vit. D —Investigations have established that RIR chicks are much more susceptible to rickets than Leghorns with low Vit. D feeds.

There seems to be every reason to believe that similar differences occur in the large farm animals, also.

II) Climatic Sensitivity

The European Cattle do not tolerate the high temperature (Heat tolerance) than zebu cattle. Crosses between these two breed types are about intermediate.

III) Infectious Diseases

A large number of experiments have been conducted with laboratory animals and poultry to determine the causes of the individual differences in resistance to infectious diseases, by exposing these experimental animals to infectious agents. By the use of suitable experimental techniques it is possible to avoid confusion between acquired, passive immunity and genetically determined resistance.

The resistance to pullorum disease—bacillary white diarrhoea (*Salmonella pullorum*) has been considerably increased in American experiments by continuous selection.

Leukosis in poultry (neural lynphomatosis or fowl paralysis and visceral lymphomatosis) responsible for great losses in poultry breeding is caused by one or several viruses. Newly hatched chicks are especially susceptible to leukosis infections. The heritability for leukosis has been estimated to be about 8-15 percent. USA workers have carried out selection for high and low mortality to leukosis.

By selection it has also been possible to increase the resistance to other bacterial and viral infections as well as to various internal parasites. Resistance to one disease does not mean necessarily resistance to other diseases.

As far as the larger animals are concerned, it has not been possible to carry out any comprehensive selection experiments for increased disease resistance. On the other hand, a number of interesting observations have been made. In cattle, suspectibility to mastitis appears to be fairly highly heritable (23-40%). Large differences among different breeds have been established for resistance to tuberculosis; zebu cattle are much more resistant than European breeds. The zebu also has greater resistance to piroplasmosis (red water) and many tropical diseases transmitted by parasites.

In pigs heritable differences have been found in the resistance to brucellosis, atrophic rhinitis.

Biological basis of Resistance:

The differences in the ability to produce antibodies, as well as the differences in the efficiency of the complement may explain part of the genetic variation in the resistance of farm animals to various diseases.

Livestock industry depends not only on animals producing economically but also on their viability. That means to say, their capacity to resist and withstand diseases and thrive. This resistance to disease is inherited. For example some breeds among the same category of animals are resistant to the diseases as noted below :

Swine: Brucellosis, erysipelas, hog cholera, dysentery.

Sheep and goats Jaagsiekte, Trichostrongyle infections,

Cattle: Mastitis, tuberculosis, bebesiosis, anaplasmosis, trypanosomiasis, foot and mouth disease, lukemia

Horses: Encephalomyelitis,

Fowl: Pullorum disease, fowl typhoid, leukosis, avian monocytosis, aspergillosis, worms.

It is quite possible to select and breed animals that are resistant to diseases. But in countries where diseases are not present, the best way to stand out any that may occur suddenly is by destruction as in the case of rabies in England and Foot and Mouth Disease in Australia.

Selection of resistant strains take a long time and one must have sufficient patience and money for this. So the best way to improve the animals is to combine selection of resistant animals with preventive inoculation. It should be brone in mind that an animal resistant to one disease may be suscepible to another and so it may not be possible to have animals resistant to all diseases. It is always better to select animals which have acclamatised themselves to the locality. And cross breeding is advocated because hybrid vigour may provide resistance.

A list of diseases that are heritable is given below. It is by no means complete since more and more heritable conditions are recognised. Fortunately most of these diseases are caused by recessive genes. But it has the great disadvantage that in heterozygous animals the bad characteristics are masked and so selection among such animals is difficult for purposes of weeding them out.

Animals with heritable diseases should not be used for breeding.

Condition	Breed	Mode of inheritance
	CATTLE	
Congenital perphyria	Shorhorns, Holsteins, Black & White Danish.	Single recessive
Hare lip	—	Not known

Condition	Breed	Mode of Inheritance
Atresia ani	—	Not known
Atresia ilei	Swedish higland.	Simple recessive
Lymphatic obstruction	Ayrshire calves (more in males than in females).	Single autosomal recessive.
Aortic aneurysm	—	Not known
Idiopathic epilepsy.	Brown Swiss	Dominant
Congenital hydrocephalus	Holsteins, Herefords. Ayrshire.	Recessive.
Congenital Achondroplasia	Dexter.	Recessive.
Congenital Cerebellar hypoplasia	Herefords, Guerenseys, Holsteins, Shorthorns,	Recessive.
Periodic spasticity	Holstein & Guerensey.	Single recessive with incomplete penetrance
Spastic paresis.	Holstein, Aberdeen Angus Danish Red, Ayrshire.	Recessive.
Neonatal spasticity.	Jersey, Hereford.	Single recessive.
Congenital Posterior paralysis.	Norwegian Red poll.	Recessive
Exophthalmus with Strabismus.	Shorthorn, Jersey.	Recessive.
Osteoarthritis	—	Recessive.
Multiple ankylosis	Holstein.	Not known
Multiple tendon contracture.	Shorthorn.	Single recessive
Reduced phalanges	—	Not known
Polydactylism	Normandy	Hereditary
Syndactylism	Holstenis.	Hereditary
Achondroplastic dwarfs	Herefords, Aberdeen Angus Holstein, Shorthorns.	Simple recessive.
Displaced molar teeth,	Calves	Simple recessive.
Agnathia	Angus and Jersey	Sex-linked recessive.
Mandibular prognathism.	Holstein, Herefords, Ayrshire, Jersey, Shorthorn.	Simple recessive
Umbilical hernia	Holstein.	One pair of autosomal recessive.
Cerebral hernia	Holstein, Friesian.	Recessive
Cryptorchidism.	Holstein.	Not known
Taillessness.	Holstein.	Not known
Muscular hypertrophy	Many breeds	Not known
Symmetrical alopecia.	Holstein.	Single autosomal recessive.

Condition	Breed	Mode of inheritance
Congenital hypotrichosis.	Guernsey (viable)	Single recessive.
	Friesian (non-viable)	Sex-linked recessive.
	Holsteins (in females)	Sex-linked semidominant.
	Hereford (partial)	Simple recessive.
Baldy calves.	Holstein.	Autosomal recessive.
Congenital Absence of skin	Non-viable.	Single recessive.
Ichthyosis	Holstein, Norwegian Red poll, Brown Swiss etc.	Single recessive.
Eye diseases:	Jersey,	Simple recessive.
Microphthalmia.		
Irideremia (absence of iris total or partial).		
Microphakia (smallness of lens).		
Ectopia lentis. Cataract.		
Opacity of cornea.	Holstein, Jersey, Swiss, Norwegian Red poll.	Hereditary
Prolonged gestation.	—	Not known
Hypoplasia of ovary.	Swedish Highland.	Autosomal recessive. with reduced penetrance.
Abortion	—	Hereditary predisposition.
Familial Polycythemia	Jersey	Simple recessive.
Chediak-Higashi syndrome (incomplete albinos).	Hereford & other breeds.	Single recessive.
Smooth Tongue.	Holstein-Friesian. Brown Swiss	
Familial Convulsions and Ataxia.	Aberdeen-Angus	Dominant
Hereditary Neuraxial edema.	Polled Hereford.	Autosomal recessive.
Congenital Osteopetrosis.	Aberdeen-Angus.	—
Congenital Parakeratosis of Calves.	Black Pied Danish.	Autosomal recessive.
Dermatosparaxia (abnormal fragility of skin).	Belgian	Recessive.

PIGS

Hemophilia	Poland	Semilethal, recessive.
Congenital porphyria	—	Simple recessive.
Congenital Atresia ani.	—	Not known

Conditions.	Breed.	Mode of inheritance.
Congenital Hydrocephalus.	Yorkshire	Recessive
Thick fore legs.	—	Simple recessive.
Inherited rickets.	—	—
Congenital Scrotal hernia and cryptorchidism.	—	Inherited.
Tail deformities	Landrace Large white.	Not known
Absence of skin.	Not viable.	Single recessive.
Dermatitis vegetans	—	Recessive, semilethal.
Trembles.	—	Hereditary,
Bent legs.	Swedish Yorkshire	Recessive,
Club feet.	Swedish and Norwegian Landrace	Recessive.

SHEEP

Goitre	Merino	Recessive
Dwarfism	Southdown	Semilethal, recessive
Cerebellar atrophy	British & Corriedale	Recessive
Osteogenesis imperfecta	Lamb	Not known
Entropion	Suffolk	Not known
Photosensitisation	South Down	Not known
Prolonged gestation	—	Not known
Lethal gray	—	Partially dominant
Impaction of abomasum & abnormalities of digestive system.	Gray.	Recessive
Microcephaly	Polwarth	—

GOAT

Congenital myotonia.	—	Not known.

HORSES

Cerebellar hypoplasia	Arab	—
Abrachia (absence of fore limbs)	—	Lethal, recessive
Congenital Atresia coli	Animal not viable, dies in the first few days.	Simple recessive
Wryneck.		Lethal, recessive
Multiple exostoses	Quarter horse	Not known
Bleeding	Thoroughbred	Semilethal, recessive
Dysplasia of hip	Dole	Not known
Cryptorchidism,	—	Dominant, sex - liked
Foal ataxia	German breeds	Recessive
Congenital Absence of skin non-viable		Simple recessive
Entropion	Foals	Hereditary

Cat

Polydactylism		Hereditary
Deafness	White	Not known

Dogs

Overshot jaws	Dachshunds	Recessive
Achondroplaia	Poodle, Scotish terrier, Bulldog, Sealyham, Basset, Beagle	Hereditary
Otocephaly	Beagle	Recessive
Cranischisis	Cocker spaniel	Recessive
Reduced tooth number	—	Incompletely dominant associated with Hairlessness
Short spine	Japanese	Not known
Hip dysplasia	German shepherd, Poodle, Labrador etc.,	Autosomal dominant Polygenic inheritance
Intervertebral disc protrusion	Long-bodied breeds eg. Dachshunds	Not known
Polydactyly	Several breeds	Not Known
Elbow dysplasia	German shepherd, Peke-Poodle, Cocker spaniel Labrador	-do-
Osteogenesis imperfecta	Toy breeds	Not known
Hairlessness	Lethal in homozygous	Dominant
Congenital lymphedema	—	Autosomal dominant
Hernia, umbilical	—	Multiple recessive genes
Ectropion and entropion	Several breeds	Not known
Microphthalmia	—	Hereditary
Cryptorchidism	Common short headed and dwarf breeds Cocker spaniel	Irregular inheritance Sex limited, autosomal recessive
Mononephrosis	Beagle	Recessive
Anterior pituitary dysfunction	—	Incomplete dominant
Hemophilia	Several breeds	Sex-linked, recessive
Kidney stones, predisposition for	Dalmatians	Recessive
Diabetes mellitus	Dachshund	Not known
Cystinuria and urinary calculi	Various breeds	Not known
Cataract	Alsatian	Simple dominant
Deafness	Various breeds, linkage with coat color-Bull terrier, Dalmatians, Sealyham	Not known

Condition.	Breed	Mode of Inheritance
Glaucoma	Fox terrier	Irregular inheritance
Retinal atrophy	Irish setter, Poodle	Simple recessive genes
Retinal dysplasia	Bedlington terrier	Recessive
Epilepsy	Keeshund, Poodle	Not known
Hydrocephalus	Bull dogs and their crosses	Several recessive genes
Tetany	Scottish terrier	Irregular recessive genes
Trembling	Airedales	Irregular
Cerebellar hypoplasia	Airedales	Familial
Hemolytic anemia	Basenji Beagle	—

POULTRY

Sticky egg	—	Recessive
Crooked neck—dwarf	—	Recessive]
Congenital loco	—	—
Congenital shaking	—	Recessive
Naked Neck	—	Sex-linked recessive.

CHAPTER 13

NEOPLASMS

(Revised by: Dr. D. Sundarasiva Rao.)

Hyperplasia and neoplasia
Differences between benign and
  malignant neoplasms
Histotogical features of malignant tumors
Causes of neoplasms
  Intrinsic or predisposing
    Heredity; Cohnheim's cell-rest
    theory: Age; Pigmentation; Hormones.
  Extrinsic factors
    Single injury; Chronic irritation;
    Chemical carcinogens; Ionizing
    radiation; Metazoan parasites;

  Viruses;
Spread of tumors
Diagnosis of cancer
Effects of Neoplasia
Classification of tumors
Connective tissue tumors
Fibroma; Fibrosarcoma; Equine
                Sarcoid
Myxoma and Myxosarcoma
Lipoma and Liposarcoma
Chondroma and Chondrosarcoma
Osteoma: Osteosarcoma; Giant-cell
                tumor
Leiomyoma and Leiomayosarcona
Rhabdomyoma and Rhabdomyosar-
                coma
Angioma; Hemangiopericytoma
Mesothelioma
Mastocytoma

Nervous tissue tumors
  Schwannoma
  Meningioma
  Glioma
  Ependymoma
Neuroblastoma and ganglioneuroma
  Aortic body tumors
Tumors of the hemopoietic tissues
  Lymphosarcoma
  Hodgkin's disease
  Myeloid tumors
  Avian leucosis complex
  Marek's disease
Epithelial tumors
  Papilloma
  Squamous cell carcinoma
    Horn cancer; Eye cancer
  Basal cell carcinoma
  Adenoma
    Sebaceous gland adenoma
    Perianal adenoma
    Sweat gland adenoma
    Primary carinoma of lungs
    Tumors of :
        Liver
        Mammary gland
        Thyroid
        Parathyroid
        Adrenal gland
        Ovary
        Testes
        Prostate
Embryonal nephroma
Adenoma of kidney
Melanoma
Canine Venereal Tumor
Adamanntioma.

In this text, tumor, new growth and neoplasm are used synonymously and to mean the definition of an autonomous growth. The word autonomous indicates that the tumor cell growth is independent of the physiological demand. The word cancer which is in current use by custom, is restricted to certain types of tumors which are malignant.

The autonomy of the neoplastic cells is characterised by two features–1. A disorder of growth that leads to unrestrained growth and cell multiplication and 2. disturbances of organising processes which normally operate— loss of tissue organisation,

A neoplasm (neo=new, plasm=thing formed or growth) is popularly known as a tumor (swelling). Though no definition is satisfactory, the following of Mallory appears to be the best. A neoplasm is a new growth of cells which 1) proliferate continuously without control; (2) bear a considerable resemblance to the healthy cells from which they arise; (3) have no orderly structural arrangement; (4) serves no useful function and (5) for the present at least, have no clearly understood cause. To these characteristics may be added a sixth namely that the neoplasm continues to grow even after the cessation of the stimuli which evoked the growth process.

Though neoplasm is also known as a tumor or swelling, all swellings are not neoplasms. For example, parasitic nodules, cold abscesses, chronic inflammations (Tuberculosis, glanders) are not tumors. These swellings subside after the causative agent is removed while neoplasms grow continuously and indefinitely

Neoplasms must be differentiated from the inflammatory and reparative processes and also from hyperplasia.

In inflammation and repair cells proliferate; but then there is a purpose for proliferation, namely to protect and replace tissues. As soon as the need is fulfilled, proliferation ceases and the growth even regresses. Similarly, the tubercular lesion is essentially a reparative process, although eventually in the process much of useful, functioning tissue may be lost. The lesion becomes healed as soon as the infection is overpowered.

We have already studied that hyperplasia results due to proliferation of cells of tissues from a definite demand. It may be through a hormonal stimulus. For example, if there is loss of liver tissue either by excision or damage by poisons, the loss is made up by proliferation of surviving liver cells. Similarly loss of renal tissue is made up by compensatory proliferation of renal tissue. Hyperplasia of erythorpoietic tissue occurs in anemias. In most pyogenic and certain acute inflammations (pneumonia for example) hyperplasia of leucopoietic tissue of the marrow occurs, resulting in leucocytosis,

Hyperplasia of thyroid occurs in iodine deficiency (goitre), Similarly hyperplasia of the mammary gland and prostate occurs in conditions wherever respective hormones are in excess. But hyperplasia in conditions described above differs from neoplasms in that the proliferation of the cells in the former is limited in "amount and duration". "It progresses only so long as the functional need or hormonal stimulus which evoked it persists". The differences are shown in the following table:-

Hyperplasia	Neoplasia
1. Arises due to a direct response to an extra-cellurlar stimulus.	1. No immediate exciting cause can be identified
2. Cessation of stimulus is followed by prompt regression.	2. Growth persists even after cessation of stimulus.
3. Proliferating cells are typical of the parent cells, i. e. differentiated and of regular size and shape.	3, The proliferating cells may be atypical, undifferentiated, anaplastic or pleomorphic.
4. Cells are organised and coordinate with the funcion of the parent tissue.	4. Cells are disorganised and are not functionally in unison with the normal tissue—autonomous.
5. Metastases or heterotopia not seen.	5. Metastases, especially in malignant tumors seen.

NOMENCLATURE AND CLASSIFFICATION OF COMMON TUMORS.

The present day nomenclature follows either a cytological or histological nomenclature. eg. chondroma in cartilage cells, fibroma in fiibroblasts etc. But this is not satisfactory with epithelial tissues and names indicating tissue and morphology also, are used. eg. Adenomatous polyp, Squamous papilloma etc.

The suffix 'Oma' is usually added to the tissue of origin to give the name of the tumour. eg, lipoma, fibroma etc.

Classifications also are varied. But from the clinical point of view, behaviouristic classification is most acceptable and is given below.

Tumors are broadly divided as benign (innocent) and malignant. By ths former term it is understood that the tumor is not ordinarily fatal. But this is not true af those that occur in important organs like heart, lung and brain. Malignant tumors are those that ordinarily cause death. The following is a brief summary.

1. STRUCTURE:

Benign tumors: In benign tumors, the proliferating cells resemble. closely the cells from which they arise. So they become differentiated like their parent tissue. For example, a papilloma of the skin consists of all the normal layers of the skin but only in an exaggerated degree. So also a lipoma consists of unmistakable fat tissue and a leiomyoma consists of adult plain muacle. So these have typical structure of adult tissue.

Malignant tumors : Malignant tumors consist of cells of varying degrees of differentiation. Some may show typical structures of the parent tissuee For example, in an epidermoid carcinoma, keratin formation is seen. In other tumors, the differentiation is so lacking and the cells are so embryonic that it is often difficult to identify source of the tumor. For example, in rapidly growing epithelial tumors, sarcomas and gliomas, it is difficult to distinguish cells arising from one tissue with those of others. This lack of differentiation and reversion to embryonic type is known as anaplasia, The more anaplastic the cells are the more malignant a tumor is.

2. MODE OF GROWTH :

Benign tumors: These are localised and the growth is expansive. Due to the pressure they exert on the surrounding tissues and organs a capsule is formed. Some benign neoplasms may lack a capsule, eg. angioma.

Malignant tumors grow by expansion; they also invade or infiltrate adjacent tissues by growing between cells along the tissue spaces. In a section, extension of growth below the basement membrane is adequate evidence of malignancy. Since the surrounding tissue is surreptuously infiltrated, a capsule is not formed. The extent of infiltration is of utmost importance to the surgeon, since one cannot accurately assess macroscopically the extent of such infiltration. So, to give the benefit of doubt to the patient and to avoid leaving behind any neoplastic cells, sufficient surrounding healthy tissue is always included in excision.

3. RATE OF GROWTH

Benign tumors grow slowly and show few mitotic figures.

Malignant tumors, on the other hand, grow rapidly and show numerous mitotic figures. The mitosis may be atypical being tripolar or even multipolar

Presence of mitotic figures is an indication of rapid cell division and so is a reliable indication of aggressive malignacy. Usually, the more number of mitotic figures seen per field, the more malignant the tumor is. Exceptions may be found. Though many mitotic figures are seen in the basal cell carcinoma, it is slow growing and relatively less maligant, being only locally invasive and ordinarily nonmetastasizing.

4. CONTINUANCE OF GROWTH

Benign tumors may grow slowly indefinitely but many may stop growing after some time. Some may even show regression spontaneously as seen in lipomas, adenomas, uterine fibroids etc.

Malignant tumors grow cotinuously and finally cause death of the host. Spontaneous regression of a malignant neoplasm though reported is very very rare.

5. METASTASIS

Benign tumors never metastasize, i. e , they do not form secondary tumors in other parts of the body

Malignant tumors always metastasize if given time. Metastasis to different tissues occurs by transport of the neoplastic cells by lymph and blood streams or across cavities by a snow drop fall.

6. CLINICAL COURSE

Benign tumors are usually harmless except in the following ways:

(a) by virtue of their position: Though benign, tumors of the brain are always serious and may be fatal.

(b) by complications: A tumor may cause severe hemorrhage as in papilloma of the urinary or alimentary tract; a pedunculated tumor of the bowel may cause intussusception; some tumors may cause ulcers which may be subsequently infected.

(c) Excessive hormone production: Adenoma of the pituitary may cause gigantism; adenoma of the islets in pancreas causes hypoglycemia.

Malignant tumors, while producing disease in the above ways, also infiltrate and metastasize; as such clinical manifestations are varied.

The tumor may recur in spite of treatment. Death is an invariable outcome of a malignant tumor.

7. HISTOLOGY

(i) **Invasiveness**: Tumor tissue beneath the basement membrane is indicative of malignancy.

(ii) **Anaplasia**: Reversion to embryonic type, due to lack of differentiation through inadequate maturation is known as anaplasia. The features of anaplastic cells are: (a) Unusually large size of nuclei—indicative of rapid growth; (b) Hyperchromasia of nuclei. (c) Multiple unequal nuclei in the cell—tumor giant cells. (d) Enlargement of the nucleolus – indicative of rapid growth. (e) Large number of mitotic figures indicating rapid cell–division (f) Nuclear : cytoplasm ratio is decreased. (g) Nucleolar : nuclear ratio is increased.

(iii) **Presence of atypical or abnormal mitotic figures**: These indicate abnormal cell division (each cell dividing into unequal number of daughter cells) and so indicate deviation towards malignant neoplasia.

(iv) **Presence of atypical blood vessels**: As the malignant tumors are growing fast, they require abundant blood supply. Very thin walled blood vessels may be present. In some places, the tumor cells may line blood sinusoids. The tumor cells may outgrow blood supply and so degeneration and necrosis are seen often in malignant tumors (which are seldom noticed in the benign variety).

(v) **Loss of polarity**: In epithelial tissues the cells are arranged in a particular order bearing a distinct relation to the neighbour. This is called polarity. eg. In a tissue of epithelium the long axis of a cell is ordinarily perpendicular to the surface. However, once malignant change occurs the cells divide haphazardly and so are arranged pell-mell with no order. This is known as loss of polarity.

DIFFERENCES BETWEEN BENIGN AND MALIGNANT TUMORS

Benign	Malignant
1. **Rate and mode of growth**	
a. Slow growth over a long period	a. Rapid growth—few days or weeks
b. Grow by expansion	b. Grow by expansion and infiltration
c. Do not penetrate basement membrane.	c. Break through the basement membrane.

d. Capsule present.

e. Mitotic figures not seen.

Continuance of growth

a. Limited

b. May show spontaneous regression.

2. **Metastases:**

a. Do not metastasize

3. **Blood supply**

a. Moderate well formed blood vessels seen.

b. Blood supply is adequate and so no degenerative changes seen.

4. **Destruction of adjacent tissue:**

a. Very little.

5. **Recurrence after removal:**

a. Do not usually recur.

6. **Structure:**

a. Cells resemble those from which they originate. Anaplasia not present.

b. Structure, therefore, typical of adult tissue.

c. Polarity of cells maintained.

d. Pleomorphism of cells not present.

7. **Nucleus and nucleolus:**

a. Nucleus and nucleolus not enlarged.

b. Shape of nucleus not altered

c. Nucleus normochromic.

d. Nucleolar : nuclear ratio not altered.

e. Number of nuclei in the cell normal.

f. Cytoplasmic : nuclear ratio unaltered.

d. Capsule not present.

e. Mitotic figures frequently seen.

a. Do not stop growing—unceasing.

b. No spontaneous regression.

a. Metastasize always.

a. Numerous thin walled blood vessels seen.

b. Blood supply not coordinated—degenerative changes and necrosis of tissues of the tumor frequently seen.

a. Abundant.

a. Recurrence is common,

a. Cells may not resemble parent cells, i. e anaplasia may be present.

b. Structure not typical of adult tissues.

c. Polarity lost.

d. Cells pleomorphic.

a. Both are enlarged.

b. Variation of shape seen.

c. Hyperchromasia of nucleus.

d. Ratio increased.

e. Sometimes multinucleated.

f. Ratio decreased.

ETIOLOGY & PATHOGENESIS OF NEOPLASIA

While defining a neoplasm, it was observed that the cause or causes are still obscure. This does not mean that we do not know the causes of all

neoplasms,   Some have been   well established.   As in inflammation, the causes
are varied for neoplasia.

While discussing repair we have observed that all cells except probably
the   nerve cells and the striated muscle cells, have retained their property of
multiplication as is evident in their regeneration while repairing damaged or lost
tissues.   This regenerative power is need based and functional.   In repair there
is no unrestrained growth, for, as soon as healing is complete, multiplication of
cells cease. On the other hand, in tumors, some intracellular irreversible
change takes place which makes the cells grow without any restraint.   What this
change is and how it is produced is still not known.   Some agents that are
detailed below may cause such a change.   Once this intracellular change is
produced, it becomes permanent and is handed down to progeny.

INTRINSIC OR PREDISPOSING FACTORS

A. Hereditory factors: In some families higher incidence of cancer
has been found.   Neuroblastoma, malignant papilloma of the colon,retinoblast-
oma and xeroderma pigmentosum are examples.   Though no responsible gene is
identified one can reasonably infer that there is distinct genetic control in these
tumors.

Experimentally, it was possible to breed a strain of mice which were
highly susceptible to neoplasms. Maud Slye had found that susceptibility to cancer
in these mice was due to a simple Mendelian recessive factor. But how far
these are applicable to larger animals and man is controvertial.

It is quite possible that though susceptibility may exist, a realization or
inducing factor may be necessary for the neoplasm to occur. For example,
among poultry Avain leucosis complex is caused by a virus But for the neoplasm
to occur, the birds must be from a suceptible strian. Because by selection, a
highly resistant strain of birds is developed in which the incidence of the tumor
is very low.

B. Age: Malignant tumors are essentially a disease of old age The period
of life at which cancer appears is called 'cancer age'. In man this is after 50 years
in cattle 8 to 10 years and in dogs 5 years. Congenital tumors like nephro-
blastoma as well as sarcomas appear in young animals.

C. Pigmentation: Melanin protects the skin from actinic rays of sun.
Thus cutaneous cancers are fewer in Negroes than in white horses. Malignant
melanoma is very frequent among gray or white horses. It is said that if these
animals are allowed to survive till old age, they will invariably die of malignant
melanoma.

It is also found that Hereford cattle the eyes of which are lacking in
melanin, suffer often from eye cancer.

D. Hormones: The breast, uterus,prostate and the thyroid are under the
active control of hormones for their normal cellular and functional activities.
Increase in the hormonal level that controls these tissues results in increased
cellular and functional activities of the respective tissues. In high hormonal
levels, frank hyperplasia results which may ultimately become neo_lastic.

Injection of estrogens for a long time causes mammary cancers in mice. Prolonged estrogen therapy in ladies has caused uterine cancer.

The basic structure of estrogens, progesterone, testosterone and adrenocorticoids is similar to that of benzanthracene and benzyprene, which are powerful carcinogens The important component appears to be the phenanthrene group that is found in the structure of all these as is evident in the following formulae.

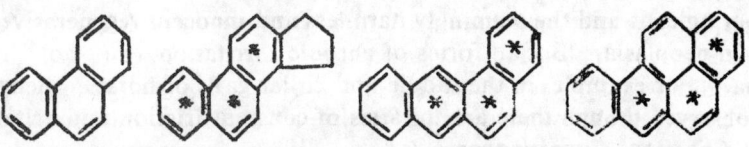

Phenanthrene   Sterol structure   Benzanthracene   Benzpyrene
common to estrogens
progesterone
testosterone
adrenocorticoids

Though it has not been proved, yet it is suspected that these hormones might have, somehow, been transformed into carcinogens. Similarly it is thought that by some metabolic error, cholic acid and cholesterol may be converted into methylcholanthrene, all of which contain the phenanthrene group.

EXTRINSIC FACTORS

A. Physical trauma : Can injury sustained only once cause neoplasia ? It has probably no role.

B. Chronic irritation : Virchow held the view that chronic and repeated irritation caused neoplasia. Examples of such association are :

a) **Cancer of lips in clay-pipe smokers :** Cancer is frequntly found on the lips of people who use clay pipes for smoking. Such people have a favourite place on lips where they keep the clay pipe. The irritation caused by constant heat of the clay pipe and its pressure on the lips, cause cancer of the ps il

b) **Kangri baskets :** In Kashmir, people keep a small bamboo-basket containing a small earthen pot with burning charcoal under the clothes, touching their abdomen to make themselves warm. This basket is known as "Kangri". Cancer of the abdomen is often seen in Kashmeres due to the continuous irritation by heat.

c) **Chutta-smokers :** In the northern districts of Andhra Pradesh, people smoke cigars (Chutta) with the burning side inside the mouth – "Reversed chutta", In such people, cancer of the palate is more frequent.

d) **Sharp-teeth :** Sharp teeth cause chronic irritation of the mouth and in such cases cancer of mouth is met with.

e) **Gall stones and kidney stones :** Stones in the gall bladder and the renal pelvis cause chronic irritation in these parts producing cancer.

f) **Yoke** : It is surmised that constant friction of the yoke in working bullocks may be responsible for horn cancer, for it was observed that incidence of horn cancer is higher in working bullocks than in cows, (which are not
worked).

In such chronic irritation, as described above, there is continuous
replacement of tissues This involves active proliferation of the cells. Eventually,
these proliferating cells get beyond the control of the restraint exerted by the
neighbouring cells and the seemingly harmless and innocent regenerative process
ends up in neoplasia, But all forms of chronic irritation do not result in
neoplasia. For example in the saddle and collar galls of horses, neoplasia is
seldom observed though these are the sites of constant friction and irritation.

C. **Chemical Carcinogens** :

(i) Carcinogenic hydrocarbons : Derivatives of Benzanthracene.

In England cancer of the scrotum was comon among chimney sweeps.
Sir Percivall Pott first drew the attention of the medical world in 1775 to the
connection between the scrotal cancer and soot. He postulated the relation of
coal and the cancer.

Cancer was also common on the thighs, scrotum and abdomen of mule
spinners.The ubricants containing petroleum products that spilled on these places
produced cancer. In fishermen of Scotland, cancer of the lips was frequent since
they used to hold between their lips the bone needle threading tarred thread.

Yamagiwa and Itchikawa of Japan produced in 1915, cancer experimentally, for the first time by applying tar to rabbit's ear.every day for 6 months.
This was an epoch-making experiment and opened the way for experimental
carcinogenesis.

Tar is complex containing many hydrocarbons. Which of these compo.
nents is carcinogenic? This was discovered by Kenneway and Cook in 1932-
By fractional distillation of tar, they isolated a hydrocarbon, benzpyrene, which
has carcinogenic activity.

It was later found that among various chemicals having a spectrum
similar to that of benzpyrene by fluorescent light, 1:2.5.6 dibenzanthracene was
highly carcinogenic. This is synthesized in pure form.

Another synthetic hydrocarbon is methylcholanthrene.This is also a powerful carcinogen. Special interest in this hydrocarbon lies in the fact that it is
synthesized from cholic acid, which is a normal constituent of bile,

3:4 Benzpyrene   1:2:5;6   Methylcholanthrene
                        Dibenzanthracene

These hydrocarbons can cause malignant neoplasms wherever applied or injected. If applied to skin they cause skin cancer. Injected into muscle they cause myosarcoma. If injected into brain, brain tumors are caused. If they are added to tissue cultures of cells, the cells become malignant. When once the cells in tissue cultures become malignant, they retain this malignancy for ever. Their progeny inherit this characteristic and we know not, yet, any method of converting these malignant cells to normal cells. If this can be achieved the problem of cancer will not exist

ii) Cyclic hydrocarbons of Naphthalene series in the aniline industry are bladder carcinogens.

iii) Inorganic chemicals; Zinc chloride when injected into the testes of roosters causes cancer of testes. Chronic arsenic poisoning causes skin cancer

Fumes from industries and exhaust fumes from automobiles are perennial hazards Of late, danger of cigarette smoking in the genesis of lung cancer has been almost conclusively accepted.

Cells differ in their susceptibility to cancer development by chemicals. Whatever may be the route of introduction the carcinogen may cause cancer always of a particular organ. For example cancer of mammary gland may be produced in mice by applying the carcinogen to the nasal mucosa. Introduction of dibenzanthracene into the alimentary tract of mice caused cancer of lungs. This may be of particular importance in man for it suggests that the site of function of a carcinogen in the body may be quite different and far distant from the site of application.

A long preparation period is required for the development of malignancy during which the cells become progressively transformed.

D. Ionizing rediation: it is a common knowledge that x-ray workers suffer from cutaneous cancers if they do not observe precautions These lesions start as chronic dermatitis, become ulcerated: and finally become malignant.

Radium introduced into the body, localises in the bonemarrow and may cause osteosarcoma and leukemia. For example, in painters of watches and clocks with luminous radioactive paints fatal leukemia was observed.

Workers in Schneeberg and Joachimsthal mines died of pulmonary cancer since the dust contained radioactive uranium.

Total body irradiation may also cause neoplasia. Victims of atom-bomb explosion died of leukemia.

Probably radiation causes changes in chromosomes by altering their composition. This may be brought about by deplolymerisation of nucleic acid and formation of hydroxyl radicles from decomposition of water (See page 123 for poison theory)

E. Actinic light rays: Among white people, the cutaneous cancers are more frequent due to the action of actinic rays of the sun. Eye cancers among cattle may be attributed to this cause.

In man is found a condition called xeroderma pigmentosa whis is hereditary and caused by a recessive characteristic. The skin is pigmented, dry and highly sensitive to sunlight which produces solar dermatitis. From this condition eventually multiple skin cancers develop. This can be prevented by protecting him from actinic rays.

F. **Metazoan Parasites**: Chronic infestation by metazoan parasites is sometimes associated with neoplasia.

i) In man, infected by *Schistosoma hematobium* is considered to be the cause of carcinoma of the urinary bladder. The parasite lives in the small veins in the wall of urinary bladder. The ova, with their spines and lytic enzymes bore through the tissues and produce inflammatory reaction resulting in ulceration of the mucosa. From these places, the ova are voided into the uirne and so reach outside where they undergo a life-cycle in a snail. The edges of the ulcers in the urinary bladder become hyperplastic and are the seats of neoplastic transformation.

ii) Infection of the rat liver by : *Cysticercus fasciolaris* the bladder worm of cat tape worm *Taenia taeniaformis* produces cysts with fibrous tissue around. This may result in fibrosarcoma. Metastases are found in lungs.

iii) Spirocerca lupi : Inhabits the walls of esophagus. A fibrous nodule forms which communicates with the lumen of the esophagus by a tiny pore, through which ova are liberated by the worms living in the nodules. Sometimes, the nodule develops into fibrosarcoma or even into osteosarcoma, by metaplastic change of the fibrous tissue into bone. Some of these tumors may even metastsize into lungs and other viscera.

iv) Eimeria stiedae causes papilliferous cystadenoma of bile duct epithelium in rabbits as a result of mechanical and toxic irritation.

v) Gongylonema neplosticum : was shown by Fibiger (1913) to cause papillomas and squamous-cell carcinomas in the stomach of rats. But it is now considered that the neoplastic change is due to metaplastic transformation of the cells resulting from vitamin A deficiency,

Some workers are of the opinion that evidence is not adequate to consider metazoan parasites as causes of true neoplasms. They feel that occurrence of neoplasms in association with the parasites may only be fortuitous.

G. VIRUSES AND TUMORIGENESIS
(Contributed by Dr. B. Sambamurthy.)

Cancer in man and animals has been ever a forbidding experience because of its incurability and fatality. Isolation, identification and propagation of tumor viruses in tissue culture and *in vitro* transformation of normal cells to a neoplastic growth by these agents, opened the flood gates of hope, that perhaps as in bacterial and viral diseases, cancer may be made preventable.

A number of viruses possessing tumorigenic and other common properties, have been listed by the International Subcommittee on virus nomenclature, 1963, under the Papova virus group.

Papova virus group. DNA viruses :

They are :—(1) Shope Rabbit papilloma, (2) Human papilloma (wart) (3) Polyoma of mice, (4) Simian vacuolating virus (SV 40); (5) Bovine papilloma; (6) Mouse K; (7) Canine oral papilloma; (8) Kilham rat papilloma, (9) Toolan H I; (10) Equine; (11) Rabbit oral (5, 7, 8, 9, 10 & 11 are candidate viruses)

The name of this group is a contraction of the first part of the names of member viruses: PA for papilloma. PO for Polyoma; VA for vacuolating agent Put together, they form the name 'Papova' virus group.

There are tumor viruses from other animal virus groups like Adeno, Pox and unclassified viruses. In the Adeno virus group, which are DNA viruses, Adeno virus types 12, 18, 7 and 31 have been credited with tumorigenic properties.

In the pox group are fibroma of rabbit, deer aud squirrel, Yaba monkey tumor virus and *molluscum contagiosum.*

Besides these there are the RNA tumor visuses consisting of the Erythroblastosis virus, Myeloblastosis virus, Lymphomatosis virus and strain RPL-12 of the Avian Leucosis Complex and Rous sarcoma virus of fowls.

Peyton Rous (1910) was the first to demonstrate the transmissibility of Rous sarcoma of fowls. Gross (1953) first discovered that filtrates from leukemic tissues of AKR mice produced parotid tumors. The bovine papillomas were known as virus induced tumors even earlier and the virus was obtained in crystalline form. The polyoma virus is used as a model virus for tumor virus research, due to its ability to induce tumors in its natural host as well as in unrelated species and transform cells *in vitro.* The biological properties of this virus as well as of other viruses of the Papova group have become the subject of intensive research to gain a break-through into the mechanism of transformation. The DNAs of the members of the Papova group of viruses were purified and found to be all double stranded DNAs. Their base ratios were determined as well as their structures Polyoma and Shope papilloma DNAs were found to be of a ring structure Polyoma has both ring structure and linear structure. The infectious agents—nucleic acids—have been extracted and found to be tumorigenic. Some viruses were found to be defective due to their inability to replicate without the co operation of helper viruses. Perhaps these latent incomplete viruses have evolved themselves to this state to adjust to their environment—virus ecology. They replicate and become a complete particle only when they chnace to get the help of the helper virus. From this phenomenon has emerged the concept of transcapsidation, where the genome of one virus and the capsid of another virus co operate to form a virion capable of replication. Another concept that has emerged is the animal viruses can act as carraiers for foreign genetic information particularly determinants of malignant transformation. This is called **partnership** Human cells transformed by SV40 are believed to be neoplastic to man. Neucleotide homology between some strains of tumorigenic viruses has been studied. Familial incidence of cancer in human beings was recorded (Steinberg 1953). Leukemic families in animals were known. Families of leukemic cattle were reported That brain filtrates from human leukemia produced frank leukemia in low leukemic mice was recorded. Vertical transmission of latent virus infection and accidental acquisition of helper or co-operating virus inducing occasional malignancy may be another possibility.

From the maze of information being accumulated it might be possible to evolve a preventive measure at least for a few of the virus induced tumors.

Some viruses have the property of inducing a wide variety of tumors on being inoculated into suitable animals. The virus after infection induces transformation of the cells of the hosts. From these transformed cells malignant cells are developed and form cancerous tissue. The behaviour of the virus-transformed cells is different from cancerous tissue Normal cells exhibit during their growth and multiplication a disciplined social behaviour called **contact inhibition** and grow in an orderly tailor-made fashion. This may be part of the genetical makeup and controlling mechanism of the host. In transformed cells this controlling mechanism of the host is lost. There is a change in the genetical makeup of the transformed cell and this change is induced by the virus which has induced the transformation. Thus transformed cells show the following properties :

1. Increased mitotic activity. 2. Lack of contact inhibition, 3. Alteration in chromosome numbers (Karyotypic changes) 4. Change in morphology. 5. Altered glucose metabolism.

These transformed cells grow autonomously in a three dimensional piling resulting in tumor formation.

It is increasingly becoming apparent that malignant transformation of a normal cell by a tumor virus, may be the result of introduction of a very small fragment of a foreign nucleic acid like that of the RNA or DNA of the tumor virus. RNA or DNA extracted from transformed cells can induce tumors when injected into susceptible animals.

Transformed cells develop foreign antigens. These antigens stimulate antibodies which endow the animal with the capacity of rejection of virus-induced homologus tumors which are free from virus. It is a peculiarity of transformed cells that the virus which has induced the tumor is absent from the very cells it has transformed. But it leaves its foot prints in the cells—viz tumor antigen. Such foreign antigens have been demonstrated in a number of cells transformed by different tumor viruses and they show a high order of specificity. Ex. Rous sarcoma. Gross mouse leukemia, **Shope papilloma** in the rabbit, SV 40 (Simian virus 40) and S E Polyoma.

Virus-transformed cells in addition have complement, fixing antigens, soluble antigens, transplantation antigens and induce neutralising antibodies. Of all viruses only Polyoma virus regularly hemagglutinates erythrocytes from guineapigs, hamsters, human 'O', mice, chicken and monkeys.

Immunology of virus induced tumors : These is no direct experimental evidence of viral etiology of human cancer but interest in tumor viruses or tumors induced by viruses has reached an all time high. The reasons are not far to be sought. Cancer has ever remained a scourge for man and animals. Though the suffering could be mitigated by suitable treatment, often with side effects cancer is neither preventable nor curable. Naturally, therefore, any hope of a preventable measure, even as a theoretical possibility. will be vigorously pursued by scientists all over the world, staking any amount of expenditure and time for such research so as to achieve the goal.

The basis for such an interest lies in :

1. That because all mammals tested so far have been found to be sensitive to one or more of the tumor viruses, it is suspected and hoped that some of the human cancers could be of viral origin.

2. That the same techniques used for the study of other animal viruses are applicable for the study of tumor viruses and so there may be a possibility for some method of immunisation.

3. So there should be some possibility of developing prophylactic agents against the tumor viruses.

4. That the fact that there are transplantation antigens in virus induced tumors has blazed a new trail for the search of common specific transplantation antigens for the tumorigenic viruses.

5. The question that when a tumor virus can induce morphological transformation of human cells *in vitro*, why does not such a change occur *in vivo?* If such a change could occur, then could it be prevented by induced specific resistance?

Among the several serological tests employed to detect viral antigens, the following tests were used to study the immune reactions in tumor virus infections. One important immunological phenomenon brought out in bold relief is the concept of immunological tolerance with reference to infection and immunity.

The tests employed can be categorised as direct and indirect.

Direct : The direct tests employed to demonstrate tumor-specific antigens are , 1. **Trausplantation tests.**

   (a) Virus induced resistance (V.I.R)
   (b) Cell induced resistance (C.I.R)
   (c) Virus induced and cell induced resistance.

2. **In vitro tests.**

   (a) Cytotoxic tests,
   (b) Inhibition of colony formation.

3. **Indirect tests.**

   (a) Fiuorescent antibody test. (F,A)
   (b) Complement fixation test, (C.F.T)

Both virus induced and chemically induced tumors contain specific antigens. In chemically induced tumors the antigens, however, differ from virus induced antigens. In the former the antigens are specific for each specific tumor whereas in the latter there is some cross reactivity. The presence of the inducing virus in some form is necessary to maintain the neoplastic behaviour of these cells. So if the virus could be eliminated from these cells by immunological means, cells can be cured of the virus and also the neoplastic property. This is a possibility Possession of specific transplanation resistance is one aspect which has promise. Complicating features are some aspects of immunological competence and incompetence and presence and absence of immunologically competent cells that may change the situation.

Neonatal thymectomy increases the frequency of onchogenesis, while thymic implantation reduces. Immunologically competent mice are resistant to polyoma infection while immunologically immature animals are susceptible. Inspite of evidence of immunological reaction in tumor virus infection, the reasons for failure to suppress or eliminate them seems to involve (a) immunological tolerance (b) specific immunological enhancement effect, (c) defective immune response and (d) low grade antigenic stimulus.

Little is known about transmission of cancers. But certain concepts have been projected. These are **vertical** and **horizontal** transmission. The genetical constitution of the animals or species seems to be an all important factor as a determinant of the liability of the host or for the frequency of occurence of cancer. Of vertical transmission of cancer, Avian Leucosis Complex in selectively inbred stock, is a good example. The virus is transmitted from hen to the chick via the egg The virus has thus an assured transmission in nature **vertically** from mother to the offspring That mouse leukemias attain high incidence, in highly inbred mice, is another example of vertical transmission. **Horizontal transmission** of the Avian Leucosis Complex is achieved in nature by other birds picking infection from the affected birds and so this is a bird to-bird transmission.

Tumor viruses can be divided into two groups on the basis of nuleic-acid core :

(1) DNA tumor viruses. Polyoma, Shope papilloma, S E. (Stewert and Eddy) Polyoma, Adenovirus types, 12, 18, 7 and 31, SV 40, SV 20, SV 38 and SV 7.

(2) RNA tumor viruses. Rous sarcoma, Avian Leucosis, Mouse leukemia virus

Peculiarities of transformation of RNA viruses . Rous sarcoma is an RNA tumor virus. It persists in the cells it has transformed. But this virus is not a complete infectious virus and hence is called "defective virus", because it cannot replicate. The complete infectious virus is produced by those cells only in the presence of another virus which helps the defective virus to become complete and infectious So the virus is called a "helper" virus. In this case the helper virus is one of the avian leucosis agents. It is also called RAS. (Rous associated virus). The change of the **defective state** to **complete state** occurs in or at the cell surface and perhaps is related to modification of function and structure of the cell membrane which may be essential for malignant transformation.

DNA tumor viruses on the other hand behave differently. The virus is absent from the very cells it has transformed to malignancy, This state of disappearance of the virus after it has transformed the cells is known as "**masking**". In some viruses like Polyoma, all evidence of virus infection in the infected cells are not detectable but in SV 40 virus-induced malignancy there is some evidence of the viral genome in a non-infectious state.

A greater understanding of the changes induced at the molecular level in the bacteriophage model has led to the knowledge of the mechanism that causes tumor induction and carcinogenesis as a consequence of tumor virus infection. But still the critical steps leading to carcinogenesis are not yet very clear.

For induction of carcinogenesis some factors are involved : the genetic constitution of the host as in leukemias of mouse and fowl is one factor, hormonal influences as in Bittner's mammary carcinoma is another, age of the host as in Polyoma, where new-born mice develop at least 20 histologically different tumors is a third factor. (Polyoma was obtained from leukemic tissues of AKR strain of mice, and was found to induce parotid tumors in new-born mice. Polyoma is widely prevalent infection of adult mice of various strains where they live in a well-balanced host-parasite relationship).

The immunologically incompetent new-born mice develop all the range of tumors to known infection with large quantities of tumor virus (Polyoma in tissue culture). The susceptibility of immunologically incompetent animals to polyoma virus is not restricted to mice alone but guiena-pigs, rabbits and hamsters also.

The DNA tumor viruses.

1. Human papilloma or wart virus has proved difficult to be propagated in tissue culture. So lack of an **in vitro** method has hampered a study of its various properties.

Human warts were known for a long time and were known to be transmissible.

2. Shope (1933) recovered the rabbit papilloma virus from the papillomas of wild cotton tail rabbits. The virus is not recoverable from the papillomas induced by the same virus in domestic rabbits. In the domestic rabbit the virus is present in a masked form. Nucleic acid extracted from papillomas of both wild and cotton tail rabbits is infectious Even when the papillomas progressed to carcinomas the virus exists 'masked'. But virus antigen is detectable in the tumors. It is thought that a part of the viral genetic material exists in an integrated state, a concept, perhaps, that developed on the analogy of a prophage.

3. Bovine papillomas were known for long to be contagious in cattle. These were shown to be induced by the Bovine papilloma virus by Creech (1929). It can transform cells from fetal bovine conjunctiva or growing skin cells

4. **The Polyoma virus** was obtained originally from filtrates of leukemic tissues of AKR mice and was first recognised as a distinct virus by Gross (1953). Susbequently Stewart **et al** (1957) isolated the virus in high titre in mouse embryo tissue cultures. They named t Polyoma because of the specturm of tumors it gives rise to. Dulbecco (1963) selected this as a model for study because the virus is able to transform cells of mouse and unrelated species *in vitro* and also induced tumors in them.

5. SV 40. Sweet and Hillman (1960) first isolated the virus from uninoculated rhesus and cynamegalus monkey kidney cultures where the virus grows without CPE. Tissue culture fluids from these when introduced into monolayers of green African monkey kidney cells induces vacuolation— hence called vacuolating agent. In new born hamsters it induces tumors. It was discovered that these monkey kidney cells which were extensively used for Polio vaccine preparation was the source of this virus and all children who took oral polio vaccine a'so took this oncogenic virus and excreted it for several weeks. Thus this virus, which

TABLE SHOWING DNA VIRUSES ONCOGENIC TO WARM BLOODED ANIMALS

Name of Virus	Source	Size uu.	Host	Tissue cultures supporting growth	Inclusion bodies	Tumors induced
GROUP—1 Papova Group						
Polyoma	Mice	43—46	Neonatal mouse, hamster. guinea-pig	Embryonic hamster, mouse, guinea-pig, human		**Mice:** By natural infection of the virus bilateral multilobular parotid tumors occur. Experimentally renal sarcoma, thymomas, mammary adenocarcinoma, epidermoid carcinoma, bone tumors, mesotheliomas, subcutaneous sarcomas, hemangioer dotheliomas, adrenal carcinomas and other miscellaneous tumors. All produced by one virus. **Hamsters:** Renal sacomas in the kidney. **Rabbits:** Tumors resemble non-malignant fibromas regressing in 1—4 months. **Rats:** Renal sarcomas. Tumors are transmissible from host to host by transplantation of cells Virus can be isolated from Polyoma tumors in tissue culture.
S. V. 40 Simian virus 40	Monkey		Rhesus monkey. Cynamegalus. African green monkey	Monkey kidney cells, New-born hamsters.		Subcutaneous inoculation into new-born hamsters induces undifferentiated sarcomas and subcutaneous nodules resembling fibrosarcoma.
Bovine papilloma	Cattle	30	Cattle	Conjunctival cells		**Cattle:** Papillomas
Shope papilloma	Cotton-tail rabbits	50				**Rabbit:** Papill. mas, carcinoma In wild rabbit papilloma virus is present in keratinised layers.

Virus	Host	Size	Cell culture / Experimental host	Inclusion	Disease / Lesions
Papilloma. (human warts)	Man	50	—	Intranuclear	Man: Verruca vulgaris ; (Human warts) Verruca plana juvenilis (Juvenile flat wart) Verruca plantaris (Plantar wart) Verruca digitata (digitate wart) Verruca filiformis (filiform wart) genital and laryngeal warts
	Dog	50	—	—	Dog: Papillomas
GROUP II. Adenotype: 7, 12, 18, 31.	Human	70	New-born hamster, rat. Hela cell, K B cells, epithelial-like cells from epidermoid carcinomas	Intranuclear	Undifferentiated sarcomas in hamsters and rats (neonatal)
GROUP III Pox Group. Fibroma	Rabbit, Deer, Squirrel	200–250	—	Intranuclear	Rabbit: Benign fibroma. Squirrel: multiple fibromatous lesions
	Yaba Monkey	—	—	Intranuclear	Monkeys: benign histiocytoma.
Molluscum contagiosum	Man	—	—	Intranuclear	Human: Multiple discrete nodules limited to the epidermal layers of the skin.

TABLE SHOWING RNA VIRUS ONCOGENIC TO WARM BLOODED ANIMALS

Virus	Host	Size	Cell culture	Inclusion	Disease / Lesions
GROUP-1 Avain Leucosis Complex Erythroblastosis virus	Fowl	89—140	Young inbred Chick embryo cells.	—	Erythroblastosis, Hemorrhagic hyperplasia in long bones.
Myeloblastosis	-do-	140	-do-	—	Myeloid leukemia
Lymphomatosis Strain RPL—12	-do-	140	-do-	—	Extravascular and intravascular lymphomatosis osteopetrosis.
RGOUP-II. Rous Sarcoma	-do-	80	Chick embryo fibroblasts.	—	Transmissible sarcoma.

can transform human and hamster cells *in Vitro* has become an object of stimulated research due to its peculiar oncogenic properties.

RNA tumor viruses:

Rous sarcoma. (RSV) : This is an RNA virus and was known earlier than most of the tumor viruses as the cause for transmissible tumor in fowl. It is an avian tumor virus, and is related to the avian leucosis viruses. After infecting a cell it produces malignant tumors in the fowl. But all the tumors induced by the virus do not contain the virus. This is a form of latency. Rous sarcoma virus in cells is a defective virus and non-infectious, requiring a helper virus, to complete its biological cycle. Non infective sarcomas, are induced by infection with RSV but yield no infective virus In one type the virus can be recovered from the tumors formed by a large dose of virus only in the early stages of the tumorigenesis but not later from the fully developed tumors This type is produced in immunologically competent birds. In the other type the sarcomas are induced with a low dose of virus in all fowls irrespective of immunological status. The second phenomenon is due to defective virus. The virus induces tumor formation at the site of injection.

Avian leucosis viruses: These are a group of filtrable agents, capable of producing a variety of neoplastic growths in the tissues of the fowl. They are all RNA viruses. They do not produce tumors at the site of injection. They are all present in the blood and organs of the affected fowls. Modern methods of study and characterisation of the viruses has enabled virologists to separate and study the biological properties of the candidate viruses of the Avian Leukosis Complex. These were named according to the Pathological changes they induce They are: **(1)** virus of erythroblastosis, **(2)** virus of myeloblastosis and **(3)** virus of lymphoblastoais. These viruses produce a complex neoplastic process, which can be individually described as myeloblastosis, visceral lymphomatosis, osteopetrosis, neurolymphomatosis, erythroblastosis and myelocytomatosis The viruses are almost univarsally distributed in the fowl. A great many cells in the fowl are infected and these cells have the potential to produce disease ultimately.

These viruses were difficult to assay due to their low pathogenicity. This was overcome in a remarkable way by the indirect tissue culture system. Tissue cultures infected with leukosis viruses induce a reristance factor (RIF) to infection with RVS virus. The discovery has helped a deeper study of epidemiology and tumor virology.

The phenomnon of immunological tolerance is the corner stone for the understanding of the natural history of the infection. In immunologically tolerant (embryonomic) fowl infected with virus through the egg, the virus multiplies throughout its life and the virus is shed through the excretions Uninfected birds pick up infection by contact. The virus is propagated through the ovum by the infected female, but not through the sperm of the iafected male though a large number of cells in the testes are infected and produce and secrete the virus. This means that the virus genome is not chromosome associated in the male. A further favourable factor for silent as opposed to explosive communication of infection is the fact that extra-cellular antibody cannot stop with intracellu'ar repl'cation of the viral genome as in the ovum. Noncytocidal virus can persist in the body indefinitely in the presence of high antibody titres in the bird.

Other tumor viruses.

  Bittner's Mammary cancer of mice: Virus induced mammary cancer in in-bred mice was one of the most fascinating developments in virology. They occur in young close inbred mice. The incidence is dependent upon the susceptibility of the animal and effects of hormonal stimulation. In strains selected from reciprocal breeding of high tumor strain mothers and low tumor strains they found that mice born to high tumor strains had high incidence of breast cancer. In the hybrids the virus appears to be transmitted by the male. The virus elicits antibodies in heterologous hosts but not in its natural hosts. The virus can be propagated in fowl fibroblasts and other cell lines.

  Friend virus: This virus strain was isolated by Dr. Charlotte Friend. This induces a particular form of leukemia. This was isolated from a Swiss mouse that has been graf ed earlier by Ehrlich ascites carcinoma. On inoculation into 3 to 4 week old susceptib'e mice, progressive enlargement of spleen and liver are noticed after a few weeks of latency. Usually animal dies of progressive disease in which abnormal white cells infiltrate the liver, spleen and other organs. Peripheral blood shows nucleated erythrocytes, abnormal white cells of the myeloid series and advanced anemia due to the destruction of erythrocytes. Sometimes spontaneous regression takes place.

  New-born rats and mice strains resistant to the Friend virus develop a myeloid or lymphatic leukemia, when inculated with Friend virus

  It is thought that the Friend virus to be able to produce 2 types of disease, actually is a mixture of 2 distinct leukemic viruses. Although this virus belongs to the broad group of leukemias, it is different from a leukemia occurring spontaneously in mice.

  Tissue culture propagation: Most of the tumor viruses can be propagated in tissue culture. The first Papova virus, to be grown on TC was Polyoma virus in mouse embryo cells and stable mouse L. cells. The virus transforms these cells into tumor cells. The virus was quantitated by Dulbecco and Freeman (1959). The virus grows slowly and requires 10 days to form plaques.

  The precise mechanism of virus induced neoplastic transformation by oncogenic viruses is still not completely understood. But the available knowledge so far gained was possible due to the available tissue culture systems and sophisticated experiments at molecular levels.

  Some of the recent trends in these studies are homology, transcapsidation, complementarity between host and tumor virus DNA. etc.

  Familial incidence of cancer in human beings, cattle and laboratory animals exist Both vertical and horizontal transmission of cancer has been established n some cases. Chance hybridisation between a latent virus in one animal and another virus in another animal may give a new dimension and property to the hybrid virus (acquisition or loss of malignancy).

Cohnheim's cell-rest theory : During embryonic life, some immature pluripotent undifferentiated cells may be trapped in some of the organs, These may later proliferate and form tumors. This is the theory of Cohnheim which says that all neoplasms arise from such "cell rests". Though this may explain the origin of teratomata and hypernephroma, it cannot explain for the occurrence of all tumors.

In summary it may be said that *for neoplasia to develop,* two factors are necessary. The first is genetical, making the animal susceptible and the second is the exciting cause, which may be a chemical, ionising radiation, virus or a hormone. By the interaction of those two factors, cancer results.

Spread of tumors : While discussing the differences between benign and malignant tumors, mention was made that benign tumors increase in size by expansion while the malignant ones infiltrate. These may be further elaborated.

Growth by infiltration : the malignant cells, actively infiltrate into the adjacent tissue spaces of the surrounding tissue and thus spread. The invasion may be due to the following factors.

(a) Rapid multiplication : The rapidly multiplying neoplastic cells push their way into the surrounding tissue by sheer pressure of increase in numbers.

(b) Motility : Neoplastic cells especially those of mesenchymal origin are actively amoeboid. Examples of such cells are the fibroblasts, R-E-cells and meningothelial cells. Cells of hemopoietic tissue, which normally are amoeboid continue to exhibit this trait when they become malignant. Amoeboid movement has been observed in tissue cultures of such cells, Motility of neoplastic cells appears to be due to deficient cohesiveness that results by decrease in calcium content of the cell membranes.

Note :—In normal tissues and in benign tumors. the cells stick to each other because of their adhesiveness. But in cancer cells this adhesiveness is lost and so are free to move about. Adhesiveness between the cells is governed by the calcium content of the cell membrane. In cancer cells the calcium content is decreased and so they do not stick to each other.

(c) Metabolites : The metabolites that accumulate in growing tumors affect the surrounding tissues. For example. the lactic acid that is produced lowers the pH of the medium causing destruction. Invasion of such damaged tissues is easy.

(d) Enzymes ; Some malignant cells produce a hyaluronidase-like enzyme which hydrolyses the inter-cellular cement substance and so renders the infiltration of the cells easy.

Routes of invasion :

1. **(a) Infiltration into tissue spaces :** When the cells are proliferating they spread first by infiltration into the neighbouring tissues. The extent of and ease with which this infiltration occurs, depends on the tissue affected. Tissues with loose structure like the muscle and parenchymatous organs are easily infiltrated. Infiltration also occurs along natural tissue planes. Tumor cells naturally grow along lines of least resistance. Tough and compact

structures like the fibrous tissue, cartilage, bone and tendons offer considerable resistance. But no structure can remain unaffected since all tissues can be invaded ultimately. The invaded tissue is finally destroyed.

b) **Intracellular invasion** of tumor cells into muscle fibres occurs. Penetrating the sarcolemmal sheaths, the tumor cells multiply and move along the fibre, destroying its substance. The sarcolemma forms a sheath around the invading and multiplying cells.

Infiltration often involves the blood vessels and lymphatics also, resulting in metastases.

2. **Lymphatic spread**; Lymphatics are often invaded by cancerous cells. Spread through lymphatics may be in two ways:

i) **by embolism**: small clumps of neoplastic cells may form emboli which are carried with lymph. This mode of spread is more common in the larger lymphatics.

ii) **by permeation**; The tumor cells extend along the lymphatics by growing along the endothelium and lumen. This is called permeation.

In both modes of spread, the neoplastic cells reach the regional lymph node and get trapped in the cortical sinuses. Subsequently by proliferation of the cells the whole node becomes a solid mass of tumor and the lymph flow is arrested. From this node, the tumor cells enter the efferent lymphatics and thus ultimately reach the blood stream via thoracic duct or other major lymphatics. Carcinomas are usually spread by lymphatics while sarcomas are less frequently metastasized by this route

3. **Blood spread**: Tumor cells may enter blood vessels in two ways (I) via thoracic duct and (ii) infiltration into the vessel. Veins and capillaries are very frequently invaded while thick walled arteries resist. The normal contraction and expansion of organs like stomach, intestines and lungs squeeze the neoplastic cells into capillaries to form emboli.

Involvement of portal veins results in metastases into liver. If the systemic veins are involved, metastases are seen in the lungs. But there is no logic in the spread of tumors. For while liver is a common site for secondary deposits, spleen and kidneys are seldom the sites. So the volume of blood coursing through the organs is not a determining factor in the establishment of metastases. Again no logic appears to be present in the occurrence of metastases in certain tumors. For example secondary tumors of the prostate are mostly found in the bone while tumors of the lungs are found in the adrenal. The causes for such distribution are still obscure.

All tumor emboli do not develop into secondary tumors. They must have a suitable "Soil" to establish themselves, proliferate and grow. Probably the inherent nature of the cells as well as the local tissue resistance govern the development of metastases of tumor cells that are spread by blood.

4. **Transcoelomic spread** Spread of tumor cells in the body cavities is known as *transcoelomic spread* The neoplastic cells that are shed into the peritoneal cavity get implanted into the serosa covering other viscera and form tumors. For example, cancer of the stomach burrows through its wall, penetrates the

serous lining and thus the cells are shed into the peritonel cavity. These may reach ovaries or rectum and grow there.

5. **Implantation** Implantation of the cells occur in the following ways:

i) **By Natural passages:** cells of tumors growing on the lining surface of hollow oragans may be cast into the lumen and may therefore be implanted in other areas. For example cells from the carcinoma of the renal pelvis may be washed down to the bladder, where they may be implanted and form a new growth. Similarly, tumors of bronchi and bowel may be implanted in other parts of the bronchial tree or bowel respectively.

ii) **Inoculation:** This is a rare hazard in surgery. While operating on a tumor, the surgeon may inadvertently implant some neoplastic cells on the edges of the wound where a new tumor may develop or may actually trans lant it in a different part of body.

iii) **Coitus:** The venereal tumor of the dog is transmitted by this method from dog to the bitch and from the bitch to the dog. The tumor cells may be implanted from the glans penis into the vagina or vice versa and so the tumor is transmitted.

6. **Spread by nerves:** The tumor cells may permeate along the perineural lymphatics to a considerable distance The nerve itself may not be affected in the beginning but later may become compressed resulting in Wallerian degeneration.

DIAGNOSIS OF CANCER

Diagnosis of cancer in animals is of very great economic importance since treatment being costly, suffering animals may be put to sleep, relieving them of much pain, misery and suffering, thereby saving the owner from needless expense on maintenance;

1. **Clinical:** Any rapidly growing tumor which bleeds profusely and does not heal must arouse suspicion of cancer.

2, **Biopsy examination or pathological** examination of biopsy material is the most reliable method of diagnosis. Anaplasia, invasion and loss of polarity are the features by which malignancy is infered.

3. **Radiology:** Radiological diagnosis of tumors in the viscera and bone is possible. But this is of limited use in Veterinary practice, since it is more applicable to small animals.

4. **Exfoliative cytology;** Cancer cells cling together less firmly than normal cells and those arising on surfaces are easily detached and cast off—exfoliated. Such cells can be collected and by suitable techniques, can bs stained and diagnosed. This technique is widely used in human medicine for the early diagnosis of uterine and bronchiogenic cancer. The test was first introduced by Papanicolou and is known as **Papa test**

5. **Chemical and serological tests:** No reliable chemical or serolo gical tests are yet available in veterinary practice. In man, in some cases like those of cancer of prostate and sarcoma of bone enzyme estimations are helpful.

EFFECTS OF NEOPLASIA

The effects of a tumor on the body depend upon its size, location, tissue from which it is derived, presence of malignancy, presence of infection and hemorrhage. The following are the effects noticed:

1. **Pressure atrophy** of the surrounding tissue occurs as the neoplasm expands. The cells suffer as fluid exchanges are interfered with. Pressure on blood and lymph vessels may also interfere with nutrition of the cells.

2. **Obstruction :** The lumen of hollow organs may be obstructed either by a tumor projectin from the wall of an organ into its lumen or, the tumor may press from out side. The narrowing or obliteration of the lumen interferes with the passage of the contents. The following complications may arise due to tumors of :—intestiness—intussuception; urinary tract—hydronephrosis; portal vein—ascites; bronchus—collapase of lungs; b.liary tract—obstructive jaundice. The obstruction of biliary tract may be due to enlarged hepatic lymph nodes, tumor of pancreas or metastatic growths, conveyed by blood and lodged in the hepatic duct.

3 **Exudate in serous cavities :** Tumor cells deposited on serous membranes cause inflammatory exud ite with high protein content. Eg malignant ascites.

4. **Destruction of tissue, blood vessels and nerves :** Invading cells destory the tissue producing functional loss. For example, destruction of hepatic tissue by cancer of liver results in fatal hepatic failure. Anemia may develop when much of hemopoietic tissue is replaced by cells of leukemia. Invasion of blood vessel may result in hemorrhage or thrombus. Thrombosis causes ischemia and necrosis of the parts supplied by the blood vessel. Invasion of nerves is followed by their degeneration and destruction. Similarly brain tumors may produce nervous disorders or death. Renal tumors bring about renal insufficiency.

5 **Infection :** Surfa e tumors may be ulcerated and subsequently infected.

6. **Malignant cachexia :** Animals suffering from malignant tumors become emaciated and weak due to starvation when the gastro-intestinal tract is affected. Cachexia may be due to complications like frequent hemorrhages. septicemia and toxemia due to infection and also due to competition of the neoplastic cells for food. The neoplasms may utilise for themselves many essential amino acids and vitamins and so the body becomes deficient in these, resulting in cachexia. Some of the catabolic products of the cancer cells such as large quantities of lactic acid may be injurious to the host cells.

7. **Anemia :** Malignant neoplasia causes a progressive anemia, which may be due to (i) malnutrition (eg. tumor of gastro intestinal tract inteterferes with proper digestion and assimilation causing nutritional anemia). (ii) recurrent hemorrhages causing hemorrhagic anemia (iii) depression of the marrow that may be caused by :

(a) **Long standing infections :** Toxins injure the hemopoietic tissue and may cau e aplastic anemia.

(b) **Secondary (metastatic) tumors** of other neoplasms may invade the bone marrow and cause atrophy of the hemopoietic tissue.

8. **Hormonal effects :** Tumors of the endocrine glands may produce increased hormones resulting in various pathological conditions. For example, a tumor of the anterior pituitary involving acidophils causes giantism. Tumor of the parathyroid results in osteoporosis and 'big head' in horses Tumor of sertoli-cells cause feminization. Tumor of adrenal gland causes Cushing's syndrome. Arrhenoblastoma in the female causes masculinisation.

9. **Death ;** A malignant neoplasm ultimately produces death of the host.

CLASSIFICATION OF TUMORS

The simplest and best method of classification of tumors is to classify them according to the tissue of origin. For example, a tumor originating from fibrous tissue is called a fibroma.

Epithelial tumors :

Benign : Those neoplasms arising from squamous epithelium are called **papillomas** while those arising from the glandular epithelium are known as **adenomas.**

Malignant : Malignant tumors are called carcinomas—the squamous-cell carcinoma arising from squamous epithelium and adenocarcinoma arising from glandular tissue.

Non-epithelial tumors

Benign :

Fibroma for tumor arising from fibrous tissue.
Lipoma for tumor composed of fat cells.
Myoma for muscle tumor.
Angioma for tumor of blood vascular tissue.
Osteoma for bone tumor.
Chondroma for tumor of cartilage.
Neuroma for tumor of nerves.

Malignant : Addition of sarcoma to the of tissue indicates a malignant tumor arising from the tissue. For example. Fibrosarcoma, Osteosarcoma, Chondrosarcoma etc.

Teratoma : This is a tumor arising out of an embryonic defect and consists of tissue developing from all germinal layers. Usually totipotential or multipotential embryonic cells give rise to teratomas. Cohnheim's cell-rest theory adequently explains the origin of teratomas.

A benign teratoma that is cystic, may arise from the totipotential cells which differentiate into ectoderum. In this tumor the cyst is lined by skin with all its adnexae (hair, sebacious and sweat glands). So it is known as **Dermoid cyst.** Dermoid cysts are more common in the ovary.

Solid teratomas are common in the testes.

CONNECTIVE TISSUE TUMORS
Fibroma and Fibrosarcoma

Fibroma is a benign tumor composed of mature fibrous connective tissue cells These cells elaborate elastin. collagen and fibroglia. By ordinary hematoxylin and eosin staining fibroglia cannot be seen.

Depending upon the amount of collagenous fibres fibroma may be **hard** (*fibroma duram*) or **soft** *(fibroma molle)*. In the former fibres p edominate while in the latter, cells predominate and fibrils are few.

Incidence; Fibroma is a very common tumor, especially in the horse and dog. In dogs, the Boston terriers and fox terriers are more often affected.

Sites of occurrence; Fibroma can arise from any place where connective tissue is present. But more often it is found in the subcutis of the head, neck, shoulder and legs.

The size of these tumors may vary, from tiny nodules to as large as those weighing 100 lbs. It is important to differentiate fibromas from chronic inflammatory granulomata like tuberculosis and actinomycosis.

Macroscopically. the hard fibroma is round and firm, On section the surface is dry and white. The soft fibromas are softer and edematous. Section reveals a pink surface.

Nasal polyps : These are polypoid growths of the nasal passages and nasopharynx. These occur mostly in the horse and cause respiratory distresss-The tumors arise from the fibrous tissue of the submucosa and gradually push on into the lumen of the nose. These are smooth, glistening and ovoid masses and are soft and slimy to the touch, being covered by nasal mucosa.

Microscopically. fibroma consists of interlacing bundles of fibrous connective tissue, which run in all directions. The nuclei of the fibroblasts are spindle shaped. Blood vessels and variable number of lymphocytes, monocytes and eosinophils are seen.

Neurofibromata arise from the fibroblasts of the connective tissue of the nerve sheaths. They are usually multiple and firm in consistency. These are mostly located subcutaneously and in man occur along the nerve trunks and is known as Recklinghausen's disease. Neurofibromas are well en-capsulated and are now supposed to be viral in origin.

Microscopically. the cells are long and spindle shaped with oval or oblong nuclei. They are arranged in whorls.

Diagnosis : Fibromas must be differentiated from keloids, which have the following characteristi s, (i) The keloids are composed of a highly vascular granulatiou tissue. (ii) They arise as a result of trauma (iii) They are infiltrated by many neutrophils. and (iv) they have large bundles of collagen.

Fibromas can be differentiated from fibrosarcomas in that the latter are characterised by (i) immature plump and rapidly growing fibroblasts which show frequently mitotic figures, (ii) having minimal amount of collagen; (iii) infiltration into the surrounding tissue and (iv) recurring after removal.

Clinical : The fibromas can be easily removed. Bleeding may be a problem while removing a large tumor.

Combinations : Lipofibroma, chondrofibroma, osteofibroma and myxofibroma may be encountered.

Fibrosarcoma : This is the malignant tumor arising from the fibrous connective tissue. The tumors may arise from places where fibromas may occur.

Fibrosarcoma occurs in horses, cattle, dogs and fowls. Though malignant, metastasis to distant organs is not a feature of fibrosarcoma But it will recur with greater speed after removal

Macroscopically, fibrosarcomas are irregular in shape and have a nodular appearance. They may be circumscribed occasionally. These tumors have abundant blood supply. Those found on the body ulcerate and emit a foul smell.

Microscopically, arrangement of cells is similar to that of a fibroma but the cells show features of malignancy like darkly staining nuclei Mitoses are rare. Malignant giant cells may also be seen Plentiful blood supply is found

Equine sarcoid, Jackson introduced the word "Sarcoid" to certain cutaneous tumors of the horses, mules and donkeys, which have the structure of a fibrosarcoma but with limited malignancy This tumor is believed to be caused by a virus which is probably similar to the virus causing cutaneous papillomatosis of the bovine and the virus of equine pox. The tumors are frequently multiple and are more often seen on the head, lower parts of legs and the prepuce.

Macroscopically, the growths are variable in size, attaining a size of man's fist. They may be pedunculated, or have a broad base They are not encapsulated. Being exposed, these tumors are subjected to trauma and ulceration with infection supervening.

Microscopically, the tumor consists primarily of spindle shaped fibroblasts with varying amounts of collagen fibres. These are arranged as bundles and as whorls (resembling the structure of a neurofibroma). The fibroblasts have elongated nuclei. Mitotic figures are seen but are fewer than in a fibrosarcoma. Eosinophils and lymphocytes are found.

In some places, the epidermis over the tumor may be ulcerated If intact, it may be acanthotic Care should be taken not to confuse sarcoid with a keloid. The latter is purely inflammatory in origin and so its collagen content, number of inflammatory cells and blood vessels are far greater than in a sarcoid.

Again sarcoid may be confused with a fibrosarcoma. Lack of anaplasia in and fewer mitotic figures of a sarcoid help in differentiation.

Malignancy : Sarcoid is locally malignant. It recurs after excision. But it seldom metastasizes.

Shope fibroma: Shope in 1932 had found that a fibroma in wild rabbits was caused by a virus. Domestic rabbits can also be infected by this tumor and the infectious agent is related to the virus causing infectious myxomatosis.

Tumors are located subcutaneously, singly or more often as multiple nodules. The tumor consists of spindle shaped fibroblasts. In nature, transmission occurs by means of mosquitoes, which mechanically carry the virus from the infected to the healthy. There is no multiplication of the virus in the vector. Affected rabbits develop resistance to infection by virus of infectious myxomatosis.

Myxoma and Myxosarcoma

Myxoma is a tumor of specialised fibrous tissue that is capable of producing mucin. Normally it is embryonic tissue that produces mucin and so mucin-producing cells are anaplastic and hence denote malignancy.

Myxoblastic change may be found associated with other tumors, giving rise to lipomyxoma, chondromyxoma and fibromyxoma. The change is supposed to be due to metaplasia. The myxomas are more frequently found in the subcutaneous, subserous and submucous tissues. Among animals this tumor must be considered to be a rare one. It may be seen in the heart and uterus of cows.

Macroscopically: the tumors are rounded and appear as a bunch of grapes. They may be encapsulated and are slimy to the touch because o mucin content.

Microscopically, the tumor consists of cells which are spindle shaped or stellate with long processes and lying in a basophilic mucinous matrix. Pleomorphism is evident and in the malignant variety mitotic figures are common.

Clinically, the tumor can be easily removed. Malignant variety may recur.

Myoxomatosis of the rabbit: This is a viral disease and affects the domestic and wild rabbits. Though mild in the latter, it is highly fatal in the domestic rabbits. In fact this condition has wiped out the entire rabbit population in certain areas.

Infection can be direct or through flies and mosquitoes. Incubation period is 3 to 11 days. The virus appears to be related to the pox - group of viruses.

Sites: The lesions are found on the lips, round the nose, the eyelids and on the skin of other parts of the body.

Macroscopically, the tumors are soft and slimy, white or pink in color. On section clear serous fluid is found to exude. Edema of lungs and bronchopneumonia may sometimes be seen. Blepharo-conjunctivitis is almost always seen. The regional lymph nodes and spleen are swollen.

Microscopically, the tumor consists of fibroblasts, reticular endothelial cells and histiocytes. The connective tissue spaces are filled with serous fluid that is rich in hyaluronic acid. Neutrophils and eosinophils infiltrate into the tissue. The stellate *myxoma cells* are supposed to be hypertrophied histiocytes which contain intracytoplasmic acidophilic inclusions. Secondary changes that occur in the histocytes and leucocytes are vacuolar degeneration, lysis, pyknosis and karyorrhexis. The epithelium overlying the myxomatous nodules shows vacuolar degeneration. Due to circulatory disturbances, the covering epithelium may be destroyed with ulcer formation. Intracytoplasmic acidophilic inclusions may be found in the epithelial cells of the epidermis also.

The course of the disease in the rabbit is 7 to 12 days. Immunity develop in those that survive.

Lipoma and Liposarcoma

Lipoma is a tumor of fat cells and so is ubiquitous. It is seen in tissues of subcutis, subserosa, mesentery and submucosa.

Incidence: Lipomas may occur singly or rarely, numerous tumors may be encountered. Lipomas are fairly common among animals, especially aged ones. Lipcsarcoma is very rare.

Macroscopically; the lipomas range in size from small nodules to big masses (one tumor in a horse weighed 120 lbs.) They are spherical and may be lobulated. They may be sessile or pedunculated and may be firm or soft. Capsule may not be evident and the tumor may merge with the normal fat tissue. On section the surface is oily and translucent and may be yellow in color. Sometimes calcification may take place. Bone may be formed by metaplasia.

Microscopically, the tumor consists of closely packed cells. The cells may be polyhedral and contain a single large fat globule or several small ones. The nucleus may be pushed to a side. Variable amount of connective tissue, divides the tumor into lobules and acts as a scaffold for blood supply. The malignant tumor is more cellular and consists of anaplastic cells with little fat.

Diagnosis: Care should be taken not to confuse fat necrosis with lipoma. In the former is found abundant infiltration by inflammatory cells. Fat-laden macrophages may be erroneously identified as fat cells.

Clinical: Lipomas may be removed easily. Sometimes the long pedunculated lipomas that may be found in the abdomen may entwine around the intestines causing strangulation. Liposarcoma does not metastasize usually.

Chondroma and Chondrosarcoma

Chondroma; is derived from cartilage cells Cartilage is formed by the proliferation of fibroblasts of the peri-chondrium and their subsequent differentiation into cartilage cells. The cartilage cells by themselves do not proliferate. Usually chondromas arise from places where there is cartilage normally. Tumors that develop on the surface of the cartilage and project under the periosteum of the bone are known as Ecchondromas, while those that remain in the interior or substance of a cartilage or bone are known as Enchondromas.

Incidence: Chondromas are one of rare tumors of animals. Dogs are the most commonly affected among animals and in these mammary gland is the most common site. The epiphyses of the long bones, the chondro - costal and chondro sternal articulations as well as the bronchi, trachea and larynx are the places mostly affected.

Macroscopically, chondromas are large in size and may be multinodular. Majority are encapsulated and may have rounded contour. They are bluish-white and on section are of translucent appearance. Some may show cysts due to degeneration. Foci of calcification may be present.

Microscopically, the chondromas may show a very cellular appearance, consisting of round or ovoid cells set in a bluish matrix. The cells are arranged singly but not in groups of four or eight as in the normal cartilage. Strands of fibrous tissue separate the tumor into lobules.

The Chondrosarcoma: is more cellular and its cells often show pleomorphism. At the periphery may be seen the immature spindle - shaped cells whilat the centre may be found fully differentiated cartilage cells. Mitosis is frequent. Hyperplastic nuclei contain many nucleoli. Blood supply is found in the connective tissue only but none in the cartilage. So large areas of the tumor may become necrosed and calcified.

Metastasis of the malignant tumors occurs through blood stream. Lungs are the frequent sites of secondary growths.

Osteoma and Osteosarcoma

Osteomas are hard tumors composed of bone and are comparatively rare in animals. They are frequently found on the head. They are usually nodular and encapsulated. Depending on its hardness, osteoma is divided into two varieties.

1. **The osteoma eburneum** or compact osteoma : This is found more often in the skull, scapulae and pelvic bones.

2. **The osteama spongiosum** or spongy osteoma : These are more voluminous than the previous variety and are found near the ends of long bones like the humerus, femur etc.

Microscopically, the structure of a bone with lamellar arrangement is noticed. Sometimes Heversian canals may also be seen. In some cases it is difficult to differentiate osteomas from exostoses, which arise from the periosteum of bones (especially those of extremities) due to inflammation and degeneration. Sometimes bone may be formed in muscles, heart and lungs by metaplasia of fibrous tissue in these places, This must be differentiated from osteoma.

Osteophytes are bony outgrowths found in the neighbourhood of a joint. **Combinations :** Osteofibroma, Osteomyxoma, Osteochondroma.

Osteosarcoma : This is also called osteogenic sarcoma. This is a highly malignant tumor and arises in the same sites where from osteomas may originate. Dog is probably the animal mostly affected by this tumor. Large breeds of dogs are more often affected.

Sites : The limbs are the sites of predilection.

Macroscopically, the tumors may be round or ovoid in shape and may replace much of the normal bone. The marrow cavity may be invaded and occupied. The tumors may be hard or soft, containing spicules of bone and cartilage They may be white, yellow or pink in color.

Microscopically, the tumor is very cellular. The neoplastic cells are plemorphic being round, spindle and polyhedral. The cells are arranged in compact masses. Tumor giant cells are present. The tumor cells possess hyperchromatic nuclei each with a single nucleolus. Numerous mitotic figures are present. Spicules of new bone may be encountered. Numerous thin-walled blood vessels are present Metastases are found in the lung mostly and in other tissues infrequently. Spread is commonly by blood stream

Diagnosis : X-ray examination reveals bone destruction and new bone formation having thick opaque shadows which indicate the tumors

Giant cell tumors : These are also known as osteoclastomas and are believed to arise from the osteoclasts. These tumors are not common among animals. As the tumor grows by expansion, absorption of the bone takes place leaving a thin cortex   When handled such bone gives an "egg-shell crackling" sensation.   Giant cell tumors are slow growing and do not metastasize.

Microscopically, numerous giant cells similar to osteoclasts are found amidst a ground substance containing oval spindle cells.   Arrangement   of the nuc'ci of giant cells are grouped in   the centre (not peripherally as in Langhan's type).

Leiomyoma and Leiomyosarcoma

Leiomyoma is common in   cow, dog and fowl. This is benign.

The tumor arises from smooth muscle wherever it is   found,   especially in   the   hollow organs—the uterus, vagina, intestines, stomach, urinary bladder and esophagus.   Leiomyomas are common in the broad   ligament   of the oviduct in the fowl.

Macroscopical'y,   the   benign tumors are smaller.   They are usually attached by a broad base though   pedunculated   var eties may be seen.   Benign tumors   are   encapsulated.   They   are   firm and lobulated and on section are pink in color.   Sometimes necrosis with calcification may be seen.

Microscopically,   the   tumor   consists of plain   muscle   bundles arranged   in   all   directions   and   planes.   These bundles   are separated by strands of fibrous tissue in which plentiful blood supply is present,   The   muscle fibres   are   spindle shaped and are arring d parallel to each other.   They have a ribbon-shaped   nucleus with rounded ends (cigar shaped).   Sometimes   short and fatter cells   may be found.   Nuclei contain a filamentous chromatin and small round or rod shared basophilic nucleoli.   van Gieson's method stains muscle yellow and white fibrous tissue white.

Leiomyosarcoma is a rare malignant counterpart.   It is more cellular and   the ce ls are anaplastic.   The cells are therefore sh rter and plumper with thicker chromatin.   Mitotic figures are frequent.   Numerous blood   vessels are found.   The tumors being invasive replace the adjacent normal tissue.   It meta-stasizes into different parts of the body with fatal results.

Rhabdomyoma and Rhabdomyosarcoma

These tumors arise from the striated muscles (skeletal and   cardiac) and are rare among animals.   Even the few reported were found mostly   in the heart. Other locations are the lungs, the tongue, sternal and neck regions.

Microscopically, the tumor consists of striated   muscle cells, which lie in different   directions.   Fibrous tissue septa may   divide the muscle bundles into lobules.   The cells are pleomorphic—some being polyhedral   while   others   are spindle shaped.   Giant cells with many   nuclei   are seen.   Cross striations are brought out by iron hematoxylin.

Angioma and angiosarcoma

Tumors composed of blood vessels are called **hemangiomas** while   those constituted of lym h vessels are called **lym hangiomas.**

Hemangioma: In animals, this is not common. Care should be taken not to confuse hemangiomas with hamartomas, which are developmental defects of the blood vascular system. This is very well exemplified by the red birth mark in man. Hamartomas consist of a collection of capillaries with a capsule. These do not grow progressively and are often multiple Sometimes these hamartomas may be enlarged and manifest hemorrhagic inflammation, edema, thrombosis and cystic changes. These changes make them appear to resemble neoplasms. Cutaneous hamartomas are particularly common in White Leghorn fowls

Hemangioma may also be confused with telangiectasis observed in dogs, cattle and cats In the latter condition, the sinusoids of the liver are widely dilated. The dilatation is always preceded by atrophy, necrosis and disappearance o hepatic cells. Since support is lacking. the sinusoids dilate Such areas of vasculara dilatation are seen extensively in the liver and the endothelium lining the dilated spaces is of the adult type. These two factors help in distinguishing it from neoplasm which usually occurs only in one place of the organ and in which the endothelium is more embryonic.

In inflammatory conditions, a rich supply of blood vessels may be present in certain areas, resembling an angioma But such lesions contain abundant connective tissue and cellular exudate.

Incidence: Hemangiomas have been described in cattle, horses, sheep, dogs, swine and fowls. Liver is more often affected Spleen, limbs, perineal region, thorax and vagina are the other parts affected Depending upon the relative presence of blood spaces and cellular tissue, hemangiomas are divided into three varieties.

1. Hemangioma hypertrophicum or solid hemangioma in which blood spaces are minimal but the cellular component preodminates.

2. Hemangioma cavernosum or cavernous hemangioma in which widely dilated blood spaces are seen.

3. Hemangioma simplex or capillary hemangioma in which the capillaries are of uniform size and in which there is minimal cellular tissue.

Macroscopically, hemangiomas vary in size. Some may weigh as much as 21 kg They are usually single but may also be multiple They are dark red to purple in color and soft in consistency. These bleed when injured

Microscopically, depending upon the kind of tumor, capillaries or blood filled spaces may be seen Usually the capillaries and spaces are lined by a single layer of endothelial cells. Sometimes there may be more layers. In others the lumen may be filled completely with the proliferated endothelial cells only The cells show various shapes—oval, polyhedral, flattened or spindle shapes The cytoplasm is clear and stains blue. The nuclei may be large and oval and contain granular chromatin.

In the malignant variety called **angiosarcoma**, the cells are anaplastic pleomorphic and polyhedral and are grouped into masses. Numerous mitotic figures are seen.

The tumor cells may sometimes resemble fibroblasts Hemorrhages are frequent and metastases are found in various parts of the body.

Clinical: Hemangiomas found externally can be removed successfully. But the malignant neoplasm is always fatal.

Hemangiopericytoma: Blood vessels have long branching cells outside their sheaths, called pericytes of Zimmermann. These are contractile but differ from muscle cells in not being able to form myofibrils. Formerly, these tumors were diagnosed as neurofibroma, neurofibrosarcoma, neurilemnoma, fibroma, fibrosarcoma, leiomyoma, hemangioendothelioma or leiomyosarcoma. The confusion arises due to resemblance of the pericytes to the cells constituting above tumors. The distinguishing feature is that in pericytoma, the spindle shaped cells are arranged always closely associated with blood-vessels.

Incidence: This tumor is found mostly in dogs though it is also reported in cows

Site Hemangiopericytoma is found subcutaneously in the trunk and limbs mostly and less often in other parts of the body.

Macroscopically, the tumors vary in size They may be encapsulated.

Microscopically, the appearance is characteristic The neoplastic cells are arranged concentrically around smaller arterioles or capillaries, giving a "finger print" appearance The cells are spindle shaped with an ovoid nucleus which has a coarsely granular nuclear membrane and a finely-meshed chromatin. Mitotic figures may be numerous. These cells are intimately associated with the vascular spaces. This association may be clearly brought out by special staining for reticulum. Sometimes the lumen of the capillaries may be occluded by sclerosis

Clinically, these tumors grow slowly and can be removed. They may recur in a few cases. Rarely metastases may be found in the internal organs.

Lymphangioma is a tumor arising from lymphatic endothelial cells and is extremely rare in animals. These tumors have been reported in cattle, dogs, horses and mules. Mostly, lymphangiomas are found subcutaneously but may also be found on the pleura, pericardium and diaphragm. These are usually encapsulated and may be lobulated.

Microscopically, the tumor consists of cavernous or cystic spaces filled with a clear fluid. Neoplastic cells resemble endothelial cells.

Mesothelioma

This is a tumor arising from the mesothelial lining cells of serous cavities, especially the peritoneum and the pleura. The mesothelial cells are derived from mesodermal elements lining the coelomic cavity and hence differ from the epithelial and endothelial cells which they closely resemble

These mesotheliomas are very rare among animals They have been reported in the ox, horse and dog. These occur in very young and new-born animals also.

Sites: Mesotheliomas are found in the thorax and abdomen arising from the cells lining the serous cavities.

Macroscopically, these tumors are pink in color and are firm. They are always nodular and multiple, scattered widely in the body cavity. A capsule may not be present.

Microscopically, the tumor consists of diffuse collections of cells resembling epithelial cells. A core of fibrous tissue is present, sometimes to such a degree as to resemble a fibro-sarcoma. The mesothelial cells have acidophilic granular cytoplasm and a big vesicular nucleus. Nucleoli are prominent. Numerous blood vessels traverse within the tumor mass. In some areas the cells may be arranged in large sheets resembling a carcinoma. Sometimes a papillary adenocarcinomatous appearance is met with. Mitotic figures may be seen. Large number of eosinophils may be present. Occasionally necrosis with calcification may be found. The tumor does not invade the blood or lymph vessels. Multiplicity or the tumor is due to transplantation or multicentric origin.

Diagnosis The tumor may be confused for tuberculosis, (Pearly disease), actinobacillosis, implantations of carcinomas, or lymphangioma.

Mastocytoma

Mast cells are found around blood vessels in the connective tissue and liver. These have no connection with the mast cells (basophils) of the blood. Tissue mast cells have a non lobated nucleus.

Tumors comprising of tissue mast cells are found commonly in the dog and rarely in cats, horses and cattle. In the dog it occurs after 6 years of age.

Sites : Mastocytoma is always found subcutaneously. The seats of occurrence are the hind limbs, groin, perineal region, scrotum, abdomen, vulvas mammary gland and gluteal region.

Macroscopically, mastocytomas are usually single, but may be multiple. They are usually small but may attain a size of 13 cm. in diameter. These tumors may be nodular and some have a peduncle. Others may imperceptibly merge with the fibrous tissue into which they may grow. The tumors may be hard and light pink or gray on section. Ulceration of the skin over the tumor is frequent. Capsule is not present. Metastasis may occur.

Microscopically, the size and shape of the cells vary according to the stage of development. In more mature forms, the cells are found with faint granular acidophilic cytoplasm and relatively small rounded nucleus. A distinct perinuclear membrane is present. Large basophilic granules which may obscure the nucleus are present.

In the immature or anaplastic form the cells are polyhedral or even angular with a faintly basophilic cytoplasm and relatively large nucleus. Mitotic figures are rare. These cells may not have the cytoplasmic granules. Eosinophils may be found scattered among the cells. Their presence is probably due to their antihistaminic nature. Neutrophils, plasma cells and lymphocytes may also be found.

Bundles of collagen may divide the tumor into lobules. Staining by Toludine blue brings out the metachromatic cytoplasmic granules clearly. This tumor has to be differentiated from the transmissible venereal tumor. The differentiating features are (a) uniform size and shape of the cells of the venereal tumor and (b) large number of mitotic figures found in it.

Mastocytoma may also be confused with a lymphosarcoma. The cells of the mastocytoma have the following distinguishing features:—(a) The nuclie are more condensed and darkly staining; (b) the cytoplasm of the cells is relatively more than in the lymphoid cells.

Clinical : mastocytoma can be removed surgically. Sometimes they may recur.

NERVOUS TISSUE TUMORS
Schwannoma

This is also known as neurilemmoma and originates from the cells of sheath of Schwann or lemmocytes of the peripheral nerves. Among animals these tumors are more common in the ox, though they were seen in other animals also.

These tumors are usually small in size, a few centimeters in diameter and are situated in the course of nerve fibres. They are almost always found internally

Microscopically, two distinct patterns of arrangement of the cells are noticed.

1   The Antony type A ; – In this type. the Schwann cels are elongated and spindle shaped with oval or cylindrical nuclei, and the fibres form interlacing bundles. The cells are arranged parallel to each other. A distinct pallisade arrangement is obtained by the alternate arrangement of the nuclei of cells and the fibres (which are nuclei free), This pallisade arrangement may take a whorl form also when they are called *varocay bodies* The fibres are derived from the neurilemma cells and are similar to collagen fibres but are distinct from them.

2.   The Antony type B –In this variety cells of varying shades are arranged in a disorderly loose manner.   Often inter-cellular vacuoles containing a watery fluid may be seen which stains blue with hematoxylin. In this type fibres are not seen

In the same tumor, the above two varieties may be found together, side by side, clearly demarcated, but never intermingling.

Effect : Since nerves are affected pressure symptoms and paralysis of the respective regions may be seen.

Clinical : Schwannomas are benign.

Meningioma

This tumor otherwise known as arachnoid fibroblastoma arises from the arachnoid fibroblasts of the brain and spinal cord. This tumor has been reported mostly in the dog but horse. cattle and cats were also found to be affected

Meningioma occurs single and by expansion causes pressure on the brain. Metastases are very rare.

Macroscopically. the tumor is white. lobulated, and encapsulated.

Microscopically. meningioma consists of spindle cells of uniform size and shape, They have elongated, oval nuclei.   The cells are arranged in whorls as in a neurofibroma but the cells of the latter are more slender than those of the former and the nuclei are more elongated. In more anaplastic type of tumors, mitotic figures may be seen. Necrosis, edema.

hyalinisation and calcification may be encountered in certain areas. Plentiful blood supply may be seen. Considerable fibrous connective tissue may be present. Besides the whorled form mentioned above the following varieties are described in tumors of man. The author encountered one such case in a bullock.

(i) **Epithelioid form**: In this type, cells resembling epithelial cells (polyhedral cells) are found in sheets or psuedoalveoli amidst vascular connective tissue.

(ii) **Psammoma form**: in this form of growths bluish calcified bodies, calcospherules, suggestive of grains of sand are dispersed in the substance of the tumor. These grains may be found in the centre of the whorls or between them. They consist of concentric laminae. Calcification of the tumor cells and the walls of small blood vessels occurs. The cells in the whorls may be replaced by hyalinised collagen, which is subsequently calcified thus giving rise to the calcospherules.

(iii) **Fbrous form**: In this variety dense fibro-collagenous tissue with or without whorl formation or sand grains is found. This structure resembles a fibroma.

(iv) **Ossified form**: In some parts of the tumor ossification with hemopoieitic marrow may be seen.

(v) **Angiomatoid form** In these tumors a rich supply of thin-walled blood vessels is present to merit its being called an angiomatoid form

(iv) **Sarcomatous form**: The growth in this category is highly cellular and anaplastic without whorl formation.

It should be noted that all the above forms may be seen in the same tumor but are not distinct entities.

Gliomas are tumors of glial tissue:

Those arising from artrocytes are called astrocytomas, while those from the oligodendroglia, oligodendrogliomas.

Gliomas are rare among animals. Among those reported, majority are astrocytomas. Most of these tumors were reported in the dog.

Astrocytoma: Besides dog, the tumors occur in the fowl, cattle, cat and horse.

Sites: Astrocytomas are mostly found in the carebrum and cerebellum though other parts of the brain may also be affected. It may occur in the spinal cord also. In all animals except the fowl, the tumor is single but in the fowl it is multiple due to multicentric origin.

Macroscopically, the tumors are soft and circumscribed. Some may imperceptably merge with the brain tissue In the fowl, these tumors are multilobular and cystic, containing a mucoid substance. Due to lipofuscin, astrocytomas are yellow in color.

Microscopically, astrocytes being of two forms either fibrillary or protoplasmic, the structure of an astrocytoma varies according to type of cell predominating. The cells may be enlarged. All forms of intermediary types are seen. These is pleomorphism, giant nuclei and multinucleated giant cells. Sometimes the cells may be arranged radially around blood vessels.

In more malignant types, the cells are highly pleomorphic, show many mitotic forms and also a few giant cells.   Such tumors are also known as *glioblastoma multiforme*.

Oligodendroglioma: These are very rare in animals.

Microscopically, the tumor cells have a faintly staining, indistinct cytoplasm containing small, regular, round and hyperchromatic nuclei.

Ependymoma, is a tumor composed of ependymal cells which line the ventricles, central canal of the spinal cord and the choroid plexus.   These tumors have been reported in the horses, cattle, and dogs

Sites: This tumor arises mostly from the fourth ventricles and central canal of the spinal cord.

Macroscopically, the tumor may be cystic, solid or papillary. It may be gray and soft.

Microscopically. the structure is variable.   The cells may be arranged in cords or may show a papillary pattern. The cells are cuboidal or columnar in the papillary form or uniform oval shaped with indistinct cell - membranes in the solid variety.   Pseudorosettes consisting of the ependymal cells arranged in a circular fashion with a clear central space are frequently seen.

Mitotic figures may be seen in more anaplastic type of tumors.

Neuroblastoma and Ganglioneuroma

Neuroblastoma is a "tumor consisting of immature undifferentiated neuroblasts", while ganglioneuroma denotes a tumor "consisting of well differentiated nerve cells and fibres" The former is a rapidly growing and malignant one while the latter is benign.   All gradations between these two varieties may be encountered.

Incidence: These tumors are extremely rare.   A few have been reported in cattle, horse, dog and fowl.   Young animals are more often affected.

Sites: Though arising from nervous tissues, these tumors are very very are in the Central Nervous System.   They are more commonly found in the adrenal medulla (where it is called *(sympathoblastoma)* and in the sympathetic ganglia of the abdominal and pelvic regions. The cervical and thoracic ganglia may also be affected.

Macroscopically: the tumors which are encapsulated vary in size. They may be white or gray in color. They are fibrous in consistency.

Microscopically; the neuroblastomas are highly cellular and consist of undifferentiated small, round, cells.   Large number of mitotic figures are present. One may find elongated cells forming "rosettes" around a central core of nerve fibres

The ganglioneuroma consists of more mature nerve cells and their fibres.   The cells show Nissl substance.   Mitoses are not seen.

Aortic body tumor

This is also known as Heart Base tumor.

Normally, the adventitia of the aorta and the carotid arteries contain a cluster of chemoreceptor cells (aortic body and caortid body respectively) that unction as chemoreceptors, which are sensitive to the carbon-di-oxide and oxygen tension and the pH of the blood. Respiration and circulation are thus regulated by these. Tumors may arise from these cells.

Incidence : These tumors are extremely rare among animals. Only in old dogs have these been reported.

Site : The tumor is located at the base of the heart, between the aorta and pulmonary artery, or sometimes encircling both (hence the name Heart Baes tumor) The rare carotid body tumor may be found at the bifurcation of the carotid arteries.

Microscopically, the tumors are reddish-brown in color and rather firm in consistency. They are encapsulated.

Microscopically, tumor cells are polyhedral in shape Cytoplasm is vacuolated or granular and acidophilic with a spherical nucleus. Connective tissue strands divide the tumor into lobules. The cells are highly invasive infiltrating into the media of the blood vessels and into lymphatics.

No endocrine function is associated with these tumors.

Tumors of the Hemopoietic tissues

Tumors of lymphoid tissues :

Lymphoid tissue comprises of cells of the lymphoid series and the histiocytic lining cells of the sinuses. The origin of these components is the primitive mesenchymal stem cell. So, neoplasms of lymphoid tissue may contain cells of these component parts in varying numbers. Hence it is that tumors comprising mostly of lining cells or lymphocytes may be encountered. Again there may be tumors that are comprised of cells from both the component parts in varying proportions (Hodgkin's Disease) It should be borne in mind that these several types of tumors are manifestations of a single entity.

Tumors of lymphoid tissue are characterised by the appearance of neoplasms in several lymph nodes and spleen at the same time. This is not due to metastasis but due to multicentric origin This makes one suspect a virus being the etiological agent, which becoming viremic affects several nodes at the same time. The obliteration of the normal structure of the nodes—follicles, red and white pulp, sinusoids etc.—by the proliferating cells is yet another argument against the contention that metastasis may be the cause for several nodes to be affected simultaneously (Multicentric).

Tumors of the lymphoid tissue are always malignant. The benign entity called **lymphoma** if such exists, is difficult to be diagnosed since hyperplasia and inflammatory conditions of lymphoid tissue resemble it. The malignant tumor has been called lymphosarcoma, lymphocytoma, leukemia, reticulum cell sarcoma, Hodgkin's Disease etc. To avoid confusion, the malignant variety is called lymphosarcoma here.

Lymphosarcoma is a very common tumor of the dog, cat, cattle and pigs. It is not so common in the horse and sheep.

Sites : The tumor primarily affects the lymph nodes wherever they are found—the peripheral and visceral lymph nodes. In the ox the walls of the forestomachs are frequently affected. Metastases may occur in any organ. No organ is immune.

Macroscopically, the affected lymph glands are enlarged. These do not exhibit the color contrast between the cortex and medulla, being diffusely pale and soft. The spleen is enlarged enormously and is very soft and friable. It is easily ruptured. Similarly liver is very much enlarged, pale, tense and easily ruptured. Bone marrow is replaced by the tumor cells and so appears pale. The metastases that occur in other organs are nodular and white. Peyer's patches stand out conspicuously.

Microscopically, the picture varies with type of cell involved that predominates.

In the lymphocytic form the neoplastic cells are the lymphocytes. These may either be normal lymphocytes or immature forms of it. The immature cell (lymphoblast) is large and polyhedral with the large nucleus placed eccentrically. The cytoplasm stains blue. Mitotic figures are present.

In the histiocytic type called the *reticulum-cell sarcoma* the cells are larger and pleomorphic with a vesicular nucleus which may show indentation or may be triangular. Mitotic figures are frequent. Reed-Sternberg giant cells may be present.

In the Hodgkin's disease type, which is supposed to be rare in animals, there is an admixture of lymphoid cells, reticulum cells, fibroblasts and eosinophils. The characteristic cell to be seen is the Reed-Sternberg giant cell with two "mirror-image" nuclei.

The neoplastic cells replace the normal tissue and so the architecture of lymph nodes is lost. In the heart, these cells may be intercellular, infiltrating between the muscle fibres, which may show degenerative changes. Some fibres may disappear finally When there is excessive proliferation of the tumor cells, they may be found in the peripheral blood, giving rise to the condition *leukemia*. In the liver, infiltration may be nodular or it may be a generalised condition. First infiltration is noticed around the portal triads. The parenchyma is slowly replaced and the remaining hepatic cells show fatty degeneration and even necrosis.

In the spleen the splenic tissue disappears gradually. In the kidney, the cells invade the interstitial spaces, crowding the renal tubules, the epithelium of which shows, therefore, varying stages of degeneration and atrophy. Utimately the tubules may entirely be replaced.

In the stomach and intestines, the infiltration may be at first, submucous, slowly extending beyond into the muscular coat externally and mucosa internally. The tumor cells separate widely the neurofibrils when they invade a nerve. In the brain, focal nodular accumulations may occur with destruction of the neurones.

Hematology : In some animals there will be leucocytosis. As bone marrow is invaded by malignant lymphocytes pressure atrophy, myelophthisic anemia and decrease in granulocytes may occur. The number of white blood cells may exceed 20,000 to 30,000 per cmm. in the final stages In some cases these may exceed 5000,000. This condition is known as *leukemic leukemia*. Conditions where there is no rise in blood leucocytes. are known as *aleukemic leukemia* or *aleukemic lymphosarcoma*. In the blood may be seen mature lymphocytes or immature lymphoblasts which have a large cellular body with an intensely bluish staining nongranular cytoplasm and a large round bluish, slightly vesicular nucleus which is eccentrically placed The haemoglobin content of the blood is reduced. Nucleated red blood cells may be seen. The coagulation of the blood is slowed.

Myeloid tumors: Myelogenous leukemia: Tumors arising from cells of the myeloid tissue are not common among animals. These tumors arise from the precursors of the granulocytes – the myelocyte or myeloblast. So a tumor of this blast cell may contain—neutrophils, eosinophils or basophils.

As myeloid tumors originate in the bone marrow, they are found there primarily. Metastases are seen in the liver, spleen, kidney and lymph glands.

Myeloid tumors are encountered in the dog, pig cattle, cat and sheep.

Macroscopically, the myeloid tumors are greenish in color. So they were called chloroma by earlier pathologists. The color fades on exposure to air but it can be brought back by applying hydrogen peroxide to the tumor. Liver and spleen may be very much enlarged, friable and somewhat paler than normal. Greyish spots and streaks may be seen on the liver.

Microscopically, the picture varies in different tumors depending on the stage of development of the cells. All grades of development from the primitive myeloblast to the well differentiated myelocytes are seen. The myelocytes are large, polyhedral or round cells with prominent rounded or oval eccentrically situated vesicular nuclei, which sometimes may be indented. The nuclear chromatin may be coarse The cytoplasm may be fairly acidophilic or neutrophilic. In more mature cells, the lobulated nucleus and characteristic cytoplasmic granules are seen. Mitotic figures are frequent. Megakaryocytes may be encountered.

In the liver, the neoplastic cells may virtually replace the hepatic parenchyma which may show fibrosis. Hepatic cells that remain show fatty degeneration. In the spleen the pulp is replaced by the neoplastic tissue. Hyaline changes of the splenic tissue may be observed.

Sometimes the tumor cell may be a plasma cell with bluish cytoplasm and an eccentric nucleus in which the chromatin granules are grouped at the periphery giving it a "clock-dial" appearance Such a tumor is called a *plasma cell myeloma*. In man this affects many long and flat bones and is known as *multiple myeloma* (Ewing's tumor). In this condition a peculiar protein, Bence-Jones protein, may be present in the urine.

Hematology The total white cell counts increase abnormally, reaching 30,000 to 40,000 per cmm. Due to pressure of the neoplastic cells on the er thropoietic tissue, anemic changes manifested by anisocytosis and poikilocytosis are noticed

In the blood smears, granulocytes number 90 to 95% of the total leucocytes. Myelocytes may constitute 5 to 10%.

Avian Leucosis Complex

The condition known as Avian Leucosis Complex was formerly described as one entity, since it was thought that the several forms in which it was manifested had the same etiology. But now it has been established that the condition could be broadly divided into two groups : Leucosis and Marek's Disease, and these are caused by different viruses.

Leucosis

This is a neoplastic disease affecting fowls of all ages, but it manifests more frequently in fowls of ages between three and eight months. Caused by a virus of RNA group, the disease spreads by contact. The virus may pass through the egg to the chick. Transmission by ectoparasites (ticks, lice,mites) is suspected though not proved. Another source of infection is the fowl pox vaccine that may be prepared from lesions harvested from fowls suffering from avian leucosis.

The disease manifests in various forms, and these were formerly known under the following names:— visceral lymphomatosis ; fowl leukemia ; big liver disease ; fowl leucosis ; lymphatic leucosis ; range paralysis ; lymphadenoma ; fowl paralysis ;   lymphomatosis ; polyneuritis : leucotic tumors ;   gray-eye ; pearl eye etc.

Leucosis is now divided into four forms. **1.** The visceral lymphomatosis or lymphoid leucosis, **2.** Osteopetrotic form. **3.** Erythroblastosis or erythroid leucosis.   **4.** Myeloblastosis or myeloid leucosis or granuloblastosis.

It is thought that lymphoid leucosis and osteopetrotic form are caused by one virus while erythroid leucosis and myeloid leucosis are each caused by a different virus.

1. Visceral Lymphomatosis: This is the most common form and the neoplastic cells are extravascular   Birds affected with visceral lymphomatosis manifest indefinite symptoms. Emaciation diarrhoea, ascites, enlargement of abdomen and paleness of mucous membranes may be noticed. Though the birds appear to feed normally they become lighter in weight.

Macroscopically, liver,   spleen and kidney are the organs frequently affected. All organs may be affected. none being immune.   These may contain nodules of the tumor or they may be enlarged due to diffuse infiltration by the tumor cells.

The liver is enormously enlarged and can be palpated from outside, "big liver" disease.   There may be focal, round, grey nodules or diffuse infiltration of parenchyma.   The liver becomes very pale and grey colored

The spleen is enlarged, even to ten times its normal size.   It may also show focal nodular or diffuse swelling.   Similarly, all other organs show round, gray nodules.   All   lymphoid tissues are affected.

Ovary is rarely affected and has granular appearance. Blood usually is aleukemic.

Microscopically, there is collection of lymphoid cells, many showing mitotic figures   The parenchyma may be virtually replaced by the infiltrating cells   The tumor cells may resemble normal lymphocytes or they may be larger immature ones. In the heart the muscle fibres may be separated by the tumor cells.

2.Osteopetrotic form: The shafts of the long bones are usually affected. They become thick, and hard resembling marble. Rarely other bones may also be affected. New bone forms from the connective tissue of the periosteum and endosteum. Eventually, the cortex becomes very much thickened and encroaches upon the marrow cavity which, therefore, becomes narrower or sometimes com. pletely obliterated. (osteosclerosis.) This may interfere with hemopoiesis and hence anemia is seen in this form.

The bones, which thus become thicker also curve anteriorly. Bilateral involvement of bones may be seen.

Some consider that osteopetrotic form is not neoplastic but an osteopathy, caused by a virus.

3. Erythroblastosis : In this condition, immature red cells crowd into the blood vessels, bone marrow and organs. The spleen, liver, lungs, and kidneys are more often affected. In this form there may be diffuse infiltration but *no nodules* form and the cells are mostly intravascular.

Macroscopically, the bone marrow appears more solid looking and is red in color. The liver and spleen may be enlarged and mottled.

Microscopically, the blood vessels are filled with the neoplastic cells These. which may also be found extravascularly in the organs, replacing the darenchyma, are the precursors of red cells and are hemoglobin free. Round or elongated cells with a clear cytoplasm and a round or oval nucleus, are noticed. In the liver, the cells first accumulate in the sinusoids from which they may infiltrate extravascularly causing atrophy and necrosis of the cord cells. In the spleen the follicles are replaced by the tumor cells.

In this condition there is true leukemia from the beginning.

4. Granuloblastosis : This is a condition in which there is tumor formation of the precursors of granulocytes in the bone marrow. Sometimes this may coexist with erythroblastosis. The proliferating neoplastic cells my invade the blood and reach liver, spleen, kidney, intestine. ovary, heart, lung and pancreas' where they may infiltrate extravascularly.

Macroscopically, the bone marrow is more solid looking, Involved organs (liver. spleen etc.) may be enlarged and show focal or diffuse masses of cells. These lesions are soft and friable.

Microscopically, the tumor cells fill the bone marrow and replace the erythropoietic cells. So myelophthisic anemia may be present. The blood vessels contain the myeloblasts or promyelocytes. In the organs these infiltrating cells may replace the parenchyma.

Hematology: There may be terminal leukemia, white cell count being abnormally high. Precursors of granulocytes are present in large numbers and some of them may contain granules. Mature heterophils may not be increased in number.

MAREK'S DISEASE—NEURAL LYMPHOMATOSIS.

Marek's Disease (M.D. for short) is the name now given to the condition formerly known as Neural lymphomatosis. The ocular form is also supposed to occur along with the neural form.

Formerly M.D. was known as Fowl paralysis, Range paralysis, Neurolyphomatosis gallinarum, Polyneuritis, Neurogranulomatosis and Avian reticulosis.

M. D. is an infectious disease caused by Herpes type virus which is a

DNA virus. All age groups of all breeds of poultry are affected But birds between 2 and 4 months of age are more often affected. In the affected flocks losses are upto 25%. Turkeys and geese may also be affected and the disease is prevalent among poultry throughout the world.

Transmission: The disease is transmitted by direct contact, through feed, water. fomites and through the egg to the chick, Air borne infection is common and important A hereditary predisposition exists. Some birds may act as carriers.

The virus:The virus could be grown on chick embryo fibroblast cultures in which the cytopathic effects seen are cytoplasmic granulation and vacuolation with .he eventual death of the cell. On chicken kidney cell culture, the CPE seen are microplaques. The CPE caused by MD virus is similar to that of H rpes type virus. The MD virus is closely cell-associated and cell-free filtrates do not have a high titre. In the affected birds, the virus is not found in the lymphoid cells but in stromal cells of bursa fabricius and in a variety of epithelial cells.

The virus has three antigens, A, B and C of which A is inherited Continued passage of the virus in tissue cultures results in loss of pathogenicity and the A antigen (by 33rd passage). Such attenuated virus can be used as a vaccine·

Inoculation of virus into chorio-allantoic membrane of 11 day old chick embryos produces pocks after five days

Strains of etiological virus: All the strains isolated, could be grouped under two main pathogenic types: (i) classic strains which induce mainly a neural di ease, occasionally with lymphomas and (ii) an acute strain which causes a high incidence of lymphomas in various organs usually with neural involvement also.

Symptoms : *Acute form* is seen mostly in young brids in which cerebral symptoms predominate.Involvement of the bronchial and lumbosacral plexuses or the nerves originating from them results in partial or complete paralysis of the wings and legs. Involvement of the vagus results in dilatation of crop and respiratory symptoms accompanied by a mucoid or purulent exudate in the passages. Involvement of the splanchnic plexus causes digestive disturbances. When the hypogastric nerve is affected, paralysis of the cloaca is seen.

Depending on the nerves affected, birds may show lameness and incoor--dination of movement. "Clutching" of toes is a characteristic symptom. The affected birds hobble or walk on the hunched toes or hopp with the affec ed legs held out stiffly in front. There may be spastic paralysis also of the legs and wings. Atrophy of the associated muscles occurs. The perching reflex may be lost and the bird lies on its side with one leg stretched forward and the other extended behind and this is a characteristic posture taken. Torticollis and drooping of tail are also observed. Capricious appetite, stunted growth and loss of condition are other symptoms noticed. Tumors may be present on the skin, particularly over the wings and breast.

In the ocular form, the iris becomes ovoid. ellipitical or pear shaped and completely obliterated. One or both eyes may be affected.

Macroscopically, the sciatic nerves, sciatic plexus and their nerves are commonly affected The affected nerve loses its glistening white appearance, becomes gray or yellowish in color with nodular swelling and loss of striation.

B'lateral involvement is not common. Sometimes vagus may be as large as a lead pencil and the sciatic may attain the size of a femur. The lumbar and celiac plexuses may stand out prominently The dorsal ganglia of the peripheral nerves extend. into the spinal cord and form tumor–like masses.

In uncomplicated cases, visceral organs remain normal. Thymus may be enlarged The bone marrow of the radius and ulna is hyperactive. There may be whitish ill-defined areas in the liver and lungs, The ovary is more often affected and shows creamy white or pinkish lobulated tumor masses. Diffuse thickening of the mesentery and bursa fabricius is common.

Microscopically: three types of lesions are seen in the peripheral nerves. In the first, there is cellular infiltration of the nerves with mature lymphocytes In the second, which is the most common, there is seperation of the nerve fibres associated with edema followed by less cellular infiltration. These cells are lymphocytes and plasma cells Infiltration with neutrophils and myelocytes may also occur. In the third type of the lesion, the nerves are infiltrated with lympho-blasts showing numerous mitoses In this condition which may be an advanced and progressive stage of the second type, all evidence of the nerve tissue is lost and it has the appearance of a neoplasm.

Myelin degeneration of the nerves is common while axonal degeneration is rare. With the electron microscope endoneural fibrosis is evident.

In the central nervous system, perivascular cuffing with lymphocytes is noticed In the visceral type, first fibrinous necrotic foci develop followed by infiltration of the lymphocytes, plasma cells and even granular cells. Finally tumor forms in these places. Such lesions are seen in the ovary frequently and in the liver and lungs less frequently. But the visceral lesions do not metastasize.

On the skin, there is gross enlargement of feather follicles, appearing as nodular elevations Skin becomes rough and tough. There is patchy infiltration of lymphoblasts, plasma cells and a few histiocytes. The feather follicular cells show PAS negative intranuclear eosinophilic inclusion bodies.

Necrosis of follicles of bursa fabricius is commonly seen, with infiltration by a few haterophils.

In the ocular form, the iris is infiltrated with small round cells and polyblast-like cells. Infiltration is also common in the choroid, retina and the ocular muscle.

Blood picture does not usually show any change except in some cases in which mild evidence of leucocytosis and lymphocytosis is encountered. Gamma globulins are decreased in the blood.

Note: It is thought that MD starts as an inflammatory process with infiltration by cells Later the cellular infiltration assumes neoplastic proportions and so this disease should not be classified under neoplastic diseases.

Laboratory Diagnosis: 1. Agar gel double diffusion technique can be used with fowl kidney cells injected with MD virus and sera from MD infected birds. 6 lines are seen.

2. Immunodiffusion test can be performed.

3. Fluorecent antibody techniaue can be utilised with feather follicular cells.

4. On histopathological examination of the affected tissues, characteristic MD cells are seen. These are large cells, basophilic and pyroninophilic. The cytoplasm is frequently vacuolated and the nucleus is obliterated by the vacuoles.

5. Presence of intranuclear inclusions in the feather follicular cells.

Differences between Leucosis and Marek's Disease.

Leucosis	Marek's Disease
1. Morphogenesis and morphology of the virus unknown.	1. Virus is similar to that of Herpes type
2. Is an RNA virus.	2. Is a DNA virus.
3. Older birds affected (3 to 8 months)	3. Younger birds affected (2 to 4 months)
4. Visceral form more common.	4. Visceral organs rarely affected.
5. Ovary rarely affected.	5. Ovary frequently affected.
6. C. N. S. less commonly affected.	6. C. N. S. and autonomus nervous system more commonly affected.
7. Lymphocytes do not show pleomorphism.	7. Lymphocytes show pleomorphism.
8. Definite serological tests absent.	8. Serological tests—agar gel double diffusion tests and immunodiffusion-test—can be performed.
9. Ocular lesions not frequent.	9. Ocular lesions frequent.
10. Neoplastic in origin	10. Inflammatory in origin.
11. No seasonal variation.	11. Seasonal variation present.
12. Not air borne.	12. Air borne.
13. Blood and bone marrow involved.	13 Blood and bone marrow not involved.
14. Cofal (complement fixing ALC factor) present	14. Cofal not present.
15. RIF (Resistance inducing factor) present.	15. RIF not present.
16. Nerves not involved.	16. Nerves involved.
17. Nodular tumors in the Bursa Fabricius present and cells are intrafollicular.	17. Bursa Fabricius shows diffuse enlargement or atrophy and the cells are interfollicular.
18. Arise From 'B' cells.	18. Arise from 'T' cells.
19. Feather follicles and skin not affected.	19. Lymphoid proliferation in feather follicles and skin present.
20. Symptoms not specific.	20 Paralysis frequently seen.

EPITHELIAL TUMORS

Epithelium covers the body as epidermis, lines the mucous membranes and ducts and forms the parenchyma of the glands. Broadly, therefore, epithelium can be divided as glandular and non-glandular. Tumors of the epithelium are classified as :-

I. Benign:

i) **Squamous papilloma**, arising from stratified squamous epithelium.

ii) **Adenoma** and adenomatous polyp arising from solid and lining glandular tissue.

II. Malignant:

i) **Squamous cell carcinoma**, arising from squamous epithelium.

ii) **Adenocarcinoma**, which is a malignant tumor of glandular epithelium.

Papilloma

This is a benign epithelial tumor projecting from an epithelial surface and is covered by squamous, transitional or columnar epithelium depending on the tissue from which it originates.

Site: Papillomas are most commonly found on the skin and on the buccal mucosa. Some may arise from the mucosa of the intestine and bladder. Very rarely, these tumors may arise from the epithelium of the ducts of glands like mammary gland.

Skin papillomas: These aye very common in all animals and are also known as warts. In calves these may be found on the shoulder, head, neck and dewlap, mostly as clusters. In horses they may be single or multiple on the shoulder, ears, around the mouth, eyelids and the scrotum. On the oral mucosa of dogs, multiple papillomas may be found and are infectious, the causal agent being a virus. Similarly multiple papillomas, which are infectious are found on the skin of calves and rabbits. Papillomas may also arise from the omasal and oesophageal mucosa of cattle.

Macroscopically; the papilloma may usually be small. Occasionally some may be large measuring 10 cms in diameter. They may be pedunculated as in the papillomas of the bladder, or have a broad base The surface may be smooth or may be rough and horny. It may be horn-like, (*cornu cutaneum*) some. times. The papillomas being benign grow outwards, away from the basement membrane but do not burrow underneath it.

Microscopically, the epithelium constituting the tumor is very much thickened. The skin papillomas have the structure of a normal skin. The epithelium grows in finger-like projections. Since its nutrient vessels are found in the connective tissue, a core of connective tissue is always present accompanying the growths· But the neoplastic change is found only in the epithelium and not in the connective tissue which is only supportive.

The infectious papillomas of dogs (verruca vulgaris) and those of cattle are caused by a virus. A vaccine conferring partial immunity against these tumors has been developed.

Clinical :—The warts may be removed, especially when single. Spontaneous disappearance of the tumors may take place.

Squamous cell carcinoma

This is a malignant tumor of squamous epithelial cells, usually of the squamous stratified epithelium

Squamous cell carcinoma is a common tumor among the cattle in India affecting the horn and the eye. It is also a common tumor of aged dogs and is fairly common in horses.

Sites: Squamous cell carcinoma is found on the skin of animals in various locations; abdomen, prepuce, anus, tail, tongue, legs, ear, eye and lips in the dog; eye, vulva, penis, hip region, base of tail, prepuce and limbs of horse: besides the eye and horn, ear, vulva and vagina of cattle. Frontal sinuses of horses and cattle have also been found to be affected.

Melanin is protective against the actinic rays of sun. Hence, in areas where pigmentation is dificient squamous cell carcinoma has been seen. Also in brand-scars which are deficient in melanin, this tumor has been seen to develop.

Macroscopically, the tumor is fast growing and has a papillary or a cauliflower appearance having a broad base. It is soft and has a grayish or pink appearance Slow growing tumors of the skin may be horny.

Tumors arising from the skin are frequently subjected to trauma, ulceration, hemorrhage and subsequent infection. So a foul smell may be exuded from such a tumor.

Microscopically, the structure may vary depending on the degree of anaplasia present. In not-so-anaplastic tumors, the structure of the adult epithelium is reproduced. In squamous cell carcinoma of the skin, all the layers of the epidermis may be found. It is the deeper cells of stratum germinativum that proliferate and these cells later become the prickle cells of the stratum spinosum. The basal cell layer is very much thickened. Tongues of these cells may invade into the dermis and differentiate into cells of various strata. A cross section of such an infiltrating column discloses the Malpighian layer (with prickle cells) externally (at the periphery) and the subsequent layers of cells are seen to lie inside this layer. So, the keratinised layer of the cells is seen in the centre. As layer after layer of kerato-hyaline is deposited concentrically by the progressive maturation of the proliferating epithelial cells, we find concentric layers of keratin forming "pearls" or "cell-nests", which are characteristic of skin cancers. Such squamous cell carcinomas are known as *epidermoid carcinomas.* (It should be remembered that cornifying epithelium is also found in rumen. vagina, tongue, esophagus and the third eye lid). The neoplastic cells are polyhedral with prickle borders and large nuclei. Mitotic figures are seen in large numbers.

In more anaplastic type of tumors the cells are found as sheets or islands without differentiation into the different layers. Here the cells are more uniform with hyperchromatic nuclei. Mitotic figures are numerous

Occasionally, the tumor cells may be so anaplastic that the cells take an elongated, spindle shape. Such cells are difficult to be distinguished from a sarcoma. But the absence of intercellular fibrils and stroma helps to diagnose the tumor as carcinoma.

In squamous cell carcinoma. the connective tissue stroma is plentiful with numerous thin-walled blood vessels. At the margins of the invading carcinoma may be found infiltration of macrophages, plasma cells and lymphocytes. Fast growing cells undergo degenerative changes showing vacuolation and karyorrhexis.

Papillary carcinoma has to be differentiated from the benign papilloma. The following characteristics of carcinoma are helpful: invasiveness into the dermis or submucosa; anaplasia of cells (which are larger with hyperchromatic, large nuclei); presence of cell nests; invasion into blood vessels and lymph channels; numerous mitotic figures and presence of inflammatory changes in the stroma.

HORN CANCER

In India, cancerous growth of the horn is very common. It is found that the incidence of the tumor is higher in bullocks than in cows. Bulls are seldom affected. The following factors are suggested to be causative :

i) Hormones : Since the tumor is never found in bulls but only in castrated bullocks some hormonal etiology is suggested.

ii) Chronic irritation: (a) Since the neoplasm is seen more in work bullocks than in cows (which are not used for draught purposes) it is suggested that the constant friction of the yoke near the base of horn may have some etiological significance.

(b) Horn cancer has been found to occur more often in animals whose horns are pared for beautification Such paring is done periodically. Injury may be caused to horn base and cancer may result.

Symptoms: usually one horn is affected. Evidently growth of the tumor causes pain since the animal shakes its head and rubs the affected horn on sharp surfaces. As the growth enlarges, it fills the horn core. The horn slowly bends. Sometimes there may be separation of the horn with the core at its base and the horn may come off leaving the horn core with the pinkish, soft tumor. It may bleed profusely.

Macroscopically, the tumor is pinkish and polypoid like a cauliflower. It is friable and easily bleeds. It fills the horn core and may invade the frontal sinus and may even corrode the cranial bones.

Microscopically, the structure is typical of squamous cells carcinoma described. Metastasis occurs into regional lymph nodes and lungs.

Cancer of the eye in cattle: Squamous cell corcinoma frequently occurs in the eyes of cattle. One or both eyes may be affected. Statistically, it was found that incidence is more in the working bullocks than in cows. It is postulated that probably actinic rays of the sun are of etiological importance since bullocks are worked in the hot sun and so are exposed to them. Probably the lower incidence in cows is because they are not worked in the sun. That actinic rays do play a part in the genesis of the tumor is supported by the fact that in Hereford breed, in which the eye pigment is deficient, eye cancer is more common. The pigment filters the actinic rays of the sun.

Macroscopically, the tumor may start as a tiny outgrowth on the nictitating membrane or the cornea. This grows rapidly and in time destroys the whole eye ball invading it.

Microscopically, four types have been described : i) *the plaque form* in which there is greater thickening of the prickle cell layer due to proliferation of the prickle cells.

ii) *The papilloma form* in which papilloma-like out-growth is seen. In some places, cells may be anaplastic, showing mitotic figures and may be grouped as islands.

iii) *The early squamous cell carcinoma* in which nests or islands of tumor cells with numerous mitotic figures may be seen.

iv) *The squamous cell carcinoma* with its typical structure.

Clinical: squamous cell carcinoma is a highly malignant tumor and metastases occur in various organs. Metastases of horn and eye cancer may be seen in the regional lymph nodes and in the lungs.

The plaque and papilloma types of eye cancer may, probably, be removed with satisfactory results since these are supposed to be benign stages.

One should be cautious in using animals affected with eye and horn cancer for breeding since a genetic predisposition is suspected.

Basal cell carcinoma

The basal cell carcinoma is also known as "*rodent*" or "*Jacob's*" ulcer or *hair-matrix carcinoma*. This tumor arises from the basal cells of the Malpighian layer of the epidermis or the basal cells of the hair matrix or of sebaceous glands. Basal cell carcinoma is locally invasive and does not metastasize.

Incidence: Basal cell carcinoma is met with in dogs, horses and cats.

Site: The head is, by far, the most common site.

Macroscopically, basal cell carcinoma occurs singly and has a broad base. It is subcutaneous, rounded and frequently encapsulated. It is firm and the covering skin may not have any hair, but is often ulcerated.

Microscopically, the tumor has varied appearance. The cells, which are small and round with round or cigar shaped hyperchromatic nuclei may be arranged in sheets or islands with plentiful connective tissue stroma. The cells may also be arranged in columns which descend into the dermis. The cells in these columns may be enlarged, with their long axis perpendicular to the columns. On cross section these columns display an "adenoid" structure but without lumina. Mitotic figures may be seen. Prickle-cells are not observed.

In some tumors, the structure so resembles the hair follicle that it is called *trichoepithelioma*. No hair is formed.

Basal cell carcinoma can be differentiated from squamous cell carcinoma as per the table given below.

Basal cell carcinoma	Squamous cell carcinoma
1. No prickle cells	1. Prickle cells present
2. Down growths of epithelium club shaped and have centrally and longitudinally arranged cells with peripheral palisading columns of cells in each club	2. Not so.
3. No pearls	3. Pearls seen
4. No metastasis (locally malignant)	4. Metastasis present

Clinical: The basal cell carcinoma is highly radio-sensitive. Since these tumors do not metastasize, prognosis is good with complete excision.

Adenoma

This is a benign tumor of glandular epithelium. So adenomas can arise from any glands (with and without ducts) in the body. But they are more frequently seen in the following locations:

Dog: mammary, sebaceous, prostate, and bronchial glands.

Horse: thyroid and sebaceous glands.

Cattle: multiple adenoma of pancreas, tumors of adrenal cortex. papillary adenoma of bladder and gall bladder.

Macroscopically, the adenomas are nodular and encapsulated. They may be firm to soft depending on the relative proportion of stroma and parenchyma. These tumors are usually clearly demarcated from tissues from which they arise. Having plentiful blood supply, adenomas may be pink in color. In hollow organs, like the intestines, stomach and bladder they may assume a polypoid shape. Adenomas may be solitary or multiple.

Microscopically, the structure of an adenoma is simple. It consists usually of a single layer of the columnar or cuboidal epithelium lining an acinus. These cells resemble the normal glandular epithelium. Should the cells show any tendency to pile up or grow in a papillary fashion into the lumen, malignancy should be suspected. The secretions do not drain as there is no duct. They accumulate causing atrophy of lining epithelium and thus resulting in cystic dilatation of glands resulting in cystadenoma. Those tumors that show intra-terminal papillary projections are called **papillary adenomas**. When a cystadenoma shows papillary projection, it is known as *papillary cystadenoma*. The adenomas are supplied with plentiful stroma. If the fibrous tissue elements also assume neoplastic character as is found in mammary tumors, the, tumor is caled a *fibro-adenoma*.

Adenocarcinoma : This is a malignant tumor of glandular epithelium. In these tumors, the cells are anaplastic, lose their polarity, are large and irregular, with hyperchromatic nuclei. The acini may contain many layers of cells, which may often show papillary projections in the lumen. The cells infiltrate under the basement membrane and into the surrounding stroma. Mitotic figures are frequent and metastases are common.

Clinical : Adenomas may be excised. Adenocarcinomas are highly malignant and require a more radical extirpation.

Sebaceous gland adenomas

Adenomas arising from the sebaceous glands are common among aged dogs.

Sites : Skin of head, neck, eye-lids, prepuce, tail and back.

Macroscopically; the tumors are small and lobulated. Greyish-yellow in color, the tumors may be greasy to the touch. A capsule may or may not be present.

Microscopically tumors consist of large polyhedral cells with a cytoplasm containing numerous fat droplets. Nucleus is small, round and central with a fine chromatin and two nucleoli. The cells are grouped into masses or islands separated by stroma forming lobules. The islands of cells may be surrounded to a depth of several layers by elongated oval cells with condensed nuclei, the reserve cells. In the septa that are seen between the lobules may be found lymphocytes, plasma cells and melanoblasts. Mitotic figures are rare. Sometimes 'pearls' may be noticed.

In the malignant variety, the cells are low cuboidal or polyhedral with a vesicular nucleus. These cells may be arranged as acini or as sheets. The cells are highly invasive and show numerous mitotic figures.

Perianal adenoma of dogs

The perianal glands are modified sebaceous glands located as a ring around the anus. Adenomas of these glands may be encountered.

Sites : The tumors are subcutaneous and may be lateral to or above the anus,

Macroscopically, these tumors are single and are multinodular. They are firm and encapsulated.

Microscopically, the tumor consists of sheets of epithelial cells arranged in lobules which are divided from one another by delicate septa containing blood vessels. The cells are large and polyhedral with faint granular acidophilic cytoplasm. The cells are large and rounded. One or two nucleoli may be seen. In the benign form mitotic figures are scarce. Acinar formation may be seen in some tumors. Rarely "pearls" may be seen.

Malignant adenomas of perianal glands may occur. The cells show anaplasia, hyperhromasia and enlargement of nuclei and nucleoli. In these tumors the reserve cells may also proliferate obscuring the glandular tissues.

Tumors of sweat glands

Tumors of sweat glands are found only in the dogs.

Sites : These tumors occur mostly on the face where sweat glands are numerous. They may also be found in other parts of the body.

Macroscopically, sweat gland adenomas may be solitary. They are small, firm and dirty gray in color.

Microscopically, the tumor consists of columnar, cuboidal or flat cells, having a glandular arrangement. The acini may have a lumen. Sometimes solid mass of cells may be found. The cells have an acidophilic, faintly granular cytoplasm. The nucleus is round or oval and is central. Connective tissue stroma is abundant, and contains large number of blood vessels, lymphocytes and plasma cells. The acinar cells become flattened due to the pressure of the accumulating secretion (sweat) giving rise to cystadenomas.

In the malignant variety, though the glandular structure is maintained, the cells lose their polarity and are anaplastic. Mitotic figures may be seen. Stroma may be plentiful.

Clinically : Metastases may be found in the regional lymph nodes and lungs in the malignant variety.

Primary carcinoma of the lung

Primary carcinoma of the lung is one of the rare neoplasms in animals. Both sexes are equally affected, though in man, incidence in the male appears to be higher.

Etiology : The following factors have a relation to the incidence of bronchogenic carcinoma in man.

1. Incidence of lung cancer is found more in urban people than in rural areas. City atmosphere has greater amount of automobile exhaust fumes as well as industrial gases. These may have carcinogenic elements. Actually, in the city of Los Angeles, the atmosphere was found to be charged with highly carcinogenic substances.

2. Tobacco smoking has been associate with lung cancer.

3. Among animals, tuberculosis, viruses (Jaagsiekte) and lung worms have been associated with lung cancers.

Incidence ; Bronchogenic carcinomas have been encountered in dogs, horses, cattle, cats, sheep and fowls. No tumors have been reprorted in swine.

Sites : The tumors may be disscrete and focal or they may also be found multifocally. The multifocal tumors may be multicentric in origin, or may be metastases.

Among animals, the large right lung is more frequently affected. Among the lobes, the diaphragmatic is more frequently the seat of neoplasia. In man hilus is a frequent site but not so in animals.

Macroscopically, the tumors vary in size and are not encapsulated. They are gray in color and may be soft.

Microscopically, the tumor cells may be of three types; (1) Squamous cell, (2) columnar or cuboidal and; (3) undifferentiated cells.

All these cells have the same origin, namely the bronchial epithelium. The bronchial epithelium as well as the epithelium of the bronchial mucous glands have the same derivation. Again, the bronchial epithelium has great metaplastic propensities and so can be converted into squmous cells.

The squamous cell type, as the name implies, is composed of squamous cells, which occur as masses. In some tumors, epithelial 'pearls' may be noticed.

In the columnar or cuboidal type of neoplasm, the tall columnar or cuboidal cells are arranged as alveoli or solid masses or as tubules. This has a characteristic adenocarcinomatous appearance. There may be papillary projection of the tumor cells into the lumen of the alveoli. The cells have an acidophilic, granular cytoplasm and a vesicular nucleus with fine chromatin. The nucleus may be flat, elongated or rounded. It may also be pyknotic. Mitotic figures are frequent. The cells may also be arranged as vertical columns, which are separated by fine stroma. The connective tissue stroma in the adenocarcinoma is scanty. The alveoli of the lung may be filled with the neoplastic cells and so *cancer pneumonia* is the name given to the condition. Some bronchogenic carcinomas produce mucin. But this is not an evidence that the tumor originates from the bronchial glands, since neoplastic bronchial epithelium is also quite capable of forming mucin.

In some tumors the cells are arranged as solid masses or as cords. But the cells are so highly anaplastic that they cannot be classified as columnar or squamous. They are large and polyhedral. Some may be ovoid, so called *oat-cell*. They may be spindle shaped also resembling a fibrosarcoma.

Though some authors describe a tumor arising from the alveolar septal epithelium, such a tumor is disputed by many oncologists, since the lining epithelium of the alveolus is not easily visible.

Infiltration into the rich peribronchial lymphatics easily occurs and so metastases are very frequent.

Clinical: Bronchogenic carcinoma is always fatal.

Tumors of the Liver

In the liver tumors may arise either from the hepatic cells or from the bile ducts. The former are called *hepatomas* or *carcinoma hepatocellulare* and the latter *carcinoma cholangiocellulare*. These tumors are not common in animals. They occur in cattle and sheep only.

Hepatoma; The liver cell carcinoma is found in adults and is usually single.

Macroscopically, hepatomas may be small or large. They usually project out on the surface as brownish or greenish nodules. The nodules may be round or ovoid and are usually clearly demarcated from the healthy area where hemorrhage has occurred. In sheep, centres of hemopoiesis may be seen in some hepatomas.

Microscopically, the tumor consists of typical hepatic cells arranged usually as columns adjoining vascular spaces. The cells are large and polyhedral with usually acidophilic granular cytoplasm but has a tendency towards basophilia. The nucleus is very large, central and pale staining Mitotic figures may be numerous. Tumor giant cells may be seen with two or more nuclei.

The stroma is fine and scanty. Numerous thin-walled blood vessels may be present in close association with the tumor cells. The new growth is devoid of bile ducts and so stasis of bile formed gives the tumor a green color. Lipoid infiltration of the cells is common.

The tumor is divided into lobules by connective tissue septa. Metastasis occurs into other portions of the parenchyma of the liver and into lungs via blood.

Cholangiocellular Carcinoma

Macroscopically, the bile duct carcinomas are small and multiple They may be round and unencapsulated.

Microscopically, the tumor consists of acini lined by columnar cells and containing mucin. In some places the cells line cyst-like spaces which may be filled with solid masses of neoplastic cells. In some tumors the cells may be so anaplastic as to be difficult to be recognised.

Since it is believed that the liver cells and hepatic duct epithelium have the same origin embryologically, it may be difficult, sometimes, to distinguish between a liver-cell carcinoma and a cholangiocellular carcinoma. The following differences aid in identification.

Cholangiocellular carcinoma	Hepatocellar Carcinoma
1. Small and multiple.	1. Single and massive
2. No encapsulation.	2. Usually encapsulated.
3. Grossly yellowish-white in color	3. Whitish or reddish
4. Early metastasis into lymph nodes- lymphatic metastasis.	4. Metastasis into parenchyma and lungs—hematogenous metastasis.

5. Cells columnar and arranged to form glandular acini.	5. Cells polyhedral and resemble hepatic cord cells.
6. Usually lumina formed,	6. No lumina formed.
7 Nucleus located basally in the cells.	7. Nucleus central.
8. Collagenous stroma	8. Collagenous stroma not as much.
9. Cells have no clo e association with blood vessels.	9. Cells have close association with with blood vessels.

Nodular hyperplasia is frequent in older animals. This condition may pose a difficulty in differentiating it from carcinoma. The following features may help 1. The cytoplasm of the cells in hyperplasia stains similar to that of normal hepatic cells. (viz. acidophilic) while the cells of the carcinoma have a tendency to basophila 2. The nulei of the cells in the benign nodule are more uniform than those of carcinomatous cells. 3. Mitotic figures are not high in the hyperplastic cells while carcinoma reveals many mitotic figures. 4. Giant cells are seen in the tumors only.

In man, cirrhosis and hemachromatosis have been associated with hepatocellular carcinoma. Among animals, liver fluke infestation is associated with bile duct carcinoma. The role of the parasite in cercinogenesis has not been established.

Tumors of mammary gland

Tumors of the mammary gland are commonest in dogs. It is strange to notice that the udder of the cow with great functional activity and development is seldom the seat of neoplasia.

In the bitches, the mammary tumors occur more commonly in older age groups. Of the five pairs of the mammary glands, the posterior two pairs are more. frequently affected. Next in frequency are the anterior two while the central pair is the least affected. The reason for posterior pair to be more frequently affected is probably because they are more prone to trauma.

Estrogens have a distinct role in the etiology. The low incidence of mammary neoplasia in ovariectomised bitches lends support to the same.

Macroscopically, the tumors are soft or firm depending upon their fibrous tissue content. They may be cystic sometimes. Those containing bone or cartilage may be hard. These tumors are grayish-white in color.

Microscopically, the tumors may be benign or malignant, In both varieties they may be simple or mixed.

If the glandular epithelium alone is affected, it is called a *simple* tumor. But if the connective tissue also proliferates, then the tumor is known as a *mixed tumor.*

Simple mammary tumors; a) duct papilloma; This is a tumor that arises from the lining epithelium of the teat canal or larger ducts. Tumors are localised and may be single.

The cells are cuboidal or columnar and these form papillary projections into the lumen of the dilated ducts which contain eosinophilic material, These must be differentiated from duct papillomatosis which are more common in the cat.

b) **Malignant duct papilloma, or Intraduct papilloma.** In this condition the cells are anaplastic and show hyper-chromasia. These are arranged in acinar or multiacinar formation. but sometimes in solid sheets also. More commonly a papillary arrangement can be noticed wherein the cells are one to several layers thick. A delicate fibrous tissue core is present in which are found blood vessels

c) **The lobular carcinoma,** The cells of the acini of mammary gland proliferate and are arranged in solid sheets or as acini with lumina. This variety is called true adenocarcinoma of the mammary gland.

Mixed mammary tumors. These are far more numerous than the simple variety. Besides the epithelial cells (either of the ducts or acini or both) the connective tissue elements and myoepithelial cells may also be involved.

a) In the benign variety, the tumor consists of papillary growth of the ductal epithelium. The acinar cells in some parts may also proliferate, It is not uncommon for the epithelium of the ducts to show metaplastic change to squamous type. Vety ofien, the myoepithelial cells, that are normally situated between the epithelial cells and the basement membrane may also proliferate to a thickness of several layers. These cells are spindle shaped with tapering ends and elongated nuclei. They can be converted into hyaline cartilage.

Stromal tissue may also proliferate and this may be transformed into myxomatous or cartilagenous or osseous tissue. Hence the tumor may be a mixture of all these tissues to merit. depending upon the tissues present, the names like fibro-myxo–adenoma or fibro-myxo-chondro-adenoma etc , Since such names are cumbersome the short "mixed-cell tumor" is now applied to such tumors.

(b) Malignant mixed tumors: In this variety, the epithelial cells are malignant. The connective tissue part may also be malignant or it may be benign. So we may find fibro chondro–osteo–adeno-carcinoma. But for exihibiting malignancy, the structure resembles that described in the benign variety. The myoepithelial cells may also show malignancy. These cells become pleomorphic and may form multinucleated giant cells.

Mitotic figures are numerous in all types of tissues. Fibrous tissue frequently shows metaplasia to cartilage and bone. Metastasis into lung, liver and kidneys is frequent. It is observed that the neoplasm affecting the anterior three pairs of glands first metastasizes into the axillary lymph node while those of the posterior two pairs into the inguinal lymph node.

Clinical. Though successful surgery has been reported in some cases mammary neoplasm of dogs must be viewed as dangerous since early metastasis occurs.

Tumors of Thyroid

The incidence of thyroid tumors follows closely that of goitre and so it is prevalent among animals suffering from that condition. Therefore it has a geographic distribution being found frequently in goitre belts.

It is rather difficult to differentiate hyperplasia of thyroid from adenoma. Actually the nodular goitre described is an adenoma, according to some.

Adenomas of thyroid are found in horses while adenocarcinoma is found mostly in aged dogs.

Adenoma: **Macroscopically** the adenoma is a small, rounded, well encapsulated tumor that is demarcated from the surrounding tissue. It is rather difficult to distinguish an adenoma from normal thyroid. But acini of the ade noma are smaller. The tumor is well demarcated from the rest of the gland and it may cause pressure atrophy of the normal tissue.

Adenocarcinoma: **Macroscopically**, the neoplasm may usually be multiple and may be unilateral or rarely bilateral. The tumors may be solid looking or may be cystic. Since the tumor is highly invasive, matastasis by way of emboli is common. The tumors are not encapsulated

Microscopically, there may be formation of acini or the cells may be arranged as solid masses. The cells are low columnar or cuboidal and have an acidophilic granular cytoplasm and an oval hyperchromatic nucleus. Mitotic figures are common. Sometimes they may form papillary projections into the lumen. Hemorrhages may be seen. Connective tissue stroma is scanty. Metastases are common and occur in the lungs mostly. They may be found in other organs also. The metastases may or may not resemble parent tissue.

Clinical: The adenomas are of no consequence unless they become toxic. i. e. when they produce hyperthyroidism and thyrotoxicosis. The adenos carcinoma is always fatal

Tumors of parathyroid

Though genuine tumors of parathyroid are not reported in animal-hyperplasia has been seen. This occurs in dogs suffering from chronic interstitial nephritis and in horses suffering from Miller's disease (big head). In these conditions osteitis fibrosa occurs. In the dog it is known as *"rubber-jaw syndrome"*.

In man tumors of parathyroid have been reported and they cause von Recklinghausen's disease (Osteitis fibrosa cystica)

Tumors of pancreas

Tumors of pancreas may either be exocrine or endocrine. Very few cases of pancreatic tumors have been reported in animals.

Carcinoma of pancreatic acini: The pancreatic new growths may press upon the common bile duct and produce obstructive jaundice.

Microscopically, the picture is one of adenocarcinoma. The cells which are low cuboidal or columnar or even polyhedral with round or elongated oval nuclei, form acini. In some areas the cells may be in the form of sheets and masses. Mitotic figures are frequent. Blood supply is plentiful and there is a delicate fibrous tissue stroma. Hemorrhages may be seen.

There may be difficulty in differentiating this tumor from islet-cell carcinoma Special stains have to be used to bring out the zymogen granules in the adenocarcinoma. Metastases are found in the abdominal organs.

Tumors of islet-cells: These are rarer still. A few adenomas have been reported in dogs and cat. Producing insulin, the tumors cause hyperinsulinism with consequent hypoglycemia.

Tumors of adrenal gland

Tumors of adrenal gland may arise from the cortex or medulla.

Adreno-cortical tumors: These may be adenomas or carcinomas.

Adenoma: Adenoma of adrenal cortex is rare among animals and is reported in old dogs, horses cattle, swine, sheep and goats. It was reported that incidence of this tumor is high among castrated goats.

Macroscopically, it is difficult to distinguish the adenoma from hyperplasia of adrenal cortex (as in thyroid adenomas) Hyperplasia of adrenal cortex is common among old dogs. The cells are similar to those of zona fasciculata or reticularis and contain lipid vacuoles A plentiful blood supply is present- Hemorrhages may be seen. These tumors do not appear to produce any recognisable effects on animals unlike in man, in whom hyperadrenalism is caused. In the case of a cow, the cortical adenoma was reported to have changed the voice to that of a bull. In man excretion of 17-ketosteroids and large quantities of oestrogen are reported. No comparable estimate of these steroids was made in the urine of animals.

Carcinoma of the adrenal cortex: This is a still rarer neoplasm. Carcinoma of the adrenal cortex was reported in cattle. Other animals are rarely affected.

Microscopically, the tumor cells are arranged as sheets or alveoli The cells are anaplastic with hyperchromatic nuclei with vacuolated cytoplasm. Mitotic figures are frequent. Blood supply is plentiful and connective tissue stroma variable. Hemorrhages, calcification and ossification may be present. Metastasis is not wide spread. The neoplasm invades locally.

Tumors of the adrenal medulla

Pheochromocytoma (Chromaffinoma). This literally means a tumor that contains dark colored cells. This tumor is supposed to arise from the cells that produced adrenaline.

Pheochromocytoma is rare and has been described in cattle, horses, sheep and dogs.

Macroscopically, the tumor may be unilateral or bilateral.

Microscopically, the tumor cells are large with central nuclei. They may, sometimes, contain lipid. Their cell boundaries are not always disinct. They may line blood spaces. Mitotic figures are few. Metastasis, occurs by blood stream into the regional lymph node, liver and lungs.

Effects on animals. Producing adrenaline, the tumor should cause symptoms of hyperadrenalism, viz. tachycardia, hypertension, hypertrophy of the heart, hyperglycemia and glycosuria

Tumors of the ovary

Granulosa cell tumor of the ovary: This tumor arises from the cells of ovarian mesenchyme. Two other closely related tumors are theca-cell tumor and luteal-cell tumor. As they often occur together in the same tumor and probably similar clinically they are grouped as one for descriptive purposes.

Granulosa-cell tumor has been seen in cows mostly. They have also been described in horses, bitches, sheep and fowls. These tumors are seen in relatively younger animals. Secreting estrogen, they cause hyperestrinism, viz. relaxation of sacro-sciatic ligament and nymphomania. Cows and mares show symptoms of continuous estrus with elevated tails. Cows bellow frequently.

Macroscopically, granulosa-cell tumor, is usually single and may be vary large. In the cow it may attain a size having a diameter of 20 centimeters. It is encapsulated, rounded and lobulated, projecting from the surface of the ovary. On section, numerous cystic spaces may be noticed together with solid areas. The tumors are yellow in color. In certain places, areas of necrosis and degeneration may be evident.

Microscopically: Structure is variable. The tumor cells are uniform, small with acidophilic, granular cytoplasm and a small centrally placed nucleus with varying chromatism. They are arranged diversely in irregular cords or columns, clusters or compact alveoli, in irregular masses or pseudo glands. The cells may be arranged in stellate or rosette-like structures reminiscent of Call-Exner bodies. These are distinctive of follicular epithelium. The lumen of many such may contain a hyaline, acidophilic material. Mitotic figures are numerous.

In tumors showing theca-cell stucture, the cells are elongated and spindle shaped. These may appear like fibrocytes and in fact the tumor may erroneously be diagnosed as a fibroma, especially if the fibrous tissue stroma is plentiful. Some of these cells may contain lipids and so the tumor may be yellow in color.

The cells of luteoma are large and contain lipid material and resemble cells of corpus luteum. The tumor is yellow in color.

Metastasis seldom occurs. Implantation of the peritoneum may occur.

Arrhenoblastoma : This is a masculinizing tumor. Though it is conjectured that the arrhenoblastoma might arise from the testicular elements of the ovary, so far, definite histogenesis has not been established.

The tumor is rare in animals being reported only in a cow and a hen.

Microscopically, the growth consists of tubules resembling seminiferous tubules, giving a glandular appearance. In some areas solid masses may be present. The cells are polyhedral.

Dysgerminoma : This corresponds to the seminoma of testes. No endocrine effects are noted. This rare tumor may be large and cystic. The cells are large and polygonal resembling seminoma cells and may be arranged as cords separated by fibrous stroma.

Tumors of the testes

Seminoma : This tumor arises from the seminiferous epithelium of the tubules of testes.

Seminoma is fairly common in dogs. The reason for its not being seen in other animals is early castration in them. Seminoma occurs more frequently in cryptorchids.

Macroscopically, seminoma is a white or gray tumor bulging from the testis. It is lobulated and may be demarcated from the healthy tissue by a fine septum. Areas of necrosis may be found in some places.

Microscopically, the cel s are arranged as sheets or islands separated by thin strands of connective tissue. Some cells may be intratubular in location. The neoplastic cells are large, rounded and uniform in size and shape. The cytoplasm is acidophilic and granular. The nucleus is large, round, central, hyperchromatic with coarse chromatin and a definite nuclear membrane and one or two nucleoli. No vacuoles are present in the cytoplasm of the cells unlike in Sertoli-cell and Leydig-cell tumors. Mitosis is very frequent.

One noteworthy feature is the presence of lymphoid cells in the stroma, sometimes in large numbers. Seminoma is a highly malignant tumor in man. But in dogs it is not so malignant, metastasizing rarely.

Sertoli-cell tumor : This is also known as sustentacular cell tumor. This tumor also is common in dogs and arises from the sustentacular cells of testes.

Macroscopically, sertoli-cell tumor is unilateral. Rarely, it may be bilateral. The tumors are firm. Cysts containing clear fluid may sometimes be seen. The tumors are white or light yellow in color.

Microscopically, the cells form tubules and so the tumor is known as tubular adenoma. The cells that are intratubular normally, may break through the basement membrane and thus be found outside the tubules as masses of cells. The neoplastic cells are spindle shaped with a clear cytoplasm and an oval or oblong nucleus In some places polyhedral cells with rounded nuclei may be seen. The tumor cells have a characteristic pallisade arrangement, their long axis being perpendicular to the basement membrane. Numerous mitotic figures may be present.

Effect on dogs : Secreting estrogen, sertoli-cell tumor is a feminizing tumor. The following clinical symptoms are noticed :— loss of libido gynecomastia; atrophy of penis and the unaffected testis; lethargy; change of voice; skin becoming blacker with greater keratin formation *(acanthosis nigricans)*, and attraction to male dogs. Since the estrogen is converted into 17-ketosteroids and excreted in urine, which gives a charactersitic smell like that of a bitch in estrus, other male dogs are attracted.

Histologically, prostate shows extensive changes. Its epithelium undergoes squamous metaplasia. Metastasis occurs into the regional lymph nodes, liver, lungs, kidneys, and spleen.

Sertoli-cell tumor has been seen in cocks, producing feminisation.

Clinical : If metastasis has not occurred, surgical removal of the testis is followed by drop in the urine levels of estrogens and disappearance of feminine characters.

Interstitial cell adenoma : Tumors arising from the interstitial cells (cells of Leydig) of testes are common among dogs and may also be found in bulls. These cells secrete androgens. But in dogs no clinical effects are noticed.

Macroscopically, the tumors occur singly. Rarely multiple tumors may be seen. These are round, nodular, yellow or brown in color and soft.

Microscopically, the tumor consists of large cells with large nuclei. The cells are larger than those of the two testicular tumors described above. The cytoplasm is foamy containing fat which is anisotropic. The nuclei are round. No mitotic figures are seen. The neoplastic cells are arranged in masses, separated by fibrous tissue trabeculae.

Carcinoma of the cells of Leydig is rare in animals.

Tumors of Prostate

Hyperplasia of prostate is common among dogs, especially old dogs. But tumors of the prostate are not so common.

Carcinoma of the prostate has been reported only in dogs. It is not associated with hyperplasia.

Macroscopically, the tumor may be small or may occupy the whole of the pelvic cavity. It may be nodular and white in color.

Microscopically, the tumor is an adenocarcinoma, with acini lined by columnar cells with hyperchromatic nuclei. Polarity is not maintained and numerous mitotic figures may be noticed. The cells may project into the lumen of the acinus to form papilliform ingrowths. In some places formation of cystic spaces may occur. Metastasis occurs rapidly and widely in various organs.

Embryonal Nephroma or Nephroblastoma

This tumor is known as Wilm's tumor in children.

Embryonal neophroma is known by many names of which adenosarcoma is the one most commonly used in the literature due to the fact that the tumor consists of epithelium-lined tubules in a solid fibroblastic tissue.

Embryonal nephroma is believed to arise from the primitive renal blastema that is pluripotent. It can differentiate into various tissues forming the kidney. vig. the renal tubules as well as the supporting connective tissue. Sometimes, in these tumors, muscle, cartilage and bone may also be seen.

Incidence Embryonal nephroma occurs most commonly in swine and fowl. It occurs even in young animals It is also seen in cattle, dogs, horses and sheep.

Site : Usually the tumor is unilateral and occurs in or on the kidney. Some have been seen in the sublumbar region also.

On the kidney, the tumor occupies a pole. It may be single or multiple and sometimes may occur at both poles

Macroscopically, the tumor may be small or big. Some attain a size of 20 centimeters in diameter, weighing 30 kgms. The smaller ones may be so small as to be buried in the renal tissue In color. the tumors may be white or gray. They are encapsulated, firm and lobulated On section they may show areas of necrosis which are soft in consistency. The healthy renal tissue may show pressure atrophy.

Microscopically, the epithelial part predominates in the swine while in cattle, the fibrous tissue part is more conspicuous. The tumor consists of tubules lined by flattened epithelial cells. These may project into the lumen. They may be found to form structures resembling glomeruli. In some places the epithelial cells may form solid masses or may line cystic spaces.   Squamous metaplasia may be seen. Mitotic figures are seen.

The fibrous connective tissue that is present in varying proportions may show different degrees of differentiation.   In some it may be adult tissue, while in others it assumes a sarcomatous type. In the latter, cells assume a round shape with an acidophilic granular cytoplasm and a rounded nucleus. (Round cell sarcoma). The tumor is provided with a well developed fibrous capsule from which septa enter the tumor and divide it into lobules   The fibrous tissue part of the tumor contains many blood vessels   The lumina of some neoplastic tubules contain hyaline material   in some tumors, muscle, cartilage and bone may be encountered.

Clinically, embryonal nephroma does not usually metastasize but grows by expansion. It may thus press on the bowel and cause obstruction.   Rarely metastases can be found in the lungs, liver and peritoneum.

If only one kidney is affected no urinary dysfunction is observed since its pair becomes hypertrophied doing the function of both.

Adenoma of the kidney :   This tumor has been reported in all animals but is rare.

Macroscopically, the tumors are small and found on the cortex of the kidney They may be single or multiple.

Microscopically, adenomas of the kidney consist of tubules. Papillary ingrowths of the epithelium may be present.   Cystic dilatations also may be seen. The cells are cuboidal.

Adenocarcinoma of the kidney :   This is also known as **hypernephroma.**   This tumor arises from renal tubular epithelium.

Macroscopically, the tumors vary in size. Some are quite large. One in a cow weighed 40 kgms They are rounded, nodular and lobulated. They may completely replace the normal renal tissue.

Microscopically, the tumor consists of tubules lined by cuboidal or columnar cells. In more anaplastic type these may be polygonal. They may form solid sheets or columns. Mitotic figures are frequent. Connective tissue stroma is scanty.

In more anaplastic type of tumors, the cells, besides being polygonal have hyperchromatic nuclei.   Some cells have clear cytoplasm and so are called "clear cells".   There are some cells showing hydropic or fatty degeneration. But these have no connection with the cells of zona fasciculata of the adrenal as erroneously thought formerly (so called hypernephroma—believed to have arisen from adrenal cell rests in kidneys).

Clinical :   Being highly invasive, metastases are found in the lymph glands, lungs and other organs.

Melanoma

Melanoma arises from specialised cells that produce melanin.   These are the dendritic cells (melanoblasts) situated in the stratum germinativum of the epidermis They are DOPA positive.

Though the origin of the melanoblasts has been subject of controversy, it appears to be now settled that these cells are neuroectodermal in origin.

The melanoblasts are also found in the choroid coat, the retina and clliary processes of the eye and in the pia-arachnoid of the central nervous system. So melanomas may also arise from these situations. They may arise from the modified lemmocytes of the tactile corpuscles.

Benign tumor is called melanoma and the malignant counterpart is called malignant melanoma.

The pigmented moles (nevi) of man are frequently the seats of melanomas. But in animals nevi are not present.

Incidence : Melanomas have been encountered in all animals. Mention has already been made that it is a very common tumor in old gray or white horses. (vide page 175)

Sites : Melanomas mostly arise in the skin. Rarely they may occur in the eye, brain and spinal cord.

Microscopically, the melanomas vary in size, from tiny specs to large tumors weighing upwards of 20 kgms. They are black or brown in color. They may be rounded, nodular, flat or pedunculated The tumors are firm and smooth. On section, a shiny, black surface is seen and the pigment soils the knife. Malignant melanomas may be soft. Cutaneous tumors are prone to trauma and, so may become ulcerated and infected.

Microscopically, it is not difficult to identify the benign melanomas Section consists of collection of pigment-laden melanoblasts amidst fibrous connective tissue. The cells are usually polvhedral or oval. Sometimes elongated. spindle shaped forms are also seen. The pigment obscures the architecture of the cells. In bleached sections, the nucleus may be found to occupy a corner as though the pigment has thrust it to a side. Mitotic figures are not present. Pigment may also be seen in macrophages.

Malignant melanoma Some malignant melanomas may not be pigmented and so are called *amelanotic melanomas.* The cells are so highly anaplastic and actively proliferating that melanin formation does not occur. The cells are larger than in the benign variety and are polyhedral with acidophilic cytoplasm. The nucleus is large and vesicular with stippled chromatin. Two or more nucleoli and numerous mitotic figures are present. Plentiful stroma with blood vessels is seen.

Clinical : Malignant melanoma is a highly invasive tumor and metastases are found in all organs.

Canine venereal tumor

This tumor is known by the following names also: transmissible lymhosarcoma, venereal granuloma, canine condyloma, venereal lymphosarcoma, infectious sarcoma, infectious lymphosarcoma, Sticker tumor and histiocytoma.

Incidence : The venereal tumor is found only in the dog and bitch. No other animals are affected. Bitches are more susceptible.

Sites   Since the tumor is transmitted by transplantation of tumor cells during coitus, it is found on the glans penis and on the prepuce in the male and in the vagina in the female.   Rarely the scrotum and perineal regions may show the tumors.   This tumor may also occur, occasionally, on the skin of other parts of the body.

Nature of the type cell :   Agreement has not been reached regarding the histogenesis of the neoplasm.   Feldman is inclined to suggest that this tumor is derived from "undifferentiated cells of lymphoid series",   Jackson favours the view that the cells are carcinomatous, while Mulligan considers the tumor a histiocytoma.   Bloom and his collegues utilising transmission, morphologic, cytologic, histochemical and tissue culture studies have concluded that the tumor cell is a mature end cell of reticulo-endothelial origin.   They suggest the retention of the name "transmissible venereal tumor"

Transmission :   The tumor can be transmitted from one animal to another not only by coitus but by injecting the intact cells subcutaneously or rubbing them on a wound on the skin or a mucous membrane.   *Cell free filtrates have no effect.*   Mere deposition of the tumor cells on an intact mucous membrane cannot reproduce the tumor.   There must be injury for the cells to develop.

At the site of implantation, the tissues of the host do not take part. The neoplastic cells alone proliferate.   After some time, the host develops immunity and so spontaneous regression of the tumor occurs and the animal is immune to further infection.

Macroscopically the tumor may be solitary or multiple, small or large, sessile, pedunculated or may spread like a cauliflower.   It is pink in color and very soft (encephaloid).   It may be ulcerated and blood-stained discharge may be noticed.

Microscopically, sections reveal a very cellular tumor, comprising of cells, uniform in size and shape.   The cells are round to polyhedral, having a finely granular acidophilic cytoplasm.   Rarely the cytoplasm may be basophilc. The nucleus is large, round, central and hyperchromatic.   Numerous mitotic figures are present. The cells may be grouped in compact masses or sheets or rows. A fine connective tissue stroma is seen   Generous blood supply is present· In places where ulceration is present, inflammatory cells are found.

Metastases, are rare.   Secondary tumors have been seen in the liver' spleen and kidneys.

Clinical :   If removed surgically, the tumor does not recur   This may be confused for a mastocytoma. In the latter are found the characteristic granu. les (stained by toludine blue) which are not found in the former.   Mitotic figures in mast-cell tumors are few while in venereal tumor they are numerous. Lastly, mast cell tumor occurs in older dogs while the venereal tumor in the younger ones.

Note: —There seems to be no connection between the venereal tumor and the heart-base tumor which is an aortic body tumor.

Adamantinoma or Ameloblastoma

Adamantinomas are tumors arising from the enamel organ. Since the enamel tissue is hard it is called adamantinoma, though the tumor itself is soft.

Incidence : Adamantinomas are frequent in animals. Cattle are more commonly affected, while the tumor has also been seen in dogs. horses, mules and cat

Site : The tumors arise from the alveolar border of maxilla or mandible. So they may be found to arise from the gums.

Macroscopioally, the tumor begins as a small soft nodule and may later widely infiltrate into the bone by expansion. They are round and lobulated. They may sometimes grow so big as to occupy the oral cavity.

Microscopically, the tumor consists of a dense fibrous stroma containing the epithelial neoplastic ameloblasts, which may be arranged in cysts or solid masses.

The type of cell is the embryonic enamel epithelial cell and is characteristic. It is columnar (mostly fusiform) with a faint basophil cytoplasm and elliptical nucleus having a fine chromatin and a single nucleolus. In the cysts, the single layer of cells are so arranged as to give a serrated edge to the external side of the cyst The cysts may contain an acidophilic debris. Or it may contain the stellate reticulum of the enamel organ.

In the solid variety, the cells are squamous and contain rarely, parakeratin pearls. This sometimes makes it difficult to differentiate from a squamous cell earcinoma.

Though the tumor cells resemble ameloblasts enamel is not produced.

Clinical : The tumor is of low malignancy and is locally destructive.

CHAPTER 14

THE CARDIOVASCULR SYSTEM

Developmental anomalies
Pericardium
    Hydropericardium
    Hemopericardium
    Pneumopericardium
    Hemorrhages
    Pericarditis
Myocardium
    Hypertrophy
    Intra and extra cardial hindrances
    Dilatation
    Cardiac failure
        Left-sided heart failure
        Right-sided heart failure
    Fatal syncope or Herztod
    Round heart disease
    Cloudy swelling
    Fatty degeneration
    Hyaline degeneration
    Infarction
    Hemorrhages

Mulberry heart disease
Myocarditis
Parasites of the heart
Endocarditis
Mächester wasting disease of cattle
Neoplasms of the heart
Diseases of the arteries.
    Hypertrophy
    Atherosclerosis
    Monckeberg's medial sclerosis
    Arteriolosclerosis
    Arteritis
        Equine viral arteritis
    Polyarteritis nodosa
    Thromboangitis obliterans
    Aneurysms
Diseases of veins
    Phlebitis
    Varicose veins
Diseases of lymphatic vessels
    Lymphangitis

THE HEART

The function of the heart is to maintain sufficient supply of blood to meet the needs of the body tissues.

"The cardiac reserve represents the ability of the heart to meet increased demands" that may arise during fever, exercise, anemia, pregnancy etc The heart is able to adapt itself to varying physiological needs and to pathological abnormalities. This ability is known as "compensation". For example, in a developing pulmonary fibrosis or chronic nephritis. when there is resistance to free flow of blood, the heart becomes hypertrophied so that the force of the heart is increased to overcome the peripheral resistance, thereby maintaining the volume flow and blood pressure in the vessels.

But in the face of an ever increasing peripheral resistance together with the increased work load put on the heart, the heart is not able to cope up with the demands and so becomes fatigued and fails. This state in which the heart is no longer able to compensate is called "decompensation". Decompensation is gradual and results in dilatation of the ventricles.

The factors that lead to cardiac failure are three fold:

1. Alteration in the return of venous blood: To a limit the heart is capable of adapting itself to increased inflow of blood. But if this condition persists, then the heart becomes fatigued and fails

If the venous return is inadequate, the heart is not capable of compensation to meet the situation and so finally fails.

267

2. **Increased resistance to out-flow:** We have already mentioned that heart compensates for increased resistance to out-flow. But in time the reserve power of the heart is exhausted and so fails. The causes of increased resistance to outflow are: hypertension, narrowing or dilatation of valvular orifices, thrombosis and arteriosclerosis.

3. **Impaired cardiac contraction**: Any disease that injures the myocardium reduces the contractile power of the heart and so the compensatory mechanism cannot work. In man the most important cause for myocardial injury is coronary vascular insufficiency.

Developmental anomalies

1. **Patency of foramen ovale:** Very soon after birth, the foramen ovale through which the right atrium communicates with the left, should become closed and blood diverted to the lungs. But sometimes this does not happen and so communication exists between the two areas. In human pathology, this condition gives rise to the "Blue babies". A small opening may not be of significance. But a large patent foramen leads to hypertrophy of the right ventricle. Patent foramen ovale has been met with in calves. Paradoxical embolism occurs in this condition.

2. **Interventricular foramina** Small openings connecting the ventricles are of no consequence. But if the openings are ½ to 1 cm. in diameter then the following functional changes are noticed:—elevation of pressure in the ventricle and pulmonary artery and hypertrophy of the right ventricle. A systolic murmur is heard.

3. **Patent ductus arteriosus :** Since in the fetus there is no necessity for blood to go to the lungs for oxidation, a shunt connects the pulmonary artery and the aorta. This is the ductus arteriosus, which should be obliterated within a few weeks after birth, Sometimes this closure does not happen, causing increased pressure in the pulmonary artery, leading eventually to hypertrophy of the right ventricle. Animals are cyanotic. Thrombosis of the ductus arterious, if it is patent, is a hazard.

4. **Persistence of the right aortic arch :** This is seen in dogs mostly and in bovines rarely. The esophagus is encircled and constricted by the ductus arteriosus of this aorta leading to obstruction of the esophagus resulting in dysphagia and then to dilatation of the esophagus proximal to the obstruction. Dilatation can be seen by x-ray examination. Puppies vomit immediately after taking food.

5. **Coarctation of the aorta :** This is narrowing of the lumen of the aorta, most commonly occurring close to the heart or between the origin of the common brachiocephalic artery and the ductus arteriosus. This leads to hypertrophy of the left ventricle as there is hindrance to flow of blood.

6. **Transposition of the aorta :** This is a serious condition in which the aorta arises from the right ventricle or sometimes from both the ventricles This is incompatible with life.

7. **Congenital aneurysm** of the aorta or pulmonary artery denotes dilatation of the respective vessels. These may cause pressure atrophy of the neighbouring structures. Fatal hemorrhage may result due to thinning of the vessel wall.

8 Pulmonic stenosis : This condition in which the pulmonary valves become fused   together resulting in stenosis, is the most common anomaly seen in dogs.   The right ventricle hypertrophies, the wall becoming twice as thick as normal.   A precordial thrill and a harsh systolic murmur are present.

9. Sub-aortic stenosis : This condition is seen mostly in Alsatians and Boxers.   A ring of fibrous tissue immediately below the cusps of   aortic valves causes   stenosis,   resulting in hypertrophy of the left ventricle.   **Clinically, the** following symptoms are observed; poor condition; inability to exercise: frequent fainting fits; tachycardia with poor pulse   volume;   systolic murmur and cardiac dtlatation in X-ray pictures.

10. Tetrology of Fallot : In some animals the   following   pathological changes may be present. **(i)** right ventricular hypertrophy, **(ii)** stenosis of pulmo- nary valves. **(iii)** defects in the interventricular septa and **(iv)** dextraposed   aorta. In man this condition also gives   rise to "blue babies".   Affected animals are stunted and the mucous membranes are cyanotic.   **Clinically,** the pulse **and** respiratory rates are increased and a murmur is heard.

11. Endocardial fibro-elastosis : This   condition may be seen in **dogs,** cattle and cats.   The endocardium of the ventricles is thickened due to increase in elastic fibres of the subendothelial connective tissue.   Death occurs soon after birth. Cause is not known.

12. Other defects are : (a) ectopia cardis in which the heart is found outside the thorax, usually in the neck region or abdominal cavity. **(b) acardia :** in which there is complete absence of the heart. **(c) diplocardia :** in which two hearts are present.   These defects are incompatible with life.

THE PERICARDIUM

Abnormal contents :

Hydropericardium : In this condition, there is excess of serous fluid. The causes are :

    (a) Cachectic diseases.

    (b) Congestive heart failure (in this condition there is increased pressure in the coronary veins and capillaries).

    (c) Renal disease.

    (d) Chronic stomach worm infection leading to hypopoteinemia

    (e) Damage to capillary endothelium as occurs in **(i)** many infections due to toxins or **(ii)** anoxic conditions.

    (f) Anemia, especially hemolytic.

    (g) Liver insufficiency—stasis of portal circulation, hypoproteinemia and general edema.

    (h) Mulberry Heart Disease in pigs.

    (i) Tumors : (i) implanted metastases c.i pericardium, (ii) lymphogenous metastases to the myocardium. (iii) primary tumors at tae base of the heart   and in the anterior mediastinum

    The fluid is straw-colored and clear. In some cases as mullbery heart disease and infectious diseases, floculi may be present.   In these conditions, due to damage of the capillary endothelium much protein flows out into the exudate.

If the fluid in the pericardial sac persists for a long time, it may become turbid and organised, giving a shaggy (bun-butter) appearance to the pericardium and epicardium.

Hemopericardium : In this condition there is accumulation of blood in the pericardial sac. If the clot completely encloses the heart, the condition is known as **cardiac tamponade**.

Causes : Trauma and rupture of heart, arota or coronary artery.

Pneumopericardium : Gas accumulates in the pericardium The sources may be :

(a) Activity of gas producing organisms, which may enter the pericardium with a penetrating body

(b) Gas may escape into the pericardium in traumatic reticulitis.

(c) From the lungs gas may enter the pericardium when a lesion involving the lungs and pericardium breaks down.

(d) In compound fractures of ribs, gas may enter from the outside.

Pyopericardium : Pus in pericardium is mostly seen in traumatic pericarditis. It may also be found in (i) tuberculosis.

(ii)   purulent pleuritis   } as secondary infection.
(iii) purulent pneumonia

(iv)   a rupturing myocardial abscess.

Serous atrophy of subepicardial fat : In animals suffering from cachectic diseases, the fat present in the grooves of the heart becomes transformed into a gelatinous mass. The fat is lost and in its place edematous fluid accumulates. The interstitial tissue also is edematous. White foci of necrosis may be seen in these lesions.

Hemorrhages on pericardium : Petechial hemorrhages are common in shock, toxemia and hypoxemia. Most common causes are the toxins of bacteria and viruses causing various diseases. Purpura hemorrhagica of horses is a condition in which patechiae of the pericardium are common. Ecchymoses and diffuse hemorrhages are common in sweet clover poisoning.

Pericarditis : Inflammation of pericardium is commonly seen in animals. It is usually due to bacterial infections and secondary to other diseases. The routes of infection may be :—

1. **Hematogenous**: occurs in septicemic conditions and other specific diseases.

2. **Lymphogenous** : from the inflammatory processes of neighbouring tissues—myocardium, pleura, bronchial or mediastinal lymphnodes.

3. **Trauma** : Directly from outside as in bullet wounds. Or by a foreign body entering through the rumen and reticulum (traumatic pericarditis).

Pericarditis in animals is found in the following conditions :—

Horses : Pericarditis is a complication of pneumonia, influenza or strangles. Streptococci are the organisms usually present.

Cattle : In bovine encephalomyelitis, pasteurellosis, contagious bovine pleuropneumonia, black quarter and coliform infection of the new born through the umbilicus.

Swine: In erysipelas, pneumonia (Mycoplasma Spp) pasteurellosis streptococcal infections, salmonellosis and hog cholera and infection by *Hemophilus suis* (Glasser's Disease)

Sheep: In pneumonia (Pasteurellosis), salmonellosis.

Dogs Rare, may be a complication of distemper and produced by secondary bacterial invaders and may also be found in leptospirosis.

Fowls: Fowl cholera and pullorum disease.

Pericarditis is classified according to the nature of the exudate.

Fibrinous Pericarditis: Pathogenesis: In the beginning, the surface of the serous membrane is dry following congestion. The glistening luster is lost. This phase is soon followed by marked exudation of serous fluid both within the subserosal tissue (thereby thickening it) and into the sac. Proliferation and desquamation of the mesothelial cells occur. Fibrin is deposited on the membranes as well as into the fluid and gives rise to the condition. "Shaggy heart" and "Bread and butter appearance"

Inflammatory cells consisting of neutrophils, lymphocytes, plasma cells and macrophages infiltrate into the subserous connective tissue. In infections by pyogenic organisms, the leucocytic infiltration into the fluid is so great that it becomes turbid and pyopericardium results.

If the cause is removed early exudate may be removed (the fluid part is absorbed through the blood and lymphatics and the solid matter liquefied by the proteolytic enzymes of the leucocytes which absorb it) mesothelium proliferates and fills in the places where it is lost and complete resolution is possible. But in severe infections with large amount o serofibrinous exudate such complete resolution is not possible and so adhesions occur. This is brought about by organisation of the exudate. This process involves proliferation and growth of granulation tissue from the subpericardial tissue. The organisation and resulttant scar tissue may be focal or diffuse. In the latter case the cavity is obliterated and the animal finally dies of decompensation. Occasionally caseation necrosis and calcification of the exudate may occur.

Suppurative pericarditis: This occurs more frequently in cattle due to bacterial infestation with a metallic foreign body passing through rumen, reticulum and diaphragm (Traumatic pericarditis). In poultry salmonellosis is accompanied by suppurative pericarditis. This condition may be a complication of suppurative pleuritis and bronchopneumonia of animals. Nutritional anemia in pigs, due to deficiency of iron and copper, is accompanied by suppurative pericarditis.

In this condition, especially in cattle of traumatic pericarditis, the pericardium is filled with pus. A thick membrane of fibrin forms on both the surfaces. Resolution in such cases is impossible and organisation supervenes, resulting in adhesion between the perietal surface and the epicardium—*cor rugosum* or shaggy heart. Sometimes the whole of the exudate is organised and so the pericardium becomes very thick.

The exudate and adhesions compress the heart, **constrictive pericarditis** though initially hypertrophy of the myocardium may occur. Due to mechanical interference and degeneration by toxins, the heart is not able to work properly, chronic venous congestion develops and the heart stops.   Death due to toxemia may supervene long before chronic venous congestion and edema develop.

In chronic constrictive pericarditis, the thickened pericardium that is tightly attached to the heart compresses it and so its normal diastolic filling is interfered with. So also contraction of the heart during systole is hampered. Since the roots of the great veins are compressed blood flow into the heart is obstructed. These result in chronic venous congestion and reduced output of the heart producing nutmeg liver, cardiac cirrhosis, splenomegaly, ascites, effusion into the pleural cavity and deposit of fibrin on the liver and spleen (sugar icing).

In traumatic pericarditis severe leucocytosis and extreme shift to the left are noticed.

Uric Acid pericarditis: In fowls suffering from visceral gout, the needle shaped salts of the uric acid and urates are deposited on the parietal and visceral layers of the pericardium.   This gives the appearance of fine frost.

The urates being irritants cause inflammation of the epicardium, with the resultant formation of granulation tissue and hence organisation of the deposits occurs,

Specific inflammations: Tubercular pericarditis is common in cattle and rare in dogs.

Myocardium

Hypertrophy: In hypertrophy of the heart muscle, there is increase in the size of the individual muscle fibres so that the walls become thicker. Usually the left side is more often affected and the ventricles suffer more frequently than the atria. Physiologically hypertrophy may occur when greater strain is put on it as in hearts of the race horses and grey hounds. The causes for hypertrophy may be intracardial or extracardial hindrance to the flow of blood.

Intracardial hindrances: Stenosis or insufficiency of cardiac valves give rise to this condition.

(a) **Aortic valves;** Hypertrophy of the left ventricle occurs due to steno-
     sis or insufficiency of aortic valves.

(b) **Mitral valves** ı Lesions of these valves   lead to hypertorophy of left
     auricle.

Ultimately, the lesions mentioned above lead to brown induration of the lungs, resulting in hypertrophy of the right ventricle

(c) Lesions of the valves of pulmonary artery produce hypertrophy of the
     right ventricle.

Extracardial hindrances: Right ventricle hypertrophies in pulmonary emphysema (heaves), chronic interstitial pneumonia, brown induration of lungs and pneumoconiosis·

Usually, hypertrophy and dilatation of the heart occur together.

The heart is enlarged with thickened walls. If the lumen of the chambers is narrowed, it is called **concentric hypertrophy**. Mere hypertrophy is called **simple** but if accompanied by dilatation it is called **eccentric hypertrophy**. Hypertrophy of the right side increases the width of the base, while the left sided hypertophy increases the length of the heart.   But bilateral hypertrophy causes the heart to be rounder.

The increased size of the cardiac muscle needs more nutrition, which may not be adequately supplied to every muscle fibre by the coronary vessels. Because of inadequate blood supply, metabolites formed in and arround the fibres are not removed and so accumulate. Due to these two factors, the muscle fibres get degenerated and in face of continued work of the heart under the circumstances that were originally responsible for the hypertrophy, atrophy of the muscle fibres results. Consequently, the compensated heart falters and is no longer able to meet the demands and so decompensation sets in, ultimately ending in heart failure. Dilatation of the heart that is a sequel to stasis in pulmonary circulation is called **cor pulmonale**.

Dilatation . In this condition, one or more chambers may undergo pathological enlargement, usually due to deficient emptying during systole. Frequently the right ventricle is affected.

Causes: Dilatation occurs in various infectious and intoxicating diseases in which myocardial degeneration or myocarditis is present and so the heart is not able to expell all the blood received into it and hence blood accumulates in its chambers dilating them.

Sudden acute dilatation is attributed to the cumulative action of toxic products on cardiac muscle in severe diseases.

Chronic dilatation usually occurs with hypertrophy in which it is a terminal lesion.

In dilatation, the heart is rounder and more globular. The walls are thinner and the papillary muscles are attenuated. Very great dilatation of the heart of man is called **cor bovinum**. Dilatation leads to congestive heart failure.

Cardiac failure : We have already noticed that the heart has great reserve powers and that it can adjust itself to increased demands placed upon it due to increased venous return and to increased resistance to outflow. In long continued states where this demand still exists decompensation sets in, especially when there is impaired cardiac contraction, resulting in cardiac failure.

Though failure of one side of the heart ultimately produces excessive strain on the other side, it is convenient to study the pathological physiology of each side separately.

Left-sided heart failure : The most common causes are :

(1) Hypertension
(2) Arotic valvular disease   } Stenosis and incompetence due to
(3) Mitral valvular disease   }   endocarditis
(4) Congenital heart disease
(5) Myocarditis
(6) Myocardial degeneration
(7) Adhesive pericarditis
(8) Nephritis in dogs.

Majority of the clinical symptoms in left-sided heart failure stem from a) diminished blood flow through the various organs and tissues of the body and (b) from pulmonary congestion caused by venous stasis. The following changes in the organs with related symptoms occur :

Lungs : Due to damming back of blood in the lungs and diminished cardiac output, there is venous congestion in lungs which is transmitted to the alveolar capillaries resulting in accumulation of the edema fluid in the alveoli. Sometimes, small capillaries may rupture resulting in small hemorrhages into the alveolar spaces (heart-failure cells are seen in such an event).

Impaired exchange of gases and reduced vital capacity of the lungs that result due to edema, cause hypoxic stimulation of the carotid sinus and the respiratory centre and so reflex dyspnoea occurs. Cough occurs due to the irritation of the respiratory mucosa by the edema fluid.

Kidneys : In the congestive left-sided heart failure, there is significant alteration in renal function. Renal excretion is impaired due to renal anoxia that results from impairment of circulation (Vasoconstriction). Salt and water are retained, raising thereby the blood volume which further adds to the workload of the heart. Further impairment in renal blood flow results in diminished excretion of nitrogenous substances—prerenal uremia. The condition is called prerenal because, the fault lies not in the kidneys, but in some prerenal or cardiac lesions. Retention of sodium and water accentuates edema, noticed first in the dependent parts of the body.

Brain : Anoxia of the brain results in increased, irritability restlessness and in far advanced cases, stupor and coma.

Right-sided heart failure: Right-sided heart failure usually occurs with the left-sided heart failure an t it is only rarely that right-sided heart failure in a pure form occurs. Because, congestion of pulmonary vessels that occurs in left-sided failure must ultimately embarrass the right side of the heart. Hence the causes for right-sided failure include those enumerated for the left side, especially mitral stenosis and incompetence, which greatly increase the pressure in pulmonary vessels.

Causes for the pure right-sided heart failure are:—

1. Myocarditis. As the right ventricle is weaker than left, it fails more rapidly when myocardium is damaged.

2. Myocardial infarction and degeneration.

3. Causes producing increased resistance for the flow of blood in the lungs—emphysema (heaves in horses) and chronic interstitial pneumonia.

4. Constrictive pericarditis: in this condition the flow of blood into the heart is blocked.

5. Hydropericardium. Here also blood is blocked from entering the heart.

6. Endocarditis, especially of tricuspid valves, producing incompetence and stenosis

In right-sided heart failure, the primary disturbance is damming back of blood in the systemic and portal venous circulations. with consequent decreased flow of blood into the left auricle from the lungs. Anoxia that is caused thereby produces renal pathology in which there is retention of salt and water increasing the blood volume. So edema results. In the horse and ox, edema is seen subcutaneously, while in the dog, ascites is manifested and in the cat pleurisy is seen.

Liver; Because of congestion the liver is enlarged. In the acute and severe phase, pulsation may be present. There may sometimes be actual hemorrhage round about the central vein. Atrophy and necrosis of the hepatic cells around the central vein are present. If the animal lives long enough (usually animals do not survive for this length) fibrosis of the central part will occur — *cardiac cirrhosis* Slight hepatic dysfunction and icterus may be evident. But clinically impaired hepatic function is not a feature.

Kidneys: In the right-sided failure, congestion and anoxia of the kidneys is more marked than in the left-sided failure. Hence renal function is much affected and there is, therefore, greater retention of salt and so the volume of blood is much increased. Due to the decreased renal blood flow there is decreased formation of glomerular filtrate and hence reduced sodium filtration occurs. The tubules reabsorb the little salt that is excreted. So along with the salt water is also reabsorbed and hence the increase in the blood volume; which together with venous stagnation produces interstitial edema, manifested in the dependent parts.

Spleen: Splenomegaly is seen due to congestion. Hemorrhages may be present liberating hemosiderin followed by organisation and fibrosis. Such areas may be calcified. Metaplasia of the reticulum to fibrous tissue causes hardening of the organ.

Other Symptoms· Due to congestion in the portal system, the stomach and intestines may also manfest venous stasis with resultant digestive troubles—diarrhoea. Cyanosis is also seen. In the horses epistaxis may occur.

In summary, therefore the pathology and manifestation in the left-sided heart failure are essentially results of stasis in pulmonary circulation, while stasis in the systemic circulation is the cause of the pathology seen in right-sided failure.

Cardiac failure may be unilateral but usually it is bilateral. Heart failure, therefore, is a syndrome of failing circulation in various organs. Hence to detect heart failure clinically or at autopsy by just examining the heart is very difficult We have to examine other organs to arrive at a correct diagnosis.

Fatal Syncope of pigs, Herztod

This is a very serious disease in pigs, which suddenly fall down and die while being driven about or while feeding or during mating. If animals could be observed while alive, symptoms of dyspnoea, paralysis and spasms can be seen.

Cause: The exact cause of the disease is not yet clear. It is probably a nutritional condition. There may be hormonal imbalance. the hormones concerned being the Thyrotropic hormone, the ATCH and the Growth hormone. A diet rich in carbohydrates and poor in proteins (feeding on potatoes) producing vitamin and protein deficiency has been cited a cause. An allergic etiology is also put forward.

Lesions: Serous cavities contain large quantity of serous fluid. All organs show passive congestion Heart shows grayish streaks and patches of hyaline degeneration with round - cell infiltration. Lungs are hyperemic and edematous. Skeletal muscles show hyaline degeneration or cloudy swelling with loss of cross striations or even atrophy. Passive congestion of the liver and severe passive congestion of the intestinal mucosa are also observed.

Thyroid is congested due to extreme dilatation of the blood vessels. The follicles show collapse due to separation of the epithelium from the basement membrane by subepithelial edema. Infiltration by lymphocytes is observed The colloid is watery and thin. The adrenals are atrophied and the immediate cause of death may be adrenocortical insufficiency. (See page 149 for fatal syncope in cattle caused by copper deficiency)

Round Heart Disease

Sudden deaths in hens under eight months of age may occur, especially in summer months.

Causes: A diet rich in carbohydrates and poor in protein is suspected. Poisoning by zinc is suggested by some workers. No evidence of infection has been seen.

Macroscopically, the heart is greatly enlarged and appears rounder, the apex losing its normal conical shape The left ventricle especially, is hypertrophied and dilated. Gray streaks or spots may be seen on the myocardium. Hyperemia of liver, kidneys, intestines and lungs may be observed. Pericardium may contain fibrinous exudate. Edema of the subcutis and lungs may be noticed. Comb is cyanotic.

Microscopically, the myocardial fibres are hypertrophied. In some areas fatty degeneration of the heart may be seen. In others small nodules consisting of accumulation of histiocytes and lymphocytes may be observed.

DISTURBANCES IN METABOLISM

Cloudy swelling: This condition is seen in different toxemias and septicemias, and in febrile conditions.

Macroscopically, the heart is slightly enlarged, has a cooked, pale appearance and is friable.

Microscopically, the muscle fibres are sightly swollen and their cytoplasm is granular, cross striations are indistinct.

Since autolytic changes occur soon after death simulating cloudy swelling care should be taken to distinguish one from the other.

Fatty degeneration: In this condition, the irritants are more severe than those causing cloudy swelling. Fatty degeneration is met with in prolonged infections; piglet anemia, toxemia, phosphorus, arsenic and chloroform poisoning, purpura hemorrhagica in horses, pyometra; avitaminosis. E.

Macroscopically, the heart is enlarged and yellowish in color. In severe fatty degeneration yellowish striping of the endocardial surface-'tigering' is seen. "Thrush-breast" heart is observed when there is mottling of subendocardial fibres. especially on the papillary muscle

Microscopically, minute droplets of fat, detected with special stains are seen in the muscle fibres, which normally do not contain any. The nuclei are degenerated.

Obesity: It is normal to find deposition of fat subepicardially. But fat is not found in the myocardium. In some cases of obesity, fat of coronary adipose tissue infiltrates between the myocardial fibres—fatty infiltration. This fat crowds out the fibres and may interfere with the action of the heart resulting in its failure.

Hyaline degeneration: Hyaline degeneration is best seen in the conditions, called white-muscle disease in calves and stiff lamb disease in lambs due to vitamin E deficiency. This may also be seen in myocarditis, in gossypol poisoning, 'Herztod' in swine and copper deficiency (falling disease) in cattle The muscle fibres are homogeneous and glassy. This is a prenecrotic lesion, terminating usually, in necrosis and calcification.

Necrosis of the myocardium is usually focal and is seen as grayish spots scattered in the muscle. This is a common lesion seen in white muscle disease of calves and lambs. Thiamine deficiency in pigs has been found to produce coagulative necrosis of the myocardium.

Hyaline degeneration and necrosis of the myocardium are seen in Foot and Mouth Disease in cattle: in Equine Viral Arteritis and in Swine Fever,

Calcification: Dystrophic calcification of necrotic myocardial fibres may occur. So this is common in white-muscle disease. In cattle poisoning by organic compounds of mercury causes hyaline degeneration of Purkinje fibres, followed by calcification.

In puppies excessive vitamin D or calcium therapy results in calcification of the myocardium.

Disturbances in circulation

Infarction of heart is relatively uncommon. This condition is due to occlusion of the coronary artery by a thrombus, atheroma or arteriosclerosis. Rarely emboli arising from cardiac vegetations (in swine) may give rise to infarction.

The sudden occlusion of a large artery may cause necrosis and deaths. But gradual obstruction of smaller branches will cause atrophy and replacement fibrosis, which being weak, is a place for dilatation to occur. This is called *cardiac aneurysm.*

Hemorrhages: petechiae under the epicardium and larger ecchymoses under the endocardium are common in cattle and sheep and are encountered in toxemias, septicemias and in death due to asphyxia.

MULBERRY HEART DISEASE
(By Dr. M. S. Kwatra)

Mulberry heart disease in pigs is a dietetic microangiopathy which is characterised by the sudden death due to heart failure, hydropericardium and typical linear hemorrhages on the heart giving it an appearance of a bunch of mulberries. The piglets belonging to Landrace breed are relatively more susceptible to this disease and the cases mostly appear on the farm during the spring season. The disease has recently been noticed in Assam. Though the morbidity rate is generally low, mortality rate is quite high. The disease commonly occurs among 3-4 month old pigs. especially in active healthy looking ones. The etiology or the cause is unknown. The disease is considered to be a dietary deficiency and addition of vit E and selenium to the diet has a beneficial effect Generalised edema is suggestive of damage to endothelium which may be due to toxins of bacterial origin. In majority of cases no symptoms are observed as the course of the disease is extremely short. Some cases may show incoordinated gait, dyspnoea, muscular tremors and weakness. The temperature is usually within the normal range. The chronic cases show nervous symptoms and poor weight gains

The characteristic gross lesions are :

1. Edema and emphysema of the lungs
2. Hydropericardium
3. Hydrothorax

The fluid is so rich in fibrin that it clots

4. Diffuse hemorrhages on the epicardium and endocard.um giving the heart a mulberry appearance.
5. Ascites
6. Hyperemia of gastrointestinal tract.
7. In animals with delayed death softening of the cerebral gyri (due to anoxia consequent to pulmonary lesions)

Microscopic Changes

A. Heart

( ) The blood vessels of the myocardium reveals generalised, hyperemia and typical fibrinoid degeneration in several blood vessels.

(ii) The myocardial fibres show necrobiotic alterations.

(iii) Hemorrhages and edema in between the bundle fibres are seen.

B. Liver sections reveal acute passive hyperemia and degenerative changes of hepatic cells in the cetrilobular areas.

C. Lungs also show passive hyperemia, enlargement of the lining cells of the alveoli, marked alveolar emphysema and edema.

D. Leucoencephalomalacia in white matter of the cerebrum is generally observed.

The mulberry heart disease appears to have no etiological relationship with 'Gut Edema Disease'.

Myocarditis

Though myocarditis is a common lesion found in many systemic diseases primary condition is rare.

Non-suppurative myocarditis: This is found as a hematogenous infection in generalised septicemias, toxemias and bacteremias.

Cattle: pasteurellosis, Foot and Mouth Disease.

Dogs: extension from valvulitis: from infective **focus somewhere** else—streptococcus or coliform septicemia:hapatitis contagiosa canis; leptospiral infecions.

Horses : infectious anemia

Extension of infection may occur from pericarditis also.

The lesions of myocarditis are usually focal and consist of pale or yellowish or grayish areas. In acute Foot and Mouth Diesase in young calves with supervening death, there is myocardial degeneration and diffuse myocarditis, With necrophorus infection, gray areas of coagulative necrosis are found.

Microscopically. there is degeneration and necrosis of the muscle fibres in the acute stage with infiltration of lymphocytes. plasma cells, macrophages and eosinophils.

In the chronic stage, there is much fibrosis since the heart muscle does not regenerate and healing occurs by granulation tissue.

Acute suppurative myocarditis: This is found in pyemia that occurs in mastitis. metritis and joint ill. The spread is by way of the coronary arteries. Direct extension from purulent pericarditis, endocarditis, pleura. lungs and bronchial lymph glands is also possible. Infection may also occur through a foreign body penetrating the myocardium through the reticulum

Macroscopically, the heart reveals abscesses with hyperemic borders. Some abscesses may be encapsulated.

Microscopically, the typical appearances of an abscess are seen with neutrophils abounding. Healing. may occur by organisation and scar formation. Sometimes abscesses may be calcified.

Specific inflammations: Tubercular myocarditis is common, especially hematogenous. Extension from neighbouring tissues is also possible. Actinomycosis of the myocardium is also met with.

Parasites of myocardium:

1. Heart worms *(Dirofilaria immitis)* are found in the right ventricle of dogs, sometimes completely filling the lumen. Depending on their number, these round worms may cause thrombosing endocarditis leading to cardiac dilatation and hypertophy or thrombosis of pulmonary arteries.

2. In the heart muscle are found the following:

   (a) *Sarcocystis tenella.* These are found universally in the hearts of cattle, sheep and pigs. The muscle fibres contain the Miescher's tubes in which are found the spores—Rainey's corpuscles.

   (b) *Toxoplasma gondii* may be present in the heart muscle fibres as pseudocysts. If these cysts rupture, a focal myocarditis develops. The lesion consists of a necrosed centre surrounded by inflammatory cells—neutrophils, histiocytes and lymphocytes. Sometimes the necrosed tissue may be calcified.

   (c) *Cysticercus bovis* in ox heart is the bladder worm of *Tenia saginata* of man.

   (d) *Cysticercus cellulosae* in pig heart is the bladder worm of *Tenia solium* of man.

   (e) *Cysticercus ovis* in the sheep is the bladder worm of *Tenia ovis* of dog and fox.

   (f) Hydatid cysts of *Echinococcus granulosus* of dog contain scolices and clear fluid.

Endocarditis

Inflammation of the endocardium is rather common in animals and is almost always caused by bacteria. Though the endocardium of all chambers and valves can be affected, it is the valvular endocardium that is mostly affected primarily. Form here, inflammation may extend to the mural endocardium.

Among animals, swine are more frequently affected.

Endocarditis is a constant lesion seen in chronic septicemic diseases, in which the circulating bacteria infect the endocardium The following organisms have been found to be causative.

Horse.*Streptococcus equi*(strangles),*Shigella equirulis*(umbilical infection) Migrating larvae of *Strongylus spp*, *Actinobacillus equili* and *Meningococcus spp*, (while hyperimmunizing with meningococci) are other causative agents.

Cattle : *Corynebacterium pyogenes* (secondary to a liver abscess or peritoneal abscess); Streptococci of intestinal origin (white scours); embolic infection from traumatic reticulitis, suppurative metritis and mastitis can occur.

Dog : Not usually observed but may exist in association with strepto-coccal infections of mouth, teeth and pharynx. Leptospirosis may also be asso-ciated with endocarditis.

Pig : *Erysipelothrix rhusiopathiae*, *Corynebacterium pyogenes* and *Streptococci*. Of these the first is most important.

Sites In the order of frequency of occurrence, these are :

In horse : Aortic valves; right auriculo-ventricular valve; left auriculo-ventricular valve and pulmonary valve.

In Cattle ; right auriculo—ventricular valve; left auriculo-ventricular valve; pulmonary valve and aortic valve.

In pig and dog : mitral valve, aortic valve; tricuspid valve and pulmo-nary valve.

The valves are more often affected because these are exposed to the circulating bacteria and to the force of blood during systole. Bacteria that are present in the blood get implanted on the valvular endocardium. The surface of the valve that is exposed to the force of blood is more often affected, viz, the auricular surface because when the blood is forced into the ventricle, it is this side that is continually brushed. Besides, it is the edges of the valve (lines of closure) which are most exposed to stress and trauma and so are more often affected. The implanted organisms grow on the injured endocardium. The toxic metabolic materials released by these bacteria damage the local cells, with the liberation of thromboplastin which converts fibrinogen into fibrin. Fibrin is a good medium for the bacteria to grow and so more and more of fibrin is formed due to the liberation of greater quantities of thromboplastin by the cells that are increasingly destroyed by the growing organisms. Thus a thrombus is formed. The enlarged valve (with the thrombus) injures the adjacent surface during its movement and during this motion the thrombi may break off and form emboli.

The lesion being chronic, the thrombus is formed slowly but progressi-vely, This is friable and resembles the head of a cauliflower and so is called vegetation. Endocarditis in which these vegetations are present is called

PART II

Special Pathology

tative endocarditis. From the basal area of the valve granulation tissue invades the thrombus which is thus organised in the deeper layers. But so long as the organisms are alive, complete organisation and healing are not possible

In swine excessive vegetations are common. But in cattle due to C *pyogenes* excessive fibrin is not common but fibrosis is frequently seen. Extension of infection from the tricuspid and bicuspid valves to the chordae tendinae makes them weak and degenerated, resulting in their rupture.

Mural endocarditis is only an extension of the inflammatory process from the valves. In cattle affected with black quarter, the wall of the left auricle is usually affected showing roughened endocardium.

In the dog. with uremia and leptospirosis, ulcerative endocarditis is common. Greenish ulcers are present in the left auricle and ventricle, pulmonary artery and aorta.

Microscopically, the lesion is a thrombus in the centre of which bacteria are seen, Leucocytes are present in the intima, Fibrosis in the chronic cases is a feature noticed

Effects of Endocarditis : The vegetations which are formed at the points of contact of the cusps of the valves prevent the closure of the valves and obstruct the lumen, thereby hindering free passage of blood through the lumen. These conditions thus give rise to valvular insufficiency or stenosis. The effect of these conditions is to cause accumulation of blood in the chamber just preceding the lesion. The following sequelae are met with in lesions of different valves

Tricuspid valve : Dilatation of the right auricle, general venous congestion and ultimately hypertrophy of the right auricle.

Pulmonary valve: Dilatation of the right ventricle, general venous congestion and hypertrophy of the right ventricle.

Mitral valve : Dilatation of left auricle, pulmonary congestion, edema and pneumonia (brown induration), hypertrophy of the left auricle, hypertrophy of the right ventricle.

Aortic semilunar valves; Dilatation of the left ventricle; hypertrophy of the left ventricle and later general venous congestion. The final outcome of these lesions is congestive heart failure. Fragments of the thrombus (vegetation) may be detached to form emboli.

The results of embolism are:

1. If the emboli are contaminated with pyogenic organisms, abscesses will be found in the kidneys (acute embolic nephritis), spleen and liver from emboli arising from left side; pulmonary abscesses from emboli arising from the right side and myocardial abscesses from emboli of coronary vessels.

2. If the emboli are bland and sterile, infarction will be caused in the kidneys and spleen from emboli arising from the left side and pulmonary thrombosis from emboli arising from the right side. Myocardial infarction may occur if emboli enter coronary circulation.

Endocardiosis in dogs:- Otherwise known as "Chronic valvular disease", in this condition the mitral valve is mostly affected. The affected valve is shrunken and distorted and so is incompetent producing gradually congestive heart failure.

The atrio-ventricular cusps are thickened, become shorter, the chordae tendinae are thickened and may sometimes be ruptured. The thickening is due to proliferation of fibroelastic tissue with abundant mucous ground substance.

MANCHESTER WASTING DISEASE OF CATTLE
(By Dr. M. S. Kwatra)

(Synonyms: Enteque seco, pasteur disease, Naalehu, Calcinosis, Calcific arteriosclerosis.)

The disease is characterized by progressive wasting, stiffness of forelegs and back and deposition of calcium in the arteries, heart muscle, lungs and kidneys. The condition has recently affected large number of Corriedale sheep and a few cattle in Ludhiana district of Punjab The disease is also found in Jamaica, Brazil and Hawaii and causes severe losses in livestock industry. The disease generally affects cattle over 15 months of age.

Symptoms: The affected animals waste away progressively; the joints of the limbs become stiff, and fore limbs are often severely affected. The animals cannot walk properly and sometimes put weight on the knees. As the disease advances, the animal is disinclined to move and stands with arched back for long periods. On auscultation heart murmurs may be heard. On exercise animals appear to be distressed Blood calcium level is increased in affected animals.

Cause: In different localities grazing on different plants containing excessive amounts of vitamin D or toxic substances causing hypercalcemia results in the calcification of arteries, heart, pulmonary tissues and kidneys. *Solanum malaecoxylon*, *Cestrum diurnum*. *Trisetum flavescens* and other plants have been found to contain toxic concentration of vit. D or some other toxic substances with similar action.

Gross Changes: The general condition of the animals is usually poor. The aorta and major arteries show plaque formation in the intimal layer. The walls of the affected vessels are hard, thick and brittle, particularly during the advanced stages of the disease In worst cases mitral valves, myocardium, lungs and kidneys may show calcium deposition. In mild cases the plaque formation is seen in the terminal portion of the aorta and in moderate cases it may extend upto the aortic arch. In severe cases whole of aorta, mesenteric,and carotid arteries are affected while in advanced cases the myocardium, kidneys and lungs also show calcific deposits. Hydropericardium, hydrothorax, hydroperitoneum may be present in cases in which the cardiac valves and lungs are involved.

Microscopic Changes; the medial layer of the aorta and arteries is intensely involved and muscular fibres show degenerative changes and deposits of calcium salts. The tunica intima shows slight proliferation of intimal connective tissue while the tunica adventitia remains, unaffected. The cardiac muscle fibres undergo calcification and degenerative changes. The coronary arteries may also reveal deposition of calcium in the tunica media and intima. In the kidneys the arterioles of the cortex and medulla are calcified. The calcification of the lungs is confined to the thickened interalveolar septa. The alveolar epithelium undergoes foetalization,

Neoplasms of the heart: primary Not common, Rhabdomyosarcoma, fibroma and myx. fibroma have been described.

Secondary: These are the commonest. The most common secondary neoplasm of the heart is the lymphosarcoma among cattle. Tumor masses may project into the pericardium or into the chambers. Diffuse infiltration of the myocardium is also seen and the neoplastic cells may be found between the muscle fibres, which undergo pressure atrophy.

DISEASES OF THE ARTERIES

Hypertrophy, Compensatory hypertrophy of the walls of the arteries is common and may affect one or all of its components. Mostly the muscle and elastic lamina are hypertrophied to withstand increased blood pressure, This is seen in renal vessels of dogs suffering from chronic interstitial nephritis and in lungs of cats

Arteriosclerosis: The exact meaning of arteriosclerosis is "hardening of the arteries" This condition is common in man, characterised by intimal thickening following proliferation of connective tissue, hyaline degeneration, infiltration of lipoids and finally calcification.

In animals there is no comparable lesion as occurring in man, The dog (with hypothyroidism) is most frequently affected among animals. Usually, the older animals are affected. One difference in the dog is that infiltration of lipoid is unusual.

In man, the lesion is described under the heads: atherosclerosis, Monckeberg's medial sclerosis and diffuse arteriolosclerosis.

Atherosclerosis: This condition affects the larger elastic arteries. Athere means in Greek a soft, mushy, gruel - like substance. In this condition such a substance is formed in the intimal layer.

The aorta and its primary branches are mostly affected though carebral and coronary vessels may sometimes be involved.

Pathogenesis: 1. The lesion commences as a focal degenerative change in the subenjothelial tissue. The mucinous ground substance of intima is increased and in these foci fine fat droplets appear. These fatty deposits consist of cholesterol, neutral fats, fatty acids and cholesterol esters. Elastic fibres disintegrate.

2. Foam cells (macrophages laden with lipoids) are found in these foci It is conjectured that macrophages transport the fats to the lesions directly from the blood

3. These macrophages cluster in the subendothelial area.

4 Subsequently the macrophages die, disintegrate and liberate the fats into the tissue spaces.

5. The fats (cholesterol crystals especially) stimulate the proliferation of the connective tissue around these foci, especially towards the luminal side. This newly formed tissue becomes hyalinised.

6. This is the plaque (atheromatous plaque), with a central debris consisting of granular, lipid-rich material and acicular crystals of cholesterol. Hemorrhages and hemosiderin granules may also be seen,

7. A well-formed plaque is supplied with numerous capillaries, which are the source of hemorrhage.

8. A few lymphocytes may be found around the lesion.

9.   Fibrous tissue increases in quantity and with further deposition of lipoids, the atheroma enlarges in size and reaches the endothelial layer, which may be pushed into the lumen.

10.   Along with the changes noticed in the intima, degenerative changes are noticed in the media.   Edema first occurs separating the muscle and the elastic fibres.   In these foci collagen is formed and scarring occurs.   The elastic fibres degenerate in the focal areas but proliferate in the adjacent areas.

11.   The fate of plaque : (1) it may be converted into a dense inflammatory scar, containing cholesterol clefts; or (ii) it may be calcified or (iii) islands of heterotopic bone may be formed in it; or (iv) the thin endothelium may be necrosed due to the subendothelial accumulation of marcrophages and fatty debris, resulting in ulceration into the lumen of the blood vessel—the atheromatous ulcer.   A thrombus may subsequently form on this ulcer.

Causes of atherosclerosis : No definite cause is attributed.   The following are some of the etiological factors suggested in man.

1.   Senility or ageing process, in which there is progressive degeneration of the walls of arteries leading to fatty degeneration of the tissue in which lipids accumulate.

2.   High blood cholesterol and lipid content.

3.   Hypertension.

4.   Intramural hemorrhages, which may be the starting points of the lesions.

5.   Probably some endocrine deficiency.

Heredity, obesity, stress, physical activity and smoking habits may be contributory factors.

Atherosclerosis as occurring in man is found only in the pig among animals.   Atherosclerosis encountered in the aorta of pigs was reported from India.   Abdominal aorta was found to be more often affected.

Maroscopically, the surface of the aorta revealed fatty streaks.   Frank fibrous plaques were not seen.

Microscopically, there was edematous swelling of the ground substance in the intima, deeper parts of which showed deposition of variable amounts of lipid.   The number of smooth muscle cells increased and marked collagenization was observed in a few cases.   It was thought that the initiation of fatty streaks in the pig was not mediated through the thrombotic mechanism, though in man microthrombi composed chiefly of platelets were supposed to be the primary initiating factor in the development of atherosclerotic lesions.

It has been mentioned earlier that among animals, atherosclerosis is more often seen in dogs, suffering from hypothyroidism with hypercholesterolemia.   The pathogenesis is slightly different.   The lesion commences in media (as against the intima in man and fowl) in the middle and outer layers of which lipoids are deposited.   Foam cells appear here and the lipoid may be found within the muscle fibres which show hydropic degeneration.   Due to the pressure of lipoids the fibrous connective tissue proliferates and replaces other structures.   This becomes hyalinised and later impregnated with iron salts.   In the adventitia lymphocytic infiltration is seen, sometimes to such an extent as to form nodules.

Along with the changes in the media the intimal elastic fibres are disrupted, destroyed and reduplicated. The endothelium shows hydropic degeneration and appears thickened. Later, there is fibroblastic proliferation that occurs around these thickened areas (plaques) of intima.

Macroscopically: the affected vessel is enlarged and less pliable and the walls are thickened. On opening the aorta, raised oval or round areas (plaques) may be seen. These are white or yellow and may be 0.1 m.m to 2 cms, in diameter.

Monckeberg's medial sclerosis : In this condition the medium sized muscular arteries are affected. The muscular tissue undergoes hyaline and fatty degeneration followed by necrosis and calcification. Sometimes heterotopic ossification occurs. This condition is found in older people but is not necessarily associated with hypertension. The disorder is considered to be related to prolonged vasotonic influences—prolonged action of epinephrine and nicotine. Hypervitaminosis D may also produce similar lesions. Medial calcification is seen in some dogs suffering from chronic interstitial nephritis.

The lesions are found only in the media and do not encroach on the intima and vessel lumen. The endothelium is intact.

Arteriolosclerosis : In this condition, there is thickening of the walls and narrowing of the lumina of small arteries and arterioles. Two types are recognised; (1) hyaline arteriolosclerosis and (2) hyperplastic arteriolosclerosis. Both forms result from hypertension. Hyaline arteriolosclerosis occurs in a slowly developing hypertension while the hyperplastic variety is a more acute condition developing due to a sudden elevation of blood pressure.

In tha hyaline arteriolosclerosis, there is homogeneous pink collagenous fibrosis and thickening of the walls of the arterioles. The cellular details of the tissue are completely lost. In the hyperplastic arteriolosclerosis an "onion skin" appearance is found due to the proliferation and concentric arrangement of the endothelial cells, subintimal fibroblasts and even the muscle cells of the media. Sometimes, there may be an admixture of the two forms. Being a reaction to increased blood pressure, arteriolosclerosis is met with in all the small arteries and arterioles, particularly those of kidney, spleen, pancreas, adrenal and small intestines.

In general, arteriolosclerosis in animals is not common and is not of clinical significance. Atheromatous plaques with calcification are frequent in the aorta of adult cattle, caused by *Onchocerca armillata* and found only at post-mortem. In dogs atheromatous ulcers may be found in infection by *Spirocerca lupi* in the migratory life cycle of which the larvae at one stage pass through the aorta.

Arteritis

The inflammation of the wall of arteries is arteritis, which may be acute or chronic.

Acute arteritis may be caused by parasites, bacteria, viruses or fungi. The routes of infection are (1) from the outside of the vessel, extending through the wall: (2) through the vasa vasorum and (3) from the lumen of the vessel.

Extension of inflammation from the adjacent tissues is common as in pneumonia, metritis and mastitis. Infection may also occur due to pyogenic bacteria, the primary lesion being elsewhere—umbilical abscess. In such cases the pulmonary vessels are the favoured places where emboli lodge and produce inflammation.

Inflammation of the intima results in the formation of a thrombus at the site—thromboendarteritis.

Equine viral arteritis : A primary acute arteritis in the horse is caused by a virus which cannot be grown on egg embryos or propagated in experimental animals. Serial passage through tissue culture attenuates it and such an attenuated virus confers strong immunity when vaccinated into horses The clinical picture is:—fever, leucopenia, conjunctivitis, rhinits with mucopurulent nasal discharges, palpebral edema, respiratory distress, depression, incoordination and edema of the limbs and abdominal wall, colic and diarrhoea, abortion in pregnant animals.

Macroscopically, petechial hemorrhages are seen in all serous membranes, in the lungs and gastric mucosa. Edema of the eyelids is prominent. All serous cavities contain excess of fluid with high protein content. The intestinal wall is thickened by edema, which is present in the mesentery and lungs also Enteritis is seen.

Microscopically, the lesions are found in the media of smaller arteries. The muscle fibres are necrosed and replaced by a hyaline fibrinoid material. There is edema of the adventitia with lymphocytic infiltration The intima and endothelium are usually intact showing no thrombosis. But the arteries of the intestines and lungs are severely affected resulting in thrombosis. Infarcts may occur in the mucosa of cecum and colon.

Chronic arteritis : This is exemplified by the arteritis of the anterior mesenteric artery in the horse due to *Strongylus vulgaris,* which by its presence causes chronic irritation. The artery is dilated, fibrosed and the wall loses its resiliency. The intimal surface becomes roughened where a thrombus forms Due to loss of elastic tissue and continuous pressure of the blood, the wall dilates and an aneurysm forms. Sometimes, the aneurysm may rupture with fatal results.

Polyarteritis nodosa; (Also known as periarteritis nodosa) This is one of the group of collagen diseases.

There is inflammation involving all the layers of the wall. Bacteria, viruses and allergy to drugs and streptococci have been thought of to be of etiological significance. Among others, sulfonamides, arsenic, iodum and desoxycorticosterone acetate have been incriminated. In some animals sarcosporidia have been suspected. Probably this parasite serves as a long acting antigen. Lesions very similar to the naturally occurring condition in man have been produced in horses by sensitising the animals to foreign protein. Among animals periarteritis nodosa is met with in viral diseases—malignant catarrhal fever in cattle, in equine infectious anemia. and in sporadic encephalomyelitis in bovines.

Small and medium sized arteries located in general musculature, myocardium, subepicardial fat and mammary gland are involved The inflammatory changes begin in the adventitia or media with edema in these places. Fibrinoid necrosis of the media occurs converting it into an eosinophilic granular material. Necrosis is followed by infiltration of neutrophils and eosinophils into the adventitia mostly and to a lesser extent into the media. After the acute phase subsides the exudate is organised. At this time macrophages and lymphocytes infiltrate. Thrombosis may occlude the lumen. Due to weakening of the wall, small aneurysms form, giving a nodular appearance externally to the affected vessel.

Thromboangiitis obliterans or Buerger's disease : This condition usually occurs in the limbs and is met with more frequently in males of Jewish race, especially in smokers. There is acute inflammatory reaction involving all the layers of the wall, resulting in inflammatory thrombosis. There is proliferation of the endothelial cells and fibroblasts. Organisation of the thrombus and subsequent canalisation may occur.

Aneurysm : An aneurysm is a localised dilatation of an artery, vein or a cardiac chamber.

Causes : The main factor involved in the causation of the aneurysm is weakening of the wall. This weakening may arise due to damage of the media. The various causes that weaken the wall are :—

1. Syphilis in man. 2. Infected embolus may cause suppuration and destruction of the media. 3. Polyarteritis nodosa. 4. Trauma. 5. **Parasites—** *Strongylus vulgaris.* 6. Infection from an abscess or tuberculous lesion may weaken the wall. 7. Congenital weakness of the walls.

Varieties of aneurysms :

1. **True aneurysm** : In this variety the sac is formed by the wall of the artery.

2. **False aneurysm** : The sac in this condition is not formed by the wall of the artery but by the surrounding connective tissue. This occurs due to rupture of the vessel by trauma.

3. **Fusiform aneurysm** is one in which a long segment of the vessel is uniformly dilated around the whole circumferance. This is mostly seen in the aorta and its branches.

4. **Saccular aneurysm** is the formation of a pouch on one side of the wall.

5. **Dissecting aneurysm** : Strictly speaking this is not a true aneurysm since there is no dilatation of the wall. In the aorta, hemorrhage occurs between the layers of media and blood circulates around within this space, dissecting the wall. Fatal hemorrhage may supervene. The condition is usually due to a degenerative lesion in the media.

6. **Cirsoid aneurysm:** These are a mass of dilated, pulsating and intercommunicating arteries and veins. usually subcutaneous in location and most are congenital. A few may be due to trauma.

7. **Arteriovenous aneurysm** : (aneurysmal varix) is an abnormal acquired communication between an artery and a vein due to simultaneous injury to both. There is pulsation in the vein since blood passes directly into it.

8. **Mycotic aneurysm** : This is due to infection by bacteria which weaken the wall, small aneurysms developing thereby. This is usually associated with vegetative endocarditis.

9. **Miliary aneurysm** This is aneurysm of minute arteries, and is usually seen in the cranium. This is congential and is also known as a *berry aneurysm*. This is a small saccular dilatation.

10. **Parasitic aneurysm** This is found in horses, in the anterior mesenteric artery due to infection by *Strongylus vulgaris*.

Sequelae.

1. Pressure atrophy of the structures around an expanding aneurysm may be seen.
2. Rupture of the aneurysm may occur with fatal results.
3. In the horse when the anterior mesenteric artery is affected colic may occur because of thrombosis of the artery or due to emboli that may emanate and occlude intestinal vessels.
4. Inflammation from the anterior mesenteric artery may spread to the neighbouring autonomic ganglia causing intestinal stasis resulting in colic.

Diseases of the veins

Phlebitis: Inflammation of the veins is usually septic in character. Infection may be;

1. In the new born animal, **umbilicus** may be infected—*Omphalophlebitis*. The usual organisms are *Shigella equirulis* in the foal and coliforms in the calves.

2. **By extension from adjacent inflammed areas** This is common in lungs (pneumonia), uterus (metritis) and the udder (mastitis). Infection passes through the thin walled veins more easily than through the thicker arterial walls.

3. **By venepuncture:** During intravenous injection if irritant chemicals are injected inadvertently outside the vein, periphlebitis and phlebitis will be set up.

4. **Foreign body:** In traumatic reticulitis, a foreign body may cause chronic phlebitis of the veins involved.

Macroscopically, the inflammed vein is enlarged, has a thickened wall with neutrophilic infiltration. Usually thrombosis develops rapidly—thrombophlebitis. An infected thrombus will get softened and disintegrated and thus septic emboli may be formed. A bland thrombus may become organised. Some times, the thrombus may be calcified—*phlebolith*. The importance of phlebitis lies in the danger of thrombosis with eventual emboli formation, causing pulmonary embolism, pyemia, septicemia or septic arthritis.

Varicose veins: Varicose veins are dilated and tortuous veins. These are not as common in animals as in man, whose leg veins and hemorrhoidal veins are most commonly affected. Stagnation of blood in the dilated vessels causes pain. In man the following causes are attributed.

11. **Exciting** causes are those that increase the pressure of blood in the veins. These may be found in the following situations.

a) Whenever there is hindrance to the return of venous blood—as in mitral stenosts, pulmonary emphysema and cirrhosis of liver; (b) Pressure on vein—by tumors, pregnant uterus, increased abdominal pressure as in straining. (c) Standing for a long time. (b) Muscular exertion as in athletes. (e) Ageing. (f) Post-inflammatory weakness of vessel wall.

There may be muscular hypertrophy followed by atrophy. The elastic tissue is replaced by fibrous tissue, leading to dilatation and in these areas thrombosis may occur.

In animals the veins of limbs are not commonly affected, The affected are the scrotal plexuses in the horse and the supramammary veins in the cow.

Diseases of lymph vessels
Lymphangitis

Lymphangitis may be non-specific or specific.

Non-specific lymphangitis: The irritant may reach the lymph vessels in two ways: (i)by extension of the inflammation of the surrounding tissues through the walls of the vessel and (ii) transport by the tissue fluid through the lumen. Hence lymphangitis is common in those vessels that drain areas in inflammation.

Again non-specific lymphangitis may be **Simple lymphangitis,** which is most commonly seen in the lungs in various inflammatory diseases.viz pneumonia in swine fever; contagious bovine pleuropnemonia; brochopneumonia in dogs and horses. The lesion in these diseases starts as bronchitis and the irritant spreads by peribronchial spread and enroute the peribronchial and perivascular lymph vessels are affected. These become very much dilated with leucocytic infiltration into their walls. Such vessels are easily seen on the surface of the lungs.

Inflammed lymphatics can be seen as thickened, reddish streaks and those of the subcuits appear as cords. These are painful to the touch. Ocelusion of the lymphatics prevents drainage of lymph and so edema develops. The nearest lymph gland that drains the area is inflamed, swollen and painful.

Purulent-lymphangitis is associated with suppurating condition of the tissue drained. There is intense infiltration by leucocytes of the wall of the vessels together with thrombosis.

Specific lymphangitis is seen as, tuberculous lymphangitis; farcy, ulcerative lymphangitis; epizootic lymphangitis; bovine lymphangitis, in John's disease and in actinomyocosis. These are described under the respective diseases in the section, 'Pathology of Specific Diseases"

Tumors of vascular system

Hemangiomas and lymphangiomas have been described.

CHAPTER 15

THE HEMOPOIETIC SYSTEM

Development of blood cells | Diseases of Lymph nodes
- Erythropoiesis
- Granulopoiesis

Terms used in describing anemias

Polycythemia

Oligocythemia

Dyshemopoietic anemias
- Porphyrinopathies
- Congenital porphyria
- Diminished stroma protein formation
- Diminished hemoglobin formation
- Toxic inhibition
- Aplastic anemia
- Myelophthisic anemia

Hemolytic anemias

Hemorrhagic anemias

Leucocytes
- Leucocytosis and neutrophilia
- Agranulocytosis
- Eosinophilia
- Lymphocytosis
- Leucopenia
- Leukemia

Diseases of Lymph nodes
- Atrophy
- Hypoplasia
- Necrosis
- Amyloid degeneration
- Hyperplasia
- Pigmentation
- Emphysema
- Inflammation-acute&chronic

Diseases of spleen
- Anomalies, Atrophy
- Hyperplasia
- Hyaline degeneration
- Amyloid infiltration
- Pigmentation
- Rupture
- Congestion
- Thrombosis and embolism
- Infarction
- Splenitis
- Splenomegaly
- Hypersplenism

DEVELOPMENT OF BLOOD CELLS

1. **In the foetus** : The primitive blood cells arise by proliferation of the endothelial cells lining the numerous blood islands. These are nucleated primitive erythroblasts, possessing basophilic cytoplasm, a large nucleus with a loose chromatin network and several *nucleoli*. These elaborate primitive hemoglobin. The intravascular formation of nucleated erythrocytes lasts upto 8 weeks in the human embryo.

2. **The hepatic phase** : From about the second month of foetal life, erythropoiesis occurs in the sinusoids of the liver and these remain active until a few weeks before birth. During this period, granulocytes begin to appear and by fourth month they are numerous.

Hemopoiesis also occurs in the spleen in which erythroblasts first make their appearance by the fourth month. However by the fourth month myelopoiesis no longer occurs in this organ. Erythropoiesis occurs till the end of the gestation period while lymphopoiesis takes place throughout life. Thymus, which is primarily a lymphopoietic organ, for a short period, produces erythroblasts.

Lymph nodes : Lymphopoiesis begins at fourth or fifth month and continues throughout life.

3. **The myeloid phase :** This begins approximately by the fifth month. In the beginning, granulopoiesis alone occurs (while liver is involved with erythropoiesis) but gradually, the bone marrow takes over the function of formation of blood cells.

The nucleated red cells, in the foetal blood gradually decrease in number and by the sixth month, none are present in the peripheral blood, which now contains only non-nucleated red cells.

Extramedullary hemopoiesis : This denotes formation of blood cells in organs other than the bone marrow. In times of need, eg. severe hemolitic anemia, the liver and to a slight degree the spleen, reassume their hemopoietic activity.

Hemopoiesis after birth : After birth, erythrocytes, granulocytes, monocytes and thrombocytes are formed in the bone marrow, while lymphocytes are formed in the bone marrow and the lymph nodes and spleen

ERYTHROPOIESIS :— Extravascular(Intravascular in poultry)

The following stages are noticed in the development of erythrocytes.

Undifferentiated stem cell
|
Rubriblast (Pronormoblast)
|
Prorubricyte (Basophilic normoblast)   } By mitotic division
|
Rubricyte (polychromatic normoblast)

Metarubricyte (Acidophilic normoblast)
|
Reticulocyte   } No more division / Only maturation
|
Erythrocyte.

In the process of erythropoiesis the following changes, in general, take place in the cells from the stage of rubriblast to erythrocyte.

1. The cell size gradually decreases.
2. The nucleus becomes progressively less sponge-like but more condensed.
3. Nucleoli disappear.
4. The size of the nucleus also decreases.
5. During metarubricyte stage, the nucleus is extruded.
6. The cytoplasm gradually turns pink from an initial blue color as hemoglobin is gradually incorporated.

GRANULOPOIESIS :— Extravascular. (Intravascular in poultry)

Hemocytoblast
|
Promyelocyte   } By mitotic division
|
Myelocyte
|
Metamyelocyte—(Juvenile)
|
Band form (stab)   } No division but maturation
|
Segmenter.

in the process of granulopoiesis. the following changes take place from the stage of promyeloblast to a segmented granulocyte:

1. The size of the cell becomes smaller.
2. The cytoplasm, which is intensely blue in the promyeloblast stage, becomes paler.
3. Granules begin to appear in the cytoplasm.
4. The nucleus which is large, becomes smaller and also becomes segmented

Erythropoietin ; For erythropoiesis. a humoral substance. the *erythropoietin*, appears to be of great importance. The granular cells of the juxtaglomerular apparatus appears to be an important source of erythropoietin, which is probably a glyco-protein, with a molecular weight between 60,000 and 70,000. It is present in plasma, urine and milk. Erythropoietin stimulates the differentiation of the bone marrow stem cells to rubriblast. It governs the rate of hemoglobin synthesis. Its secretion is controlled by oxygen content of renal arterial blood. Hypoxia is a stimulus for erythropoietin secretion. This is the reason for the polycythemia found in high altitude disease.

Anemia of chronic renal disease may mostly be due to decreased erythropoietin production by the damaged kidneys.

Androgens, adrenal corticoids, thyroxine and growth hormone stimulate erythropoiesis. The first three probably act directly by stimulating erythropoietin production, while he last has a direct effect on the marrow, stimulating erythropoiesis. Estrogens depress erythropoiesis, probably by competing with erythropoietin production or by competing with it in its action on stem cells.

Bone marrow : Macroscopical examination of bone marrow indicates:
   (a) Hyperplasia :—denoting increased activity when it is red, cellular and opaque or
   (b) Hypoplasia,—decreased activity when it is gelatinous and yellowish.

As the animal grows older, the red, active hemopoietic marrow is substituted by a yellow or white, fatty and inactive marrow. But in conditions of extreme necessity, this inactive yellow marrow can be converted into red, active marrow.

TERMS USED IN DESCRIBING ANEMIAS AND DISEASES OF THE BLOOD

Anisocytosis denotes variation in size of erythrocytes. In cattle blood slight anisocytosis is normal.

Annulocytes are erythrocytes that have a narrow rim of hemoglobin surrounding a large central pale area. These are also known as **Pessary cells.**

Basophilia This indicates that the erythrocytes instead of taking a normal red stain take a bluish or pale-bluish stain. That means to say that ribonucleic acid, which takes a blue stain, is still retained. This condition denotes incomplete maturation and is met with in anemias. It also indicates lack or deficiency of hemoglobin.

Basophilic stippling or punctate basophilia. In this condition the erythrocyte contains blue staining granules scattered throughout—the remnants of

RNA. This is met with in conditions in which there is acute and intense erythrogenesis as in anaplasmosis of bovines and hemonchosis in sheep and in lead poisoning.

Cabot rings are bluish thread like rings in the erythrocytes and are nuclear remnants. These may be found in hemolytic and toxic anemias and are found in lead poisoning.

Crenation means at normal notching of the erythrocytes. These may be seen in delayed drying

Heinz bodies are refractile inclusions found in the erythrocytes of horses that undergo phenothiazine therapy. These are supposed to be associated with denatured protein and are seen in hemolytic anemias. Their presence indicates erythrocyte injury.

In man Heinz Bodies are noticed following treatment with primaquine, acetanilid. sulphanilamide, phenylhydrazine, phenacetin, sodium nitrite, sodium chlorate- para-amino-salicylic acid, nitrofurantin.

Heinz Bodies are not visible if the blood smear is fixed in methyl alcohol. They can, however, be seen in blood stained with supravital dyes like brilliant cresyl blue. A drop of blood is mixed with 3 or 4 drops of 5% of the dye in saline. The preparation is taken on a slide and ringed with paraffin, a cover slip is placed and then examined under oil immersion. The bodies are seen as blue bodies.

Howell-Jolly bodies are remnants of nuclear material and appear as single or double spherical bluish bodies situated eccentrically usually. Normal blood of cats and horses may contain upto one percent of erythrocytes with these bodies. They may also be seen normally in young pigs and dogs. In the bovine H-J bodies must be distinguished from Anaplasma marginale. The anaplasma is uniform in size while H-J bodies vary in size. These are seen in anemias and lead poisoning

Hyperchromasia (Hyperchromic erythrocytes) indicates intensity in staining of erythrocytes. This is not due to increased hemoglobin content but to increased thickness of the cells.

Hypochromasia (Hpyochromatic cells) indicates decreased intensity in staining of erythrocytes which may be due either to thinness of the cells or to decreased hemoglobin content

Leptocytes are thin erythrocytes with larger surface without increase in volume.

Macrocyte is an erythrocyte the diameter of which is larger than normal, having therefore, higher Mean Corpuscular Volume.

Megaloblast is an immature cell in the erythrocyte series comparable to prorubricyte stage seen in animals due to Vitamin B_{12} or folic acid deficiencies.

Meniscocytes (Drepanocytes) are crescent shaped erythrocytes characteristic of sickle cell anemia. This abnormal shape is due to the insolubility of hemoglobin S in its deoxygenated form Doubly refractile masses consisting of rodlike particles are formed which deform the erythrocytes.

Microcyte is an erythrocyte the diameter of which is smaller than normal

Normocyte { cells of normal size and staining
Normochromic { with normal intensity

Ovalocytes are elliptical erythrocytes. These are normal in the camel family. Some ovalocytes may be seen in advanced anemia with poikilocytosis.

Pappenheimer bodies(siderotic granules)are purplish coccoid granules seen at the periphery of erythrocytes in anemias due to impaired heme synthesis (sideroachrestic anemias). The granules contain iron in ferritin.

Poikilocytosis: denotes variation in shape.

Polychromasia: or **polychromatophilia** denotes the staining of the erythrocytes with many colors, red blue and intermediate color

Spherocytes: are not seen in animals. In man these occur as congential abnormality. These are dome shaped and are thicker than normal,

Target cell is one with a central rounded area of pigmented material surrounded by a clear ring without pigment outside of which is the pigmented border of the red cell (resembles bull's eye). These are more commonly seen in the dog's blood and probably are artifacts.

Polycythemia:

This is an increase in the circulating erythrocytes and the blood picture is normochromic and normocytic

Polycythemia may be

(a) **Relative:** There is reduction in the total blood volume and so increased concentration of normal number occurs whenever there is excessive fluid loss as in hemoconcentration in dehydration due to continued vomiting, diarrhoea, sweating, shock and collapse.

(b) **Absolute;** There is increase in the total number of red cells, while the blood volume remains normal.

i) *Primary:* — Polycythemia vera—a tumor of erythropoietic marrow, was reported among six dogs, a cat, a cow and 14 heifers (it was found to be familial in these heifers) Evythropietin levels are low.

ii) *Secondary:*— 1) Neonatal—cause obscure.

                         2) Physiological (a) Permanent—as in high altitudes

                              (b) Temporary:— Splenic contraction in sporting dogs and racing horses.

iii) *Pathological.*— Compensatory increase in prolonged anoxic states eg., cardiac and pulmonary disease. It was found in Tetrology of Fallot, encountered in dog, cat, cow and horses.

Increase in erythropoietin production causing polycythemia was found in man in the following conditions: neoplasm of the liver; cerebellar hemangioblastoma, pheochromocytoma; hydronephrosis; cysts, adenoma and carcinoma of kidney; adrenal adenoma and uterine fibroid.

Oligocythemia: This is decrease in the quantity of erythrocytes in peripheral blood.

Anemia: is reduction in the quality and or number of erythrocytes below normal. Now the two terms are used synonymously.

Relative oligocythemia: There is increase in total blood volume (with

normal number of erythrocytes) and there is ultimately reduced red cell concentration; eg hemodilution.

Absolute: oligocythemia and anemia: In normal animals production of
erythrocytes by hemopoiesis is equal to destruction. So the condition under
'Anemia' can be conveniently grouped as.

    a) Production low, but destruction normal
        Dyshemopoietic anemias
    b) Production normal but destruction excessive
        i) Hemolytic anemias   ii) Hemorrhagic anemias

DYSHEMOPOIETIC ANEMIAS

In this are grouped all those types af anemias in which there is defect
in the formation of erythrocytes. The defect may lie in the formation of stroma
protein or in the formation of Hb etc. These are described as follows.

Porphyrinopathies : Porphyrins are necessary for the normal synthesis
of heme. If certain enzymes are lacking, then heme is not synthesized and excessive amounts of porphyrins are found in the urine (porphyrinuria — urine turns
red on exposure to light) and in the body — porphyria.

Congenital porphyria occurs in the bovines and pigs. This is a heritable
disease due to simple Mendelian recessive gene. The pigment is photosensitive
and so when deposited in the teeth (dentine) it takes a red color known as "pink
tooth". In the bone it is responsible for the condition. "Osteohemochromatosis."
In the kidneys, the pigment is deposited in the tubular epithelium and interstitial
tissue. When found in urine, it imparts a red color after exposure to light.
Affected animals suffer from photodynamic dermatitis if exposed to sun light—
photosensitisation, but if protected from direct sun, no harm seems to occur.

The following deficiencies have been found to cause anemia. After
restitution of the deficient hematinic, regenerative forms appear. "Reticulocyte
shower"

A. Diminished stroma protein formation Blood picture is **macrocytic**
and **normochromic or hypochromic.** Bone marrow is megaloblastic showing
numerous megaloblasts and giant metamyelocytes.

    a) Dietetic deficiency of Extrinsic factor: Cobalt—cyanocobalamin-
Vit. B_{12}.

Vit. B_{12} deficiency is not seen in animals other than ruminants. In
ruminants this is synthesized by ruminal microorganisms provided cobalt is
available. But in other animals. the vitamin is ingested as it is, and so deficiency
is seen only in ruminants in areas where the pasture is deficient in cobalt. Vitamin
B_{12} is necessary for the synthesis of RNA and DNA. B_{12} and Folic acid deficiency causes arrest of maturation of prorubricytes and metamyelocytes. These
have larger nuclei than normal Depressed DNA synthesis causes delayed nuclear
maturation but hemoglobin synthesis is not affected and so continues. When
hemoglobin synthesis reaches a certain concentration in the erythrocytes, the
nucleus leaves them and so macrocytes result giving rise to macrocytic anemia.

    b) Dietetic deficiency of Folic acid : Folic acid is required for the
maturation (especially the nuclei) of the erythroblasts. In its absence maturation is slowed down and so macrocytic anemia results.

c) **Deficiency of the intrinsic factor** : The intrinsic factor which is secreted by the gastric mucosa and which is supposed to be in the nature of an enzyme, helps in the absorption of the macromolecular Vit B_{12}. In its absence (gastric diseases) Vit B_{12} may not be absorbed and so will result in anemia.

d) **Failure to store the Erythrocyte Maturation Factor:** Erythrocyte maturation factor (Hematinic principle) is now known to be Vit. B_{12} which is normally stored in the liver. Hence is the efficiency of liver extracts in anemia. In diseases of liver the storage of this vitamin does not occur and so anemia results.

e) **Failure to use the Erythrocyte Maturation Factor:** The EMF may either be not utilised or may not be mobilised from the liver resulting in macrocytic type of anemia. This is known as *Achrestic anemia*. (Achrestic means failure to utilise). The bone marrow is megaloblastic and this differentiates it from aplastic anemia.

f) **Hypopituitarism:** Anterior pituitary seems to exert a potent influence in erythropoiesis, directly or through the Thyroid which influences metabolism of carbohydrates, releasing needed energy.

B. **Diminished hemoglobin formation:**

Blood picture is **normocytic** } Becoming **microcytic and**
and **hypochromic** } **hypochromic**

a) **Dietetic deficiency of iron:** can occur due to

i) **Deficient intake** : Milk of sows is poor in iron and so **piglet** anemia develops if rooting is prevented.

ii) **Defective absorption:** Excessive phosphorus and phytic acid, form insoluble complexes of iron, which are excreted through the feces.

iii) **Increased requirement** : In young, growing animal and pregnancy.

b) **Dietetic deficiency of copper** : Copper acts as a catalyst in the utilisation of iron in hemoglobin formation.

The deficiency is seen in piglets, cattle and sheep. Essentially this is an iron deficiency anemia. In iron dificiency, decreased hemoglobin synthesis leads to retention of the nucleus beyond normal number of cell divisions. Some cells undergo additional mitosis and so microcytic erythrocytes result in iront deficiency.

c) **Dietetic deficiency of ascorbic acid:** Vit. C is dietary reducing agent and so facilitates the reduction of Fe^{+++} to Fe^{++} state which is easily absorbed. Vit. C is also required for the synthesis of folic. acid and for its conversion into the more active folinic acid.

d) **Dietetic deficiency of pyridoxine**: Pyridoxine is required for the utilisation of iron in hemoglobin synthesis. So deficiency of pyridoxine results ultimately, in anemia resembling iron deficiency.

e) **Dietetic deficiency of Nicotinic acid:** Nicotinic acid is concerned in the synthesis of pyridine nucleotide which takes part in cell respiration. So, deficiency of Nicotinic acid interfers with the respiration of immature red cells. This is noticed in dogs and pigs.

f) **Dietetic deficiency of Riboflavin:** Riboflavin is concerned in the metabolism and arrangement of amino acids of the protein of hemoglobin molecule and so is useful in hemoglobin synthesis. This condition is met with in dogs.

g) **Deficiency of Thyroxine:** Thyroxine along with vitamin C is required for the conversion of folic acid to folinic acid. Thyroxine is necessary for the metabolism of carbohydrates and fats (So required for energy production).

In Myxedema, the secretion of intrinsic factor is depressed (and so absorption of vitamin B$_{12}$ is poor). So a normocytic or macrocytic anemia may be encountered.

C. Toxic inhibition Here the marrow appears to be normal and active but is unable to utilise the hematinics.

Blood—Normochromic: Microcytic.
No regenerative forms

Examples:

i) **Chemical Poisons:** Nitrogen mustard (which is cytotoxic).
Folic acid antagonists—antimetabolites—6 mercaptopurine etc.
Streptomycin, chloromycetin—Antibiotics and sulphonamides.
Metals—Bismuth, Arsenic and Gold (by injection).
Others—Benzol, hair dyes, insecticides.

ii) **Chronic interstitial Nephritis:** In advanced cases there is uremia which suppresses erythropoietic cells. Probably erythropoietin is not produced in the kidneys in this condition.

iii) **Oesophagastomiasis:** This worm causing pimply gut depresses absorption and so may cause various deficiencies, resulting in anemia.

iv) **Chronic infections:** In chronic infections like Tuberculosis, Brucellosis and Rheumatic fever (in man) a normocytic normochromic anemia is noticed. In these conditions there appears to be some abnormalities in hemoglobin synthesis for there is increased excretion of coproporphyrins. There is hypoferremia with reduction in serum iron binding capacity. Along with these, hypercupremia is noticed.

Absence of regenerative forms indicates impaired erythropoiesis. It is suggested that during infections and inflammation there is great demand for iron by the tissues affected and so it is side tracked to these areas instead of to the bone marrow and so hemoglobin is not formed and anemia results.

v) **Ionising radiation:** The hemopoietic system is highly sensitive to radiations, the leucopoietic being the most. After exposure there is lymphopenia with the spleen and lymphoid tissue becoming soft and shrunken Decrease in granulocytes is much more sooner than development of anemia since granulocytes are short lived. Hemorrhages occur due to thrombocytopenia and damage to the vascular endothelium.

D. Aplastic anemia: This occurs due to aplasia of bone marrow, wherein there is utter inactivity. The anemia seen is **normochromic** and **normocytic.** No regenerative forms are present.

Aplastic anemia may be divided into.

i) **Primary or idiopathic :** rather rare.

ii) **Secondary:**

c) **Exhaustion:** *Due to chronic hemorrhages*—Gastric and intestinal ulcers
                  (rare in animals ; Blood sucking worms, Neoplasms;
                  Deficiency of vit. C. K, and prothrombin.

b) **Toxic:** ionising radiation, Chemical poisoning—same as those
                  detailed under the heading "Toxic inhibition"but in a higher
                  dose and exposed for a longer duration.

e) **Metabolic:** Another form of aplastic anemia occurs in baby pigs
                  that are born of sows which suffer from protein malnut-
                  rition during pregnancy. This can be prevented by
                  feeding sows, during pregnancy, high protein diet
                  containing Vitamin B_{12}, folic acid and iron. Once the
                  symptoms are seen in the baby pigs, no treatment is of
                  any avail.

   E. Myelophthisic anemia: There is replacement of bone marrow by other
tissues. Since in this disease immature forms of granulocytes are found in the
peripheral blood, it is also known as **leuco-erythroblastic anemia.**This condition
is found in :

1. Secondary metastasis of other tumors—lymphatic leukemia in dog and cat.
2. Osteodystrophies—where the myeloid tissue is replaced by connective
   tissue and.
3. Primary tumors of the reticulo-endothelial system—Nieman-Pick Disease;
   Hodgkin's Disease etc.

<h2 style="text-align:center">HEMOLYTIC ANEMIAS</h2>

   In this condition intravascular destruction of erythrocytes occurs.
Anemia is **normochromic and macocytic** becoming **hypochromic and microcytic** as the iron stores are used up. Many regenerative forms are seen. Bone
marrow is active usually in this type of anemias, while erythrocytes show increased hyptonic fragility and spherocytosis

   We have already studied the normal breakdown of hemoglobin. Hemoglobin breaks up into heme and globin. The iron of the heme is stored by the
RE cells for future use. The pigment part is excreted as cholebilirubin and
urobilinogen. The protein moiety is broken down in the liver into amino acids
which are used again in the synthesis of hemoglobin. In some hemolytic anemias
their breakdown of hemoglobin occurs at a faster rate. So there is

a) jaundice—with increased bile pigments in the blood, feces and urine and
b) increased stroage of iron in the form of hemosiderin crystals.
   In other types of hemolytic anemias, there may be hemoglobinuria.
   Causes of the hemolytic anemias may be classed as:

1. **Abnormal auto-antibodies** the presence of which may be
   a. Primary or idiopathic, or
   b. Secondary due to
   i) Malignant disease—lymphatic neoplasms; ovarian tumors, gastro-
      intestinal carcinoma.
   ii) Collagen diseases—disseminated lupus erythematosis.
   iii) **Viral diseases** - infectious mononucleosis
      In this condition spherocytes, spontaneous agglutination of erythro-
   cytes and hemoglobinuria may be found.

2. Abnormal iso-antibodies: Due to the presence of hemolysins in the plasma. produced by (a) incompatible blood transfusion (b) injection of blood products (c) pregnancy—blood group antigens of the foetus pass to the dam which does not possess these antigens. Icterus neonatorum that develops in such a condition has already been studied. (Page 179)

3. Toxic (Partly also toxic dyshemopoietic)

A. Chemicals: i) Copper poisoning This condition is seen only in sheep in which excess of copper released suddenly in stress, produces hemolysis resulting in jaundice and hemoglobinuria.

Copper is a poorly excreted element and so if there is continuous ingestion of unduly large amounts of copper liver becomes loaded with this element. Such poisoning can occur by

a) ingestion of fodder treated with copper-containing insecticides and fungicides.

b) too heavy a dose of water containing copper sulphate given as a preventive and curative for stomach worms.

c) ingestion of large quantities of salt lick containing $CuSO_r$.

d) eating forage contaminated by copper from mines and dumps.

e) too much of supplemental mineral mixture containing copper sulphate.

f) eating forage that contains large quantities of copper due to soil having greater concentration of this element,

Stress can be brought about by: 1. Transport 2. Starvation after good feeding. 3. Excessive exercise, especially if unaccustomed. 4. Sudden stoppage of food. 5. Drenching 6 Exposure to cold 7 Loss in body weight.

Lesions seen are icterus, yellow and friable liver (which may be shrunken in later stages, distended urinary bladder(excretion of blood-colored urine). smooth kidneys, which are dark brown in color and a dark colored swollen spleen. Hemoglobinuria and dyspnoea are noticed clinically.

ii) Onion poisoning Occasionally. fatalities occur in cattle and sheep fed onions. in regions where they are extensively grown. The toxic principle is n-propyl disulphide. The symptoms are hemolytic anemia with hemoglobinuria and icterus. The carcass smells of onions.

iii) Poisoning by castor seeds: Ricin in castor seeds produces hemolysis and so ingestion of large quantities of castor results in hemolytic anemia.

iv) Phenothiazine poisoining (Drug sensitivity) : Phenothiazine, a good anthelmintic, sometimes even in threapeutic doses, has been found to be hemolytic. especially in horses. Cattle are also susceptible though to a lesser extent than horses. The symptoms are hemolytic anemia with hemoglobinuria Other lesions are hepatitis and nephritis.

v) Naphthalene used as moth balls may be accidetally ingested by pet animals and hemolytic anemia results.

iv) Lead also may produce acute hemolytic anemia.

vii) Hypersensitivity to certain drugs like sulphanilamide, quinine, paraminosalicylic acid and some anti-pyretic drugs may result in hemolytic anemia.

viii) **Snake venoms :** Snake venoms contain a lecithinase which acting on lecithin converts it into lyolecithin which is highly hemolytic.

B. **Post-parturient hemoglobinuria :** Also called – post parturient hemoglobinemia. This is found only in dairy animals, usually after parturition. This condition is associated with hypophosphatemia and so in such animals, signs of phosphorus deficiency are seen besides anemia : eg. pica, shifting lameness, decreased productivity, lordosis (curvature of the spinal column).

Lesions include i) a pale, slightly enlarged liver having centrilobular necrosis due probably to thrombosis of portal vein capillaries by "ghost corpuscles". ii) Black colored kidneys—due to deposition of Hb. iii) Dropsy of serous cavities, iv) Ecchymosis. v) Edema of lungs.

C. **Infections :** hemolysis occurs in infection by :

   a) Protozoa — Anaplasmosis, Babesiosis, Hemobartonellosis ; Eperyth-
       rozoonoses, **Ehrlichia canis.**
   b) Bacteria—Leptospirosis, Clostridia; Streptococci and Staphylococci.
   c) Viruses – Equine infectious anemia; feline infectious anemia.

D. **Hypersplenism** is found in some dogs in which there is severe anemia, macrocytic or normocytic in type   Lesions include splenomegaly and icterus.

E. **Cold hemoglobinuria in calves.** Ingestion of excessive quanities of cold water by calves (rarely in older cattle) resulted in a mild disease, characterised by intravenous hemolysis and hemoglobinuria, associated with cardiac insufficiency and pulmonary edema   Spontaneous recovery usually occurred.

HEMORRHAGIC ANEMIAS

In this condition extravascular destruction of erythrocytes occurs. We have noted, that normally there is a balance between blood production and blood loss. But in hemorrhagic anemia blood loss is greater than production. The bone marrow can rally round to meet the situation only if necessary basic ingredients, most important of which is iron, are available in sufficient quantity.

In cases where there is a balance between blood loss and production, the picture is one of normocytic, normochromic anemia with many regenerative forms. In cases where this balance is maintained with difficulty, i. e., where the bone marrow is working at a fast rate, macrocytes will be found. In due course as iron stores become depleted, the picture turns to one of microcytic hypochromic anemia, with numerous regenerative forms. Ultimately, when the bone marrow becomes exhausted and is no more able to cope up (aplastic stage) a normochromic and normocytic anemic picture is seen but without any regenerative forms.

Therefore 1. the amount of blood lost. 2. the rate at which the blood is lost and 3. the diet controlling the balance between blood loss and production determine the nature and type of anemia that develops.

The various types are : A: **Acute hemorrhagic anemia** due to injury; sweet clover poisoning; Warfarin poisoning; bracken fern poisoning.

Sometimes, in scarcity and famine conditions, horses and cattle may ingest, large quantities of bracken fern when poisoning may occur. In this condition there is an acute thrombocytopenia which is the direct cause of hemorrhages.

At necropsy one finds hemorrhages in the gastro-intestinal tract under the mucosa (resulting in ulcers), in the myocardium, in the liver and kidney.

Bracken fern is a cumulative poison and there may be terminal bacteremia due to granulocytopenia, resulting in bacterial embolism, hemorrhages and infarcts of heart and kidneys. Bracken fern contains thiaminase which destroys thiamine (Vit. B_1) and so in single stomached animals it may produce thiamine defic ency. But in ruminants which can synthesise thiamine this is not a problem. There is some other unknown factor responsible for the hemorrhage in them. Acute hemorrhagic anemia may be seen in ulceration of stomach in pigs, bleeding abomasal ulcers in cattle, cocidiosis in poultry; bovine enzootic hematuria ; hemonchosis and anylostomiasis.

B. Chronic hemorrhagic anemia

Due to blood sucking worms; Hemonchus, Fasciola, Bunostomum, in cattle and sheep; Strongyles in horses, Ancylostomes in dogs.

Ectoparasites—ticks, lice and fleas

Protozoa—Coccidiosis in dogs

Hemorrhagic diseases—Chronic bovine hematuria

In gastrointestinal ulcers and vascular tumors.

C. Purpura and hemorrhagic diseases

Purpura is accumulation of blood, under the skin due to spontaneous rupture of the capillaries. Hemorrhages result even due to mild damage.

Purpura is a syndrome but not a disease. The causes may broadly be divided under ;

1. **Vascular disorders ;**

i. **Purpuric infections :** Symptomatic purpura. This is found in various diseases characterised by petechial hemorrhages, eg. Hemorrhagic septicemia, Anthrax etc. Cause is injury to the vessels—capillaries and venules—by the toxins. In viremic diseases the endothelium is directly damaged due to the multiplication of the virus in the endothelial cells eg. Infectious canine hepatitis and hog cholera

ii. **Allergic purpura or purpura hemorragica :** This is a symptom of post-infectious toxemia as in Strangles. It is also seen in fistulous withers, poll evil and emphysema of guttural pouches. Besides the petechial hemorrhages noticed on the mucous membranes, edema of subcuits, peritoneal cavity and muscles is also seen. In this condition there is no thrombocytopenia. The defect appears to be injury to the vascular endothelium due to development of an allergy, resulting in increased capillary permeability.

iii. **Congenital purpura :** Purpura may develop in the foetus. The mechanism is suggested to be similar to the one found in erythroblastosis foetalis. Iso-agglutinins formed against platelets in the mother pass into the foetus via placenta and produce thrombocytopenia.

iv. **Senile purpura :** This is not seen in animals but is sometimes seen in old men and in very under-nourished people. The vessels of the skin are easily injured as there is no subcutaneous fat and the skin is very much atrophied.

v. **Vitamin C deficiency :** In avitaminosis C, there is no thrombocytopenia. Hemorrhages occur due to increased capillary permeability and capillary fragility since cement substance of capillary wall is not synthesized. However this condition may not be met with in animals since vitmin C is synthesized in their gut.

II. Impaired Clotting mechanism:

A) Thrombocytopenia:

a) Idiopathic or primary thrombocytopenia; cause is unknown. Probably auto-antibodies against platelets are present.

b) Secondary Thrombocytopenia:

1. Damage to the bone marrow:

Ly Chemicals— i) Nitrogen mustard, benzol, urethane antimetabolites

ii) Individual sensitivity to therapeutic doses of—sulphanilamide quinine, gold salts, Oxytetracycline, Streptomycin, P. A. S, soda salicylate , ergot organic hair dyes, D. D. T. etc.,

iii. Animal toxins—snake venom, extensive burns.

iv) By Physical agents —ionising radiation, heat stroke.

v) Infections: in septicemias, occasionally.

2) Myelophthisic replacement; In Leukemias.

3) Hypersp'enism—destruction of thrombocytes.

4) Aplastic Anemias—In this condition due to causes already described there is complete atrophy of the bone marrow and so there is no production of platelets.

5) Bracken fern poisoning;—already dealt with

B. Other Coagulating defects

i) Hemophilia is a condition in which coagulation of blood does not occur after an injury and so in some cases ends fatally. It is an inherited defect.

Two types of hemophilia, A and B, are recognised. Type A is due to absence or reduction of anti-hemophilic globulin (A H G). In this, coagulation time is prolonged. This is sex-linked, conditioned by a recessive gene and the defect is evident only in the males passing through the females. The condition is met with in dogs and swine But in swine it does not appear to be sex-linked So it is found in both sexes.

Type B (found in man) is due to deficiency of the Christmas factor or factor IX. It is also sex-linked

ii) Prothrombin deficiency:

This is mostly due to impaired formation.

1) Liver diseases—In hepatic-disease the following, that are necessary for clotting mechanism are not synthesized; fibrinogen, factor V, prothrombin, factor VII and factor IX. Deficiency of bile that may occur in hepatic disease may lead to deficiency of vitamin K.

2) Deficiency of Vit K. Animals usually do not suffer from vit. K. deficiency as it is synthesized in the intestines. Only fowls with short intestines may be affected. Pigs medicated with sulpha drugs and antibiotics may also suffer as micro-organisms are eradicated by these, and Vit. K synthesis stops.

3) Impaired absorption of Vitamin K ; For absorption of vitamin K, bile salts are necessary. So if there is deficiency of bile due to hepatic disease or

as in obstructive jaundice, vitamin K cannot be absorbed and so prothrombin and factor VII cannot be synthesized.

Similarly, in diseases of intestines, absorption of vit. K may be interfered with eg. as in colitis, sprue etc.

4) Poisoning by dicoumarin and Warfarin: Sweet clover disease: Sweet clover contains coumarin, which is converted into dicoumarol (which is 3.3 methylene–bis-4 hydroxycoumarin) This is a powerful anticoagulant. It probably antagonises the activity of vit. K and so depresses the formation of prothrombin, factor VII, factor IX and factor X. Poisoning occurs among cattle and sheep in which extensive hemorrhages are seen—under the subcutis, on the serous membranes and in the viscera. Anemia results.

Swine may also suffer but horses appear to be refractory, probably because the detoxicating activity against dicoumarol is well developed.

Warfarin, which is chemically similar to dicoumarol is used as a rodenticide and so may accidentally be eaten by pets. Its action is similar to dicoumarol and causes extensive hemorrhages.

iii) Presence of Circulating anticoagulants, a) Heparin is a powerful anticoagulant, producing this effect by preventing the conversion of prothrombin into thrombin. Heparin is produced by mast cells. In anaphylactic shock in dogs, large amounts of heparin are liberated resulting in bleeding.

b) Some snake venoms are also anticoagulants and so bites by such snakes may result in fatal bleeding.

C. Unknown Etiology:—

i) **Mouldy corn poisoning in cattle and swine:** Corn spoiled by mouldy growth, if consumed by cattle and pigs, produces among other conditions, acute hemorrhages in various parts of the body, together with necrosis of the hepatic parenchyma and renal epithelium. Abortion in pregnant cows may be noticed. Lesions include centrilobular necrosis, cloudy swelling and fatty degeneration of the renal tubular epithelium with glomerular atrophy and necrosis of some tubular epithelium.

ii) **Epistaxis in horses:** In some families of horses bleeding from nose occurs during strenous exercise. It is due to a non sex-linked recessive character. The walls of blood vessels are very thin and so rupture whenever distended during great exertion (as in racing)

SECONDARY EFFECTS OF ANEMIA

The secondary effects and symptoms noticed in anemia are mainly the results of anoxia, which leads to; 1. Hyperplasia of hemopoietic tissues, evidenced by the regenerative forms, which are larger and more fragile and less efficient than the normal erythrocytes.

Extra medullary hemopoiesis may be observed in the liver and spleen,

2. Dyspnoea and tachycardia.
3. Fatty degeneration of the parenchymatous organs.
4. Rapid fatigue, due to incomplete metabolism.
5. Compensatory hypertrophy of the heart, in the early stages. If decompensation sets in, C. V. C. and resultant hydropericardium and ascties may occur.

6. Edema—due to damage to capillary endothelium, which becomes more permeable.

7. Petechiae.

Pallor of the skin, glossitis, anorexia, flatulence, constipation, diarrhoea, vomiting, albuminuria, fever and splenomegaly are other symptoms, noticed in anemia.

LEUCOCYTES

Increase in number : This may be due to :

Leucocytosis : which is a temporary phase, useful to the animal and reversible.

Laukemia : A cancer of the leucopoietic tissue and so is progressive, irreversible and fatal.

Decrease in the number of circulating leucocytes is called *Leucopenia*.

Leucocytosis : Neutrophilia is increase in the number of neutrophils in the peripheral blood. When bone marrow is stimulated, immature neutrophils may be found in the peripheral blood, in greater numbers than normal. This phenomenon is called "shift to the left". To measure this shift to the left Schilling has proposed his hemogram in which the four following stages of neutrophils are estimated.

1. *Myelocytes*
2. *Juveniles* in which the nuceus is indented.
3. The *band* form or *stab* in which the nucleus is curved or bent and
4. The *segmenter*. The first 3 are immature forms.

This shift to the left is of two types.

a) **Regenerative reaction :** in which because of increased activity of the bone marrow immature forms are a little more in number than mature forms.

b) **Degenerative reaction :** in which immature forms are far more in number than the mature ones. In this condition there is a depression in the maturation of leucocytes in the bone marrow and this denotes a very severe infection with unfavourable prognosis. In severe infections, neutrophils contain large toxic granules called "Dohle's Bodies". Dohle's Bodies are aggregates of rough endoplasmic reticulum. Some neutrophils may show vacuoles in the cytoplasm. These are supposed to be due to leakage of hydrolytic enzymes released from ruptured lysosomes under the influence of bacterial toxins.

Causes of Neutrophilia : Physiological : may be found in new-born animals, in pregnancy, during exercise and with high protein diets.

Pathological 1. Acute infections : especially by cocci and also by leptospira, psittacosis organism, poliomyelitis virus, small pox virus, *E. coli* and *Actinomyces bovis*.

2. **Metabolic :** as in uremia, diabetic coma, burns.
3. **Poisoning by**
    Chemicals – lead, digitalis, mercury
    Organic—foreign protein; epinephrine
4. Acute hemorrhages and hemolysis
5. After surgical operations

6. Malignant diseases—leukemia and any other rapidly growing tumor. Ruptured immature neutrophils are called "basket cells".

Agranulocytosis is a condition in which there is almost complete disappearance of the granulocyte series of leucocytes from the peripheral blood. There will, therefore, be concomitant leucopenia.

The causes are either total suppression of leucopoiesis or inhibition of maturation of the granular series in the bone marrow. The causes include :

(1) toxic chemicals and drugs acting on the leucopoietic tissue—eg. benzol, arsenical preparations, barbiturates, amydopyrine.

(2) bacterial toxins—toxins of *Staphylococcus aureus, Streptococcus hemoliticus, Streptococcus viridans.*

(3) X-ray irradiation.

Myelopoiesis is totally destroyed in the Viral Feline Panleucopenia and so agranulocytosis is a symptom of that disease.

Usually there is hyperplasia of the stem cells of the bone marrow in the early stages. In chronic and prolonged cases myeloid hypoplasia supervenes.

Eosinophile leucocytosis or eosinophilia

This is seen in ; 1. Allergic diseases—asthma, hay fever, serum sickness.
2. Parasitic infections—trichinosis, helminthiasis
3. Skin affections—eczema, scabies
4. Following recovery from acute diseases
5. Chronic eosinophilic myositis of dogs
6. Following splenectomy
7. Following administration of certain poisons and drugs: arsenic, copper, sulpha drugs, chlorpromazine, digitalis
8. In certain diseases of hemopoietic system: chronic myelocytic leukemia, Hodgkin's disease
9. Following mild irradiation

Lymphocytosis

Usually absolute increase in lymphocytes is rare though relative increase is common,

Causes : 1. Certain viral infections—Mumps, influenza
2. Bacterial infections: usually chronic infections—Brucellosis, Tuberculosis
3. Thyrotoxicosis
4. Lymphatic leukemia
5. In convalescence
6. Adreno-cortical insufficiency
7. Following vaccination

Lymphocytes that are damaged during preparation of smear are called "smudge" cells.

Monocytosis

1. In protozoal diseases : Trypanosomiasis, Malaria, Kalaazar
2. During convalescence following acute diseases
3. Rickettsial affections : Typhus

4 Hodgkin's disease

5. Chronic bacterial diseases; Tuberculosis, Brucellosis

6. Monocytic leukemia

Leucopenia This term denotes reduction in the number of leucocytes in the peripheral blood. Usually all leucocytes are affected equally.

Causes:

A. **Diminished production:** 1. Bracken fern poisoning—already discussed,

2. Viral diseases- Rinderpest, Distemper, Infectious Canine Hepatitis, Mucosal Disease

3. Certain bacterial infections—Brucellosis, Typhoid, Tuberculosis

4. Protozoal infections - Kalaazar

5. Fungal diseases—Histoplasmosis

6. Ricketisial diseases—Tick-borne fever

7. Cachectic states and starvation

8. Metabolic disturbances as in hypothyroidism and hypop'tuitarism in which there is general lowering of body metabolism and so activity of bone marrow is decreased.

9. Chemical and physical agents that produce hypoplasia of the marrow Eg. benzol, ionising radiation, X-rays, urethane, nitrogen mustard, etc. described earlier under anemias.

10. Hemopoietic disorders: Anemias—Aplastic—Myelophthisic.

11 Of unknown cause cirrhosis of liver, primary splenic neutropenia.

B. **Increased destruction:** By 1. Physical agents—large doses of ionising radiation

2. Loss of leucocytes in large numbers in pus and in inflammatory exudate.

3. By bacterial toxins—toxins of *Clostridium welchii* and Pasteurella as in pasteurella pneumonia of sheep.

4. By protozoa —*Theileria parva* (East-coast fever)

5. Destruction by leucocytic antibodies, especially when amidopyrene is administered antibodies are formed which destroy leucocytes in the presence of the drug.

6. Hypersplenism:- found in certain diseases, like tuberculosis, Hodgkin's disease - cured by splenectomy.

C. **Altered distribution.** 1: In anaphylactic shock the leucocytes are trapped in the sinuses of the liver, spleen, and lungs.

2. In stress, liberated cortisone of the adrenal cortex produces eosinopenia and lymphopenia.

LEUKEMIA

This is a primary neoplastic disease of the bone-marrow and other reticulo-endothelial tissues. Comparatively leukemia is not so common in animals as in man

Again in animals, the neoplastic condition affecting lymphopoietic tissue is more common. The neoplastic cells often flood the blood, when it becomes *eukemic* At other times, when the blood picture is relatively normal the condition is known as *aleukemic leukemia.*

Neoplasia of granulopoietic tisssue is called *granulocytic leukemia* while if the lymphocytic type is involved it is known as *lymphatic leukemia*. Similarly, *monocytic leukemia* is also met with. Granulocytic leukemia is further divisible into neutrophilic, eosinophilic, and basophilic depending upon the type of cell involved.

In the circulating blood immature forms of leucocytes or "blast" cells are frequently encountered It is quite often impossible to determine to which type of cell this 'blast' cell is a precursor. And so by examining a blood smear alone it may sometimes be difficult to determine the type of leukemia. Examination of bone marrow smears and the presence or otherwise of hepatomegaly, splenomegaly and involvement of lymph nodes help in the diagnosis. If the lymph nodes are not affected but if the bone marrow smear reveals presence of numerous myeloblasts, and if there is, in addition, hepatomegaly and spleno-megaly, often a diagnosis of myeloid leukemia can be arrived at. On the other hand, a quiescent bone marrow with involvement of all the lymph nodes is indication of lymphatic leukemia

Leukemia is always fatal.

In animals lymphocytic leukemia is variously known as lymphocytoma, malignant lymphoma, lymphomatosis or lymphosarcoma. The last named term is usually applied to the condition wherein neoplastic masses of cells are found internally as in other sarcomas. Involvement of all the lymph nodes suggests a multicentric origin. Metastases are found in almost all the organs in which the normal structure may be completely changed and replaced.

Other infrequent conditions seen *are the reticulum cell sarcoma and the giant follicular lymphoma arising from the reticulum cells.*

Incidence of different types of leukemias in animals :

Cattle : Lymphatic leukemia is the tumor most commonly met with in bovines next to squamous cell carcinoma. Tumor masses are found in almost all the organs and all the lymph nodes are invariably enlarged. Metastatic tumor tissue is found more frequently in the liver, heart, walls of the abomasum, oma-sum, reticulum, uterus and ureter.

Blood picture may reveal anemia. In the blood leukemia will be moderate with immature blast cells. Neutrophilia may be present. The affected lymph nodes show central necrosis,which may incite sterile inflammatory reaction.

A solitary case of granulocytic leukemia was reported in a four month old calf.

Symptoms : Enlargement of the superficial lymph glands—prescapular, sub-maxillary and precrural—is the first symptom noticed followed by pallor of the mucous membranes, laboured breathing and a gradual loss in general condition. Digestive symptoms may be exhibited if the liver and abomasum are involved. Paraplegia is found if the spinal cord or brain are affected.

Sheep : Compared to bovines, incidence of leukemia is not as frequent. Among the different types lymphocytoma is more common and all the viscera are affected. Subcutaneous lesions are more common in this species.

Swine : Lymphomatosis is common in swine, deposits occurring even in long bones. Reticulum cell sarcoma resembling Hodgkin's Disease is seen.

Horses: In the horse also generalised lymphomatosis is found though not very frequently.

Dog: Mostly lymphocytic leukemia is seen.   A few cases of granulocytic and monocytic leukemias are described. In the lymphocytic variety, generalised lymphadenopathy is met with.   Splenomegaly and hepatomegaly are often observed. Leukemic stage is usually absent, and so the blood examination is often misleading. Moderate neutrophilia is constantly observed. Anemia is common. Biopsy is very useful in diagnosis.   Hodgkin's disease has been reported among dogs.

Cat: Lymphomatosis is common in the cat and appears as a generalised condition.   Lymphadenopathy is not a feature.

Plasma cell myeloma

This tumor found rarely in man and known as Ewing's tumor, is of rarer occurrence in animals.   This is found n ostly in the bone marrow and some times in other organs.   Multiple in olvement of bone may result in fractures. Blood picture may or may not show the neoplastic cells.   Normal peripheral blood does not have plasma cells.   Perssons suffering from this tumor excrete Bence-Jones protein in their urine.

DISEASES OF THE LYMPH NODES

Fowls do not have any lymph nodes

Function: 1. Production of lymphocytes.

2.   The reticulo-endothelial system of the lymph nodes have the following functions:

a) Phagocytosis of foreign particles and worn out blood cells.

b) Rarely, extramedullary hemopoiesis in severe anemia

c) Probably production of normal and antibody globulin by plasma cells

Atrophy: Atrophy of lymph nodes is associated with

a) Some viral infections

b) Ionising radiation

c) Excessive doses of adrenal cortical hormones and sex hormones (lymphopenia also occurs) "Alarm reaction"

d) Senility

e) Starvation and

f) Chronic wasting diseases

Hypoplasia of lymph nodes is caused (together with degenerative changes) by infection and toxic agents or hormonal mechanism. In "Alarm reaction" of stress diffuse dissolution of lymphocytes is seen.

Necrosis : Necrosis of the whole or a part of a lymph node may occur when infectious agents grow locally, In anthrax and erysipelas, necrosis of the lymph node drain ng the affected area occurs.

Macroscopically, the necrotic areas are dry, and circumscribed. In some infections, gas bubbles may be present.

Amyloid degeneration; In general amyloidosis, amyloid may be found in the lymph nodes.   Deposition of amyloid starts in the germinal centres and spreads outwards.

Hyperplasia of the lymph nodes is an usual reaction to subacute or chronic type of irritants and is met with either as a general or a local phenomenon in such diseases like canine distemper, chronic enterritis or chronic pneumonia.

Macroscopically, the affected nodes are enlarged, whitish-gray and firm but not fibrosed or calcified. Follicles are prominent.

Microscopically there is great enlargement of the germinal centres with a zone of mature lymphocytes surrounding them.

If the underlying disease is removed, hyperplasia subsides.

Pigmentation

Exogenous: Exogenous pigmentation of the lymph nodes is most common in the pulmonary and mesenteric nodes.

Anthracosis: Coal dust in the bronchial nodes is common in animals especially dogs that live in industrial areas and pit ponies. The coal particles are found in the macrophages of the medullary cords.

In ruminants mesenteric and other nodes develop a grey exogenous pigmentation of the medulla probably due to some pigments ingested with feed. In tatooed animals, the granules of the pigment used for tatooing are found in the regional lymph nodes.

These exogenous pigmentations are not of clinical importance.

Endogenous pigmentation: Hemosiderin is the most common endogenous pigment and is found in lymph nodes draining areas where hemorrhage has occurred.

Macroscopically, such nodes are brownish in color.

Microscopically, brown amorphous crystals of hemosiderin are found in the reticular and sinusoidal macrophages.

Bile pigments may be found in the hepatic lymph nodes. Melanin is found in the superficial lymph nodes of old grey horses.

Emppysema: In association with intestinal emphysema of pig, emphysema of the mesenteric lymph nodes is seen.

Emphysema of the bronchial nodes is common among cattle suffering from pulmonary interstitial emphysema.

Macroscopically, the nodes are enlarged, soft and puffy. The cut surface looks like a sponge.

Microscopically, vesicles are found in sinuses and the sinus endothelial cells become macrophages and even giant cells. These cells occurring as clusters cause pressure atrophy of the lymphoid tissue.

Circulatory disturbances; Hemorrhages are seen in lymph nodes in severe infectious diseases, hemorrhagic diathesis, local trauma and passive venous congestion.

Macroscopically, reddened areas are noticed, which may be diffuse, focal or even petechial

Hemolymph nodes must not be confused with hemorrhagic lesions.

Inflammation.

Lymphadenitis is inflammation of the lymph nodes. This may be non-specific, local or general. Functioning as a filter, the lymph node naturally is affected by any irritant that may be present in the area it drains. The following are the irritants that may cause non-specific lymphadenitis:

(a) Irritant chemicals, (b) Soluble toxins from trauma and burns, and (c) bacteria. Depending on the nature of the exudate, lymphadenitis may be acute, serous, hemorrhagic, suppurative or chronic.

Acute serous lymphadenitis : This condition is common in the nodes draining lymph from acutely infected or inflammed areas. In some septicemic diseases the nodes throughout the body may be affected eg. anthrax, pasteurellosis, swine erysipelas, hog cholera; salmon disease Mesenteric nodes may be affected by the absorption of irritants from the gastro intestinal tract.

Macroscopically, the affected node is enlarged, moist and reddened.

Microscopically, hyperemia and edema are noticed. Due to proliferation of the lymphatic parenchyma and reticulo-endothelial tissue, the lymph sinuses are filled with lymphocytes, mononuclears (derived from the RE, system) plasma cells and a few neutrophils.

Hemorrhagic lymphadenitis occurs when the irritant is stronger than in the serous variety. The best example is anthrax. The exudate in the gland is mixed with blood. Microscopically lymph sinuses contain large number of erythrocytes.

Suppurative lymphadenitis: Pyogenic bacteria cause suppurative lymphadenitis. The common organisms producing this are : *Streptococcus equi* in horses (strangles) *Corynebacterium ovis (caseous lymphadenitis* in sheep).

Macroscopically pus may be found in the nodes

Microscopically, the predominant cell of the greatly infiltrating leucocytes is the neutrophile. There is necrosis and liquefaction of the parenchyma and several small purulent foci may be present which may coalesce to form a big abscess.

Chronic lymphadenitis: the affected nodes are large, hard and dry. This is seen in Johne's disease (mesenteric lymph nodes),

Microscopically, there is hyperplasia of the R.E. system with numerous endothelial cells becoming rounded, swollen and cast off into the lymph sinuses that are much distended. To this picture is given the name of *"sinus catarrh"* Macrophages predominate. Reactive hyperplasia of the lymph nodules is also present. Fibrosis that occurs is the cause of hardness.

Specific lymphadenitis :

Lymphadenitis is a characteristic lesion of the following diseases.

(a) Tuberculosis
(b) Glanders   ⎫ granulomatous lymphadenitis with
(c) Actinobacillosis   ⎬ caseation and calcification.
(d) Johne's disease   ⎭
(e) Salmon poisoning in dogs
(f) Strangles in horses
(g) Caseous lymphadenitis in sheep
(h) Bovine lymphangitis and lymphadenitis caused by *Pasteurella pseudotuberculosis rodentium.*
(i) Brucellosis in guinea pigs
(j) Tularemia in rodents

(k) Epizootic lymphangitis.

(l) Helminthic larvae--Pentastoma and other helminthic larvae in mesenteric lymph nodes of cattle; lungworm larvae in the bronchial nodes.

Neoplasms — Primary benign tumors of lymph nodes are not common. But primary malignant neoplasms—lymphosarcoma—are common.

Secondary tumors that are common in the lymph nodes are: carcinoma, malignant melanoma and occasionally sarcoma, which invade the lymph vessels.

DISEASES OF THE SPLEEN

Functions of spleen :   1   Production of lymphocytes.

   2.   Through the reticulo-endothelial system.

   (a)   Phagocytosis of foreign particles.

   (b)   Phagocytosis of effete erythrocytes.

   (c)   Conversion of hemoglobin to bilirubin and storage of iron.

   (d)   Extra-medullary hemopoiesis.

   (e)   Production of antibodies by plasma cells.

   3.   Blood vascular system.   The spleen is a "great reticulo-endothelial sponge" (Boyd) and so holds a large amount of blood.   The speed of blood flow is controlled by the presence of sphincters, muscular trabeculae and a muscular contractile capsule.   Blood cells can be "sequestrated" in the red pulp, so that the macrophages can act and destroy them

Anomalies : Accessory spleens are acquired and are found scattered in the gastrosplenic omentum.   These are implanted pieces, produced by traumatic rupture of the spleen.

Doughnut spleen is a circular spleen with a hole in the middle, rarely seen in the horse.   Sometimes a primitive lobulated spleen may be encountered.

Aplasia and hypoplasia of spleen may sometimes be met with.

Atrophy of the lymphoid tissue is similar to that seen in the lymph nodes, (causes are also similar).

In swine, cats and dogs, there is a form of atrophy of spleen resulting from induration due to chronic stasis of blood.   The spleen is much reduced in size and the capsule is shrivelled.   The parenchyma is scanty and Malpighian corpuscles are not visible.

Hyperplasia :   Focal hyperplasia of the spleen is common in old dogs and is characterised by round, soft and grey projecting 'nodules – *nodular hyperplasia*. These nodules comprise of newly formed hyperplastic lymph follicles, which do not contain the central arteriole.

Hyaline degeneration is seen in the walls of the arterioles.

Amyloid infiltration occurs as a part of generalised amloidosis.   Two varieties are recognised:

(1) Focal: "the sago-spleen" in which the central arteries of the Malpighian corpuscles are affected.   The involved foci are prominent, pale and translucent standing out against a red back ground, like boiled sago.

(2) Diffuse: "the Bacon spleen" in which the arterioles and fibres of the reticulo endothelial system are affected.   The organ is enlarged with rounded edges and the cut surface is smooth and translucent.

Of these the focal is more common.

Pigmentation: Hemosiderosis is not of much consequence, unless in excessive amounts, since normally hemosiderin is stored in the spleen in the reticular macrophages. On destruction of larger number of erythrocytes, as in hemolytic anemias, there may be increased amounts of this pigment in the spleen.

Rupture of the spleen is common in dogs due to automobile accidents or due to sharp blows on the abdomen. Such rupture may divide the spleen into two. Healing may take place and scars be visible postmortem.

Splenomegaly due to congestion, amyloid disease, hyperplasia or tumors may predispose the spleen to traumatic rupture. Hemorrhages into the peritoneum may occur with fatal results.

Circulatory disturbances; Due to wide variations in size, congestion of the spleen is difficult to interpret.

Acute Congestion. This is common in acute infectious diseases and in acute bacterial intoxications as in enterotoxemia.

Acute passive Congestion: In euthanasia of pet animals by barbiturates acute congestion of the spleen is noticed since the barbiturates relax the smooth muscle and when the smooth muscle of the splenic trabeculae and capsule is relaxed blood fills into the organ.

Paralysis of the splanchnic nerve results in relaxation of the splenic musculature and hence passive congestion results.

In cardiac failure, acute congestion of the spleen may be noticed as a part of general venous congestion.

Macroscopically, the spleen is very much enlarged and soft. On section the cut surface bulges and dark blood oozes.

Chronic passive hyperemia: This is not common among animals. The cause are;

1. Partial or complete obstruction of venous return

a) Thrombosis or pressure by cysts, tumors and absceses on veins draining the spleen.

b) Torsion of the stomach and spleen in dogs.

c) Torsion of the splenic ligament in the pigs.

2. Cirrhosis of liver, leading to congestion of portal vein.

3. Lesions of heart and lungs giving rise to general chronic venous congestion.

Macroscopically, the spleen is moderately enlarged and firm due to increased fibrous tissue in the pulp and in the trabeculae.

Microscopically. there is progessive induration of reticular stroma and trabeculae. The pulp cells and follicles are gradually replaced. Hemosiderin accumulates in the phagocytes.

Thrombosis: Thrombosis of the splenic veins is rare but occasionally met with in the following conditions;

In cattle ;   traumatic reticulitis and portal thrombosis,

In horses;   extension of infections from parasitic abscesses.

Embolism may involve splenic artery and its branches. Emboli originate from the valvular vegetations and cause infarction.

Infarction of the spleen is common. If the splenic artery is occluded by an embolus the whole organ may undergo infarction.

In hog cholera, occlusion of the follicular branches of the splenic artery by proliferated endothelial cells results in hemorrhagic infarcts. The base of the infarct is red in the beginning but later turns pale with the diffusion of hemoglobin.

Hemorrhages;— In dog, due to automobile accidents, if rupture does not occur, blood may collect under the capsule and form hematomas.

Splenitis ; The inflammation of the spleen may be acute or chronic.

Acute Splenitis is a common feature of acute generalised infectious diseases such as salmonellosis, anaplasmosis, infectious anemia of horses, eperythrozoonosis and swine erysipelas. The infective organisms grow in the spleen.

Macroscopically the spleen is enlarged dark and soft. The pulp is fluid If infeetion is by pyogenic organisms, abscesses may be found.

Microscopically, necrosis of the pulp and neutrophilic infiltration may be seen in the sinusoids which are congested Certain amount of proliferation of lymphocytes is present and there is reaction of germinal centres. In the red pulp proliferation of the reticular cells and macrophages may be found. Along with the above changes, plasma cells proliferate (for production of antibodies)

Chronic splenitis occurs in such chronic diseases as tuberculosis. glanders, actinomycosis. pyemia pseudotuberculosis of sheep and histoplasmosis. The spleen is enlarged, firm and tough. Abscesses or granulomatous foci that are specific features of the diseases are seen.

Splenomegaly : Enlargement of the spleen is found in many different kinds of diseases. Since spleen is an organ of ant body production, the reason for such enlargement is quite understandable. There is marked hyperplasia of the reticuloendothelial system as well as of the white pulp with diffuse infiltration by neutrophils. The following are some of the diseases in which enlargement of spleen may be noticed It should be noted that absence of splenomegaly does not rule out the presence of the diseases.

Horses: Equine infectious anemia; metastatic melanoma; Tuberculosis; salmonellosis, Anthrax.

Cattle; Anthrax; Salmonellosis; Babesiosis; Anaplasmosis; Theileriasis Lymphocytoma; Acute congestion in bacteremic and toxemic conditions.

Pigs; Erysipelas; Salmonellosis; Eperythrozoonosis; acute congestion; Torsion.

Dogs ; While using barbiturates and chloroform; Histoplasmois; hemangiomas and hemangiosarcomas, lymphomatosis, myelogenous leukemia.

Fowls; Spirochetosis; lymphoid leucosis.

Splenomegaly is also seen in the following conditions.

a) in congestive heart failure,the spleen is enlarged due to stasis of blood and resultant fibrosis. b) when bone marrow is destroyed extramedullary hemopoiesis occurs in spleen. known as myeloid metaplasia with resultant splenomegaly (c) in conditions in which the histiocytes of the liver are saturated there by stimulating the splenic histiocytes to undergo hyperplasia and (d) in hepatic fibrosis in which condition the antigens from the intestines that are not detoxicated enter the general circulation and thence to spleen in which plasma cell production is stimulated for antibody manufacture.

Hypersplenism : This is a pathological condition in which there is excessive activity of the phagocytes. This condition may exist with or without splenomegaly.

In this condition there is excessive hemolysis (resulting in hemolytic jaundice). leucocytosis (resulting in leucopenia) and thrombocytolysis (resulting in thrombocytopenia and so purpura). The causes may be: 1. Depression of marrow function or inhibition of the maturation of cells of marrow by spleen through some hormonal influence.

2 Hypersequestration in which there is increased stasis in the sinuses leading to increased fragility of the erythrocytes (consequent on loss of plasma and erythrocyte potassium). Such red cells are easily phagocytised.

3. Antibody formation against the erythrocytes, leucocytes and thrombocytes leading to destruction of these elements.

Splenectomy relieves the clinical picture and the blood cells increase in number to the normal levels.

Tumors : **Primary neoplasms** of the spleen are rare. Fibrosarcomas, lymphosarcomas, myeloid leukemia, Hodgkin's disease; follicular lymphoma)Splenoma), cavernous angioma, reticulum-cell sarcoma, leiomyosarcomas and hemangiosarcomas may be seen. Of these the hemangiogenous tumors are more frequent.

Secondary metastases are not common since spleen is a "poor soil" for the growth of the tumors.

CHAPTER 16

THE RESPIRATORY SYSTEM

Anomalies
Nose
  Congestion
  Epistaxis
  Acute rhinitis
  Atrophic rhinitis of swine
  Rhinohyperplasia or Bull nose
  Tumors of nasal cavity
Diseases of larynx and trachea
  Roaring
  Laryngitis
Diseases of Bronchi
  Bronchostenosis
  Bronchiectasis
  Acute tracheo-bronchitis
  Chronic bronchitis
Lungs
  Emphysema—Heaves
  Vascular disturbances
  Hyperemia—active and passive

Edema
Hemorrhages
Thrombosis and embolism
Infarction
Inflammation—Pneumonia
  Lobar
  Bronchopneumonia
  Necrotic, gangrenous and
        aspiration
  Verminous
  Interstitial
  Mycotic
  Pulmonary adenomatosis
  Maedi
  Tumors of the lung
Diseases of the pleura
  Pneumothorax
  Pleuritis
  Tumors of pleura

The main function of the respiratory system is the exchange of oxygen and carbon-dioxide between the blood and environmental air. As such to satisfactorily. discharge this function, respiratory system depends on the work of the heart. Diseases of the circulatory system are often accompanied by some abnormalities of the respratory system.

Anomalies : *Cleft palate or palatoschisis* is a fairly common defect seen in the new-born animals. In this condition there is an abnormal connection between the nasal cavity and the mouth and hence milk passes into the lungs. So the animals do not survive long, dying of pneumonia and starvation.

NOSE

Congestion : Congestion occurs whenever animals are exposed to cold air. The blood vessels in the nasal passage dilate so that the air breathed in may be sufficiently warmed Secondary bacterial infection may result in inflammation and edema.

Epistaxis is hemorrhage from the nasal cavity.

Causes : 1. Trauma; 2. Convulsive expiration; 3. Parasites—*Oestrus ovis*; 4. Erosion of the vessels by pathological processes in the nasal cavity—Neoplasms; 5. Compression of the jugular veins by too tight collars in the working horses; 6. During certain infectious diseases Eg. Glanders, Anthrax, Purpura; Infectious bovine rhinotracheitis; Malignant catarrhal fever, Septic metritis; 7, Idiopathic—familial in certain race horses; 8. Neoplasms—hemangioma; 9. Uremia; 10. Poisoning by nitrates, bracken, sweet clover or mercurials

Acute rhinitis or coryza or acute nasal catrrrh

This is acute inflammation of the mucous membrane of the nose. In man this is the common 'cold'.

Causes : 1. **Irritants :**
   (a) **Physical :** Dust, foreign bodies like chaff, pollen.
   (b) **Chemical :** Irritating gases and smoke.
2 **Parasites :** *Linguatula serrata* in dog, larvae of *Oestrus ovis* in sheep.
3. **Fungi :** *Aspergillus fumigatus.*
4 **Bacteria :** *Sphærophorus necrophorus, Bordetella bronchisepticus, Pseudomonas aeruginosa, Streptococci, Staphlococci.* May be found in the course of infections by *Mallcomyces mallei* (Glanders), *Pasteurella, M. tuberculosis, A bovis, A. lignieresi* and *Cryptococcus.*
5. **Viruses :** May be found in infections by viruses of :
   a. Rinderpest   c Equine influenza   e. Laryngotracheitis (fowl)
   b. Swine Influenza   d Canine distemper   f Fowl pox

Macrosopically, the mucous membrane is swollen and congested. Dry at first, a mucous discharge occurs subsequently, which turns mucopurulent later.

Microscopically, hyperemia, inflammatory exudate with inflammatory cells and hydropic degeneration of the epithelial cells (goblet cells) are seen. Extension to sinuses results in sinusitis.

Infectious sinusitis of turkeys is of economical importance to the turkey raisers Caused by a virus, there is an acute inflammation of the mucosa, resulting in a thick mucous exudate The infra orbital sinus is more often affected. Usually there is closure of the opening of the sinus cavity into the nasal cavity thereby preventing escape of the exudate

Chronic rhinitis : This is usually a sequel of the acute variety. There is ulceration of the mucosa, which in some places, may be thickened and congested.

Atrophic rhinitis of swine

This disease, especially of weanlings, is infectious. A virus was incriminated to be the cause, while Trichomonads were secondary invaders. At different times it was thought that *Pasteurella multocida, Hemophilus influenza suis, Fusiformis necrophorus* and a *Mycoplasma hyorhinitis* (P.P.L.O.) were of etiological importance. Though osteitis fibrosa (general osteodystrophy) is produced in swine with lesions and symptoms similar to those of Atrophic Rhinitis by feeding with calcium deficient diets producing secondary hyperparathyroidism thereby, characteristic lesions of Atrophic Rhinitis with atrophy of turbinate bones could be produced only in those pigs exposed to nasal installation of nasal washing from natural cases Though a specific organism could not be attributed to be the cause of the disease, yet it is proved that it is infectious in nature and not just a specific effect of diet.

The disease starts with slight catarrh and swelling of nasal mucosa due to irritation. Progressive affection results in dyspnoea, anorexia and finally death due to inanition occurs. The disease is chronic and animals grow rather poorly. Lameness and fractures are often encountered. In many animals there is deformity of the snout which bends to a side or upwards due to excessive bone resorption around the skull sutures.

Deviation of the snout is due to uneven resorption around the skull sutures. So there should be growth on one side while resorption occurs on the other so that the snout bends on the side of resorption due to growth pressure.

In the early stages there are foci of congestion of turbinate bones, which may be depressed. In 2 to 4 weeks, the turbinate bones are absorbed leaving a strip of hard tissue   When the nose is sawn transversely in front of the second upper premolar, a characteristic appearance is noticed. Normally the turbinate bones completely fill the cavity at this level. But in atrophic rhinitis, a large cavity is seen. A mucopurulent exudate forms, which is discharged through the nose. Intermittent epistaxis may be present. The parathyroids may be larger in size.

Microscopically, there is rarefaction and disappearance of the turbinates. This is of the nature of osteolysis rather than osteoclasia No evidence of osteoclastic activity is present. The nasal septum also disappears leaving a fibrous band. Infiltration by lymphocytes and a few neutrophils is seen. It is postulated that "inflammation of the nasal mucosa is not a cause of atrophic rhinitis". In some places there may be metaplasia of the epithelium of the nasal mucosa to squamous type. The lamina propria is diffusely infiltrated with mononuclears There is hyperplasia of the parathyroids with increased secretion of parathyroid hormone, which is responsible for the bone lesions.

The condition can be prevented by a dietary calcium level of 1.20 per cent and a phosphorus level of 1.00 percent. Due to increased calcium intake, zinc in a concentration of 100 parts per million was recommedded to be fed to pigs.

**Rhinohyperplasia or Bullnose is a chronic suppurative inflammation of young pigs, said to be caused by *Spherophorus necrophorus*, but other organisms such as *Streptococci, Pseudomonas aeruginosa*, Staphylococci and *Corynebacterium pyogenes* have also been incriminated.

Clinically, the nose is enlarged posterior to the cartilaginous rim

Lesions include. caseous or suppurative necrosis involving the bones of the nasal cavity, its lining and the skin adjacent to the nares. Being a slowly developing condition, fibrosis is predominant and this with edema is responsible for the swelling noticed. Extension of the inflammation to the sinuses and regional lymph nodes (lpmphadenitis) is observed. Exaudate obstructing the nasal passage and distortion of the maxillary and nasal bones are responsible for the dyspnoea seen.

Nasal granuloma—see under "Diseases caused dy Helminths".

Rhinosporidiosis : *Rhinosporidium seeberi* produces a chronic rhinitis in animals, characterised by the formation of polypi—some sessile while others are pedunculated and cauliflower-like. These are soft and jelly-like to the touch, covered by the nasal epithelium.

Microscopically, the fungus is present in loose fibrous or myxoma'ous tissue. The lesion bleeds easily. A few lymphocytes and epithelioid cells may be seen. (See under "Diseases caused by Fungi")

Tumors in the nasal cavity; Nasal polypi are soft moist masses. These have already been studied, (vide page 227)

The following tumors are encountered.

Benign	Ma'ignant
Osteoma	Osteosarcoma
Chondroma	Chondrosarcoma
Myxoma	Myxosarcoma
Fibroma	Adenocarcinoma
Angioma	Squamous cell carcinoma

One interesting tumor is the adenocarcinoma arising from the glands of ethmoid and olfactory mucosa, which occurs in epidemic form in some countries Horses and cattle are affected. The tumors are highly malignant and destructive and metastasise rapidly.

DISEASES OF THE LARYNY AND TRACHEA

Roaring or laryngeal hemiplegia; In normal health the arytenoid cartilages are drawn outwards during inspiration to allow ingress of air. The important muscle that operates this is the *Cricoaryrtenoideus*. If for any reason there is injury to and degeneration of the nerve supplying this muscle, then the cartilage cannot open and so will stand in the way of air passing freely into the wind pipe.

In horses a condition is noticed, in which there is hyaline degeneration and fibrosis of the left Cricoarytenoideus muscle together with demyelinization and Wallerian degeneration of the left recurrent laryngeal nerve that supplies the muscle.

The cause for the paralysis of the nerve is obscure. One theory is that it is subjected to repeated trauma by the pulsation in the aorta as the nerve circles round aortic arch where the nerve is situated during its course.

Other causes are lead poisoning and pressure on the nerve by aneurysms, enlarged lymph nodes, abscesses, tumors, oesophageal divericula and other traumatic conditions.

When the nerve is paralysed, and the muscle is degenarated and replaced by fibrous tissue, the arytenoid cartilage does not open out during inspiration and so air cannot enter the treachea freely and this condition is accentuated when the animal is exercised, and a noise is heard by brushing of air with the arytenoid cartilage. This is therefore known as *Roaring*.

Laryngitis

A mild catarrhal laryngitis is met with which may progress to chronic form if the cause perisists.

Cause; Usually an extension cf infection from the nasal cavity or the pharynx in infectious diseases. Eg. Distemper in dogs; Strangles and influenza in horses; Infectious laryngo-tracheitis in poultry; (Bacteria and viruses)

2. Irritant vapours, chemical irritants.
3. Mechanical injury,—kicks, bites, grass awns; injuries while passing probang or stomach tube.
4. Excessive barking, clergyman's throat.

5. Specific diseases; Tubercu'osis, g'anders and actinomycosis.

Macroscopically, the lesion is a swelling of the mucous membrane of the larynx and trachea, which is hemorrhagic and dry at first, later becoming coated with a mucoid exudate that may turn mucopurulent.

Croupous or membranous laryngitis ; The laryngeal mucosa is coated with greyish fibrinous deposits as in *laryngotracheitis of fowls* A membrane consisting of fibrin and leucocytes is formed over the necrotic mucosa. Death in fowls is due to asphyxia

BRONCHI

Bronchostenosis is a narrowing of the bronchial lumen due to obstruction or peripheral pressure

Cause : 1. Aspiration of foreign bodies; 2 Accumulation of exudate; 3 Parasites within the lumen. 4. Inflammation of the wall of the bronchus producing alteration in it—the exudate infiltrating the wall reduces the diameter of the bronchi. 5. Pressure from outside the bronchial wall—abscesses, tumors, enlarged lymph nodes and exudate of the pleural cavities; 6 Spasm of the muscles of bronchi – as in allergy (Asthma)

A partial closure of the bronchi or bronchioles results in balooning of the lung involved, since air that enters during inspiration is not expelled during expiration and so is trapped. Repeated inspirations will therefore result in ba-looing of the alveoli.

Complete obstruction of a bronchus results in collapse of the lung.

BRONCHIECTASIS

This is dilatation of the bronchus.

Causes: 1. This usually follows a chronic inflammation of the bronchi, in which the elastic tissue; the musculature and even the cartilages may be destro-yed. Due to loss of the elastic tissue, contractile power of the bronchus is lost and so the bronchus dilates. At the place of dilatation, exudate accumula-tes thereby still further dilating the bronchus.

2. In chronic pneumonia, the contraction of the fibrous tissue exerts a pull thereby widening the walls. The dilatation is facilitated by a weakening of the wall in bronchitis, which, may also be found.

3. In bronchostenosis, air accumulates during inspirations and so causes dilatations of the bronchi below the level of obstruction When the bronchi are completely closed resulting in atelectasis, there is an elastic pull on the bronchial wall due to negative pressure in the pleural cavity.

Macroscopically, two forms are recognised A. The saccular, which is less common, is an out-pouching of the bronchial wall; usually results due to loc-alised necrotising foci in bronchitis, found in cattle and sheep in lungworm infection.

B. The cylindrical variety, which is more common, especially in cattle, is a uniform dilatation of the bronchi.

Macroscopically, the wall of the affected bronchi shows variable infiltra-tion by chronic inflammatory tissue. The musculature, cartilage and the lining ep.thelium may disappear in varying degrees.

The affected lung is collapsed and is carnified. There may usually be pleural adhesion.

Results: The course is chronic and unfavourable. These animals cough persistently and become debilitated. Complications include abscesses with metastases, bronchiolitis with emphysema, bronchopneumonia, and secondary amyloidosis.

ACUTE TRACHEO-BRONCHITIS

This condition is usually encountered along with upper respiratory diseases More often it is a condition seen in pneumonias.

Though bronchitis means inflammation of the bronchial epithelium, frequently the inflammation spreads to the wall of the bronchus and from there to the lung tissue—expanding peribronchitis—or the infection may spill into the alveoli from the terminal bronchioles thereby resulting in bronchopneumonia.

Causes:

1. **Inhalation of irritants** —dust, feed, industrial fumes, medicaments, smoke
2. **Infections**:   a) Bacterial. — Pasteurellosis
                     b) Viral;—   Ranikhet disease, Infectious bronchitis of
                                       fowls, Infectious bovine rhinotracheitis.
                     c) Parasites;— Lung worms.

Macroscopically; the mucosa is thickened, reddened and covered by an exudate which may be catarrhal, fibrinous or purulent.

In aspiration of foreign material, a gangrenous bronchitis is seen in which there is extensive necrosis of the mucosa which sloughs. The wall of the bronchus may also be destroyed.

Microscopically, there is congestion and infiltration by inflammatory cells in which neutrophils predominate. There may be increased secretion of mucus and in severe cases the epithelium may be destroyed. The lumen contains mucus, leucocytes, dead epithelial cells, lung worms and their ova.

Results; Recovery; bronchiectasis; abscess formation; bronchopneumonia; chronic bronchitis.

CHRONIC BRONCHITIS

Causes; 1. Mild, continuous irritants—smoke and dust; 2. Chronic venous congestion—as in heart disease; 3. Chronic infection of upper respiratory tracts—chronic sinusitis; 4 Bronchiectasis 5. Most common cause in animals is lung worm infection, tuberculosis and lung abscesses.

Macroscopically, the bronchial mucosa is thickened and has a velvety feel. Sometimes it may be congested but more often is pale and edematous. The exudate is mucoid or mucopurulent and in cases of worm infection, it is mixed with worms and their eggs. The bronchi may also be dilated.

Microscopically: there is infiltration by lymphoid cells. The ciliated epithelium is lost and replaced by a cuboidal variety The mucous glands may show atrophy There is hyperplasia of peribronchial glands which now resemble goblet cells. There is increased fibrosis of the walls,producing polypoid projections into the lumen —*bronchiolitis obliterans*. Lymphoid follicles may be formed in the walls of the bronchi.

Results : 1. Bronchopneumonia; 2. Atelectasis; 3. Bronchiectasis; 4. Emphysema—putting greater strain on the right side of heart—chronic venous congestion will result ultimately.

LUNGS

Atelectasis : The failure of the alveoli to open and contain air is called atelectasis. So the alveoli become collapsed. This condition may be congential or acquired.

Congenital atelectasis: Animal is born dead and has not breathed.
Causes; 1. Obstruction of the bronchi by mucus or inhaled liquor amnii.

2. Damage to the respiratory centre that may occur in injury to the brain.

The lungs are dark and reddish-blue in color due to dilatation of the alveolar capillaries The lungs are firm to the touch and sink in water since there was no aeration. The alveoli are collapsed and the epithelium lining the alveoli is cuboidal. Sometimes the alveoli may contain fiuid.

Acquired atelectasis (Pulmonary collapse)

Cause : 1. **Obstruction ;** The lumen of the bronchus is obstructed and the air in the alveoli is resorbed. The cause of the obstruction may be

    a) **In the lumen :** foreign bodies, pus, mucus, masses of parasites.

    b) **On the wall**—tumors, abscesses, enlarged lymph nodes, cysts.

2. **Compression ; Extrapulmonary**—hydrothorax, pneumothorax, hydro-pericardium; abdominal distension as in tympany of rumen and ascites.

Macroscopically, there is never a total collapse but only focal. The affected lung tissue is dark or reddish-blue in color and is depressed from the level of the surrounding healthy lung. The affected part sinks in water and is leathery in consistency. Pleura is thickened and wrinkled.

Microscopically, the alveoli are devoid of air. They may appear as small or elongated clefts or the walls may lie in apposition with each other with no lumen visible. Due to the absence of pressure of alveolar air, the capillaries become dilated and engorged with blood. In later stages the alveolar epithelium may be desquamated while the tissue of the interlobular septa may proliferate.

EMPHYSEMA

Emphysema is increased air in the lung. This is divisible into:

1. **Acute alveolar emphysema;** The alveoli are greatly distended and some-times may rupture, orming "Vesicles".

Causes; a) **Compensatory** – In pneumonia or atelectasis when affected portions of lungs cannot dilate other healthy parts dilate to a greater extent to fill the space created by the expansion of the chest cavity.

    b) **Over exertion**—in coughing and struggling over-ventilation occurs.

    c) Feeding on lush pastures.

    d) Allergic or toxic agents—feeding mouldy forage or mouldy sweeet potatoes. Parathion poisoning causes acute emphysema.

2. **Acute interstitial emphysema** often accompanies the acute alveolar emphysema. In this condition air collects in the interlobular space beneath the pleura and other interstitial tissue of the lung. This is seen more often in cattle and sheep.

Causes : a) Condition that produces dyspnoea—pneumonia.

b) Bellowing — in oestrum or when separated from calf ; common in kosher-killed animals.

c) Perforation of lung by mechanical means—foreign body through the rumen and reticulum.

d) Forced breathing as in old hunting dogs.

e) Pulmonary strongylosis.

In these cases, alveolar emphysema, when it occurs, is so severe that the alveoli rupture and air escapes into the interstitial tissue of the lung, especially in the inter-lobular septa. In severe conditions air may escape via the thoracic inlet into the subcutis of the neck and may accumulate there along the spine, from the pole to the base of the tail.

3 Chronic alveolar emphysema : Commonly called 'Broken wind' or 'Heaves' in horses, in which animal it is more often seen.

Causes; 1. Working horses immediately after a heavy meal — digestive organs are distended and so prevent the expansion of diaphragm during inhalation. So animal makes violent respiratory efforts;

2. Dusty and mouldy foods—causing coughing.

3. Obstruction of the bronchi—as in bronchiolit s.

4. Allergy to some sensitising agents (pollen, dust or fungi) may play a part The above conditions cause severe and continuous cough.

So during violent cough and inspiratory movements alveolar walls are subjected to undue pressure. This process, repeated for months and years produces atrophy of the walls and their subsequent rupture. Pressure also interferes with capillary circulation — so the nutrients and oxygen supply to the alveolar wall is diminished with resultant fatty degeneration of the alveolar epithelium, degeneration and disappearance of elastic fibres and the inter - alveolar and the inter-infundibular septa, resulting in rupture of the alveolar walls. Hence the adjacent alveoli become confluent—"bullae"

In bronchiolitis with obstruction, air passes through the pores of Kohn and so adjacent alveoli become balooned—collateral ventilation.

Owing to their diminished elasticity and permanent distension, alveoli expand with difficulty at each inspiration and also contract less easily on expiration and consequently continue to increase in size and finally rupture.

Besides the mechanical stress that produces emphysema as detailed above, it may primarily be also due to inflammation of respiratory bronchioles, alveolar ducts and alveoli. These structures becomes necrotic and weakened thereby becoming balooned due to the pressure of air that may be trapped.

Macroscopically, affected lung is voluminous and is pale due to decreased blood circulation. Lung pits on pressure easily and the indentations of the ribs are seen clearly.

Microscopically, the alveoli are over distended. Their walls are atrophied and then rupture of some alveoli with confluence of neighbouring alveoli give rise to giant alveoli.

Sequelae Hypertrophy and dilatation of right ventricle—chronic venous congestion.

Vascular disturbances of the lung

Active Hyperemia is commonly seen in some acute general infectious diseases or in acute pneumonia.

Passive Congestion occurs generally in older animals.

Causes: 1. **Cardiac lesions'** a) Myocardial weakness; b) Lesions in the mitral valve; c) Chronic pericarditis.

2. **Extrathoracic Lesions** Bloat—intra—abdominal pressure increases

3. **General vascular dilatation** : Shock.

Macroscopically, the lung is larger in size, dark red in color and is firm to the touch. On section blood oozes.

Microscopically, alveoli contain red cells and macrophages that have engulfed the red cells (Heart failure cells)

Sequelae : 1. Edema of lung—brown induration ; 2. Bronchopneumonia-

3. Dyspnoea; 4 Generalised hypoxia—fatty changes in the liver and kidneys:

Hypostatic congestion; In animals in moribund conditions blood accumulates in lung on the side on which the animal reclines. This is because the heart is too weak to maintain sufficient blood pressure.

EDEMA

Usually precedes pneumonia.

Causes : 1. Passive congestion of lungs—in lesions of myocardium and mitral valve

2. Toxic material that increases the permeability of the capillary endothelium—ANIU (Alpha Naphthyl Thiourea) poisoning; in shock

3. Hypoproteinemia

4. Acute anemias

5. After inhalation of smoke,phosgene,chlorine,ammonia,nitric oxid

6. Mulberry heart disease of swine

7. Bacterial toxins (exo or endo, produced locally or elsewhere)

8. Allergy (vaccination with strain 19 against Brucellosis and in helminthic infections)

9. Intravenous administration of fluids increasing venous pressure in pulmonary circulation.

Macroscopically, lung is large and firm. On section edematous fluid drips from the cut surface. Trachea and bronchi contain froth—due to churning action of the tracheal air on the protein-containing fluid. Edema of the inters; titial tissue is present. The alveolar septa stand out prominently.

Microscopically the alveoli and bronchi contain a pink stained homogeneous material. The pink staining capacity is proportional to amount of the protein present. The greater the amount of protein present, the pinker will be the fluid stained.

The edematous fluid is a good medium for the growth of microorganisms and so pneumonia is a frequent sequel.

Hemorrhages Blood in sputum due to pulmonary hemorrhages is called *hemoptysis* This is rare and must be distinguished from the blood present in lungs by inhalation while the animals are slaughtered.

Causes: 1.   Erosion of the blood vessels and rupture into a bronchus.
2. Extreme over exertion—as in young horses.
3. Extreme cardiac action as in death due to asphyxia.
4. Injury from a foreign body.
5. General hemorrhagic conditions
   Bacterial diseases—Pasteurellosis, Anthrax
   Viral diseases—Hog cholera
   Defects in coagulating ability — hemophilia and bracken fern poisoning
   Toxins—Uremia in which injury to capillary endotheliem occurs.

   Macroscopica'ly, hemorrhages may be in the nature of petechiae, ecchymoses or even hemato.ysts. Iulmonary hemorrhages are dangerous because air is displaced and hypoxia may result. Blood in bronchi and trachea foams.

   Microscopically, blood may be found in alveoli, bronchioles and bronchi.
THROMBOSIS, Often in pneumonia, due to extension of infection to the blood vessels thrombosis may occur and is frequently observed in septicaemic diseases—Pasteurellosis and Hog cholera.
EMBOLISM ; The strategic position of the lung is well suited for the arrest of emboli. Emboli emanate from;
   A. Thrombi occurring in
      1. Heart-worm infection in dog. 2. Mesenteric veins in horses (*Strongylus vulgaris* infection) 3. Uterine veins and pelvic veins (metritis and mastitis). 4. Posterior vena cava—extension of infection from hepatic abscess. (No 3 and 4 more common in cattle)
   B.   Ascarid larvae
   C.   Tumor cells—metastasis in malignant tumors,
   D.   Fat — not so common in animals as in man in whom it arises from crushing fractures of bones.

   Sequelae ; Abscesses, thrombosis, infection and parasitic cysts.
INFARCTION ; For the infarction of lungs to occur there must be damage to both pulmonary and bronchial circulations.

   Macroscopically, the affected areas appear dark-red, firm and solid looking. They bulge on the cut surface. They are cone shaped with base at the pleural surface and by careful dissection it is possible to disclose the thrombus and embolus at the apex of the wedge.

   Microscopically, the whole area of infarction, including the alveoli, capillaries and the septa, is filled with blood. Later necrosis sets in the alveolar walls.

INFLAMMATION OF THE LUNGS

   Inflammation of the lung is called pneumonia. Pneumonitis—literally meaning inflammation of the lungs—is used by different authors to convey slightly different meaning. Runnels uses it as a synonym with pneumonia; Smith and Jones apply the term for "any inflammatory disease of the lungs" preferring the use of the term pneumonia to "apply to one of the acute infectious inflammations with copious exudate filling the alveoli". Jubb and Kennedy differentiate the two terms by saying, "in pneumonitis the reaction is largely confined to the wall of the alveolus and in pneumonia it is the alveolar lumen which reveals the most obvious changes". In other words they reserve the term pneumonitis to the specific disease in which alveolar wall is the primary site of pathology. To avoid confusion, the term pneumonia alone is used in these descriptions.

Pneumonia is a very common disease found in animals except probably the cat, in which it is rarely met with.

In man, a specific type of pneumonia called the *lobar or croupous pneumonia* caused by *Diplococcus pneumoniae* is met with, in which whole lobes may be affected, characterised by a fibrinous or croupous exudate in the alveoli. Differing from this condition is another type, called the *Catarrhal or lobular or bronchopneumonia* which is patchy and in which only parts of a lobule or only a lobule are affected, characterised by a catarrhal exudate of the alveolus

In animals, it is the lobular pneumonia that is frequently seen. There is no condition similar to the lobar pneumonia of man. Most varieties in animal-may start as a lobular pneumonia but end up as a lobar variety. So all gradations may be met with in the same animal and, what more, the same etiological factors may give rise to these different grades of pneumonia.

Causes 1. By far the most common causes are the bacteria, the viruses, fungi and parasites. Again pneumonia may also be a lesion found in many specific diseases.

Bacteria :

Corynebacterium equi
Streptococcus equi   } **HORSES**
P. mallei

Mycoplasma mycoides
Pasteurella multocida
Corynebacterium pyogenes
Actinobacillus lignieresi   } **CATTLE**
Staphylococci
Streptococci
Mycobacterium tuberuclosis

C. Pyogenes
Pasteurella multocida
Staphylococci; C. ovis   } **SHEEP**
Streptococci: E. coli.

Streptococci; E. coli
Bordetella bronchisepticus
Staphylococci   } **DOG**
Klebsiella

Streptococci; C. pyogenes
Haemophilus suis   } **SWINE**
Pasteurella multocida

Pasteurella multocida
Coliforms   } **CATS**

Viruses

Equine infectious pleuropneumonia, Equine influenza, Calf pneumonitis, Canine distemper, Ranikhet Disease, Sheep pox.

Fungi : Blastomyces, Coccidioides, Histoplasma, Actinomyces, Aspergillus, Cryptococcus, Mucormycosis.

Parasites : *Metastrongylus apri* in swine; *Dictyocaulus viviparus* in cattle; *Protostrongylus rufescens* and *Dictyocaulus filaria* in sheep and goats; *Ascaris lumbricoides* var *suum* in pigs.

2. **Irritants** :   Inhalation of dust, pollen, foreign bodies, smoke, hot and cold air, anaesthetics, war gases, medicinal agents.

Routes of infection :

1. **Through the respiratory passages—Bronchogenous** : This is by far the most common route.   Some infectious agents like *Aspergillus fumigatus*, *Streptococci*, *Staphylococci*, *E. coli* and *Corynebacterium pyogenes* and some viruses (Eg. viruses of Ranikhet disease and viral pneumonia of pigs.) may invade the lung through this route.

2. **Through the blood vascular system—Hematogenous** : Blood stream may carry bacteria (Salmonella, Pasteurella) and parasitic larvae to the lungs.

3. **Through penetrating wounds** : Wounds by bullets, knives, pitch forks etc., may penetrate the lungs from the exterior, usually carrying bacteria and producing pneumonia. Similarly, foreign bodies penetrating through rumen, reticulum and diaphragm may set up pneumonia.

Predisposing causes : Conditions, called predisposing factors make the animals more susceptible to diseases of respiratory system. These are fatigue, exposure to cold air, long travel by train or ship, severe hunger, malnutrition; chronic undernutrition, parasitism, exposure after dipping in winter months, cardiac weakness and recumbency for a considerable time.

The predisposing factors produce rhinitis and laryngitis by virtue of their lowering the resistance with superimposed infection.   These affect or even destroy the ciliary movements and so predispose the organ for infection. Edema of the lung, that results due to cardiac lesions and other factors is an ideal condition for pneumonia to occur since the edematous fluid is a good medium for the organisms to develop and so cause pneumonia. In the recumbent animal "Hypostatic Pneumonia" may ultimately result.

Some etiological agents appear to become pathogenic only when the pulmonary tissue is suitably altered by other agencies, eg: in swine, virus of porcine pneumonia requires *Haemophilus suis* : in swine, *Pasteurella multocida* paves the way for a pleuropneumonia-like organism.

Some organisms, which are natural inhabitants in the upper respiratory tract become virulent and pathogenic, when the vitality of the animal is lowered eg. *Bordetella bronchisepticus* produces pneumonia in dogs in Canine distemper infection The virus of Distemper *per se* may not be able to cause pneumonia. Similarly pasteurella will cause pneumonia in Hog cholera.

Varieties of pneumonia

It has already been observed that in animals lobar pneumonia as it occurs in man is not met with. So the clear-cut classical stages described in human pathology are not seen in animals in entirety. Since a description of the stages of human lobar pneumonia helps in the understanding of the genesis and development of pneumonia the same is given below.

1. **The stage of congestion :** This is the early stage in which there is active heperemia and edema of the alveoli.

Macroscopically, the lungs are congested and swollen. These still float in water On section, blood tinged-fluid escapes.

Microscopically, the capillaries on the alveolar walls are dilated and filled with blood. Alveoli contain a little serous exudate and often a few red blood cells. Depending on the irritant, this may develop within a few minutes (chemicals) to a few hours (infectious agents).

2. **Stage of red hepatisation :** Macroscopically, the affected portion of the lung is quite conspicuous being readily discernible from the healthy. A distinct line of demarction is found. The affected part is red, and consolidated, solid looking resembling liver—**hepatization.** Portions of the affected parts sink in water, since all air is replaced. Over this area the pleura is inflammed and dull red in color. A membrane may form. Lymphatics are obstructed by fibrinous plugs. The pleural fluid is increased. The peribronchial and perviascular lymph-hatics are dilated with protein-rich fluid.

Microscopically, the alveoli reveal a fibrinous exudate containing erythrocytes, polymorphonuclear leucocytes and desquamated epithelial cells. Dilatation of lymphatics and widening of septal cells are observed.

Develops in two days.

3 **Stage of grey hepatisation :** Microscopically, the lung is still consolidated and sinks in water. The color is less red than the previous stage and some parts are grey like grey granite The rednes is due to persistence of capillary hyperemia or the hemorrhagic nature of the exudate.

Microscopically, the alveolus appears to be less filled than in the previous stage. Fibrin can clearly be seen and strands may be found to pass from one alveolus to another through the pores of Kohn Erythrocytes have almost disappeared from alveoli. The greyness of the affected tissue in this stage is attributed to i) ischemia of the alveolar capillaries due to pressure of exudate on them, ii) increased infiltration by leucocytes iii) thrombosis in the alveolar capillaries and iv) lysis of red blood cells

The liquefaction of the exudate commences and the nuclei of polymorphs become blurred and less distinct.

4. **Stage of Resolution :** At this stage liquefaction and removal of the exudate take place. The liquefied material may be absorbed via lymphatics or veins or may be expectorated.

Microscopically, the exudate is disappearing What remains is granular; polymorphs are either absent or the few that remain are degenerated. A number of macrophages derived from alveolar epithelial lining as well as from the blood are in evidence. The epithelium, most of which has died and was desquamated is regenerated. Thus the lung returns to the normal state of functional activity.

In animals, though the classical lobar pneumonia described above is not met with, yet, in infection by *Pasteurella multocida* in cattle, sheep and swine a croupous type of pneumonia is encountered. But, even then, the quantity of fibrin in the alveoli is not as much as in the human condition Infection is usually by inhalation, the earliest lesion produced being a bronchiolitis From this place the infection spreads rapidly. If the bronchiole is obstructed., the flooding exudate infects the surrounding parenchyma. And through the pores of Kohn, the edema fluid may infect the neighbouring lobule, which is not directly served by the obstructed bronchiole. Infection may also be direct and continuous from the bronchiole to the alveolus In a third manner, the infection may take place through the wall of the bronchus, namely by peribronchial path-way, when infection passes on to the alveoli lying immediately adjacent to the bronchiole by contiguity. Lastly, the infection may spread rapidly via lymph and blood.

Usually the apical, cardiac and intermediary lobes and the anterioventral portions of the diaphragmatic lobes are affected,

Due to differance in age of the lesions in different areas, one may see different stages of the disease in the same animal and hence we may see different colored areas. Again, thrombosis of pulmonary vessels, producing infarction may add difference in color of some areas.

Contagious pleuropneumonias of cattle and goats are of this lobar variety in which the interlobular septa are dilated and prominent due to a great outpouring of plasma and fibrin into them. It is the dilated septa that give the "marbling" effect to the lung in these areas.

Bronchopneumonia

This is the commonest type of pneumonia found in animals.

Causes;

Calves—is enzootic, caused by a virus.

Sheep—is enzootic? (Virus?)

Swine—i) Virus plus *Hemophilus suis*—swine influenza

    ii) *Salmonella cholerae suis*—hematogenous

    iii) In Hog cholera—*Pasteurella sp* plus *Hemophilus sp.*

Foals—*Corynebacterium equi* (may also be a complication of Strangles)

Primarily, infection starts as a bronchitis and bronchiolitis from where it may spread to the alveoli as described earlier viz. by flooding with exudate when there is occlusion of the bronchi, by direct extension from bronchi and bronchioles to alveoli and by peribronchial path-way. Compared to the lobar variety, extension to different parts is slow. The anterior and vantral parts of the lungs are more commonly affected because the bronchi to these parts take off vertically and so infection gravitates. Also the respiratory movement of these parts is less due to limited rotatory action of anterior ribs. As the infection spreads, fresh foci are set up and hence different stages of the disease are noticeable within the same animal

The *lesions* are patchy in distribution. One or several lobules may be affected which are red and firm, sinking in water. Areas adjacent to the affected show "compensatory emphysema".

The lymphatics are swollen and become easily visible. The lymph nodes are hemorrhagic and swollen. The bronchi contain hemorrhagic exudate. Ths pleura over the affected area shows inflammation with fibrinous exudate.

Microscopically; one or the other of the four stages described under the lobar variety may be seen.

Sequelae: 1. Death – due to i) Toxemia; ii) Hypoxia and iii) Cardiac failure.

2. Atelectasis—when bronchiole is still obstructed, even after resolution of the alveolar exudate.

3. Suppuration and abscess formation—if the causative agent is pyogenic. Common in dogs, caused by *Bordetella bronchisepticus* and in foals by *Corynebacterium equi*.

4. Gangrene—this occurs when there is superimposed infection by saprophytic putrefactive bacteria. Usually occurs in aspiration pneumonia.

5 Septicemia when the infective organisms enter the blood, causing inflammation in other parts of the body eg. arthritis, meningitis etc.

6 Incomplete resolution—resulting in organisation of the alveolar exudate. Fibrosis of the lungs and pleura may result—*carnification*. (becoming like flesh).

OTHER TYPES OF PNEUMONIA

Necrotic, Gangrenous and Aspiration pneumonia

Causes : 1. Faulty drenching in cattle and careless passage of stomach tube in horses.

2. Inhalation of irritant drugs, oils, anesthetics or feed (lambs and pigs)

3. Aspiration of i) milk or gruel (in pail – fed calves)
ii) ingesta—in paraylsis of throat in parturient paresis of cattle.

4. Hematogenous—from gangrenous lesions elsewhere in the body. eg gangrenous metritis or mastitis.

5. Penetration of sharp foreign bodies through rumen and reticulum.

6. Direct Infection by *Spherophorus necrophorus*.

The irritants produce severe inflammation and extensive thrombosis of the blood vessels resulting in necrosis. The leucocytes and bacteria produce liquefaction of the necrotic material, resulting in cavitation. Putrefactive organisms produce gangrene.

Macroscopically, there is extensive consolidation of the anterior and ventral portions of the lung with foul smelling exudate. The affected parts are greenish or black in colour and sometimes large cavities are seen. The area around these lesions shows congestion and intense reaction.

Gangrenous pneumonia may be a sequel of choke (esophageal obstruction) Sequelae—Death in all cases.

Metastatic suppurative pneumonia

This may be acute or chronic This is due to embolic deposition of pyogenic organisms from lesions somewhere else in the body. eg Suppurative metritis, mastitis, lesions in Strangles, navel ill and heart valve lesions. (Vegetative endocarditis)

Microscopically: because the pathogen is hematogenous, there is uniformity in the distribution and size of the lesions. The diaphragmatic lobe is mostly affected and a number of foci may be found beneath the pleura. The lodged organisms produce inflammation with suppuration.

There may be several abscesses scattered with a zone of acute inflammation in the parenchyma surrounding them. There may be a capsule, the thickness of which indicates the duration of the process, (older the lesion, thicker the capsule.) Encapsulated abscesses may be organised, leaving a depressed scar.

Verminous pneumonia

Pneumonia in animals is caused by many species of parasites. They are

Parasite	Habitat
Cattle—*Dictyocaulus viviparus*	Bronchi
Paragonimus westermanii	Bronchi and cysts in the lungs
Sheep—*Dictyocaulus filaria*	Bronchi
Protostrongylus rufescens	Bronchi
Mullerius minutissimus (capillaries)	Alveoli and blood vessels
Paragonimus westermanii	Cysts of lung and bronchi
Swine—*Metastrongylus apri*	Bronchi
Ascaris larvae	Migration
Horse—*Dictyocaulus arnifieldi*	Bronchi
Cat— *Aleurostrongylus abstrasus*	Bronchi
Dog— *Angiostrongylus vasorum*	Pulmonary arteries
Paragonimus westermanii	Cysts in lung and bronchi

The infection by strongyles, whose habitat is in the lungs, starts by the ingestion of infective larvae with feed and water. The larvae enter the portal circulation and finally reach the capillaries on the alveolar walls. Here they pierce the capillaries and enter the alveoli where they develop and become mature in about a week to 10 days. The mature worms then go up to the bronch and settle.

The pathology produced by these "lung worms" may, therefore, be described in two stages. In the first stage, the larvae enter the alveoli, develop and mature and in the second, they settle down in the bronchioles and bronchi. There may not be any clear-cut demarcation of the two stages, which may therefore overlap. In the place where the larvae enter the alveolar walls are found microscopic necrotic foci surrounded by an infiltration of neutrophils, eosinophils and macrophages. Thus thickening of the alveolar walls occurs. A few giant cells may also be seen. In massive infection, there may be severe hemorrhages.

When the parasites have become mature, they migrate, into the respiratory bronchioles and bronchi where they set up inflammation on the walls. A thick mucous exudate containing polymorphs, eosinophils and macrophages forms at the site, often occluding the bronchi. The epithelium of the bronchi and bronchioles becomes hyperplastic and thickened. Weakening of the bronchial walls because of pressure and inflammatory process results in bronchiectasis. Obstruction of the bronchioles and bronchi results in emphysema, since during inspiration air enters but which is not exhaled. So repeated trapping of the air results in emphysema. In places where there is complete occlusion of the bronchioles, atelectasis develops as the air in the alveoli is resorbed.

The lesions are found mostly in the diaphragmatic lobes and consist of focal wedge-shaped areas of emphysema and atelectasis. Secondary bronchial infection may complicate the picture.

Macroscopically, the lesion is one of bronchopneumonia. The exudate in the bronchi contains the worms with their embryonated eggs along with neutrophils, macrophages, lymphocytes and desquamated epithelial cells. The alveoli may contain besides the inflammatory cells, worm ova.

Mullerius capillaris, a parasite of sheep and goats lives, not in the bronchi, but in the pulmonary blood vessels. The lesions produced are nodular and consist of granulomatous process—viz. fibrosis, giant cells, mononuclears and eosinophils. When the worm dies the nodule may become calcified.

Paragonimus westermanii, a fluke. is found in the lungs of cattle, sheep and dogs (also in man). These are usually found in pairs in inflammatory cysts in the lungs Most of the cysts have communication with the bronchi for the passage of eggs. Those not having such outlets become organised.

Larvae of Ascaris in their itinerary to complete their life-cycle pass through the lung, where they may cause severe damage and in some cases even death, if infection is sufficiently heavy. As in the strongyle worms, the larvae reach the lung via blood. From the capillaries, the larvae enter the alveoli through their walls thereby damaging them in the process. In heavy infection there may be massive hemorrhage when the larvae pass through capillaries. While the lung tissue is thus damaged, bacteria, normally found there migrate to the damaged areas and produce inflammation Thus the combined action of the parasitic larvae and bacteria causes death of the animal.

Sometimes liver flukes may be found in the lungs, but this is an aberrant location. A nodule is formed with a fibrous capsule and the parasite dies as it cannot thrive in a strange locality.

Interstitial pneumonia

This is a condition in which the alveolar septa are affected. Though some exudate may be found in the alveoli, by far most of the changes, prolferative, are found in the alveolar septa. The following characterise this variety.

i) Thickening of the alveolar septa by
   a) exudate—serous or fibrinous
   b) infiltration by leucocytes and
   c) formation of new connective tissue

ii) Proliferation of the epithelial cells of the alveoli

iii) Formation of hyaline membrane in the alveoli and over the alveolar ducts and

iv) Almost complete absence of neutrophils in the exudate.

Causes are many and varied.

1. The usual cause are viruses, which may be the primary cause as in swine influenza and feline pneumonitis.

2. Or this condition may be pulmonary manifestation of generalised septicemic diseases like Erysipelas, Leptospirosis, endocarditis of the right side, Salmonellosis etc.

3. Psittacosis group of organisms.

4. Larval migration of ascarids and other intestinal parasites (not usually important)

The condition may be acute or sub-acute in intensity. Injury to the capillary endothelium is responsible for the fibrinous exudate in the the alveolar septa. Bacteria and parasitic larvae injure the endothelium directly and so make it more permeable. In the case of the viruses, increased permeability occurs in an indirect manner. The viruses affect the alveolar epithelium which proliferates and so becomes thicker preventing oxygen transfer. Hypoxia injures the capillary endothelium which therefore, becomes more permeable. This edema is partly responsible for the thickening of the septa.

Macroscopically, the lungs may be pale or reddened and do not collapse when the thorax is opened. As a whole, the lung appears edematous, with fluid dripping from a cut surface. The interlobular septa are also widened. Usually interstitial pneumonia is the seat of secondary bacterial invasion and resultant bronchopneumonia.

Microscopically, alveoli may contain serous or serofibrinous exudate, which is also present in the alveolar and interlobular septa. There is fibrous thickening of the septa and metaplasia of the alveolar epithelium, which is flat and pavemental normally and not easily seen, becomes cuboidal and prominent. These are called **Cells of tripier**. This change is said to be *Fetalisation* or *epitheliolisation*. Furthermore, the cells may proliferate, become rounded and desquamated, forming macrophages with phagocytic properties. In some cases, giant cells also form by the fusion of the macrophages thus formed from the alveolar epithelium or by the proliferation of their nuclei without the division of the cytoplasm. Predominance of these cells gives rise to the "giant-cell" pneumonia.

Lymphocytes and macrophages infiltrate into the alveolar septa and around the small blood vessels.

One characteristic feature of interstitial pneumonia is the presence of hyaline membranes, which are found covering the alveolar epithelium. The membrane is thought to be derived from and composed of the plasma proteins. Due to vascular damage and increased permeability of the capillary endothelium, rpofuse exudate forms, which nearly resembles plasma and hence the formation of the membrane.

Complete resolution may occur with alveoli regaining their normal histological structure and physiological function. The hyaline membrane may be removed by the macrophages. If resolution does not occur, healing by fibrosis takes place.

The fibrinous exudate is organised if it is not resorbed and remains longer than 2 to 3 weeks. This organised tissue is shrunken and grey and is like flesh—carnification.

Mycotic pneumonia

Inflammation of the lung caused by a variety of fungi is called mycotic pneumonia. Various fungi that may invade the lung and cause pneumonia are Aspergillus, Blastomyces, Mucor, Coccidioides and Cryptococcus. Of these, the most common is *Aspergillus fumigatus*.

Aspergillus fumigatus is ubiquitous and is found everywhere in nature. It is surprising that more cases of pneumonia are not caused in man and animals with the wide spread distribution of the spores of this fungus.

Though man and domestic animals are susceptible to Aspergillus and stray cases of pneumonia are encountered, but by and large, fowl is the most frequently affected.

The spores of the fungus are found in the poultry litter and mouldy grain. Infection is by way of the respiratory tract. Spores may be inhaled with the infected dust or may be aspirated from the mouldy grain. Since this affection is very common in brooder houses, it is called "Brooder pneumonia".

Lesions are found in the trachea, bronchi, and air sacs. Sometimes, larynx also may be affected. White or greenish, thick cheesy material is found in the affected areas.

The spores after entering the terminal bronchioles and alveoli, grow by budding and formation of septate hyphae. This produces a local inflammatory reaction, a nodular bronchopneumonia and there is infiltration by polymorphs and macrophages. This focus expands and more and more lung tissue is affected.

Macroscopically, the lesion is nodular. Evacuation of the central caseous material into the bronchus results in a central cavity. In lesions that are not progressive, increasing number of macrophages and epithelioid cells may be present with a fibrous capsule.

Microscopically, septate hyphae of the fungus with the inflammatory cells may be found around a central caseous material. In some cases, foreign-body giant cells may be found at the periphery. In lesions that are not progressive and are being obliterated, the hyphae are very short and the lesion resembles that caused by actinomyces—the *"asteroid body"*.

Pulmonary adenomatosis

This is a disease of domestic animals, characterised by hyperplasia and hypertrophy of the alveolar epithelium, giving it a glandular or adenomatous appearance.

The disease has been described in sheep (in various parts of the world and in India), in cattle, pigs, horses and man. The causes are:—

Sheep; A filterable virus is supposed to be the cause, though it is not yet conclusively proved. A *mycoplasma* may also play an etiological role.

Cattle: Food allergy or some intoxication is thought to be the cause: feeding on mouldy foods.

Man: Nitrogen peroxide from silos is supposed to be the etiological agent.

Pigs and horses: A pleuropneumonia-like organism and toxins of *Crotalaria dura* respectively are supposed to cause a condition similar to pulmonary adenomatosis in these animals

The disease in sheep is called Jaagsiekte (Jaag=drive, siekte=sickness) or driving sickness In Afrikans since fits of coughing are produced when affected animals are driven (exercised). The disease may occur sporadically or as an outbreak The incubation period appears to be about 7 months. After the onset of symptoms, the course is slow, death occurring in 2 to 3 months. Natural transmission is by droplet infection.

The disease can be differentiated from other pneumonic conditions by the absence of rhinitis in this condition.

The lesion starts as a small nodule, resembling that produced by *Mullerius capillaris*. But the former is softer and more friable. The nodules that are discrete to start with may coalesce to form bigger nodules and so finally, the lung is bulky and heavy, showing the imprints of the ribs. Emphysema of unaffected lungs is also noticed. Bronchial lymph nodes may frequently show metastases.

Microscopically, the characteristic change noticed is hyperplasia and hypertrophy of the alveolar epithelium. The normal pavemental epithelium, not easily discernible, becomes cuboidal or columnar with rounded nuclei. There may be papillary projection of epithelium into the lumen of the alveolus. In some places, there may be desquamation of the epithelium into the lumen. The alveoli seldom contain any exudate found in other pneumonias. The bronchial epithelium may also show papillary projection into the lumen. The bronchial lumen may contain desquamated epithelium as well as alveolar septal cells. Lymphocytes and plasma cells may infiltrate in the stromal tissue.

The thickening of the alveolar epithelium prevents the normal exchange of gases and so hypoxia is produced, causing dyspnoea. Due to decrease in the pulmonary capillary bed, dilatation of the right side of the heart may result. Fibrosis of the lung usually supervenes since the condition is not resolved completely and early.

In cattle lesions similar to the above may be seen but the condition is more acute and so fibrosis is not usually observed.

In man, pulmonary adenomatosis is common in those working near corn silos, where nitrogen peroxide is liberated. This is, therefore, called "silo-filler's disease". Death is quick occurring in six weeks after initial exposure,

In the human disease, bronchiolitis fibrosa obliterans is more in evidence due to the organisation of the bronchiolar exudate. Grossly these lesions appear as numerous uniformly distributed, firm miliary nodules.

Maedi

This means dyspnoea in Icelandic language. Maedi is found in Iceland and probably in Holland and North America. A condition similar to Maedi was reported in India.

The incubation period is long, upto 3 years. The disease is contagious and occurs as an outbreak. A virus is supposed to be the causal factor.

Maedi affects older sheep, usually those over two years. Affected animals show progressive debility, dyspnoea and inanition and may linger on for three to six months. In final stages acute pneumonia may supervene (due to pasteurellosis).

Macroscopically, the lungs do not collapse, when thorax is opened, but feel rubbery. They are greyish-blue in color, larger and very heavier than normal. The alteration in lungs is uniform in the absence of any complications. There is hyperplastic lymphadenitis and so the nodes are swollen. Except in terminal stages, consolidation, as found in other types of pneumonia is seldom met with. So, on section, no exudate oozes and the cut surface is dry.

Microscopically, the lesion is chronic interstitial pneumonia and the reason for the increased density of the lung is due to increase in the thickness of the alveolar walls. There is infiltration of walls by lymphocytes and reticuloendothelial cells — macrophages and short reticular cells. The alveolar lumen may contain de quamated epithelial cells. There is hypertrophy and hyperplasia of the bronchiolar epithelium. Peribronchiolar and perivascular infiltration with lymphocytes is marked. There may be increase in the smooth muscle of the respiratory bronchioles, which may give a false appearance of fibrosis. No healing is observed.

Inclusion bodies, μ to 3μ in diameter, bluish-grey with Giemsa's stain are found in the cytoplasm of the alveolar macrophages.

Jaagsiekte	Maedi
1. Alveoli show adenomatosis	1. Adenomatosis is not seen
2. Inclusions absent	2. Inclusions present
3. Lesions focal, not diffuse	3. Lesions diffuse
4. Lymph nodes not affected	4. Lymphadenitis present
5. Course shorter	5. Course longer

TUMORS OF THE LUNGS

In the lungs primary tumors are rare. Those reported are classified as having the following cell types: squamous cell, columnar cell, mixed cell and undifferentiated cell. The exact sites of origin are still under dispute, whether from the bronchial epithelium (though most of the tumors are believed to arise from this site) or the alveolar epithelium or from the epithelium of the mucous glands.

Both benign and malignant tumors occur. A few connective tissue tumors have also been described. Lymphocytoma is the commonest.

DISEASES OF THE PLEURA

Congestion and edema (hydrothorax) are common in acute poisonings and passive congestion.

Pneumothorax is the presence of air in the pleural cavity. This may be due to entry of air into the pleura by the following ways:

1. From the lungs; when some bullae rupture,
2. Through the chest wall—by piercing sharp objects.

Pneumothorax causes atelectasis of the lungs and this is of **great importance** in the horse in which the right and left cavities communicate.

Pleuritis; Pleuritis is inflammation of the pleura. It is also known as **pleurisy** Mostly pleuritis is secondary to pneumonia, through primary infection of the pleura may occur. The infection may be by several routes:

1. **By direct extension** from the underlying lungs or mediastinal glands or esophagus.
2. **By blood stream** in septicemic diseases.
3. **Introduction through the thorac wall**—trauma by knives, bullets etc
4. **Introduction from the rumen** via reticulum and diaphragm.
5. **Through the esophagus**—when sharp bones or pins may penetrate the pleura and convey bacteria or in the horses when choke occurs.

Pleurisy may also be a condition noticed in specific diseases like swine erysipelas and contagious bovine pleuropneumonia. When pleura is involved in pneumonia, the condition is known as *pleuropneumonia*. But it is not neceassry that pleura should be affected whenever lungs are. It is also possible to have pleuritis when the underlying lungs are healthy as in Black Quarter.

Bacteria that reach the pleura are readily spread due to respiratory movement. Bacterial growth damages the mesothelium and the endothelium of the superficial blood vessels, resulting in an inflammatory reaction, characterised by congestion and thickening of the membrane. The friction, between the parietral and swollen visceral surfaces causes intense pain during respiration The pain is relieved on the formation of inflammatory exudate which separates the two surfaces.

The exudate may be serous but more often is serofibrinous or fibrinous. Usually it is copious and may press on the lung, producing pain. Pyog ic bacteria may produce purulent exudate when the conditions is known as *empyema*.

If the exudate is purely serous, it may be absorbed. But if serofibrinous or fibrinous, complete absorption does not take place, organisation sets in producing "adhesions" in which the parietal and visceral surfaces get sewn, causing pain during respiration.

If saprophytic organisms gain entry to the pleura along with foreign bodies, gangrenous pleuritis will occur.

Tuberculosis of the pleura is most common in cattle, frequent in dogs and cats but seldom found in other animals.

Tumors of the pleura are rare. They may be (a)Primary—mesothelioma, a malignant tumor or (b) secondary,—metastasis of melanoma

---

THE DIGESTIVE SYSTEM

The Teeth

Caries is decay of teeth in which the enamel is decalcified followed by softening and discoloration. Caries is rather rare in domestic animals. This occurs occasionally in pet dogs with imbalanced and inadequate diets. The affected teeth have, usually, one or more depressed areas, which are brown or black in color The organic acids, especially lactic acid, that are formed due to the action of bacteria on carbohydrates, dissolve the salts of the enamel. Then, the same acids corrode into the less stronger dentine, which contains in its structure 30% of protein. The damage to the dentine is deeper and most widespread. The opposed surfaces of adjacent teeth may be more frequently affected "*Enamel Flecks*" are yellow stained spots on the enamel in the early, incipient stages The affected teeth are shaky and are very painful and so interfere with mastication.

Causes include disturbances in calcium and phosphorus metabolism as well as dietary deficiencies of these minerals.

As we have already seen, caries is frequently seen in fluorine poisoning.

MOUTH AND PHARYNX

Malforamations :Clefs of the mouth occur due to failure in embryonic processes. The most common is palatoschisis or hare lip. In this condition the ingesta is likely to enter the respiratory passages

Stomatitis : This is diffuse inflammation of the mucous membrane of the mouth But when confined to particular parts of the mouth it is known as:

Gingivitis for inflammation of gums.

Glossitis for inflammation of the tongue.

Lampas for inflammation of the palate.

Cheilitis for inflammation of the lips.

If pharynx alone is affected, inflammation of that part is known as **Pharyngitis**. If tonsils alone are affected, then the condition is known as **tonsillitis**.

Stomatitis is a common affection noticed in animals and more often it is a symptom of some other disease. Again, it may be a primary affection or may occur as secondary to other associated diseases viz. gastritis or infectious diseases.

Cause: 1. Physical (a) **Trauma** by awns, thorns, burrs, wood pieces, glass pieces; sharp bits, irregular sharp teeth; sharp edged feeding utensils

(b) **Heat**: Hot drenches, (c) eating frozen foods

2. Chemical ; caustic alkalies, corrosive acids, fertilisers.

3. Deficiency of vitamins; a) Hypovitaminosis A; especially in fowl
b) Niacin deficiency : Black tongue in dogs—(necrotic stomatitis)

4. Microorganisms ;

i) BACTERIA; *Actinomyces bovis* ; *Actinobacillus lignieresi* ; *Spherophorus necrophorus* ; *Pseudomonas aeruginosa* ; *Corynebarcterium pyogenes* ; *Streptococci and Staphylococci*:

ii) FUNGI; *Monilia albicans* and *Oidium pullorum* in poultry,

iii) VIRUSES : Foot and Mouth disease; Rindespest: Virus diarrhoea—Mucosal disease ; Infectious Canine Hepatitis ; Contagious ecthyma ; Vesicular exanthema ; Fowl pox, Blue tongue in sheep,

Macroscopically, the lesion starts as catarrhal inflammation of the mouth and pharynx with reddening and swelling of the mucosa, which is covered by small, whitish spots, raised than the surrounding. "*Aphthous stomatitis*" is the name given to the condition. These spots may later develop into small crusts or into ulcers.

Thrush: is found in birds, This is the name given to a condition in which grey or yellowish thick material tenaciously gets attached to the mucous membrans.

Vesicular stomatitis is the condition in which vesicles or blebs or blistere containing fluid are formed on the mucosa and seen in Foot and Mouth disease,

Infectious Vesicular Exanthema and Infectious vesicular Stomatitis.   Rupture of the blisters results in the formation of erosions, which subsequently heal.
Catarrhal and vesicular stomatitis may develop into ulcerative variety.

Fibrinous and necrotic stomatitis; are seen in infection by *Spherophorus necrophorus* Very severe irritants may cause gangrenous stomatitis.

Fowl pox produces diphtheritic stomatitis and pharyingitis in which a grayish membrane is found.

Sequelae; 1. Starvation as prehension and mastication are prevented.

2. Spread of infection to other parts—esophagus, lungs, stomach etc.,

Tumors of the mouth and pharynx are common. Most common neoplasms in the dog and calves are the infectious papillomata, occurring as clusters on the lips and gums. These are supposed to be viral in origin (verruca vulgaris).

Epulis is a fibroblastic tumor, usually occurring in the gums consisting of dense fibrous tissue with varying amounts of epithelium and a few giant cells

Carcinoma, sarcoma, fibroma and melanoma are other tumors occasionally seen.

SALIVARY GLANDS

Pathological processes are very rarely found in the salivary glands of animals because (1) the salivary secretions have some anti-bacterial properties: (2) there is good flushing by the secretions and (3) the glands are in fairly well protected situations.

Foreign bodies are occasionally found in the ducts especially of the parotid and submaxillary glands. These are usually, awns, slivers of wood and kernels of grain, causing inflammation. Sometimes these may produce obstruction and consequent dilatation of the ducts.

Dilatation of the salivary ducts may occur when the flow of saliva is obstructed by foreign bodies, inflammatory exudates etc.,

When the dilatation occurs as cyst on the floor of the mouth it is called a *ranula*, which is smooth and rounded, containing a clear fluid and which can be easily ruptured.

Sialoliths are salivary calculi and are common in horses. These are formed by the precipitation of minerals around nuclei of foreign matter in the ducts Salivary calculi are usually single, and sometimes may be very large; preventing the flow of saliva. These calculi produce stasis, distension of ducts and finally atrophy of the gland.

Inflammation of the salivary glands; *Sialadenitis* or *parotiditis* is very rare in animals and may be due to traumatic injury or due to infection by bacteria. Stasis of saliva in the ducts facilitates infection. This may be associated with strangles in horses, mastitis in cattle and distemper in dogs.

Macroscopically, the glands are swollen and red. Abscesses may be found in glands and sometimes cystic dilatations may occur. Inflammation of the salivary glands may cause atrophy of the gland.

Neoplasms of salivary glands are not common in animals.

ESOPHAGUS

Choke; is obstruction of the esophagus, occurring in horses and cattle, but more common in the former.

Causes : 1. Impacted masses of feed due to : improper chewing, bad teeth and rapid gulping of dry feed.

2. Lesions of esophagus—stenosis or diverticulum—repeated choking may occur,

3. Old age.

4. In cattle, large objects of food—beet root, carrot, apples, potatoes, fetal membranes, sticks, wire, (large bones in dogs).

5. Enlarged lymph nodes—mediastinal and cervical.

6. Enlarged thyroids.

7. Neoplasms of adjacent tisssue—especially thymus, thymoma in new-born animals.

In the horses choke occurs in the thoracic area while in cattle and dogs the pharynx is obstructed. Choke may be complete or incomplete. In complete choke feed will be returned and water will come out of the nostril when animal is watered. Inhalation of the feed will cause secondary foreign-body pneumonia. In cattle, complete obstruction will cause dangerous tympany.

Because of pressure, ischemia and resultant necrosis develop leading to gangrene

Infection may spread to the surrounding tissues—cellulitis or to the lungs—gangrenous pneumonia. Sapremia or toxemia that occurs is the cause of death in fatal case .

Partial obstruction will give rise to dilatation of esophagus above the obstruction,—the *esophageal diverticulum.*

Sequelae : 1. Death due to gangrenous pneumonia, bloat, cellulitis or asphyxiation. 2. Esophageal diverticulūm. 3. Rupture of esophagus.

DILATATION (ECTASIA) The dilatation of esophagus may be *fusiform or cylyndrical* of which the former is the more common.

Causes : 1. Accumulation of food proximal to a stenosed area.

2. In ruminants accumulation of food during regurgitation on the distal side of stenosis.

3. Trauma from horns etc. rupturing the muscular coat.

4. Idiopathic, due, probably, to relaxation of the esophageal muscles consequent on nervous lesions.

The food gets accumulated in these areas producing pressure.

Sequelae : 1. Rupture due to pressure of the food.

2. The food may become decomposed and produce softening of the epithelium, inflammation, ulceration gangrene and death. 3. In ruminants, bloat.

ESOPHAGITIS

Inflammation of the esophageal mucosa is rare in animals because of the thick and resistant condition of the mucosa.

Causes : Trauma—probang; stomach tube, foreign bodies.

Chemicals—corrosives

Parasites—bot-fly larvae in horses, and hpoderma larvae in cattle.

Persistent vomiting, in dogs and pigs.

Avitaminosis A in Fowls.

Macroscopically, the mucosa is red and swollen. In the catarrhal variety, the exudate is mucous. The ulcerative variety is met with in conditions caused by trauma (Stomach tubes), in virus enteritis and in Mucosal Disease in cattle.

In the fowl, thallium sulphate poisoning produces esophagitis.

When pyogenic bacteria enter the place of obstruction (in choke) which had become necrotic due to pressure suppurative esophagitis occurs.

Sequelae : Usually recovery occurs. Stenosis may result if severe.

NEOPLASMS : Neoplasms of the esophagus are not common. Of those that are encountered, the connective tissue tumors of the dog must be mentioned. In the thoracic portion of the esophagus are found fibrosarcomas and osteogenic sarcomas that have some connection with *Spirocerca lupi* infection. The osteogenic sarcoma is evidently a metaplastic manifestation of the fibrosarcoma. Metastases of these tumors are sometimes found in the lungs and other tissues.

Carcinoma in cat and horse and papilloma in cattle were other tumors met with.

OBSTRUCTION OF CROP IN BIRDS

Causes : 1. Atony or paralysis of wall leading to stasis of food.

2. Ingestion of large quantities of dry grain which swell in the crop and form a hard mass.

3. Foreign bodies like wire etc.

The stagnated food gets decomposed, gas accumulates and inflammation sets in.

Sequelae : 1. Rupture due to distended food and gas or due to penetration of a foreign body.

2. Death because of (a) asphyxia, due to comperssion of trachea, (b) heart failure due to pressure on heart, (c) starvation since food does not enter the proventriculus (d) intoxication due to absorption of toxins from fermented foods.

INGLUVIITIS : (Inflammation of crop)

Acute catarrhal ingluviitis

Causes : 1. Trauma by foreign bodies.

2. Chemical agents : phosphorus, fertilisers.

3. Toxins from decomposed food.

4. Infectious diseases.

5 Parasites— *Acuaria sp.*; *Capillaria sp.*

Lesions include cogestion, edema and tympanites. Diphtheritic ingluviitis is found in fowl pox.

FORE STOMACHS OF RUMINANTS

TYMPANITES OR BLOAT : Normally anima's get rid of gases produced in the rumen by eructation. Bloat (or accumulation of gas) can therefore occur when the gas is produced at too rapid a rate than can be eructated or when the eructation mechanism is faulty.

Bloat may be acute or chronic. The chronic variety occurs whenever there is any hindrance to eructation in the esophagus, either within or without— pressure by tumors, foreigh bodies, enlarged lymph nodes, abscesses, constrictions or diverticula. Or the lesion may be in the rumen causing decreased contractions of the ruminal wall as in atony, serosal adhesions, paresis, diffuse lymphomatosis.

Acute Tympanites ; This may be due to choke in esophagus. Or it may also be due to sudden changes of feed or to excessive feeding on legumes that are wet with dew or rain.

There are various theories propounded to explain bloat but none of them are quite satisfactory. The following are some of them.

1. Some legumes contain HCN, which is toxic, causing paralysis of the ruminal or reticular musculature and so inhibits eructation.

2. Some legumes contain Phosphatase which with arsenates accelerates fermentation so that large quantity of CO_2 is produced.

3. H_2S, CO_2 and CO produced in large quantities casuse paralysis of ruminal muscles.

4. If fed excessively on green plants only, which do not contain sufficient stiff fibres, the mucosa of the rumen is not adequately scratched to elicit the reflex contraction of the musculature.

Saliva which has important antifoaming properties plays a significant part in the prevention of bloat. Mechanical stimulation of cardia, especially by roughages, increases the rate of secretion of saliva. But with ingestion of young succulent legumes, too little saliva may be secreted and so foaming is not counteracted and bloat results. Hence hay or straw must be fed liberally when animals are fed with succulent legumes. Mucin in saliva prevents formation of froth. But ruminal mucinolytic bacteria may destroy salivary mucin thereby producing bloat. Polysaccharides produced by capsulated ruminal bacteria may be another etiological factor in bloat.

Interference with the nerve pathways that are responsible for the eructation reflex may also lead to tympany. The receptors for this reflex are in the reticulum and the afferent and efferent nerve fibres are in the vagus nerve. Any lesions in this nerve may, therefore, lead to bloat, since the reflex that leads to eructation is interrupted.

Bloat is of two varieties : the dry and frothy. The former is less harmful since in this condition the gases can be more easily got rid of by eructation. On the other hand, in the frothy bloat, the gas is trapped as small bubbles in the fluid forming a foamy mass, which is not easily eructated. The following are supposed to produce frothy bloat.

1. Saponin found in plants is incriminated. Saponin is a good saponifying agent.

2. Water-soluble proteins of the legumes are capable of forming froth. This is probably of greater importance than No. 1, since bloat was observed even in animals fed with low-saponin plants.

3. Changes in surface tension and viscosity. It was found that (i) feeding times; (ii) kinds of feed and (iii) amounts of water, change the viscosity and surface tension. Those increasing the viscosity and lowering the surface tension produce froth. Normally, in rumen due to bacterial activity, fatty acids are produced. These increase the surface tension. If the production of these fatty

acids is decreased then their protective action is not available and so surface tension will be lowered favouring froth production. This is the theory behind the use of vegetable oils in the treatment of bloat.

Distended rumen compresses other abdominal organs and causes passive congestion since the pressure on thin-walled veins impedes circulation. Along with this, there is forward thrust on the diaphragm, pressing on the lungs, which become smaller and sometimes atelectatic. The result of this is hypoxia and ultimate asphyxia and death.

In animals that die of bloat, besides congestion of the abdominal viscera, one may notice hemorrhages on the serous membrane of the lungs, on the pericardium, on the tracheal mucosa and in the lymph nodes of head and neck. Blood is tarry, as in Anthrax. The bronchial lymph nodes may be hemorrhagic. Liver is pale. Sometimes the rumen or diaphragm may be ruptured. If the animal is dead for some hours, the ruminal epithelium peels off.

Sequelae; If quickly relieved, acute bloat can be cured. If not, death may supervene due to asphyxia.

Impaction of the rumen and reticulum

This is a common condition in cattle, the important feature of which is that the rumen stops functioning, the musculature does not contract and so the food ingested stagnates.

Causes: 1. Overfeeding with large amounts of highly fermentable carbohydrate feeds.

2 Penetration of the wall of the rumen or reticulum by sharp objects—wire, nail etc,

3 Lack of water.

4 Defective mastication and insalivation due to defects in teeth or lesions of the tongue.

5. Paresis of rumen which may occur due to injury to vagus by pressure from abscesses, tumors, tubercular nodules, swollen lymph glands and ruminal displacements.

Lack of exercise and debility are supposed to predispose the animal to atony of rumen. Tight packing of the rumen leaves no room for bacterial growth and normal ruminal fermentation and digestion. This leads to weak contractions of the ruminal and reticular walls and so the food does not get propelled. The stagnated food becomes putrified with the liberation of foul smelling gases. Anorexia develops and regurgitation stops.

In some animals, diarrhoea may be present if the putrid ingesta finds its way into the intestines causing enteritis. In mild cases, if the primary cause is removed, normal state may be regained. But in severe cases, toxemia will cause death At necropsy, the rumen will be found to contain hard, caked, undigested food, with evil smelling odor.

The pathogenesis of atony and impaction of the rumen after ingestion of large quantities of carbohydrate-rich feeds is as follows; The carbohydrates are fermented by gram positive organisms, notably *Streptococcus bovis*, with the formation of lactic acid, resulting in lowering of pH of the ruminal contents to as low as 4 or 4.5 from a normal 5.5 to 7.5. Due to the production of lactic acid the osmotic pressure of ruminal contents increases and so fluid is drawn into the rumen from the blood and so hemoconcentration, anuria, dehydration and circulatory collapse result.   As the pH of the ruminal constituents falls, the motility of the rumen decreases and there may even be complete stasis.   At the lowered pH, the normal microflora of the rumen are destroyed, the lactobacilli and streptococci thrive and the salivary secretion ceases so that the buffering action of the saliva is absent.   Absorption of the lactate causes acidosis.   Histamine may also be produced in the rumen which on absorption is toxic.

In such an atonic rumen, in which the normal microflora are lost, *Fusiformis necrophorus* and fungi of the family *Mucoraceae* (those belonging to the genera *Mucor*, *Rhizopus* and *Absidia)* invade the ruminal wall producing ruminitis and ulcers.

In animals that die of acute atony, the contents of the rumen and reticulum are thin, porridge-like and bulky.   The cornified epithelium is soft and peels off easily, exposing hemorrhagic areas underneath.   The blood is dark and thick. Lungs show bleeding into the alveoli and bronchi.   Heart musculature is flabby. In animals that survive for three days and more, demyelination of the nervous system may occur.

Complications : Enteritis, peritonitis and ketosis.

Traumatic reticulitis

This is a very common condition in older cattle.   These animals ingest and swallow, along with their feed, a wide variety of sharp objects like needles, nails, pieces of fencing wire and screws. Contraction of the rumen and reticulum during pregnancy may aid in the development of the condition

The sharp object pierces the wall of the reticulum during its contractions. Usually, it pierces the antero-ventral wall, which is near the diaphragm. The passage through it is usually slow and so the track formed by the moving nail is thickened by a dense fibrous wall. Piercing the diaphragm, the foreign body may enter the pericardium and even the heart, producing inflammation enroute. Or sometimes, it may take a downward slope and pierce the chest wall near the xyphoid cartilage forming an abscess there

At the point where the object pierces the reticulum, a localised peritonitis is formed and so adhesion of the reticulum to the diaphragm at this place occurs.

The sharp object may sometimes penetrate the lungs or the liver or the spleen.   If not contaminated, only mechanical injury to the affected parts is seen. Traumatic pericarditis with serofibrinous or purulent exudate may be the result when the pericardium is affected.   Other sequelae of traumatic reticulitis are : vagus indigestion, when the ventral branch of the vagus is affected by the inflammatory and scar tissue formed by the penetrating foreign body ; diaphragmatic hernia that may occur due to weakening of the diaphragm by lesions produced by the foreign body ; abscesess of liver and spleen and death due to rupture of left gastroepiploic artery.

At necropsy : The thick-walled track followed by the foreign body and adhesions of the reticulum to the diaphragm may be clearly seen.   Along the track may be found abscesses and fistulae connecting one abscess to the other.

Fibrinous pericarditis with hypertrophied myocardium may be seen if foreign body has entered the pericardium.

Ulcers of the fore stomachs may be seen in cattle occasionally and are usually due to *S. necrophorus* infection. These ulcers may be found in animals that are started on heavy grain feed and also in calves kept on milk. Chemical ulceration due to corrosives may occur, though rare.

Ulcers are seen in Viral diarrhoea and Mucosal disease.

Hepatic abscesses are considered to be complications of ruminal ulcers.

THE STOMACH

MALPOSITIONS :

Diaphragmatic hernia of stomach may be met with in dogs and cats involved in automobile accidents. The diaphragm is ruptured and the stomach enters the thoracic cavity.

Abomasal displacement : Abomasum may be displaced from the normal position either to the left or to the right. But the left-sided displacement is more common, in which it comes to lie between the rumen and left abdominal wall. The greater curvature of the body of the abomasum which is more mobile slips under the ventral ruminal sac.

Causes : This condition is met with more frequently after parturition. It is suggested that during pregnancy the rumen may be lifted by the expanding gravid uterus and the abomasum may slip under the rumen. After parturition, when the uterus recedes, the rumen is dropped to its normal position, when it traps the the abomasum. Atony of the abomasum due to feeding large quantities of concentrated feeds is a contributory cause for the codition to continue without any tendency at correction. Atony may be caused by the inhibitory effects of high fat or protein feeds. Post-parturient diseases like milk fever, mastitis, metritis and ketosis may cause atony of the abomasum. Abomasal displacement has been met with in cases treated surgically for chronic indigestion. In such cases the incisions made are the weak spots where the abomasum may slip through. Violent activity like jumping in estrus, may be a cause in non-parturient cases. A hereditary predisposition may exist.

Symptoms manifested are vague and are like those of chronic indigestion— anorexia alternating with voracious appetite, abdominal pain and ruminal tympany. Animal loses weight rapidly, is listless, dull and has a tucked–up–appea— rance. The dung is scanty but soft (not caked as in chronic acetonemia). Mild ketonuria is present. Frequent abnormal tinkling abomasal sounds may be heard at the level of paralumbar fossa.

Due to pressure and compression the normal function of the abomasum is interfered with. Because of the abnormal position (consequent on the displaced position of the abomasum). the function of the esophageal groove may be affected.

Diagnosis : Displacement of abomasum must be differentiated from chronic acetonemia, traumatic reticulitis, vagus indigestion, diaphragmatic hernial pyelonephritis and lymphomatosis. Laparotomy may be needed for diagnosis.

Treatment consists in correction of the condition by **(1)** rolling the anima, gently. If this is not successful, **(2)** the animal is operated on the left or both flanks, the rumen raised, the abomasum released and replaced in its original position.

TORSION of the stomach may sometimes be seen in old dogs. There is twisting of the stomach around the esophagus. This is due to sudden movements (jumping, rolling etc.) especially when the stomach is full. With the esophagus as the pivot, the heavy stomach rotates clockwise. The twist closes both the openings of stomach and so gastric tympany develops with resultant dyspnoea In some cases the stomach may rupture. As the blood vessels are compressed, there may be congestion and hemorrhage The contents of the stomach are blood stained.

RUPTURE OF STOMACH ; This is common in the horse and is usually due to tympanites and dilatation. Sometimes trauma and violent gastric contractions may be the causes.

ACUTE DILATATION OF THE STOMACH is due to excessive amounts of food or gas accumulating in the stomach.

Causes ; 1. Overeating; especially with grain in horse.

2. Excessive fermentation when easily fermentable foods are eaten.

The following factors act as accessory causes ; Obstruction of the pylorus, (by rags or sacking n calves), reflex closure of the pylorus, atony of the gastric musculature, incomplete mastication due to poor teeth, diseases of the stomach wall and hard work immediately after feeding, external compression of the pylorus by lipoma in horses and lymphocytoma in cattle.

The accumulated feed and gas in the stomach cause such a stretching of the gastric wall that contractions are interfered with. And this stretching causes severe pain also. Dilatation of the stomach causes vomition. Loss of fluid may result in fatal dehydration and alkalosis. Rupture will result due to stretching of the muscle fibres, when death will occur due to shock.

The stomach which is dilated 4 to 5 times its normal size presses upon other organs: pressure on diaphragm and lung results in dyspnoea and congestion. Sometimes the diaphragm may rupture. Obstruction of the pylorus due to a tumor or cicatricial constriction and wind sucking may cause recurrent dilatation with resultant hypertrophy of the muscular wall of the stomach.

GASTRITIS

Inflammation of the stomach is a fairly common condition in animals. It may be primary or may be secondary to some other infections, as in canine distemper, viral diarrhoea, swine erysipelas. Gastritis may be acute or chronic Causes may be the same for both but of different severity and acting for different lengths of time.

    A. **Physical :** 1. Overfeeding. causing dilatation of stomach, is always accompanied by gastrtis.

2. Feeding with frozen roots causing bloat, produces gastritis.

3. Feeding very coarse material (eating bedding in horses and dogs.)

4. Faulty dentition, preventing mastication is another cause of gastritis.

5. Foreign bodies may traumatise the gastric mucosa.

6. Spoiled, mouldy and fermented hay and silage.

7. Too sudden changes of feed may also be responsible.

B. Chemicals ; 1. Caustic and corrosive chemicals like mercury, lead, copper, arsenic and phosphorus.

2. Toxic plants.

3. Uremia; (Excretion gastritis is caused by the excretion of the poisoning material through gastric glands.)

4. Feeding easily fermentable foods liberates irritating substances which produce gastritis.

5. Feeding heavily fatigued animals has the same effect as number 4 above, since in such animals, the feed is not easily digested, stagnates, ferments and so produces irritation.

6. **Stress** ; In stress, adrenaline is produced in large quantities which is responsible for gastritis. This is seen in nervous dogs and in calves separated from their mothers

C. Bacterial ; In calves — enterotoxemia, colibacillosis ; pigs—erysipelas, vibrionic dysentry, salmonellosis, colibacillosis.

D. Viruses; Pig— hog cholera; transmissible gastro-enteritis in baby pigs Cattle : rinderpest, mucosal disease.

E Fungi ; Mucormycosis, moniliasis and aspergillosis cause gastritis in many animals.

F. Parasites ; Stomach worms—Trichostrongylus, Hemonchus, Ostertagia, larval paramphistomes in ruminants. Larvae of Habronema and *Gasterophilus equi* in horses. In pigs *Hyostrongylus rubidus, Ascarops strongylina* and *Physocephalus sexalatus.*

In general, close confinement and insanitary conditions where bacteria thrive contaminating feeds and feeding utensils are predisposing factors.

Gastritis may be acute or chronic. Acute gastritis may be catarrhal, fibrinous, suppurative, hemorrhagic or necrotic, depending upon the cause and their severity. By far the most common is the catarrhal and to a lesser extent, the hemorrhagic.

Catarrhal gastritis: Macroscopically, there is increased thickening and reddening of the gastric mucosa. A thick mucous exudate will be found covering the mucosa, which in some places may show ulceration. If severe, there may be hemorrhages and the gastric contents may be blood stained.

Microscopically, usual characteristics of inflammation—hyperemia, exudation and leucocytic infiltration in the mucosa are noticed. Some of the gastric glands may be damaged and lost.

Acute hemorrhagic gastritis is also common. Due to hemorrhage, the mucosa is bright red in color and the gastric contents are blood stained. Digested blood imparts a brownish coloration to the contents Besides, the changes described under cattrrhal variety are seen. Hemorrhagic gastritis is met with in acute infectious and intoxicating diesases like pasteurellosis, braxy, uremia, leptospirosis (in dogs) and in caustic chemical poisoning.

In gastritis, food does not get digested, motility of the gastric wall is retarded and irritation may produce pain and vomition.

PARASITIC GASTRITIS

Special mention must be made of parasitic gastritis, since this is very common in animals. The parasites that produce gastritis are:

Pigs:—*Hiostrongylus rubidus, Physocephalus sexalatus, Simondsia paradoxa, Ascarops strongylina.*

Cat:—*Gnathostoma spinigerum.*

Cattle:—*Hemonchus contortus, Ostertagia ostertagi, Trichostrongylus axei,*

Sheep—Same as above

Horses:—*Trichostrongylus axei, Gasterophilus equi larvae; Habronema larvae.*

The strongyles are blood suckers and they produce minute injuries on the mucosa The larvae may burrow into the mucosa for completion of their life cycle and thereby cause damage to the glands and epithelium. Heavy infestation, besides causing anemia, will produce cattrahal gastritis. Gasterophilus in the stomach may produce ulcers while Habronema larvae live in granulomatous nodules which may be infected by secondary bacteria and which produce abscesses

CHRONIC GASTRITIS in animals has usually the same causes as the acute but operating for a longer time. Sometimes it m.y be secondary to chronic gastric dilatation and hepatic cirrhosis. Partial anemia in the former and passive hyperemia in the latter decrease the local resistance thereby facilitating infection

This condition is usually of a hypertorphic type with thickening of the gastric wall. The mucous membrane is thickened and covered with tenacious, viscid glassy mucus. Mouths of the glands may be occluded resulting in retention cysts.

Microscopically. there is desquamation of the epithelium with increased interstitial connective tissue and this is the cause for exaggeration of mucosal foldings Hyperplasia of gastric glands and hyperplasia of muscle fibres with infiltration of inflammatory cells and hypertrophy of basal lymphocytic nodules are other features seen. The mucosa may be thrown into polypoid folds giving rise to *polypoid gastritis.*

GASTRIC ULCERS are common among animals. Calves are more often affected. Usually gastric ulcers run an acute course, heal promptly and seldom become chronic as in man So most often these ulcers are seen postmortem. Small superficial defects are known as *erosions* and these are common among ruminants Gastric ulcers of animals rarely perforate.

The exact causes of gastric ulcers are still obscure. The following are some of the causes cited.

1. **Trauma :** This appears to be the commonest cause of abomasal ulcers in calves. In early weaning of calves (before fourth week after their birth) when they are put on roughages before the fore-stomachs have attained their full functional development and before rumination has started, the coarse plants irritate the tender gastric mucosa producing gastritis which results in ulcer formation.

2. **Infections;** Erosions are common in the abomasum of cattle in rinderpest, mucosal disease, pox and bovine malignant catarrh.

3. Circulatory disturbances; In intense hyperemia (stasis) focal hemorrhages by diapedesis, that may occur through vascular nerves, the involved epithelium, because of disturbed nutrition, undergoes autolysis or even nercrosis and then is acted by the gastric juice; which digests it. Because the hemoglobin is digested by acid the area becomes brown in color. Such vascular stasis and hemorrhages are found in many infectious diseases—foot and mouth disease, canine distemper, rinderpest, rabies, purpura hemorrhagica and in uremia in which toxemia is present.

4 Nervous effects; The sympathetic and the vagus nerves control the secretion and blood circulation of abomasum. Any factors that disturb these nerves may produce stress, hemorrhges and increased secretion of glands. Stress in pregnancy and abomasal displacement may cause such disturbances and so it is that abomasal ulcers are common in post-parturient dairy cows.

Vasoconstriction produced by adrenaline in stress is thought to be another factor. Vagus indigestion causes atony of abomasum and ulcers are thereby caused there.

5. Obstruction of pylorus in calves and adult cattle is another cause of gastric ulcers. In calves pylorus may be obstructed by foreign bodies or coarse feed. In adult cattle pyloric obstruction occurs in traumatic reticulitis and in abomasal displacement.

6. Chemicals: Corrosives and escharotics may act either directly on the mucosa or indirectly through their action on the nerves. Grazing on pastures heavily fertilised with nitrogen is associated with gastric ulcers.

7. Parasites: In horses larvae of *Gasterophilus* species and *Hebronema megastoma* cause gastric ulceration.

8. Neoplasms of the stomach, especially lymphocytoma, are accompanied by ulceration.

9 Nutrition: Nutritional hepatic dystrophy (due to deficiency of vitamin E) in pigs is associated with gastric ulceration. Occurring in the young pigs, there may be heavy mortality due to massive hemorrhage.

10. Fungi : Mucor sp and Monilia sp. cause gastric ulcers in pigs.

Macroscopically, the erosions (in cattle) are of the size of a millet and usually affect the mucosa superficially. Slowly, by the action of the gastric juice the erosions may enlarge and become deeper to form ulcers. Sometimes blood vessels may be eroded and fatal hemorrhages may result. But this is not so common in animals as in man. The ulcers are usually demarcated, having raised borders and so have a punched-out appearance. The base of the ulcers may be the propria or submucosa or muscular coat or in some cases be even the serous coat. When the serous coat is the base, perforations are likely to occur.

Miroscopically, the erosion is covered by an exudate consisting of mucus, fibrin and inflammatory cells. The epithelium is desquamated and the sub-mucosa may be infiltrated by large number of leucocytes.

Fate of the ulcers: Ulcers develop very rapidly. The abomasal ulcers develop in three to four days. Perforation, when it occurs, may happen in six to seven days.

Most of the gastric ulcers heal by the formation of granulation tissue, which fills from the base of the ulcer. The resulting scar tissue (which contracts) is covered by the epithelium of the mucosa. Destroyed glands are not regenerated. Healed gastric ulcers have a star-like appearance.

Hemorrhages and perforation resulting in fatal peritonitis are the complications seen in some ulcers. Stress that may occur in lactation may be one of the causes of perforation of abomasal ulcers in cattle. Progressive anemia and emaciation may result in some cows.

Neoplasms : Most common is the lymphosarcoma of the wall of the abomasum in the bovine. Other tumors seen are : Leiomyoma, adenocarcinoma.

THE INTESTINE

Congenital anomalies : Atresia of the intestine may be found as a hereditary defect among cattle.

In pigs and foals **atresia** of rectum is an inherited lethal characteristic.

In cattle, **imperforate anus** at birth is a semi-lethal character.

Persistent omphalomesenteric duct is known as Meckel's diverticulum, which is sometimes met with in pigs and horses and is probably hereditary. This may produce complications since it may become obstructed or inflammed or even ruptured resulting in peritonitis.

Stenosis of the intestines sometimes occurs congenitally.

CIRCULATORY DISTURBANCES

Acute passive hyperemia occurs in conditions where there is sudden obstruction to the venous out flow as in prolapse, torsion, hernia or intussuception.   The involved part is dark-red in color, and swollen. There may be effusion into the peritoneal cavity and edema of the intestinal wall. Hemorrhages are common. Due to acute passive congestion, the following may result : enteritis, necrosis, gangrene, rupture, peritonitis and death.

Chronic passive congestion is common in liver diseases or obstruction of the portal vein or may be part of the general venous congestion (with lesions in the heart and lungs). The veins are dilated and stand out prominently. The intestines become predisposed to enteritis. The intestinal wall is thickened and edematous and ascites is also found. In long standing cases, there is fibrous thickening of the wall and atrophy of the glands.

Hemorrahages are common on the mucosa and serous surface of the intestines. The causes are; — blood sucking parasites, hemorrhagic diseases, enterotoxaemia (in sheep) and specific infections in which this is a lesion.

Infarction : If circulation is obstructed either by embolus or from pressure from the out side, infarction may result.   Putrefactive organisms normally resident in the bowel invade the infarcted part producing gangrene.

Thrombosis : Thrombosis of the intestinal vessels is common in the horses and is mostly due to larvae of *Strongylus vulgaris*.

The infective larvae burrow into the mucosa and travel along the intestinal arteries, crawling against blood flow along the intima and reach the anterior mesenteric artery where they settle. At that place the endothelium is damaged and a thrombus is formed. This thrombus may become organised and canalised. Complete organisation of the thrombus is prevented by the penetration of larvae which produce irritation. The damage caused by the penetration of larvae as well as their continued presence so weakens the wall of the artery that an aneurysm may develop. Due to the replacement of the elastic tissue by fibrous tissue the resilience of the arterial wall is curtailed and so it cannot contract when stretched and so dilates (aneurysm). The wall may enlarge gradually and sometimes may rupture. Calcification of the fibrosed wall may sometimes occur but this still more weakens the wall.

Of various branches of the anterior mesenteric artery, it is in the right branch that thrombus more often occurs. This branch supplies the ventra colon and so thrombosis of the vessel produces ischemia of these parts. Ischemia of the parts results in atony of the musculature leading to decreased peristalsis and so to stagnation of and impaction by the ingesta. Gas may be produced if fodder is succulent. Accumulation of gas and impaction give rise to colic.

Emboli from the thrombus may occlude some branches of the artery resulting in infarction, gangrene, shock and death.

Mechanical obstruction

Causes : Congenital : Atresia and imperforation have already been described.

Acquired : (a) **Stenosis** may be due to pathological lesions: hematoma, neoplasms, abscesses, and chronic inflammatory scars, displacements like torsion, volvulus. intussuception, hernia.

(b) **Impaction :**

   (i) Foreign bodies—Bone, stones, cartilage, rubber ball, rags, golf balls.

  (ii) Hair balls—in cats.

 (iii) Impacted undigested coarse food, especially in the horse; sudden changes in feeds and faulty dentition are accessory factors.May also be found in dogs—coproliths.

 (iv) Impacted meconium in new born animals.

  (v) Parasites—masses of round worms in pigs and fowls and tape worms in sheep.

 (vi) Enteroliths.

(vii) Neoplasms—lipoma in horses.

Obstruction causes weak peristaltic movements of the bowel above the points of obstruction, resulting in dehydration of the contents at that place. Spasms with violent contractions of the gut above the place of obstruction causes intense pain (colic) and sometimes rupture may occur. Vomition is an usual symptom of intestinal obstruction in dogs and cats.

Macroscopically, the place where obstruction has occurred is found to be distended. The contents are hard, which pressing on the mucosa may cause necrosis and erosion. Ultimately stenosis may develop at this part. Rupture and peritonitis leading to death may result if the obstruction is not relieved.

TORSION is a twisting of intestines on its axis.

VOLVULUS is a twisting of the bowel on itself as occurs when it passes through a tear in the mesentery. These conditions seen in horse more frequently may also be met with in other animals.

Causes : 1: Violent movements as in rolling and struggling.

2. Violent peristaltic movements.

3   Foreign bodies—sand or enteroliths, by their weight make the part heavy and aid in its winding around other parts.

4. Gas. Accumulation of gas makes the part bulge and twist round other viscera.

Torsion occurs more often in the small intestines, which have a long mesenteric attachment. In the horse, the right colon is fixed by ligaments and so torsion occurs in the left and transverse colon. In the cattle torsion of cecum is more common.

The changes that occur in torsion are — acute passive congestion leading to edema, hemorrhage, gangrene, peritonitis and death.

Macroscopically, the affected portion is swollen and darkened in color. The wall is very easily torn. Peritonitis may be evident in some

INTUSSUSCEPTION is telescoping of a portion of intestines into another, usually the anterior into the posterior, and occurs mostly in the jejunum and cecum in dogs and cattle. Along with the portion of intestines, its mesentery also is dragged along and so there is compression of the thin-walled veins resulting in acute passive hypermia.

Macroscopically, the affected part is dark-red or bluish and swollen. Usually gangrene and peritonitis supervene terminating in death.

In some stray cases, the invaginated portion may be sloughed off, healing occurring by granulation tissue Epithelium covers the scar. But at the site of scar. circular stenosis may form. Death in volvulus and other intestinal displacements is due to : 1. Acute anemia which may occur due to extensive hemorrhages into the intestine and peritoneal cavity. 2. Asphyxia and heart failure due to compression of lungs and heart by pressure on the diaphargm by excessive gas formation 3. Rupture of intestines and stomach leading to peritonitis and absorption of toxic products. 4. Toxemia due to absorption of toxins from decomposed food and bacterial growth

INCARCERATION of the intestine is trapping of the intestine internally, from pressure on its external surface. Incarceration may occur due to.

(i)   adhesion of the intestine to other abdominal organs.

(ii) the loop of intestine may pass through the epiploic foramen of of Winslow

(iii) occasionally a persistent urachus may cause incarceration.

(iv) Similarly an adhesion to the uterus may cause this condition.

(v)   when the bowel passes through a fissure of the mesentery, congenital or acquired, incarceration may supervene

The changes in these conditions are similar to those found in acute passive congestion viz. stagnation of the intestinal contents followed by venous stasis due to non-return of venous blood as the thin-walled veins are compressed; edema, infarction, gangrene and peritonitis. Ultimately rupture, shock and death occur.

PROLAPSE OF THE RECTUM : Sometimes the rectum protrudes through the anus. The causes are straining, irritation, abdominal pressure, diarrhoea, increased peristalsis and constipation.

Macroscopically, the rectum, bright-red in color, will be found hanging through the anus. It may be edematous and soon becomes gangrenous. The changes are similar to those found in incarcerated intestines. If not attended to early, the prolapsed rectum will be pecked by fowls or injured by swine. Due to swelling, fecal matter cannot be voided. Antemortem prolapse can be distinguished from the post-mortem prolapse by the absence of congestion in the latter.

HERNIA of the abdominal organs is the protrusion of the abdominal viscera through a natural or artificial opening. Hernia of intestines is commonly seen in domestic animals, especially the pig and horse.

The intestines may pass through a natural opening, the internal inguinal opening which is patent in the males. The umbilicus, if not healed is another site of hernia. Other causes are trauma when the abdominal muscles may rupture or even the diphragm may tear resulting in the intestines passing through the opening Violent straining during parturition or defecation may also be another cause.

Depending upon the location, hernia may be *external* or *internal* (diapharg-matic, pelvic).

Among the external, are;—

The **ventral**, when the abdominal muscles are ruptured. This is common in horses (spontaneous in pregnant mares) and occasionally in cattle. Causes; Trauma:- Horn injuries, kicks, automobile accidents, laparotomy and castration scars. In pregnant ewes this may due to muscular degeneration of nutritional origin.

The **umbilical** when the bowel passes through a congenital or acquired defect of umbilicus and seen in foals. calves and pups.

The **inguinal** when the bowel passes through the internal inguinal ring. This is not so common in animals as in man, because of horizontal position: seen in colts and pigs.

Scrotal hernia: the intestines slide into the tunica vaginalis along the inguinal canal in contact with the spermatic cord. The testes may undergo thermal atrophy when in contact with the intestines.

Femoral hernia may develop when the omentum and intestines pass through the femoral triangle along the femoral artery and so the bowel is found on the inner surface of the thigh.

The **perineal hernia** may occur in old dogs due to violent straining in cases of enlarged prostate.

External hernia consists of; — (i) a hernial sac formed by the parietal peritoneum and the covering skin, (ii) a hernial ring which is the opening in the abdominal wall and (ii) the hernial contents.

If the hernial contents can be returned into the abdominal cavity it is called a *reducible hernia*. But if it cannot be so returned is is called *irreducible*. The causes of the latter are;— (i) adhesions between the visceral mass and the hernial sac. (The adhesions arise due to inflammation of the peritoneum) (ii) accumulation. of ingesta in the loop of intestines making it too bulky to be reducible and

(iii) venous stasis, due to incarceration, whereby the volume is so increased that, the bowel cannot be reduced.

If the hernia does not have a parietal peritoneal covering of the viscus, it is called a *false hernia*. In such cases, opening of the skin will reveal the bowel. The condition is called *eventration*.

Strangulated hernia is one in which the blood supply is cut off by the pressure of the hernial ring through which the intestines pass. If not relieved in time, this condition is fatal since infarction, gangrene, peritonitis and shock will develop within 24 to 36 hours.

ENTERITIS

Enteritis is the term denoting inflammation of the whole of the intestinal tract. But usually it is applied to the inflammation of the small intestines. The inflammation of the colon is called *colitis,* that of cecum *typhlitis* and of rectum *proctitis.*

Enteritis is very common in domesticated animals and fowls and is of immense economic importance.

Since enteritis occurs along with gastritis (the same irritants causing gastritis passing on to intestines produces enteritis also) gastro-enteritis is a frequent condition met with.

Causes are many and varied and they include, bacteria, viruses, protozoa, rickettsia, helminths, fungi, chemicals, disturbed metabolic processes as in rumiants, venous congestion as in portal hypertension and congestive cardiac failure, toxins of clostridia, coliforms and spoiled or mouldy feeds and avitaminosis. In enteritis, the whole length of the bowel may not be affected, inflammation localising only at one part or the other.

Based on the nature of the exudate and the changes produced in the intestinal tract, enteritis is classified, as catarrhal, hemorrhagic, fibrinous, suppurative and necrotic.

Acute catarrhal enteritis is the mildest of inflammations of the intestinal tract, occurring in a diffuse manner throughout the bowel.

Causes include mild irritants like foreign bodies, sand, coarse feeds, bites of parasites (hook worms), chemicals and drugs, *Vibrio coli* (causing winter diarrhoea in cattle). Acute catarrhal enteritis may be noticed in :

  i) Enteritis in sucklings—scours in calves, lambs, foals and piglets caused by *E. coli*, pasteurella, salmonella, proteus, vibrios and streptococci.

In calves and lambs avitaminosis A is a predisposing factor while in young pigs deficiency of animal proteins and trace elements predispose them to infections. In such a state, the organisms are able to gain a foothold and thrive causing the disease.

  ii) Viral Diarrhoea–Mucosal Disease in cattle.

  iii) Enterotoxemia in sheep.

  iv) Virus gastroenteritis in pigs.

  v) Salmon poisoning in dogs.

  vi) B W.D, infectious cloacitis, pullet disease and ornithosis in fowls.

  vii) Oral antibiotic therapy may cause enteritis in two ways : (1) these may themselves be irritants or (2) they may so alter the intestinal flora that there is over growth of other bacteria (*Staphylococci, Proteus* sp; *Pseudomonas* sp) and fungi *(Candida albicans)* which are normally kept under restraint and so enteritis results.

Macroscopically, one should be able to distinguish this condition from the normal hyperemia that occurs during active digestion. The mucosa is reddish in color and slightly thickened, covered with a mucinous exudate. The Peyer's patches are prominent being hyperplastic, outlined by a zone of hyperemia.

Microscopically, the edema of the intestines is due to exudate with leucocytes in the lamina propria and to a little extent in the submucosa. Hyperemia is evident by the engorgement of the blood vessels. Goblet cells are numerous arising from metaplasia of the epithelial cells and produce large amounts of mucin. The tips of villi may be reddened and edematous.

The intestinal contents are watery, consisting of mucus, fibrin and desquamated epithelial cells.

Sequelae : When cause is removed inflammation may subside and the bowel returns to normal. But if irritant persists, the condition may develop into the chronic state.

Chronic catarrhal enteritis may develop from the acute condition or more usually it may arise gradually as in Jhone's Disease, intestinal helminthiasis, chronic venous congestion (due to congestive cardiac failure) and cirrhosis of liver.

Macroscopically, the wall of the intestines is greatly thickened. The mucosa is smooth (covered by thick mucus) and thickened due to infiltration by macrophages, plasma cells and lymphocytes. This infiltration makes the mucosa corrugated due to infoldment. The corrugations are at right angles to the length of the intestines ; *(Chronic polypoid enteritis)*

Microscopically, the characteristic appearance is the presence of numerous macrophages, plasma cells, lymphocytes and connective tissue cells in the lamina propria and even in the sub-mucosa. The intestinal glands are atrophied due to the pressure of the infiltrating tissue. Sometimes retention cysts due to closure of the mouths of glands are found. The mucosa is covered with tenacious mucus.

Hemorrhagic enteritis : This is a more severe form of catarrhal enteritis, characterised by the presence of erythrocytes in the exudate. Always patchy in distribution, this is mostly seen in septicemic, bacterial and viral diseases e. g. Anthrax and Rinderpest. This condition may also be found in uremia of dogs ; coccidiosis, poisoning by arsenic and croton oil, enterotoxemia, in vitamin B deficiency in dogs and pigs and in colibacillosis. Continuous feeding of dogs with horse meat causes an anaphylactic condition manifested by hemorrhagic enteritis.

Macroscopically, there is infiltration of blood in the intestinal wall which is thickened and the intestinal contents are blood stained. Blood found in the anterior portion of the intestines is digested and so is brown in color while in the posterior portion it is bright red.

Microscopically, red blood cells may be found in the exudate of the mucosa. The villi may show necrotic changes and thrombosis of some enteric vessels is evident.

Sequelae: Being very severe, death usually occurs. If treated in time, prognosis may be favourable.

Fibrinous enteritis : This is of the diphtheritic type and occurs in cattle, pigs and cats, and rarely in horse and fowl.

Causes:

Bacteria; *Salmonella cholerae suis* (necrotic enteritis in swine), *Escherichia coli*

Chemicals; salts of mercury and arsenic.

Parasites: *Echinostomum*, a fluke in turkeys

Macroscopically,the characteristic finding is the presence of strands of fibrin on the mucosa of the intestines. The wall of the intestine is edematous. In more severe conditions, a thick, grayish or whitish-gray membrane may be covering the intestinal mucosa, which is hemorrhagic and edematous   The inflammation may extend into the submucosa and petechial hemorrhages may be seen. The mesenteric lymph glands are swollen, hemorrhagic and juicy.

Macroscopically, the membranous exudate consists of strands of fibrin containing in its meshes varying number of neutrophils and desquamated epithelial cells together with mucosa. Mucous membrane shows edema, hyperemia and infiltration by neutrophils. Coagulative necrosis of the epithelium occurs in some places, which along with the exudate forms the false membrane which is adherent to the intestine.

Sequelae ; Being very a severe condition death is a common sequel. In those that are able to withstand, recovery with complete healing occurs.

Suppurative enteritis is not common and may result due to infection by pyogenic organisms of wounds caused by helminths.

Causes; Pyogenic organisms ; *Streptococci, Salmonella & Shigella.*

Macroscopically, the exudate contains pus.

Microscopically, the exudate contains, besides mucus, desquamated cells neutrophils and bacteria.

Necrotic enteritis ; Necrosis of the intestinal epithelium and underlying tissues occurs.

Causes ; Severe irritants; Chemical — croton oil, mustard gas, wood preservatives; insecticides.

Bacterial—Necrophorus organism, *Salmonella.*

Viral—Rinderpest, Viral Diarrhoea—Mucosal Disease, Hog cholera;

Protozoa—Coccidiosis, Histomoniasis.

Vitamin deficiency—Niacin deficiency in swine.

Macroscopically, patchy necrotic areas are seen. The necrosis of the mucosa extends into the sub-mucosa also. Fibrin may be found on the necrotic mucosa. When the necrotic material is removed, a red, raw, bleeding surface is seen. The mesenteric lymphatic glands are swollen and juicy. In hog cholera, the characteristic lesion is the ''button ulcer'', which is a spherical ulcer in the mucosa of the colon This is circumscribed with sharp edges.

Microscopically,besides hyperemia, exudate and cellular infiltration, necrosis of the epithelium of the mucosa is seen. The ulcer reveals a demarcated zone of necrosis in the mucosa and sub-mucosa.

Actually, these button ulcers are tiny areas of infarction that arise by the occlusion of small arteries by the swollen and proliferated endothelium.

Sequelae : The condition is mostly fatal. If the condition is one of niacin deficiency, restitution of the deficiency may cure the condition.

ENTEROLITHS; sometimes stones are found in the large intestines of horses. They are formed of triple phosphates which are deposited concentrically, layer after layer, over a nucleus of sand or a metal piece or an undigested vegetable fibre.

Genesis ; When animals are fed on wheat or bran which are rich in magnesium phosphate, intestinal calculi can occur. Normally magnesium phosphate is dissolved by the gastric juice and then absorbed in the intestines. On the other hand, when excessive amounts are fed to an animal, and that too to one suffering from chronic catarrhal gastritis in which gastric juice is not secreted, much of the magnesium phosphate reaches the intestines in an undissolved state. This combines with ammonia that is formed from the decomposition of protein (which is also abundant in wheat and bran) to form triple phosphate. This triple phosphate crystalises around foreign bodies like a grain of sand, a piece of metal or undigested plant fibre Enteroliths do not form in the small intestines because (i) the movement of the food is too rapid there to allow the deposition of salts and formation of calculi and (2) bacterial decomposition of proteins to form ammonia does not take place there. The following may be contributory factors: disturbance in the colloid protection of dissolved salts; change in bacterial flora, with altered fermentation conditions and sluggish intestinal movements that occur in the dilatation of the bowel or in relaxation of intestinal muscle met with in feeding with bran. Enteroliths may sometimes attain a large size, some may weigh as much as 20 lbs. and are usually round and smooth.

Phytobezoars ; These are food balls (phyto=plant, bezoar=concretion)

These arise from plant fibres and awns which are impregnated with triple phophate and rol ed into balls. These have a velvety surface, are light in weight and are brown in color. They may also be found in the crops of birds

Trichobezoars (Piliconcretions)—Hair balls.

Hair balls are found mostly in the rumen. Animals having itching skin conditions (animals infected with mange or lice) may lick each other when the loose hair may be swallowed. Similarly, calves kept together, suck and lick each other's ears, tails etc. swallowing hair. The hair is rolled into balls during ruminal contractions. Mucus of rumen may form a smooth coat over such balls.

The enteroliths, phytobezoars and trichobezoars are usually of no consequence unless they obstruct the passage, when, sometimes, even rupture may occur. Cattle may regurguitate a food ball into the esophagus which may be choked More often these concretions are found only at postmortem.

COLI GRANULOMA IN FOWLS (Hjarre's Disease)

A granulomatous condition of the fowl intestine and liver, caused by a mucoid strain of *E coli* is reported in many parts of the world including India. Usually adult fowls are affected.

Macroscopically, a large number of grayish - white nodules varying in size from a millet seed to a hazel nut are found projecting from the serous surface of the intestines. These are distributed diffusely from the duodenum to the ceca. The large intestines are free of the lesions. In some places, the whole circumference of the bowel may be involved.

Microscopically, the lesion is a granuloma involving all the structures of the bowel wall, with desquamation of the mucosa and fibrous thickening of the serosa.

A typical nodule consists of the following structures from within out-wards: a central structureless mass which is calcified in some places; peripheral to this is a zone of caseo-necrotic material with cells in varying stages of degeneration and necrosis; peripheral to this is a zone of granulation tissue, with epithelioid cells and a few giant cells interposed in between it and the previous zone. E. *Coli* could be demonstrated in and isolated from the lesions. Similar lesions are found in the liver.

The following are the differences between the lesions of coli granuloma and tuberculosis. The lesions of coli-granuloma are single while those of tuberculosis form conglomerates; lesions of coli granuloma are not found in spleen and bone while tuberculous lesions are found in those situations; in tuberculous lesions the acid-fast organisms can invariably be found while in coli-granulomatous lesions only *E. coli* are seen.

Neoplasms of the intestines: Lymphocytoma is the most common neoplasm met with in animals. Masses of the neoplastic cells are found in the wall of the bowel.

Other tumors met with are: — adenocarcinoma, lipoma, leiomyoma, sarcoma and papilloma.

Anal glands of dogs may sometimes be inflammed and may become purulent if infected by pyogenic organisms. There may also be swelling of the glands due to retention of the secretion. These swellings may cause constipation and so need to be manually evacuated.

Adenoma of the anal glands has been met with.

COLIC

Colic means pain in the abdomen. This is a symptom manifested by animals suffering from diseases of various organs. Animals suffering from colic have an anxious look, lie down and get up frequently, roll on the ground, look towards the flank often, have polypnoea and tachycardia. Though all animals may suffer from colic it is the horse that is most often affected.

Colic is not a term used by the pathologist. But for the convenience of the students, the conditions in which colic is a symptom are listed below (after Cohrs)

Diseases of the stomach : acute and chronic dilatation of the stomach (caused by overfilling of the stomach or by pyloric stenosis), gastritis, gastric ulcers, gastric parasites.

Diseases of the intestines; Volvulus; torsion; intussception, stenosis, displacement, obstructions, impaction, retention of meconium, enterolithiasis, foreign bodies, thrombosis and embolism of mesenteric vessels, enteritis, parasites.

Acute peritonitis.

Diseases of the liver and bile passages : impaction of gall stones, acute hepatitis; sudden enlargement of the liver due to hemorrhage, rupture of the liver:

Diseases of the urinary organs : acute nephritis; renal abscess, pyelitis, displacement with obstruction of the ureter, cystitis, urethral obstruction.

Diseases of the genital organs; torsion of the uterus, uterine contractions associated with the movements of the foal, labour pains during normal parturition or abortion, enlargement of the prostate causing retention of urine.

Diseases of other organs Certain diseases of the esoghagus (dilatation, stenosis, displacement, contractions), irritation of the rectum and the surrounding tissues by parasites (*Oxyuris, Gasterophilus*), equine myogl. binuria associated with paralysis, hunger and extreme exhaustion.

Of all horses examined postmortem, colic has been responsible for 34 to 50 percent of deaths. The following are the most common causes of colic among horses: volvulus of small intestines, volvulus of large intestine, other forms of displacements, primary gastric dilatation, obstruction of small intestines, cecal impaction, impaction of large intestines, thrombosis and embolism of mesenteric vessels, and enteritis.

Chronic or recurrent colic, which is not so frequent may also occur and is due to embolic thrombosis, chronic impaction of the cecum, chronic dilatation of the stomach and intestines, obstruction causad by adhesions, old incarcerations, tumors, stenosis and inflammatory new growths; stones in the bowel and parasites.

LIVER

In addition to being the largest organ in the body, liver also discharges the greatest number of functions. Hepatic cells are among the highly specialised cells in the body. At any one time, 25% of the blood in the body flows through the liver.

THE FUNCTIONS OF LIVER ARE :

1. **Secretion of bile;**—Bile contains pigments and bile salts. Bile pigments are not useful to the body. On the other hand, retention of these (hyperbilirubinemia) is toxic to the body.

Bile salts are formed in the liver from cholesterol, and excreted as sodium salts of taurocholic and glycocholic acids after conjugation with taurine and glycine. These bile salts play an important role in digestion, especially of the fats.

The bile sults;— (a) activate pancreatic lipase and amylase;

(b) aid in the emulsification of fats in the intestines.

(c) aid in the absorption of fats.

(d) aid in the absorption of fat soluble vitamins (especially vitamin K).

(e) act as cholagogues and

(f) maintain a stable pH in the Intestines.

Bile contains mucin and related substances which act as stabilisers for the fat emulsion in the bowel.

Though not strictly antiseptic, bile renders the intestines uncongenial for the bacteria to thrive and so is bacteriostatic.

2. **Protein matabolism;** (a) Amino acids are deaminised.

(b)   Uric acid is converted into allantoin.

(c)   Highly toxic ammonium salts are detoxified by converting them into urea.

(d)   The non-nitrogenous residues obtained after deamination of amino acids, are converted into glucose and ketones which are used by the body.

(e)   From Amino acids are formed :
     i)   Plasma proteins (albumin, globulin, fibrinogen, prothrombin).
     ii)   Tissue proteins and
     iii)   Protein reserves stored in the liver.

3.   **Carbohydrate metabolism :**   Glycogen is synthesized and stored in the liver. Excessive carbohydrates ingested are converted into lipids and stored in the fat depots. With the assistance of pancreas, liver maintains a constant level of blood glucose. In times of need, gluconeogenesis from proteins and fats occurs in the liver.

4.   **Fat metabolism :**   We have already seen how bile salts assist in the absorption of fats. Fats, that are characteristic of animals. are also synthesized from fatty acids and glycerol by liver

With the assistance of choline, liver is able to transform the depot fats into tissue fats (phospholipids) so that the tissues can utilise them.

5.   **Erythropoiesis :**   In the bird. liver is the site for erythropoiesis. In other animals, during fetal life, erythropoiesis occurs in the liver In these animals under certain circumstances (in severe anemias) erythropoiesis takes place in the liver even in the adult, i.e., extramedullary hemopoiesis.

6.   **Iron metabolism :**   The reticulo-endothelial cells of the liver are capable of destroying the red blood cells and the minerals released (Fe. Cu and Co) are stored in the liver for use again by the body.

7.   **Detoxication :**   Some toxic substances, especially putrefactive products from the alimentary tract are detoxified by the liver by conjugation while bacterial toxins and hormones produced in excess of requirements are inactivated Many drugs used therapeutically are also made harmless by the liver, eg. morphine, barbiturates, phenol, camphor.

8.   **Vitamin metabolism and storage :**   Failure of bile excretion due to hepatic damage interferes with the absorption of fat-soluble vitamins—A, D, B. and K. Vitamin A is stored in the liver and Vitamin K is utilised there for the formation of prothrombin and so these functions will be interfered with in the diseases of liver. Some members of the vitamin B group, especially Thiamine, Riboflavin and Niacin are partly metabolised in the liver where they may also be stored.

It is, therefore, obvious that with severe disease of the liver, a great many vital procesess will be affected. The following are the more important pothological conditions met with :

1.   Jaundice due to retention of bile.

2.   Bleeding may be due to (i) failure of prothrombin formation; (ii) lack of absorption of Vitamin K; (iii) lack of formation of fibrinogen.

3.   Hypoglycemia due to impairment of glucose metabolism: glycogen is not stored in the liver nor is it released into the blood. This condition makes the animals weak and irritable.

4. Hypoproteinemia—due to failure to synthesize plasma proteins. Animal becomes emaciated and generalised edema develops.

5. Anemia due to iron and protein deficiencies; liver stores iron and so in liver disease iron stores are deplete i

6. Toxemia due to failure in detoxication of proteins and intestinal toxins.

7. Renal failure—Heptorenals syndrome—in severe hepatic injury, the toxins that are not detoxified are excreted through kidneys, which are affected by these toxins resulting in renal degeneration. Renal function suffers—uremia develops.

8. Pyrexia—the heat regulating center is affected by the toxins since they are not detoxified by the injured liver.

Liver Function tests

Several tests have been evolved to measure different functions of the liver. It must be remembered that liver has a great reserve power and it has enormous ability to recover from injury. So the tests are not adequate, clinically, to evaluate the correct state of the health of liver and hence it is not wise to put too much reliance on these tests. Since the functions of the liver are carried out by the activity of enzymes, inadequacy or absence of one particular enzyme may affect one function and so a decrease of one function does not mean that other functions are affected.

Postmortem changes : Postmortem decomposition of liver occurs rapidly since gas-forming organisms (Cl. welchii) invade from the intestines which are close by. Liver, particularly rich in nutrients, is a good medium for the growth of these bacteria. Gas bubbles form in the blood vessels. The parenchyma and the blood vessels adjacent to the bowel are stained by hemoglobin blush-black. The presence of gas gives the liver a foamy appearance—"Foamy-liver". Imbibition of bile by the liver tissue surrounding the gall bladder is noticed.

DEGENERATIONS

Cloudy swelling is common in the liver :—

Causes : i) Bacterial toxins :-seen in all infectious diseases.

ii) Poisons : (a) Chemicals : salts of heavy metals—arsenic and lead.

(b) Plant toxins —glucosides, saponin.

(c) Drugs—carbon tetrachloride used as an anthelmintic

iii) Viruses.

iv) Hypoxia.

Macroscopically, the liver is enlarged and the capsule is tense. Consistency is softer Borders are rounded. The organ has a dull, cooked appearance. On section it bulges at the edges. Lobular marking are indistinct.

Micoscopically, the cells of the liver are swollen and have a pale granular cytoplasm due to swelling of mitochondria. The nuclei may be indistinct.

Sequelae : Recovery of the cells occurs if cause is removed. But if it continues the condition may progress to fatty degeneration or necrosis.

Fatty change in the liver is common and sometimes may be of severe degree.

As explained in general pathology, all visible fat in the liver is due to fatty infiltration. The liver is too sick to metabolise the fat brought to it from the depots.

The main causes are :—Toxins, poisons and anoxia

Poisons : Inorganic—phosphorus, arsenic, antimony.

Organic : chloroform, carbon tetrachloride, tannic acid, tetrachlorethylene alkaloids of phytotoxins, Aflatoxin, senecios.

Anoxia : Chronic venous congestion.

Nutritional: inadequate choline; Metabolic — Diabetes mellitus in dogs and cats, deficiency of thyroxine and anterior pituitary hormones.

Macroscopically, the liver is enlarged, has a smooth surface and is pale or yellowish. On section it bulges on the cut surface and fat droplets are seen on the blade. Very fatty livers, as in pregnancy toxemia of sheep, float in water. They are friable.

Microscopically, the hepatic parenchymal cells contain fat droplets, either as a single large globule or as multiple small globules If single and large the nucleus may be thrust to a side. The sinusoids are compressed and so appear anemic. Usually the distribution of the lesions in the liver may be diffuse or zonal. In chronic venous congestion it is in the periphery of the central vein. But on the other hand, in poisoning, when poison is brought through the portal vein, the fatty changes are found at the periphery of the lobule

Sequelae : If the cause is removed early, the condition can be completely corrected, But in continued presence of the pathogen hepatic fibrosis—portal cirrhosis—will eventually result.

HEPATITIS

Essentially hepatitis is an alterative inflammation of liver in which the various degenerative processes like cloudy swelling, fatty degeneration and necrosis are caused by irritants which also produce inflammation. Besides, in liver these degenerative changes are accompanied by lmphocytic and exudative infiltrations typical of an inflammatory reaction. Hepatitis is classified as *alterative inflammation* because the inflammatory process is caused by the same etiological agents that also produce degeneration and so alteration in the parenchymatous cells is produced.

Hepatitis may be either infectious or non-infectious or toxic. The latter may again be acute or chronic. The chronic variety is usually called **Cirrhosis**. **Infectious hepatitis ;** This is found in various infections as detailed below.

A. Conditions in which the liver is only or primarily affected

   (i) Infectious Canine Hepatitis (Rubarth's disease)

  (ii) Blackhead in turkeys (Histomonas affections)

 (iii) Wesselsbron disease of sheep, a viral disease found in South Africa

  (iv) Leptospirosis.

   (v) Viral hepatitis of ducks.

  (vi) Viral hepatitis of poultry.

B Conditions in which liver is also affected along with other organs;

   (i) Necrobacillosis

  (ii) Suppurative conditions

 (iii) Tuberculosis

  (iv) Histoplasmosis

   (v) Toxoplamosis

(vi)   Coccidioidomycosis
(vii) Rift valley fever
(viii) Salmonellosis
(ix)   Coligranuloma of poultry
(x)   Pasteurellosis
(xi)   Brucellosis
(xii) Glanders
(xiii) Actinomycosis
(xiv) Botriomycosis

Routes of Infection : Infection to the liver may be conveyed through several routes. The following are the more important.

1. **Portal vein** : Ingested organisms enter the portal vein and so are conveyed to the liver.

2. **Hepatic artery** : Organisms when present in the blood as emboli or in a bacteremic state, reach the liver.

3. **Umbilical vein of the new born animals:** When the umbilical vein is contaminated, organisms grow well in the partially coagulated blood which acts as a good medium and reach the liver. *S. necrophorus* and pyogenic bacteria are the commonest organisms involved producing hepatic necrosis and abscesses respectively.

4. **Bile ducts** : Infection may ascend from the duodenum. Obstruction of bile ducts causing stasis may be another source for infection.

5. **By direct extension** from neighbouring organs as in traumatic reticulitis.

Acute toxic hepatitis ; This is characterised by necrosis, which is usually preceded by degenerative changes like cloudy swelling and fatty degenaration.

Hepatic necrosis is conveniently classified, as per anatomical distribution into focal, centrilobular, midzonal, peripheral, diffuse and paracentral necrosis.

Focal necrosis In this variety, numerous microscopic necrotic areas are seen scattered in the liver and may be found in any part of the lobule.

Causes ; 1. Viral—as in equine viral rhino-pneumonitis in the foetus,

2. Bacterial—in bacteremic or septicemic affections—Johne's disease, Salmonellosis, Tularemia, Listeriosis in new-born.

3. Obstruction of biliary passages.

4. Due to parasitic migration.

Focal necrosis of the liver is not of much consequence since the function of the liver is not affected. Healed lesions show some scarring but this also disappears after some time.

Centrilobular necrosis or Periacinar necrosis.

In this condition, the cells nearest the central vein are affected.

Causes; Anoxia ; i) Acute hemorrhagic anemia; ii) Low atmospheric pressure; iii) Congestive cardiac failure; iv) In Shock— due to reduced blood pressure, reduced oxygen tension and reduced volume flow.

Toxins: Blood borne especially: Carbon tetrachloride. (See page 137)

Macroscopically.the liver is enlarged and paler than normal. In severe cases, the organ may be redder due to increased quantity of blood. The lobular markings are exaggeratèd. This is due to the difference in color at the center and periphery. When congestion of the central part is present, the periphery is paler due to degenerative changes in the cells. On the other hand, if necrosis of the cells in the center occurs, then the center will be pale while the peripheral cells are darker.

Microscopically, the cells round about the central veins have disappeared, blood taking up their places. Away nearer the periphery the cells may show fatty degeneration and those beyond these cells cloudy swelling. Infiltration of the periportal connective tissue by lymphocytes is seen after some days.

Sequelae : Single affection may heal by regeneration. Repeated attacks, however, will result in fibrosis which will ultimately reduce the size of the organ.

Pseudolobulation with proliferated bile ducts and resulting nodulation is the ultimate result found in frank post-necrotic cirrhosis.

Mid zonal necrosis : This lesion that is found in *yellow fever* of man affects the hepatic lobe, mid-way between the periphery and the central vein. This condition is not seen in animals.

Periportal necrosis : In this condition the cells adjoining the portal tract become necrotic and so the toxins should have been conveyed by the portal vein. This is more commonly seen in phosphorus poisoning. Accompanying inflammation of the portal triads results in cirrhosis, similar to portal cirrhosis.

Massive necrosis or Acute yellow atrophy : In this condition, there is necrosis of considerable number of the cells in a lobule. This may be a severe manifestation of various types described above.

Since whole parenchyma of the lobule is dead no regeneration occurs, the reticulum and fibrous frame work collapse and there is post-necrotic scarring.

The liver is yellow because of fatty degeneration and necrosis and smaller in size due to loss of parenchyma.

Causes :   Virus :–in man

    Poisons :–Carbon tetrachloride, chloroform, phosphorus.

    Dietetic :–Deficiency of sulphur-containing amino acids, Tocopherols and
           Selenium.

    Sequelae :–Death

Paracentral necrosis is a peculiar type of wedge-shaped necrosis occurring only on one side of the central vein, but not around it and extending up to the periphery. This type is encountered in Rift-valley fever and in uremic conditions.

Saw dust liver : In well-fed young cattle, at postmortem, focal necrosis of the liver is common. The animals do not manifest any symtoms while alive. The foci of necrosis may be few or many, and appear to the naked eye as though saw dust is sprinkled on the liver.

Microscopically, the lesion consists of hepatic cells which have undergone coagulative necrosis and infiltration by lymphocytes and neutrophils. These spots are evidently scars resulting from inflammatory reaction. It is conjectured that the irritant is borne by the portal veins from the gut and it is for this reason that the lesions are found nearer the portal areas.

CIRRHOSIS

Cirrhosis of the liver is chronic hepatitis characterized by fibrosis, degeneration and hyperplasia of hepatic cells. The stimulus for the fibroblastic proliferation is some irritant, chronic and severe enough to produce degeneration and necrosis of the parenchymatous cells.

The irritant may reach the liver through (a) The portal vein (b) hepatic artery and (c) bile ducts.

Based on the route of infection, the cirrhosis is classified as follows :

1. Portal or nodular cirrhosis ;

Causes : Usually, the causes of portal cirrhosis are the same as described under acute focal toxic hepatitis. But frequently one may not be able to ascertain the cause. It should be noted that the irritant is mild and acting for a long time. In this context mention must be made of toxic plants and chemicals. Among the toxic plants known to cause cirrhosis are :—

Crotalaria saggitalis in horses; plants of *Senecio* family in horses, cattle and sheep; *Atalaya intermedia* in horses; *Amsinckia intermedia* (tar weed) in horses, swine and cattle; plants containing high selenium content—in horses. (Wheat, loco weed).

The following chemicals are found to produce this condition :—

Pitch in tar paper, repeated exposure to chloroform, carbon tetrachloride and phosphorus.

Long continued intestinal toxemia is another cause of this condition.

Pathogenesis :

When the irritant is conveyed via the portal veins, changes are noticeable first at the periphery of the lobules—area next to the portal tract. Due to the action of the irritant the following changes take place : degeneration of the hepatic parenchyma, stimulation of the connective tissue in the interlobular septa to proliferate, infiltration of lymphocytes and macrophages into the islands of Glisson: Depending upon the severity of the irritant necrosis of the hepatic tissue may also occur. Along with the new connective tissue, new blood vessels are formed. These irregular blood vessels anastamose with the network of the portal vein as well as with the branches of hepatic artery. Thus arterio-venous shunts result and so ischemia of some parts of the liver occurs leading to further hepatic necrosis.

Along with these changes, hyperplasia of the surviving cells takes place, replacing those that are destroyed. But the connective tissue, which is young and cellular in the early stages becomes mature and fibrous, then contracts, interfering with blood circulation. The decreased blood supply interferes with the proliferation of the hepatic cells and so hyperplasia does not progress further.

In the new fibrous tissue, in the portal areas especially, new bile ducts are formed. These are not functional, lacking in outlets and so stasis of bile occurs.

As the fibrous tissue grows into the liver lobule, the hepatic cells become atrophied due to pressure and lack of nutrition The central vein becomes narrowed (due to the pressure of the fibrous tissue) impeding the out-flow of blood, thereby rendering the irritant to stay longer in the liver. Growth of the fibrous tissue into the lobule divides the parenchyma into small islands of hepatic cells—pseudolobulation.

If irritant enters the liver through the hepatic artery, changes of damage are first noticed in the tissues of portal canal and inter- lobular connective tissue. The features here are: lymphocytic infiltration and proliferation of the connective tissue which slowly encroaches into the lobules producing changes described above.

One noteworthy feature is that when once the fibrous tissue is stimulated to proliferate, this proliferating fibrous tissue itself becomes an irritant. So, even if the original irritant is removed or destroyed, cirrhosis progresses with more and more fibrous tissue formation until the condition terminates fatally.

Macroscopically, the liver is hard and firm. The surface is uneven and nodular. In the early stages the organ may be large. But as the condition progresses. due to atrophy of the parenchyma and contraction of the fibrous tissue, the liver may be reduced in size. The color of the organ is tawny or yellowish-gray and it is to this color that the name "Cirrhosis" was first applied. The color is due to the stasis of bile in the liver.

The architecture of the liver is lost (the normal marking of the lobule disappears) and the hyperplasia that is present gives nodularity to the organ (Hobnail liver) The nodules lack a central vein and are usually greenish in color due to the stasis of bile, which cannot be excreted since the newly formed bile ducts lack on outlet. Stasis leads to deposition of the bile pigment.

On section, the liver cuts with difficulty due to the dense fibrous tissue formed. While cutting a peculiar grating sound can be heard.

Microscopically, the characteristic picture is the increase in fibrous tissue—within and around the lobules. In the portal area small new bile ducts and inflammatory cells (lymphocytes and macrophages) are present. Pseudolobulation is evident. Central veins in some lobules are either absent or are placed eccentrically.

The parenchymatous cells show various stages of degeneration—cloudy swelling, fatty degeneration and even frank necrosis.

Regeneration of surviving cells is evident in some places, giving rise to the nodules noticed macroscopically. These regenerating young cells are plump, robust and stain more intensely.

Multinodular or Atrophic or Gindrinker's or Laennec's cirrhosis. This is portal cirrhosis of man and merits description here briefly, since in the dog, a similar condition is met with, though not due to similar etiology but to toxins absorbed from the intestines.

Though the exact causes of cirrhosis are still obscure, it is thought that deficiency of Vitamin B complex and lipotropic factors; especially in drunkards, produces this condition. Lack of Vitamin B complex and lipotropic factors results in a highly fatty liver, the fat globules literally occupying the cell cytoplasm pushing the nucleus to a side. Along with this infiltration. there is proliferation of the fibrous tissue which is infiltrated by chronic, inflammatory cells. The bulging cells, pressing on the sinusoids produce ischemia resulting in necrosis of the parenchyma. New capillaries form and invade the lobule and connect the central vein with the portal vessels. The penetrating fibrous tissue divides the parenchyma into smaller lobules. Some surviving cells proliferate and form nodules

(Hobnail). Contracting fibrous tissue makes the liver smaller and hence "Atrophic cirrhosis" results.

Biliary cirrhosis (Monolobular or hypertrophic cirrhosis)

In man this type of cirrhosis occurs consequent on obstruction and infection of the biliary tract. The causes are:

a) Cholangitis—the inflammatory exudate and the desquamated cells clog the bile ducts.

b) Pressure on the bile ducts from without—tumor of the head of pancreas.

c) Stone in the common bile duct.

d) Stricture of the duct.

e) Obstruction of biliary passages by flukes (*Chlonorchis sinensis*) and ascarids.

Macroscopically, the liver is enlarged and the surface is either smooth or finely granular. It is greenish in color.

Microscopically, connective tissue encircles individual lobules (hence monolobular). The bile ducts may be dilated and tortuous. There is great infiltration of the connective tissue with chronic inflammatory cells. Newly formed nonfunctional bile ducts are also found. Hepatic cells reveal degenerative changes. Jaundice is a constant symptom. Ascites is not common.

Biliary cirrhosis in animals is rare because cholangitis and cholangiostasis do not occur in them. Liver flukes that inhabit the bile ducts do not cause extensive cirrhosis but only a local fibrosis.

Effects of Cirrhosis

1. **Due to disturbance in portal circulation**

    A. Ascites; due to

    i. Increased hydrostatic pressure in portal veins—flow of blood through liver is hindered due to compression and distortion of the portal and hepatic veins as well as sinusoids by the regenerating nodules. The effect is more in portal cirrhosis since the number of such nodules is greater in this condition than in the biliary type.

    ii Decreased colloid osmotic pressure—since there is decreased production of plasma proteins, particularly albumin.

    iii Hormones are not inactivated by a damaged liver In health, the liver inactivates the mineralocorticoids of the adrenal and the anti-diuretic factor of the posterior pituitary. But if these are not inactivated more of sodium chloride is reabsorbed and with it more of water is also reabsorbed, resulting in conservation of more fluid in the body and so ascites results.

    B. Varicosity of esophageal veins—sometimes resulting in rupture and so hematemesis occurs.

    C. Splenomegaly.

    D. Gastroenteritis—result of C. V C. of abdominal viscera.

    E. Caput medusae in man—this is dilatation of the cutaneous veins around the navel and is seen distinctly in white skinned people.

11. Loss of inactivation of hormones, toxins etc.;

i. Estrogens normally are inactivated in the liver in the male. But in hepatic cirrhosis this does not occur and so gynecomastia and testicular atrophy occur.

ii. Toxins—exogenous or endogenous—are normally detoxified by the liver. If this is not done, the toxins affect the brain, producing degenerative changes resulting in "walking disease" in horses.

iii. Jaundice—due to pressure on the bile capillaries by the compressed cord cells (by fibrous tissue). So there is obstructive jaundice resulting in digestive disturbances.

iv. Bleeding due to deficiency in production of prothrombin.

v. Anemia—since iron and Erythrocyte Maturation Factor cannot be stored.

vi. Vitamin A deficiency since Vitamin A cannot be stored in the liver.

Other forms of Cirrhosis

Pericellular cirrhosis; In this condition the fibrous tissue invades the parenchyma and encircles individual cells. This picture may be seen in the far advanced stages of the multi and monolobular cirrhosis. In well developed aflatoxicosis pericellular cirrhosis is often found.

Pigment cirrhosis: This is the fibrotic condition of the liver that is found in hemochromatosis (bronzed diabetes of man). The macroscopical and microscopical appearances are similar to mild portal cirrhosis with nodulation. The large amounts of hemosiderin deposited in the hepatic cells seem to irritate the organ causing cirrhosis.

Glissonian cirrhosis: Correctly speaking this is not a true cirrhosis since the liver as a whole is not affected. Inflammation of the Glisson's capsule (the result of regional peritonitis)extends to the adjacent liver parenchyma. Though macroscopically resembling portal cirrhosis,microscopically the fibrosis is seen to extend from the capsule to a short distance beneath it.

Cardiac or central or congestive or stasis cirrhosis In chronic venous congestion resulting from cardiac lesions the cells round about the central veins suffer—degeneration and necrosis due to pressure and hypoxia; As the hepatic cells disappear, a relative increase in the fibrous tissue is evident. Later on there may be diffuse fibrosis and alteration in the architecture in some cases. This may giverise to atrophy and granular appearance of the organ.

Parasitic cirrhosis: In this variety, the irritant enters the liver through the bile ducts. The cause is usually a chronic obstruction of the bile ducts by flukes or other parasites. In swine mature ascarids invading the bile ducts cause biliary obstruction.

In this condition the changes are localised and are usually restricted in animals to fibrosis of parenchyma for a short distance around the biliary passages. Cirrhosis may spread out due to the penetration of bile into the tissues that surround the bile ducts.

Macroscopically, the liver appears larger, hard, firm and greenish in color. The surface is smooth. The bile ducts are hard and stand out due to extensive calcification—'clay-pipe' appearance.

Microscopically, the fibrous tissue is found encircling the bile ducts and the individual lobules (hence perilobular or monolobular cirrhosis). In this type there is formation of a large number of new bile ducts that are nonfunctional. Also there is infiltration of a large number of lymphocytes into the fibrous tissue. The bile ducts may be completely occluded by the flukes, blood stained exudate, and debris.

Because of the obstruction in the bile ducts, jaundice and deposition of bile pigment in the liver are present.

Besids the flukes and round worms that obstruct the bile ducts, other parasites damage the liver and produce cirrhosis during their larval migratory phase. Wherever the larvae lodge, chronic inflammatory changes arise with resulting fibrosis. In milder infections, the lesions are usually diffuse white spots, one to three cms. in diameter. In heavy infections, advanced fibrosis may be encountered. In pigs the scars produced by ascarid larval migration are depressed while those of *Stephanurus dentatus* larvae are elevated.

In infections by schistosoma species, dense white zones of fibrosis develop around intrahepatic portal branches. The lesions produced are similar to portal cirrhosis with nodularity.

Microscopically, the ova or their remnants may be seen in the fibrous tissue.

Kupffer's cells contain brownish pigment granules.

Abscesses of the liver: Due to the entrance of pyogenic bacteria abscesses may be found in the liver. These bacteria enter the liver by way of portal veins and hepatic arteries mostly. Infection may also occur from the umbilical vein in the young animals. Though the primary umbilical site may heal, the liver may still have abscesses. In the adult and older cattle, infection may occur from traumatic reticulitis.

In countries where cattle are fattened for slaughter, abcesses are frequently encountered in the liver. The cause is *S. necrophorus* gaining entry through the portal vein. In these animals, highly concentrated grain appears to produce ruminal disturbances, resulting in ulcers. Through these ulcers *S. necrophorus* enters the portal vein and ultimately reaches the liver, where it produces first coagulative necrosis. Subsequently this lesion becomes liquefied by the R. E. cells and then gets encapsulated. Later these abscesses heal leaving fibrous scars. In time, these scars may disappear.

Neoplasms: Tumors of the liver may be primary (arising from the liver parenchyma and bile ducts) or secondary (metastases of tumors located somewhere else)

Primary: Hepatomas arising from the hepatic cells or tumors arising from the bile duct epithelium are now considered as not rare in animals. They may be benign or malignant. The benign tumors are more common.

Another primary tumor of the liver that is found in the dogs is the hemangioma causing, sometimes, fatal hemorrhage. Primary fibroma may also be found in the liver.

Secondary: Metastases of lymphocytoma and pancreatic carcinoma are mostly seen but metastases of any malignant tumor may be found in the liver. In the cow metastases of uterine carcinoma are common. The neoplastic cells of lymphocytoma may form nodules, or may infiltrate dffusely, replacing the parenchyma gradually.

Mammary gland carcinoma in the dog metastasizes in the liver from the secondary tumors in the lung.

CHOLANGITIS : In liver fluke infection, cholangitis is met with, caused by the irritation of the spines on the cuticle of the parasites as well as the toxins liberated by them. The lumen of the bile ducts is dilated and its wall is thickened due to fibrous tissue proliferation around it. These ducts stand out as thick cords. In some cases, due to calcium deposition, these may feel hard also.

Microscopically, the mucosa is thickened and forms papillary projections into the lumen. Infiltration of the walls by macrophages, lymphocytes and eosinophils is common. The lumen contains parasites, cell debris and some mucus. The fibrous tissue that proliferates around the walls of the bile ducts may extend to a short distance into the parenchyma of the liver.

Occlusion of the bile ducts may give rise to obstructive jaundice.

CHOLECYSTITIS : This is rare in animals. Infection is usually ascending. from the duodenum. Stasis of bile by the presence of foreign bodies, parasites, concretions or by pressure on the biliary duct by pancreas are other causes since the retained bile is itself an irritant. *E. Coli* and *Salmonella* are frequently found.

Usually the catarrhal variety is noticed with congested mucosa and increased secretion of mucus by the glands.

CHOLELITHIASIS : Gall stones or choleliths are not as common in animals as in man. These are found mostly in cattle. The gall stones may be found in the gall bladder or bile ducts but unlike in man, bile ducts are more often affected because of frequency of parasitic involvement. They may arise in the bile passages of the liver also.

Gall stones are composed of a mixture of cholesterol, bilirubin, bile salts, calcium and an organic matrix. These may be dark brown or yellowish-green in color. There may be numerous small stones or a few large ones in the gall bladder. The larger ones may be faceted due to rubbing against one another. They are light and friable.

Etiology : Almost always gall stones occur as a result of cholescystitis. The dead cells or bacteria or mucus may form the nuclei around which are deposited cholesterol, bile pigments and bile salts. Sand particles and food materials that may reach the gall bladder through the bile duct from the duodenum during violent peristalsis may also form nuclei of the stones. Cholesterol is normally held in solution by loose combination with bile salts. This combination may be easily broken up. In cholecystitis, the bile salts are rapidly absorbed, leaving the cholesterol which is precipitated.

Sequelae : Most of the gall stones are "silent". That is, they cause no symtoms, being observed at autopsy only. But some may cause colic, nausea and dyspepsia.

If the bile passages are obstructed, obstructive jaundice may occur. If the obstruction of the bile duct is complete, rupture of gall bladder may occur sometimes.

Exocrine disorders : In animals, diseases of the pancreas are not common.

Acute pancreatic necrosis (acute hemorrhagic pancreatitis, necrotising pancreatitis; acute hemorrhagic necrosis).

This condition may sometimes be met with in dogs, cats, swine and horses. Ruminants are believed to be not affected. The essential lesion is necrosis of pancreas by its own enzymes. How this is brought about is obscure. The proteolytic enzymes are the most important. How and where trypsinogen, secreted in the pancreas, is activated to trypsin, to produce this acute condition is not yet clear. Probably occlusion of the duct (by parasites) or injuries or circulatory disturbances or regurgitation of bile or bacterial infection (via blood or from the intestines by ascending infection) may be the causes.

The ezymes escaping out of the pancreatic tissue digest the surrounding peripancreatic fat first and the pancreatic parenchyma subsequently. The fats are hydrolysed with the liberation of fatty acids, which form calcium soaps in the tissues round about the pancreas. Entering lymph channels the lipase may produce fat-necrosis in different and distant organs, even as far away as anterior mediastinal region.

Macroscopically, in fatal cases, there is a small quantity of fluid in the abdominal cavity. Hemorrhages may be present in the omentum. In the mesentery and around the pancreas, whitish areas or nodules of fat necrosis with an inflammatory zone surrounding them are found.

The pancreas is swollen, and soft, yellowish or slightly hemorrhagic. The lesions may be widespread or localised. If limited to a little area, encapsulation may occur. On section, yellowish-grey, soft (pus like) areas of necrosis may be visible.

Microscopically, one finds necrosis of the parenchymatous cells and fat, edematous swelling, infiltration by a few leucocytes, hemorrhages and thrombosis of vessels. Crystals of fatty acids and bluish calcium soaps are found in the necrotic area. Foreign-body giant cells are seen at the periphery.

Results : Death in acute cases after manifesting severe abdominal pain and cardiovascular collapse in shock. Chronic inflammation may result if the episodes are repeated and chronic fibrosing pancreatitis results with atrophy of the organ which is nodular (in cats and dogs). In the horse and in some dogs, on the other hand, the organ is enlarged due to great increase in the scar tissue. Pancreatic fat will reveal a granulomatous reaction.

Steatorrhoea occurs due to loss of the pancreatic juice. In this condition the feces is fatty and foul smelling.

Diabetes mellitus with glycosuria may be seen in dogs

Neoplasms ? Tumors of the pancreas are not common. Even in the few that are described, the exocrine tumors (tumors of the acini) are more common. The acinar neoplasm is an adenocarcinoma usually, while that of islets of Langerhans is an adenoma.

ENDOCRINE DISORDERS

Diabetes mellitus : It is pertinent to review here the carbohydrate metabolism since diabetes is essentially its derangement.

Carbohydrates are absorbed as glucose, which is converted into glycogen and stored in the liver. When needed by the tissues, muscles in particular, glycogen is converted into glucose—6—phosphate and then oxidised releasing energy. The by-products—CO_2 and water—are eliminated. These processes are regulated and controlled by various hormones. The most important of these is insulin.

Insulin : This is a hormone produced by the Beta cells of the islets of Langerhans. It is a protein, having a molecular weight of 6000. It has 51 amino acids arranged in two chains and]these contain 17 different amino acids.

Insulin has the following functions; helping in the storage of glycogen in the liver, facilitating the entry of hexoses across the cell membrane into the cell (muscles especially); stimulation of hexokinase for formation of hexose—6—phosphate and inhibition of activity of hepatic glucose—6—phosphatase and thus preventing overproduction of glucose.

Conditions may arise when

1. Insulin may not be adequately secreted due to necrosis of pancreas. The causes for necrosis have already been described.

2. Insulin may not be liberated into the circulation though synthesized by the Beta cells. Cause is unknown.

3. Diminished production of insulin due to "work-exhaustion". This occurs when insulin-antagonists act for a long time. Under this category must be mentioned.

   a) Insulinase, a proteolytic enzyme which destroys insulin.

   b) Glucagon and epinephrine are anti-insulin by virtue of their capacity to stimulate hepatic phosphorylase and produce glycogenolysis and hyperglycemia.

   c) Growth hormone (S. T. H.) This antagonises

      i)   the effect of insulin on hexokinase ;

      ii) the ability of insulin to transport glucose across the cell membrane.

      iii) by stimulating insulinase and

      iv) by probably stimulating the release of glucagon.

   d) Thyroxine—This increases the metabolic rate and gluconeogenesis.

   e) Adrenal corticial hormones—antagonise by gluconeogenesis and supporting the action of growth hormone.

The above antagonists first stimulate the islets of the pancreas, which become hyperplastic, releasing excess of insulin to arrest the hyperglycemia produced by them. In time, the cells become exhausted and atrophied.

The modern concepts of diabetes in man are ;

1. Diabetes may be present from birth as an inherent defective carbohydrate metabolism and this will be manifested later as diabetes due to various causes, viz, stress due to pregnancy, ACTH therapy, Cushing's syndrome, overeating, streptococcal infections and acromegaly. This error is due to an inherited recessive Mendelian factor.

2. Insulin in diabetes may be in the body as inactive complexes.

3. There may be autoimmunity so that patient's antibodies against insulin may inactivate the insulin in the body.

4. In diabetes there is inability to store sugar as glycogen and so it accumulates in the blood leading to hyperglycemia.

The following changes in the metabolism of carbohydrates take place in insulin deficiency. Though carbohydrate is transported in the form of glucose, it should be first converted into glycogen in the muscle cells before it can be metabolised to CO_2 and water (by way of TCA cycle). For this, therefore, glucose has to be transported across the cell membrane, to enter into the cells.

In the absence of insulin, normal quantities of glucose molecules are unable to move across the cell membrane and so it is not utilised and hence blood glucose level rises—hyperglycemia. When this is above the renal threshold (in dog normal is 160 to 180mg. per 100 ml) renal tublues [are unable to completely reabsorb the glucose of the glomerular filtrate and so *glycosuria* also results. Now because of glucose in the urine its osmotic pressure rises and this prevents the reabsorption of the water by the tubules and so polyuria also results giving rise to increased thirst, *polydypsia* and *dehydration*.

Glycogen stores of the liver are depleted due to glycolysis. So sufficient amounts of pyruvic acid and oxaloacetic acids are not formed to combine with active acetate formed from the fats. So this active acetate accumulates, condenses and forms ketone bodies, which in excess produce ketonemia and ketonuria. Being acidic, the ketone bodies neutralise the alkali reserve resulting in acidosis' which terminates in air hunger and coma.

Since tissues are unable to utilise glucose (except nerve cells and red blood cells which do not require insulin for glucose utilisation), catabolism of proteins and fats takes place as source of energy. Since fat of fat depots has to move into liver for phosphorylation (without which it cannot be utilised in the tissues) fatty infiltration of liver occurs. Ketone bodies are therefore formed in excess (due to catabolism of excess of fats) and so ketonemia occurs. (Normally small quantities of ketone bodies are produced but these are metabolised in the body.) So ketonemia gives rise to ketonuria. The breath and urine have the characteristic sweet odor. These keto acids interact with sodium and potassium salts and so these bases are lost in the urine and acidosis develops. Acidosis, dehydration and ketonemia give rise to coma.

Protein is catabolised to amino acids from which glucose (gluconeogenesis) and fatty acids are formed. Glucose cannot be utilised and so is lost in the urine. Hence body weight decreases. Excess of amino acids are deaminised in the liver and so there is elevation of blood and urine non-protein nitrogen. With the depletion of carbohydrates, fats and proteins, body loses weight inspite of consuming considerable quantities of food.

Diabetes mellitus may be found in dogs and cats. In dogs it is a disease of older animals, especially in females, due to chronic pancreatitis. For some unknown reason, such dogs develop cataract in the eye.

In lambs diabetes is seen in those that are overfed on carbohydrates. Glycosuria is met with in enterotoxaemia in sheep.

Macroscopically, the animal is emaciated and dehydrated. The liver is highly fatty. The pancreas may either be normal or show pancreatitis and necrosis with fibrosis. Lipemia is evident with the serum appearing white.

Microscopically, lesions are not very constant and conspicuous, Necrosis and hyalinisation of Beta cells have been noticed. Vacuolation of the Beta cells and the epithelium of ducts is present and is due to glycogen infiltration. Similar glycogen infiltration of the epithelial cells of the Henle's loops, and the distal convoluted tubules of the kidney is noticed. The liver cells are loaded with fat (foam cells).

Thr retinal and vascular lesions of man are not met with in animals.

Hyperinsulinism : This can occur in dogs with (1) excess of insulin injections or increased production of insulin by a tumor of Beta cells. In this condition, glucose is removed from the blood by (a) glucose oxidation by insulin-sensitive tissues, (b) deposition of glycogen in the liver and (c) by lipogenesis, resulting in *hypoglycemia*. The nervous system which is dependent primarily on glucose for energy suffers and its dysfunction is manifested by incoordination, dizziness, muscular weakness, tremors, loss of consciousness and convulsions.

2. Moderate hypoglycemia (50 to 60 mg. percent) activates the sympathetic nervous system. and so epinephrine is released. This brings about glycogenolysis in liver. Similarly glucagon may be released which also causes glycogenolysis in the liver. Hypoglycemia also stimulates the release of ACTH which in its turn causes the production of glucocorticoids which raise blood glucose level by gluconeogenesis and suppressing the peripheral utilisation of glucose,

Glucagon : This is a polypeptide hormone secreted by the Alpha cells of the islets and contains 29 amino acids in a single chain. Its function is quite. opposite that of insulin, namely to produce glycogenolysis of liver glycogen. The release of glucagon is brought about by hypoglycemia.

Glucagon action is to increase the activity of liver dephosphorylase kinase which activates phosphorylase and this causes glycogenolysis leading to elevation of blood sugar level. This activity similar to that of epinephrine. But glucagon does not cause glycogenolysis of muscle glycogen since it has no effect on muscle phosphorylase, while epinephrine acts both on liver and muscle phosphorylase.

THE PERITONEUM

Ascites or Hydroperitoneum is edema of the peritoneum and is common in dogs and cats but may also be encountered in sheep and ¦cattle.

Causes: 1. Portal obstruction—due to hepatic lesions—cirrhosis, hydatidiasis. fascioliasis, neoplasia (secondary) and pressure upon the vein by neoplasms, abscesses and enlarged lymph nodes.

2. General chronic venous congestion—cardiac valvular disease or pulmonary lesions.

3. Urinary obstruction in male cattle and sheep with or without rupture of bladder.

4. Hypoproteinemia — gastro-intestinal trichostrongylosis and Johne's Disease in which there is protein loss.

5. Cachectic diseases :- anemia and starvation—in which general dropsy develops.

6. Increased capillary permeability—due to histamine release in shock or due to toxins as in Edema Disease of pigs.

7. Lymphatic obstruction—by neoplasms.

8. Carcinomatosis—primary (malignant mesotheliomas) or secondary (extensive implantation carcinomatosis).

Hemorrhages into peritoneum are common in all animals and may be due to trauma of organs or sweet clover disease. Small focal hemorrhages are common in acute toxemias (enterotoxemia) and infectious diseases – anthrax, hemorragic septicemia and infectious canine hepatitis. These hemorrhages are found on the serosa of the diaphragm, stomach and intestines.

Hemorrhages in the peritoneum are also seen in the course of certain parasitic diseases such as distomiasis, *Strongylus edentatus* infection etc.

PERITONITIS is a very common condition in most of the domestic animals and may be localised or generalised.

The pathogens may be bacteria :- *E. coli*, *Streptococci*, *Staphylococci*; *Corynebacteria*, *Clostridia*; *Pasteurella* group, Anthrax in pigs.

Viruses—of bovine encephalitis

Helminths

Chemicals—introduced for medication

Endogenous—Bile and pancreatic juice.

Routes of entry : 1. Externally, through surgical wounds or from truama.

2 By blood stream as in bovine viral encephalitis.

3. By rupture of an abdominal organ.

4. Extension through the walls of stomach, intestine or uterus when their mucosa is inflammed.

5. Through osteum abdominale of an infected oviduct.

6. From an infected umbilicus.

7. By way of lymphatics from scrotal infection and infection of abdominal wall.

8. Direct extension from an infected kidney

The irritant first produces a serous inflammation which later becomes fibrinous or fibrinopurulent. The fibrin is helpful in localising the inflammation by forming adhesions. Being a very large absorptive surface of the body, toxins are speedily absorbed from the peritoneum damaging other parenchymatous organs.

In the condition known as Glasser's Disease in swine, a diffuse serofibrinous peritonitis is seen.

In visceral gout of birds, *uric acid peritonitis* occurs characterised by the deposition of urates on the serous membrane which consequently shows inflammatory changes.

Tuberculosis of the peritoneum is very frequent in cattle, less frequent in dogs and rarely met with in other animals.

One of the protective mechanisms of nature is the mobilisation and movement of the omentum which covers and sticks to the area of inflammation thereby

restricting its spread.   But this has its own drawbacks, since adhesions may form
between it and the inflammed parts.   Fibrin that forms, if not removed within
6 to 10 days, is organised, thereby inhibiting the movements of the intestines
and   impeding the digestive process :s

Neoplasms : the primary tumors, mesotheliomas (malignant) arise from the
serosa and are common in the young and newborn animals.

The secondary tumors are metastases from the liver or uterus.

Transcoelomic implantation of ovarian tumors found in women is not
common in animals.

CHAPTER 18

THE URINARY SYSTEM

Kidneys
  Functions
  Errors in renal function
  Postmortem changes
  Anomalies
  Circulatory disturbances
  Hydronephrosis
  Nephrosis – chemical
    Mercury poisoning
    Oxalate nephrosis
    Sulphonamide nephrosis
    Endogenous toxic nephrosis
    Lower nephron nephrosis
    Nephrocalcinosis

Suppurative nephritis
Pyemic nephritis
Pyelonephritis
Non-suppurative nephritis
Interstitial nephritis
Glomerulonephritis
Neoplasms
Urinary bladder
  Anomalies
  Chronic bovine hematuria
  Cystitis
  Neoplasms
Urethra
  Obstruction
  Urolithiasis
  Urinary casts.

THE KIDNEYS
Functions of the kidney :

1. Excretion of metabolic end products, especially end products of nitrogen metabolism—urea, creatine, creatinine, ammonium salts etc.—so as to maintain a standard chemical compsition of the blood.

2. Regulation and maintenance of acid–base balance of the extra-cellular fluids.

3. Selective reabsorption and thereby conservation of substances useful to the body : sodium chloride, glucose etc.

4. Maintenance of standard extra-cellular body fluid volume by excretion of water or its reabsorption, whenever indicated.

The above functions are performed by changes in the rate of excretion of the constituents of plasma.

The following extra-renal factors interfere with the functions of the kidney.

1. **Hemoconcentration** : In this condition the viscosity and osmotic pressure of blood are increased, resulting in decreased blood flow in the glomeruli. So there is decreased filtration of fluid resulting in **reduced urine**.

2. **Low blood pressure** : Causes :-Shock and cardiac decompensation.

Due to decrease in the effective filtration pressure there is complete stoppage of urine formation (Anuria)

3. **Obstruction to the out-flow of urine**: When urinary passages are obstructed by calculi or tumors, back pressure develops in the urine, which opposes the filtration pressure in the glomeruli and so urine is not filtered. Obstruction of urinary passage may also occur in cystitis.

The following intra-renal factors may affect functions ;

A. Injury to the glomerular filter ;— This may cause (i) increased permeability of the glomerular capillary endothelium, thereby facilitating the passage

377

of larger protein molecules into the capsular filtrate, and so albuminuria results
or (ii) reduction in the filtering surface as in (a) acute inflammation and (b) fibrosis. In condition (ii) the capacity of the glomerular filter is reduced, resulting in
reduced urine formation—anuria or oliguria.

B. Injury to the tubules :   This results in alteration of the tubular functions:
(a) selective reabsorption does not take place and so the essential substances
needed by the body like sugar etc. are lost; (b) water from the glomerular filtrate
is not reabsorbed, resulting in polyuria and dehydration; (c) tubules may be
blocked, thereby obstructing the formation of urine—anuria, uremia; (d) the
filtrate formed in the Bowman's capsule may be completely reabsorbed by the
lymphatics and veins resulting in anuria, and uremia; (e) substances that are
normally eliminated selectively by the tubules are retained, eg. creatinine.

C. Alteration in the circulation of kidney :   This may cause reduction in
tubular function due to decreased blood and oxygen supply to the tubules. This
is brought about by (a) narrowing of the arterioles when blood supply to the organ
is reduced and (b) fibrosis of the renal capsule in which condition the renal parenchyma is compressed so much that the total capillary bed is decreased and so the
blood supply is reduced.

Mention must be made of the peculiarity of the blood supply in the kidney.
The afferent vessels of the glomeruli arise from the arteriae rectae.   The efferent
arteriole (having smaller diameter than the afferent) on emerging from the tuft,
breaks up into capllaries, which surround the tubules and are the nutritive vessels
for these structures. Hence, it is natural to find atrophy and disappearance of the
tubules if glomeruli are destroyed. And it is to be known that when once
destroyed regeneration of glomeruli does not take place. Once lost they are lost
for ever (but tubules can re enerate very well).

ERRORS IN RENAL FUNCTION

PROTEINURIA : This is the presence of proteins(albumin mostly) in urine.
Ultimately this condition will lead to hypoproteinemia and so to generalised
edema (renal edema) finally.

Causes are;- (a) increased permability of the glomerular capillaries.
(b) tubular injury.   (c) inflammatory reaction.

Albuminuria is met with in the following conditions: congestive heart failure,
glomerulonephritis, renal infarction, nephrosis and amyloidosis.

GLYCOSURIA : This is the presence of glucose in urine and may be found
in diabetes mellitus, enterotoxemia in sheep due to *Clostridium welchii* type D,
following intravenous injection of large quantities of dextrose solution and in
njection of adrenocotricotropic hormones.

KETONURIA is the presence of ketone bodies in the urine and is met with
in diabetes mellitus, acetonemia of cattle, pregnancy toxemia in ewes and in
starvation.

ANURIA means complete urinary failure and **Oliguria** is a condition of
reduced excretion of urine, These are brought about in the following manner.

(a) Glomerulonephritis : In this condition due to (i) swelling of the capillary
endothelium and (ii) infiltration of inflammatory cells, the capillaries of the

gomeruli are compressed and so blood flow through them is blocked. So urine is not filtered.

(b) **Cloudy swelling and fatty degeneration of the tubular epithelium:** In these conditions, the pressure within the kidney is raised so much by the swollen cells that it obliterates the blood vessels. The tough, inelastic capsule of the kidney does not permit any expansion of the kidney and so the pressure in the organ is passed on to the vessels, compressing them.

(c) **Stagnation of the secreted urine:** If the urine formed is not evacuated from the kidney due to obstruction, the back pressure thus exerted will oppose the filtration pressure thereby preventing the formation of urine.

(d) **Extensive destruction of tubular epithelium:** In this condition there is diffusion of the urine filtered by the glomerulus into the lymphatics and veins.

(e) **Extreme dehydration ;** Sufficient fluid is not present to be excreted.

(f) **Low general blood pressure :** This has already been dealt with above.

POLYURIA : This is increased amount of urine passed.

      Causes :- (a)   Diabetes mellitus.
              (b)   Diabetes insipidus (posterior pituitary involvement).
              (c)   Moderate injury to tubular epithelium and so water is not reabsorbed.
              (d)   Chronic interstitial nephritis.
              (e)   In hypercalcemia and hypomagnesemia.

PYURIA : signifies pus in the urine and is found in suppurative inflammation of the kidney or some other part of the urinary system.

HEMATURIA is blood in the urine which is therefore colored red. On centrifugation or standing of urine, the erythrocytes settle down leaving a clear supernatant fluid. This condition is due to hemorrhage from any part of the urinary apparatus—glomeruli to urethra.

      Causes:- (1) Diseases of the urinary organs; acute nephritis; pyelonephritis' cystitis, chronic bovine hematuria; urethritis; renal infarction
      (2) Traumatism
      (3) Chemical irritants: cantharides; turpentine; carbolic acid;
      (4) Calculi
      (5) Acute septicemic conditions :- H.S. Anthrax
      (6) Neoplasms—Carcinoma of the bladder or kidney.
      (7) Parasites : *Dioctophyma renale.*

HEMOGLOBINURIA · This is the presence of free hemoglobin in the urine which is brown or coffee colored. Since there is hemolysis, red cells cannot be sedimented either by standing or centrifugation. This condition arises due to hemoglobinemia, hemoglobin escaping through the glomerular filter. Hemoglobinuria is found in

      (a) Protozoan infections—Babesiosis in cattle, horses, sheep and dogs.
      (b) In certain infective diseases : Streptococcal septicemia; infection by *Closiridium,* leptospirosis
      (c) Chemical poisoning—potassium chlorate poisoning, chronic copper poisoning in sheep

(d) conditions of unknown etiology–post-parturient hemoglobinuria in cattle.

UREMIA : Uremia is a toxemic syndrome resulting from renal insufficiency. It is associated with urea retention but is not caused by that condition. It is due partly to the retention and toxic action by non-protein nitrogenous substances including urea, creatine, uric acid, ammonia etc., and partly to the development of an acidosis.

The term 'azotemia' means an increase of non-protein nitrogenous material in the blood. This may be due to extra-renal factors such as dehydration, rapid break down of proteins or increased metabolism. Azotemia due to renal cause is uremia and can be differentiated from azotemia due to the extra-renal causes.

Uremia is a fairly common condition met with in significant derangement of renal function. It is estimated that five percent of all dogs that are examined at autopsy have some degree of uremia. This condition is more common in males than in females in all species ef animals.

Urea is formed in the liver by the breaking down of amino acids. This is not toxic as is evidenced by injection of urea intravenously without any untoward effects However, the blood urea level is a good index of the toxemia that develops in the uremic condition. The uremia that develops may not be of renal origin at all. It may be:

(i) **Post-renal** when the urinary tract is obstructed by calculi. In post-inflammatory strictures, carcinoma of bladder, prostatic enlargement and congenital defects there is retention of urine, which with back pressure in the kidney opposes filtration pressure and so anuria results.

(ii) **Pre-renal** found in :

(a) Lowered blood pressure due to shock, trauma and intestinal hemorrhage. There is decreased glomerular filtration pressure. Waste products of protein metabolism which are thus retained, aggravate the condition.

(b) Diarrhoea, vomition and intestinal obstruction and excessive sweating. In these conditions there is salt deficiency, dehydration and electrolyte imbalance.

(c) Fever, large infarcts, gangrene, diabetes, high protein intake. In these conditions there is increased protein destruction.

The renal lesions resulting in uremia may be :

(a) Glomerulonephritis; b) Chronic interstitial nephritis (small granular contracted kidney, c) Toxic tubular necrosis and d) Extensive amyloidosis. These lesions cause the following disturbances in the Physiology of the kidney: (a) decrease in glomerular filtration resulting in retention of urea, phosphates and sulphates, (b) decrease in tubular reabsorption resulting in the loss of water and electrolytes. (c) decrease in tubular secretion, resulting in the upsetting of the balance in potassium, H–ion, ammonia and creatinine, leading to hyperkalemia (producing cardiac inhibition and death) and acidosis. (d) Probable decrease in the detoxifying mechanism of the kidney so that the accumulated toxins act on the hemopoietic system and cause anemia.

Normally, the blood urea nitrogen (BUN) is less than 14 mgm. per 100 ml. of blood. But in uremic states, this is well above this figure. Besides the non-protein nitrogenous substances (uric acid, ammonia, creatinine, urea, amino-acids) other substances, such as sulphates and phosphates of potassium and chlorides are retained. These deplete the alkali reserve and so acidosis develops. This acidosis may be due to (i) reduction of glomerular filtration of the acid substances or (ii) loss of base especially sodium. In the damaged tubular epithelium "ion exchange" does not take place With the rise in phosphates, the calcium content of the blood is diminished.

Lesions met with in uremia are :

1. Toxic degenerative changes in the parenchymatous organs: ulceration of the mouth and stomach accompanied by hemorrhagic gastro-enteritis. The toxins are presumed to be excreted through the alimentary tract, causing the inflammatory changes. (Excretion gastritis—See page 347).

2. Injury to the neurones : by the toxic materials retained in the blood.

3 Deposition of calcium urates and urea on serous membranes : These minerals produce trauma of the serous membaranes and joints and so inflammation in these places occurs.

4. Dyshemopoietic anemia due to

(a) Decreased intake of iron and Vitamin B_{12} as animals suffer from anorexia or vomition. (b) Toxic suppression of hemopoiesis in the bone marrow.

5. Hyperplasia of parathyroid : In chronic nephritis, especially in the dog with extensive tubular damage, there is phosphate retention. This produces osteodystrophy. The following processes are involved. The damaged tubules reabsorb the phosphate in discriminately thereby elevating the plasma phosphorus level Excess of the phosphorus, when being excreted through the intestines combines with calcium, forming insoluble calcium phosphate. So calcium is not available for absorption and hence hypocalcemia results. This hypocalcemia in turn stimulates the parathyroid resulting in hyperplasia and increased parathyroid hormone production. This in turn produces (i) resorption of bone by osteoclasts, thereby releasing calcium and phosphorus and (ii) increased renal tubular excretion and diminished reabsorption of phosphorus

These are attempts by nature to retain the plasma calcium and phosphoru levels within physiological limits. But the damage of the kidney in chronic renal disease is irreparable and so a vicious circle is established, with more and more phosphorus retention and consequent hyperparathyroidism and osteoporosis resulting. This is clinically known as osteodystrophy. Nature tries to strengthen the softened bone by the cheaper fibrous tissue, which becomes hyperplastic and this is more evident in the bones formed by intra-membranous ossification, namely those of the head and jaw. Therefore these bones become very soft and pliable and can be bent. Hence they are known as "Rubber nose" and 'Rubber jaw" respectively. Histologically, some degree of osteodystrophy can be found in all the bones.

6 Polyuria : due to excretion of large amounts of calcium and phosphoru osmotic polyuria and attendant polydipsia result.

7. Metastatic calcification : Hyperparathyroidism with consequent hypercalcemia will result in metastatic calcification of soft tissues, especially gastric mucosa, larynx, trachea, lungs, visceral pleura etc..

8. Degeneration of liver with icterus.

9.   **Cardiovascular lesions** :   Found mostly in dogs.

(a) Increase in pericardial fluid with small amounts of fibrin

(b) Necrosis of the endocardium of the left auricle and the intima of the aorta and pulmonary artery to their first few centimeters. Later, the necrotic areas become fibrosed and calcified in those animals that survive

(c) Cardiac hypertrophy (especially of left side), medial hypertrophy of arterioles and capillaries. These changes are attributable to hypertension that develops due to (i) resistance to flow of blood through the kidney in chronic interstitial nephr tis(leptospiral infection) consequent on fibrosis or (ii) production of excessive amounts of renin by the damaged kidney, which releases hypertensiu (a vasoconstricter) acting on plasma hypertensinogen.

10.   **Terminal pulmonary edema**

POSTMORTEM CHANGES IN THE KIDNEYS

As in the liver, in the kidney also, postmortem autolysis occurs rapidly after death So if autopsy in not conducted immediately after death, it will be difficult to differentiate between antemortem degenerative processes and postmortem autolysis. The changes are observed in the cells of tubular epithelium in which the nuclei and brush borders disappear and the Altmann's granules become clumped together. In birds there is hyperchromasia of the nuclear wall and karyorrhexis.

More rapid autolysis of the kidney parenchyma may occur if prior to death there are already degenerative changes in the renal tubules. This is especially true in "Enterotoxemia" of sheep, in which the kidney is very soft—"pulpy-kidney". To the naked eyes, the cortex appears opaque and grey in this postmortem change. Gas bubbles may occur from bacterial putrefaction.

Hypostatic congestion, imbibition of hemoglobin and pseudomelanosis ar other changes noticed postmortem.

ANOMALIES OF DEVELOPMENT :

Agenesis—absence of one or both, of the kidneys may be met with.

Hypoplasia, in which the kidney is smaller, is more often seen. In such animals, the other kidney shows compensatory hypertrophy.

Persistent lobulation (normal in fetal life) may be seen in dogs, sheep and swine.

"Horse-Shoe kidney" is seen in all species of animals. This results from fusion of the kidneys at the posterior poles.

Duplication of one kidney may be seen in pigs and so in such animals three kidneys may be noticed.

Cysts ; Cysts in the kidneys are the most common congenital defects. One or more cysts may be found. Kidney with numerous cysts is known as *congenital polycystic kidney*. These arise due to lack of continuity between the nephron and the collecting duct and so urine formed in the nephron is not evacuated but collects to form a cyst.

Cysts may also be formed (acquired) after birth especially in chronic interstitial nephritis, in which the fibrous tissue compresses the tubules with resulting dilatation (and accumulation of urine) of the proximal part.

CIRCULATORY DISTURBANCES

HYPEREMIA : *Active hyperemia* is noticed in acute nephritis and in genera-lised acute septicemias and bacterial intoxications. *Passive hyperemia* is a feature found in generalised passive congestion. In these conditions the kidney may be slightly enlarged but due to unyielding nature of the tough capsule and the dense renal parenchyma, spectacular changes are not found. Usually, congestion is more evident in the medulla.

INFARCTS of the kidney are common due to occlusion of the branches of renal artery and are anemic in type. Renal infarcts are very common in cattle, especially cows, the commonest causes being thrombosis of the uterine veins after parturition and ulcerative thrombotic endocarditis caused by *C pyogenes* and *Streptococci*. In pigs infarcts are seen in erysipelatous endocarditis. They are wedge shaped, with base of the wedge towards the cortex and apex towards the pelvis. If the condition is not septic, healing by scar tissue will ensue, with pitting on the surface.

EDEMA of the kidney is not common because there is no place in the kidney for the fluid to accumulate. The capsule is inelastic and the parenchyma is firm. In acute interstitial nephritis some inflammatory edema may be observed.

DEGENERATIVE PROCESSES IN THE KIDNEY : Degenerative renal lesions are known as **Nephrosis**. This term is applied to necrotic lesions of the kidney also. Though formerly distinction was made betwen nephrosis (degenertive changes) and nephritis (inflammatory), as per our concept of alterative inflamm-ation, the distinction is no longer tenable.

The degenerative processes affecting the tubules mainly, include cloudy swelling, fatty degeneration and even necrosis. It is the highly functional and specialised epithelium of the proximal convoluted tubules that is greatly suscep-tible to the irritants. Next in order that are affected are the Henle's loops and the distal convoluted tubules

HYDRONEPHROSIS : In this condition there is dilatation of the renal pelvis due to obstruction to free flow of urine. The obstruction produces stasis of urine, which with its back pressure causes atrophy of the renal parenchyma.

The obstruction may be any where in the urinary tract—from the urethra to the renal pelvis. For hydronephrosis to develop, the obstruction must be partial leading to gradual stasis. If it is complete, atrophy of the corresponding kidney results. Hydronephrosis may be congenital or acquired. Among the causes for the acquired are :- (a) calculi; (b) hemorrhagic cystitis; (c) enlargement of the prostate in the dog; (d) tumors of renal pelvis, ureter, bladder and urethra ; (e) compression of the ureters by surrounding inflammatory tissue, neoplasms, gravid uterus, ovarian or uterine masses ; (f) displacement of the bladder in perineal hernia.

Hydronephrosis may be unilateral or bilateral. Extreme degree is observed only in the unilateral affection with partial obstruction proximal to the bladder. Here the whole kidney may be converted into a bag with a paper-thin capsule. In the not-so-severe cases, there is atrophy of the tubules, some of which may be widely dilated. The cortex will be thinner and grayish in color.

Sometimes, the fluid in the kidney may be pus due to supervening pyelonephritis, when the condition is known as **pyonephrosis**.

Results : In bilateral—death due to uremia

In unilateral – in uncomplicated cases, hypertrophy of the other kidney.

Pyonephrosis if infection supervenes as in pyelonephritis.

TOXIC NEPHROSIS : The toxins are conveyed to the kidney by blood The toxins are:-

hemical : Inorganic : salts of ; mercury (used as calomel; mercuric perchloride; mercurial fungicides ; uranium (uranium nitrate); chromium (potassium dichromate); copper (copper sulphate used to eradicate molluscs, parasites and fungi); bismuth (in B.I P.P. as dressing for wounds); cadmium, arsenic and phosphorus

Organic : Carbon tetrachloride and tetrachlorethylene *(anthelmintics)*, insecticides containing chlorinated hydrocarbons, oxalates and oxalic acid *(from plants)*: sulphonamides, turpentine, iodoform, cantharides, phenol.

Mercury poisoning is common in cattle, horses and swine. Mortality is high. This is a chronic type of cumulative poisoning, a single dose of acute poisoning not being common. The kidneys are enlarged, pale and bulge on cut surface.

Microscopically, in the acute disease there is coagulation necrosis and desquamation of the epithelium of the proximal convoluted tubules. Granular and hyaline casts are formed. There is moderate dilatation of the lumen of the tubules Infiltration of edema fluid and lymphocytes is found in the intertubular tissue. If death does not occur within about a week, regeneration of the epithelial cells occurs. These new cells are flat and dark staining. A notable feature is the deposition of calcium in the basement membrane, on the necrotic epithelium and the debris. By the third week, regeneration is complete (See page 133).

Oxalate nephrosis occurs in dogs and cats by swallowing ethylene glycol (used as an antifreeze) and in sheep that eat plants rich in oxalic acid or oxalate. The kidneys are slightly enlarged and are greyish brown in color. The doubly refractile, rectangular crystals of calcium oxalate are deposited in masses in the tubules, especially in the proximal convoluted tubules The tubular epithelium reveals vacuolar degeneration. There may be focal necrosis but damage is mild. Anuria is explained by the obstruction of the tubules by crystals.

Sulphonamide nephrosis may be met with in calves subjected so sulphonamide therapy but with inadequate sodium bicarbonate and water. In the acidic urine, the crystals are deposited in various parts of the kidney.

The epithelium of the proximal convoluted tubules and the Bowman's capsule undergo hydropic degeneration. Swelling and proliferation of the epithelium of the distal tubules is also evident. Anuria may result due to mechanical blockade of the tubules.

Endogenous toxic nephrosis occurs in ketosis, icterus and degenerative changes found in severe septicemic and toxic diseases. There is at first cloudy swelling and then early necrosis of the tubular epithelium.

Among the above varieties, special mention must be made of *Cholemic nephrosis*, seen in the jaundiced animals. The nephrosis is believed to be due to the action of the substances retained in the bile on the tubules. This condition, therefore, is more pronounced in obstructive jaundice. In these conditions the kidneys are swollen, pale and opaque and are brown in color due to the deposition of blood pigment in the renal tubules. The pigment is found intracellularly.

Hepato-renal syndrome   When there is damage to and necrosis of liver parenchyma, severe degenerative changes and necrosis may be found in the renal parenchyma also. Normally liver detoxifies various toxic substances formed some where in the body   But when it is damaged, this function of the liver is deranged and so the toxic substances reach the kidney and damage it.

Anoxic nephrosis : When adequate supply of oxygen is lacking, degeneration of renal tubules may result. Anoxia may be found in (i) urine retention producing increased intra-renal pressure; (ii) shock—due to severe burns; (iii) intestinal obstructions. The kidney is pale and soft.

Lower nephron nephrosis—the crush syndrome—hemoglobinuric nephrosis;

When large quantities of hemoglobin are excreted through the kidneys, lesions are found in the lower portions of the nephron, viz, the ascending loop of Henle and the distal convoluted tubule. This condition, therefore, occurs whenever there is hemoglobinemia as in (1) excessive hemolysis—by poisons or protozoa or incompatible blood transfusion; (2) severe burns; (3) equine Azoturia and (4) in crushing injuries—automobile accidents and air raids. In acidic urine the pigment is precipitated, which blocks the tubules resulting in anuria. Selective damage seems to be caused in the lower nephron with hemoglobin casts in these places. Tubular epithelium contains fine particles of hemoglobin and may show hyaline droplet formation.

Macroscopically, the kidney is swollen and pale with reddish streaks in the medulla.

NEPHROCALCINOSIS

Three types of disturbances in calcium metabolism are met with. White streaks or spots may be seen macroscopically denoting the places of calcium deposition. Calcium is deposited as phosphates and carbonates,

   1. Dystrophic calcification. (Primary epithelial calcification): In this condition the epithelial cells of the tubules are affected. As explained earlier (page 142) this occurs whenever there is necrosis of the cells and so may be encountered in nephrosis. Actually calcification is met with in sub-acute mercury poisoning. Due to poisoning by the mercury salts, the tubular epithelial cells are degenerated and necrosed and calcium salts are deposited on them.

   2. Calcium casts : When the urine is concentrated and inspissated in the distal convoluted tubules, the ascending loops of Henle and the proximal part of the collecting tubules, the calcium in the urine is precipitated on albumin casts These calcium casts may destory the tubular epithelium, which may also, on the other hand, be calcified. Later there is inflammatory reaction in the interstitial tissue, evidenced by infiltration of histiocytes and fibrosis.

3.   Deposition of calcium in the interstitial tissue: This condition is seen in dogs between the ages of 6 months and 4 years and is usually found in such affections in which there is reaction in the interstitial tissue.   Therefore, this is seen in leptospirosis.   The process starts with the deposition of calcium salts in the basement membrane of the tubules and later it may extend to the interstitial tissue   The capillary vessels may also reveal calcium masses in their walls· Sometimes this condition may be found in association with deposition of calcium in other organs

INFLAMMATION OF THE KIDNEYS

The following classification of nephritis, among animals, appears to be the simplest :

1.   Suppurative :—   i. Pyaemic—hematogenous
                      ii Pyelonephritis—urinogenic— ascending
2.   Non-suppurative :—   i. Interstitial
                          ii. Tubular (already dealt with under nephrosis)
                          iii. Glomerulonephritis.
3.   Specific :—   Tuberculosis. etc.

PYAEMIC NEPHRITIS OR EMBOLIC NEPHRITIS is a focal suppurative nephritis arising from infection with pyogenic organisms which are blood borne.

Causes are :—   Cattle—coliforms in calves.

Corynebacterium pyogenes and Streptococci in adults.

Foals :   Shigella and Streptococci

Swine :   Streptococci and Staphylococci

Infection may occur as a secondary to suppurative processes elsewhere—the umbilical vein, mammary gland, uterus, the pericardium or the lung. In the foal, the infection may be acquired in utero

The bacteria may reach the kidney in clumps or in the emboli from cardiac vegetation. These are arrested in the glomeruli or in the intertubular capillaries where abscesses are formed

Macroscopically, there may be numerous tiny abscesses, literally studding the kidney. These may be visible on the cortex through the capsule. Abscesses in the cortex are circular while those in the medulla are elongated. All the abscesses are of the same size, being of same age.

Microscopically, abscesses with leucocytic infiltration are found. Bacterial emboli are found in the glomerular loops and in capillaries between the tubules

Sequelae: Mostly terminate with death.

PYELONEPHRITIS :   This is a term used to indicate the inflammation of all parts of the kidney—involving the pelvis and parenchyma of the kidney.

Usually, this condition is met with in cows, but may also be found in sheep and swine

Infection in most of the cases, is thought to be an ascending one from the lower regions of the urinary tract. For such an infection to occur stasis of urine is an essential predisposing factor. When stasis occurs, the bacteria, especially the motile ones, ascend to the pelvis.

Causes of stasis :   In young animals - anomalies of the urinary system— pervious urachus in calves and kinking of ureters in pigs.

In mature animals :—In pregnant females gravid uterus presses upon the bladder and ureters; in males (dogs especially) enlargement of the prostate and obstruction by urinary calculi. Tumors, abscesses and fibrosis along the urinary excretory tract may prevent evacuation of the urine. Similarly, cystitis and ureteritis may also hinder speedy emptying of urine.

The causative orgnisms :

Cattle : *Corynebacterium renale* mostly and *C pyogenes* sometimes.

Sow : *Coliforms, Proteus, Streptococci. Pseudomonas* and *Salmonella.*

C. renale appears to have selective affinity for the pyramids of the pelvis and is the organism seen in most of the cases in the cows. The condition occurs more frequently in the post-parturient period because at that time there may be infection of the uterus and vagina from which infection spreads to the pelvis of the kidneys by way of the short and broad urethra. Infection may sometimes be hematogenous also.

At first there is pyelitis from where infection spreads to the kidney parenchyma by way of large uriniferous tubules.

Macroscopically, bladder is enlarged and ureters are dilated. The pelvis is widely dilated with pus; beneath the capsule on the cortex are visible irregular spots, grey in color, indicating tiny abscesses. Gray streaks may be found in the medulla in the early stages. The calices are widely dilated and filled with purulent material containing calcium particles. The walls of these are red and ulcerated. The papillae are either absent or it they are present are dirty-grey in appearance with erosions and a zone of hyperemia around. The renal pelvis is filled with purulent material which contains triple phosphates. The ureters are thickened and their mucosa is roughened. Due to increasing accumulation and retention of urine, back pressure may produce pressure atrophy of the parenchyma so that in some instances only a thin rim of cortex may be found.

Microscopically, the lesion is a purulent process with neutrophils and a few lymphocytes, which may be found as streaks among the tubules. Glomerular loops and Bowman's capsules are filled with leucocytes and bacteria. Tubules may contain cell-casts as well as bacteria. The walls of the collecting tubules, as well as the interstitial tissue, may become necrosed and this is demarcated from the healthy tissue by a dense zone of leucocytes and hyperemia. The epithelial lining of the pelvis may be necrosed and there may be leucoytic infiltration underneath.

Urine is bloody and contains pus.

The condition is fatal.

NON-SUPPURATIVE NEPHRITIS

INTERSTITIAL NEPHRITIS : Among animals, this is the most common type of nephritis seen. In dogs, it is an important disease. Bloom believes that at least 55% of all dogs autopsied have some form or other of interstitial nephritis. The condition is found more frequently in older and male dogs. The disease is sometimes observed in horses, swine, sheep and cattle.

In a normal kidney, no tissue is found between the tubules (interstitial space). But in interstitial nephritis there is infiltration of inflammatory cells and exudate as well as proliferation of the fibrous tissue.

Causes ; In most of the cases, the cause is obscure. Usually the condition is associated with retention of urine. Lesions of the lower regions of the urinary tract causing hindrance to free passage of urine are conducive to the development of the condition. Though infection is mostly hematogenous, ascending infection is not ruled out.

Possibly tubular damage is a starting point and a factor in many cases, damage occurring while toxins are excreted by the kidney. Interstitial nephritis has been seen in the following conditions:

Dogs   Leptospiral infection; infectious Canine Hepatitis; metritis; pyometra; bronchopneumonia and other respiratory affections; cystitis; prostatitis, chronic peritonitis and chronic infections.

Cattle and swine : Leptospiral infection. In pigs this condition is a metastatic infection and the organisms responsible are. organism of erysipelas, corynebacteria, *E. coli* and non-hemolytic staphylococci and streptococi, besides leptospira. Though the condition is frequent in these animals, it is benign and is encountered at autopsy in normal looking animals,

Fowls : The condition is usually seen in infectious diseases notably *Pullorum disease*. It is focal in distribution.

Young calves : Pneumonia and enteritis (White spotted kidney).

The disease may be acute, subacute or chronic. Again it may be diffuse or focal.

DIFFUSE INTERSTITIAL NEPHRITIS

Macroscopically, in the acute type, the kidney may be of normal size or slightly enlarged. The capsule strips off easily. The cortex shows a mottling of red and gray. The gray areas are the places of infiltration by inflammatory cells and are present in the cortex and may also occur in the outer medulla. In the normal yellow kidney of cat these changes are usually not easily observed.

The subacute and chronic types of interstitial nephritis merge imperceptably. The kidney is smaller in size, pale-grey in color, is hard and cuts with difficulty. The thickened capsule peels with difficulty and when stripped some portion of cortex is torn. The cortex is shrunken and very narrow. The surface is uneven due to irregular contraction of the fibrous tissue—the "Small granular contracted kidney".

It is not easy to distinguish kidney with chronic glomerulonephritis from that of chronic interstitial nephritis. The following differences may, however, be noticed.

Chronic glomerulonephritis	Chronic interstitial nephritis
1. Surface finely granular	1. Surface coarsely granular
2. Inflammatory process begins in the glomeruli	2. Inflammatory process begins in the interstitial tissue
3. Involvement of interstitial tissue secondary	3. Involvement of glomeruli is secondary
4. Diffuse involvement of glomeruli.	4. Glomeruli not involved or only few affected.

Similarly difficulty in differentiating between the kidney with interstitial nephritis and that with pyelonephritis may be felt. But the lesions in the latter are more irregular and asymmetrical and lesions may also be found in the pelvis and bladder.

Retention cysts of varying numbers and sizes are present   Sometimes they may be so many as to resemble the polycystic kidney.

Microscopically, the acute diffuse interstitial nephritis presents a picture that is essentially a true inflammatory reaction, consisting of exudation   infiltration and proliferation, affecting the interstitial tissue.   The infiltration, consists mainly of lymphocytes and plasma cells, with fewer number of neutrophils. The infiltration of the leucocytes is found mostly in the cortex and outer medulla and is diffuse and wide spread. Heaviest infiltration is found at the cortico-medullary junction Early in the lesions fibroblastic proliferation is evident, especially in a case of leucocytic infiltration. Glomeruli, however, are found to be normal The epithelium of the tubules shows degenerative changes (especially in Canine leptospirosis), which may be so severe and extensive as to cause death by uremia. These changes are more pronounced in the proximal convoluted tubules.

The acute episodes may pass off with resolution and healing. But the condition may also progress to the chronic phase, with scarring, especially with repeated acute attacks.

The gradual increase in the fibrous tissue in the chronic diffuse interstitial nephritis produces atrophy and disappearance of tubules, by pressure. Some of the tubules may show cystic dilatation in the parts proximal to constriction by the fibrous tissue. Granular and hyaline casts are found in such dilated tubules. As the fibrous tissue increases, the leucocytes diminish in number but do not completely disappear denoting thereby that the inflammatory reaction still exists. The Bowman's capsule may show rings of fibrous tissue around them. Eventually, the fibrous tissue replaces much of the renal parenchyma. Collagen that forms may contract, producing the granularity on the surface. In some places hyaline changes can be seen in the fibrous tissue and some glomeruli may be represented by hyaline nodules only.

In the cortex may be found hyperplastic and hypertrophic changes in those tubules, which are still normal, with tall columnar epithelium, having an "adenomatoid" appearance but which is purely compensatory. Calcium may be deposited, especially in the glomeruli. The walls of blood vessels become thickened and the lumen narrowed.

As in the liver, after a certain stage of fibroblastic proliferation, the fibrous tissue appears to continue to proliferate and is self-perpetuating even though the causative factors may no longer be present. The stimuli for such proliferation appear to be (1) the constituents of urine that may leak out of the damaged tubules; (2) the degenerative changes in the tubules due to encircling fibrous tissue and (3 the alteration in the blood vessels and circulation caused by the fibrosis.

FOCAL NON-SUPPURATIVE INTERSTITIAL NEPHRITIS

The causes for this condition may be those enumerated for the diffuse variety. But the lesions are focal and sparse. Incomplete resolution of diffuse interstitial nephritis may result in focal type, while coalescence of numerous foci in the focal interstitial nephritis may be the cause for diffuse variety.

The best example for focal interstitial nephritis is the "White spotted kidney" seen in calves, which do not seem to suffer much of this condition and show no symptoms, since, when they grow old, complete resolution and obliteration of the lesions ensue. It is only incidentally that the lesions may be observed in calves that die of some other disease. *E. Coli* is incriminated to be the etiological factor in these animals.

Macroscopically, the lesions are small. Pin-point-sized, grey-white, circumscribed areas that are found scattered on the cortex under the capsule and also deep into it, are seen on section. Outer medulla also may show these lesions. Some foci may bulge on the cortex and are clearly noticeable through the capsule.

Microscopically, there is inflammatory reaction of the interstitial tissue only while the glomeruli are free. There is edema of the interstitial tissue with infiltration by moderate number of lymphocytes and plasma cells Fibrous tissue proliferates causing degeneration of the tubular epithelium and later atrophy and disappearance of some of them. As in the diffuse variety, some tubules may be dilated and cystic. Granular and albuminous casts are present in the tubules.

Clinical correlation ; In the early stages of interstitial nephritis, may be found marked albuminuria, casts and either oliguria or polyuria.

In the chronic stage is found polyuria with low specific gravity, because the damaged tubules are not able to concentrate the glomerular filtrate.

With progressive loss of renal parenchyma, due to replacement by connective tissue the renal function fails and uremia supervenes with ultimate death. In such animals there may be calcification of laryngeal mucosa, left atrial endocardium, pleura and pulmonary artery (due to metastatic calcification that occurs consequent on secondary hyperparathyroidism that develops on phosphate retention occurring in this condition)

GLOMERULONEPHRITIS

In glomerulonephritis, the glomeruli are chiefly affected. This condition is not as common in animals as it is in man. It may sometimes be found in dogs, cats, swine, horses and mink. Horses used for antisera production suffer from this disorder. Acute glomerulonephritis may sometimes be seen as an enzootic among minks vaccinated against distemper.

The exact cause of glomerulonephritis is not yet known but it is believed to be an allergic (antigen-antibody) reaction to foreign proteins The glomerular capillaries become sensitised to the foreign protein and get damaged due to deposition of antigen-antibody complex. Two mechanisms by which the antigen-antibody reaction is brought about in the kidney have been identified. In one, there is production of antibodies by the animal to the glomerular basement membaranes. In this variety, deposition of antibody, complement and fibrin beneath the endothelium on the basement membrane is demonstrable by microimmuno—fluorescent technique. In the second variety, there is deposition of antigen-antibody complex (which is not of renal origin) in the basement membrane beneath the epithelium. In this condition it is thought that the antigen antibody complexes circulating in the blood react with the cells releasing histamine, which increases the permeability of the glomerular capillaries, thereby facilitating deposition of such complexes on the basement membrane.

Glomerulonephritis is often seen as a sequel to bacterial or viral diseases elsewhere in the body.

The primary changes are in the glomeruli. The tubular and interstitial lesions are secondary to the glomerular affection since their blood supply depends on the efferent arteriole emanating from the glomerulus. Among animals though acute diffuse glomerulonephritis as occurring in man is not observed, focal glomerulonephritis is met with in various acute septicemic infections such as acut⁻ swine erysipelas and infections by *Coliforms, Leptospira, Streptococci, Staphylococci, Pasteurella* and *Salmonella*.

Acute glomerulonephritis

Macroscopically, both the kidneys are enlarged and pale. The capsule which peels easily, is tense. On the cortex, red dots indicating the congested glomeruli, are seen. Sometimes small hemorrhages are also observed on the cortex. On section, the kidney slightly bulges on the edges.

Microscopically, there is hyperemia of the glomerular capillaries, soon followed by proliferation of the endothelial and epithelial cells This proliferation blocks the capillary lumens and so glomerular ischemia results. Infiltration by inflammatory cells aggravates the ischemic condition. Hence the efferent arteriole becomes ischemic, and so nutrient blood supply to the tubules is diminished. So now there is "increased cellularity" of the glomeruli. The capsular space, thus appears completely occupied by the swollen tuft, leucocytes, precipitated protein and a few erythrocytes. The subepithelial basement membrane becomes thickened. Electron microscopical examination reveals swelling of the basement membrane and deposition of electron dense material between the membrane and endothelium, within the membrane and between the membrane and epithelial cells. The changes in the basement membranre are evidently due to the increased permeability of the membrane. Fibrin-thrombi forming in the golmerular capillaries may cause hemorrhages. Collagen-like material may also be deposited between the capillaries. The ischemia that results is the cause of degenerative changes in the tubular epithelium which may contain hyaline droplets. Casts of protein, leucocytes and erythrocytes are seen in the tubules.

Sequelae : The condition may heal and resolve in mild cases. If not, it may progress to subacute and chronic phases.

Subacute glomerulonephritis: *Macroscopically* in the subacute glomerulonephritis we find the "Large white kidney". The kidney is enlarged, pale and smooth with non-adherent capsule Cortex may reveal a few hemorrhages. The capsule is tense. The cortex is wider than normal and is yellowish in color and so there is distinct color contrast between it and the medulla.

Microscopically, the proliferation of epithelial and endothelial cells is more pronounced. The proliferation of the epithelial cells of the parietal layer of Bowman's capsule results in a crescent-shaped tissue, several cell-layers thick— "epithelial crescents".

Due to deposition of collagen-like material between the crescents and the tufts, adhesions develop. Hence the tufts are obstructed and subsequently destroyed.

The tubular epithelium undergoes fatty degeneration, which progresses to hyaline droplet degeneration and necrosis. The tubules reveal casts of protein, leucocytes and necrotic epithelial cells. The interstitial tissue is edematous and contains infiltrated inflammatory cells together with some amount of collagenous tissue. The basement membrane becomes diffusely thickened and permeable to proteins.

Chronic glomerulonephritis: The subacute glomerulonephritis may imperceptably merge with the chronic phase. It is not necessary, it must be understood, that all cases of acute glomerulonephritis should culminate into chronic form. Only a small proportion may do so.

Microscopically we find in this stage, a shrunken and contracted kidney with a finely granular surface. The capsule is adherent and when removed some of the cortex is peeled off. On section, the cortex is found to be narrower and markings are obscured. Small retention cysts, due to obstruction of the tubules, are seen

Microscopically almost all the glomeruli are found to be affected and most of them are fibrosed. Some show hyaline changes while quite a few are atrophied and may disappear altogether. Still others may show adhesion between the tufts and the capsular epithelium. The inflammatory changes in the interstitial tissue is more pronounced. Greater number of lymphocytes and more n arked fibrosis are evident. In the scarred tissue, many tubules have disappeared. Arteries show thickening due to proliferation of the media and intima and so are narrowed. Some of the tubules which are still connected to functional glomeruli, are dilated

Clinical Correlation

Acute stage :

Urine ; Oliguria—due to ischemia of the glomeruli no blood is available for filtration.

High specific gravity of urine : some of the still normal and functional tubules concentrate the urine. Albumin, erythrocytes, leucocytes, renal cells and casts seen.

Blood ; Raised B U N content of blood.

Chronic stage : Urine ; Polyuria with low specific gravity—later urine decreases in volume, albuminuria, casts, red and white cells.

Blood ; Progressive anemia. B U N abnormally high.

Filtration rate and renal plasma flow are reduced.

MAIN TYPES OF NEPHRITIS IN DIFFERENT SPECIES OF ANIMALS

Species.	Type.	Cause	Frequency of incidence.	Source.
Cattle.	Pyelonephritis	C. renale C. pyogenes	Common	Associated with metritis and retained placenta following parturition

Species.	Type.	Cause	Frequency. of incidence.	Source.
Cattle	Pyemic	*C. pyogenes* *Staphylococcus*	Rare	
	Specific	*M. tuberculosis*	Rare	Miliary
Calves	Focal interstitial.	*E. coli.*	Common	From umbilical infection White spotted kidney.
Sheep.	Focal interstitial.	—	Fairly Common	—
	Pyelonephritis.	*Streptococcus* *Staphylococcus*	Rare	Associated with calculi.
Horse.	Glomerulonephritis	not known (antigen-antibody reaction?)	Rare	Associated with recent streptococcal infections of the upper repiratory tract.
Foal.	Pyemic.	*Shigella equirulis*	Common	Associated with joint ill
Pig.	Focal interstitial	*E. rhusiopathiae*		
		E coli, Corynebacteria, nonh-molytic strepto & staphylo,	Common	Associated with vegetative endocarditis.
	Pyelonephritis	*E coli* *Streptococcus* *Staphylococcus*	Common	Probably an ascending infection.
Dog.	Interstitial	*Leptospira canicola*	Very Common	
		L. icterohemorrhagiae	Not so Common	Associated with rats
	Tubular.	Poisons. P, As, $HgCl_2$	Common	—
	Pyelonephritis	*Streptococcus* *Staphylococcus*	Rare	Associated with calculi
	Specific.	*M. tuberculosis*	Common	Miliary.
Cat.	Interstitial	not known	Seen frequently	Leptospirosis occasionally.
	Glomerulonephritis.	not known	Rare	
	Specific.	*M. tuberculosis*	Common	
Fowl.	Interstitial.	*S. pullorum.*	Common	Associated with pullorum disaese.

NEOPLASMS

Primary neoplasms of the kidney are not very common. Two neoplasms deserve special description They are "Embryonal Nephroma" and "Hypernephroma". These have been described under "Neoplasms" (Vide pages 261 and 262)·

Secondary tumors may be found more often than the primary. The most common is the lymphosarcoma. Masses of the neoplastic cells may be found in the kidney as discrete nodules of varying sizes or the cells may infiltrate diffusely into the intertubular tissue. Then it may present a problem to distinguish these neoplastic cells from inflammatory infiltrating cells. But a careful examination reveals that in the neoplasm only the lymphoblastic cells are present while in the inflammatory lesion, besides the smaller lymphocytes, other inflammatory cells are also present.

DISEASES OF THE URINARY BLADDER

Anomalies : Persistent **urachus** may sometimes be seen in foals but very rarely in other animals. Urachus is the tube that connects the bladder to the umbilicus in the foetus. Just before birth, this is severed from the umbilical cord and becomes obliterated But in some instances, it is still patent after birth and is said to be "pervious". This is a problem needing surgery, since infection of bladder may occur through a pervious urachus.

Diverticula of the bladder wall and division of the bladder into several cavities are other anomalies rarely observed.

Rupture of the bladder may occur due to(a) Trauma—automobile accidents, gunshot wounds and faulty catheterisation (especially when bladder is full) or (b) obstruction of the urethra—by calculi, enlargement of prostate, neoplasms of urethra or inflammatory debris.

Prolapse of the bladder may occur through the urethra into vulva in cows, mares and sows due to straining during parturition. The short, broad urethra facilitates this condition. If not corrected, necrosis and gangrene of the bladder will supervene.

Perineal herniation of the bladder may occur in male dogs during straining in prostatic enlargements.

Calculi in the bladder form due to causes described under the subject "Urolithiasis". (See later) The calculi may be small and numerous and have a smooth surface. Or they may be large, when due to mutual rubbing against each other during contractions of the bladder, smooth facets develop.

Calculi are of consequence only when these occlude the urethral passage resulting in stasis of urine and attendant sequelae:- Viz, cystitis catarrhal or fibrinous developing into hemorrhagic); rupture of the bladder or uremia.

BOVINE ENZOOTIC HEMATURIA OR CHRONIC BOVINE HEMATURIA

This is a chronic disease of cattle that is observed in all parts of the world. It occurs as an enzootic only in certain localities In India this is mostly seen in the hilly regions of Darjeeling, Ooty, Kodaikanal. Garwhal, Kumaon and other Himalayan regions.

Though the exact cause has not yet been definitely established some toxic principle in the plants ingested may ultimately prove to be the cause.

Various agents have been incriminated from time to time. These are:—

Coccidia; liver flukes; piroplasms; schistosomes; *Aspergillus kamala;* high oxalic acid content of the food; deficiency of minerals in the soil (calcium, phosphorus, manganese etc.), feeding on bracken fern and a virus. In Russia and Poland, plants of the genus *Ranunculaceae* have been found to damage the kidney and bladder

The course of the disease is chronic and may last for months or one or two years. The early symptoms are the presence of blood in the last few drops of urine. As the disease progresses the blood content increases and in the later stages, pure blood may be passed. Secondary anemia supervenes with degenerative changes in various organs of the body. Animal becomes emaciated and finally dies

Macroscopically, the lesions may be found on the mucosa of the bladder. In the early stages, petechiae or ecchymoses may be found, which enlarge and become confluent as the condition progresses. The mucosa becomes thickened and red cauliflower-like tumor masses develop on the walls.

During excretion, the irritant produces inflammation of kidneys which therefore manifest hemorrhages. Later due to mechanical obstruction to passage of urine cystic dilatation may be found.

Microscopically, in the early stages may be seen hemorrhages on the bladder, followed by hyperplastic proliferation of the epithelium. Metaplasia to squamous or columnar (mucin producing) epithelium is frequent. The hyperplasia may be of neoplastic dimensions and in some cases may be a precursor to carcinoma (precancerous stage).

The capillaries also proliferate and growths similar to hemangiomas are observed. These are of two varieties. One is arranged as cavernous hemangioma, with thin walled, large dilated blood spaces. The second is capillary hemangioma in which masses of endothelial cells invade the surrounding structures, even into the muscular coat. Lesions of glomerulonephritis, tubular degeneration and interstitial nephritis may be seen.

INFLAMMATION OF THE BLADDER—CYSTITIS

Causes : The most important predisposing cause is retention of urine, especially if associated with trauma. Hence paresis of the bladder is a chief contributory cause.

Infections by bacteria in the majority of cases is by the ascending route and so this is more frequent in the female animals with short urethra—cow, sow and mare. (from suppurative endometritis and vaginitis). In some cases, infection may be descending also (from suppurative nephritis, and pyelonephritis).

Pervious urachus in the foals and calves is yet another route of infection while catheterisation is also found to be a factor in the causation of cystitis.

Infection may also occur by expansion from neighbouring organs. The infective organisms are : *Escherichia coli, Corynebacterium renale* (especially in cow and sow). *Proteus vulgaris, Staphylococci* and *Streptococci*

Cystitis may be acute or chronic

Macroscopically, acute catarrhal, fibrinous, purulent or hemorrhagic types may be found. In the catarrhal form, the mucosa, is thickened, edematous and reddened. The urine is cloudy. In the hemorrhagic form due to hemorrhages in and on the wall of the bladder, the urine is coloured red, while in the purulent form the urine is opaque. In the fibrinous variety, flakes of fibrin are present in the urine.

Microscopically, the epithelium is degenerated and desquamated. There is leucocytic infiltration of the mucosa and submucosa, and form dense sheaths around blood vessels. In the purulent variety the infiltration of leucocytes is heavy extending even to the muscular coat. Congestion of the blood vessels is seen. In severe conditions, ulceration of the mucosa supervenes.

In *chronic catarrhal cystitis,* (usually due to calculi) the mucosa is very much thickened due to fibrosis and frequently there is epithelial desquamation. There may be hypertrophy of the muscular layer with infiltration by lymphocytes.

Sometimes one may encounter chronic polypoid cystitis in cattle, in which the mucosa is thrown into folds. In this condition, there is dense infiltration of the proliferated connective tissue by mononuclear leucocytes.

In the dog, a follicular form is frequently seen in which the mucosa is studded with small, grey nodules, resembling lymph nodes consisting of aggregations of lymphocytic cells

Neoplasms of the bladder are seen occasionally. The most common is the metastatic lymphosarcoma. in which the neoplastic cells infiltrate the wall of the bladder, forming diffuse or nodular thickening. The primary tumors include:. leiomyoma, papilloma, transitional-cell carcinoma and squmous-cell carcinoma-

The Urethra : Affections of the urethra are rare since the lining mucosa is sufficiently tough to withstand infections.

Obstruction of the lumen by calculi is the most common cause of *urethritis.* In this condition, if the calculi are not removed, necrotic or hemorrhagic urethritis may result. When healing from urethritis takes place, stenosis of the urethral lumen may sometimes occur.

UROLITHIASIS

Urolithiasis means the presence of calculi in the urinary system. They may be found in the urinary tubules, in the ducts of Bellini (microconcretions), in the pelvis of kidney (nephrolithiasis), in the ureters, in the bladder (cystic calculi) or in the urethra.

Urinary calculi are found in all animals and they are of greater importance in the ox, since the stone may be arrested at the sigmoid curve and cause fatal obstruction. The groove in the *os penis* of male dog is a frequent site for the lodgement of the calculi.

The calculi differ in their chemical composition in different species of animals. This difference is largely governed by the pH of the urine. In alkaline urine (herbivores) the calculi are mostly carbonates and phosphates of calcium, magnesium and ammonium. In the acidic urine (carnivores and omnivores) oxalates, urates, xanthine and cystine predominate.

With infection the nature of the urine may change depending upon the infecting organism. Staphylococcal infection renders the urine alkaline while infection by coliforms renders it acidic, So even in herbivores, in coliform infection of the urinary passages, oxalates and urates may be found.

The following types of calculi are common among animals:

Horses : Calcium carbonate; triple phosphate (ammonium-magnesium phosphate); magnesium phosphate; magnesium carbonate.

Ox and other ruminants : Phosaphates and carbonates of calcium, magnesium and ammonium; silicates; oxalates.

Pigs : Triple phosphates; carbonates of calcium and magnesium; magnesium phosphate; oxalates; silicates

Dogs and Cats ; Oxalates, urates, uric acid, cystine, triple phosphates, calcium carbonate and phosphate.

The uric acid calculi as uratic calculi (composed of ammonium or sodium urates) are small, hard and brown in color. They show concentric rings on section These are common in Dalmatian dogs.

The oxalate calculi are very hard and the surface is rough and spiny (mulberry calculus). So it produces severe irritation, inflammation and hemorrhage and hence these stones are dark in color as they contain blood. These may also show concentric rings, but not as well defined as in uric acid calculi. These are mostly found in acid urine and may be solitary and large in the bladder.

Phosphate stones consist usually of calcium phosphate, magnesium phosphate or triple phosphate and are white, smooth, chalky and easily broken. These are formed in alkaline urine and may contain small amounts of urates, oxalates and carbonntes.

Xanthine stones are brownish-red, often laminated concentrically and are easily broken.

Cystine calculi are small, irregular, friable and yellow, becoming greenish on exposure to air. These are more common in cats.

Size : The size of the calculus depends on its location. The renal calculi may be microscopic in the tubules. The pelvic calculi attain to a size upto 8 cms and are hard and ovoid in shape.

The cystic calculi may vary from a grain of sand to that of a tennis ball. They may be smooth or rough, circular or ovoid and laminated or smooth on section. Some calculi develop smooth facets by the rubbing of one with the other.

They may be white or if containing blood, dark.

Formation of calculi . For the formation of the calculi there must be a nidus or nucleus around which salts may be deposited. Such a nidus may be organic material that may be found in the following conditions ;

a) casts
b) bacteria
c) leucocytes and
d) degenerated cells

} From the injured nephrons in nephritis

e) keratinised desquamated epithelial cells
f) mucoproteins

} In vitamin A deficiency and in st.lbestrol administration

Causes : How urinary calculi form is not very clear. The following factors singly or in combination, may be the causes :

1 **Vitamin A deficiency**: In vitamin A deficiency, the transitional epithelium of the urinary tract undergoes metaplasia into keratinised stratified squamous epithelium. The keratinised epithelial cells get exfoliated and may form the nidus of calculi

2. **Infection** : Infection of the urinary tract by *Streptococci*, *E Coli* and *micrococci* may occur when formation of calculi may be facilitated because the exudate and bacteria may not only form the nidus but the reaction of the medium may be suitably altered for the deposition of salts   But it is not always possible to detect infection in urolithiasis.

3. **Concentration of salts** : The organic and inorganic salt content of the food and water has an influence on formation of calculi. If the feeds consist of concentrates, with inadequate water, then calculi may be formed. Also if drinking water contains a high precentage of minerals, formation of calculi is facilitated, because in these circumstances, the mineral concentration of urine is increased. Hypervitaminosis D may cause hypercalcemia and so hypercalcuria· Curtailment of water, excessive sweating and ingestion of plants with high oxalic acid content increase the salt concentration of urine.

An inborn error in the metabolism of these salts, probably predisposes formation of calculi.

4. **Deficient green feed:** It was found that when animals were maintained on dry concentrates without alfalfa or green forage, incidence of urinay calculi was greater This was due to the excretion of muco proteins in the urine in the absence of the green fodder.   These mucoproteins act as nuclei for the calculi.

5. **Sulfonamide medication** : Formerly, in the early days of sulfonamide therapy, calculi were common. If sulfa dugs were used without sodium bicarbonate and sufficient water acetyl salts were formed and precipitated in the renal tubules, pelvis and ureters, forming calculi.

6. **Hormones** ; (a) For fattening lambs diethly stilbestrol pellets are implanted under the skin. This has a metaplastic action on the urinary epithelium, which is transformed into keratinised squmous epithelium.   The desquamating cells. form nuclei for the deposition of mineral salts.   The calculi that form in the bladder may cause obstruction of the urethra.

(b)   In man tumors of parathyroid, in which there is hypercalcemia, are associated with urolithiasis

7. **Prolonged confinement** : In man when patients are confined to bed for long periods with inability to move their limbs, the bones are decalcified and phosphatic calculi are formed in the bladder.

Sequlae : Calculi are harmful in two ways :

1. They may irritate the urinay passages and cause inflammation.

2   Obstruction of the passages may occur.

The results depend upon the place of obstruction.

a) If the obstruction is in the urethra, there is retenion of urine with attendant uremia and dilatation of the bladder. In some extreme cases the bladder may rupture with fatal results.

    b) if the obstruction is in the ureter :

        i) Atrophy of the corresponding kidney if obstruction is complete.

      ii) Hydronephorsis if the obstruction is partial.

URINARY CASTS

Casts are supposed to be products of albuminous exudate into the tubules from the blood vessels. Presence of casts in the urine is known as *clyindruria*. Casts may also contain tubular epithelial cells that may become swollen, destroyed and desqumated. Usually urine containing casts gives positive test for albumin.

Presence of casts in urine, in most cases, is indicative of some type of nephritis and is of great diagnostic value.

Casts have been classified according to their microscopic appearance as follows :

Hyaline casts are pale, often colorless, homogeneous and cyindrical in shape. They have straight parallel sides and rounded ends. These casts dissolve in acetic acid (but fatty casts do not), Hyaline casts are found in congestion and in inflammation of the kidney.

Granular casts may be either coarsely granular or finely granular. They may be curved or straight with rounded or broken ends. The granular material is composed of albumin, fat, epithelial cells or disintegrated leucocytes or erythrocytes.

Epithelial casts contain epithelial cells of the tubules and indicate acute nephritis. They may be yellowish due to imbibition of blood pigment. When the condition changes to subacute or chronic, these casts may become fatty or waxy.

Waxy casts are more opaque than the hyaline casts and are yellowish in color. These are found in chronic nephritis, but seldom in acute. They are invariably found in amyloid degeneration of the kidney.

Blood casts are found in acute nephritis, renal hemorrhage and in acute congestion of the kidneys. They are formed of erythrocytes and so may have a reddish color.

Fibrin casts are found in cases where there are hemorrhages. They may be yellowish due to altered blood pigment.

Pseudocasts have no connection with renal disease but occur due to conglomeration of various substances on mucous threads. Urates and phosphates may aggregate together to resemble casts. The following are examples of pseudocasts :

Fatty casts : These contain numerous fat globules which can be stained by Sudden III or Osmic acid. The fat is probably derived from the degenerated epithelial cells. The fat globules are not soluble in acetic acid

Pus casts consisting of pus cells that have been kneaded together are found in renal suppuration. They may contain a few fat globules also.

Cylindroids are not casts but consist of mucus and some fat globules. They are striated.

Bacterial casts occur due to proliferation of bacteria in stagnant urine but are not true casts.

CHAPTER 19

THE NERVOUS SYSTEM
By Dr. J. L. Vegad

General concepts
Reaction of nervous tissue to injury
  Neurons
  Neuroglia

The cerebrospinal fluid in neuropatho-
                logy
Paths of infection
Blood-brain barrier
Congenital anomalies
Distrubances of circulation
Hydrocephalus
Disturbances in growth

Calcification
Traumatic injury to nervous system
Necrosis
Inflammation
  Fibrinous
  Suppurative
  Lymphocytic
Specific inflammations of brain and
         spinal cord

Parasitic encephalomyelitis
Cestodiasis (Gid; Sturdy)
Cerebrospinal nematodiasis–
                (Kumari)
Wobbles
Toxoplasmosis

'Allergic' encephalitis
Meningitis
  Pachymeningitis
  Leptomeningitis
Myelitis
The peripheral nerves
  Degeneration (Wallerian degene-
                ration)
  Regeneration
  Neuritis
Marek's disease
Congenital myoclonia of pigs
Epilepsy
Sway back (Enzootic ataxia)
Tumors

GENERAL CONCEPTS :

The general laws of pathology remain fully applicable to the nervous system but certain peculiar characteristics of the nervous tissues have to be recognized' For example, when the central nervous system is attacked, the functional distur_ bances in turn may be widespread, and could affect the entire body. This becomes all the more important when we realize that once a neuron is destroyed, it cannot be replaced, The functional distrubances of the neurons (nerve cells) in the brain are therefore of supreme importance because they govern so many vital activities of different organs. When irritants alter the colloids of the nerve cells and their fibres, the functions of the nervous system may be directly disturbed, indirectly affecting the activities of the organs. The animal may not only be affected mentally, but its body movements, its gland secretions and its reflexes are also affected. Various activities of the brain can be upset by diseases. These are usually manifested by loss of consciousness, nervous dep ression, and increased nervous irritability. Another important feature of the central nervous system is that it is often possible to diagnose, from a careful study of symptoms such as paralysis or other disturbances of function, as to in what particular portions of the brain or spinal cord lesions may be located. Actually, many symptoms noticed in diseases of other systems or in generalised diseases are due to either malfunction of or injury to the nerve tissue. In diseases of the nervous system the following symptoms are exhibited :—muscular tremors, ataxia, convulsions, dullness, coma, fever, anorexia. sleepiness, mania stupor and paralysis. Though cytological and macroscopical changes can be demonstrated in the nervous system in some of the diseases showing the above symptoms, in some others, functional disturbance is not accompanied by morphological changes. For example, in Tetanus, though profound functional disturbances take place, yet no macroscopical changes occur in the nervous system

Loss of consciousness :- results from the effect of variuos toxic agents upon the brain. Complete loss of consciousness is said to be the state of **coma.** In this animal lies outstetched and motionless, its reflexes are gone, the pupils are dilated, respiration is slow and irregular, heart beat weak, and skin cool. It usually ends fatally. **Nervous depression** results from pressure upon the brain, such as the one brought about by hemorrhage within the cranial cavity, a brain tumor, or even collection of fluid within the ventricles. There is loss of feeling, sleepiness, and muscular incoordination. A good example of this functional disturbance is equine encephalomyelitis. **Nervous excitement** results from congestion and inflammation of the brain and its coverings. There is delirium and mania, and even convulsions. The best example of this functional disturbance is rabies.

Disturbances in nervous functions also affect muscles in two ways. In the first ease. there is increased activity (**spasm**), and in the other loss of contractility (**paralysis and paresis**). In muscle spasms there are sudden, violent, involuntary contractions. They may be continuous (**tonic spasms**), or intermittent (**clonic spasms**). When the spasms are mild and are confined to groups of muscles, they are called **tremors**; a very good example of this being epidemic tremor (avian encephalomyelitis) of chicks. If the muscle spasms are widespread and involve the whole body, including the limbs, they are called **convulsions.** They are often seen in puppies infested with ascarids. When tonic and clonic spasms alternate, and are accompanied by loss of consciousness, they are termed **epilepsy.** In the second form of neuromuscular functional disturbance **paralysis** (the complete immobility of a muscle) and **paresis** (the incomplete loss of motion), the underlying cause is the defective innervation of the muscle. The defect may lie in the motor centres or in the conduction paths. It prevents the flow of motor impulses and immobility results. **Hemiplegia** is the paralysis arising in the brain cortex and in the peripheral nerves and is unilateral. A bilateral paralysis of the posterior parts of the body and hind limbs resulting from injury to the cord is called **paraplegia.**

REACTION OF NERVOUS TISSUE TO INJURY ;

Nervous tissue is very susceptible to injury. Further, when a neuron is destroyed, it cannot be replaced. When the nervous system is sick and fails to perform its functions, the structural changes may be of three types: 1. macroscopic alterations. 2. microscopic alterations, or 3. the changes may be of a biochemical nature, and nothing may be visible even microscopically. It would be helpful to examine separately the reactions of the neurons and the glial cells to injury.

Neurones: These are the cells that are responsible for carrying the impulses. The neurone consists of a cell body and one or more processes—the axon or dendrites. Degenerate neurones are normally found in healthy brain but more so in the young animals. Since autolysis occurs very early after death in the nervous system it is essential to know how to distinguish between autolytic and degenerative changes. The following are the changes noticed postmortem:—

Imbibition of large amounts of fluid giving a spongy, wet appearance to the tissue; Neurones and glia shrink, leaving a clear space between the cell and the surrounding parenchyma; Shrinkage and condensation of the nucleus (P. M. pyknosis); Fragmentation, fading and disappearance of the nucleus and cytoplasm and Nissl granules; Axis cylinders, which are normally unstained, take stain diffusely.

The neurones being highly specialised are easily susceptible to injury by hypoxia or toxic materials, evidenced by degenerative and necrotic changes. The following reactive changes are noticed.

1. **Shrinkage** characterised by cells becoming very irregular, nucleus pyknotic, clumping and condensation of Nissl substance and tortuousness or sclerosis of the processes. This is seen in senility and chronic infections.

2. **Swelling of the nerve cells** : Here the cytoplasm stains very faintly and only the cell outline may be discerned with fragmentation of the processes. This is a reversible condition and occurs in severe intoxications and systemic infections.

3. **Vacuolation** of the nerve cells may be seen in toxic conditions and in viral encephalomyelitis.

4. **Chromatolysis** : In this condition the Nissl substance becomes fine and dispersed and later may disappear. The nucleus may be eccentric. This change is seen in injury to the axon. It is suggested that the Nissl substance being a ribonucleoprotein is actively involved in the synthesis of axoplasm and so chromatolysis may denote an exhaustion phase. When axon regenerates and is repaired, Nissl substance is restored.

In viral infections and other severe intoxications chromatolysis occurs when the Nissl substance disappears. The cytoplasm also shows degenerative changes: swelling and rounded contours. Unlike in the axonal trauma, in viral infections and injury to central axons chromatolysis is irreversible since the neurones are destroyed.

5. **Neuronophagia**: when the nerve cell dies, microglia and oligodendroglia invade the cell and remove it by phagocytosis.

6 **Satellitosis** : Normally every neurone has one or two oligodendroglia near them. They are called satellite cells in this location. Whenever a neurone is damaged, oligodendroglia and microglia crowd around such cells, without actually invading them, and this phenomenon is known as "satellitosis". Both satellitosis and neuronophagia are indications of necrosis of a neurone.

The following are some of the causes that bring about the degenerative changes described above :—

Inoroganic salts : Lead, arsenic.
Organic : Anesthetic agents.
Metabolic : Toxic products of uremia.
Infectious agents : Neurotropic viruses.
Nutritional deficiency : Deficiency of B_1, Copper, Cobalt.
Vascular : Interference with blood supply—causing anoxia.

Effects of vaccination : Allergic encephalitis occurs after vaccination with vaccines containing brain tissue. There is destruction of myelinated tracts in the white matter

Toxic agents : a) In liver disease, toxic agents, exogenous or endogenous, are not detoxified and so these pass on to the brain producing degenerative changes

b) In mercury poisoning, neural degeneration and necrosis of the brain and demyelination of the nerve tracts extending to the spinal cord are noticed.

c) In *lathyrus* posioning there is degeneration of neurones in the spinal cord accompanied by gliosis and ultimate atrophy of the spinal cord.

d) In poisoning, by chlorinated hydro. arbons, there is Nissl degeneration and necrosis of neurones, especially of the ganglia and brain stem.

Astrocytes, which are star shaped, are the supporting cells found throughout the central nervous system. These cells react to injury and proliferate and this process is known as "gliosis", which may be uniformly diffuse or may be focal.

Apart from the function of forming a supporting matrix, astroglia probably actively participate in the transport of fluids and solutes between the blood vessels and the nerve cells.

When nervous tissue is destroyed, repair does not take place by fibrosis The fibrous tissue from the adventitia of the blood vessels may repair sometimes. But, if a cavity arises due to softening and absorption of the brain, it is not filled in but a cyst is formed with a thin fibrous capsule around which the astrocytes proliferate. Inclusion bodies may be present in the astrocytes also. (gemistocytes in Canine Distemper).

Oligodendroglia are cells containing dark round nuceli and are found mostly in the white matter, in long rows between the fibres. They may also be found in small numbers as satellite cells around the nerves and blood vessels. Though their exact function is obscure, oligodendroglia are supposed to be connected with the maintenance of myelin sheaths. In places where medullated nerve fibres are destroyed oligodendroglia are found to disappear. The cells invade the dead neurones and engulf them (neuronophagia).

Microglia are mesodermal in origin (while astroglia and oligodendroglia are of neuroectodermal in origin) and so belong to the reticulo-endothelial system. They are the phagocytes of the central nervous system and are found in both white and gray matter. With ordinary stains, only their oval dark nuclei are noticed. Their branching cytoplasmic processes require special stains to be seen.

Microglia are amoeboid and phagocytic and become hypertrophied and proliferate. Hypertrophied cells that engulf the dead tissue are rounded and the cytoplasm is foamy containing lipids and are called "*Gitter cells*" (from *Gitter-zellen,* German), "*Compound granular corpuscles*" or *fat granule cells*". Often these gitter cells migrate to the perivascular spaces (space of Virchow-Robin).

The meninges : The coverings of the central nervous system consist of the dura and pia-arachnoid. While the dura in the cranium is attached to the cranial periosteum, in the spinal column it is separated widely from the vertebral periosteum. Pia closely follows the brain and is separated from the arachnoid by the subarachnoid space. The space contains the cerebrospinal fluid and has the spongy network of arachnoid trabeculae.

The meninges are composed of cells that are mesodermal in origin. Their reaction to injury is comparable to that of other tissues elsewhere in the body Reaction to injury is by inflammation and fibrosis, unlike that found in the response of glial tissue.

Blood vessels :— The blood vessels of the brain have some peculiarities. These are : (1) The arterioles and venules are very thin walled, devoid of elastic and muscular tissue. (2) Veins do not have valves. (3) The blood vessels acquire a meningothelial sheath as they pass through the subarachnoid space and a second outer sheath derived from the pia. So a perivascular space is formed between these sheaths—the space of Virchow-Robin, which is continuous with the perineural and interstitial space of the C. N. S. It is in this space that the cells accumulate and give rise to "*Perivascular Cuffing*". Depending on the nature of the pathogen, the cells vary. In baterial infections, neutrophils predominate while in viral, lymphocytes and in allergic encephalitis macrophages, lymphocytes, plasma cells and eosinophils are found. In infections of the brain, inflammatory exudate collects in the space of Virchow-Robin. The adventitial wall is the source of macrophages (in some instances) and the fibrous tissue elements that compose the capsule in abscesses that arise in some bacterial infections.

Cerebrospinal fluid is found in the ventricles, spinal cord and the subarachnoid space and probably serves the function of lymph (which is absent in the C. N. S.). Its function is not only to serve as a medium for metabolic exchange but also to serve as a protective cushion for the delicate C. N. S.

Cerebrospinal fluid is formed by secretion and dialysis from the blood by the choroid plexus in the lateral ventricles. From the lateral ventricles the fluid passes into the third ventricle through the foramina of Monro and from there, through the aqueduct of Sylvius into the fourth ventricle. From the fourth ventricle, the fluid passes through the foramina of Luschka into the subarachnoid space. Mention has already been made that the subarachnoid space is continuous with the perivascular space. So the fluid is found in this place. Since perineural space also communicates with the subarachnoid space, the fluid is also found in the perineural space.

Most of the fluid is drained into the venous sinuses found in the dura through the activity of arachnoid villi, which project into these venous sinuses. Some fluid may also diffuse into the blood vessels from the space of "Virchow-Robin".

Normally, the cerebrospinal fluid is colorless and clear with Sp. gr. never exceeding 1009, pH. is 7.4 to 8. Leucocyte count is very low, fewer than 10-12 cells (mostely mononuclears) per cubic millimeter being present. The globulin content doses not exceed 10mg/100ml. But in infections,the C. S. F. may show considerable change. It may become cloudy or even bloody and the globulin content may be very high, 130 to 1500 mg / 100 ml. and the leucocyte count also being very high. Normal sugar content is 35 to 70 mg%.

Paths of infection :

— Infection may reach the central nervous system by way of the blood stream or the lymph stream, or it may pass along the axis cylinders of motor or sensory nerves.

Blood-brain barrier :

It is now well established that the cerebral blood vessels differ from other vessels in their permeability. It has been demonstrated, for example, that when, large molecule dyes are injected intravenously they appear in the reticuloendothelial cells of liver, spleen and lymph nodes, but not in the microglia of brain, which also belong to the reticuloendothelial system, and are the brain histiocytes. Thus, blood-brain barrier is an obstacle that exists between circulating blood and the brain which effectively prevents a wide variety of toxic substances of large molecules from reaching the brain. The blood-brain barrier, thus is essentially a defence mechanism against noxious agents. In pathology, its importance lies in the fact that it successfully prevents the entry of most bacteria and viruses from the circulating blood into the brain. And, it is only when certain toxic substances, including bacteria and viruses, break down the barrier that they are able to enter into the brain tissue and set up the infection, or induce other pathological changes. Histologically, the blood brain barrier is composed of the vascular endothelium of the cerebral blood vessels, the basement membrane, and the perivascular glial membrane which is the close application of the cytoplasmic foot processes of innumerable astrocytes to the endothelium of the capillaries.

CONGENITAL ANOMALIES :

Anencephaly	is the absence of most of the brain, and is seen in most animal species.
Acrania	is the complete failure of cranial development.
Amyelia	is the absence of spinal cord.
Cranioschisis	is congenital fissure of the cranium.
Encephalocele	is the protrusion of meninges, alone or with part of the brain, through a defect in the cranium.
Exencephalus	is the absence of cranial vault exposing the fully developed brain.
Microcephaly	is the presence of an abnormally small brain.
Meningocele	is hernia of the meninges, which protrude through an opening of the skull or spinal column.
Rachicele	is hernia of the spinal cord.
Spina bifida	is a congenital defect in walls of spinal canal caused by lack of union between the laminae of the vertebrae. Through this defect the spinal cord may herniate. The condition has been repor-

ted in cattle, dogs and sheep. Sometimes there is no protrusion and swelling to indicate the defect. This means the condition is hidden, and is then called **spina bifida occulta** (hidden.)

DISTURBANCES OF CIRCULATION

Hyperemia :

Acute general active hyperemia is present when bacterial or viral diseases involve the entire central nervous system (rabies, viral equine encephalomyelitis, and hog cholera).

Acute focal active hyperemia is seen in the vicinity of abscesses, tumours and infarcts.

Chronic general passive hyperemia : This occurs when there is a passive hyperemia due to lesions in the heart or lung or an obstruction to the flow of blood from the brain such as thrombosis of both jugular veins. Histologically, there are increased number of glial cells throughout the brain and spinal cord, which indicates chronicity of the condition.

Chronic focal passive hyperemia : This occurs when a tumour or abscess presses upon a vein, or a thrombus forms within a vein causing a reduction in the flow of blood from a local area of the brain.

Anemia :

General anemia : This occurs when anemia involves the entire individual. This may be seen in parasitic anemia as in gastrointestinal parasitism, excessive hemorrhage and in anemias associated with deficiency of iron. copper and the vitamin B complex.

The brain and spinal cord are whiter than normal, and the blood vessels contain decreased amount of blood and are therefore less prominent. Histologically, areas of liquefactive necrosis as well as gliosis and neuron degeneration may be present. These result from oxygen deficiency.

Local anemia or ischemia : This occurs from a deficiency of arterial blood in a local area of the brain or spinal cord, the two main causes being thrombosis and embolism.

Thrombosis and embolism of cerebral arteries are rare in animals and may occur in the brain and spinal cord. Emboli may be (1) detatched vegetations from the cardiac valves or may arise from lesions of the lungs, left atrium or coronary artery, (ii) clumps of bacteria, (iii) tumor cells, (iv) Parasites (larvae of ascaris, onchospheres of tapeworms, young trichinella etc.) or (v) agglutinated erythrocytes.

Thrombus can arise from lesions of cerebral vessels (atheroma) or can occur in diseases that damage the vascular endothelium :— trauma causing fractures of the skull. invasion of the vessel wall by neoplastic cells, abscesses and hog cholera. If collateral blood supply is inadequate infarction results. The infarcted area is finally liquefied, a cyst being formed

If the blood supply is not adequate to maintain the nutritive and oxygen requirements of the area, **infarction** occurs. The infarcted area is pale or red depending upon the blood supply. Infarction ends up in liquefactive necrosis of the involved area.

Hemorrhage

Petechiae are common in acute septicemic diseasees (Anthrax, Hemorrhagic septicemia, Hog cholera, Leptospirosis) or in infections by pyogenic organisms. These also occur after thrombosis or in degeneration of the vessel walls or in general hemorrhagic diseases as bracken fern poisoning

Rupture of an artery will give rise to large areas of hemorrhage with clots causing apoplexy. Rupture may occur in injuries—automobile accidents, gun-shot wounds, diseases of wall of blood vessels (atheroma) with hypertension as in arteriosclerosis, chronic-nephritis, bursting of an aneurysm as in parasitic aneurysm in horses.

The first symptom in cerebral hemorrhage is shock, later passing on to coma and terminating in death. Animals that survive the first shock suffer from some degree of paralysis due to pressure on and damage to neurones. Hemorrhages may be found subdurally. They may also occur in the substance. of the brain. When hemorrhage is present in the ventricles, the cerebrospinal fluid may be blood tinged.

The blood clot in the brain first contracts separating the serum which is absorbed. The clot that remains is liquefied and a cyst is formed with a clear fluid—the "apoplectic cyst". The capsule of the cyst is formed by the neuroglia.

Hyperemia of the brain and meninges together with petechial hemorrhage and edema are found in the following conditions: (a) Electrocution, (b) Lightning stroke and (c) Sunstroke

Edema of brain : Edema of brain may be focal, caused locally by local lesions such as :

(a) Neoplasms ; (b) Trauma accompained by hemorrhage and laceration; (c) Meningitis ; (d) Focal necrosis ; (e) Cerebral and meningeal hemorrhages.

Generalised edema of brain may be found in; i) Diffuse meningitis; ii) Viral encephalitis ; iii) Enterotoxemia caused by *Clostridium welchii* iv) Lead poisoning; v) Poisoning by organic mercury compounds; vi) Shock; vii) ANTU poisoning; viii) Sunstroke ; ix) Causes that give rise to general edema of the body ; x) Salt poisoning in pigs.

Macroscopically the brain appears more moist and heavy. The gyri are widened while the sulci are narrowed. Swollen gyri that press against the skull appear flattened.

On section. the gray matter appears wider while the internal white matter is softer. The ventricles appear narrowed.

Microscopically, the white and gray matter appear to have a loose texture and the interfibrillar space is widened. Neurones and glia appear swollen. Edematous fluid accumulating around the space of Virchow–Robin (Perivascular space) widens these areas.

Note :— Swelling of the brain occurs as a post-mortem change (Autolytic change) especially in the young animals and so care is necessary in interpreting the condition.

HYDROCEPHALUS is the condition on which there is abnormal accumulation of cerebrospinal fluid in and around the brain.

In animals this is a congenital condition due to some error in development, obstructing the pathways of fluid passage. Vitamin A deficiency during intrauterine life may cause internal hydrocephalus in calves and pigs.

If the accumulation is in the ventricles, the condition is called *internal hydrocephalus*. But if fluid accmulation occurs in the sub-dural space or pia-arachnoid, it is cailed *external hydrocephalus.*

Internal hydrocephalus can arise whenever there is obstruction to the free passage of the cerebrospinal fluid. The obstruction can occur at the foramen of Monro, the aqueduct of Sylvius or the foramina of Luschka. Cysts, (hydatids and coenurids) tumors or inflammatory exudate are the usual causes for the blockade. Congenital narrowing of the lumina may also be a contributory cause.

Due to pressure of the accumulating fluid the ventricles dilate and the adjoining nerve tissue atrophies The cranium is greatly enlarged causing foetal dystocia. If hydrocephalus develops before the cranial sutures fuse, the cranial bone may grow to a large size.

External hydrocephalus results due to either too much fluid formed and not rapidly drained by the arachnoid villi or to hindrance to the drainge of normaily produced fluid as occurring in congenitally constricted tentorial aperture. It is usually the result of rupture of the thin dorsal wall of the third ventricle which allows the fluid to escape into the subarachnoid space between the cerebral hemispheres and the cerebellum The external hydrocephalus is called the "Communicating hydrocephalus" which is less common than the internal variety. The accumulated fluid exerts pressure on the surface of the brain. So there is a general atrophy of the brain and widening of the sulci between the convolutions. Usually. the exrernal hydrocephalus is "acquired". The result of hydrocephalus is pressure atrophy of the surrounding nervous tissue causing depression, incoordination, ataxia, and death.

DISTURBANCES IN GROWTH :

Aplasia : Aplasia of portions of the brain and spinal cord are observed in youug animals.

Hypoplasia : This is relatively more common than aplasia; certain important examplesbeing-congenital posterior paralysis in calves and swine, spastic paresis in cattle and cerebellar hypoplasia in pigs, dogs, cats, lambs. goats, and calves.

Cerebellar hypoplasia: This anomaly is seen in calves and cats mostly but may also be found in other animals. In the Jersey calves and cats, this defect is inherited. Animals may usually die shortly after birth. Those that survive for a short while, show locomotor disturbance and incoordination. At necropsy, eerebellum may be found to be rudimentary or even absent. Cerebellar hypoplasia was encountered in calves born of cows which were affected with Virus diarrhoea-Mucosal Disease while pregnant. Similarly modified hog cholera virus causes, (when used as a vaccine, in the dam) cerebellar hypoplasia in the fetal pig.

Microscopically, molecular and granular layers are reduced in size, There is a relative reduction in Purkinje cells.

Hypertrophy : This may result from increase in size of the glial cells, microglia showing the greatest degree of hypertrophy. The neuron does not increase in size-

Hyperplasia : This results from an increase in the number of glial cells. Glia, especially the microglia, increase in number under conditions of hypoxia. Hyperplasia of the neurons does not occur.

Metaplasia : This does not occur in the nervous tissue proper. It may occur in the connective tissue of the meninges and blood vessels, in which case cartilage and bone may be found.

Atrophy ; Atrophy of the cerebrum may occur in hydrocephalus. Pressure atrophy also occurs in the vicinity of tumors, abscesses, haematocysts, and depression fractures of the skull.

DISTURBANCES IN CELL METABOLISM ;

Cloudy swelling : The neurons and the glia undergo cloudy swelling as a result of hypoxia or irritation produced by toxic substances or infectious agents. The cells become larger, cellular outline more round, and cellular structures indistinct

Fatty degeneration ; This appears as fat droplets in the cytoplasm of the neurons.

Hpdropic degeneration : It is a continuation of cloudy swelling: In this droplets of edematous fluid are observed in the cytoplasm of the neurons and glia

Amyloid infiltration . This is uncommon in the central nervous system of most domestic animals.

Glycogen infiltration : This does not occur in the central nervous system.

Pigmentation : In cattle and sheep, melanin is most frequently encountered in the pia mater of the anterior one-fourth of the brain, Focal areas of melanin may be found in other portions of the meninges and even within the brain and spinal cord.

Calcification : This is more commonly found in the meninges than in the brain and spinal cord proper. It occurs in the presence of dead tissue and faulty circulation, examples being abscesses, infarcts, parasitic lesions, sites of old hemorrhage and in necrotic neurons.

Traumatic injury the Nervous system : The brain being soft, is susceptible to shock that emanates from impact, especially from fast moving objects.

A sudden blow on the cranium may result in fracture of the cranial bones which may not be depressed, On the other hand a blow on the vertebral column results in fracture or dislocation. Fracture of the skull causes considerable damage to the meninges and brain. Hemorrhage may occur and nerve fibres disrupted. Hemorrhage aggravates the condition by the pressure of the accumulated blood on the brain tissue.

Concussion occurs when the skull receives a sharp blunt blow suddenly, not accompanied by fracture. There is loss of consciousness. The condition is not fatal and no morphological changes are present. Recovery is the rule. Lesions may consist of small hemorrhages in the brain and under the skin at the site of injury

Laceration : In this condition there is discontinuity of the tissue and usually occurs in automobile accidents. Blunt objects may cause laceration and *contrecoup* lacerations occurs on the brain on the side opposite to that on which the injury is struck. This is due to striking of the brain on the skull on the opposite side, since normally the brain is smaller than the craninm and is slightly movable. In such places hemorrhages are common. Penetrating wounds, usually caused by gun-shot wounds. are followed by severe hemorrhage. Fractures are also common in such injuries. Penetrating wounds are usually followed by secondary infections and are fatal.

NECROSIS

1. Coagulative necrosis :

This involves the neurons and the glia   The causes are severe injury to the cells brought about by hypoxia, chemical poisons, bacterial toxins and viruses.

No changes are seen macroscopically. Microscopically, the cells are swollen and become more globular in shape. The Nissl substance may eventually disappear (chromatolysis or tigrolysis), the cytoplasm stains more intensely with eosin. and the nucleus shows pyknosis, karyorrhexis, or karyolysis. Microglia accumulate around the necrotic neurons, the process being known as **satellitosis**. When the microglia phagocytose the necrotic neuron, the process is called **neuronophagia**

2. Liquefactive necrosis :

This is the most common type of necrosis encountered in the brain and spinal cord compared to coagulative and caseoue types because the nervous tissue contains little coagulable albuminous material but is rich in lipoids In fact, necrosis of the brain is almost always liquefactive in nature. It so happens that when necrosis occurs in the nervous system, the autolytic enzymes released from lysosomes of the dead cells. cause disintegration of myelin into a liquid mass that consists mainly of lipoid

Infarction is one of the common causes of liquefactive necrosis.   It may also occur when the central nervous system is invaded by pyogenic bacteria. The lysosomal enzymes released from neutrophils induce liquefaction of myelin, neuroglia, and other structures, and this is known as *encephalomalacia*.   Softening of gray matter is known as **poliomalacia** and that of white matter **leucomalacia**. Encephalomalacia is commonly seen in the following conditions :

Deficiency of Vit E. in young chickens (crazy chick disease); Mouldy corn poisoning in horses (Cornstalk disease); Acute pancreatitis in all animals; Ante—natal copper deficiency in lambs (sway back);   Cobalt deficiency (Enzootic marasmus); Enterotoxemia in lambs; Mulberry heart disease in swine; Vitamin B deficiency (Chastek paralysis) in fur-bearing animals and in calves and sheep, when it is called cerebro-cortical necrosis; Blue tongue in sheep; Rift valley disease of Kenya; Distemper of dogs; toxoplasmosis and lead poisoning; Infarction due to an embolus consisting of tumor cells or parasites or a piece of a thrombus or due to thrombosis of an artery; poisoning by mecuric salts.

The lesions seen are:-thickening of the blood vessels, endothelial hyperplasia and liquefaction of brain substance. Thrombosis and hemorrhage may be found in some cases. Around the area, there may be proliferation of capillaries and the formation of a capsule by the cells of meninges. Astroglia proliferate and surround the area of encapsulation. The involved tissue undergoes liquefaction and a serous fluid is present.

3 Caseous necrosis :

The cause of this type of necrosis is infection of the brain by **Mycobacterium tuberculosis.** The necrotic area is seen as a dry, crumbly, yellowish - white mass. It may even contain areas of calcification. Microscopically, all cellular or architectural structures are lost, and the necrosed area is surrounded by a zone of inflammation.

Necrosis of nerve fibres of the peripheral nerves, the tracts and central nervous system is first indicated by fatty degeneration of the myelin sheaths of the nerve fibres affected. This change occurring in the brain and spinal cord is called demyelination. Ultimately the axon may disappear, but demyelination alone can render the nerve fibre non-functional. If demyelination alone has taken place, regeneration is possible with restoration of function.

Gangrene :

This could occur if brain is invaded by saprophytic microorganisms, as if the case of traumatic injuries of the skull or as septic emboli from areas on gangrene in the lungs.

INFLAMMATION

Terminology :

Encephalitis   — is inflammation of the brain
Myelitis   — is inflammation of the spinal cord
Encephalomyelitis — is inflammation of the brain and spinal cord
Meningitis   — is inflammation of the meninges
Pachymeningitis — is inflammation of the dura mater
Leptomeningitis — is inflammation of the pia mater
Meningoencephalomyelitis — is inflammation of the meninges, brain and the
                              spinal cord.
Polioencephalitis — is inflammation of gray matter in brain
Poliomyelitis — is inflammation of gray matter in the spinal cord.

The same general laws of inflammation apply to the brain and spinal cord as elsewhere. Since there are no mucous membranes, catarrhal inflammation does not occur. Serous inflammation also probably does not occur; if it does it resembles edema. Hemorrhagic exudates are rarely met with and fibrinous inflammation is limited practically to the meninges. Purulent, lymphocytic and proliferative inflammations are the types which are regularly encountered in the central nervous system.

Fibrinous encephalitis, myelitis and meningitis :

These are seen in cattle and sheep during pasteurella infection of the central nervous system They are characterized by cardinal signs of inflammation and increased fibrin content in the sub-arachnoid and Virchow-Robin space.

Suppurative (Purulent) encephalitis, myelitis and meningitis:

These are observed in all species of animals There are the usual cardinal signs of inflammation ; the principal constituent of the exudate being pus. The inflammation may be focal or diffuse.

The pyogenic organisms responsible are staphylococci, streptococci, corynebacterium, pasteurella, listeria, and pleuropneumonia-like organisms.

Routes of infection : 1. By direct extension from suppurative conditions of the middle ear, nasal passage, cribriform plate or from meninges.

2. Through blood stream (in septicemic diseases) and lymphatic vessels accompanying nerves (Listeriosis).

3. Infection of penetrating wounds of the skull. Suppurative myelitis will result due to infection of the wound made while docking the tail.

The lesions may be microscopic and consist of focal collection of neutrophils and lymphocytes. The accumulation of pus causes pressure and destruction of the local tissue. If an important area in involved, severe effects follow. The abscesses do not have well developed capsules as mesodermal cells that form it are few. Astroglia proliferate and form a poorly defined capsule around the cerebral abscess.

Listeriosis (Listerellosis)

This is the most frequent cause of a purulent reaction in the brain of farm animals. Infection by **Listeria monocytogenes** produces suppurative meningoencephalomyelitis in cattle, sheep and goats. How actually this organism reaches the central nervous system is not known. The disease is characterized by the presence of multiple microabscesses which contain the organism.

Lymphocytic meningencephalomyelitis :

This is the most important form of inflammation of central nervous system in animals. In this the lymphocyte is the principal constituent of the exudate : the cells being trapped in the Virchow-Robin spaces as they leave the vessels— **perivascular lymphocytic infiltration** or **perivascular cuffing**. This type of inflammation is caused mainly by viruses.

The viruses may be (1) Neurotropic. That means to say, those that affect almost only the nervous system. Examples of these are the Rabies in dogs and Borna disease in the horses.

(2) Organotropic. The viruses that affect other tissues may also infect the nervous tissue by chance. Examples : Canine distemper; Hog cholera; Epidemic tremor of fowls ; Malignant catarrhal fever and Rinderpest.

Encephalitis is said to be due to allergic causes, when no other obvious etiological factors can be discerned. Under this category are listed those conditions that result after vaccination (Post-Vaccinal encephalitis) with vaccines containing nerve tissue. Others that may produce encephalitis are the causal agents of psittacosis and ornithosis, PPLO, rickettsia and trypanosomes.

Routes of entry: 1. Blood stream—Hog cholera.

2. Nerves—Rabies.

3. Neurolymphogenous—infection ascending along the lymph pathways of cerebrospinal nerves.

The virus entering a neurone kills it. Demyelination of the nerve fibres may be present. Petechiae also may be found.

Macroscopically, no gross lesions of significance are noticed. Hyperemia and edema of the pia-arachnoid may be the only lesions seen. Occasionally localised areas of softening may be found.

Macroscopically, the following are the characteristic features noticed :

1. **Congestion and hemorrhages.**

2. Cuffing of blood vessels : i.e. accumulation of lymphocytes in the space of Virchow-Robin. Later plasma cells and macrophages may also be found in these places. This lesion is found both in the white and gray matter.

3. Edema.

4. **Gliosis :** diffuse proliferation of the astrocytes throughout the brain giving the tissue a dense and celluar appearance.

5. **Satellitosis**—appearance of scavenger cells or "Hortega cells" or "Gitter cells" around the necrotic area. These are the microglia and oligodendroglia which remove the dead neurones and debris by neuronophagia. The gitter cells contain lipid and are the only cells seen in the area where neurones were situated previously. Some times in more chronic cases, the microglia increase in number, their nuclei become elongated and their cytoplasam contains iron deposits and such cells are known as "Rod cells"

6. **Neuronophagia.**

7. Rarely proliferation of blood capillaries may te seen.

8. Inclusion bodies may te found in the neurones or astroglia in a **number** of diseases ; rabies, canine distemper, infectious canine hepatitis, Borna disease.

SPECIFIC INFLAMMATIONS OF BRAIN AND SPINAL CORD :

Rabies :

It is an acute viral disease of domestic animals characterized by a very severe lymphocytic inflammation of nervous system. There is diffuse and severe meningoencephalomyelitis. A characteristic feature of the disease is the presence of intracytoplasmic inclusion bodies (Negri bodies) in the cells of the hippocampus and cerebellum.

Pseudorabies :

It is an infectious viral disease of cattle, pigs, dogs. and cats. In pigs, there is diffuse lymphocytic meningoencephalomyelitis which does not seem to occur in cattle.

Hog cholera (swine fever) encephalitis :

In hog cholera 80 to 90 per cent of the animals suffer from an acute diffuse lymphocytic meningoencephalomyelitis.

Canine distemper encephalitis :

The virus of canine distemper also produces a typical diffuse lymphocytic meningoencephalomyelitis. However, not all the affected dogs develop lesions in the central nervous system.

Infectious viral equine encephalomyelitis :

This is an acute viral disease of horses and mules and is also characterized by a typical diffuse lymphocytic meningoencephalomyelitis. It terminates fatally in about 50 per cent of the cases.

Borna disease :

This is an acute diffuse viral meningoencephalomyelitis of the horses that occurs in Europe, especially in Germany. It has not been described in India.

Louping ill :

This is an acute diffuse viral lymphocytic meningoencephalomyelitis of sheep in Scotland, England, and Ireland.

Avian viral meningoencephalomyelitis or **Epidemic tremors :**

It is an infectious viral disease of chicks. Microscopically, the disease is characterized by a diffuse lymphocytic inflammation of the entire central nervous system.

Ranikhet disease (Pneumoencephalitis of poultry) :

It is a viral disease of chickens which is also characterized microscopically by diffuse meningoencephalomyelitis.

Chronic meningoencephalomyelitis :

The only important example of this in domestic animals is that of tuberculous meningoencephalomyelitis. The lesions in the central nervous system consist of a central area of caseous necrosis which may be partially calcified. This indicates that a generilized tuberculosis is present.

PARASITIC ENCEPHALOMYELITIS

Myiasis :

Hypodermo bovis larvae may be found in the fat of the vertebral canal in cattle. At times they may invade the spinal cord or the brain. The larvae of *Oestrus ovis* have also been reported to invade the brain.

Cestodiasis:

Tapeworm cysts are found in the central nervous system of the domestic animals, examples being the cysts of *Multiceps multiceps*, *Taenia pisiformis* and *Taenia echinococcus*. The ova of these tapeworms are ingested by the animal. In the intestine, the hexacanth embryo comes out, pierces the intestine, and is carried by blood stream to various places in the body. Some of the larvae, particularly those of *Multiceps multiceps*, reach the brain where they encyst. The path of migration of the larvae in the brain is macroscopically visible as red streaks due to the persence of hemorrhage.

The larval stage of *Multiceps multiceps*, a dog tapeworm is known as *Coenurus cerebralis*. It causes a rather uncommon disease of the central nervous system of sheep, known as '*gid*' or '*sturdy*'. The symptoms depend on whether the bladderworms are located in the brain or spinal cord. In both the places, they form cysts reaching 50 mm in diameter or more. Each cyst is filled with clear fluid and contains even up to 500 scolices. As the cyst enlarges there is pressure atrophy of the surrounding nervous tissue. So the convolutions may be flattened and cortex becomes thinned Even the cranial bones may be subjected to pressure atrophy and some may be punctured even. The chronic irritation induces a chronic lymphocytic meningitis, encephalitis, or myelitis, depending on the location of the parasite. In severe infection, an acute diffuse lymphocytic meningoencephalomyelitis is produced, and the animal dies. Cysts usually involve the lumbar portion of the spinal cord, which results in such symptoms as inco-ordination and paralysis of the posterior extremities. Death of the larvae will result in calcification of the cyst.

Nematodiasis :

Various nematode larvae are found in the central nervous system. Strongyle larvae are found in the horse. The larvae of ascarids, strongyloides, hookworms and microfilaria of *Dirofilaria immitis* may be found in the capillaries or in the nervous tissue of the brain, spinal cord or meninges. Their lesions are characterized by a chronic lymphocytic inflammation.

Cerebrospinal nematodiasis (Neurofilariasis; Kumri);

This is found in sheep, goats and horses, and is caused by *Setaria digitata*. n horses it is known as Kumri (Hindustani for 'weakness of the loin') in our country, whereas in sheep and goats it is known as 'lumbar paralysis'. Besides India, the disease also occurs in Srilanka, Burma, Korea and Japan

Setaria digtiata is a natural parasite of cattle. But when the microfilaria find-entry into heterologous hosts, like sheep and goat, they wander away to the central nervous system. Symptoms are basically neuroparalytic and include motor weakness, inco-ordination and loss of balance. Severe cases exhibit par-esis of one or all limbs, the hird limbs being most freqnently and severely affected. The cases may terminate fatally, or recovery may follow.

Macroscopically, lesions are found in the brain and spinal cord and the severity depends on the number of parasities present Narrow tortuous tracks of hemorrhage and softening may be found den ting the path taken by the parasite,

Microscopically, the lesions consist of a central space (where the parasite was) surrounded by a degenerated and necrotic tissue The necrosis is liquefac-tive in type. Hemorrhages may be present in this area Due to damage by the larvae the axis cylinders in the affected area are swollen and degenerated and appear enlarged and fragmented. The myelin sheath becomes swollen and distor-ted, accompanied by glial proliferation.

Lymphocytes, eosinophils and microglia infiltrate around the area. Perivas-cular cuffing of nearby vessels is observed. The larvae may not be visible in these lesions as they might wander off. Careful microscopic examination of the tissue and cerebrospinal fluid is neceesary to see the larvae.

The condition known as "Wobbles" among young horses and mules may be a manifestation of cerebrospinal nematodiasis, Animals one to two years of age are affected. Suffering animals move with difficulty and on motion sway (or wobb'e) from side to side. They may fall frequently and show difficulty in rising. Trauma of the spinal cord at the cervical region is suspected to be the cause- by some authors. Weakness inherited genetically is also suggested. Lesions are found only in the cervical portion of the cord where tracts of hemorrhagic necrosis and bilateral symmetrical areas of malacia are found

Microscopically, the following are noticed. liquefactive necrosis, hemorrhage, degeneration and demyelination of the peripheral nerve tracts, perivascular cuffing, satellitosis, neuronophagia and gliosis.

In some cases, though parasites may be found in the central nervous sytem neither symptoms nor lesions are observed.

Toxoplasmosis :

Toxoplasma gondi, a protozoan parasite is found in the central nervous system of domesticated animals. Lesions consist of a central area of coagulative necrosis, surrounded by microglia and neutrophils. There is also lymphocytic meningitis, lymphocytic perivascular cuffing, and gliosis

'Allergic' encephalitis (Postvaccinal encephalitis) :

This sometimes occurs in dogs following rabies vacccination, It occurs 2 to 3 weeks after vaccination, and is characterized by a lymphatic meningoenceph-alomyelitis. There is motor paralysis of one or more limbs, which may later involve most of the body. Death is the usual outcome.

MENINGITIS

Pachymeningitis : The inflammation of the dura mater is usually secondary to infection of the middle ear or adjacent bone. It may be surpurative or non-suppurative In the suppurative variety which is more common, local abscesses may be found on the dura and the peridural spaces. Subsequently chronic fibrosis may develop when the dura is thickened with local adhesions Infection may spread to the arachnoid causing leptomengitis.

Leptomeningitis is the inflammation of the pia-arachnoid. When associated with inflammation of the brain, which is usually the case, the condition is known as meningoencephalitis.

Leptomeningitis may be suppurative or non-suppurative. The causes are:-

1 Extension from adjacent tissues—as in viral encephalitis (swine fever, rabies etc).

2. Mechanical injuries—fractures.

3. Batcterial infection from neighbouring areas—middle ear, nasa cavity and sinuses (Usually *Streptococci* and *Staphylococci* : other bacteria that may cause leptomeningitis are :— *Listeria, C. pyogenes, Pseudomonas,* coliforms, Pasteurella ; Chronic meningitis is produced by *Toxoplasma, Mycobacterium tuberculosis* and *Cryptococcus* In swine *Leptospira pomona* causes non-suppurative meningitis).

4. Hematogenous infection in septicemic conditions—navel ill; enz otic pneumonia, Colibacillosis, purulent pneumonia, metastasis from infections such as mastitis, metritis or peritonitis.

5. Parasitic invasion as in multiceps infection in sheep.

6 Hemorrhagic meningitis is seen in acute lead and copper poisonings. Due to the movement of the C. S. F. the inflammation is usually diffuse.

Hyperemia is severe in meningitis. The causative organisms grow on the surface of the pia-arachnoid and in its spaces. Injury to the blood vessels is responsible for the inflammatory exudate. In suppurative meningitis, the exudate which is yellow or greenish accumulates in the pia-arachnoid space. Pus may also be found in the spinal fluid. When it accumulates in the lateral ventricles, the convo'utions tend to be flattened. Due to gravi'y the inflammatory fluid collects at the base of the brain. Suppurative inflammation is characterised by the infiltration of neutrophils while mononuclears (lymphocytes and macrophages) predominate in the non-suppurative variety.

Examination of the spinal fluid collected from a lumbar puncture gives valuable information as to the nature of infection.

Myelitis is inflammation of the spinal cord. Usually myelitis is found along with encephalitis when the condition is known as encephalomyelitis Myelitis may be suppurative or non-suppurative and the lesions are comparable to those of the brain. Sometimes non-suppurative myelitis may occur without any attributable cause. In such cases trauma (automobile accidents) are believed to be the cause. Fractures of spinal column and protrusion of inter-vertebral disc may be other causes.

Macroscopically, there may be congestion of the pia, petechiae on and inside the spinal cord and in advanced cases, softening of the nervous tissue.

Microscopically, congestion, infiltration by inflammatory cells and degenerative changes of the nerve cells are seen. The nerve cells are swollen. Nissl substance disappears and the nucleus assumes an eccentric position. Degenerative changes of the nerve fibres are also observed.

THE PERIPHERAL NERVES

Degeneration : When a nerve cell undergoes degeneration due to action of an irritant, the degenerative process also affects the nerve fibre of that cell. This is known as descending degenaration. Degeneration can also begin in the nerve fibre and progress towards the nerve cell (ascending degeneration). **Microscopically,** both axis cylinder and myelin sheath are simultaneously involved (total degeneration). Loss of the myelin substance is called **demyelination.** Lipoid of the myelin can be stained and demonstrated by Marchi's method.

When a nerve fibre (axon) gets severed from its cell body, the distal part of the nerve fibre undergoes characteristic degenerative changes known as **Wallerian degeneration.** The axis cylinder disintegrates and disappears, the myelin sheath (medullary sheath) also degenerates and is transformed into a chain of lipoid droplets which can be stained black by Marchi's method. The cells of the sheath of Schwann proliferate and get converted into phagocytes which remove the remnants of axis cylinder and the lipoid droplets. Similar change occur in the proximal part up to the first node of Ranvier.

Regeneration : In a degenerated nerve fibre there are also attempts at repair. Nerve fibres in the central nervous system cannot regenerate, but the peripheral nerves regenerate fairly rapidly. Schwann cells play a leading role in the healing of nerves. If the sheath of Schwann (neurilemma) is intact, the Schwann cells proliferate and arrange themselves in both proximal and distal ends in the form of a tube. Along this tube new axis cylinders grow and unite the two severed ends. They fail to heal the gap if it is more than one inch. In such cases, the gap is filled in by granulation tissue which originates from the three connective tissue coverings of the nerve and its bundles. The Schwann cells proliferate at both ends. In case of amputation, the axon fibrills coil up and form a nodule called *amputation neuroma* which is covered by fibrous tissue.

When a peripheral nerve is cut degenerative changes occur in the neurons. These changes are called *Nissl's Degeneration* in which the cells become enlarged and the nucleus is pushed to a side (eccentric) Chromatolysis of Nissl's substance occurs after breaking up. When regeneration of the nerve fibre starts the neuron tends to return to normal. Nissl granules reappear, nucleus takes up a central position and the cell becomes smaller. Repair in a nerve fibre is a prolonged process requiring 10 to 12 months for complete healing. Repair of the nerve fibres of central nervous system, lacking a sheath of Schwann, does no occur.

Neuritis is inflammation of the peripheral nerves Inflammation of the nerves is usually accompanied by degenerative changes.

Causes : 1. Trauma.
2. Toxins—bacterial mostly. Neuritis occurs in infectious diseases as in strangles, protozoal—dourine, viral—rabies and distemper
3. Chemical poisons—lead, mercury, arsenic, alcohol.
4. Plant poisons—*Lathyrus sativus.*

5   Nutritional deficiency—deficiency of the members of vitamin B group.

6.   Allergic factors.

7.   Viruses—Marek's disease, Ranikhet disease.

Macroscopically, the nerve may be swollen and reddened. More often no naked eye changes are noticed.

Microscopically, degenerative changes, even leading to Wallerian degeneration are found. Edema and infiltration by inflammatory cells of *interstitial connective tissue* can be seen. The exudate may be serous (serous neuritis) or purulent (purulent neuritis). The latter variety may destroy the nerve completely.

Gross evidence of any abnormality may be completely absent. In some cases, however, nerves may reveal mild congestion or they may be unusually swollen, soft or flabby. **Microscopically**, inflammatory changes are present in the connective tissue and degenerative changes in the nerve fibres. All degrees of changes are met with, the more severe forms leading to destruction of the nerve fibres and Wallerian degeneration in the distal portion of the fibres.

Marek's disease :

It is a lymphoproliferative disease of the domestic fowl which has an unusual predilection for peripheral nerves. It occurs in classical and acute forms. The classical form is characterized by peripheral nerve enlargement, and paretic and paralytic symptoms. The nerves that are commonly affected are the brachial and sciatic plexuses, coeliac plexus, abdominal vagus and intercostal nerves. Microscopically, the nerves are infiltrated with lymphoid cells, which include primitive and activated reticular cells, lymphoblasts and small, medium and large lymphocytes

CONGENITAL MYOCLONIA OF PIGS (TREMBLES)

New born pigs are affected by this condition, which is characterised by clonic convulsive movements due to hyperirritability of muscles. After a few weeks the symtoms may disappear and the pigs may then thrive. In severe cases they die of inanition as the piglets are unable to suckle properly.

Cause : Definite cause is not known. The following are suspected : heredity factors, hypothyroidism, virus infection, defective care and deficiencies of the sow during pregnancy. Many workers feel that the last two may be important

Macroscopically, no visible changes may be noticed. Edema, thickening and hemorrhage in the cerebellum may be noticed. Congestion and hemorrhage may be observed in the brain, lymph glands, liver, kidney, lungs, spleen, thymus and ocular muscles.

Microscopically, there may be delay in the myelin sheath formation in the spinal cord. Ganglion cells of corpus striatum may reveal changes of shape and vacuole formation. Vasculitis affecting small arteries (sometimes obliterating them) is seen in various organs.

EPILEPSY

Epilepsy is a sudden brief (petit mal) or prolonged (grand mal), loss of consciousness usually preceded by convulsions.

Symptomatic epilepsy : may occur in animals due to organic brain lesions such as neoplasms or inflammation or trauma; disturbances in brain metabolism due to visceral pathology or metabolic diseases or poisons; cerebrospinal nematodiasis ; verminous infestations or profound toxemias.

True or Idiopathic epilepsy is an inherited condition in Brown Swiss cattle and Cocker spaniels. The inheritance is through a recessive factor. Between attacks the animals are perfectly well and the condition persists for life.

Symptoms : A true grand mal epileptiform seizure is manifested by an early period of alertness, followed by a state of tetany, which gives way after a few seconds to a clonic convulsion with padding, opisthotonus. champing of jaws and salivation. The clonic convulsions are followed by a period of relaxation. The convulsion may spread from the initial area to the rest of the body, which is referred to as Jacksonian epilepsy. The animal is unconscious throughout the seizure. Evacuation of the bladder or bowel or both is common during the seizure. The animal may quickly regain its normal state after the seizure or act dazed or uncoordinated for a few minutes The temperature may be elevated or normal. The pulse is frequent and respiration rate is increased. The blood, cerebrospinal fluid and urine are normal.

The attcks are always recurrent and the animals are normal in the intervening periods.

Symptomatic treatment may be administered. Sedatives are usually resorted to.

Control : The animals should not de used for breeding.

SWAYBACK OR ENOZOOTIC ATAXIA

The disease is seen in new born lambs in certain parts of the world. The symptoms noticed are severe ataxia, locomotor disturbance, paralysis and inability to walk. Affected animals may be blind and so are unable to move. Death may also be due to broncho-pneumonia (exposure).

Swayback is attributed to a deficiency of copper. The ewes which are maintained on a copper deficient diet or grazed on lands with molybdenum-rich grasses may manifest anemia and produce "Steely" wool. Lambs of such ewes show demyelination (see page 149) and suffer from "Swayback".

Macroscopically, lesions are not prominent in mild cases. But in severe cases, cavities containing gelatinous material may be found in the white matter due to liquefactive necrosis with secondary internal hydrocephalus. The lesions are bilaterally symmetrical. Flattening of cranial bones occurs due to cystic degeneration and increased intracranial pressure.

Microscopically, diffuse symmetrical destruction of the white matter in the cerebrum is noticed, which is liquefactive necrosis. There is destruction of descending myelinated tracts. Gitter cells are numerous in the area. There is reduction in the cytochrome oxidase activity of neurones.

Tumors (Neoplasms)

Primary tumors of the brain and spinal cord are rare. However, tumors of the central nervous system are most common in dogs and least common in the pig and sheep. Primary tumors include those of the neuroglia (gliomas astrocytoma, oligodendroglioma), nerve cell and fibres (neuromas), ganglion cells (ganglioneuroma), ependymal cells (ependymomas), and of meninges (meningiomas). Central nervous system is also prone to secondary tumors, which are metastatic, their primary sites being the lung or some other organ.

The brain tumors are of limited malignancy, metastases not occurring elsewhere. Pressure on the brain by the developing neoplasms produces various symptoms depending upon the part of the brain involved and the functional disturbances in turn are dependent on the neurons of that part which are responsible for them. Death is the invarible outcome.

CHAPTER 20

THE REPRODUCTIVE SYSTEM
(Revised by Dr. P Rama Rao)
THE FEMALE GENITAL SYSTEM

Developmental anomalies
  Freemartin
  Arrests in the Mullerian duct system
    White heifer's disease
  Intersexes
Ovary
  Developmental anomalies
  Disturbances in growth
    Aplasia
    Hypoplasia
    Polyoogonia
  Hemorrhage
  Perioophoritis
  Oophoritis
  Cystic ovaries
    Follicular cysts
    Leutin cysts
  Tumors
Bursa
  Bursitis
  Adhesions of bursa
Fallopian tubes
  Malformations
  Hydrosalpinx
  Salpingitis
  Pyosalpinx
  Tuberculosis
  In birds
Uterus
  Malformations
  Rupture
  Malposition
    Torsion
    Hernia
    Prolapse
  Circulatory disturbances
    Hyperemia
    Hemmorrhage
    Thrombosis
  Atrophy
  Hypertrophy
  Metritis

Acute catarrhal (endometritis)
Chronic non-suppurative
Acute suppurative
Chronic suppurative
Sclerotic
Perimetritis
Tuberculosis
Brucellosis in pigs
Mucometra/Hydrometra
Uterine abscesses
Endometriosis
Mummifiication of fetus
Maceration of fetus
Abortion
Cervix
  Malformation of cervix
    Double External Os
    Hypoplasia of cervix
    Tortuosity of cervical canal
    Cervical dilatation and diverticula
    Prolapse of cervical rings
  Cervicitis
  Mechanical injuries
  Cysts
  Neoplasms
Vagina
  Developmental anomalies
  Cysts
  Rupture
  Vaginitis
    Granular
    Vesicular
    Epivag
  Pneumovagina
  Tumors
Oviducts of birds
  Abnormal eggs
Mastitis
  In bovines
  In ewes
  In pigs

421

CHAPTER 20

DEVELOPMENTAL ANOMALIES

Freemartin

The bovine freemartin is a genetic female born cotwin with a normal male with which it has exchanged whole blood. The structural modifications of female genitalia are supposed to result from the influence of androgenic hormones produced by the male foetus.

The gonads are undifferentiated. Ovaries are small. The Mullerian duct system is not differentiated fully. There are usually portions of tubular system, which do not develop and often the uterus is small, and incomplete. The vulva has frequently long tufts of hair. The clitoris is quite prominent the vagina. is fairly developed and the cervix is usually absent. One feature of the reproductive tract, which is very useful in distinguishing this condition from severe cases of aplasia of Mullerian ducts is the presence of seminal vesicles. Epididymis may be present or absent. The histologic appearance of the gonad is one of a quite undifferentiated structure. There are small tubular structures resembling primitive seminiferous tubules with lining cells similar to the sertol. cells. There are interstitial cells, which in the new born freemartins resemble fibroblasts. In the freemartin which is allowed to live the age of one or more years, the interstitial cells develop and resemble luteal cells in the ovary or Leydig cells of the testis. In the older animals, these develop into multiple large masses of orange or tan colored masses which resemble both interstitial cell tumor or corpora lutea on gross examination. Most freemartins do not develop ovarian follicles. Endometrial glands are present and produce fluid resulting in cystic distension of vestigeal remnants.

The seminal vesicles are usually small and have abundant fibrous stroma. The epithelium resembles that of seminal vesicles of a castrated bull.

Arrests in the Mullerian duct development

These defects are of significance only in cattle and swine.

White heifer's disease

This condition is seen more commonly in short horn cows due to arrest in the Mullerian duct system and consists of a number of abnormalities.

Depending on the intensity of arrest in the development it may be classified into three groups.

Group A — Hymenal constriction; absence of anterior vagina, cervix, uterine body; cystic dilatation of uterine horns. Presence of well marked Wolffian bodies and occasional submucous vaginal channels.

Group B — Uterus unicornis, the abnormal horn being present as a flat muscular band. Hymenal constriction may or may not be present.

Group C — Essentially hymenal constriction and the rest of genitalia comparatively developed. If constriction is complete, gross uterovaginal distension results.

In addition to aplastic or hypoplastic defects of Mullerian ducts, there are common anomalies which results from failure of fusion of caudal portions of the ducts. Complete failure of fusion results in double vagina and double cervix. The more common failures of fusion occur in or adjacent to the cervix. The anterior vagina may be partitioned by a dorsal septum in conjunction with a double cervix. A dorso-ventral band may be present across the external os (double external os), the cervix and vagina being properly fused. The failure of fusion may involve only a part of the cervix, chiefly the caudal part, so that there is one uterine body and bifurcated cervical canal with duplication of external os. The cervix and uterine body may be completely divided, a condition known as **uterus didelphys**.

Intersexes. The intersex is an individual with congenital abnormality, where the diagnosis of the sex is confused. Intersexes may be of two types; 1) True hermaphrodites in which gonads of both sexes are present; 2) Pseudohermaphrodites having gonads of one sex only but possessing reproductive organs with some characteristics of the opposite sex. Male and female pseudohermaphrodites depending on the gonads present are recognised.

Pseudohermaphroditism is very common in goats and studied in detail from genetic point of view. It is caused by a recessive gene The incidence of hermaphroditism in Sannen breed of goats is high and mostly present as male pseudo-hermaphroditism.

Intersexes are common in pigs but not to the same degree as in goats. In bovines intersex seems to be a rare condition.

THE OVARY

Placed in the abdominal cavity, the ovaries are well protected from exogenous causes.

The functional activity of the ovary is under the control of the anterior pituitary through the two hormones, Follicle Stimulating Hormone (FSH) and the Luteinising Hormone (L H) Hence any pituitary endocrine disurbance affecting the gonadotropin levels affect the ovaries considerably. Placenta also elaborates certain hormones which have considerable effect on the ovarian function. Ovary produces two hormones, estrogen and progesterone, which have very important functions on the development of tubular female genital organs and control of estrus cycles.

DEVELOPMENTAL ANOMALIES.

Supernumerary and accessory ovaries are very rarely seen in cow. Supernumerary ovary is an extra gonad which is entirely separate from the normally placed gonad and appears to arise from a separate analogue. An accessory ovary is situated near the normally placed gonad and may be connected to it giving the impression that it developed from a normal ovary.

Anomalies in the position of the ovaries occur and vary according to the length of the broad ligament. It is not unusual to find cows with a short broad ligament on one side resulting in the ovary being located closer to the body wall than normal In cows with uterus unicornis the gonad of the affected side is generally located near the body of the uterus. On the other hand in bitches with uterus unicornis, the ovary may be located some distance from the uterus. In such cases, the ovary may be overlooked during routine ovariohysterectomy.

DISTURBANCES OF GROWTH

Aplasia or absence of ovaries is occasionally seen, especially in swine and sheep

Hypoplasia of the ovary is considered to be due to the failure of migration of primordial germ cells from the yolk sack to the developing gonad during embryonic stage. Thus the developing gonad (i.e. the ovary) becomes devoid of germinal epithelium, which is the precursor for the follicular system Both ovaries may be affected, or sometimes only a single ovary or a part of the ovary may be affected In the Swedish Highland breed af cattle, hypoplasia is determined to have been caused by a single recessive autosomal gene.

The ovaries are small and rudimentary in the form of a thin band with wrinkled rough and irregular surface similar to that of new born calves.

In ovarian hypoplasia gonads consist of predominantly medullary tissue. Tunica albuginea is thick and covered by low cuboidal epithelium. Follicles are completely absent The stroma is dense and made up of thick fibrous tissue with several anovular cords of Type I and Type II, blood vessels and rete tubules.

Polyoogonia is a condition in which each follicle, which normally contains only one ovum, may contain several ova without disturbing the function of genital organs.

HEMORRHAGE occurs : 1, during ovulation (of no consequence) 2 while enucleating corpus luteum manually in the cow, 3. while expressing ovarian cysts in the treatment of sterility and 4, in hens affected with avain leucosis complex, pullorum disease and fowl pest.

PERIOOPHORITIS

Usually chronic in nature and is more common than oophoritis. It is more commonly a localised serositis which is seen as red fibrin and serosal tags containing numerous leucocytes attached to the surface of the ovary. Fine bursal adhesions may result but are too delicate to interfere with ovulation.

Serosal granulomas may occur in bovine peritoneal tuberculosis or porcine brucellosis or in setariasis in bovines. These appear as reddish nodules or tags. These infective granulomas remain strictly localised to the surface of the ovary and do not penetrate its substance.

Grossly the ovarian surface is shaggy and often encapsulated with adhesions of bursa

Microscopically, the tunica albuginea is thickened by fibrous tissue with occasional lymphocytic and plasma cell infiltration.In tuberculosis, ovaries may be covered by tuberculous granulomatous tissue with Langhans' type of giant cells In Setariaris, granulomatous foci with sections of larvae along with neutrophils and surrounded by mononuclears, epithelioid cells and fibrous capsule may be seen.

OOPHORITIS : (Inflammation of the ovary) is rare in animals. Sometimes abscesses may be found in or about the capsule. They are usually the result of an ascending infection from the oviduct or uterus.

Peritonitis may involve the ovaries when adhesions may occur. This is especially true in the case of bovine tuberculosis of the peritoneum and brucellosis of swine, when infective granulomata may be found on the surface of the ovaries.

CYSTIC OVARIES

The ovaries contain one or more cysts of varying size. Though met with in all animals, cysts of the ovary are more frequent in the cows, sows and mares.

Cysts are found in high milk yielding cows more frequently and so endocrine disturbance is the main cause. Great enhancement in the milk yield by selection has resulted in more and more cows suffering by this disorder.

Follicular cysts : In this condition, the graffian follicle does not rupture as it should normally and so liquor folliculi accumulates and so the cysts enlarge upto as much as 11 cms. in diameter.

The granulosa cells which are normally of several layers degenerate as atretic bodies. In most of the follicular cysts a single layer of granulosa cells are left appearing as a string of pearls lining the antrum. In some, even the single layer of granulosa cells disappears leaving the membrana propria to line the antrum. The cumulus oophorus and the ovum degenerate. However, the glandular cells of Theca interna continue to be secretory adding to the estrogenic pool.

The incidence of cystic ovary is high in high lactating animals immediately after parturition. Defective feeding and faulty animal husbandry practices may be some of the contributory factors. There may be an involvement of hereditary predisposition.

The cause of the cystic ovaries is considered to be failure of release of LH or failure of release of LH in sufficient quantities to cause ovulation. It may also be due to imbalance of FSH and LH.

Changes have been noticed in the basophil cells of Hypophysis which are believed to elaborate the gonadortophic hormones. The cells assume a bigger and bizarre shape, with large nuclei and large nucleoli. The basophilic granules are at first large and numerous. Later, the cytoplasm becomes clear, homogeneous and later even becoming acidophilic. Side by side with these changes, the acidophils become hypertrophied and densely laden with granules which produce growth hormone and prolactin.

The following extra-ovarian lesions are observed :
1. Relaxation of the sacro-sciatic ligament.
2. Uterus enlarged and edematous
3. Cervix—enlarged with patent os
4. Endometrium—cystic endometrial hyperplasia, "swiss-cheese" type. The hyperplastic glands secrete excess of mucin, producing retention cysts.
5. Vagina and vulva are edematous.
6. In the dog mammary tumors and uterine fibroids
7. Enlargement of the thyroid gland and so hyperthyroidism
8. Increase in the width of the zona fasciculata of the adrenal gland.

Clinically, nymphomania (persistent sexual desire) is observed in cows and bitches.

Lutein cysts: Normally, after ovulation, corpus lutuem forms from the proliferation and luteinization of the cells of theca interna and the follicular epithelial cells.   A small central cavity is present.   But in lutein cysts, there is abnormal accumulation of fluid in this cavity.   These are more common in cows and sows than in other animals.

The cause is probably non-release of adequate quantities of luteinising hormone by the anterior pituitary.

Increased production of Progesterone by these cysts renders the uterus susceptible to infection, pyometra being the usual sequel.

The ovary is large, round or oval and soft.   The corpus luteum is not discernible on the surface as in the case of normal one.   The cyst has a narrow internal lining of yellowish-brown luteal tissue and contents are opalescent, light yellow and gelatinous.

Microscopically, the cyst wall comprises of three layers surrounding the central cavity containing homogneous contents.   The inner layer consists of a thin band of loose connective tissue separating the adjacent luteal tissue from cystic contents.   The middle layer has varying thickness of lutein tissue.   The outer layer consists of concentrically arranged dense bands of connective tissue merging with ovarian stroma.

Intra-follicular luteinisation

Formation of luteal tissue within the follicles even before rupture is called as intrafollicular luteinisation.   Usually the ovaries are of medium size.

This condition is noticed only on histopathology.   The follicle is located in the centre of ovarian stroma.   Tunica theca is thick and lined by several layers of granulosa cells.   In the centre there are irregularly arranged lutein cells with large vacuolated cytoplasm and large nucleus having scanty chromatin.

It may probably be due to hormonal disturbance between FSH and LH and requires further investigation.

Embedded corpus luteum :

Small encapsulated yellowish-brown lutein tissue is located in the middle of cortical stroma and associated with endometritis.

The embedded corpus luteum has lutein cells of normal appearance but the fibroblastic proliferation forming irregular masses is conspicuous

The embedded CL seems to be significiant by way of inhibitting the oestrum resulting in anoestrous condition.

Small and Sclerosed ovaries :

This is the commonest abnormality noticed in ovaries of buffaloes.   The ovaries are small with smooth surface.   Neither follicle nor corpus luteum is apparent on the surface.   The cut surface reveals dense stroma with no developing follicles.

Microscopically :   The surface epithelium is usually absent.   Tunica albuginea is thickened with dense fibrous tissue.   The cortex is reduced in thickness. Developing follicles are completely absent and a few atretic follicles may be seen. The stroma is dense with thick fibrous strands runnig in different directions. Aggregates of thick walled, closely packed capillaries are seen in the stroma.

Different views are put forth for the development of this condition Subactivity or inactivity could be due to hypofunction of thyroid as evidenced by low blood levels of thyrotropic hormone in buffaloes with small ovaries. An imbalance of gonadotrophic hormones is suggested. Nutritional error might have a significant role.

Epoophoron : These consist of intercommunicating, short, closely packed. acinar structures in the loose connective tissue of mesovarian attachment at either poles of ovary. The lumen is narrow and lined by cuboidal to columnar epithelium with large lightly stained nucleus. These have the origin from anterior mesonephric tubules.

Rete ovarii : Consists of tubular net work of anastomosing canals, separated by thick bands of connective tissue at the hilus. These tubules are lined by low cuboidal to columnar epithelium with round or elongated lightly stained nucleus. At times the epithelium may be hyperplastic and assume adenomatous appearance. Rete ovarii is seen prominently in the old ovaries. The origin is from mesonephros.

Anovular cords : These are seen in the ovaries of bovines as scattered or in groups in the ovarian stroma. Anovular cords are probably originated either from groups or nests of epithelial cells which never had oocytes from normal follicles in the early stages of development replacing the follicles.

Three types of anovular cords are recognised.

Type I anovular cords : These are elliptical and surrounded by a thin layer of PAS positive membrane. The cords are filled wth 3 to 4 rows of irregularly arranged epithelial cells with no ovum. The nucleus of the cells adjacent to basement membrane is oval with diffuse chromatin. The cytoplasm is stained light and contains a net work of very thin eosinophilic fibrils.

Type II anovular cords : These are elliptical or round and slightly larger than type I cords. The cells are arranged in one or two layers of cells. The lumen contains a moderate amount of PAS positive material.

Type III anovular cords : These are larger than Type II, the diamater reaching 200 microns. There are two layers of epithelial cells with PAS positive amorphous substance in concentric layers in the lumen. The connective tissue around the anovular follicles is arranged in several circular layers and in these epithelioid and eosinophilic cells are often found.

Presence of anovular cords is directly proportional to the severity of hypoplasia of ovaries. In very severe cases Type II and Type III forms are seen. In less severe cases Type II and Type I cords and in early cases of hypoplasia Type I cords are predominantly seen.

Folliculoids : These are seen in ovaries on microscopic examination and are present in one or both the ovaries of aged bovines. Two types of folliculoids are recognised.

1) Trabecular type
2) Colloid type

Trabecular type : These have distinct connective tissue PAS positive capsule with the invaginations of septa into the lumen dividing it into smaller cavities. The septa are lined on either side by single or double layers of granulosa-like

cells.   These cells are elongated having large vesicular nucleus and scanty cytoplasm.   Several rossette-like structures consisting of eosinophilic irregular bodies
surrounded by radially arranged single or double layers of cells are present in the
cavity.   These structures have a resemblance to Call-Exner bodies, characteristic
of Granulosa-cell tumor but the origin and morphogenesis of rossettes is different

Colloid type : A few irregular shaped PAS positive colloid bodies are
characteristically seen in the lumina of solitary folliculoids. The cellular elements
are few. Two types of colloid bodies are seen.   One type is large, irregular in
size and shape with a laminated appearance.   The other type is small and spherical with homogenous structure.   Both the types are surrounded by a single layer
of granulosa-like cells.

The common association of anovular cords with folliculoids and their close
morphological resemblance suggest that the anovular cords might be precursors
of the folliculoids. Probably under constant stimulation of gonadotrophin, particularly in aged animals, the anovulatory follicles proliferate to form folliculoids.

Parovarian cysts : These occur frequently in most species of domestic animals in the vicinty of ovary in mesosalpinx and vary in size from a few mm. to
1 cm. or more in diameter.   The cysts are lined by a single layer of cuboidal
epithelium. The wall of the cyst contains smooth muscle.

Parovarian cysts arise from the remnants of either Mullerian or Wolffian
ducts.

TUMORS OF THE OVARY :   These are comparatively rare in animals.

Of the primary tumors, cystadenoma and cystadenocarcinoma are the most
frequent and met with in the bitch and hen   These may be unilateral or bilateral
and may be unilocular or multilocular, the cavities containing clear fluid and
lined by either cuboidal or low columnar cells.   Sometimes cilia may be noticed
in these cells.   More frequently, there may be papillary projections from the
lining, filling the cavity.   Usually, such papilliferous tumors are malignant.
Peritoneal implantation of the ovarian carcinoma may be observed. Metastasis is
by way of lymphatics.

Tumors consisting entirely of the epithelial cells mentioned above are known
as solid carcinomata and are seen in fowls and rarely in bitches.   Transcoelomic
spread occurs in this tumor.

Other tumors, called endocrine tumors of the ovary that may be rarely seen
are :- granulosa-cell tumor, theca-cell tumor   arrhenoblastoma and dysgerminoma   Of these, the granulosa-cell tumor is more often encountered in the
cow and bitch.

The following secondary tumors may be found; lymphosarcoma; mammary
tumor of the bitch, intestinal carcinoma of the cow

Teratomas and dermoid cysts may be found in the ovaries of animals and
birds

AFFECTIONS OF BURSA

Bursitis is generally due to the extension of inflammation from the peritoneum
or from the infundibular end of the oviduct.   The ascending infection from the
oviduct is common in cases of retained placenta and septic metritis. Perimetritis

may also contribute to bursitis. Excessive pressure during enucleation of corpus luteum may cause inflammation of ovarian bursa

Adhesions : Occur as a consequence of bursal inflammation or due to hemorrhage caused by ovulation or enucleation of corpus luteum. The bursal adhesions cause infertility by interference with the transport of gametes. The adhesions may be in-between the membranes.

DISEASES OF THE FALLOPIAN TUBES

The disease of the Fallopian tubes (oviducts) are not common in animals. Affections of the oviduct are of importance since the ovum is transported to the uterus via these tubes and any disease of the tubes, therefore, will interfere with pregnancy and reproduction.

Malformations : absence of fallopian tubes or segmental aplasia may be met with. Accessory tubes, reduplication of the tubes are other malformations seen.

Of importance in animals are the following conditions of the oviducts :- hydrosalpinx salpingitis and pyosalpinx. These conditions are more important in the cow and sow.

Hydrosalpinx denotes a cystic dilatation of a part of the oviduct, containing clear fluid This condition arises due to some obstruction in the oviduct. It is usually a result of salpingitis in which, occlusion of the lumen may arise.

Based on gross changes and histopathology two types are recognised :
1) Hydrosalpinx simplex and
2) Hydrosalpinx follicularis or multilocularis.
It may be seen affecting one or both the tubes.

In the simple form, the fallopian tube is considerably distended, elongated and tortuous forming several coils in the mesosalpinx. The wall is thin, translu-cent and distended with varying amounts of clear fluid. The distension is more often located in the ampulla.

In the follicular form, unlike the simple form, the tube is distended with a little fluid but is hard, tortuous and irregularly beaded. On cross section the lumen presents a multilocular appearance.

Microscopically, the mucosal folds in the simple form are considerably atrophied and lined by low cuboidal to columnar epithelium devoid of cilia. The lamina propria and muscular coat are thin.

In the follicular form, the fibrous septa are usually thin, but in some places, thickening is marked. The trabeculae are lined by low cuboidal or flat epithelium. Infiltration of lymphocytes, a few plasma cells and eosinophils are seen in the lamina propria.

Salpingitis : This is the most common disease of the oviduct and which is usually not diagnosed while the animal is alive. This is of great economic importance since salpingitis is one of the causes of sterility.

The organisms that are incriminated are : *Streptococcus viridans, Staphylococcus aureus,* M, *tuberculosis* and *Brucella suis.*

The organisms may enter the oviduct
1. By way of the blood stream—in generalised infection as in tuberculosis.

2.   Through the osteum abdominale—spread of peritonitis (descending)

3.   Through the osteum uterinum—extension of endometritis (ascending)

Sometimes irritants may be introduced by uterine insufflation or surgical operations

Macroscopically, there may not be visible changes in the tubes except for slight enlargement and congestion of the mucosa. In the milder forms, there may not be any exudate in the lumen. In more severe cases catarrhal or fibrinous exudate may be present, consisting of dead cells and debris

Microscopically, there may be mononuclear infiltration besides congestion of the mucosa. desquamation of epithelium and proliferation of stromal elements. Plasma cells are particularly abundant. These microscopical changes may be evident even in normal looking (grossly) tubes

The mucosa of the fallopian tubes does not possess much of regenerative capacity and so when once the epithelium is lost it is not restituted.

Sterility occurs in salpingitis for the following reasons :- (1) The ciliated epithelium and contractile muscle necessary for transport of ovum, are destroyed, preventing the movement of the ova to the uterus. (2) The inflammatory exudate is toxic to the spermatozoa, causing their death, (3) Exudate or proliferating cells may occlude the lumen of the tubes. (4) Fibrosis in chronic salpingitis may cause occlusive stenosis.

Pyosalpinx ; This is pus in the salpinx and occurs in suppurative salpingitis, which is usually a sequel to suppurative metritis. Pus accumulates in some seg. ments of the tube due to occlusion of the lumen in certain places by inspissated exudate or inflammatory thickening or by chronic granulation tissue.

The wall of the oviduct is infiltrated by neutrophils, lymphocytes and plasma cells, which are also found in the exudate that collects in the lumen. Metaplasia of the epithelium to squamous variety is common.

Pyosalpinx invariably ends in sterility.

Tuberculosis : Tuberculosis of fallopain tubes is common in cows. Two varieties are seen   1. *Caseous tuberculous salpingitis*, in which the tube is very much thickened and swollen. The mucosa is much thickened and caseated. The lumen contains the disintegrating tissue masses. Adhesions may be present between the ovary and the tube and between the individual coils of the tube.

2. *Nodular tuberculous salpingitis* in which miliary tubercles are found in the mucosa, as a result of generalised tuberculosis.

In birds, salpingitis is a common affection.

Causes : Primary, bacteria :— *Salmonella pullorum*, (descending infection from ovary), *E. coli.*

Parasites : *Prosthogonimus macrorchis*

Secondary : Pullorum disease, vitamin A deficiency.

Macroscopically, the oviduct is swollen and distended. The body cavity contains masses of yolk material and fibrin strewn about, which may cause loops of intestines to become adherent. The mucosa of the oviduct is red and edematous. The cloaca and parts round about are soiled.

Microscopically, catarrhal, fibrinous, hemorrhagic or purulent salpingitis may be encountered. Ascending infection may cause ovaritis or peritonitis.

Neoplasms : The most common neoplasm in the oviduct of fowls is the leiomyoma. It is also frequently met with in the right mesosalpinx

THE UTERUS

Malformations : The following malformations may sometimes be seen :— Aplasia, hypoplasia, duplication of cornu, longitudinal division of the uterus by a septum.

The malformations of the uterus are due to the failure of development of Mullerian duct system. The conditions are bilateral in the case of infantile genitalia and freemartins and unilateral in the case of uterinum unicornis. Sometimes cystic dilatations are noticed in the undeveloped segments.

Rupture of the uterus may occur during parturition due to violent contractions or due to obstetrical manipulation in dystocia. Rupture may involve only the mucosa in which case healing will occur. If the whole wall of the uterus is involved. death may supervene, due to (1) hemorrhage, (2) or inflammation of the uterus spreading to the peritoneum : or (3) entry of the placenta into the abdominal cavity.

Rupture may occur in prolonged dystocia and torsion due to weakening of the wall Another rare cause may be over distension of the uterus with introduced fluids

Malpositions : **Torsion** of the uterus is most common in the cow, especially during the terminal stages of pregnancy. Minor twists are self corrected. The condition assumes importance only if the twist is 180° or more. The veins of the broad ligament and ovarian ligaments are compressed while acute hyperemia occurs in the arteries. If the condition is not corrected the dam will die of gangrene, sepsis and peritonitis. As the cervix is tightly closed in the twist, parturition cannot take place unless the disorder is corrected. The uterus is liable to rupture easily in this condition as the walls become weakened and friable.

Hernia : Displacement of the uterus, especially uterus in advanced pregnancy, through a ruptured diaphragm into thoracic cavity may occur in dogs and cats (possessing sufficiently long broad ligaments) as a result of automobile accidents

Displacement of the uterus in abdominal and ventral hernias is also met with. In the bitch perineal hernia is also seen. Herniation into the inguinal and femoral canals may occur, when it is called a **metrocele** if a peritoneal lining of the sac is present.

Prolapse of the uterus through the vulva is most common in the cows but may also be seen in other animals. It may be due to strong uterine contractions for expelling the fetus, the placenta or the exudate. Forced traction during distocia, post-parturient hypocalcemia and retained placenta are the predisposing causes. The sequelae are similar to those found in intussusception viz. acute congestion, hemorrhage, necrosis, infection, gangrene and death. Sows and poultry may injure everted uterus.

CIRCULATORY DISTURBANCES

Physiological hyperemia and edema of the endometrium are found during estrus.

Hemorrhage occurs during estrus, analagous to the menstrual discharge. The source of the blood is capillaries of the endometrium.   This is more often found in heifers and bitches.   Ecchymoses on the serosa and musculature are normal in heifers during estrus.

Pathological :   Acute hyperemia is present in metritis.

Chronic general venous congestion is found with cardiac and pulmonary lesions hindering normal blood flow through these organs.

Hemorrhage can occur during parturition and dystocia. One of the common causes is manual intervention in dystocia when rupture of the uterus with hemorrhage may take place.

Torsion of the uterus as well as prolapse may also be responsible for hemorrhages.   Hemorrhage from the arteries of the broad ligament, especially in the sow, is a surgical hazard during cesarean operation, if too much traction is applied on the uterus. In dogs metrorrhagia may be the result of hormonal disturbances.   There may be severe lowering of folliculin.

Lastly, massive hemorrhages may be met with in cattle in sweet clover poisoning.

Thrombosis of the uterine vein may be observed in septic metritis, torsion or prolapse.   The affected veins are dilated and tortuous and do not collapse at death.

ATROPHY OF UTERUS :   Causes;- 1. Senility

2.   Oophorectomy after a normal full growth of uterus is attained.

3.   Hypopituitarism :   (a) due to wasting disease or   (b) primary lesion of the pituitary.

HYPOPLASIA may be observed in oophorectomised young animals, the uteri of which have not attained full size.

HYPERPLASIA :   Hyperplasia of the endometrium is observed in all species of animals but is more often met with in the dog,   It is also known as *cystic hyperplasia* of the endometrium.

The cause appears to be increased estrogen and or progesterone secretion, under the influence of which the endometrium undergoes hyperplasia.

This condition has been noticed in animals having granulosa-cell tumor, papillary cystadenoma and persistent corpora lutea and ovarian follicular cysts.

Feeding on pasture legumes—lucerne and clover—containing substances having estrogen activity has been found to cause hyperplasia of the endometrium in the ewe and cow.

Though a variety of organisms, including *E. coli, Staphylococci* and *Streptococci,* have been isolated from the affected uteri, they must be considered as secondary invaders of the already existing hyperplastic endometrium.

Macroscopically, endometrium of both the horns contains cysts of different sizes some microscopic and some others as big as 4 or 5 mm. in diameter.   The cysts may completely fill the lumen giving it a "swiss-cheese" appearance.   They contain clear fluid. The lumen of the horns may also contain, in some cases, mucus or pus (if infection by pyogenic organisms has occurred) which may flow out of the vulva.

Microscopically, no inflammatory changes may be present in the uncomplicated cases but only cyst formation of the glands which are increased in number and irregularly distributed (unlike the normal orderly arrangement). These cysts contain a single layer of epithelium enclosing clear watery fluid. In some animals there may be plasma cell infiltration of the lamina propria. The endometrium may show thickening due to polypoid proliferation. In those animals in which infection has taken place, there will be pus in cystic glands and neutrophilic infiltration of the lamina propria. In this type, the uterine horns show alternating constrictions and dilatations which appearance mimics pregnant uterus

Clinical features : 1. Abnormal uterine bleeding.

2 Disturbances in estrus cycle : (a) irregularity (b) longer or short duration (c) diminished or enhanced characteristics of the different phases. 3. Sterility; 4. Aborton 5. Long or pregnant lactation. 6. Development of secondary infection, characterised by high leucocytic count with a shift to the left.

INFLAMMATION OF THE UTERUS

Metritis is inflammation of the uterus and is found in all animals. If the inflammation is restricted to the endometrium alone the condition is known as *endometritis*. On the other hand, if the whole thickness of the wall is involved *metritis* is the term used. Inflammation of the serosa is known as *perimetritis*.

ENDOMETRITIS

Causes 1. Infection

(a) *Trichomonas fetus*, *Vibrio fetus*, *Brucella*, weaker strains of pyogenic *cocci* and *coliforms* Infection may occur during coitus or during artificial insemination or during manual handling of the uterus for therapeutic purposes.

(b) A severe metritis might have subsided leaving a low grade inflammation of the endometrium.

2 **Irritants :** Introduction of too hot fluids or too irritating chemicals into the uterus, thereby injuring the delicate mucosa.

Macroscopically, no gross lesions are evident. An increased secretion of tenacious mucus may all that may be visible. The mucosa may be swollen, red and rough instead of having a smooth surface and covered with fragments of necrotic material.

Microscopically, in the mild catarrhal variety there may be slight but diffuse infiltration by lymphocytes, plasma cells and macrophages. The blood vessels are engorged.

More severe forms invariably involve all the layers of the wall and must be considered as metritis

Clinical consideration : Though mild in appearance, endometritis must be attended to promptly or else conception may not occur. The inflammatory exudate being toxic is lethal to the ovum whether fertilised or not. Besides, the condition may progress to the chronic stage when permanent sterility may supervene.

Causes : Metritis is caused by bacteria, infection being facilitated by the following, which may cause an initial injury: parturition, dystocia, mechanical injuries by the obstetrical instruments, projecting fetal bones after embryotomy, excessively warm irrigating fluids, chemical antiseptics and disinfectants introduced into the uterus.

The organisms that invade the injured uterus are *Corynebacterium pyogenes*, *Streptococci*, *Staphyylococci*, *E. coli*, *Sphenophorus*, *Clostridium sp.*

Tuberculosis and other chronic granulomatous infections may also invade the uterus.

The parturient uterus, with its lochia is a good medium for the growth of bacteria. Retained placenta and albuminous exudate are also ideal for the propagation of microorganisms and so infection is common in those animals in which lochia is plentiful and involution is delayed. The latter condition may result due to weakened and injured uterine musculature in prolonged dystocia.

Nutritional deficiencies and disturbances and hormonal and endocrine disturbances are considered as predisposing factors by preparing the soil for infection.

Routes of infection

1. **Ascending infection** from the vagina. This is the most important route.
2. **Descending infection** from the abdominal cavity through the fallopian tubes.
3. **Through lymphatics** from the peritoneum.
4. **By way of blood** : this is important in tuberculosis.

Depending upon the virulence and nature of the organisms, metritis may be acute catarrhal, acute suppurative and chronic suppurative.

Acute Catarrhal Metritis : This is acute endometritis and has been described above. This condition is difficult to be differentiated from an uterus during estrous cycle.

Chronic non-purulent endometritis can be met with in which there is heavy infiltration of leucoytes, mostly plasma cells, in the mucosa, which, therefore becomes thickened. This thickening is not uniform because of the fixed position of the outlet ducts of the gland and so assumes a polypoid appearance. This is known as *Chronic Polypoid Endometritis*.

Because of the proliferation and the subsequent contraction of the superficial connective tissue, the mouth of the glands may be closed rendering them cystic. This condition is known as *Chronic Cystic Endometritis*.

If the causes of the endometritis is removed, the leucocytes are replaced by fibrous tissue which on contraction produces atrophy of the glands and the mucosa and so *Chronic atrophic endometritis* results.

In cows the mucosa in chronic endometritis may undergo degeneration and subsequently be calcified and the condition then is known as *endometritis calcificans*.

Acute suppurative metritis : This condition which usually arises from infection by pyogenic organisms is a frequent complication of dystocia, retained placenta or abortion.

Microscopically, the mucosa of the uterus is very much reddened, thickened, rough and is covered by a purulent, often reddish exudate. This may freqnently contain shreds of disintegrated fetal membranes. The uterine wall is thickened and friable. The mnscle fibres atrophy and disappear or show Zenker's degeneration. Subserosa is edematous and infiltrated with leucocytes. Serous coat also shows inflammatory changes.

Microscopically, there is infiltration of the endometrium by large number of neutrophils. After several days, macrophages, lymphocytes and plasma cells infiltrate the endometrial stroma.

Infection extending into the uterine-veins results in thrombosis of those vessels.

Chronic Suppurative metritis : Pyometra: Pyometra literally means pus in the uterus. But usually this term is applied to chronic suppurative metritis. This condition is seen in dogs, cats, cows and swine.

In cattle, pyometra is encountered as a result of retention of placenta. The placenta putrifies since it is a very good medium for bacteria to thrive. Incomplete involution may be an associate factor.

The pathogenesis in the dog is different. In this animal, increased progesterone activity seems to be the prime cause.

Pyometra arises due to infection of the hyperplastic endometrium met with in "Pseudopregnancy" of bitches, which is caused by persistent corpus luteum releasing large amount of progesterone.

Corynebacterium pyogenes, E coli, Proteus sp. and Staphylococci have been isolated from the uterine exudate.

In the cow *Trichomonas fetus* infection is a common cause.

Macroscopically, in the cow, thin cream-like pus may be discharged through the vulva, soiling the tail and the perineum When the animal lies down, due to pressure on the abdomen, pus which may stagnate due to gravity during a standing position, may flow out. The uterus is dilated, and involution may not be complete.

In the bitch, the exudate is always retained as the cervix is completely closed. The horns are dilated and thin-walled and contain chocolate-colored fluid. The abdomen is enlarged as in a full pregnancy. The serosal surface may show congested vessels and evidence of inflammation. The mucosa may be thickened irregularly and in some places it may be ulcerated and hemorrhagic and covered with necrotic shreds of membarane appearing as though bran is sprinkled. Retention cysts may sometimes be seen.

Microscopically, in the cow the appearances are similar to those of endometritis. viz, congestion of the blood vessels, infiltration by inflammatory cells especially neutrophils and lymphocytes and plasma cells. These cells accumulate under the epithelium, leading to its purulent softening and separation of the necrotic area of tissue and these appear as bran-like material.

The condition in the dog is more acute and so greater infiltration of neutrophils and lymphocytes occurs. There is hyperplasia of the endometrium, producing pseudostratification or papillary proliferation. Sometimes squamous metaplasia may be noticed.

Extragenital lesions :   These are found in the dog and cat and may be due to the 'toxic' effect on other organs as well as to periodical bacteremia that may occur.   The lesions are :   1 Anemia due to depression of the bone marrow.

2. Extramedullary leucopoiesis, especially in liver, spleen, kidneys, adrenals, lungs and lymphnodes.

3. Lesions of the kidney—glomerulonephritis, tubular degenration, hemorrhages in the medulla, infarcts, pyelonephritis.

4. Congestion and degenerative changes in the liver.

5. Adrenals—necrosis of the cortex and hemorrhages in the medulla.

6. "Sinus Catarrh" of the lymph nodes

7. Intense leucocytosis.   The total white cell counts vary from 30,000 to 160,000 per c.mm.   An extreme shift to the left and toxic granulation of neutrophils are found.

Sclerotic metritis is characterised by complete destruction of endometrium as a result of severe chronic endometritis.   A thick dense connective tissue layer replaces the endometrium.   The foci of infection in the connective tissue layers is responsible for purulent exudate in the uterine cavity.   The uterine caruncles and endometrium are destoryed resulting in permanent sterility.   The uterus on rectal examination appears hard and firm and the cervix thickened.   Cow is usually anestrus and the corpus luteum is found deeply embedded.

Perimetritis and Parametritis are characterised by varying amounts of adhesions between uterus and broad ligaments, with other pelvic and abdominal organs.   The adhesions are reusltant of severe metritis, douching with strong irritant solutions, perforation of rectum with leakage of its contents, torsion of uterus vaginal and cervical lacerations curing diffcult birth, excessive bleeding, following enucleation of corpus luteum or vigourous massage of infected uterus. The condition may also be due to peritonitis or tuberculosis of the genital organs.

In women uterus may be infected with organisms producing gas gangrene while abortion is induced.   The muscles and peritoneal coat are involved with inflammatory changes and gas formation.   This condition is known as **Physo-metra.**   Among animals a similar condition may be met with in infection of the uterus by the bacillus of Black Quarter.

Necrobacillary metritis that may occur in cows due to puerperal infection by *Spherophorus necrophorus* is always fatal.

Tuberculosis of Uterus : Infection may be descending from tubercular pertonitis or it may be hematogenous as occurs in generalised tuberculosis.   Two forms are seen :   *(1) Disseminated miliary tuberculosis* in which the tubercles are found uniformly in the mucosa.   *(2) Diffuse caseating endometritis* in which the body or the cornua are diffusely thickened.   The lumen contains large quantity of serous or purulent exudate containing large caseating masses.   The mucosa is very much thickened and caseated.

Actinomycosis :   Incidence of actinomycosis is not common.   Large swellings with extensive pelvic adhesions are produced.   Prognosis is generally poor·

Brucellosis of the uterus of pigs : Some aged pigs reveal pea sized, miliary yellowish-white nodules in the uterine mucosa, caused by *Br. suis* These nodules may be single or may occur in groups.   The mucosa which may be raised may sometimes be ulcerated. In the centre, the nodules contain a little pus. Histologically the nodules are granulomas. Animals having these lesions give positive test for brucellosis.

Mucometra/Hydrometra

The two conditions are considered together as the difference is probably only in physical properties and depends on the degree of hydration of mucin, which in turn may be related to the relative activity of estrogenic hormone.

The accumulation of thin or viscid fluid in the uterus is concurrent with the development of endometrial hyperplasia or is proximal to an obstruction of the lumen of the uterus, cervix or vagina. In the first instance the amount of the fluid may be several litres and the greater the volume of the fluid the less viscous it is. Small amounts of mucin gives the mucosal surface a gummy stickiness. In cows with cystic ovaries, the large volumes of fluid is usually associated with functional cysts of the follicles.

In the second instance, that of obstruction to the lumen, the volume of fluid depends on the site of obstruction. The fluid is slightly cloudy and watery.

Animals with mucometra are sterile. If affected uterus becomes infected, an intractable pyometra results.

An abnormally long and tortuous cervix may result in a form of mucometra caused by the retention of uterine secretion.

Microscopically : The endometrium is thin and lined by a single layer of cuboidal to low columnar cells. The uterine glands are reduced in number. The endometrial stroma is edematous.

Uterine abscesses are usually man made lesions. These are commonly present on the dorsal wall of the uterus at the anterior extremity of the body and produced either by an insemination pipette or some instrument used for uterine medication.

These are of variable sizes containing thick yellow pus.

Endometriosis : Means presence of endometrial glands and stroma in places other than endometrium. If these are seen between the muscle bundles of myometrium, the condition is called **endometriosis interna or adenomyosis**. On the other hand, if these are seen in places other than uterus such as mesosalpinx, ovary, cervix or intestinal serosa, it is called **endometriosis externa**.

Endometriosis is occasionally noticed in bitches and bovines. The endometrial glands show changes in response to ovarian activity and often there are hemorrhages. In women these appear as chocolate coloured cysts. Histogenesis of endometriosis is still controversial

Mummification of fetus :

Mummification of a dead fetus is seen occasionally in any, but usually in multiparous species and most commonly in the cow. In the mare it is typically one of twin fetuses which is mummified.

A prerequisite for mummification is absence of infection unlike in maceration The fluids are reabsorbed and the membranes become closely applied to the desiccated fetus. The whole fetus becomes brown or black and rather leathery, moist on the surface with sticky mucus without odour.

The time required for complete mummification depends on the size of the fetus but probably requires as long as 6 to 8 months. In uniparous animals, the mummified fetus is usually retained indefinitely or if aborted may only be delivered into vagina. In the case of multiparous animals, it may be delivered along with viable fetuses. Animals which had and recovered from mummified fetus usnally breed normally on subsequent occasions

In bovines, haemic mummification is seen while in mare and sows it is or paperaceous type.

Maceration of fetus : Depends on the presence of infection in the uterus. If the early embryo succumbs to uterine or embryonic infection maceration is usually followed by resorption from the uterus or expulsion along with a small amount of purulent discharge. If the fetus is about three months, complete foetal maceration does not occur and bones resist maceration. These may be discharged or be retained in the pus of pyometra indefinitely, often near the cervix.

Advanced uterine lesions accompany the macerated foetus. The uterine wall is thickened and the reaction within it varies from the acute exudative inflamm· ation of pyometra to more or less complete sclerosis and replacement by graaul· ation tissue in long standing cases.

ABORTION

Expulsion of a dead fetus prior to the normal full gestation period is called abortion. Abortion is mostly due to infection of the fetus, placenta or the uterus since these conditions cause death of the fetus.   A dead fetus is a foreign body and so is expelled from the uterus .

The following are the causes :

1. Specific infections :

A. Brucellosis : *Brucella abortus, melitensis* and *suis* affect the cow, sheep goat and pig. *Brucella ovis*, a new species, affects only the sheep   These strains can be differentiated by means of biochemical methods.

Bovine brucellosis : The typical abortion occurs at about the 7th month of gestation in the cow   The organism has special affinity for the pregnant endothelium A few weeks after abortion or parturition, the organism can no longer be detected in the uterus.

Routes of infection :

1. Alimentary canal—ingestion of feed or water contaminated by fetal membranes, fetus or uterine discharge.
2. Vagina—coitus, artificial insemination.
3. Conjunctiva.
4. Skin.
5. Contamination of healthy udder from an infected one during   milking.

The organism produces abortion in the following manner :

1. First placentitis is produced by the invading organism.
2. Sero-purulent exude accumulates between the endometrium and chorion.

3. Edema and infiltration of the chorion by macrophages, lymphocytes and plasma cells.
4. Necrosis and hyalinisation of chorio-allantois.
5. Thus the membranes become separated from the uterine endometrium.
6. Fetus dies.
7. Severence of blood supply to fetus.
8. Dead fetus is a foreign body and so expelled—abortion.

In milder cases, a live fetus may be born, which is usually weak and may succumb soon after. In the chronic cases, there is fibrotic adhesion of the placenta to the endometrium, resulting in retention of placenta. In these cases, calves may be born alive.

In the aborted fetus may be found croupous or catarrhal pneumonia; edema of the pericardium, umbilical cord and skin and serosanguineous exudate in the serous cavities. Suppurative or hemorrhagic gastro-enteritis is present. Hyperplasia of lymph nodes and spleen are prominent lesions.

Brucellosis of swine has a course similar to the bovine but is more acute and severe, abortions occurring between the 2nd and 3rd months of pregnancy. Coital infection is more frequent.

Brucellosis in dogs. Brucella canis causes abortion in dogs. This is a highly infectious disease and transmission may be by contact, through infective discharges and also by venereal transfer. In male dogs epididymitis, testicular atrophy and complete sterility may be caused.

No fever is noticed. Abortion occurs between the 7th and 9th weeks of gestation.

Ovine Brucellosis (Br ovis): Infection is probably by ingestion.

In rams after an initial bacteremia and mild systemic reaction, the organisms localise in the epididymis causing sterility. Semen is of poor quality and contains leucocytes and brucellae. There may be acute inflammation of the scrotum with edema. The condition may become chronic with enlargement of epididymis, thickening of the scrotum and atrophy of testes.

In ewes abortion may occur in late pregnancy or still births may also result due to placentitis. There is purulent exudate on the placenta and edema of allantois. There may be elevated, firm, yellowish-white plaques in the inter-cotyledonary areas and the cotyledons are enlarged and edematous.

Brucella melitensis abortion in goats : Infection is by ingestion. Abortion may occur, but sometimes live kids may be born. Viable kids are infected and infection persists in a latent form and at maturity clinical symptoms are manifested. In the goats an acute systemic reaction develops and later localisation of the organisms in the placenta causes placentitis and thus abortion results. After abortion the uterine infection persists for over 5 months and the mammary glands remain infected for many years. In some cases, spontaneous recovery may occur.

B. Vibriosis : Vibrio abortion occurs in cows and sheep and infection is by ingestion in sheep and coitus or artifical insemination in cows.

Vibrio causes acute catarrhal endometritis, cervicitis and vaginitis.

The pathogenesis of abortion is similar to that of brucellosis, the initial lesion being a placentitis followed by exudation, necrosis, vasculitis, separation of the placenta from the endometrium, death of the fetus and abortion.

In the cow abortion occurs between the 5th & 7th months of pregnancy while in the sheep at 2 months. Usually placenta is not retained.

In the fetus may be found edema of the subcutis, serofibrinous pleurisy, peritonitis, pericarditis, fatty degeneration of the liver and kidney and hemorrhages in the renal cortex. Infection causes repeat breeding.

C. Trichomoniasis :

Trichomonas fetus is transmitted to the cow through coitus, the bull harbouring the flagellate in the mucous membrane of the penis, terminal portion of urethra and prepuce.

In the cow, within three days after infection, vulvitis and vaginitis develop, from where infection spreads producing cervicitis and endometritis and placentitis. There is copious greyish-white thin exudate and abortion will occur within about 16 weeks of pregnancy. Sometimes the dead fetus may be macerated in the exudate. Or the fluid may be absorbed (if not infected by other bacteria) and the fetus may be *mummified*. If pyogenic organisms invade the uterus *(Corynebacterium, Staphylococci* and *Streptococci,)* pyometra will result. In chronic infection with fibrosis, placenta may be retained. Infection causes repeat breeding.

D. Listeriosis :

Listeria monocytogenes, which primarily affects the brain, may sometimes infect the pregnant uterus and cause abortion in cattle and sheep, The organisms become septicemic in the fetus and cause its death. Abortions usually occur during the last trimester of pregnancy.

The fetus shows hemorrhage in the kidneys; anasarca; areas of necrosis and granulomas in the liver, spleen, lungs and kidneys; catarrhal gastro-enteritis, cardiac vegetations and hemopericard.

E. Epizootic bovine abortion :

A virus of the family psittacosis lymphogranuloma group is found to cause abortion in cows and ewes (In sheep it is called enzootic abortion).

An arthropod vector is probably important in transmission Animals rarely abort more than once, due probably to development of immunity. In an outbreak 75% of affected may abort.

Abortion occurs during the last trimester.
The virus causes death of the fetus. The characteristic lesion found in all the organs is a focal inflammation consisting of neutrophils, lymphocytes and macrophages. Injuring the vascular endothelium, the virus is responsible for petechiae found on the skin and internal organs.

Macroscopically, the fetus shows subcutaneous edema, hemorrhages on the conjunctiva, on the mucosa of ventral surface of the tongue and on the tracheal mucosa. Skin at the groin shows erythematous patches. Body cavities are filled with serosanguineous fluid. The pathognomonic lesions are found in the liver, which is enlarged, friable, pale red to reddish-orange in color and has a coarsely granular surface (due to chronic venous congestion). Placenta is not retained.

Microscopically, the liver may either show changes consequent on chronic venous congestion (dilatation of the central vein and sinusoids and consequent pressure on the hepatic cells causing necrosis) or granulomatous lesions in the hepatic capsule, in the portal triads or in the adventitia of the central veins. Meningitis, focal encephalitis and mild degenerative changes in the kidney, pancreas and lung may be noticed. There is hyperplasia of the reticulo-endothelial tissue of the spleen, thymus and lymph nodes and so these organs are enlarged. Infiltration of the adventitia of meningeal and parenchyma cells of the brain by pleomorphic mononuclear cells, arranged concentrically is a characteristic appearance.

The granulomas wherever they are seen (liver, kidney, spleen, lymph node) consist of central necrotic areas surrounded by neutrophils, epitheloid cells. and lymphocytes, surrounded by fibrous tissue.

Diagnosis : Is difficult because of inconstancy of lesions. It is difficult to isolate the virus. Serology is not useful. Only symptoms must guide one to arrive at a diagnosis.

F. Leptospirosis in cattle

In cattle various strains of Leptospira produce abortion, after 6th month of pregnancy. The placenta is avascular with collapsed blood vessels. Cotyledons are atonic, yellow-brown in color and leathery. No inflammatory infiltration is noticed.

In the fetus are found edema of the subcutis, peritoneum, umbilical cord pericardium ; focal interstitial nephritis with round-cell infiltration, glomerulonephritis, infiltration of eosinophils into the cortex and round-cell infiltration into the periportal tissue of the liver.

Abortion is due to fetal death.

G. Abortion in mares by Salmonella abortus equi :

The organisms produce a purulent hemorrhagic placentitis. Allanto-chorion is edematous and exhibits hemorrhages and necrosed areas with a wall of hemorrhagic reaction separating it from the surrounding tissue. Abortions occur late in pregnancy. Infection is followed by development of immunity.

H. Equine viral abortion

Two viruses of horses are incriminated : (1) The virus of equine rhinopneumonitis (influenza). The fetus shows edema of the subcutis, jaundice and edema of the lungs. The lungs are heavy and voluminous. Liver shows focal necrotic areas. Visible under the capsule are grayish-white foci. Such necrotic areas are seen in the spleen and lungs also. Petechiae are found throughout the body. Abortion occurs in the 9th or 10th month. Acidophilic nuclear inclusions are found in the bronchial and alveolar epithelial cells.

(2) The virus of equine arteritis brings about the death of the fetus. The fetus shows hemorrhages in the splenic capsule and respiratory mucosa.

I. Abortion caused by the virus of Infectious Bovine Rhinotracheitis

Abortion in cows may be caused by the IBR virus when the animal suffers from respiratory affection caused by this virus. Vaccination of cows with the IBR virus vaccine also brings on abortion.

Symptoms: Abortion occurs during the last trimester of pregnancy. No symptoms are noticed prior to abortion. There may be history of vaccination by IBR vaccine or of a respiratory affection. Fetus is usually decomposed when aborted since abortion occurs only 24 to 36 hours after its death. Animals which abort do not become sterile.

Macroscopically, fetus shows petechiae on the heart. Serous cavities contain serosanguineous fluid. There is edema of lungs and placenta.

Microscopically, there is focal necrotising hepaititis and placentitis. Renal cortex shows hemorrhagic necrosis. No inclusion bodies are seen.

Diagnosis : 1. By symptoms.

2. Affected animals are serologically positive.

3. Virus isolation from cotyledons.

J. Mycotic abortion : Cattle and sheep

Abortion in cattle may occur due to infection by fungi of the following species : *Aspergillus, Absidia, Mucor* and *Rhizopus*. The infection is a secondary one, the primary lesions being in the lungs, abomasum (ulcers) and the intestines. Infection is by the blood stream.

Abortion in affected cows occurs during the later half of the gestation period between 6th and 8th months and placenta is retained.

Macroscopically, the fetus may show only cricumscribed grayish plaques on the skin resembling ring worm lesions. Internal organs are free

In the cow, lesions are present in the placenta. The Chorion-allantois is thick and leathery. Infection occurs first in the placento nes which show necrotic plaques and the fungus can be demonstrated in these locations,

Microscopically, the typical lesion consists of focal collection of inflammatory cells macrophages predominating. Extensive necrosis of the placentomes occurs. In the uterine wall, the intercaruncular areas show red patches covered in places by a thin yellowish-grey pseudomembrane. Thrombosis and perivascular necrosis occur in these places and hyphae are found both in the tissues and over the mucous surface. Some degeneration of circular muscle is noticed and small arteries are hyalinised throughout the uterine wall. Hyperemia and hemorrhages are common in the affected area. Separation of the placenta from the cotyledons causes death of the fetus.

2. **Abortion may occur in infections by various organisms which first produce metritis**, followed by placentitis and abortion or birth of weak fetus. The following are noteworthy :

Cattle : *Salmonella sp; Corynebacterium pyogenes; Streptococci; Staphylococci; M. tuberculosis; Actinobacillus; Pasteurella*

Mares : *Streptococcus zooepidemicus; Klebsiella genitalium; Shigella equirulis; E. coli.*

Ewes : Virus of Ovine Abortion; *Solmonella abortus ovis.*

3. **Poisoning on administration of Ergot**: Ergot being an ecbolic produces violent contractions of the uterine muscle resulting in abortion.

4. **Neutralisation of the effect of progesterone by estrogens**: Progesterone maintains pregnancy while estrogen terminates it by inducing uterine contraction.

5. Poisons : (a) Chlorinated naphthalenes (which are anti-Vitamin **A** and so may produce metaplasia of the uterine epithelium, infection and separation of the placenta); (b) Purgatives; (c) nitrates through ingestion of plants containing large quantities of this chemical,

6. Faulty nutrition of the mother: Deficiencies of minerals and vitamins.

7. Vaccination of mother during pregnancy against bacterial and viral disesses.

8 Severe and acute septicemic diseases of the mother : Leptospirosis, dourine, viral diarrhoea, hog cholera, erysipelas, infectious rhino-tracheitis In these diseases. abortion may frequently occur.

9. Hereditary predisposition.

10. Torsion of the umbilical cord (rare).

11. Traumatic injury to the placenta (very rare).

NEOPLASMS of the uterus : In the domestic animals neoplasms are neither common nor important as in man.

The most common uterine tumor is the lymphosarcoma, a local manifestation of a generalised condition. In the uterus the neoplastic cells may aggregate as nodules or may diffusely infiltrate the organ.

Adenocarcinomata of the uterus are encountered with metastases in the lungs and liver

In the bitches, uterine fibroids (leiomyomas) are most frequent.

CERVIX

Malformations

These occur more frequently in the cervix than in other parts of the reproductive tract.

Varying degrees of persistence of the median wall of the Mullerian ducts which are destined to develop into cervix result in the formation of a complete or partial duplication of cervix.

Incomplete double cervix occurs much more frequently than a complete duplication and usually involves the portion of the cervix adjacent to vagina. In the case of both incomplete or complete double cervix, if the insemiration is done pregnancy may occur but it may result in dystocia.

Absence of external os may be commonly encountered. In this case, the expulsion of uterine secretions cannot occur resulting in accumulation of fluids in uterine horns— hydrometra.

Double external os : Presence of a dorso-ventral band adjacent to external cervical os giving an impression as though two cervical openings are there. It may not interfere with conception or pregnancy but may cause dystocia occasionally. The fetal membranes may be caught on this dorso-ventral band This condition is inherited and conditioned by a single recessive gene with low penetrance.

Hypoplasia of cervix : The cervix may be very small and there may be deficiency in number of cervical rings. Such a cervix is usually defective in protecting the uterus against bacterial invasion from vagina.

Tortuosity of cervical canal : Extreme degrees of tortuosity of the cervical canal may be a cause of inferility in heifers. There may be S-shaped kink and insemination pipette cannot be inserted into cervix.

ABORTION DISEASES OF CATTLE : CAUSES AND DIAGNOSIS

Disease	Clinical features	Abortion		Lesions		Laboratory diagnosis	
		Rate	Time	Placenta	Fetus	Isolation of etiological agent	Serological
Brucellosis (Br. abortus) reported in India.	Abortion and repeat breeding	High, upto 90% in susceptible cattle	Late abortion, after 6 months	Placentitis, necrosis of cotyledons	Pneumonic lesions, fetal diarrhoea may be present.	Culture of material in 10% Co_2 tension. Chromogenic on potato medium.	A) *Individual tests* 1) Blood plate agglutination with colored antigen. 2) Serum agglutination (tube agglutination 1 in 40 positive). 3) Vaginal mucus agglutination test. 4) Semen agglutination in bull. B) *Herd tests* 1) Milk ring test 2) Whole milk and whey plate agglutination test.
Trichomoniasis (Tr. foetus) Reported in West Bengal. Uttar Pradesh Madya Pradesh	Repeat breeding due to embryonic death. Post-coital pyometra in 10% of the cows.	Moderate 5 to 30%	Early, within 2-4 months	Placenta is edematous Uterine discharge with flocculent material, clear and serous.	Fetus slightly macerated.	1. Cultural examination of fetal stomach. uterine exudate and preputial washings	Serological test with vaginal mucus not popular.

Disease	Effects	%	Time	Placental lesions		Diagnosis	Serology
Vibriosis (*V. foetus*) No confirmed reports in India.	1. Repeat breeding 2. Prolonged and irregular estrus cycles	Low, 5% to 20%	5—6 months	Placentitis, semi-opaque, small thickenings; patechiae, localised avascularity and edema	Flakes of pus on visceral peritoneum	2, Demonstration of the organisms in the above. Culturing of organisms under 10% CO_2 tension. Pathogenic strains are catalase positive.	Cervical mucous agglutination tests (1/50 positive) Mucus to be collected after 40 days of suspected service
Leptospirosis (*Lep. pomona*) Other strains are incriminated in A.P.	Repeat breeding due to endometritis; Abortion may occur during febrile condition; later due to fetal death	25—30%	After 6 months	Avascular placenta, atonic, yellow brown cotyledons- Brown gelatinous edema between allantois and amnion.	Fetal death common	Culturing fetal stomach, placenta and uterine exudate	Serum agglutination and microagglutination lysis test. against all leptospiral antigens. (21 days after febrile condition) Positive cases in brucellosis- free animals reported in A.P.

ABORTION DISEASES OF CATTLE : CAUSES AND DIAGNOSIS

Disease	Clinical features	Abortion		Lesions		Laboratory diagnosis	
		Rate	Time	Placenta	Fetus	Isolation of etiological agent	Serological
Mycosis— *Mucor, aspergillus* and *absidia* species.	Abortion	6-7% unknown abortions	3-7 months	Necrosis of maternal cotyledons. Adherence of necrotic material to chorion causes—soft yellow cushion-like structure, small yellow raised leathery lesions on inter-cotyledonary areas.	Small raised grey but soft lesions or diffuse white areas on skin resembling worm lesions.	Direct examination of cotyledon, fetal stomach contents for presence of hyphae; cultural examination can be undertaken.	Gel diffusion test using the srum from aborted cases and furgal mycelium as antigen.

Cervical dilatation and diverticula ; Dilatation and diverticula usually occur in heifers at the level of third and fouth cervical rings. The cervical canal is usually very small anterior to the defect so that it may be difficult to insert insemination pipette. With age tenacious mucus tends to accumulate in the area of the defects.

Prolapse of cervical rings : is a condition which usually develops with age following repeated parturition. Lacerations and hemorrhages which occur during parturition, results in the formation of excess fibrous stromal tissue, enlargement of cervical rings, vascular embarrassment and occasionally squamous metaplasia of the affected rings. The first and sometimes the second cervical rings prolapse into the vagina.

Cervicitis is the inflammation of the cervix and normally follows abnormal parturition such as abortion, premature birth, dystocia, retained placenta, post partum metritis, pneumovagina and vaginitis The organisms responsible for cervicitis are the same as those of metritis.

Cervicitis, always occurs whenever metritis or vaginitis is present, since cervix is located between these two. Causes iuclude :

    1. **Mechanical injuries :** (a) during parturition,
                         (b) copulation and
                         (c) phooka — criminal stimulation of the
                                  vagina or os for higher milk yield.

    2. Diseases of uterus and vagina

Cysts ; Retention cysts of the cervix are seen in cows. These are usually small. Bigger ones may partially occlude the cervical canal.

Neoplasms Squamous cell carcinoma may be encountered.

VAGINA

Developmental abnormalities :

Double vagina due to persistent median septum along the vaginal passage.

Median vertical bands connecting the floor with the roof at the hymenal border. This condition is more common.

Heterotopic vulval opening may be located in the inguinal region behind the udder instead of below the anus in the perineum

Cysts Dilatation of the Gartner's canals (which are remnants of Wolffian ducts) may be noticed in the cow poisoned with highly chlorinated naphthalenes and in those having ovarian follicular cysts. Multiple cysts are located on the floor of the vagina as parallel rows. These contain a thin clear fluid and are lined by simple cylindrical epithelium.

Rupture of vagina may occur during parturition or during coitus (especially in sows). Infection of the rupture may result in abscess, phlegmon, gangrene and peritonitis.

Vaginitis and Vulvitis

Causes include Physical—Trauma; Chemical; Nutritional deficiencies. Bacterial and Viral agents that are the same as for metritis and abortion

Appearances are similar to those of inflammations of other mucous membranes.

Granular vaginitis is otherwise called *nodular venereal disease*. This disease is described as one of the causes of infertility and is believed not to cause abortion.

The causative agent is not clearly established, Many organisms have been listed : hemophilus, pleomorphic rods and viruses. The incidence of the disease is highest in naturally served herds.

Raised orange-red areas about 3 mm. in diameter are noticed in the posterior part of the vagina upto the urinary meatus. The lesions are most commonly seen in the region of the clitoris, below the lips of the vulva. Sometimes lesions extend up to the dorsal commissure. The raised areas or granules are isolated and in severe cases may coalesce. The granules are lymphoid follicles or lymphoid accumulations. The epithelium over the granules is easily injured and bleeding may occur.

"Vesicular venereal disease". "vesicular vaginitis", "Coital exanthema" The cause is supposed to be a virus, which is considered to be the same causing infectious rhinotracheitis and infectious keratoconjunctivitis in cattle, an epitheliotropic virus.

This is a highly contagious disease, frequently transmitted by coitus. The course of the disease, is about 10 days. Recovery is the rule with transient immunity. **Incubation period is I to 3 days**

Macroscopically, pustular lesions may be found only in the vagina and vulva. Early fever and leukopenia may be seen during viremic phase. Starting with hyperemia of the mucosa of this part hemorrhages may be observed later in the submucosal lymphoid follicles. The mucosa is covered with thick mucus. Within 24 hours, there is mucopurulent discharge from the vagina. There may be pustules over the lymphoid follicles. Rupture of these lesions results in ulceration. Extension of infection to cervix and uterus results in cervicitis and metritis. ·

Microscopically, the epithelial cells undergo hydropic degeneration and acidophilic intranuclear inclusion bodies are found in them, since the virus is epitheliotropic. Neutrophilic infiltration is present near these lesions. In the lamina propria, infiltration by lymphocytes and plasma cells may be found together with edema and hyperemia

Resolution occurs in about 8 days.

Specific bovine venereal epididymitis and vaginitis (Epivag) is a chronic viral disease of cattle, transmitted by coitus and is found in Africa. The disease is characterised by mucopurulent vaginal discharges in females and causing permanent adhesions of fallopian tubes. In the bulls the disease causes swelling of epididymis.

Pneumovagina is common in mares and is due to the deformities or injuries to the vulva and its suspensary apparatus It is also due to the vice of crib biting In this condition, the mare makes an inspiratory effort holding something hard in its mouth. Because of this effort, and because of the already existing negative pressure of the uterus, and due to the inability of the vulva to keep back the external air from entering the vagina, air, enters the vagina and causes balooning of the vaginal wall. Along with air, urine and dung also gain

entrance into the vaginal cavity. This contamination causes vaginitis, cerviciti-
and endometritis. **Klebsiella** can generally be isolated from the exudates of the
vagina.

Tumors : Fibromas may be found, which may be soft or hard, pedunculated or sessile.

Leiomyomas are also seen. Some have fair amount of collagen fibres when
they are known as fibromyomas. These are comparatively harder than the pure
leiomyomas.

Transmissible venereal tumor of the bitch is frequently seen in the vagina.

THE OVIDUCT OF THE BIRD

The oviduct is divisible into five portions.

The infundibulum : This is thin walled and sucks in the ovum released
from the ovary.

2 **The magnnum** is very glandular and has a good muscular wall The yolk
during its passage through has a rotatory motion due to the spiral arrangement
of the outer muscular layer. It is in this part that the yolk gets its albumin
layer.

3. **The isthumus :** Here the two egg membranes are added to the albumin
covered yolk.

4. **Shell gland or uterus :** Here the shell is formed.

5. **The vagina :** This is the terminal portion. Between shell gland and the
vagina is a sphincter.

The formation of a normal egg depends upon :

1. The speed with which it passes through the several portions of the ovi-
dnct and 2. The normal function of various parts. So abnormal eggs may be
encountered if :

i) the yolk is held too long in one section—excessive amounts of the substance that is normally added, will be deposited.

ii) the ovum is driven back by anti or reverse peristalsis. In this condition
more or less of the substance already deposited will be added.

iii) there are disturbances of secretion of the parts. In this condition
more or less of the substances may be added.

iv) if the movement of the ovum is too rapid, the quantity of the material
to be deposited, may be too thin or even no material may be added.

v) if there is any inflammatory exudate in the oviduct. This will also b
enclosed in the egg formed.

The following types of **abnormal eggs** may be encountered,

1. **Double yolked eggs:** In this condition two yolks enter the oviduct simulta
neously and so are enveloped in a single layer of albumen, membrane and shell.

2. **Ovum in ova :** In this condition there is fully formed smaller egg within
a large one. A fully formed egg, due to reverse peristalsis reaches anterior parts
and comes in contact with another ovum and gets attached to it. Then this combined mass while moving downwards is enclosed with albumin and shell.

3. **Yolk-less eggs :** In this variery, the fully formed egg does not contain
yolk. The explanation offered is the yolk that is released does not enter the

oviduct due to closure.   But at the same time stimulates the oviduct to produce various layers of albumin, membrane and shell.

Or, a nidus of exudate may be coated by albumin etc. to from a yolk-less egg.

4. **Soft shelled eggs** : This variety is also known as *leathery eggs*. In this a shell is lacking.   The causes are various and include deficiency of calcium, vitamin D or there may be disturbances of secretion of calcium by the   shell gland. This disturbance may be due to:

(a) poisons – chemicals such as Zinc sulphide used as sprays on trees.

(b) infective agents as in infectious bronchitis; or

(c) too rapid peristalsis at that part.

5. **Layered eggs** : In this the various layers are duplicated. This is due to forward and backward movement of the egg mass so that various substances are deposited again and again

6. **Egg concretion:** When there is some inflammatory process of the oviduct the egg may be retained and so more and more clayers of albumin, membrane and calcium are deposited. There may be deposition of fibrin also and the whole mass may be dehydrated and calcified to form a huge mass of concretion.

7. **Foreign bodies within eggs** :   Eggs may contain various foreign bodies These may be :

a) Feed, feathers, feces, small pieces of wood etc. that have reached through the cloaca, upper parts of the oviduct by antiperistalsis and get incorpo rated into an egg.

b) Parasites — *Prosthogonimus macrorchis* which is found in the Bursa of Fabricius normally, may go up the oviduct and be incorporated in an egg. So also *Ascaridia galli* may go up the oviduct from rectum and be incorporated in an egg.

c) Blood clots may be found when hemorrhage occurs due to rupture of vitalline membrane,

Abnormal location of eggs

Sometimes eggs, in various stages of development may be found in the abdo minal cavity. They may be there because :

1. Yolk does not get sucked by the infundibulum and so is found in the abdominal cavity.

2   By antiperistalsis, partially formed egg may be thrown out through the infundibulum into the abdominal cavity.

3. The oviduct may rupture so that the egg is found in the peritoneal cavity.

The eggs and their contents acting as foreign bodies produce peritonitis and death. Sometimes, if secondary infection were to occur, the organisms multiply quickly since the egg contents are good medium, death resulting.

Egg bound condition in the fowl is that in which the egg is lodged tightly in the oviduct or cloaca and is not laid. This gives rise to local irritation, inflammation, peritonitis and death if not relieved in time.

This condition may by due to :

(a) narrowing of the oviduct as a result of inflammation, thereby making it difficult for the egg to be laid.

(b) Far too large an egg like a double yolked egg.

(c) Paralysis of the muscles of the oviduct. In this condition the egg is not propelled further towards the cloaca.

THE MAMARY GLAND

MASTITIS (mammitis)

Mastitis or inflammation of the udder, may theoretically be caused by trauma of various kinds. But by far the most common causes are the infectious agents. All domestic animals suffer from this condition, but it is in the cow that mastitis is of importance because of the economical loss the owner may suffer.

Bovine Mastitis : Causes : The bacteria that have been found to cause mastitis, are *Streptococcus agalactiae* : *Streptococcus dysagalactiae*; *Staphylococcus aureus and albus*; *Corynebacterium pyogenes*; *E. coli*; *Pseudomonas aeruginosa*; *Pasteurella multocida*; *Brucella abortus*; *Mycobacterium tuberculosis*; *Actinomyces bovis*; *Actinobacillus lignieresi* ; *Nocardia*; *Mycoplasma* and *Cryptccoccus neoformans*.

Though formerly *Streptococcus agalactiae* was the commonest organism causing mastitis, after the advent of antibiotic treatment *Staphylococcus aureus* has been found to be the major cause of mastitis in cows.

In India, *Staphylococcus aureus* and *pyogenes* have been isolated from large number of mastitis cases. In some herds, on the other hand, gram negative organisms (*E. coli* and *Aerobacter aerogenes*) have been isolated.

Since the pathology of *Streptococcus* mastitis has been well studied, the same is described here When once udder becomes infected with *Streptococcus agalactiae*, it is said that it never becomes free of this organism. Though some kind of equilibrium develops between the udder and the organism, at times acute exacerbations may occur when the organisms multiply and increase in great numbers.

In a herd all cows are not equally affected. The route of infection appears to be through the teat canal. Wounds that occur in cow pox or those caused by suckling calves, facilitate infection. Contaminated cups of milking machines, milkers hands and farm utensils are other sources of infection.

The teat canal is lined by the same type of epithelium that covers the teat, but this epithelium seems to secrete a type of smegma, (rich in fatty acids) and this inhibits the streprococci.

The development of mastitis can be described under three phases : (1) **The invasion phase** in which the bacteria are able to enter the teat orifice and be present in the teat canal and cistern; (2) **The infection phase** in which the organisms are able to overcome the resistance and multiply and lastly (3) **The inflam.matory phase** in which the organisms invade the udder.

When the streptococci invade the epithelium of the ducts inflammation results and due to the rapid development of the granulation tissue beneath the epithelium it is thrown into folds of polypoid thickening.

The organisms that penetrate the interstitial tissue cause edema and infiltration by neutrophils which destory some of the organisms. The lymphatics in the stroma become widely dilated due to infiltration by leucocytes that migrate from the regional lymph node. The epithelium of the acini becomes vacuolated and desquamated. Streptococci are numerous in the ducts Milk being a good medium for the growth of bacteria infection is much more serious in a lactating udder than in a dry one The exudation process gives rise to pathological fibrosis and involution of the acini.   Subsequently macrophages and fibroblasts increase in number while neutrophils decrease. There may be stagnation of secretion in the smaller ducts and at this stage the udder may be firm and indurated due to the inflammed interalveolar tissue and retained secretion. When acute stage passes off and the damage caused is slight regeneration of the acini may occur. But if there is large scale destruction, regeneration is not possible and so the acini collapse and are replaced by granulation tissue. Interstitial spaces are infiltrated by lymphocytes.   Such a gland is reduced in size and becomes hardened in consistency—"the shrunken quarter".

The acute systemic symptoms are due to the action of the bacterial toxins that diffuse into the general circulation.

Macroscopically, one or more quarters may be affected The secretion may be serous or may contain floculi and sometimes it may be purulent.

The gland is swollen and slightly hard.   On section, it dose not possess the silky pink color of normal udder but is red or white. Lobulation is distinct. When fibrosis has set in it can easily be seen surrounding tne lobules and the ducts.

Mastitis caused by *Streptococcus dysagalactiae* is more severe than that caused by *Str agalactiae* and more destructive leaving non-functional udder.

Streptococcus uberis produces a mild and chronic mastitis

Staphylococcus aureus usually affects younger animals; especially after, parturition.   Infection is supposed to be contagious and through the teat canal. There may be a peracute and fulminating type or more commonly a chronic type. Treatment is not very satisfactory. Differing from the stretococcal variety, in this type organisms persist in the interstitial tissue producing the granulomatous lesions described hereunder.

In acute cases mortality may be high due to toxemia and the udder is hard, swollen and very painful. Secretion of milk is very little and that too blood stained. Uninfected quarters are also swollen because of the action of the toxin (alpha) that has diffused into them. Gangrene may supervene when infection spreads to the blood vessels causing thrombosis. The udder then becomes cold and greenish or blue. Pitting edema is seen in the flank, inguinal region and the ventral aspect of the abdomen anterior to the udder. Gas may be present in the affected gangrenous area producing crackling sound on pressure. In such cases death may supervene or if the animal survives, the udder is totally lost.

In less severe cases necrotic foci are found surrounding which is a zone of leucocytes and this in turn is enveloped by fibrous tissue This granulomatous lesion is known as *botriomycosis*, in the centre of which can be seen gram positive cocci. The udder tissue contains numerous such granulomata. Fibrosis ultimately occurs resulting in the shrunken quarter described under streptococcal mastitis.

Staphylococcus pyogenes produces a very acute type of mastitis accompanied by severe systemic disturbances and fever. Animals die in a few days. Gangrene may supervene.

Corynebacterium pyogenes is the cause of the so called "Summer Mastitis" affecting both immature and lactating glands. The organism being pyogenic large amount of pus is produced, resulting in abscesses. There may be fistula discharging the pus to the exterior as well as large scale necrosis and sloughing. The latter are evidently due to thrombosis. Fibrosis with loss of function results in those animals that survive.

Coliform organisms, *E coli* and *Aerobactor aerogenes,* produce, sometimes, an acute inflammation of the udder. Infection is supposed to be by blood stream though galactogenic infection cannot be ruled out. The affected quarter is hot, painful and edematous. Clotted milk, sometimes blood tinged, may be seenc Infection may subside, with involution of the quarter or it may become chroni· with acute exacerbations developing later. Sometimes severe general toxic symptoms may be noticed with death following due to the potent toxins.

Gangrenous mastitis : In severe cases of mastitis caused by virulent strains of organisms thrombosis of the mammary vessels occurs resulting in infarction and gangrene. *Staphylococcus aureus* and *E coli* with *Clostridium welchii* produce this condition. Usually all the four quarters may be affected. The udder becomes cold, and bluish within 3 to 4 days after infection. In many cases death may supervene.

Cryptococcal mastitis is a surgical hazard that may be encountered in repeated intra-mammary infusions. *Cryptococcus neoformans* produces an acute inflammatory reaction. The gland becomes hard. One or more quarters may be affected. The milk turns to a watery, flaky secretion. The gland is fleshy and interlobular septa are distended with edema. There is large-scale destruction of the glandular tissue and the alveolar and ductal epithelium is liquified to form a viscid, mucoid material. In sections, the double refractile fungus can be seen in large numbers, some of which are found englufed by the histiocytes. In some isolated chronic cases'granulomatous nodules with interlobular and intralobular fibrosis occurs together with infiltration by histiocytes and lymphocytes.

Organisms may be found in the supra-mammary lymph gland. Metastases, in some cases, may be found in the lungs.

Brucella mastitis :

In infection by Brucella, udder is the reservoir of the organism, which is excreted through the milk. The udder may not show any changes or only scattered lesions may be observed which are not very characteristic.

Histologically the lesion is a granuloma which is intralobular. It consists of lymphocytes, plasma cells, histiocytes and a few Langhans' type giant cells.

In some cases, instead of the granuloma, an interstitial mastitis is seen with infiltration of lymphocytes, histiocytes, plasma cells and fibrous tissue. Regional lymph nodes may or may not be swollen.

Mycoplasmal mastitis : This is identified to be caused by several strains of *Mycoplasma sp.* Cows of all ages are affected. All four quarters may be involved, with sudden drop in milk yield Milk becomes abnormal grossly. There is cessation of lactation and the animal will not be useful again for dairy purposes. Faulty milking machines and unsterilised teat syringes, contamination of teats during milking and inhalation are the sources of infection

The condition is a purulent mastitis and the organism may also invade the blood and then affect the joints and other tissues causing systemic symptoms. There is arthritis with swelling of the joints and lameness.

Milk is thick and cheesy and may be tinged with blood. There may also be clots or granular material in some specimens. The udder is swollen and later it gets atrophied. Treatment so far has not been of any avail.

The following organisms also may cause mastitis : *Pseudomonas aeruginosa; Pasteurella multocida; Nocordia asteroides; Candida sp,* and *Mycobacterium tuberculosis.*

Mastitis ine wes

The organisms responsible for mastitis in ewes are :—*Staphylococcus aureus, Pasteurella hemolytica; Corynebacterium pyogenes; Streptococci and Collforms,*

Infection is usually ascending through the teat canal. Injuries made by suckling lambs provide a route of entry for the organisms.

Staphylococcus aureus produces a more acute disease than in the cow. Morbidity is 25% and mortality is greater than in the cows, being 25 to 50%. There are severe systemic disturbances with intense edema of the udder, which may extened upto the belly and gangrene may supervene, when the udder assumes blue color and so the condition is named "blue bag". There is serous or sero-fibrinous or purulent exudate in the acini, interlobular septa and in the interacinar septa. Alveolar exudate containns desquamated epithelial cells and leucocytes. Large scale necrosis may occur with abscess formation. Abscesses may rupture on the skin. Ultimately, the gland becomes fibrosed and functionless. Mastitis caused by other organisms is less severe than that produced by *Staphylococcus aureus.*

Contagious agalactiae of goats and sheep :

This is a disease primarily of goats but slightly infective to nearby sheep also The causal agent is *Mycoplasma agalactia.*

Though the udder is mostly affected, the disease may run a septicemic course in which mortality is heavy (10 to 33%). Later, infection localises in the eyes, the joints and the udder.

Infection is probably by ingestion, though it may also occur by way of teat-canal and the conjunctiva. The infective agent is eliminated in the secretions and discharges.

Both adult goats and kids are susceptible. Pregnant animals may abort. If live kids are born, they may be found infected.

The lesions noticed are :

Mammary gland : The inflammation commences in the interstitial tissue, with fibrosis. Later acini may be involved and they may be atrophied as the fibrous tissue increases and encroaches on the acini. Ultimately the udder is completely fibrosed and so lost. Organisms are voided in the milk.

Eye : In about half the number of cases mucopurulent conjunctivitis and keratitis complicated by ulceration are noticed.

Joints : Mostly the carpal and tarsal joints are affected showing arthritis and periarthritis, manifested by lameness. The peri-articular tissues appear swollen due to inflammatory edema.

Mastitis in pigs.

Staphylococcal mastitis. S*tphylococcus aureus* causes sporadic cases. The condition is chronic with the formation of large fibrous nodules in the gland. The nodules may open out through sinuses and then pus containing small granules may be discharged. The udder is lost.

Mastitis caused by Coliform bacteria : This is usually found in sows heavily fed with concentrates and in those which farrow in unhealthy pens. Heavy losses are encountered.

The udder is swollen, hard, discolored and painful. There may be severe systemic reaction. Tne sow does not get up and does not allow piglets to suckle. So the piglets die of starvation. If they suckle, they suffer from enteritis. There may be concomitent metritis. The udder is lost and even if the pig survives, it will not be useful for breeding.

Mastitis caused by C. pyogenes.

This is very frequent in pigs. The infection may be *primary* or it may be *secondary* from metastatic involvement from a focus somewhere else. In the udder there are absecesses with central collection of greenish pus. Fistulae may be present opening out on the skin from the abscesses. Such animals are no more useful for breeding.

Mammary neoplasms : As descrebed earlier, tumors of the mammary gland are common only in the canine species in w ich the mixed tumor (fibrochondro-adenocarcinoma etc.) is frequently seen.   (pages 255, 256.)

THE MALE GENITAL SYSTEM

Testes	Seminal vesicles
Cryptorchidism	Prostate
Hypoplasia	Hypoplasia
Hematocele	Penis and prepuce
Hydrocele	Posthitis
Testicular degeneration	Balanoposthitis
Orchitis	Phimosis and Paraphimosis
Epididymis	Neoplasms
Spermatic cord	
Funiculitis	

TESTES

Cryptorchidism : In the fetal life the testes are located in the sublumbar region, from where they descend into the scrotum during later stages of pregnancy. Sometimes the testes do not descend and so are retained in the abdominal cavity. This condition is knowu as cryptorchidism.

Cryptorchidism may be unilateral or bilateral. The bilaterally cryptorchid animal is sterile The condition is more often seen in swine and horses. In horses it is believed to be hereditary, being a sex-linked dominant factor.

Because of the higher temperature in the abdominal cavity, the testes do not develop normally Spermatogenesis, therefore, is arrested. The interstitial cells of Leydig may be normal or hyperplastic. Testes are smaller and softer than normal.

Cryptorchid stallions sometimes develop increased sexual urge (Satyriasis) and become vicious and difficult to control. In dogs, cryptorchidism appears to predispose the testes to tumors. Seminoma and Sertoli-cell tumor are met with more often in such animals.

Hypoplasia of testes :

This condition, though seen in all animals is more frequent and of greater importance (economic) in the bull In the Swedish Highland cattle, hypoplasia of the testes is a problem, being hereditary. Hormonal disturbances, vitamin deficiencies and some poisons may be the other causes.

Hypoplastic testis is smaller in size and is harder in some cases. **Microscopically** the picture, varies with different degrees of hypoplasia. In severe conditions the tubules are narrowed and have only one layer of cells, which do not show mitotic activity, denoting cessation of spermatogenesis. The basement membrane may be hyalinised and thickened. Increase in the peritubular connective tissue is responsible for the hardness. No change may be noticed in the cells of Leydig which may be increased in number.

In milder cases varying degrees of spermatogenesis may be found, some showing giant spermatids.

Hematocele is the presence of blood in the tunica vaginalis. Apart from trauma locally, this condition is found whenever there is hemoperitoneum. In leptospirosis and Infectious Canine Hepatitis hemolysed blood is seen in tunica vaginalis

Hydrocele is the condition in which clear serous fluid accumulates in the tunica vaginalis. This condition may be an accompaniment of generalised edema or ascites. It may also be due to local inflammation as a result of trauma, when the fluid may be turbid or blood stained.

TESTICULAR DEGENERATOIN

Degeneration of seminiferous epithelium is the commonest type of bull infertility encountered. This condition may be unilateral or bilateral.

Causes : The causes of testicular degeneration are many and it is not possible to pin-point the primary cause, because the factors responsible for degeneration may have ceased to exist before the degeneration is noticed.

Failure of thermo-regulatory mechanism which is responsible for the maintenance of testicular temperature lower than that of the body is proved to be one of the influencing factors for testicular degeneration. Excess of scrotal fat, short Cremaster muscle, inguinal or scrotal hernia, periorchitis and dermatitis and edema of the scrotum, interfere with thermo-regulatory mechanism.

Physical : Excessive heat, freezing, trauma, hematoma and laceration of scrotum may cause testicular degeneration. The condition may also be noted in the animals recently transported.

Localised and systemic infections are also common causes of testicular degeneration. Fever, toxemia, inflammatory changes of tunica vaginalis, orchitis, inflammation of the scrotum and epididymitis, contribute to the testicular degeneration.

Belgian workers have found an enteric virus to be a cause of this condition.

Nutritional factors : deficiency of : Vit. A, phosphorus, protein and energy requirements.

Avitaminosis A causes testicular degeneration through inhibition of release of gonodotrophins

Vascular lesions : torsion of testes, testicular tissue biopsy, spermatic cord compression (while controlling hemorrhages), inflammation of testicular artery, especially in the horses due to migrating strongyl larvae may cause testicular degeneration.

Hyaline degeneration of arteriolar walls and thrombosis of testicular arteries which are associated with age also cause testicular degeneration.

Obstructive lesions of the head of the epididymis interfere with the flow of spermatozoa and secretions of the tubular system of testes. The back pressure thus caused produces degeneration of the germinal layers of the seminiferous tubules.

Toxic substances that cause testicular degeneration include a number of chemicals (arsenical dips), metals, rare earth salts, and radiation.

Injection of Cadmium chloride causes testicular degeneration through lesions of vascular endothelium resulting in thrombosis

The degeneration caused by highly chlorinated napthelenes is generally reversible. Spermatogonia-B and late spermatogonia-A are highly sensitive to radiation.

Hormonal imbalance of the FSH and LH and improper administration of hormones.

Auto immunisation with autologus or isologus spermatozoal materials.

Macroscopically, the gross lesions include soft and flabby consistency in normal sized or smaller testes. The tunica albugenia does not bulge on the cut surface. In chronic cases fibrosis of the testes may be recognised. In cases of extensive fibrosis the testis is firmer and smaller. Calcium deposition may be noticed as yellowish white flakes in the degenerated products of the tubules

In acute degeneration deposition of calcium is observed in the connective tissue.

Microscopically, the tunica albugenia is condensed, thickened and wrinkled Sometimes it becomes impossible to differentiate the hypoplasia from testicular degeneration. Histological changes vary with the severity and stage of degeneration. The degenerative changes may not necessarily involve the tubules uniformly. The entire length of some tubules may be affected, while in others only partial. During early stages of degeneration, failure of maturation of spermatozoa, and degeneration of spermatids, are evident. Many spermatids are observed to be necrotic. Some of these give rise to multinuclear phagocytic giant cells. During advanced degeneration cytoplasmic vacuolation and nuclear pyknosis are observed in the precursors of spermatids.   As the condition further advances denudation of layers of germinal epithelium occurs leaving the basement membrane exposed.   Lastly even the very resistant sertoli cells may become denuded.   The tubules then collapse and are replaced by connective tissue.

In some cases when the degeneration is not rapid, along with the changes in the germinal epithelium, thickening and hyalinisation of the basement membrane and an increase in the interstitial connective tissue are noticed.   Finally the the tubules become replaced by dense hyaline connective tissue.   In the lumen of stenotic tubules large polyhedral mononucleate and binucleate cells with granular eosinophilic cytoplasm are observed.   These may contain within their cytoplasm a golden yellow pigment—"wear and tear" pigment.   The nature of these cells is not known but are probably altered spermatogonia with an unusual capacity for survival.

Stenosis of a portion of tubules leads to stagnation of spermatozoa and tubular secretions and ultimately calcified foci desquamate and the basement membrane fragments.   Contact of degenerated sperm, with the connective tissue causes granulomatous reaction.

The semen picture is of great diagnostic value in testicular affections.   The semen volume is not generally affected but the density may tend to be poor.   The sperm mobility is affected.   Sperm count may be low.   The number of abnormal sperms increases ranging from 30% to 50%.   Detatchment of heads, variation of the size and the shape of the heads are common abnormalities. Presence of proximal protoplasmic droplets, looped middle piece and tail, tight coils of middle piece and tail, may be other abnormalities encountered.

ORCHITIS

Inflammation of the testes occurs more frequently in sheep, cattle and swine.
Causes :

1. **Trauma :** This is more common in rams, in which the testes are pendulous.

2. **Bacteria :** A variety of bacteria are responsible. Infection may be hematogenous or may be extension from the lower genital organs through the epididymis.

The most common bacteria causing orchitis are :- *Brucella abortus* in bulls, *Brucella suis* in boars.

Salmonella abortus equi and P. *mallei*—in equines.

Corynebacterium pyogenes, Corynebacterium ovis—in rams.

Orchitis produced by *Brucella abortus* in bulls. is mostly an acute condition with swelling of the scrotum, which is hot and painful.   Since the testis is located compactly in tunica albuginea, a tough fiibroelastic membrane, swelling of the testis to any appreciable degree does not occur.   It is only the accumulation of the inflammatory exudate in the tunica vaginalis and the scrotum that is responsible for the swelling seen.   The exudate is fibrino-purulent and sometimes hemorrhagic   Adhesion of the parietal and visceral layers of tunica vaginalis may occur.   Due to pressure by the tunica albuginea, and the action of the pathogen, necrosis of the testes occurs with suppuration and abscess formation. Abscess may open out on the scrotum.

Microscopically, there is mycrocyst formation, degeneration and desqumation of the tubular epithelium. Infiltration of lymphocytes, macrophages. and plasma cells occurs in the interstitial tissue and the organisms can be seen in large numbers in the epithelial cells and necrotic areas In many cases, an accompanying epididymitis is present.

In chronic cases, in the initial stages, miliary tubercle-like granulomas form (as in mastitis) in the tubules and intertubular connective tissue. Because of pressure *of the infiltrating tissue* and action of the pathogen, degenerative changes and atrophy of tubular epithelium occur along with fibrosis and shrinkage. Finally the testes are hard, shrunken and very much smaller in size.

In the boar, the characteristic appearance is the formation of multiple abscesses, consisting of a central necrotic and caseated material surrounded by epithelioid cells, Langhans' type giant cells, plasma cells. lymphocytes and a connective tissue capsule.

Salmonella abortus equi produces an acute suppurative orchitis in donkeys and horses The seminiferous tubules rupture and the sperms spill into the interstitial space, and so a foreign body reaction is set up. General systemic and febrile reaction is usually seen in this condition.

Pasteurella pseudotuberculosis rodentium produces suppurative orchitis in rams. Pathogen is transmitted from the rodents by tick bites.

Tubercular orchitis : The lesion is usually a secondary hematogenic infection from a primary focus somewhere else. Lesions noticed may be calcified miliary nodules in which the typical histology of a tubercle nodule may be noticed or it may be in the form of a diffuse caseating lesion radiating from the rete testes. Very often the epididymis is also affected.

Straus's test used for diagnosis of diseases caused by *Malleomyces mallei* (glanders), *Malleomyces pseudomallei* (melioidosis). *Cryptococcus neoformans* (epizootic lymphangitis), *Corynebacterium ovis* (ulcerative lymphangitis and ovine lymphangitis), *Brucella abortus* and *Pseudomonas aeruginosa*, is essentially a suppurative orchitis and peri-orchitis in a male guinea-pig that results in 2 or 3 days following an intraperitoneal injection of a small amount of the culture of the organism.

Tumors : Described under the chapter "Neoplasms". (Page 259)

EPIDIDYMITIS

Inflammation of the epididymis is usually seen along with orchitis.

Of special importance is the epididymitis in rams caused by *Brucella ovis*. In this condition no lesions may be found in the testes

Usually the tail of the epididymis is affected, where the lesion produced is a granuloma. The pathogen incites an inflammatory reaction at the place of its localisation with inflammatory edema, infiltration by lymphocytes and macrophages. Neutrophils may appear a little later. The epithelium first shows papillary hyperplasia and later undergoes hydropic degeneration accompanied by fibrosis of the interstitial tissue, rendering the organ to become hard Epididymis is enlarged to 3 or 4 times.

Occlusion of the lumen by the debris and exudate results in **spermatocyst** formation. Should this cyst rupture into the tunica vaginalis, a foreign body reaction is set up with dense adhesions between its visceral and prietal layers.

Though primary orchitis is not found in this condition, secondary degenerative changes with calcification may be seen in the seminiferous tubules due to stasis of sperm.

Spermatic cord : The inflammation of the spermatic cord which is usually seen after castration is called **Funiculitis**. This may be acute and necrotizing as seen in the pig or it may be chronic—**scirrhous cord**—in horses and cattle. It is common in the pig because in this animal contamination can occur very easily due to the proximity of cord to the ground and to the general insanitary conditions of the environment.

In the scirrhous-cord, there is excessive formation of granulation tissue in the stump of castrated cord due to infection, usually by staphylococci. (*S. aureus*). Abscesses with thick walls may be present in this tissue. Leucocytic infiltration is common. A characteristic appearance is the presence of granules in the inflammatory tissue consisting of colonies of bacteria surrounded by a zone of clubs and inflammatory cells as in actinomycosis. This is known as **Botriomycosis**. In the spermatic cord and testes of horses, verminous granulomas caused by wandering larvae of *Strongylus spp* may occasionally be found.

SEMINAL VESICLE

Segmental aplasia of ampulla and seminal vesicle usually occurs in association with segmental defects of the epididymis.

Duplication of seminal vesicle on one or both the sides may be occasionally seen.

Seminal vesiculitis : Inflammation of seminal vesicle is usually rare but this condition is of serious concern when it occurs in a bull which is used for artificial insemination. In these cases, the pathogenic organisms may be transmitted to a wide population.

Causes : *Brucella abortus, Corynebacterium pyogenes, PLV* and *Mycoplasma* are some of the organisms incriminated.

Macroscopically, seminal vesicles are enlarged and tender on palpation in acute stages. There is a tendency for loss of lobulation.

Microscopically, the acute stage of the disease is characterised by infiltration of alveoli and interstitial tissue with neutrophils. In chronic cases, lymphocytes, plasma cells and histiocytes become numerous in the interstitial tissue.

Not all cases of seminal vesiculitis can be diagnosed by mere rectal palpation. In some cases, the seminal vesicles may not be noticeably enlarged. For clinical diagnosis of vesiculitis, palpation of the vesicles is useful if they are considerably enlarged. If they are not enlarged, massage of the vesicles and stripping of the ampullae will force inflammatory cells into the urethra and exudate will drip from the penis. Semen collected following this procedure has a marked increase in leucocytes and there may be clumping of exudate so that the sample will have the appearance of curdled milk.

THE PROSTATE

Hyperplasia of the prostate : This condition is seen only in dogs. Dog above five years of age and especially housebred, are commonly affected.

Causes : The precise cause of prostatic hyperplasia is obscure. It is thought that excessive testosterone may produce this condition which yields to castration. Forced retention of urine in housebred dogs in sometimes thought to be one of the factors.

Macroscopically, the gland is very much enlarged, with either a smooth or nodular surface. Sometimes the biloted appearance of the gland is lost. Fluctuating cysts are palpable under the capsule.

Microscopically, the picture is one of hyperplastic adenoma in which the acinar cells are increased both in size and number. The supporting tissue is also increased. The epithelium is tall and is frequently thrown into folds as papillary projections into the lumen. There is always infiltration of lymphocytes and plasma cells in the interstitial tissue Some acini may be cystic with increased amount of secretion which pressess upon the epithelium, flattening it. The interlobular connective tissue may be increased. Bladder may show compensatory muscular hypertrophy.

Clinically, due to pressure on the rectum by the enlarged prostate constipation may be produced. Difficulty in micturition is attributed not to the pressure on or narrowing of urethral lumen, but to paresis of the bladder resulting from pressure of the enlarged gland on the parasympathetic nerves.

PENIS AND PREPUCE

Inflammation of the prepuce is called **posthitis** and that of glans penis **balanitis.** Usually both occur together as **balanoposthitis.** In the dog it is a common condition. Cause may be trauma or bacteria. There is catarrhal exudate with infiltration of leucocytes into the degenerated epithelium. Mucosal lymph follicles may be enlarged. In other animals balanoposthitis is associated with various organisms inculding *Pseudomonas aeruginosa*;*Corynebacterium pyogenes* and *C. renale*. This condition is also met with in bulls that cross cows suffering from "Infectious pustular vulvovagintis". In the last named tondition (known as infectious pustular balanoposthitis) pustules form on che preputial lining and glans penis, giving them a granular appearance. Infection does not extend into urethra. Edema of penis and prepuce may cause paraphimosis

Phimosis is a condition in which the penis cannot be extended from the prepuce, due to inflammatory swelling.

Paraphimosis is the opposite condition in which the extended penis due to inflammatory enlargement, cannot be withdrawn into the prepuce.

Neoplasms : In the bull, transmissible fibropapilloma is encountered. These are multiple and cauliflower-like.

In the horse squamous cell carcinoma may be met with

In the dog transmissible venereal tumor is common.

CHAPTER 21

THE MUSCULOSKELETAL SYSTEM

THE SKELETAL SYSTEM

The following terms indicate certain anomalies and abnormalities in the skeletal system.

Abrachia :—Absence of anterior limbs.

Amelia :—Absence of limbs. The scapula and pelvic girdle may be intact or rudimentary.

Apodia :—Absence of posterior limbs.

Micromelia :—All parts of limbs are present but are of smaller size.

Perodactyly :—Absence of some of the toes.

Adactylism :—Absence of all the toes in a limb.

Brachydactylism :—Abnormal shortening of toes.

Polydactylism :—More number of digits; seen in horse and pig.

Syndactylism :—Fusion of toes, seen in cattle and pigs.

Prognathism :—Having a long jaw,—pig-mouth condition in horse.

Brachygnathism :—Having a short jaw. —parrot-mouth in horse.

Lordosis :—Is the curvature of the spine with a ventral convexity due to heavy loads or heavy abdominal organs; terminal parts of the thoracic spine and the lumbar spine are involved. The spinous processes rub against each other and so periostic osteophytes develop.

Kyphosis :—Abonrmal curvature and dorsal prominence of spine—hump-back. This is rare in animals

Scoliosis:—Abnormal lateral curvature of the spinal column—may be congenital and sometimes inherited. May be due to diseases of bones like achondroplasia, osteodystrophy.

Torticollis :—Wry neck—twisting of the neck with an unnatural position of the head.

Various factors govern bone formation. Briefly these are :—

1. **Minerals:** Sufficient amount of calcium and phosphorus must be supplied in the food in correct proportion and the intestinal tract must be healthy and of correct pH for their absorption.

2. **Proteins :** Sufficient amount of protein must be fed for the formation of the ground substance.

3. **Vitamins :** Vitamins A, D and C control bone formation. Vitamin A is necessary for the proper stimulus and bone growth. Its deficiency produces inanition and growth rate is retarded. It is concerned with the metabolism of endothelial cells and so is required for the proliferation of endothelial cells of the capillaries, for their transformation into osteoblasts and for the erosion and removal of the calcified cartilage. Vitamin D controls absorption and utilisation of calcium and phosphorus How these are brought about is not clear. Vitamin C controls the formation of osteoblasts and so controls deposition of osteoid

4, **Endocrines :**

(a)**Parathyroid** controls calcium and phosphorus metabolism. The action of the parathyroid hormone is two fold (1) It increases the phosphate diuresis and (2) it produces hypercalcemia through its action on the osteoclasts which withdraw calcium from the bone.

(b) **Anterior Pituitary:** The growth hormone influences the growth of connective tissue, especially bone. Gigantism occurs where there is increased secretion of growth hormone and the bony growth is enormous.

(c) **Thyroid :** Thyroxine controls the metabolism of carbohydrates and fats and so energy production is under its control. Indirectly therefore, bone formation is influenced by the thyroid as energy prduction is controlled by it. In hypothyroidism there is retardation of endochondral bone formation and osteporosis which occurs due to negative metabolism balance.

(d) **Gonads and Adrenal cortex:** Bone, growing and mature, is affected by estrogens and androgens. Their mode of action is not clear. These hormones accelerate the epiphyseal closure and maturation of the bone. In deficiency of these hormones there is disproprotionate elongation of immature long bones.

The following three types of cells are found in the bone :—

1. **Osteocytes** are the ordinary bone cells that are found in the lacunae. These are old cells that cannot divide.

2. **Osteoblasts** are bone producing mesodermal cells and line the deep layer of periosteum, the endosteum and the Haversian canals These cells, like fibroblasts, have great power of proliferation and produce alkaline phosphatase. Osteoblasts secrete precursors of collagen and mucopolysaccharides. The latter act as the cement substance and in this is embedded collagen. These form the matrix of the bone called osteoid

3. **Osteoclasts** are the phagocytes af bone and are multinucleated Foreign body giant cells can be formed from them. These are under the control of parathyroid and under its influence remove bone.

Phosphatase : The alkaline phosphatase found in bone is formed by the osteoblasts   It is believed that this enzyme splits the organic phosphate compounds liberating excess of phosphate which upsetting the local calcium phosphate balance leads to the precipitation of calcium salts.   It is in this manner that mineralisation of the osteoid (the organic matrix of the bone) is believed to take place.   Evidence is now available that the phosphatase of bone is concerned in the elaboration and secretion of protein—the organic matrix.

Osteodystrophy denotes a disturbance in the growth of bone. Osteodystrophies may be acquired or congenital.   Causes are many and varied.

1.   Lack of minerals and vitamins (Rickets and osteomalacia etc.)
2.   Excessive hormones—gigantism, acromegaly. osteoporosis.
3.   Unknown causes.

RICKETS

This is a condition seen in growing young animals in which there is a failure of adequate calcification of bones.   Similar condition in adult mature animals, in which growth of bones has stopped, is known as *osteomalacia*—literalty meaning softening of bones.

Essentially, rickets is a deficiency disease—deficiency of calcium, phoshporus or vitamin D. The defiiciency of these may arise in several ways.
Deficiency of calcium :

1.   **Deficiency of calcium in the diet :** Inadequate calcium in diet may not occur in animals.

2.   **Improper balance of calcium and phosphorus :** Excess of phosphorus in the ration (feeding too much of bran etc.), may combine with calcium and form a relatively insoluble $Ca_3(PO_4)_2$ which is excreted in the feces.

3.   **Failure of absorption of calcium :** Calcium is mostly absorbed as $CaH PO_4$ and for this the medium must be acidic.   If the intestinal contents are excessively alkaline, calcium cannot easily be absorbed.

4.   **Formation of insoluble complexes :** (a) Oxalates and phytates present in some green leaves and grains respectively may form insoluble compounds in monogastric animals and are lost in the feces.

Excess of oxalic acid in leaves and excess of lactic, tartaric, and malic acids in silage bind calcium in large quantities.   Similarly acid breakdown products of proteins bind calcium if food is too rich in protein.   In too coarse a food greater amounts of hippuric acid is formed from cellulose. Too rich or poor fat reduce the utilisation of calcium.   Reduced body movement may also be a contributory cause.

(b) When sulphur is fed to chicks as a coccidiostatic, it combines with calcium to form insoluble compounds which are lost in the feces.

5.   **Steatorrhoea :** In dogs, fatty acids from fats that are not assimilated may combine with calcium, forming calcium-soaps which are lost in the feces. This is common in man in *Coeliac disease.* Since Vitamin D absorption is conditioned by absorption of fat, in steatorrhoea, Vitamin D also is not absorbed and this still further affects the absorption of calcium, since vitamin D is not only necessary for calcium absorption but it also increases calcium absorption by the intestinal mucosa.

6. Renal Disease : In nephritis, phosphorus is not excreted as it should be and so accumulates in the blood and body. The excess phosphate ions are excreted through the intestinal tract, where they combine with calcium to form insoluble compound which is lost in the feces.

7. **Increased requirements in growing animals** : If adequate quantities are not allowed in the ration deficiency may arise in growing animals in which the needs for calcium and phosphorus are great.

Deficiency of Phosphorus :

1. **Inadequate amounts of phosphorus in diet** : In certain parts of the world soil is deficient in phosphorus and so animals maintained solely on the plants from such soils develop phosphorus deficiency. which is clinically manifested as osteophagia (pica). (See page 145)

2 **Change of reaction of intestinal contents** : As explained above an acid medium is required for the absorption of $CaH PO_4$. But if the reaction changes absorption cannot occur.

3 **Steatorrhoea** : Vitamin D is necessary for absorption of phosphorus also. In steatorrhoea Vitamin D is not absorbed and so is not available to the body and so phosphorus is not absorbed.

4. **Formation of insoluble complexes** : Excess of calcium, iron and aluminium form insoluble phosphorus compounds and so phosphorus deficiency results.

5. **Increased requirements** : In growing animals, if adequate quantity of phosphorus is not allowed in the ration, deficiency will result.

Deficiency of Vitamin D :

Insufficiency of vitamin D may occur in young animals.

Causes : I. **Steatorrhoea** : As mentioned earlier. vitamin D is not absorbed in the absence of fat.

2. **Diseases of liver** : If sufficient bile is not secreted, absorption of Vitamin D may be interfered with.

3. **Deficiency of sunlight** : Since vitamin D can be formed in the skin by the action of ultroviolet rays on ergosterol, deficiency of sunlight may cause deficiency of vitamin D. Smoke and smog in industrial places filter the ultraviolet rays and so rickets may supervene.

The essential defects in rickets are :

1. The cartilage cells are not calcified.
2. The cartilage cells are resistant to degeneration.
3. Failure of the blood vessels to invade and corrode the cartilage.
4. Overgrowth of the osteoid on the persistent and growing cartilage.
5. Increased growth of fibrous tissue in the osteochondral zone and
6. Absence of calcification of osteoid.

While describing the normal formation of bone it was mentioned that the cartilage cells nearest the diaphysis should become degenerated, must be calcified and then the capillaries nearby should invade and corrode the cartilage, on the remnant trabeculae of which the osteoblasts that arrive there form the organic matrix. the osteoid, over which calcium salts are deposited. In rickets, due to the deficiencies enumerated above these events do not take place. So the cartilage cells persist and grow and the zone of cartilage is wider and longer. Therefore chondrocostal and osteochondral junctions are widened and enlarged. Since osteoid is not calcified, the osteochondral zone is softer than normal. So, on pressure and due to weight the bones bend.

Symptoms noticed are stunted growth, bowing of the limbs. ("bow legs"). "Pot-bellied" appearance of the abdomen, kyphosis and scoliosis, enlargement of ends of bones and joints; bending of knees and fetlocks; overextension of pasterns with overgrowth of hooves, deformity of the pelvic bones (which later may cause maternal distocia), cranium is more dome shaped and the fontanels are wide ; "Rickety-rosary". (enlarged chondrocostal joints appearing as a string of beads) and crooked sternum in birds. Teeth may be poorly formed and irregular; jaws cannot be closed. Shortening of bones results In rachitic dwarfism.

Macroscopically, the epiphyseal cartilages are abnormal, wider and soft and so can easily be cut.

Microscopically, epiphyseal cartilage is wide and the osteochondral junction is irregular Tongues of surviving and resistant cartilage cells appear to be arranged in a disorderly and crooked manner Osteoid which is pink staining is abundant while the blue staining bony trabucule are few and widely separted.

Overgrowth of fibrous tissue occurs at the osteochondral zone and in the marrow. There is therefore, reduction of myeloid cells.

Restitution of the deficiencies corrects the disorder. But deformities persist eg. "bow legs" remain,

Diagnosis : X ray shows enlargement of epiphyseal plate, enlargement of epiphyseal line. and bending of bones. There may be decrease in serum calcium and phosphorus and increase in alkaline phosphatase.

OSTEOMALACIA

This condition is othewise known as adult rickets and occurs in animals in which endochondral ossification has ceased. Like rickets, this condition is characterised by failure of calcification of matrix and appearance of excess of osteoid. Atrophy of the bone substance is also present due to excessive resorption in bone consequent on negative mineral balance. Because of either excessive demands of the body for or deficiency in the dietary intake of the minerals, to meet the demands of the mineral requirements by the body, bone resorption occurs.

The medullary cavity is enlarged. Compact bone becomes spongy, In extreme cases all that is left is a membranous sac covered over by the periosteum containing traces of bone.

Because of lack of mineralisation the skeleton becomes soft and fragile and so fractures and deformities occur. The articular heads of some of the bones may sometimes separate. Kyphosis and lordosis are frequently seen together with narrowed pelvis.

The causes for osteonalacia ave similar to those of rickets :

i) deficiency of calcium and phosphours, ii) deficiency of Vitamin D and iii) chronic nephritis in which phosphorus excretion is diminished.

Osteomalacia may also be seen in pregnancy when maternal calcium is drained to the fetus and in high yielding cows in which large amounts of calcium are excreted through the milk.

Microscopically, the following may be observed : active resorption of bone by osteoclasts, reduction in size and number of the trabeculae of spongiosa and presence of excess of osteoid

Osteodystrophia fibrosa : This is a condition of the bone dependant upon disturbance in metabolism and frequently occurs in animals It is noticed mostly in animals imbalanced in calcium and phosphorus contents.   Normally, the Ca:P ratio in food should be 2:1.   But if this ratio is reversed and becomes 1 : 3 or wider, osteodystrophia fibrosa results.

In the horses, osteodystrophia fibrosa is known as "Bran Disease" or "Big Head" or "Millers' Disease" and is common among horses maintained by millers.   Since bran is a cheap by product in the milling of wheat, horses of the millers were maintained exclusively on bran which has a high phosphorus content   This phosphorus combines with calcium of the food and forms insoluble $Ca_3 PO_4$ in the intestines and is excreted in the feces.   Therefore sufficient amount of calcium and phosphorus is not available to the body and so hypocalcemia results.   This in its turn causes osteomalacia. Hypocalcemia stimulates the parathyroid, which becomes hypertrophic, producing excess of parathormone. This hormone acting on the bone (through the osteoclasts) decalcifies it producing osteomalacia.   All bones are not equally affected.   This affection is first noticed in the facial bones.   The bones that are most active are affected.   The bones become soft as calcium is withdrawn from them.   So as to strengthen the bones, fibrous tissue proliferation occurs.   Since soft bones bend and twist irritation is produced and this causes inflammation to occur which ultimately is responsible for fibrosis.   This fibrosis is most apparent under the periosteum hence the bone appears larger than normal.   The bone marrow may also be replaced by the fibrous tissue.   The facial bones of the horse appear swollen and hence is the name "Big Head".

To start with, the disease is manifested by abnormal gait, stiffness and shifting lameness. These symptoms are later followed by anorexia, anemia and cachexia.   The anemia is myelophthisic in origin—due to the replacement of marrow by the newly formed fibrous tissue.   There may be swelling of the jaws, dyspnoea due to narrowing of the nasal passages, difficulty in mastication, loosening and loss of teeth.   Fractures are common.

Microscopically, large masses of fibrous tissue are seen in which are found remnants of bony trabeculae. Sometimes the fibrous tissue may be of such proportion as to resemble a fibroma.

Osteitis Fibrosa Cystica : (von Recklinghausen's disease) This condition is a form of osteodystrophia fibrosa The essential nature is decalcification of bone, substitution by fibrous tissue and formation of cysts.

Csuses are :

Hyperparathyroidism :

    A. **Primary :** As in a tumor of the parathyroids.

    B. **Secondary :** i) Dietory calcium insufficiency, ii) Dietory phosphorus insufficiency,   iii) chronic renal disease.

It has already been noticed that the function of parathyroid is two fold.

    i) to facilitate excretion of phosphorus in the urine.

    ii) to remove calcium from bone through the mediation of osteoclasts.

The parathyroids are sensitive to blood calcium level. Any decrease in blood calcium stimulates the parathyroids and hyperplasia occurs with increased production of parathyroid hormone. Acting on the bone, this hormone is responsible for withdrawal of calcium through the activity of osteoclasts.

Normally glomeruli filter phosphates some of which are reabsorbed by the tubules. But in renal disease the phosphate excretion is much reduced and so the phosphate level of the blood rises – hyperphosphatemia. To compensate for the rise and to keep the Ca:P ratio constant, calcium is withdrawn from the bones. Besides, the retained phosphate is excreted through the bowel where it combines with calcium and forms an insoluble compound and so is lost to the body resulting in hypocalcemia, which stimulates parathyroid liberating excess of hormone and this withdraws calcium from the bone to maintain normal blood calcium level.

The function of parathyroids is therefore homeostasis, to maintain the optimum Ca : P level in the blood.

Under the circumstances described above when calcium is removed from the bones, they become soft and weak and so to strengthen them, there is fibrous tissue proliferation. This change, though found in all the bones, is more prominent in the bones of the head. In the dog, the lower jaw becomes so soft that it is as pliable as rubber. "Rubber-Jaw syndrome". Since the newly formed connective tissue is poorly supplied with blood. degeneration, softening and cyst formation occur—hence the name Osteitis fibrosa cystica.

Microscopically, osteoid and fibrous tissue are more in evidence. Attempts at formation of new bone in some places is evident by the presence of osteoblastic activity. In others, osteoclasts are seen nibbling away spicules of bone. Cysts of varying sizes and hemorrhages are seen.

OSTEOPOROSIS

In this disorder, there is reduction in the bony matrix. But what is present is fully mineralised (whereas in rickets and osteomalacia tissue matrix is formed but inadequtely mineralised). The bones become porous and brittle, as in this condition destructive proeesses exceed the productive in the remodelling of bone. Biochemically, the blood levels of calcium and phosphorus are normal.

Causes :

1. **Senility** : causes not fully known Probably due to decreased osteoblastic activity or decreased sex hormones (see below).

2. **Lack of protein** as in loss of protein (renal disorders) or decreased production as in liver disease or defective absorption due to intestinal disorders. Protein is essential for the formation of osteoid, without which bone cannot be formed.

3. **Deficiency of vitamin A**.

4. **Deficiency of Vitamin C** : osteoblasts and osteoid are not formed.

5. **Local pressure** on bones may cause atrophy :– for example tumors, *Coenurus cerebralis,* hydatid cysts and pulsating arterial aneurysms in contact with vertebrae.

6. **Disuse**: For proper healthy condition of the bone to be maintained exercise is necessary. If a part is immobilised for a long time bone of the part

becomes thinner and porous due to increased activity of the osteoclasts and inctivity of osteoblasts (due to lack of normal stimulus of stresses and strains).

7. **Loss of nerve supply** to the part results in paralysis and so the part cannot be moved and osteoporosis will result.

8 **Deficiency of trace elements**—copper deficiency in dogs, manganese deficiency in pigs and zinc deficiency in fowls.

9. **Hyperthyroidism** : Osteoclastic activity is probably increased.

10. **Hyperparathyroidism** : Increased resorption of bone occurs.

11 **Cushing's Syndrome** : Excess of glucocorticoids probably suppresses the osteoblastic activity. Bodies of vertebrae are severely affected.

12. **Lack of either androgens or estrogens** : In human pathology, osteoporosis is frequently observed after menopause. The sex hormones appear to have some, influence over the osteoblasttc activity.

13. **Poisons**—For example lead posioning in sheep and goats causes osteoporosis.

Macroscopically, the bones appear lighter and thinner—atrophied. The cortex is thinner but the marrow cavity is wider. Bones become brittle and so are prone to fractures.

Microscopically, the bony trabeculae are thinner with decreased number of osteoblasts. Osteoclastic activity denotes destruction of bone.

Alkaline phosphatase of serum is normal Can be diagnosed by lzuka's test·

PULMONARY OSTEOARTHROPATHY

This condition is known as Marie's disease in humans and found in all animals, though of greater incidence in dogs. The lesions are found in the lungs. The bones of the limbs are affected, while the joints are frequently not.

There is formation of new bone, mostly under the periosteum, which is therefore pushed out (Periosteal hyperostoses). As the osteophytic formation is not even, the bony surface is rough. Osteophyte means a bony excrescence. The articular surfaces are free Joints may be swollen due to periarticular proliferation

In the lung, foci of new bone formation are seen.

Causes : The exact causes are not known. This disease is noticed in the following conditions :

1. Chronic disease of heart and lungs—bronchiogenic carcinoma, bronchiectasis, emphysema, chronic tuberculosis and congenital heart disease.

2. When there is interference in the vascular supply to the extremities.

3. Passive congestion of the affected parts.

4. Neoplastic condition of the lungs.

5 *Spirocerca lupi* infection.

6 *Dirofilaria* infection.

It is surmised that anoxia, probably with some obscure toxins, is the causative factor. It is also thought that the skeletal changes are the result of reflex vasomotor disturbances in limbs secondary to circulatory disturbances in the lungs. A familial predisposition is noticed in man.

OSTEITIS AND OSTEOMYELITIS

Inflammation of the bone is called **osteitis,** and that of periosteum is **perio-stitis**. Inflammation of bone marrow is known as **osteomyelitis**. Inflammation of vertibrae is **spondylitis.**

Osteitis and Osteomyelitis may be acute or chronic. Acute purulent osteo-myelitis is always caused by bacteria which gain entry into the bone in the following ways.

   1 Direct :
      (a) through compound fractures
      (b) gunshot and other wounds.
   2 By lymph vessels in draining neighbouring purulent areas such as :
      (a) purulent arthritis
      (b) purulent periostitis
      (c) gathered-nail wound
      (d) suppurative otitis media.
   3. By blood stream from a suppurative lesion elsewhere and in pyemia.

This condition is not so frequent in animals as in man and the organisms that cause it are *Pyogenic bacteria S necrophorus, Erysipelothrix rhusiopathiae, Salmonella* and *Cryptococcus neoformans.*

Acute periostitis may be non-suppurative. usually caused by trauma (con-cussion) and is seen in horses as "sore shins" (due to working on hard roads).

Macroscopically, in periostitis, the usual inflammatory reaction is seen in the periosteum, hyperemia with purulent exudate accumulating between the cortex and periosteum. The exudate may separate the periosteum from the bone and necrosis of the cortex results. Periosteum may be ruptured, liberating the pus into the nearby tissue. Since periosteum is in continuity with the endosteum and medulla, pus may pass on to these structures. In such an event, necrosis of the bone occurs due to separation of both periosteum, and endosteum on which the nutrition of the bone depends.

In suppurative osteomyelitis, pus is found in the medullary cavity and it may burst through the cortex But more often, such a drainage is difficult and the condition progresses to a chronic stage.

In the young growing animals abcesses are found at the chondrocostal joints and in the epiphyseal plates.

The necrosed bone is separated from the healthy bone by the action of osteoclasts and a *sequestrum* is formed. Osteoblasts nearby are active and produce new bone, which forms a case, as it were, around the sequestrum and this is known as *involuctum.* Pus is discharged to the outside from the sequestrum through small openings in the involucrum called *cloacae.*

Sequelae :

   1. Pathological fracture due to extensive destruction of the bone.
   2. Chronic osteomyelitis.
   3. If suppurative osteomyelitis is extensive and present for a long time amyloid degeneration may occur.
   4. Resolution and healing with timely treatment.

5. Suppurative arthritis may occur due to extension of infection to the neighbouring joint; metastatic abscesses.

6. Death due to pyemia and septicemia.

Chronic osteomyelitis :

Causes : 1. May be a sequel to osteomyelitis.

2. Repeated injury or concussion, in horses especially.

3. Bacteria—of low virulence—*Actinomyces, Brucella, Mycobacterium tuberculosis, Salmonellosis*

4. Fungi—*Coccidioidomyocosis.*

In the case of chronic trauma and concussion **exostosis** results. This is the formation of granulation tissue of the bone, Just as fibrosis occurs in chronic inflammation of soft tissues, so also, in the bone, chronic inflammation results in the formation of new bone. Essentially this is a result of chronic ossifying periostitis.

In the horse special names are given to exostoses occurring in certain locations

1. **Ring bone,** if the exostosis is found on the 1st or 2nd phalanx. This is a painful condition causing lameness.

2. **Splint :** Exostosis at the end of metacarpal or metatarsal bones; not usually painful and so no lameness seen

3. **Spavin :** Exostoses on the medial portion of the distal tarsal bones, causes lameness as the bony growth pinches the cunean tendon, which passes over it.

The exostoses or osteophytes have the structure of a compact bone, but do not have haversian system.

The lesions produced by bacteria in chronic osteomyelitis are granulomas. **Microscopically.** in chronic osteomyelitis, centres of pus are surrounded by granulation tissue and inflammatory cells. consisting mostly of mononuclears and a few giant cells. Due to activity of the osteogenic layer of the periosteum new bone is formed and so the shaft is thickened and marrow narrowed—**osteosclerosis** In actinomycosis and tuberculosis there is rarefaction of bone—**rarefying osteitis.** In tuberculosis, there is extensive destruction of bone with the formation of caseous material but new bone is not formed.

FRACTURES

A fracture is a break in the continuity of a bone and is usually due to trauma.

Varieties of the fractures :

1. Simple fracture : Fracture of bone without an opening over the overlying skin.

2. Compound fracture : Fracture with an opening on overlying skin.

3 Comminuted fracture : The bone is splintered into many pieces

4. Impacted fracture : When one fragment of a fractured bone is firmly driven into the other.

5. Greenstick fracture : Here one side of the bone is broken while the other is intact as occurs when a green stick is bent.

6.   Pathological fracture :   The fracture is not due to trauma only but due to some bone disease existing. Eg. osteosarcoma.

7.   Articular fracture :   When joint surface of a bone is involved.

8.   Depressed fracture :   In the skull where the involved bone is depressed below the surface

9.   Linear fracture .   Here bone is split lenghthwise.

10.   Transverse fracture :   Fracture at right angles to the axis of the bone.

11.   Multiple fracture :   Here are two or more lines of fracture of the same bone but not communicating with each other.

12.   Oblique fracture :   Break extends in an oblique direction.

Healing of fracture :

Along with fracture of bone, there is hemorrhage as the blood vessels near- by are torn and ruptured.   Moreover, the capillaries of the haversian canals also contribute to the hemorrhage.   Because of ischemia (due to cessation of local circulation) bone cells die and these incite an inflammatory reaction.   The accu- mulated blood clots and in twenty four hours this clot is invaded by fibroblasts and capillaries from the periosteum and is organised.   This fibro-vascular tissue is strong enough to keep the two broken ends together and is known as **a soft tissue callus.**   (Callus, Latin for a hard substance).

Osteoblasts dervied mostly from the deeper layer of the periosteum invade the blood clot along with the capillaries and within 4 or 5 days trabeculae are formed around central spaces which become Haversian canals. This is the osteoid laid down by the osteoblasts.   This osteoid is well formed by the end of second week   Osteoblasts are also formed by metaplasia of the fibrous tissue.   Later calcium salts are deposited on the osteoid to form bone   The newly formed bony tissue unites the two ends of the fractured bone and is known as **provisional callus.**

The callus formed by the periosteum and located sub-periosteally is called *external callus.* that present in the medullary region is called *internal callus* and that between the ends of the shaft the *intermediate callus* or *in-line callus*

The callus formed is larger than the outlines of the bone and so bulges on the periosteal side   In the beginning there is no orderly arragement of the trabeculae and haversian systems.   Later the provisional callus is removed by osteoclasts and remodelled by osteoblasts into regular bone   This is called **defi nitive callus**   It may take several months for this definitive or hard callus to form.   Finally, during the remodelling processes, excess of the callus is removed.

If the gap between the two ends of a broken bone is too wide, the fibro- blasts of the provisional callus may become cartilage cells by metaplasia and this is later converted into bone—endochondral ossification.

Factors that interfere with healing

1.   **Non-alignment of the two ends of the bone :**   Due to this, deformity, excessive callus formation and displacement may occur

2.   **Infection**   This is common in compound fracture, leading to necrosis and osteomyelitis, which retard the process of healing

3   **Deficiency of calcium, phosphorus, vitamin D and proteins:** These may occur in dietary deficiency, starvation, metabolic or infectious diseases eg. renal

disease, malabsorption diseases due to gastro-intestinal pathology, parathyroid disorders ; excessive loss of protein as in albuminuria or heavy stomach worm infections.

4. **Presence of foreign bodies** hinder normal and rapid healing. These may be bullets, muscle, fat or clothing.

5. **Fragments of necrotic bone :** This is more common in comminuted fractures where, the necrotic bone acts as a foreign body, producing inflammation and preventing healing.

6. **Inadequate immobilisation : A false joint** or pseudoarthrosis may occur if the fractured ends are not firmly immmobilised. The provisional callus is not sufficiently mineralised and so permits bending at the fractured area.

7. **Senility :** In older animals, healing is slow due to decreased vascularity and retarded metabolic processes.

8. **Pathological :** presence of osteodystrophy or neoplasms prevents healing of fractures

Tumors of bones :

Primary Fibroma, myxoma, lipoma, chondroma, osteoma, chondro sarcoma, fibrosarcoma, osteogenic sarcoma, and Giant-cell tumor

Secondary: Metastatic carcinoma and sarcoma from other parts of ths body.
Chondrodystrophia foetalis (Achondroplasia).

In certain breeds of cattle, the Dexter and Norwegian Telemark, some calves do not attain full development of cartilage growth. So they become dwarfed. The exact cause is not known, though endocrine deficiency is suspected. But the condition is known to be hereditary, transmitted by a lethal factor.

Since endochondral ossification is retarded and even stopped, the bones of the limbs and extremities are shortened. There is brachygnathia and so the animals have a bulldog appearance of the head (Bulldog calf). The skin is thick and much folded due to subcutaneous edema. The calf has a vaulted skull. Growth of the skull being by intramembranous ossification, the head is disproportionately large Such calves are usually aborted between their third and eighth months of intrauterine life.

JOINTS

Inflammation of the joints is called **arthritis**.

Inflammation of hip joint is called **Coxitis** while that of sitfie joint **gonitis**. Arthritis may be acute or chronic.

Acute Arthritis :—

Causes :

1. Contusion or strain in which there is stretching of the joint capsule.

2. **Bacteria :** Routes of infection may be (a) via blood stream (b) by extension from neighbouring tissue and (c) by puncture wounds.

Trauma usually produces a *serous type of* inflammation in which there is increased production of synovia, distending the joint capsule. The condition is mild showing a slight hyperemia of the articular cartilage and the synovial membrane.

Bacterial arthriris may be conveniently classified as :
1. Non-suppurative and
2. Suppurative.

In non-suppurative arthritis, there is acute serous or serofibrinous exudate.
Causes are :—

Erysipelothrix rhusiopathiae (serous polyarthritis) in sheep and pig
Hemophilus influenza suis in pig.

Macroscopically, exudate contains yellowish flakes,which are often compressed into flat structures which float in the joint fluid   The synovial membrane is thickened and studded with hemorrhages,

Microsccopically, hyperemia and neutrophilic infiltration are common. Articular cartilage may be eroded.

Suppurative Arthritis: This condition is usually associated with Navel ill. The bacteria localise in the joints because of the rich blood supply there and also probably to the weak defences in that region,

The following organisms are incriminated.

Species of animal affected	Causative bacterium	Nature of lesicns
Calf	1. *E. coli*	Septicemia and acute arthritis, cloudy synovia. Many organisms in the joint,
	2. *Corynebacterium pyogenes*	Purulent arthritis with destruction of joint. Organisms may be found in pure culture or mixed with other organisms.
Colt	1. *Shigella equirulis*	Swollen joints
	2. *Streptococci*	Purulent exudate
Sheep	1. *Staphylococci*	Purulent exudate with
	2. *Corynebacterium*	joint destruction
Swine	1. *Streptococci*	
	2. *Brucella abortus*	Purulent exudate

Infection may be
a) Primary— penetrating wounds of joints.
b) Secondary—i) Extension of suppurating process from neighbouring lesions ii) metastatic lesions in pyemia.

Macroscopically, all the symptoms of an acute inflammation are seen notably swelling of the joint.   White, yellow or green pus may be present in the joint depending on infecting organisms. In mycoplasma infection it is thin and colorless.

In suppurative arthritis the articular cartilage is destroyed and infection may spread to the underlying bone Suppurative osteomyelitis, necrosis and caries of bone result   Particles of disintegrated bone are found in the pus, like grains of sand. Sometimes, the pus may be discharged through a break in the skin resulting in an open joint. The articular cartilage may be inflammed and eroded. Synovial fluid which is increased is purulent. There may be inflmmation of periarticular tissue.

Microscopically, there is infiltration by neutrophils.

If the condition becomes chronic, there is excessive fibrosis of the joint.

Sequelae : The condition has unfavourable prognosis in young animals. Due to pain, they will not be able to move about and in chronic condition, fibrosis and ankylosis of the joint will result.

Chronic arthritis

Causes :   Sequel to acute arthritis.

            Primary

            (a)   Chronic traumatic.

            (b)   Bacterial Tuberculosis in ox and pig, Fowl Cholera in
                  fowls

Chronic serous arthritis : Due to destruction of the articular cartilage, there may be fibrous adhesion between the two articular surfaces. Subsequently the two bones may fuse together producing **ankylosis of the joint.**

Tuberculous arthritis is characterised by the granulomatous inflammation. It is manifested in 3 forms :

i) **Miliary form** in which miliary nodules are found in the synovial mem-brane, The neighbouring tubercles may coalesce and project into the joint cavity as 'pearls'. This form is seen in pigs

ii) **Infiltrating tuberculosis** is seen in cattle, characterised by diffuse tuberculous granulation tissue, containing epitheliod cells and giant cells (Chronic organ tuberculosis.)

iii) **Caseating tuberculous** synovitis with caseation but without specific granulation tissue

In chronic stages of Fowl cholera *Pasteurella avium* may get localised in the joints and tendon sheaths, where there is an accumulation of a cloudy or cheesy material, thus giving the structures a swollen appearance.

Myocaplasmal arthritis in swine

Mycoplasma granularum is a common cause of arthritis in 100 to 200 pound swine. Heavy muscling, genetical background and stress act as predisposing factors.

The disease is usually an acute one, with sudden onset of lameness. The course runs for 3 to 10 days, Subsequently flare ups cause longstanding chronic arthritis.

Macroscopically, in the acute form, there is increased serosanguineous synovial fluid in the femuro-tibial, coxo-femoral, cubital or scapulohumoral joints. The synovial membranes are swollen, hyperemic and discolored but the joint capsule and articular surfaces appear normal.

Microscopically, hyperplasia of synovial lining cells, villous hypertrophy and extensive mononuclear infiltration are noticed.

Degenerative arthropathy Osteoarthritis—deformans

This condition should be distinguished from the conditions of joint resulting from arthritis. In arthropathy, no inflammation occurs initially, but is an ageing process　Normally, the young cartilage is white and translucent.　As it ages it becomes opaque, yellowish and less and less elastic.　To start with, th cartilage cells undergo hydropic degeneration and fatty changes.　The fibrils of the cartilage become visible.　Subsequently fissures form　on the cartilage followed by fibrosis of the ground substance.　The cartilage soon becomes separated and eroded, exposing the bone underneath.　The older cartilage has lost its power of regeneration and growth and so repair does not occur.

The cartilage being avascular depends on the synovial fluid for its nutrition So any changes that may occur in the synovial fluid as a result of ageing process may contribute to the degeneration of the cartilage.　When ulceration of the cartilage occurs, the bone is exposed and subjected to stress and becomes sclerosed and hard (eburnation).　Granulation tissue grows from the exposed bones of the two articular surfaces and thus fills the articular cavity.　This tissue subsequently becomes ossified resulting in ankylosis　At the margin or edges of the joint are formed periostitic exostoses.

The synovial membrane becomes fibrous and thickened　The villi become thickened, fibrous and long and contain fatty tissue.　Occasionally near the area of degeneration, some cartilage cells proliferate and form into small nodules which may be calcified following degeneration.　These nodules may become detached into the articular cavity.

Causes are obscure　The following are noteworthy :

1.　Probably it is an ageing process.
2.　Repeated trauma—as in concussion sustained by working horses on hard roads; sprains.
3　Obesity
4.　Faulty circulation.
5　Absorption of products of faulty digestion

Ring bone :　This is a condition of degenerative arthropathy affecting the inter-phalangeal articulation of horses, resulting in ankylosis and lameness.

The articular cartilages may be destroyed resulting in ankylosis due to union of the articular ends of the bones by granulation tissue which becomes ossified. So ankylosing arthrosis results—*articular ring bone.*

More often there is chronic inflammation of the periosteum and the ligamental apparatus due to repeated concussion and this results in **periarticular ring bone,** in which the exostoses occur as a ring round about the ends of bones. Some times these periarticular exostoses may fuse bridging the joint and fixing it.

Spavin : This is arthropathy of the tarsal joint affecting its distal and medial parts. Ankylosis may result.　The condition first starts with degeneration of the cartilages of the second and third tarsal bones.　Subsequently other tarsal and metatarsal bones may be involed

The normal white or bluish cartilage undorgoes degeneration, becoming opaque and fibrous.　It breaks down and ulcerates.　Granulation tissue from the exposed bone grows and fuses with that growing from the opposite end.　When this becomes ossified, the joint becomes ankylosed　No periarticular changes may be noticed　So such a condition is known as **occult spavin.**

In some cases, the synovial membrane may become thickened due to irritation. The fibrous layer of the articular capsule proliferates and then becomes ossified, resulting in large exostoses, which can be easily seen on the internal and medial aspects of the hock joint. These exostoses pinch the cunean tendon and so pain may be caused, resulting in lameness.

Ankylosing spondylosis : In this condition. the small vertebral articulations become ankylosed.

In the dog it is due to the protrusion of the nucleus pulposus irritating the periosteum and the ventral spinal ligament, resulting in exostoses which may subsequently fuse and join the vertebrae.

In bulls also this condition is met with in those that are used for stud. Due to frequent trauma attendant on their work. there is constant irritation. The lum. bosacral region is more often affected causing paralysis or ataxia.

Protrusion of inter-vertebral discs

This condition is met with in man and dog.

Normally, the intervertebral disc consists of a central *nucleus pulposus* which is semisolid mucoid connective tissue. This is enclosed in a thick fibrous covering, the *annulus fibrosus*. Due to violent trauma and degenerative changes in senility, there may be a rupture in the annulus, from which the nucleus pulposus escapes and becomes displaced. The susceptibility of the disc to degeneration is inherited. Usually two forms of displacement occur: (i) dorso-lateral prolapse of the nucleus pulposus into the spinal canal. (ii) ventral prolapse beneath the spinal ligaments. In this variety due to formation of osteophytes ankylosng spondylosis results.

In the chondrodystrophic breeds (Dachshunds, Pekingese, French bull dogs), at a very early age, the nucleus pulposus becomes cartilagenous, which later becomes degenerated and calcified. So the nucleus pulposus, which is normally a gel and so able to withstand shocks and transmits pressures uniformly to the annulus fibrosus, becomes transformed into a cheesy mass which crumbles easily. This material transmits pressure to localised portions of the annulus, which also undergoes degeneration. Its lamellae become hyalinised and later split.

In other breeds, the above changes occur in mid or later life.

The displaced nucleus pulposus presses upon the spinal cord producing nervous lesions. The protrusion of the disc may occur at any level, but occurs more frequently in the lumbar region or in the posterior thoracic region. Complete paralysis of the posterior region may be noticed. Pressure on the spinal cord may produce hemorrhage and necrosis in the involved area. Wallerian degeneration of the nerves may be noticed in the spinal nerves arising from the affected region as well as demyelination of nerve tracts

Symptoms include : (a) pain with exaggerated reflex movements which may be intermittent or occur over long or short periods. (b) Partial paralysis of the limbs (c) Violent reaction to stimuli—spastic type. (d) Rapid progressive paralysis and early death due to respiratory failure.

BURSITIS

Inflammation of the bursa over the joint is bursitis. This is of frequent occurrence in animals. Examples of these are :

Hygroma of the carpal joint in cows and ' capped elbow or hock joints of horses.

Causes :

1   Trauma, especially, if repeated.

2. Over-use

3   Infection (brucella infection in cows produces Hygroma and in horses "Fistulous withers." and "Poll evil").

Macroscopically, the inflammation may be serous, serofibrinous or purulent. Trauma produces serous type and one example is the serous bursitis of hock joint in the horses. This is called Bog spavin. Here the joint is filled with serous fluid

Pole evil is the inflammation of the bursa between ligamentum nuchae and atlas and axis

Fistulous withers is the affection of bursa between the ligamentum nucnae and the thoracic spines. The inflammation is a suppurative granulomatous one in which fistulae open on the surface of the skin

Causes may be traumatic, parasitic *(Onchocerca cervicalis)* or *Brucella abortus* and *Actinomyces bovis*. The suppurative and granulomatous reaction is attributed to the two organisms, infection occurring hematogenously

Navicular disease : This is bursitis and arthritis involving the distal sessemoid or navicular bone in the horse. Usually the fore limb is affected.

First there is serous inflammation of the lining membrane of the podotrochlear bursa, with hyperemia. This is followed by erosion and ulceration of the articular cartilage, over which the flexor tendon passes. Due to the changes in the cartilage, the tendrills of the tendon become frayed and ultimately rupture of the tendon may occur. Later the bone is inflamed, becomes rarefied and may fracture.

Infectious synovitis : This is a chronic disease of chicks. Morbidity and mortality are low. The characteristic lesion is a purulent synovitis of the leg joints

Causes : *Mycoplasma synoviae* and *Mycoplasma gallisepticum.*

Infection occurs in chicks, 12—14 weeks of age, by ingestion and the incubation period is 24 to 80 days.

Symptoms seen are :— emaciation, retarded growth, pale, comb, distended hock joints and swollen foot pads and lameness.

Macroscopically, in the early stages is found a creamy exudate in the synovial membrane of the joints, especially those of the hock and foot. This material becomes caseous as the disease progresses. The surface of the affected joints becomes yellow or orange.

In the early septicemic stage, the spleen, liver and the kidneys may be swollen.

Microscopically, brain may show gliosis and degeneration of purkinje cells. In the liver and spleen there is proliferation of the reticular cells of the reticuloendothelial system. Bile duct proliferation may also be seen.

With timely treatment using antibiotics the disease can be cured.

Sometimes a focal infiltration by mononuclears and necrosis of the myocardium and fibrinous pericarditis may by noticed. Thymus and bursa af Fabricius may be atrophied due to degeneration of lymphoid tissue,

DISEASES OF SKELETAL MUSCLE

Atrophy :

Etiology :

1. **Senility :** In old age there is gradual atrophy of all muscles. But cattle and swine are slaughtered young and so atrophy is not seen in them. In dogs, milch cows, horses and ewes which are allowed to grow old, atrophy may be observed. It is likely that there may be under nutrition or the animal is not able to metabolise available nutriments in old age. The muscle cells are not able to assimilate the nutrients and so catabolism exceeds anabolism.

2. **Disuse :** This is seen in fractures of bones when the parts are immobilised for long periods and the muscles therefore are not utilised. Disuse of the limb may also occur due to pain as in rupture of a tendon, acute arthritis, ankylosis and diseases of the bones and muscles.

3. **Starvation :** Sufficient food is not available to make up for the catabolism that takes place.

4. **Atrophy of wasting diseases, cachexia and malnutrition :** In chroni wasting diseases like tuberculosis and Johne's disease; in debilitating conditions like neoplasia; in cachexia and in malnutrition, the food digested is either not effectively metabolised or is not used by the body and so atrophy results.

5. **Denervation :** When a nerve is injured or severed, the muscles supplied by it become paralised and atrophied. Examples are; a) Atrophy of laryngeal muscles when the recurrent laryngeal nerve is injured (Roaring). b) Atrophy of supraspinatus muscles when the suprascapular nerve is injured. (c) Atrophy of muscles in lesions of the central nervous system—poliomyelitis, protrusion of intervertebral disc, tumors etc.

6. **Pressure :** Continuous pressure on the muscle, producing ischemia locally and interfering with movement will cause atrophy The cause of pressure may be tumors, abscesses, cysts, ill-fitting, collars and saddles as well as infiltrating lymphoid cells in neoplasia of these cells.

Macroscopically, the muscle which is normally pink, loses this color and turns pale, grey or brown. It is firmer due to replacement by fibrous tissue. Due to uneven atrophy of different muscles, disfigurement may occur. Skeleton becomes prominent.

Microscopically, the size of the muscle fibres is reduced. Sarcoplasm may become so reduced and in some places may even disappear, that the sarcolemmal nuclei become prominent. There may be deposition of "wear and tear" pigments at the poles of nuclei giving the muscle a brown color—'brown atrophy' The cell nuclei may proliferate and fill the empty sheath.

As the etiological factors for atrophy also cause degenerative and necrotic changes, cloudy swelling, fatty degeneration and coagulative necrosis may be encountered.

In later stages there may be infiltration of fat in some areas *(atrophia lipo-matosa)* and fibrosis in others.

Inflammation of muscles :

Myositis : Inflammation of the muscle is called myositis. This may be acute or chronic.

The routes of infection and causes are :

1) Trauma. 2) By direct extension from lesions of neighbouring arthritis, osteitis or periostitis. 3) In pyemia, hematogenously. 4) By parasitic infection.

Acute myositis may be non-suppurative.

Best example of acute non suppurative myositis is Black quarter in cattle and sheep. In this condition, the organisms, *Clostridium chauvoei* causes inflammation and necrosis of the muscles with production of gas. The muscle fibres are torn by the gas bubbles. Local hemorrhage is present and the area is black due to formation of black iron sulphide. Regional lymph nodes are actutely congested and parenchymatous organs show fatty changes. Serous cavities contain blood stained serous fluid.

Microscopically, there is necrosis of the muscle, infiltration by neutrophils and clumps of the anaerobe.

Suppurative myositis : Hematogenous infection may occur from other foc as in Strangles and Glanders. Infections may occur in lacerating and penetrating wounds or by extension from adjoining areas.

The usual changes of suppurative inflammation are found, viz. abscesses or phlegmon Microscopical appearances are typical of any other suppurative inflammation with a great outpouring of netrophils. The muscle fibres undergo liquefaction following coagulative necrosis.

Sequelae : As there is loss of muscle tissue, healing is by means of fibrous tissue proliferation and scar formation. If severe, septicemia may result.

Chronic myositis : Examples are found in infections by Actinomycosis and Aetinobacillosis. The muscles of the tongue, cheek and throat are affected. The lesions consist of chronic suppurative myositis in which the "Sulphur granules" are noticed in a mass of inflammatory granulation tissue. There is infiltration by large number of lymphocytes, neutrophils and plasma cells. Muscle fibres are destroyed.

Parasitic myositis : The following parasites are found to infect the muscles of animals.

i) **Toxoplasma.**

ii) **Trichinella spiralis** is fonnd in man, pig and other animals. The larvae are encysted in many muscles, especially those of diaphragm, intercostal muscles and tongue. The cysts are parallel to the muscle fibres which undergo granular degeneration of the sarcoplasm. Intense infiltration by eosinophils, plasma cells, histiocytes and lymphocytes occurs. Sarcolemmal nuclei proliferate. The encysted larvae may be alive for as long as 20 years.

iii) **Sarcosporidia** are present in the skeletal and cardiac muscles of all species of animals. No specific disease has been attributed to these parasites. though light infections cause no perceptable symptoms, heavy infections may be responsible for lameness, weakness, paralysis, emaciation and sometimes even death. Parasitised muscle fibres are destroyed by the parasite and the adjacent cells undergo pressure atrophy.

iv) Cysticercus, (Measles).

a) Cysticercus cellulosae : The bladder worm of *Taenia solium*, a tape-worm of man, infects the muscles of pig The muscles of the shoulder, neck, diaphragm, tongue, intercostals, abdominal and cardiac muscles are affected. Heavy infection may result in fatal anemia and cachexia.

b) Cysticercus bovis is the intermediate stage of tape worm *Taenia saginata* of man, found in the muscles of cat le. All muscles may be a affected but especially those of tongue, mastication and heart are more often infected.

c) Cysticercus ovis is the intermediate stage of dog tape worm *Taenia ovis* and is found in the muscles of sheep.

Diseases of unknown origin

Eosinophilic myositis :

A: Cattle and Sheep :

Very rarely, at slaughter, yellowish green areas may be noticed in the lingual, oesophageal, cardiac and diaphragmatic muscles of cattle and sheep.

The green color fades on exposure to light.

Microscopically, large numbers of eosinophils, histiocytes, plasma cells and lymphocytes, are found between the muscle fibres and in tissue spaces. Though extensive degeneration of the muscles may not be noticed, in some places necrosis and invasion of muscle fibres by the eosinophils are observed. In more chronic cases fibrosis is evident.

There is some suspicion that the condition may be a manifestation of allergy and sarcosporidia are mentioned in this connection as the sensitising factor.

B. Dogs.

Clinically, the condition involves the masseter, temporal and pterygoid muscles chiefly. Other muscles may be affected. The muscles of mastication are enlarged bilaterally so that opening of the mouth is painful and mastication is interfered with The eyes bulge out, resulting in keratitis and corneal ulceration, since eyelids cannot close completely. German shepherds and Alsatians are more often affected. As the animals cannot eat, they die finally, of inanition.

The local tonsillar and mandibular lymph nodes are also swollen. Blood picture reveals high eosinophilic count upto as much as 90%. Temporary remissions may occur but are followed by repeated attacks and the animal finally dies.

Macroscopically, the affected muscles which are swollen and hard to the touch, show grey and red streaks and white and yellow spots Hemorrhage is present The regional lymph nodes are congested and swollen.

Microscopically, there is heavy infiltration of eosinophils, lymphyocytes, plasma cells and macrophages into the muscle, producing atrophy, and hyaline necrosis, vacuolar degeneration and lysis of muscle fibres. Hemorrhage is common. The necrotic muscle is removed by macrophages and fibrosis follows. In the liver is found periportal lymphocytic infiltration.

Cause : Unknown.   Some kind of allergy is suspected ... Probably a nutritional problem, Vitamin E deficiency, may have something to do with the condition

White muscle disease :   Stiff-lamb disease.

This is essentially a coagulative necrosis of the muscles due to various causes.   The disease occurs in calves and lambs and can be produced in rabbits and guinea-pigs.   The clinical picture is classified into three main types.

1. **The stiff type ;**   The head is carried low and has a drooping posture. Animal experiences difficulty in rising and walking.   While walking, the gait is stiff.   The weight-bearing and active muscles, for example, muscles of the croup and quarters. diaphragm, heart and intercostals, are mostly affected. In lambs this is the form encountered and the animals are always recumbent and do not like to move.   On forcible movement, they have stiff gait and wobble. (Stiff-lamb disease)

2. **The respiratory type :** Here the muscles of respiration (diaphragm and intercostal muscles) are affected and the animal may show symptoms of respiratory distress.

3. **The cardiac form :**   In this type, animals show considerable weakness, inability to stand, rapid pulse and low blood pressure.   Since the heart is affected and weakened, exertion brings on respiratory distrees and even death. In animals with cardiac involvement alone, sudden death occurs without any other symptoms.

Macroscopically, those muscles, which are continuously active, viz , diaphragm and intercostal muscles, show the changes. The muscles are bilaterally affected and are pale like fish flesh   The whole muscle bundle may not be affected but only a part of it will show the change.   The muscles become hard and wooden. The paleness is due to loss of myoglobin which is excreted in the urine. The change in color is also due to changes in optical characterirtics of the muscle protein when it becomes coagulated.   Pneumonia, edema, hydrothorax, C.V.C of liver and hydropericardium, will be found when heart is involved. Heart shows yellowish or grey streaks or patches.   The left ventricle is more often affected

Microscopically the muscle fibres are swollen with loss of striation and with widespread hyaline degeneration.   This progresses to coagulative necrosis. Fibres are fragmented and may completely disappear. Marked sarcolemmal proliferation is present Some fibres may be calcified, Infiltration by macrophages and lymphocytes is seen.   Similar lesions may be found in the heart.   In some places healing by fibrous tissue is evident.

One noteworthy feature is that the nerves and C. N. S. are normal without showing lesions.

Clinically the serum glutamic oxaloacetic transaminase (SGOT) level will be more than 300 units while the normal is less than 100 units.

Causes :

1. **Vitamin E deficiency :** Vitamin E is an antioxidant and in its absence, oxidation in the muscles is increased to 400 times the normal and so degeneration and necrosis occur.

Vitamin E deficiency may occur in the following manner.

a) Dietetic deficiency.

b) Feeding too much of cod liver oil. The unsaturated fatty acids in th 4od liver oil antagonise Vit E.

2. **Selenium deficiency :** Selenium is required in minute quantittes. In its absence muscle necrosis occurs. Selenium deficiency can occur in the following manner.

a) Deficiency in the soil;animals that are grazed on fodder grown on soils deficient in selenium suffer from the diease.

b) Excess of sulphur, used in fertilsers, inhibits the uptake of selenium by plants,

3. **Vitamin B deficiency :** It is found that Thiamine deficiency, especially, produces cardiac necrosis. Deficiency can occur in animals when the ruminal flora are not active to synthesize the vitamin as occurs in cobalt deficiency.

4 **Abnormal ruminal fermentation :** Some toxic products are probably produced in the rumen that cause muscle necrosis

5. **Deficiency of choline** produces muscle necrosis in rabbits (experimentally)

6. **Vitamin A deficiency :** Probably vitamin A deficiency produces this disease in swine

7. **Multiple deficiencies :** Lastly, in starvation as occurs during drought and malnutrition, multiple deficiencies of vitamins and minerals (phosphorus) may occur and muscle necrosis may be encountered. Similar lesions are seen in hypothyroidism.

AZOTURIA (equine myoglobinuria, Monday-morning sickness; Paralytica hemoglobinuria)

This disease in the horses, literally means "nitrogen in the urine."

Azoturia is found to occur suddenly in horses going to work after complete rest for a few days but maintained on full work-rations. The animals suddenly stop, sweat, shiver and show great suffering from pain in the lumbar region. The affected muscles, which are those of gluteal, lumbar and femoral regions, are swollen and board-like. Soon the animal passes coffee colored dark-brown or black urine since it contains large quantities of mygolobin. Animals lie down and soon die. Those that survive, are weak and it takes a long time for them to recuperate and for the atrophied muscles to regain their normal state.

Pathogenesis

In normal muscle contraction, muscle glycogen is converted into pyruvic acid ($CH_3COCOOH$) Due to inadequate oxygen,only 1/5 of this is oxidised to CO_2 and H_2O to liberate energy The rest is converted into lactic acid ($CH_8CHOH\ COOH$) which is converted into glycogen in the liver and used again When the animal is at rest but well fed, the muscles are well stored with glycogen. When it is put to work suddenly much of this glycogen is converted to lactic acid in the muscles and large amounts of this stimulate extreme contraction of the muscles, which become hard (board-like) In the contracted state of muscles, blood circulation is poor and so oxygen supply is reduced. Under this hypoxic condition more of lactic acid is formed (from pyruvic acid) which still further contracts the muscles and so greater curtailment of blood flow occurs leading to still greater reduction of oxygen supply.

Thus a vicious circle is established, the net result being that the muscles do not get sufficient amount of oxygen and nutrition, and so necrosis results. Necrosed muscle liberates myoglobin which is excreted in the urine. Large masses of myoglobin in the urine appear to produce renal blockade, renal ischemia and lower nephron nephrosis, wherein the epithelium of the distal convoluted tubules as well as that of Henle's loops are degenerated, some of which become necrosed and desquamated. Renal vasoconstriction, that may be caused by the same factors responsible for the hemoglobinuria, produces renal ischemia. This condition causes degenerative changes in the tubules and so anuria and fatal uremia result

Macroscopically, the affected muscles are swollen, pale and have increased amount of interstitial fluid.

The affected kidneys are swollen and on section the cortex is brownish and medulla has reddish streaks.

Urine shows granular reddish casts and a few hyaline casts.

Microscopically, the changes in the muscle are those of Zenker's degeneration, in which the muscle becomes a homogeneous hyaline mass without striation. The fibres may be fragmented There may be disappearance of all the constituents of the muscle fibre excepting the sarcolemma and fibrous stroma This is the cause of atrophy noticed in surviving animals. In animals that survive regeneration may occur, but it is a very slow process

In the kidneys, lesions are found mostly in the tubules. The epithelium of the proximal tubules may be degenerated and all stages from cloudy swelling to necrosis are encountered Desquamation of epithelium occurs. Similar changes may be noticed in the epithelium of Henle's loops and distal convoluted tubules. The lumens of the tubules may contain, besides the desquamated cells, masses of myoglobin. These form granular pigmented casts. A few hyaline casts may also be found.

Death is due to renal insufficiency leading to uremia.

TUMORS

Primary :

Rhabdomyoma, rhabdomyosarcoma, lipoma, liposarcoma, fibroma, fibrosarcoma; myxoma

Secondary

Metastases of carcinoma and sarcoma, are not frequent since muscle does not afford a suitable 'bed' or 'soil' for them to grow. The following may be found occasionlly:

Lymphosarcoma, adenocarcinoma, melanoma, and angiosarcoma.

THE ENDOCRINE GLANDS
(Revised by Dr. A. Rajan)

The endocrine glands are ductless glands and are of vital importance to the body since, through their hormones, they act as chemical regulators. These glands have widespread and specific influence on various procesess connected with metabolism, growth and reproduction. For example, the parathyroids con. trol the metabolism of calcium and phosphorus, the thyroid that of iodine, adrenal cortex that of sodium and pancreas (Islets of Langerhans) that of carbohydrates. The anterior pituitary controls the growth. Gonads and the anterior pituitary control repoduction.

General outline of hormonal relation :

The anterior pituitary controls the activities of most of other glands and is called "the conductor of endocrine orchestra". Through neurohumoral pathways via the hypothalamus, the anterior pituitary is connected to the central nervous system and so to a certain extent, the endocrine system is under nervous control. In most cases, the anterior pituitary doses not act directly on tissue cells but through the mediation of another endocrine gland which is called. "Target gland". The hormones affecting these target glands are "tropic" in character. Direct influence of tropic hormone on tissue cells is exemplified by the action of lactogenic hormone of the anterior pituitary on mammary glands in which milk secretion is stimulated. (Trophic is also spelled as tropic).

Adrenal medulla has no tropic hormone for its stimulation, but it is under the direct control of central nervous system, eg liberation of adernaline in fright.

In the case of parathyroids, neither tropic nor nervous influences are present since they respond directly to the body tissues. eg. hyperplasia of parathyroids in hypocalcemic states.

The hormones of the target glands have an inhibiting action of the anterior pituitary. This is a *"feed back mechanism"* or *servomechanism* This may be brought about by the action of the hormones of the target glands directly on the pituitary or indirectly through inhibition of the "releasing factors" in the hypothalamus. Hypbthalamus stimulates the anterior pituitary through certain factors called 'relasing factors' to produce the corresponding hormones by it For example, corticotropin releasing factor or harmone for ACTH (CRF), thyrotropic hormone releasing factor (TRF) and similar releasing factors for STH.LH and FSH Hypothalamus may be stimulated for the release of the factors by the hormones of the target glands or by stimuli arising from higher centers. This servomechanism acts as a safety valve so that over-production of tropic hormones is prevented and thus the energy metabolism is kept under control. Therefore, continuous administration of an excess of target gland hormone will produce atrophy of the corresponding tropic-hormone-producing gland.

Hypothalamus produces some inhibitory hormones like prolactin inhibiting harmone (PIH) in mammals and growth hormone inhibition hormone otherwise called SOMATOSTATIN

Compensatory atrophy : This may be seen in a gland if there is prolonged administration of its own hormone: eg. tumor of one adrenal will produce atrophy of its pair.

Exhaustion atrophy : If there is continuous stimulation of a target gland by its tropic hormone, exhaustion atrophy of the stimulated gland occurs (after an initial period of hyperplasia and over-production by this gland).

An animal may be affected by diseases of the endocrine glands and such disease processes may be the result of :

1. **Hypofunction** of the gland. This may be due to :
    (a) congenital hypoplasia
    (b) destruction of the tissue by
        i) Disease processes ; inflammation, tumors
        ii) Atrophy—of unknown origin.
        iii) Thrombosis and embolism.

2) **Hyperfunction :**
    i) Tumors
   ii) Non-neoplastic hyperplasia.

THE THYROID

Thyroid is situated in the cervical region, and consists of two lobes connected by a narrow isthmus. The gland is composed of follicles (the thyroid follicles) which are closely apposed to each other. These are supported by a delicate connective tissue stroma. The gland is provided with a rich blood supply, A few lymphocytes are present.

The follicles are lined by a layer of epithelial cells. The gland is very labile. That is to say, it is susceptible to various influences and demands for thyroxine, and hence the histological structure varies with function, breed, age, season, metabolic activity etc.

At stress, when thyroxine is most needed, the epithelium assumes a tall columnar shape and the colloid is resorbed But at rest, the colloid accumulates and the epithelium assumes its normal size and shape (involution).

The anterior pituitary through its thyrotropic hormone controls the function of the thyroids as follows : (1) The thyrotropic hormone stimulates the trapping of the iodides (2) Two iodinated tyrosine molecules are coupled to form the tetraiodotyrosine under the influence of the thyrotropic hormone. (3) Thyroxine after formation is combined with a protein to form thyroglobulin (molecular weight 680,000). This compound is hydrolysed by protease, when needed, under the influence of the thyrotropic hormone and thus free thyroxine is liberated into the blood.

Hypothalamus, which is the controlling center for emotional activity releases thyrotropic hormone releasing factor (TRF) which acting on the anterior pituitary causes the production of the thyrotropic hormone. Thyroxine may inhibit production of the thyrotropic hormone by "feed back mechanism" or "servomechanism". This check is needed, for otherwise, due to continued stimulation of the gland, exhaustion .rophy of the thyroid may result. Emotional disturbances through hypothalamus may play a part in the production of thyroid hyperactivity. (See exophthalmic goitre later).

Goitrogenic substances : In the synthesis of thyroxine, iodine must first be absorbed by the gland as iodide, which is later converted into an organic compound by combination with tyrosine. Those substances, therefore, that inhibit these two processes, viz uptake of iodide by the thyroid and its subseqent conversion into an organic compound, prevent the formation of thyroxine. A low level of thyroxine is a stimulus for the production of thyrotropic hormone which causes hypertrophy and hyperplasia of the follicular epithelium. This is manifested as goitre. So these substances that are responsible for the production of goitre are called goitrogenic substances. Examples of these are :—Phenothiazine; thiocyanates, cabbage, linseed, soya bean, sulphonamdies.

The uptake or trapping of iodine by thyroid is blocked by thiocyanate and perchloride while thiourea, sulphonamides and phenothiazine may prevent or nhibit the synthesis of thyroxine. So these substances are known as *antithyroid substances.* Ultimately these substances may also produce goitre.

Functions of the thyroid and thyroxine.

The function of the thyroid through its hormone thyroxine, is to maintain a high rate of metabolism in the animal (that means, it raises the basal metabolic rate, BMR).   The rate of metabolism is affected by the hormone acting at one or more points in the citric acid cycle.   Since the enzymes that are involved in this cycle are located in the mitochondria, it is postulated that it increases the permeability of the cell membrane or the membranes of the mitochondria. The hormone increases also the number of mitochondria.   So more metabolites are exposed to more enzymes of the mitochondria,   thereby increasing the meta- bolism.

When thyroxine is administered, a number of enzymes, especially cytoch- rome oxidase, cytochrome C and succinic oxidase increase in the tissues. Oxy- gen utilisation is stimulated.   Thyroxine causes increased utilisation of carbohy- drates, increased catabolism of proteins and increased oxidation of fats as shown by loss in weight.   Protein synthesis is enchanced by thyroid hormones, through the mediation of mitochondria.

Central nervous system requires thyroxine.   If it is deficient the nerves are permanently damaged in the young growing animal and so   it is lethargic, dull and stupid. Increased thyroxine stimulates the activity of central nervous system and so the animal is jumpy, nervous, irritable and hyperactive.

Thyroxine causes hepatic glycogenolysis and so hyperglycemia is produced. It also affects the normal reproductive functions. In hypothyroidism, androgens may not be produced and so thyroxine is required for libido.   In the female the litter size may be reduced, milk production   is lowered and cysts form in the ovary leading to sterility. It is necessary for estrus and so in its deficiency, silent heat occurs.

Lower environmental temperature causes release of the thyrotropin releas- ing factor (TRF) of the hypothalamus and so thyroxine is released which acting on the thyroid causes thyroxine production.   This raises the basal metabolic rate and so temperature of the body is raised.

Therefore,   thyroid controls the body metabolism and so is necessary for proper physical, sexual and mental development and function.

Cysts :   Cysts may be ultimobranchial duct cysts or thyroglossal duct cysts or parenchymal cysts.   The former are seen either attached to the thyroid gland or embedded in the gland and are seen in the dogs The Cyst contains clear fluid and histologically the presence of ciliated columnar epithelial lining differenti- ates it from parenchymatous cysts

Atrophy : This may occur secondary to lesions in hypophysis or due to certain unknown causes. Glands become smaller. Sections show fatty meta- plasia of interstitial tissue. Follicles become smaller and they are filled with basophilic colloid and corpora amylaceae. Connective tissue stroma is promi- nent and there is condensation of reticular fibres

Hypothyroidism This occurs in areas where there is incidence of endemic goitre due to deficiency of iodine in the water and soil. Due to deficiency of thyroxine, the basal metabolic rate is abnormally low and so all the vital proces- ses are slowed down   The condition found in young growing children is called **cretinism.**

Cretinism may be sporadic or endemic.

Sporadic cretinism occurs in young of healthy parents. There appears to be a genetic defect by which there is an inability to produce thyroid hormone, Low blood level of thyroxine stimulates pituitary to produce thyrotropic hormone in excess with resultant formation of goitre.

Endemic cretinism occurs in areas where the incidence of goitre is common in man and animals (due to iodine deficiency).

Symptoms in man : The cretin is a dwarf physically, sexually and mentally. Growth is arrested, bones are brittle, abdominal muscles are flabby leading to pendulous abdomen; skin is dry and cold; lips and face are swollen (mouth is half open always) and the tongue is large. The patient is extermely lethargic and has a vacant, idiotic look and the gonads are ill developed. "What was intended to be created in the image of God has become what has been called the pariah of nature, and all for want of a little iodine" (Boyd)

Calves usually are either born dead or die within a day or two Animals have myxedema and alopecia, The fetal placenta is retained. There is swelling of the throat (goitre) and this may be so large that fetal dystocia may develop. Growth of cranial, body and limb bones is arrested Eruption of teeth and second dentition are retarded. Deafness, idiocy and hypoplasia of the pituitary may be noticed. Similar symptoms are seen in colts, lambs, piglets and kids.

The young pigs show deficiency of hair, cyanosis of skin and subcutaneous tissue, edema of skin, shortening of neck (thick neck) and limbs.

Hypothyroidism in adults is called **myxedema** and occurs in man at about the age of 40 years and is more frequently encountered in females.

Causes . 1) Thyroidectomy. 2) Following an earlier severe hyperthyroidism. 3) Atrophy and fibrosis of unknown etiology. 4) Sequel to thyroiditis 5) Radio iodine therapy. 6) Hypopituitarism—decreased production of thyrotropic hormone.

Clinically, the patient is lethargic, heavy and has no inclination to move. She is cold and feels cold. The skin is dry and rough and hair is lost. The face is puffed up and broad. The basal metabolic rate is low and heart rate lower than normal (bradycardia) Serum cholesterol level is high. In the subcutaneous and other connective tissues, there is an accumulation of mucoid or myxomatous substance, which gives the puffed up appearance to the face. Females are frigid and become sterile while males are impotent The thyroid gland is atrophic and hard. In some places it is just a mass of fibrous tissue. The glandular parenchyma is widely separated by fibrous tissue. Lymphocytes infiltrate the tissue and may also form, in some places, nodules.

Myxedema does not appear to occur in animals other than dogs, in which the symptoms are obesity, alopecia, thick skin and lethargy. Serum cholesterol level is very much increased, but serum protein-bound iodine is low.

Hypothyroidism in cattle is manifested by sluggishness with agalactia, silent heat, retained placenta, still births and a tendency to purulent endometritis.
In general, we find the following changes in hypothyroidism:

Decrease in : BMR; oxygen utilisation in the liver, kidney and muscles; number of mitochondria; blood flow; cardiac output; blood pressure; cardiac rate (so bradycardia); nervous function and myelination (so animal is dull, stupid; sluggish and sleepy); gut motility; absorption of glucose; phagocytic activity of leucocytes; serum proteins (so myxedema develops due to decrease in colloidal osmotic pressure of blood); egg production in fowls.

Increase in ; body weight; circulation time (due to weak heart) susceptibility to infection.

Other changes noticed : constipation; hypophagia; weakness and hypotonia of muscles; skin is dry and brittle; loss of hair; thickened skin; dermatitis and retarded feather development

GOITRE

Goitre is non-inflammatory and non–neoplastic enlargement of the thyroid.

I.   Parenchyamatous goitre : This is also called the hyperplastic goitre or goitre of cretinism and is congenital.   This condition occurs in areas the soil of which is deficient in iodine and this results in iodine deficiency in animals and man of the locality   As explained under cretinism, there is increased production of thyrotropic hormone with resultant hypetrophy and hyperplasia of the thyroid epithelium

Macroscopically, the affected thyroid gland is enlarged, meaty, and firm. In very severe cases, cysts may be present   The color is darker than normal due to increased vascularity

Microscopically, hypertrophy of the follicular epithelium is seen to start with.   The cells become tall and plump and so encroach into the lumen that is either reduced in size or even obliterated.   So there is reduction in colloids Later, there is hyperplasia of these cells and so one finds papillary projection into the lumen, completely filling it.   There may be formation of new follicles in the midst of the newly formed clusters of the epithelial cells.   Later when exhaustion phase supervenes, the epithelial cells are degenerated and desquamated and fibrosis occurs.   Himalayan goitre is characterised by prolonged hyperplasia and mild involutionary changes.   There is pronounced stromal hyperplasia and when there is involution the stromal reaction persists indicating previous hyper-plasia.

Microscopic picture may simulate a papillary carcinoma but in the absence of anaplasia of the cells and invasion of the basement membrane this can be ruled out.

II Colloid goitre :

This is also called *simple goitre* and is the more frequently seen form among animals. especially dogs.   The thyroid glands are much swollen.

Causes : 1. Low levels of iodine in soil and water. 2. Excessive demands-adolescene, pregnancy. 3. Diseased conditions interfering with assimilation of iodine—gastro -enteritis. 4. Ingestion of goitrogenic substances; thiouracil, soya beans

The glands are enlarged   Cut section is transluscent.

Microscopically, the acini are widely dilated and contain watery, faintly staining colloid.   The epithelium is flattened   Colloid is deficient in iodine and thyroxine.   Sometimes neighbouring acini may coalesce to form cysts   Leakage of the colloid into the interstitial space may occur due to rupture of cysts.

Since this condition is supposed to be an involutionary phase of hyperplastic goitre, here and there some acini may show papillary projections of the epithelium.

Usually no other symptoms except difficulty in swallowing and dyspnoea due to pressure of the enlarged gland is noticed. This is not congenital.

III Nodular or adenomatous goitre :

Nodules of varying sizes, microscopic to several centimeters in diameter, are found very often in old and senile horses, dogs and cattle. Some consider that this is an outcome of alternating hyperplasia, hypertrophy and involution affecting the gland. On section, the nodules are translucent and may contain cysts or vesicles filled with gelatinous colloid.

Microscopically, the picture is variable. All gradations may be found in different nodules, from the picture of a colloid goitre to that of hyperplastic goitre. But in a nodule, the picture is constant. In some nodules dilated acini filled with colloid and having flattened epithelium may be present In other places, acini containing papillary projections of the epithelium obliterating the lumen, may be found Retrogressive changes, leading to necrosis with subsequent softening and liquefaction are responsible for cyst formation (*pseudocyst*). The connective tissue which is increased undergoes hyalinisation. Calcification (*calcareous goitre*) and metaplasia into bone (*osseous goitre*) of the connective tissue may also be encountered.

Cystic goitre results due to formation of the cysts by the confluence of smaller colloid-filled follicles. These are lined by epithelium. (Pseudocysts are just spaces without an epithelial lining).

Hemorrhage into follicles due to erosion of vessels causes *hemorrhagic goitre.* The conglutination of the fibrin produces *rubber colloid.*

Is nodular goitre neoplastic? This question has not been conclusively settled. Some consider the nodules as adenomas while others as a hyperplastic involution process only. The regularity of structure in the nodules and want of evidence of expansion by absence of compression of the adjacent tissue make one consider goitre as non-neoplastic.

In some cases, the nodules are inert and functionless. But in some others thyroid hormone may be actively secreted when **toxic adenoma** is applied to the condition. This condition is met with in horses and dogs with signs of hyperthyroidism.

A hereditary predisposition (*endogenous factor*) to goitre is suggested. But there should be some exogenous factor (like deficiency of iodine etc.) to precipitate the overgrowth Sexual excitement and pregnancy are also factors to be considered. Fluorine containing compounds have goitrogenic effects.

VI Exophthalmic goitre (*Grave's disease. Basedow's disease*) Primary thyrotoxicosis

This disorder of the thyroid gland found in man is probably not seen in animals. Women are more often affected than men.

Causes : The exact cause of this condition is still obscure. Iodine deficiency is definitely not a cause Some kind of shock, probably psychic, connected with sex, is suggested as being causative. There is genetic predisposition of the condition. Lesions of the anterior pituitary is suggested by some. A long acting thyroid stimulator (LATS), probably a gamma globulin, for the formation of which lymphocytes and plasma cells may play a part, is suggested to be of etiological significance by some workers.

Clinically there may not be enlargement of the thyroid. The following are the characteristic symptoms: 1. Exophthalmus 2. Tachycardia 3 Muscular tremors 4. High B. M. R. and other symtoms of hyperthyroidism—weight loss, sweating etc.

Macroscopically, the gland is meaty, darker in color and on section is not translucent but fleshy.

Microscopically, the follicular epithelium is tall columnar and hyperplastic. Papillary projections into the acini are often seen. Colloid is scanty in the acini and is thin and watery, vacuoles may be present in the colloid near the epithelium suggesting the resorption of colloid by the cells. Throughout the stroma is found lymphocytic infiltration, sometimes to the extent of lymphoid follicle formation with active germinal centres. Increase in vascularity is marked

Other lesions : 1. Lymphoid hyperplasia in thymus, Peyer's patches, lymph nodes and tonsils. 2 Lymphocytosis. 3 Myocardial degeneration and fibrosis. 4, Muscular weakness. 5. Increased amount of fat and water in the eye and the extra-orbital muscles are swollen and firm. Retraction of eye lids occurs so that the sclera are visible These lesions are supossed to be due to the action of an *exophthalmus producing substance* (EPS) which is distinct from the thyroid stimulating hormone but produced by the pituitary. Ultimately, exhaustion atrophy occurs with symptoms of myxedema supervening.

In general, we find the following changes in hyperthyroidism.

Increase in : BMR; size of the skeleton in the young; oxygen utilisation in the liver, kidney and muscles; oxidation of enzymes; number of mitochondria (with swelling of mitochondria); blood flow and cardiac output; blood pressure; cardiac rate (and so tachycardia); blood volume (due to vasodilatation); nervous function (so animal is alert, quick, irritable, anxious, wakeful, restless and fatigued); motility of bowel (so diarrhoea and polydipsia); glucose absorption;

Decrease in : body weight; fat depots; serum cholesterol level.

Other changes seen : Muscle weakness and tremors; mobilisation of calcium from bones and increased loss of this element in urine and feces (but blood Ca level is not altered) and hence fractures occur easily and frequently.

THYROIDITIS may be classified into.

Infectious thyroiditis {	Acute {	Non-suppurative Suppurative.
	Chronic {	Specific Non-specific
Thyroiditis due to physical agents {	Trauma Irradiation	
Due to undetermined etiology {	Hashimoto's struma Subacute granulomatous (Quervain's thyroiditis) Riedel's struma	

Thyroiditis is less common in animals. But in generalised infections it may also get involved. Poor development of the reticulo-endothelial system, small arterial branches and smallness of the organ have been attributed as reasons for this. As an extension of a subcutaneous suppurative inflammation, lesions may spread to the thyroid.

Hashimoto's struma (struma lymphamatosa). It is relatively common in dogs and there is a genetic predisposition. Incidence is high in Beagle breed of dogs. The cause is an autoimmune reaction. Thyroglobulin which is a secluded antigen leaks out into the blood due to defect in the basement membrane and stimulates antibody production. These antibodies are being formed by plasma cells and lymphocytes which infiltrate the gland. Three specific antigen-antibody systems have been described: (1) Microsomal. (2) Thyroglobulin and (3) An altered colloid

Gross lesions are not characteristic. Histologically there are multiple isolated lymphoid nodules with well formed germinal centres replacing the parenchymal tissue. The interstitial tissue and the parenchyma show dense streaks or collections of lymphoid cells, plasma cells and large mononuclears. Larger or oval oxyphilic cells are occasionally seen in the follicles (Askanazy cells).

Subacute granulomatous (Quervain's) Thyroiditis. The disease follows an acute respiratory infection caused by a virus. It is associated with epithelial necrosis, disappearance of epithelial cells and infiltration by histiocytes, mononuclears, fibroblasts and multinucleated giant cells.

Riedel's struma: The gland is very firm, hard and is adherent to the surrounding tissue. Histologically there is severe fibrosis and diffuse, moderate infiltration with lymphocytes and mononuclear cells. The involvement is unilateral compared to the Hashimoto's struma. There is no giant cell reaction.

Thyrocalcitonin: This is a hormone produced by the parafollicular or 'C' cells of the thyroid. These cells are independent of the thyroid secreting follicular epithelium. The 'C' cells are larger and paler and are not incontact with the thyroid colloid. The cells are argyrophilic, stain metachromatically with toludine blue after hot hydrolysis. Cytoplasmic granules can be stained by strong basic dyes after oxidation with performic acid.

Thyrocalcitonin is a single chain polypeptide, consisting of 32 amino acids. Its molecular weight is 3000. It is now synthesised.

The function of hormone is to reduce bone resorption and to maintain Ca level thereby inhibiting osteoclastic activity. Hypercalcemia stimulates the release of the hormone. Its action is opposite that of paratharmone. It is thought that in postparturient hypocalcemia (milk fever) there is sudden release of thyrocalcitonin. This is used as a therapeutic agent in hypercalcemic state and in demineralising bone disease. There is significant rise in thyrocalcitonin level in the blood in thyroid medullary carcinoma.

Calcitonin enchances the excretion of sodium, phosphates and calcium.

Neoplasms of the Thyroid

Primary tumors.

Adenoma in the horse while adenocarcinoma and adenoma in the old dogs are more common. These tumors are found in areas of endemic goitre.

Microscopically, the adenoma resembles the parent tissue and may be diffi-cult to differentiate.   The neoplastic acini are smaller with very little colloid and their cells are hyperchromatic.   A connective tissue capsule encloses the tumor.

The cells of the adenocarcinoma are highly anaplastic and it is difficult to identify them.   Acinar formation is minimal or may not be present but only solid sheets of cells which are cuboidal or cylindrical in shape may be present. Sometimes papillary arrangement is noticed.   The nuclei are central, large and hyperchromatic.   Nucleoli are large and mitotic figures are numerous   Where acini are found colloid is absent in them.   Capsule may not be present.

Metastases are common in the malignant variety and secondary deposits are found in all organs, especially in the lungs

Secondary tumors of the thyroid :

Metastases from mammary tumors and lymphosarcomas are common in the thyroid.

THE PARATHYROID GLANDS

After birth, two pairs of parathyroid glands are present.   In the dog one is situated on the thyroid at its antero-lateral aspect and the other is found in the areolar tissue just anterior to the thyroid.   Unless looked for carefully, the parathyroids are usually missed being translucent and merging with the areolar tissue   In size, these are small. and weigh about 0.1 gm.

Histologically, the gland cells are of two types

1   **Chief cells or water-clear cells** which are vacuolated.

2   **Oxyphil cells**, the cytoplasm of which is granular and stains red.

The cells are arranged as islets or clusters in a vascular fibrous stroma.

Function : The main function of the gland, through its hormone, parathor-mone, is to maintain the calcium-phosphorus balance.   This is brought about in two ways.

1.   by influencing the renal excretion of phosphorus by depressing the renal resorption of phosphate, causing thereby phosphate diuresis and

2.   by regulation of the osteoclastic activity in the bone—demineralisation of bone.

Parathyroid hormone helps in the absorption of calcium and magenesium from the intestines.

Parathormone is a protein having a moleculer weight of 8500.   It has a single chain of 84 amino acids.

Hyperparathyroidism :

May be primary or secondary

Primary hyperparathyroidism is due to an adenoma, in which increased parathormone is secreted.   Changes consist in osteopathy, nephrocalcinosis and urolithiasis. The osteopathy has already been described under renal osteodystrophy (Page 467)   In nephrocalcinosis there is diffuse calcium deposition in the renal tubular epithelium.   (Page 386).

Secondary hyperparathyroidism is found in :

1. Inadequate intake of calcium in diet; or imbalance of calcium : phosphorus ratio.

2. Hypocalcemia due to vitamin D deficiency.

3. Steatorrhoea

4, Chronic renal failure—phosphorus excretion is interfered with.

5 Pregnancy and lactation—calcium is side-tracked to the fetus and milk respectively.

In these conditions there is a lowering of blood calcium or rise in the level of blood inorganic phosphorus leading to hyperplasia of the parathyroids. Osteitis fibrosa cystica and Rubber Jaw Syndrome are the sequelae in these conditions which have already been studied (Pages 467 and 468).

THE PITUITARY GLAND

The pituitary gland or the hypophysis is located in the sella turcica. The gland is divided into adenohypophysis and the neurohypophysis.

Adenohypophysis: This arises from Rathke's pouch, which is an invagination of the pharynx. This glandular part is divided anatomically into :—

    i)   Pars distalis – anterior lobe,

    ii) Pars intermedia– the intermediate lobe.

    iii) Pars tuberalis, which surrounds the infundibular stem.

Neurohypophysis : arises as a downward growth from the floor of the third ventricle

Structure : Pars distalis is composed of three types of cells, as distinguished by the staining affinities of their granules.

    1. The Chromophobes, which do not have stainable granules and are supposed to be immature cells. These form 50% of the cells.

    2. The Acidophils. forming 40% have granules taking up acid stains (red with eosion). These are PAS negative.

    3. The Basophils, forming 10% have granules taking up basic stains (blue with hematoxylin). The basophils are again divided into the larger **beta cells** and the smaller **delta cells**

Neurohypophysis : The neurohypophysis is connected to the hypothalamus by nerve tracts. This contains modified glial cells called "pituicytes".

Hormones : The adenohypophysis produces the following hormones.

By acidophils :

    1. Growth hormone—STH

    2. Prolactin.

    3 Adrenocorticotropic hormone—ACTH (Doubtful)

By basophils :

    1. Thyroid stimulating hormone ; T.S.H. (Thyrotropin)

    2. Follicle-stimulating Hormone – F.S.H.

    3. Leuteinising hormone – L.H. (In the male it is called Interstitial-cell stimulating hormone, I C.S H.)

These are glycoproteins and contain a carbohydrate. The thyrotropin is believed to be secreted by the beta cells while the gonadotropins (FSH and LH) by the delta cells.

The intermediate lobe secretes the melanophore stimulating factor, M. S. F. or hormone (M.S H.) See page 175.

The neurohypophysis produces : 1. The oxytocin and 2. The Antidiuretic hormone—A.D.H, or Vasopressin These hormones are produced in the hypothalamic nuclei and then pass along the axons of neuro-secretory cells to be discharged into the small blood vessels of pars nervosa, which is outside the blood-brain barrier.

The Somatotropic Hormone (STH): STH is specific. It acts on all the cells (ectodermal, mesodermal and endodermal) and influences the metabolism of proteins, carbohydrates and fats. This is a complex protein having 396 amino acids. *Its functions are :* (1) increasing the growth of soft and osseous tissues. This is brought about by increasing the cell permeability to amino acids thereby favouring a build up of the muscle mass of the body and nitrogen retention. Growth hormone (STH) acts by stimulating the production of Somatomedins in the liver and kidney and these are then released into the circulation These somatomedins act on the epiphyseal cartilage of bones, and help in the transport of amino acids into cells for growth purposes It brings about protein synthesis by (i) interfering in protein breakdown, (ii) inhibiting coversion of amino acids into urea and (iii) accelerating protein synthesis from amino acids. (2) Probably STH causes the release of glucagon, which in its turn raises the glucose level of the blood by glycogenolysis in liver. (3) It probably prevents the entry of glucose into the cell and also inhibitis the action of hexokinase So through (2) and (3), STH is antagonistic to insulin and so is 'diabetogenic'. This hormone has profound effect on lactation, which can be induced and enhanced by STH injections.

Prolactin : This is a protein with a molecular weight of 32,000. This hormone also is necessary for intiation of lactation Besides it has a corpus luteum stimulating effect.

Adrenecorticotropic hormone (ACTH) : This is a protein having 39 amino acids The regulation and output of this hormone is intimately associated with the hypothalamus and a servomechanism exists. The adrenal steroids act on the hypothalamus influencing the amount of corticotropin releasing factor (CRF) discharged. Besides external stress stimuli like hemorrhage, temperature, toxins and emotional states influence the release of ACTH by affecting the release of CRF

ACTH stimulates the adrenal cortex but not the medulla. It increases the parenchyma of the cortex as well as its hormones.

ACTH stimulates the phosphorylase activity of the adrenal cortical cells thereby bringing about a series of enzyme-catalised reactions which lead to production of energy needed for corticoid biogenesis.

Thyrotropic hormone (TSH). This a glycoprotein with a molecular weight between 10,000 and 28,000. Its functions are (1) to control the uptake of iodide by the thyroid (trapping), (2) to control the formation of iodine-organic compound and later thyroxine and (3) release of thyroxine from the thyroglobulin compound. TSH also causes the hypertrophy and hyperplasia of the chief cells of thyroid.

Follicle stimulating hormone(FSH) This is a glycoprotein, with a molecular weight varying between 29,000 and 67,000 among various species of animals, Its secretion is under the control of hypothalamus with a servomechanism Its release is also influenced by environmental conditions such as changing seasons and the length of daylight. Probably visual stimuli through hypothalamus influence it production

The function of FSH is to control spermatogenesis in the male and development of ovarian follicles in the female. It does not stimulate the Leydig cells or cause the production of estrogen.

Luteinising or Interstitial cell Stimulating Hormone (ICSH). LH is in the female while ICSH is in the male. It is a glycoprotein with a molecular weight varying in different species of animals, from 40,000 to 100,000.

LH output is also regulated by the hypothalamus and there is a servomechanism.

The function of the LH is to control the development of the ovarian folli cle towards maturation, cause the production of estrogens and ovulation (if FSH has already acted) It may also control the development of corpus luteum. In the male ICSH stimulates the cells of Leydig for secretion of testosterone, causing steroid synthesis.

Oxytocin : Is a peptide containing 9 amino acids (with two sulphur molecules.) It acts on (1) the smooth muscle of the uterus and (2) on the myoepithelial cells of the mammary gland, causing their contraction and hence is useful in the contraction of the uterus (at the end of parurition) and in ejection of milk respectively

Antidiuretic hormone (ADH) or Vasopression, This is also a peptide having 9 amino acids with two sulphur molecules. Its main function is to promote reabsorption of water from the urine by the distal convoluted tubules and collecting tubules. This is brought about by increasing the size of pores or channels.

Regulation of ADH production depends on the water content of the body. Increased water content of the blood inhibits the release of ADH and so water is not absorbed by the renal tubules and so excess of water is got rid of. Whenever there is decrease in body water or increase in the electrolytes in blood, ADH is released and this mediates the reabsorption of water from urine, increasing the blood and body water, which dilutes the electrolytes and thus matintains homeostasis The osmoreceptors are located in the supraoptic and paraventricular nuclei.

Pain favours ADA release leading to oliguria while ethyl alcohol inhibits ADH release and so we find diuresis after drinking alcoholic beverages.

Pituitary is subjected to various types of morphological alterations due to normal physiological processes and one should know these to interpret effectively the pathological processes.

Ageing is characterised by accumulation of herring bodies and wear and tear pigments in pars nervosa and proliferation and hyalinisation of sinusoidal reticulum in pars distalis causing splitting up of parenchymal acini. (Herring bodies are globules of amorphous acidophilic colloid-like material).

Aplasia : Aplasia of the hypophysis has been observed in Holstein-Friesian cattle, controlled by an autosomai recessive gene. Histologically there is evidence of failure of differentiation of acidophils.

Cysts : Cysts are relatively common in dogs. They may be developmental cysts (Rathke's cleft cyst; Craniopharyngeal cyst: Evagination cyst) or acquired cysts (common in cattle and horses due to nutritional deficiency of Vitamin A or due to senility) Craniopharyngeal end of Rathke's pouch may persist resulting in the formation of pharyngeal hypophysis, a cystic non-glandular structure

Cysts are common in short muzzled dogs and cross breeds. Cysts may be single or multiple and may be unilocular or multilocular. These are seen commonly in the periphery of pars distalis and pars tuberalis. Cysts are lined by cuboidal, columnar, ciliated columnar or squamous epithelium. Though cysts may be encountered only occasionally, they cause functional disturbances.

Atrophy may be caused by the pressure of cysts or Tumors. Atrophy may be present to a great extent without signs of hypophyseal deficiency. Basophils are more sensitive to pressure than acidophils.

Pituitary lesions in disorders of other organs :

Hypothyroidism : Adenohypophysis is enlarged and the large Beta basophils are hypertrophied. The granules of these cells become finer and disappear eventually. In severe cases, these Beta cells become vacuolated giving a spongy appearance to the medulla, where these cells are found in large numbers.

Gonadal deficiency : Gonadotropin-producing basophils become degranulated and are transformed into hypertrophic aminophils, a change characterised by enlarged nucleus, thinness and irregularity of nuclear membrane.

Cystic ovaries : This has already been considered.

Stress : In chronically stressed animals basophils of both types store granules This accounts for gonadal inactivity in such animals. The acidophils are hyperactive and may be completely degranulated or a few granulated cells persist adjacent to sinusoids

Hyperpituitarism : This condition is manifested by overgrowth and proliferation of bone. In man this is called "Gigantism" in young growing individuals in whom the ossification of bone has not yet stopped and "acromegaly" in adults in whom no more growth occurs.

Gigantism : Gigantism is due to increased secretion of somatotropin in the young. The individual grows very tall and the skin and subcutaneous tissues show fibrous hyperplasia. Since STH is diabetogenic, glycosuria is a symptom. As this condition occurs in adenoma of the acidophils and since the neoplastic growth produces pressure atrophy on the basophils impotence in the male and amenorrhoea in the female are observed. If the patient lives beyond the age of epiphyseal fusion, acromegaly results.

Acromegaly is the condition that is observed in the adult. Since no growth in the bone is possible, the bones become thicker and broader. The hands and feet are abnormally large (akros = extremity; megale = enlarged) and the fingers are crooked and knotty. The facial bones become long and thick, especially the jaw, resulting in prognathism. Viscera are enlarged (splanchnomegaly or macrosplanchnia) and fibrous hyperplasia of skin and subcutaneoue tissue is common. Nose, lips and ears become large. Kyphosis is also seen in some. Impotence in the male and amenorrhoea in the female are other symptoms. Diabetes mellitus occurs due to diabetogenic action of the hormone. Eye lesions may be noticed due to pressure on the optic chiasma by the tumor.

Cushing's syndrome : Basophil adenoma may produce the disorder. Conversely, hyperfunction of the adrenal cortex may cause changes in the pituitary. Details of this condition will be studied under adrenal gland.

Hypopituitarism: Since the pituitary is enclosed on three sides by hard bone, even slight enlargement of some part will cause pressure atrophy on others and so corresponding decrease in the activity of the cells results.

Causes of pituitary hypofunction : 1, Pressure by: (a) Tumors, (b) Cysts: 2 Inflammation, sclerosis; 3. Infarction, necrosis. 4. Hydrocephalus—bulging of the floor of the ventricle. 5. Abnormal development. 6 Tuberculosis.

Hypofunction in the young:

Pituitary dwarfism or infantilism : Not seen in animals.

In children, though stature is small, yet the build is proportional. Examples are the midgets of circuses These subjects are normal mentally. Sexual development may be retarded.

Symmond's disease (Pituitary Cachexia) or Sheehan's Syndrome; This is found only in females, due to postpartum necrosis of the pituitary consequent on thrombosis following hemorahage. Hence severe hypopituitarism develops. The characteristics are ;— severe cachexia, loss of sexual function; weakness, low metabolic rate, loss of hair and pigmentation, mental apathy and drowsiness. microsplanchnia and extreme dehydration and emaciation.

Since trophic hormones are not secreted, there is atrophy and fibrosis of the thyroid, adrenal, ovaries and parathyroids together with the symptoms and lesions consequent on the deficiency of the hormones secreted by these glands and structures.

This condition was described in dogs.

Frohlich's Syndrome—Dystrophia adiposogenitalis : This develops probably due to the pressure by a tumor or hydrocephalus and is mostly found in ladies. The characteristic features are:

(1) Obesity : There is disproportionate and excessive accumulation of fat on the abdomen, buttocks and thighs, while other parts are thinner. (2) Genital hypoplasia and decreased sexual function. (3) Idiocy or mental retardation. (4) Thin skin and hair. (5) Reduced sweat secretion.

In man it is feminising with the characteristic distribution of fat as in a female.

Diabetes insipidus : Normally under the influence of the antidiuretic hormone of the neurohypophysis, 80% of the water in the glomerular filtrate is reabsorbed by the epithelium of the Henle's loops and distal convoluted tubules. But if the secretion of the ADH is interfered with due to failure of the hypothalamic-hypophyseal system, reabsorption of water from the glomerular filtrate does not occur and so large quantities of urine with low specific gravity are passed and this condition is known as *diabetes insipidus.*

Lesions of the pars nervosa or any causes that injure the hypothalamus will produce diabetes insipidus. The secretion of ADH by the pars nervosa is under the control of stimuli from the hypothalamic nuclei The causes that may produce this situation are :

1. Trauma—surgical or fractures. 2. Pituitary tumor or metastases from bronchogenic carcinoma or mammary carcinoma, 3. Meningitis:—pressing on the stalk, 4. Encephalitis

CHAPTER 22

Tumors of the pituitary

Chromophobe adenoma : This tumor was reported very rarely in horse and dog The tumor consists of cells grouped with an alveolar arrangement. The cells have non-granular cytoplasm. Not producing any hormone, no endocrine disturbances are noticed due to this tumor. Secondary hypopituitarism may develop due to pressure atrophy of the acidophils or basophils, manifested as Frohlich's Syndrome and visual disturbances

Acidophile adenoma : This tumor comprises of cells having an acidophilic granular cytoplasm. Results are gigantism or acromegaly.

Basophile adenoma consists of cells having a basophilic granular cytoplasm Result is Cushing's Syndrome.

Carcinomata of the above three types of cells may be noticed. But these are only locally invasive.

THE ADRENAL GLANDS

The adrenal consists essentially of two glands, differing in structure and function. The cortex, which is external is of mesodermal origin, being derived from the urogenital ridge, from which the gonads and the urinary organs also are derived and so is closely related to them. The medulla, on the other hand, is derived from the neural crest from which the sympathetic nerve cells also arise.

Structure

The cortex : The cortex is conveniently divided, histologically, into three zones. The zone of cells lying just under the capsule and arranged as small nests of cells is called *Zona glomerulosa.*

A zone wider than the above and situated beneath it, and in which the cells are arranged as parallel cords, is called *Zona fasciculata.* The zone of cells lying between *Zona fasciculata* and the medulla (the innermost zone) in which the cells are arranged as interlacing cords and containing sinusoids of reticuloendothelial system, is called the *Zona reticularis.*

The cells of all the zones appear similar, polygonal in shape and containing lipids, which give the yellow color to the gland. The cells are also rich in vitamin C content which is probably useful in the synthesis of hormones.

The adrenal cortex is under the control of the anterior pituitary. The adrenocorticotropic hormone (ACTH) controls the form and function of the cortex ACTH may increase the production of cortical hormones and the parenchyma of the cortex is also increased. Decrease or increase of ACTH is followed by atrophy (regressive transformation) or hypertrophy (progressive transformation) of the adrenal respectively.

When ACTH is lacking, atrophy of all zones occurs. Differentiation of the cells is lost. The cytoplasm and nuclei become smaller and the storage capacity of the lipid is lost. The capsule becomes thickened and the fibrous tissue of the gland is increased. Continuous ACTH administration causes hypertrophy and hyperplasia of the cortex, which, therefore, becomes broader. The cytoplasm and nuclei of the cells become increased in size and the storage capacity of the cells for lipid is increased.

The medulla consists of nests of large cells containing brown granules, nerve fibres, sympathetic ganglion cells, and a rich vascular supporting tissue. The chief cells are large, do not contain any fat but have granules staining brown with chromic salts and so are called chromaffin cells. It is thought that these granules indicate adrenaline content of the cells.

Hormones : The three zones of the cortex probably produce three different physiologically active hormones. But in chemical composition, all have the basic sterol nucleus.

On the basis of physiological activity, three groups of steroids are recognised.

1. Glucocorticoids. These hormones are secreted by the Zona fasciculata The most important glucocorticoids are the *hydrocortisone* or *cortisol* (compound F) and *cortisone* (compound E). The former is more potent and active physiologically The functions of the glucocorticoids are :

1. Conversion of amino acids into glucose—gluconeogenesis from protein—antianabolic

2. Inhibition of peripheral glucose utilisation and so increasing the blood glucose level (hyperglycemia). This is probably brought about by inhibiting the hexokinase reaction and so antagonises insulin action.

3. Decreasing tissue stores of glycogen, especially in the liver

4. Cause catabolism of proteins leading to negative nitrogen balance So urinary excretion of nitrogen and uric acid is increased. Anabolism of proteins is discouraged and so growth in the young ceases, wounds heal more slowly and because of this there is inhibition of antibody production.

5. Because of breakdown of proteins, there is greater blood level of amino acids. And these are converted into glucose as mentioned above. While the permeability of the membrane of extrahepatsc cells is decreased for amino acids the permeability of the hepatic cells is increased and so there is increased formation of plasma and liver proteins Hence there is greater deamination in the liver.

6. Shifting of body fat stores. Circulating fatty acids are more and these are used for energy. Some are converted into glycogen of the liver, sparing glucose.

7. Sodium retention and potassium diuresis So continued administration of cortisol leads to edema, metabolic alkalosis, hypokalemia and hypochloremia.

8. Suppression of ACTH production.

9 In excess, due to potassium depletion, muscular weakness.

10. Prolonged use causes increased elimination of Ca, P & N leading to osteoporosis and fractures

11. Suppression of connective tissue response to injury, that means to say, it is anti-inflammatory : suppression of the activity of the fibroblasts, depression of vascularisation and granulation tissue formation and intercellular ground substance. All these are probably related to protein catabolism Due to altered permeability of the capillaries, there is decreased exudation of plasma into the tissues

12. Reduction and even absolute disappearance of eosinophils in the blood

13 Causing lysis of lymhocytes in the blood and lymph nodes.

14. Decreasing the secretion of pepsin and HCl.

15.   Decreasing the hyaluronidase activity.

16   Interference of antigen-antibody reaction.

17. / In man euphoria is produced—that is a sort of well being is experienced by man in such painful cases as cancer and in depressive states.

Control of production : The cortisol level of the blood inhibits the ACTH releasing factor of the hypothalamus Cortisone may also directly inhibit the pituitary from producing ACTH. The hypothalamus may also be influenced by impulses coming from sites of injury or burns and by impulses coming from the cortex of the brain and by "stressors". Hypothalamus, as already explained, is influenced by the epinephrine levels.

Mechanism of acton of glucocorticoids : These stimulate the critical enzyme production by activating a DNA-dependent synthesis of RNA which in turn accelerates the formation of specific enzymes in the cells.

Therapy with glucocorticoids or ACTH : ACTH can be used only if the adrenal cortex is healthy and responsive: Synthetic glucocorticoids like *prednisone* etc, are 3 to 5 times more potent than cortisol and prolonged use of these cause adrenal atrophy

DOG : Primary adrenal cortisol insufficiency has been met with in dogs and they manifest: anorexia, diarrhoea, asthenia, polydipsia, azotemia, hyponatremia, cardiovascular effects, eosinophilia, dehydration, anuria, hair loss, emesis hyperkalemia. Cortisone 1 mg/lb body weight daily cured the symptoms. *Arthritis* in dogs may be usefully treated with cortisone in the dose mentioned above, Acute conditions respond while chronic ones do not *Asthma* may be treated with an initial dose of 1 mg per pound body weight and later maintained with reduced doses.

Purulent dermatitis may be treated with the combination of cortisone and an antibiotic.

Ottitis externa shows moderate response to cortisone therapy. In *eye conditions*, cortisone is beneficial in inflammation DO NOT use when there are ulcers of the cornea. After healing of the ulcers cortisone can be used to alleviate inflammation if still present.

HORSE : Glucocorticoids have been useful in lameness of horses. It is particularly useful in the acute stages provided adequate rest is given after treatment. It is not so useful in chronic cases. Injection into the tendon sheath at 50 to 250 mg. of cortisol, a total of 4 injections, each 2 to 3 days apart has given good results. This therapy is useful in acute carpitis, gonitis and in metacarpophalangeal arthritis.

In laminitis in the early stages, 50 to 150 mg can be given intravenously with advantage

For treatment of enlargement of bursa, capsule or sheath, aspirate the fluid with a sterile needle and syringe and then with the same needle but a different syringe inject cortisol and if bacterial infection is present (as evidenced by turbidity or discoloration) add an antibiotic.

BOVINES: Theoretically ACTH can be given in ketosis but intravenous injection of calcium-boro-gluconate is so effective that none probably thinks of treatment with ACTH.

Shipping fever : A condition of hypoadrenalism can arise due to stress of transport, cold, hunger, fatigue, lack of water and food, fright and infectious agents. This stimulates the production of ACTH, which in its turn stimulates the cortex which finally becomes exhausted. Infectious organisms may then invade and cause pneumonia. In such situations one can treat with cortisone and antibiotics

Note : Do not use cortisone in the later stages of pregnancy for abortion may occur and the palcenta may be retained

SWINE : In animals that are unable to get up with arthritis, 50 to 75 mg. of cortisone per day enables them to get up and move about.

The following points may be remembered in cortisone therapy :

1. Prolonged cortisone therapy inhibits ACTH and so atrophy of the adrenal may result. Hence, if you stop cortisone abruptly, non-functional adrenal may result. So discontinue slowly and gradually.

2. Because of its anti-inflammatory effect cortisone is not useful in specific diseases. It is indicated as a replacement therapy in Addison's disease only.

3. Be careful to check if no infectious disease in present. Because if one is present, the organisms will quickly spread and overrun the body since the defences of the body are diminished

4' The natural walling processes of diseases as in tuberculosis are destroyed by cortisone If you have to use it, then include a specific antibiotic against the infection.

5. Cortisone or ACTH may be useful in chronic diseases as in arthritis.

6. When you desire slow healing, you can use cortisone as a topical application. For example, in surgical or traumatic wounds of the prepuce you would like to have a slow healing instead of a rapid one resulting in strictures. In such cases you can use this topically.

7. Remember that synthetic glucocorticoids are metabolised slowly and so they are longer acting than cortisone. So their dose should be suitably adjusted.

II Mineralocorticoids :

These hormones, supposed to be secreted by the cells of Zona glomerulosa, control the electrolyte and water balance The chief hormone is the aldosterone. Desoxy-corticosterone acetate (DOCA) is the synthetic hormone having similar properties.

The functions of aldosterone are :—

1. Promotion of excretion of sodium and water by the renal distal convoluted tubules.

2 Promotion of excretion of potassium. So in aldosterone administration there will be muscle weakness due to loss of potassium.

3. Regulation of the extracellur fluid (ECF) volume. Aldosterone promotes reabsorption of sodium—so osmotic pressure of serum increased—supraoptico—hypophyseal system stimulaled—more of ADH released-promotes reabsorption of water from urine—lowers the osmotic pressure of serum—ADH production inhibited—ECF volume kept in balance

In aldosterone therapy, due to increased reabsorption of water (along with sodium)extra-celluar fluid is increased—so blood volume is increased, thereby increasing the cardiac output and blood pressure.

In deficiency of aldosterone, sodium, chloride, bicarbonate and water are lost—extracellular fluid is decreased, thereby minimising the cardiac output and lowering the blood pressure - this leads to failure of circulation, culminating in shock, coma and death. Retention of potassium is toxic and the acidosis due to loss of bicarbonate and sodium is harmful to the body aiding in   the genesis of shock and death.

Anterior pituitary has no control over the production and activity of aldosterone.

Mode of action of aldosterone : Aldosterone first stimulates   the genes or DNA molecules and so specific   messenger RNA is synthesised, which in its turn acts to form the specific enzymes that are necessary to supply the energy for sodium transport

Control of production of mineralocorticoids: 1. The sodium and potassium content of blood directly stimulates the adrenal cortex.

2. Aldosterone stimulation can be induced by sodium deficiency or hemorrhage in which conditions there is lowered blood pressure. In this condition the juxta glomerular cells (J.G. cells) situated on the afferent arteriole of the glomerulus act as stretch receptors. So when blood pressure is lowered, renin which is a proteolytic enzyme is released. This enzyme splits angiotensin from alpha serum globulin. Angiotensin acts on the cells of zona glomerulosa and aldosterone is released and this acts on the distal tubule cells of the kidney and sodium is resorbed. Along with it water also is resorbed—hypernatremia causes thirst and so more of fluid is ingested—this with reabsorbed water increases the blood volume thereby increasing the blood pressure. This acts as a servomechanism since the J G. cells are stretched and renin release is shut off and no more aldosterone is released ultimately.

Clinical use : Mineralocorticoids may be useful in the treatment of chronic interstitial nephritis in the dog, because in this condition there is potassium intoxication, sodium loss and uremia. 2 to 5 mg. of DOCA intramuscularly for several days with glucose, fluids and antibiotics is indicated when potassium is eliminated and symptoms will abate

III Sex hormones : These hormones are produced mostly by the Zona reticularis, are anabolic and play a part in protein synthesis. Most of these hormones are antagonistic to the glucocorticoids in this respect. Most of these hormones are androgens (masculanising) and a few only estrogens (feminising).

Medulla   The hormones of the medulla are adrenaline and noradrenaline, which are produced in the ratio of 4 : 1.

The main functions of these two hormones is the maintenance of blood pressure and changes in carbohydrate matabolism and other adjustments so as to meet stressful conditions. Noradrenaline is responsible primarily for circulatory adjustments while epinephrine is for metabolic changes necessary to meet emergencies. Both are powerful stimulants of the heart and increase its force, frequency, and amplitude of contraction So blood pressure is raised. Coronary arteries are dilated. Skin vessels are constricted. The rate and depth of respiration is increased.

Adrenaline :   1.   It elevates blood glucose level by (i) glycogenolysis of liver glycogen, (ii) breaking down muscle glycogen into lactic acid which in the liver is synthesised into glucose (latic acid can be utilised by cardiac muscle to yield energy); (iii) by stimulating ACTH release which is its turn through gludo-corticoids causes gluconeogenesis and glyconeogenesis ;   (iv) depression of utilisation of carbohydrates in the tissues.

2.   It raises oxygen consumption and so **BMR** is raised.

3   From fat of depots, free fatty acids are released and these are utilised for energy production.

4   The output, rate and force of heart is increased.   Therefore systolic blood pressure is increased.

5.   Cutaneous arterioles constricted.

6.   Visceral arterioles dilated.

7.   Coagulability of blood is increased.

8.   Dilatation of respiratory passages.

9.   Rate and depth of breathing increased.

10.   Erection of hair.

11.   Increased sweating.

Noradrenaline :   1.   Generalised vasoconstriction. So both systolic and diastolic blood pressure is increased. 2. Increase in sweating. 3. Erection of hair.

Adrenaline acts on the phosphorylase of liver and activates it for rapid gly-cogenolysis. In the muscles it is able to convert the glycogen into latic acid but not into glucose since muscles do not have phosphatase to split the glucose-6- phosphate and hence blood glucose level is raised by virtue of muscle glycogenolysis. The lactic acid has to go to the liver for conversion into glycogen and glucose (some of it can be metabolised by the   heart since it is rich in lactic dehydrogenase).

Reaction of Adrenal glands to Strees :   Adrenal cortex reacts to stress by depletion of lipids. Three major patterns occur based on variation of lipid content.   1. Focal lipid depletion with alteration in cell type; 2 focal deple-tion of lipid with alteration in cell type and degenerative changes and 3 lipid reversion. In the first type, cells are hypertrophic and compact   In the second, degenerative changes are characterised by cytoplasmic degeneration. Solid cords of cells in zona fasciculata break up to from pseudotubules   and tubules with lumina formation. The lumen of the tubules contain detached cells, pale staining fluid or erythrocytes.   Occassionally infarcts affecting mid zona-fasciculata may be seen.

Lipid reversion is local depletion, i.e. lipid is absent in outer fasciculata, scanty in reticularis but abundant in the remaining fasciculata. This picture is characterstic of conditions in which the adrenal recovers from stress.

Adrenal in systemic infections :

Acute infections stimulate adrenal cortex and cause lipid depletion   The inflammatory reaction in the adrenal when there is primary granulomatous inflammation is modified due to influence of cortisol and is predominently caseating and organisms are abundant. Generally, the adrenal is   spared of infectious process.

Colloid formation : Hyaline eosinophilic bodies having a laminated appearance varying in size from 2 to 25 μ consisting of phospholipid materials are seen as inclusions in the adrenal cortical cells. Their significance is not known These have also been reported in patients treated with spironolactone (Aldosterone antagonistic). *Note* : The adrenal glands in mammals can be divded broadly into two groups on the basis of lipid content of the cortex:—glands rich in lipid in man, monkey, rat and rabbit, and glands poor in lipid, in cattle, sheep, camel and horse. A prominent zona glomerulosa rich in lipid is separated from the rest of the cortex by a thin layer of compact cells, the sudanophobic zone. Next to this zone is zona fasciculata, rich in lipid and beneath this zone the zona reticularis, which is poor in lipid. Mitotic activity is greatest in the zona reticularis.

Senile changes in the adrenal are charecterised by narrowing and flattening of zona glomerulosa, thickening of the capsule, formation of irregular conglomeration of lipids in the zona fasciculata and hyalinisation of thickened arterioles. Arteriosclerosis and atherosclerosis are common in the capsular arteries associated with ageing. Striking presence of thick longitudinal muscle bundles of the the adrenal vein indicates systemic hypertension

Myeloid metaplasia in association with extramedullary myelopoiesis in other organs is common. Myeloid elements arise by metaplastic differetiation of reticuloendothelial cells lining the cortical sinusoids.

Dystrophic calcification following necrosis is common in the cat and dog. Extensive necrosis and calcification may cause symptoms suggestive of adrenocortical insufficiency.

Hypocorticalism can be classified into :

Primary hypothalamic-	Primary adrenal.
Symmond's Disease.	Addison's Disease.
Sheehan's syndrome.	Waterhouse-Friderichsen syndrome (Adrenal hemorrhage)
Iatrogenic (Steroid induced)	Contralateral adrenal atrophy
Congenital adrenal hypoplasia	Congenital adrenal hypoplasia
(Anencephalic)	Adrenal cysts.

Congenital hypoplasia of the adrenal gland has been reported in dogs and a hereditary predisposition has been indicated. Adrenal hypoplasia associated with maldevelopment of hypophysis is relatively frequent in calves.

Hypofunction of adrenal cortex
1. Acute hypofunction (Extirpation), rare
2. Chronic hypofunction (Addison's disease).

The manifestation of Addison's disease may occur only if there is bilateral destruction of the glands. This condition is exceedingly rare in animals, but may rarley be met with in man.

Causes : The causes are those conditions causing atrophy and destruction of the adrenals. These may be : 1. Tuberculosis, 2 Atrophy, 3. Amyloid disease, 4. Secondary tumors, 5. Histoplasmosis, 6 Hypopituitarism, 7. Drug allergy.

Symtoms : 1. General weakness;anemia; 2 Low blood pressure, feeble heart, 3. Brown pigmentation of skin, 4. Vomiting and diarrhoea, nausea, loss of appetite, slowed absorption, 5. Atrophied thyroid, 6. Loss of water along with sodium leading to hemoconcentration, 7. Acidosis due to loss of bicarbonate, 8. Inability of liver to convert amino acids to glycogen at usual rates.

These symptoms are explainable by the absence of cortical hormones. The mineral and glucose metabolisms are deranged, leading to elevated potassium level and lowering of sodium; hypoglycemia, hypotension, higher blood urea and anemia Pigmentation is due to increased melanin production as tyrosine is side tracked to the skin when adrenaline is no longer synthesised from this amino acid by the destroyed adrenals. Another explanation is that with low glucocorticoid level, production of ACTH is stimulated The first 13 amino acids in the structure of ACTH and MSH are the same and hence their actions are also similar so far as melanin production is concerned Hence high ACTH levels cause increased formation of melanin giving rise to the bronzed coloration of the skin. Lymphoid tissue of thyroid and lymph nodes is increased. In anterior pituitary there is decrease of basophils and increase of chromophobes. Atrophy of thyroid and heart is noticed

Adrenocortical hyperplasia Cortical hyerp asia is relatively common, particularly in dogs Hyperplasia may be of zona glomerulosa, zona fasciculata or zona reticularis It may be focal or diffuse and may or may not be associated with clinical manifestation. Hyperplastic reaction of zona fasciculata is a manifestation of adaptation syndrome in stress reaction. Progressive and regressive transformations have been described in stress reaction. The former is characterised by cellular hypertrophy, increased formation of new cells and storage of fat, while the regressive transformation is characterised by depletion of fat and hypofunction of cortex. Regressive transformation is seen in acute diseased conditions. In severe degree of progressive transformation, there may be formation of accessory cortical nodules (extra-capsular extrusions), There is proliferation of reserve cells in the sudanophobic zone and transformation of these cells into zona fasciculata type In the hyperplastic zone acini formation and secretory activity may be seen and this is a morphological expression of severe hyperactivity, wherein the gland takes recourse to exocrine type of secretion Accessory cortical nodule formation is characterised by formation of nodules on the capsule of the adrenal Histogenesis of the cells which form the nodule is from subcapsular blastema The reserve cells in the subcapsular blastema proliferate and invade the capsule which gets split up and later the proliferating cells get encapsulated by fibrous tissue. The small size of the lesion, absence of encapsulation and acini formation are all features of cortical hyperplasia while adenoma is characterised by large-sized nodules, encapsulation, acini formation and the type cells are columnar, lipid-depleted cells

Results of hyperfunction of adrenal cortex—Cushing's syndrome :

This condition seen in man, is due to excess of circulating hydrocortisone which again may be due to basophile adenoma of the anterior pituitary Zona fasiculata is increased and the adrenals are yellow in color Hypercorticalism may be observed in cases of tumors of non-endocrine tissue (ectopic ACTA syndrome), unassociated with regulation of adrenocortical function, like bronchial carcinoma, pancreatic carcinoma and thymoma. These tumors produce ACTH-like substances and induce Cushing's syndrome.

The effects are : 1. Painful adiposity of neck and trunk (limbs not affected) the buffalo type of obesity ; moon face. 2. Wasting of muscles and weakness 3. In females and pre-adolescent males, hairsuitism 4. Amenorrhoea in females. 5. Osteoporosis and kyphosis. 6 Peculiar striations on the abdominal wall. 7 Atrophy of the skin 8. Hypertension, 9. Hyperglycemia and diabetes, 10, Sodium retention, polydipsia, polyuria, urine with low Sp. gl. 11. Slow wound healing 12. Depression of lymphoid tissue and so lymphopenia. 13. Susceptibility to infection. 14. Pot belly. 15. Thinning and atrophy of skin 16. Dermatitis.

Hyalinisation of the basophile cells of the anterior pituitary together with disappearnce of basophile granules is a frequent finding in Cushing's syndrome Adrenogenital syndrome; Adrenal virilism :

A syndrome in which "little girls become little boys and little boys little men" In this condition there is an excess of androgens—masculinising hormones In the female fetus, if the excess of hormone occurs during the first few weeks of intrauterine life; pseudo-hermaphroditism results. But if it occurs later in females or in boys, precocious puberty results

Clinically the following symptoms are noticed 1. Rapid growth with great musclarity in children (infant Hercules) 2. Hairsuitism. 3 Virilism. 4 In girls—enlragement of clitoris, harisuitism on chest. 4. Impotence in boys (testes are atrophied). 6. In women: amenorrhoea, deep voice; hairsuitism on face and body. 7. Urinary 17 ketosteroids increased. 8. There may be deficiency of hydrocortisone—so hypoglycemia

Neoplasms : Tumors of the adrenal cortex : Adenomas may be seen in old cows and rarely in dogs and horses

The tumors may press upon and protrude into the aorta nearby. These tumors have thin capsules and are yellow in color due to fat content.

Microscopically, the neoplasm consists of cells of normal adrenal cortex. Encapsulation may be present. Cysts and areas of calcification may be seen.

Adenocarcinomas with limited metastases have been reported Affected animals show virilism, increased libido and weakness.

Pheochromocytoma : Medullary chromaffinoma :

Tumors of the adrenal medulla have been seen in dogs, cattle, horses and sheep and are usually solitary and unilateral These tumors may attain a size of 10 cm in diameter. The tumors are friable and brown in color.

Microscopically, the cells are pleomorphic and are large and ovoid, with a large central vesicular nucleus. The cytoplasm is granular and acidophilic. Mitoses are not common The cells may be arranged in rosette formation Cells take a dark color with chromium salts and hence the name pheochromocytoma (a tumor of dark staining cells).

Pheochromocytomas in animals are benign and do not metastasise. In most cases, no clinical symptoms are noticed. Rarely symptoms associated with hyperepinephrinism and hypernoradrenalism may be noticed These are paroxysmal attacks of hypertension with tachycardia, sweating, pallor and glycosuria

THE THYMUS GLAND

Thymus is an unpaired organ situated in the thoracic cavity, in the anterior mediastinum, in close association with the pericardium and the great veins at the base of the heart.

The endocrine gland is of vital importance in the young animals. Histologically thymus consists of lymphocytes and an epithelial reticulum. The gland undergoes involution, which commences in animals attaining maturity, viz., between 4 and 6 years in cattle, between 2 and 2½ years in horses, between 1 and 2 years in goats, sheep. pigs and dogs ; betweed 6 months and 1 year in cats and between 10 and 15 years in man.

Though considered to be a lymphoid organ, the thymus does not contain any germinal centers The epithelial reticulum probably acts as a blood—thymus barrier, blocking the penetration of antigens.

Functions : The exact functions of the thymus are not yet clear. But from the mass of experimental data, the following appear to be significant.

1. Thymus is an important site of lymphopoiesis in the embryo and the new born It plays a very significant role in specific resistance. Thymectomised animals when stimulated with an antigen after whole body irradiation do not produce detectable antibody, but thymus graft or injection of thymus cells alters the picture with a significant antibady rise. Clones of thymic cells are believed to migrate to the lymphoid tissues in Payer's patches. spleen and other lymph nodes, colonise and enable these organs in later life to give an antibody response to specific antigenic stimulation. The antibody making cells of adult life originate from the clones of antibody making cells of the thymus.

2. Thymus produces a hormone which has an immunotrophic effect, making those cells that have the immunologic potenital, immunologically competent.

Thyroid, adrenal and gonads antagonise the action of thymus. Involution of thymus begins at the time of sexual maturity. After castration or adrenalectomy, the thymus is enlarged and its involution delayed.

Status thymolymphaticus : Sometimes, men die suddenly without any obvious cause and in such people it was found that the thymus and the lymphoid tissue throughout the body were enlarged. In these cases, the hyperplasia of the thymus and other lymphoid tissues is accompanied by the underdevelopment of adrenals and the cardiovascular system, leading to lowered resistance of the patient, who succumbs to various infections. Adrenal insufficiency in such persons renders them unable to adapt themselves to stress This condition has been reported in dogs, cats and cattle.

Myasthenia gravis : This is a very peculiar disease of man, characterised by great muscular weakness. The muscles of the face are more severely affected. All the voluntary muscles are involved. No histological changes are noticed e ther in the muscles or in the motor end plates. Lymphocytic infiltration may occur later.

The cause is still obscure but the condition is associated with hyperplasia of and presence of lymphoid germinal centers in the thymus, which appear to produce some kind of antibody, which prevents the transmission of impulses at

the motor end plate.   This is an example of auto-immune mechanism.   Anti-
nuclear antibodies are demonstrable in the blood of such   patients.   This condition has not been met with in animals.

Tumors :   The neoplasms of the thymus are known as thymomas, which are
rare among domesticated animals.   Unlike in man, these tumors are not accompanied by myasthenia gravis   These tumors have been described in goats
cattle, sheep, dogs, horses, pigs and rabbits.   Most of the tumors were encountered among adult animals.

Site :   The tumors are found in the thoracic cavity, anterior to the heart.
They may extend to the neck

Macroscopically, thymomas are encapsulated, soft in consistency and gray
ish in color.   Many tumors may contain cysts and hemorrhagic foci.

Microscopically, the tumors consist of broad sheets of epithelial cells. These
cells are derived from the epithelial reticulum of the gland.   The cells may sometimes be arranged as whorls or rosettes.   Large number of lymphocytes are
present

Clinical .   Usually thymomas do not metastasize.   Some may spread to the
lungs, pericardium and regional lymph nodes.

Lymphosarcoma or Hodgkin's disease may sometimes be met with in the
thymus and these are usually seen in younger animals.   They are similar to such
conditions occurring elsewere in the body.

THE PINEAL GLAND

This is a tiny gland placed above the posterior extremity of the third ventricle.   In structure it is composed of epithelial cells in a loose connective tissue
stroma   Some gila cells and lymphocytes may be found.

The pineal gland secretes a hormone, *Melatonin* (5-methoxy-N acetyltryp-
tamine).   Serotonin which is found in the gland is transformed into melatonin
by the action of the enzyme, hydroxyindole-O-methyl transferase (HIOMT) which
is found in large quantities only in the pineal gland   Melatonin antagonises the
action of the melanocyte stimulating hormone (MSH) of the posterior pituitary
Melatonin acts on the brain to depress the rate of gonadal maturation and to
interfere with subsequent gonadal function and cyclicity.   Pineal hyperfunction
is associated with delayed puberty while hypofunction with precocious puberty.
In man complete destruction of the gland by tumors cause cachexia, trophic
disturbances, adiposity, premature development of genital organs, premature
spermatogenesis and growth of interstitial cells.   Melatonin inhibits thyroid
hormone secretion rate and the secretion of adrenal steroids.

Tumors :   Three types of tumors may arise from the pineal gland.

1.   **Pinealomas :**   These consist of islands of large epithelial cells, with
acididophilic cytoplasm and enclosed in a fibrous stroma   Among the large cells
are scattered small cells. believed to be lymphocytes   These tumors have been
described in a goat, horse, dog, silver fox and in a cow.

2.   Gliomas comprising of glia cells.

3.   Teratomas, from the totipotential cell rests.

CHAPTER 23

THE SKIN AND ITS APPENDAGES

Terms used in studying skin lesions
Anomalies
Alopecia
Congenital Ichthyosis
Acanthosis nigricans
Eczema
Dermatitis
  Serous
  Acute vesicular
  Seborrhoea
  Impetigo
Folliculitis
Injury by ionizing radiation
Urticaria
Laminitis
Cysts
  Epidermoid
  Dermoid
  Sebaceous
  Sudoriferous
Calcinosis circumscripta
Tumors.

The following terms are used while studying skin lesions.

Acanthosis is thickening of the epidermis due to hyperplasia of the cells of Malpighian layer. This condition may or may not be associated with hyperkeratosis and parakeratosis

Balooning degeneration is characterised by intracellular edema (hydropic degeneration) which is the early stage in vesicle formation. The cells are swollen and prickles disappear (acantholysis) and the cells become isolated from one another. This is seen in viral diseases.

Bulla (Bleb) is a space, containing fluid situated intra—epidermally or sub-epidermally These are larger than vesicles.

Dyskeratosis is a faulty development in which the cells of the Malpighian layer undergo abnormal, premature or imperfect keratinisation. The changes may suggest developing malignancy since the following characteristics of malignancy are found : hyper-chromatism, loss of polarity and large number of mitotic figures.

Erosion is loss of superficial epithelium. It is also called **excoriation** and is usually caused by continuous discharges.

Fissure or Rhagades is a deep linear defect in the epidermis, often extending to the dermis and occur in dry, crusty skin in which the elasticity is lost.

Hyperkeratosis is the abnormal thickening of the stratum granulosum.

Lichenification is thickening of the skin in irregular areas with exaggerated markings of the skin and is seen usually in chronic dermatitis.

Makula is a discolored spot of the skin, which is not elevated above the skin and may be seen in hemorrhages and focal hyperemia.

Pachyderma is thickening of the skin, all the layers of which are affected. The individual cells are normal. The condition is due to nonspecific dermatitis. There may be hyperplasia of the connective tissue. This is also known as elephantiasis and may be seen in the hind limbs of horses and scrotum of old dogs.

Actinomycotic pachyderma is a special form, often seen in the ears of swine. The ear is very much enlarged and hard. This is an actinomycotic granuloma.

Papule or pimple is a small circumscribed solid elevation of the skin resulting from an infiltration of the deep corium.

Parakeratosis is a condition in which the keratin layer is imperfectly formed but in which the nuclei of the horn cells are retained. The stratum granulosum is reduced in size In this condition, there is production of dandruff.

Pseudoepitheliomatous hyperplasia is a severe acanthosis in which there is deep downward growth of rete pegs, resembling a carcinoma. The lesion is seen at the margins of burns, indolent ulcers and other chronic focal inflammations.

Pustule is a vesicle filled with pus.

Scales are bran-like, thin flakes, consisting of imperfectly keratinised superficial layers of the epidermis and are usually seen in chronic dermatitis.

Spongiosis is inter-cellular edema of the epidermis This is seen in inflammation. A severe spongiosis will result in vesicles or bullae.

Ulcer is a break in the continuity of the epidermis, exposing dermis and so is deeper than erosion.

Vesicle is a small bulla in the epidermis containing serum, plasma or blood, covered by a thin rim of epithelium and raised above the surface of the skin. These may coalesce to form bullae.

Wheal (Urtica) is a sharply circumscribed, flat. edematous elevation of the skin and is found in urticaria Several may coalesce to form large plaques.

Anomalies

Epitheliogenesis imperfecta is an inherited skin defect in calves, piglets, pups, lambs, kids, and foals. due to an autosomal recessive character. The skin fails to develop around the feet, nose and ears. Infection occurs and septicemia results quickly.

Hypotrichosis congenita : (congenital alopecia) This is congenital absence of hair and is seen in calves, foals and dogs (Dachshunds). Such animals are easily susceptible to cold and sunburns. (Congenital hairlessness in calves is considered to be a Mendelian recessive trait).

Alopecia : This is lack of hair, wool or feathers. The old hair falls out and further growth does not occur. This is a symptom of many skin diseases such as dermatitis, eczema, mange etc.,

In fowls congenital lack of feathers is known as *apennosis* and may be found throughout the body or in some parts.

Causes :

1. **Inflammatory skin** disease : There is irritation and itching to relieve which animals rub themselves. Due to friction, hair is denuded. This is usually seen in mycotic infections or infestation by ectoparasites.

2 **Chemical agents** : (a) **Thallium poisoning** : In chronic poisoning dermatitis is produced in which the fibres break and the shaft is weakened. (b) In chronic poisoning by arsenic and selenium, alopecia is noticed

3. **Following severe febrile disease** : This is noticed in sheep, especially those that have recovered from blue tongue. There may be a partial or complete loss of wool.

4 Iodine deficiency in the new born : Animals born of mothers fed on iodine-deficient food, are hair-less.

5 Endocrine imbalance: Hyperestrogenism in male dogs (testicular tumors produce symmetrical alopecia).

In cryptorchidism (either unilateral or bilateral) bilateral alopecia develop due to deficiency of testosterone.

In bitches, hypoestrinism produces alopecia on the posterior part of the abdomen, inside the thighs and under the tail.

In hypothyroidism (due to atrophy of the thyroid gland, either primary or secondary to disease of pituitary) bilateral alopecia, hyperpigmentation of the skin (diffused or localised), followed by hyperkeratosis are noticed

6. Vitamin deficiency : In Biotin deficiency, alopecia is believed to be produced.

Vitamin C deficiency in calves causes loss of hair and cracking of thickened skin

Vitamin E deficiency in cats produces loss of hair on the head and extremities as well as dryness of the skin.

7. Plant poisons: When sheep and goats feed on *Tamarindus indica;Senecio* and *Chrysocoma tenuifolia,* alopecia develops.

8. Injury to nerves: Peripheral nerve injury is followed by alopecia,

9. Psychic disturbance- : In man *alopecia areata* is believed to be due to psychic nervous stimuli There have been many cases of alopecia coincidental with business failure, death or departure of beloved persons and the ordeals of getting married or divorced. Capillary blood supply is not affected.

10 Other Causes : In chronic interstitial nephritis in dogs partial alopecia develops. Partial alopecia and hyperkeratosis are found in cattle and dogs suffering from functional disorders of the liver such as cirrhosis.

Feeding of excessive whale oil, palm oil and soy oil as milk replacers causes alopecia. The fibres break easily Alopecia may de due to the variation in the blood supply to the hairs or due to variation in the nutritive quality of the blood supplied.

Inherited symmetrical alopecia is sometimes met with among cattle and is due to a single autosomal recessive character. Calves begin to lose hair between 6th week and 6th month of their postnatal life. The condition commences on the head, back and hindquarters and progressively extends to other parts of the the body, It is symmetrical. Affected parts are completely bald.

Congenital Ichthyosis : Ichthyosis means scaly skin of fish. Some calves are born which have a skin, usually devoid of hair, but which is composed os hard, horny plates, with fissures seperating them The fissures follow the normal skin folds, Affected animals do not usually live longer than one or 2 days.

Macroscopically: there is severe hyperkeratosis and some acanthosis This condition is supposed to be hereditary, conditioned by a simple recessive lethal gene

Acanthosis nigricans : This condition in dogs results from hormonal imbalance—(due to decrease in thyroid uptake). In this condition there is hyperkeratosis with increased pigmentation. This is seen mostly in dogs. The lesions are bilateral and are small, poorly circumscribed patches, blacker than normal and found on the skin of the abdomen, axilla, innerside   of the thigh, inguinal and circumanal region.   The skin is thickened and folded and usually devoid of hair

Microscopically, there is elongation of the dermal papillae. Congestion may be noticed. Prickle cell layer may be increased in thickness with prominent rete pegs   Pigment is not confined to the basal layer but is   present in cells of all layers.   The glands and hair follicles   may be atrophied. The cause of acanthosis nigricans is obscure.   This condition is found in association with certain visceral tumors like adeno-carcinoma of the liver. Dogs affected with sertoli-cell tumor and hypoplasia of the pituitary show this condition.

ECZEMA

This is an inflammatory skin condition (dermatitis), charcterised by vesicle formation, infiltration by inflammatory exudate, watery discharge and development of scales and crusts   These are the several manifestations of an illunderstood condition.   The term eczema is a very general term and is not very descriptive   There is no valid reason to retain in   since *dermatitis* can be better described specifically according to the nature of exudate; as serous, papular, suppurative, necrotic or parasitic.

Eczema may. however, be applied to the allergic condition of the skin though in animals the nature of allergin cannot usually be determined.   The allergins may be exogenous or endogenous.   The exogenous allergins may be chemicals (medication of the skin), fungal or parasitic. The endogenous allergins may be ingested proteins or they may be formed in the   intestines due to over eating or stasis of food as in constipation; or they may be digested parasites.   In some animals there may be inherited predisposition. Continued sweating may also predispose the animal

Macroscopically, in the acute or moist eczema, vesicles, bullae and infiltration by inflammatory cells can be noticed First the vesicles start as spongiosis and later edema of the dermis may develop. Pustules arise which may later dry up. Vesicles may rupture causing weeping on the surface.   Dermis may show inflammatory changes.

The chronic or dry eczema : there may be scratching and rubbing leading to thickening of the skin with scale formation. lichenification and formation of scabs and fissures

Microscopically, there is hyperkeratosis and acanthosis. The pegs are prolonged. The dermis shows fibrosis and infiltration by lymphocytes.

DERMATITIS

Severe inflammation of the skin is called dermatitis.

Causes of dermatitis include the following :

1.   Physical :
   a)   Pressure—exemplified by decubital ulcers.
   b)   Trauma – abrasions, scratches etc.
   c)   Beta irradiation.
   d)   Photosensitization.
   e)   Heat and cold – if excessive
   f)   Sunburn—in unpigmented animals especially.

g) Exessive wetting causing maceration of the stratum corneum. This is easily invaded by baceria.

2. Chemical irritants—acids, bases, thallium poisoning.
3. Bacterial—Swine erysipeleas, Anthrax,Tuberculosis of skin, *Streptococcal* and *Staphylococcus aureus* infections.
4. Fungal—various types of ring worm, moniliasis *(Candida albicans,* Epizootic lymphangitis).
5. Parasitic—Nematodes, *Stephanofilaria*; ectoparasites;
6. Viral—Pox (Variola), contagious ecthyma and in Foot and Mouth disease, rinderpest, mucosal disease etc
7. Allergy—allergic dermatitis of horses
8. Dermatitis due to nutritional deficiencies—as in deficiency of B vitamin complex in pigs.
9. Protozoa—Leishmaniasis.

Lesions : These vary with cause.

Serous dermatitis : This is the mildest type seen and occurs in sunburns and mild friction due to ill-fitting saddle and harness as well as mild chemicals, heat and cold.

Macroscopically, the skin is red and raised due to spongiosis. Sweat and sebum may be secreted in large quantities.

Microscopically, hyperemia, edema and infiltration by a few leucocytes are seen.

Result : If the cause is removed, the condition resolves quickly and no permanent damage is done. If the cause is more intense and persists, acute dermatitis will follow :

Acute Vesicular Dermatitis :

Causes :

1. More intensive sunburn
2. Stronger chemicals   } Than in the previous variety
3. Hotter and colder applications.
4. Photosensitisation
5. Sepecific diseases—pox

Macroscopically, besides redness and heat of the locality, there is edema of the dermis leading to swelling of the part. Blebs form in the epidermis containing clear fluid. The roof of the vesicle is formed of stratum corneum.

Microscopically, the lesion starts as an erythema, (morbid redness of skin), with edema in the dermis and infiltration by lymphocytes and histiocytes. With increasing exudate, spongiosis occurs Hydropic degeneration of the prickle cells occurs, which may rupture due to pressure from fluid inside. Due to coalescence of adjoining ruptured cells, vesicles from, which later may enlarge to form bullae. As the fluid accumulates in this vesicle, the superficial stratum corneum is pushed outwards to form the roof of the vesicle and this layer ultimately dies. If the pressure is great, rupture of the stratum corneum will occur, revealing a red base consisting of hyperemic stratum germinativum. Intact stratum corneum prevents infection. Presence of leucocytes in the vesicles changes them to pustules. Later the pustules may rupture and a crust is formed consisting of the coagulated exudate that forms at the site. Beneath the crust infiltrating neutrophils protect the area while epithelium regenerates and healing occurs

Seborrhoea is increased secretion of sebum and is usually met with in dermatitis. The common examples are the greasy heel of horses and seborrhoea in dairy cows. In the former, there is excoriation and soreness on the back of the pastern and may be due to standing in unhygienic barns The part appears greasy and is sensitive to the touch, resulting in lameness Similar condition of the legs may also occur in cattle that stand in mud.

In the seborrhoea of dairy cows, that have newly calved the lesions are commonly found in the groin between the udder and the thighs or in the udder itself between the two halves There is voluminous sebum secretion and the skin may become necrosed. Bad smell exudes. The affected skin may be shed.

Impetigo : This is pustular dermatitis usually caused by *Staphylococci* and rarely by *Streptococci*

Infection may occur through bite wounds in pigs (bites by dogs or other pigs). Lesions are mostly found on the face—on the snout, ears and over the eyes.

In Canine distemper lesions are found under the belly.

In sheep wounds caused by thorns are infected and dermatitis develops.

Dirty litter and humid places are predisposing factors. Skin that is continuously drenched by discharge, either from a wound or a natural opening becomes soft and easily infected.

On the udder small pustules are found at the base of teats, Other parts of the teat and udder may also be affected. Spread to other animals occurs during milking This condition is noticed in association with staphylococcal mastitis.

Lesions : **Macroscopically**, initially erythematous patches may be observed which soon become vesicles and pustules. A scab forms on removal of which a red weeping area is seen. Healing occurs quickly.

If dermis is also affected, a diffuse inflammatory area is seen—**phlegmon**.

Microscopically, the pustule contains serofibrinous exudate with neutrophils The dermal vessels are congested.

Folliculitis is the inflammations of hair follicles. Inflammation of sebaceous glands is called **acne**. **Boil or furuncle** is an abscess of the hair follicle. **Carbuncle** is a cluster of boils situated close to each other, opening on to the skin through several pores.

Causes ; Usually *Staphylococci*; *C. Pyogenes* may also be found. Contributing causes are sweating, contamination by filth ; decreased vitality.

It may also result from ringworm, mange and as a complication of distemper.

Lesions ; The earliest lesions are small raised papules, which may either subside or progress to pustules or abscesses Discharge that occurs at the roots of hairs dries to a scab, removal of which results in loss of hair, which, however, regenerates. Local lymph nodes may be enlarged.

Microscopically, the follicle contains inflammatory exudate with leucocytes.

Injury by ionsing radiation : Though, comparatively speaking, skin is radioresistant, it is radioresponsive. The beta particles appear to be more pathogenic. The radiant energy seems to alter the permeability of the cell membrane after injuring it, so that the passage of fluids, electrolytes, oxygen and nutrients is so altered that vacuoles appear in the cytoplasm.

The skin may be severely exposed in therapeutic use of ionising radiation. Within 24 to 48 hours, erythema occurs. This may fade and return within 10 to 21 days due to vascular changes. In radiation, vascular endothelial cells are damaged which become swollen. Thrombosis occurs obstructing the vessels. The fibrous tissue on the walls of the blood vessels becomes hyalinised.

In severe exposure, dermatitis with edema develops, followed by loss of hair (epilation) and necrosis of the epidermis. A highly keratinised, dry and scaly epidermis results—parakeratosis. If the hair follicle is damaged, hair may not grow again. So also, sweat and sebaceous glands may be destroyed. If the necrotic area is shed and ulcer forms, secondary infection may occur by saprophytes leading to gangrene.

The changes noticed in the dermis, following damage to the epidermis are : hyperemia, edema. hyalinisation of the fibrous tissue and infiltration by lymphocytes and neutrophils.

If the dermis is not affected, complete healing may occur. But if there is wide spread necrosis with involvement of the dermis, scar tissue may be formed. But this scar is very weak and the epidermis is thin and inadequately keratinised. Carcinomata are supposed to arise from such scars of old X-ray burns.

Urticaria : This is an allergic condition and is characterised by the appearance of wheals (or urtica) on the skin.

Causes : Primary : Insect bites, contact with nettles and caterpillar hairs, unusual food, larvae of warbles; or drugs, eg Penicillin.

Secondary : This occurs along with other diseases in which allergic manifestations are present Eg.: Strangles in horses and Erysipelas in swine, Chronic fowl cholera (wattle disease) in fowls.

The lesions are those found in an allergic reaction. First there is erythema due to vascular dilatation, followed by exudation.

LAMINITIS (Founder)

The inflammation of the sensitive laminae of the hoof is called laminitis This occurs usually in horses but may sometimes be seen in cattle.

In horses. the onset is sudden and acute with high temperature, rapid pulse and accelerated respirations. The animal shows distress and does not like to bear weight on the affected limb.

In the hoof, acute congestion with venous stasis is seen Rarely in very severe condtion, after some days, suppuration may occur. As the foot is enclosed in a very tough and unyielding hoof the pressure of the increased blood and edema on the sensitive tissue causes extreme pain which may be severe enough as to cause systemic disturbances Hemorrhages into the tissue may occur. There may be separation of the sensitive laminae from the inner laminae after a few days in some cases Due to weight of the body, the *os pedis* takes a more perpendicular position and it may actually, sometimes, pierce the sole. Due to the abnormal position of the *os pedis*, weight is borne by the heels and so the toe grows exceedingly long. As the keratogenic lamellae are affected and some destroyed, formation of the hoof is uneven.

There is evidence to show that the basic defect is not vascular but degenerative changes in the "keratinogenic layers" of the hoof. These degenerated lamelle do not produce sufficient keratohyaline. There is loss of eleidin granules and "Onychogenic fibrills". Probably some toxic factor destroys some substances important in the formation of hoof. The following hypothesis explains the pathogenesis of this condition. The absorbable toxins, indol and phenol, are formed from tyrosine and tryptophan. These toxins become bound to the free sulphate ions of blood and so the balance between the blood serum sulphur and the sulphur in chondroitin sulphuric acid is disrupted. Hence the collagen loses its strength and elasticity. Consequently the keratinous material loses its supporting tissue, laminitis developing thereby.

Causes :

Laminitis is met with in the following conditions in the horse.

1. After excessive ingestion of concentrates, causing enteritis and diarrhoea.
2. Drinking very cold water after the animal is over-heated.
3. Following the use of irritating purgatives like aloes.
4. Concussion on hard roads—long drives.
5. Standing for a long time.
6. Chronic laminitis develops in fat ponies that are kept at pasture without much exercise.
7. In certain septicemic conditions—metritis, retaind placenta, mucosal disease. (In cattle)
8. In cattle certain feeds are supposed to cause allergic laminitis. Cotton seed cake, mouldy hay and barley have been incriminated.

In cattle and sheep that are overfed with concentrates, laminitis might develop.

It is reported that treatment with methionine was very effective, one treatment being sufficient in acute and subacute cases. Medication was continued in chronic cases till symptoms abated.

Sequelae: If the condition subsides in 72 to 96 hours no permanent damage results. But if it persists permanent damage to hoof ensues manifested by dropped sole, convexity of the hoof, elongated toe and concentric rings on the hoof parallel to the coronet.

Non-neoplastic cysts

Epidermoid cysts : These are mostly found in the dog. The cysts are rounded or oval and located subcutaneously and so are easily movable. But if found in the dermis, they become fixed. These cysts enlarge slowly.

Epidermoid cysts may be solitary or may occur as clusters or sometimes be even generalised. Their size may vary from a faw millimeters to a few centimeters in diameter. The cysts are well demarcated, with a thin wall. The contents are gray or brownish and may be semi-solid or dry in consistency and contain hair or wool.

Usually, epidermoid cysts are benign and can be removed surgically. Rarely do they ulcerate and become infected.

If the contents escape into the tissues, a foreign body reaction is set up. Occasionally, epidermoid cysts may manifest malignant transformation.

Microscopically the cyst wall consists of a collagenous capsule surrounding squamous epithelium. Contents consist of keratin, deposited concentrically. The wall of the cyst does not have any skin adnexa. The cyst probably arises by the occlusion of the mouth of hair follicles with resulting trapping of the epithelium, which continuously desquamates keratin into the lumen.

Dermoid cyst : This is similar to the epidermoid cyst but differs from it in that the wall of the dermoid cyst contains skin appendages. The dermoid cysts are located at the junction of the dermis and the subcutis and so are mobile They are soft and round or oval. Occasionally, a cyst may communicate with the surface through a tiny pore. The cyst contains keratinised, greasy substance together with hair and desquamated cells. Calcium and cholesterol may be deposited in the contents

This is a benign lesion and has no connection with congenital "Dermoid", of the eye or the teratomatous "Dermoid" of the ovary.

Sebaceous cyst : Dilatation of sebaceous gland or its duct gives rise to a sebaceous cyst, which contains the greasy sebum and cholesterol.

Sudoriferous cyst: Occlusion of the ducts of sweat glands gives rise to cysts which have a thin capsule, lined by a single layer of columnar or cuboidal epithelium and containing a watery fluid.

Calcinosis circumscripta : These are commonly seen among dogs as raised elevated or bulging masses under the skin, one to ten cms. in diameter. On section, white, granular masses are present separated by thin connective tissue septa On incision, chalky paste-like material may be enucleated.

Microscopically, the lesion consists of masses of granular material, staining blue with hematoxylin, evidently a calcium salt, surrounded by granulation tissue with lymphocytes, plasma cells, foreign body giant cells and macrophages

TUMORS OF THE SKIN

Epidermal : Squamous cell carcinoma, Papilloma, Melanoma (Benign and Malignant). Basal-cell carcinoma.

Dermal : Mastocytoma, Fibroma, Fibrosarcoma, Histiocytoma, Myelocytoma, Venereal tumor of dogs, Keloid of horses, Equine sarcoid.

Adnexa : Hair matrix tumor, Basal-cell carcinoma, Adenoma and adenocarcinoma of perianal glands, Adenoma and adenocarcinoma of sweat glands Adenoma and adenocarcinoma of sebaceous glands.

In the subcutaneous tissue: Hemangioma, Hemangioendothelioma, Lipoma Lymphangioma

CHAPTER 24

THE EYE AND EAR

THE EYE

Though detailed anatomy of the eye is beyond the scope of this book, a brief description is necessary to understand the pathological conditions that may be met with.

The eyeball is located in the bony cavity of the skull known as the orbit It is protected by eyelids, which have skin on the outside and startified squamous epithelium (the conjunctiva) lining the surface that comes into contact with the eyeball. The borders of the eyelids have eyelashes containing hair. Just behind the eyelashes are a row of tiny sebaceous glands, the Meibomian glands, the secretion of which serves to lubricate the eyelashes, preventing their adhesion. Movement of the eyeball is brought about by various muscles that are attached to the orbit and the eyeball.

Membrana nictitans or the third eyelid is situated near the inner canthus and consists of elastic cartilage covered by conjunctiva.

The conjunctiva is a membrane with a fibrous tissue stroma covered by epithelium and lines the inner surface of the lids and the third eyelid and is also reflected on the eyball—on the sclera and cornea The conjunctiva also contains small glands similar to lachrymal glands.

Lachrymal glands, producing tears, are situated below the supraorbital processes. The duct draining the tears is situated near the nasal canthus and opens into the nostrils.

Harderian glands are located in a cavity below the membrana nictitans. They secrete an unctuous material that is useful to lubricate eyelids and membrana nictitans

The **sclera** or sclerotic coat is a fibrous membrane lined externally by the conjunctiva. Though it contains many blood vessels, normally they are not seen, but come into view when they become "injected". The functions of the sclera are : (a) to protect the vascular tissues underneath and (b) by affording rigidity to the eyeball, its shape is maintained constant. Otherwise, by the pressure of the eyelids and muscles, the shape is likely to be altered and vision will be affected.

The **cornea** is the transparent portion of the eyeball It consists of 5 layers: (1) stratified epithelium, which is continuous with that of the sclera; (2) a structureless elastic layer—the Bowman's membrane; (3) the substantia propria consisting of transparent connective tissue fibres; (4) Descemet's membrane, which is very elastic and hyaloid This is deposited by the endothelial cells that line the posterior surface of the cornea This membrane is very tough and is useful when ulceration of cornea occurs. Though the substantia propria may be ruptured, it is this membrane that prevents the herniation of the iris and (5) the endothelial cells that line the Descemet's membrane This forms the anterior boundary of the anterior chamber that contains the aqueous humor. The cornea has no blood vessels and so it derives its nutrition from (1) aqueous humor; (2) from the capillaries of the sclera surrounding it and (3) from tears.

During wakeful hours, oxygen supply is obtained by the cornea directly from the air and when closed by eyelids from the blood. Vitamin A is necessary for the health of the cornea. In its deficiency xerophthalmia results.

The **choroid coat** is a pigmented, vascular membrane that lines the sclera. The retina is situated in its internal face.

The **iris** is a muscular structure with an opening in the centre. This opening is called the pupil. The iris functions as the diaphragm in a camera by virtue of the muscle fibres.

The **ciliary body,** which is a ring of tissue is joined to the iris in front and to the choroid behind. This structure consists of the ciliary processes which secrete the aqueous humor and the ciliary muscle, which helps in the "accomodation" of the eye.

Dilatation of the pupil is known as **mydriasis.** This can be brought about by various drugs: atropine, hyocyamine and stramonium (all of which are parasymptholytic); cocaine, adernaline, and amphetamine which are sympathetic stimulants

Mydriasis may be seen in the following conditions : hypertension, injury to the third cranial nerve (oculomotor), strychnine poisoning and in the later stages of chloroform anesthesia

Constriction of the pupil is known as **myasis.** This can be brought about by pilocarpine, physostigmine and ergotamine.

Contracted pupil is the tell-tale appearance in morphine addiction.

Myasis may also be met with in the following conditions : keratisis, ulceration of the cornea, inflammatory conditions of the uveal tract and meningitis.

The **lens** is a peculiar structure composed entirely of epithelium. It has neither stroma nor vascular tissue In front it is bathed by the aqueous humor and is nourished by it. Actually the anterior surface of the lens forms the posterior boundary of the anterior chamber. Its anterior surface is in contact, partly, with the iris. Its posterior surface fits into the depression of the vitreous—the hyaloid fossa. The lens has a tough hyaline capsule outside, which is impermeable to bacteria and leucocytes.

Aqeous humor : The aqueous humor (intraccular fluid) is secreted by the ciliary process into the anterior chamber from where it is drained at the filtration triangle into the veins through the spaces of Fontana.

The functions of this fluid are :

1   It nourishes the lens and cornea which are avascular.

2.   It serves as a refractive medium.

3.   It helps in the maintenance of the shape of the globe For proper vision it is essential that the shape of the globe is not altered.

4.   It helps in the removal of waste products of metabolism.

There should be a constant turnover of the fluid to achieve the above functions or otherwise the fluid may become cloudy and the nutrition of the lens may be affected leading to cataract.

The **vitreous body** is a jelly-like, transparent structure lying between the posterior surface of the lens and the retina and is enclosed in a delicate capsule, the hyaloid membrane

Of the structures enumerated above, the cornea, aqueous humor, the lens and the vitreous, constitute the refractive media and so to a large extent perfect vision depends on their perfect transparency Should they become cloudy or opaque due to pathological processes (to be described hereafter) vision will be impaired and blindness may ensue.

Perfect vision is also dependent upon the health and proper function of the photosensitive nervous receptors in the retina and optic nerve.

The **retina** is connected to the brain by the optic nerve. It santerior (inner) surface is in close association with the hyaloid membrane of the vitreous and the outer surface related to the choroid

The histological structure of the retina is very complex and consists of ten layers The rods and cones of the last layer (i. e. the layer in contact or near the choroid) are the most important in vision

The **optic nerve**, which is the second cranial nerve, enters the globe at the optic papilla and the fibres get distributed to the retina

Branches of the third, fourth, fifth and sixth cranial nerves innervate different parts of the eyeball and the fibres get distributed to the lachrymal glands,

Congenital anomalies of the eye

Anophthalmia congenitus This is complete absence of one or both eyes. Instances of this condition have been reported in foals, pups, calves and piglets. Histological examination may reveal remnants of some ocular structures

Microphthalmia is a condition in which one or both eyes are small. Most of the cases have been reported in swine (result of hypovitaminosis A of the dam) and dogs (a hereditary condition caused by sub-lethal genes)

Cyclops is a condition in which there is only one eye due to fusion of the two orbits and is seen in monsters

Ankyloblepharon is a condition in which both the eyelids are fused together.

Strabismus is squint of human beings. In animals the condition is bilateral with the two eye globes turning inwards. Squint is seen in Siamese cats in which it is congenital. This has also been seen in calves of beef breed and in collie dogs

Entropion is turning in of the eyelids and is a hereditary congenital condition seen in sheep, dogs and foals.

Ectropion is turning out of the eyelids. Usually the lower eyelid is affected. This is also a hereditary condition.

Coloboma is a congenital anomaly due to failure of the closure of embryonic ocular cleft and so the eylids, ciliary body, lens or iris may be affected showing fissures or gaps in their continuity In the eyelids a small wedge shaped portion may be missing or even larger area may be lacking. Similarly small areas of iris and lens may be absent.

Dermoids of cornea : These tumors occur as a congenital condition due probably to a sublethal factor. The cornea of one or both eyes is partly covered by skin containing hair. Actually the cornea at the place affected is replaced by the skin Because of constant friction and irritation by hair, conjunctivitis, pannus formation, keratitis or ulceration may result.

Congenital anterior synechia is a condition in which there is adhesion between the iris and the posterior surface of the cornea.

Congenital opacity of cornea may be met with, sometimes, as a result of anterior synechia In certain breeds of cattle (Holstein-Friesian, Swiss, Norwegian, Red poll) hereditary opacity of cornea has been met with

Microphakia — the lens is small and is spherical. it may be opaque or transparent.

Displacement of the lens or luxation of the lens *(ectopia lentis)* may be met with in some animals. The dislocated lens is opaque. In certain breeds of dogs (Sealyheam and terriers) it is a developmental defect.

Cataract is a condition in which the lens becomes opaque. In some animals it is a congenital condition.

Congenital aplasia of the retina and hypoplasia of the *optic nerve* may be met with in calves and they are born blind.

EYELIDS

Trichiasis is the turning in of the eyelashes and therefore cornea is irritated by the hair. Keratitis and attendant lesions may be encountered.

Blepharitis is inflammation of the eyelids. This may be a part of generalised dermatitis. But it is usually a complication of (a) distemper in dogs and cats, (b) trauma, (c) conjunctivitis or (d) inflammation of the lachrymal glands. Complications of blepharitis are : (i) **ankyloblepharon**, when adhesions between eyelids take place, (ii) **Symblepharon**, where there is union between the conjunctiva lining the lids and that covering the eye balls.

Hordeolum or stye is the inflammation or even abscess formation of the follicles of an eyelid and is a very painful condition.

Chalazion is the abscess formation of the Meibomian glands. Sometimes a cyst may form in these glands.

Edema of the eyelids : This usually results due to (i) trauma, (ii) infection, (iii) conjunctivitis, (iv) allergy and (v) fracture of the orbital ring. The following condtions are accompanied by edema of eyelids—*Horse :* Influenza, pink eye, purpura and allergic conditions

Cattle : malignat catarrhal fever, distomiasis, stomach worm affection, traumatic pericarditis and penetration of a foreign body into the eye like husks etc.

Pig :gut edema, chronic hog cholera.

Dog and Cat : distemper, allergy, stings by nettles and ants.

Neoplasms : Squamous cell carcinoma which may arise from the conjunctiva or the skin of eyelid or from the membrana nictitans is very common among cattle

Papillomata, basal- cell carcinoma and angiomata (cavernous or **capillary**) may be found. Adenoma of the Meibomian glands may also be met with.

ORBIT.

Exophthalmos : Protrusion of the eyeball

Enophthalmos sinking of the eyeball into the orbit.

Orbital cellulitis : is the inflammation of the orbit. Is rare.

Causes :   Nutritional—in cats.
                  Foreign bodies
                  Extension from periodontitis of posterior molars in dogs.
                  Orbital contusion complicated by fracture of the ring of the orbit.
                  Extension of ophthalmitis.

This condition is usually suppurative in nature. Non-inflammatory edema with slight exophthalmos occurs in edema disease and Mulberry heart disease of swine, congestive edema of head, purpura hemorrhagica and urticaria.

Lachrymal apparatus:Dacryoadenitis is the inflammation of the lachrymal glands and is rarely met with as a complication of conjunctivitis or trauma. There is diffuse congestion and enlargement of the gland, causing protrusion of the membrana nictitans. An abscess may form rupturing on the upper eyelid. The ducts are dilated with inflammatory exudate.

Xerophthalmia due to vitamin A deficiency has already been studied (Page 157)

Occlusion of lachrymal canal : This may be congenital or acquired. In doge of certain breeds (Sealyheams and Poodles) there may be congenital absence of the puncta.

The canal may be occluded in the following conditions. (a) entropion, (b) inflammatory swellings found in conjunctivitis and rhinitis, (c) atrophic rhinitis,in swine, (d) miller's disease in horses and (e) neoplasms of the nasal passage.

Adenoma of the lachrymal gland is very rare.

CONJUNCTIVA

Conjunctivitis is commonly met with among animals and the causes are :—

(a) **Bacteria** : Some bacteria can penetrate the intact conjunctival mucosa—*Brucella, Listeria* and *Pasteurella tularensis.*

Usually, the conjunctival mucosa is free of bacteria either due to the flushing action of the tears or to the bacteriostatic property of the lysozyme.

Secondary infection by *Staphylococus aureus, Pseudomonas aeruginosa* and *E. Coli* may occur following trauma or debilitating viral diseases (distemper in dogs).

(b) **Chemicals :** Disinfecting fluids, lime in white washing material, irritant gases such as formalin vapour and sulphur, smoke, acids, alkalies, sheep dips, parasiticides, skin dressing, iodism.

(c) **Foreign bodies** Awns, oak husks, mud, dust and sand.

(d) **Parasites :** *Thelazia lachrymalis* in horse, *T rhodesii* in the ox, *T. callipeda* in dog and *T. leesi* in camel.

(e) **Allergy** to pollen, horse serum etc.

Symptoms : There is congestion of the conjunctiva and increased production of tears, which flow over the face as the lachrymal cannal may be closed due to swelling of the membrane. The tears are clear at first but soon become turbid and thick due to the presence of leucocytes and mucoid material. It may also contain flecks The eye lids may be glued by the sticky material.

Infections by pyogenic organisms as occur in distemper of dogs and periodic ophthalmia of horses produce purulunt conjunctivitis. Croupous or diphtheritic conjunctivitis is mostly encountered in fowls. In cattle infection by *S. necrophorus* causes croupous conjunctivits. In this condition there is gray or chocolate colored membrane covering the eyeball.

Infection may spread to the cornea and keratitis may, therfore, result. In the purulent and croupous varieties, keratitis and ulceration of the cornea arvery commonly seen.

CORNEA

Pannus is a condition in which vascular granulation tissue is found betweee the corneal epithelium and the Bowman's membrane. Calcification of the granulation tissue may sometimes occur.

Keratitis. The causes of keratitis are the same as detailed for conjunctivitis.

Symptoms inculde photophobia and blepharospasm (in which the eyelids are tightly closed). The cornea may, in the initial stages, show edema of the epithelium followed by opacity due to infiltration by leucocytes. In the acute stage ulceration may occur. Vascularization of cornea may be seen.

Corneal ulceration occurs during acute or chronic conjunctivitis. It may a so occur as a result of suppurative conjunctivitis or due to trauma (thorns, nails, barbed wire, cat scratches, horn gores). Nutritional imbalance due to deficient proteins and vitamins and impaired nerve supply may also cause ulcers. Virulent organisms can cause ulcers due to the activity of their toxins.

If the Descemet's membrane is also perforated, aqueous humor is lost. There may be prolapse of the iris through the rupture (**Staphyloma**), followed by dislocation of the lens. Secondary infection of an ulcerated cornea can infect the whole globe (panophthalmia) and the eye will be completely lost.

Healing of a corneal ulcer is slow and is similar to healing of an open wound. The scar tissue contracts and a tiny scar, which is opaque. is left, which never completely disappears. Depending on its density, the corneal scar is known as nebula, macula or leucoma.

Infectious keratoconjunctivitis in cattle : Pink eye : This condtion may occur as an epizootic in many parts of the world and is common in summer and autumn. Probably flies transmit the disease from animal to animal. The causative organism is *Moraxella bovis* which is gram negative and is found in the tears. An endotoxin that causes necrosis of the skin is produced by this organism. Blood stream is not infected.

Symptoms include conjunctivitis, copious lachrymation, photophobia and blepharospasm A slight elevation of temperature accompanied by anorexia occurs. Within one to two days following the onset of the above symptoms, corneal opacity develops in the centre followed by ulceration in two more days. Ulceration in the young expands and vascularization may occur. The cornea may become completely opaque. There may be purulent discharge from the eyes. As the condition subsides, opacity decreases and complete recovery occurs in three to five weeks. Recovery is followed by immunity which lasts for a year. Immunity is local and may be due to peristence of the organism in the conjunctival sac

A mild conjunctivitis is also caused by the virus of Infectious Bovine Rhinotracheitis characterised by congestion of the conjunctiva with increased lachrymation which is serous. Cornea may become slightly cloudy. Cornea is not ulcerated

Infectious keratoconjunctivitis in sheep—contagious ophthalmia: In sheep this is also known as pink eye, *caused by Rickettsia conjunctiva* and occurs as outbreaks in various parts of the world in summer. Though the condition is mild and not fatal, animals become temporarily blind and become weak and undernourished as they cannot graze properly and adequately.

Symptoms include conjunctivitis, keratitis, increased lachrymation, blepharospasm and opacity of the cornea with vascularization. The discharge which is watery at first becomes purulent subsequently. Recovery starts within three to four days and by the tenth day is complete. In some animals, certain amount of opacity of the cornea remains. Recovery is followed by partial immunity and the recovered animals are carriers for a year.

Among goats, the condition is mild.

Infectious keratoconjunctivitis in fowls

This condition is found in Africa, Russia and Denmark among fowls. A rickettsia is said to be causative. It is known as *Ricloasia conjunctivae.*

Birds under three months are usually affected, the incubation period is 4 to 6 days. The condition is transmissible

Symptoms : Usually one eye is affected. In a few days the other eye is also affected due to spread of infection. The eyelids are glued together and opacity of the cornea develops. If no secondary infection occurs, recovery in 2 weeks is the rule. Rarely growth of the bird may be retarded. Morbidity is 20% to 30% while mortality is less than 1%

No organisms are seen in the conjunctival sac. Inclusion bodies are seen in the conjunctival epithelium. Since antibiotics are not useful in treatment a virus is suspected to be the cause.

THE LENS

Because of the peculiar structure of the lens, the changes that could occur in this structure are of limited range. These are : (a) changes in its position and (b) degenerative or metabolic changes in which the transparency of the structure is altered.

Luxation of the lens : The lens is anchored by the zonula of Zinn or the suspensory ligaments to the ciliary body. If these ligaments are ruptured, the lens may be displaced into the anterior chamber or into the hyaloid fossa or into the vitreous If it is displaced into the anterior chamber, opacity of the cornea occurs due to the pressure of the lens on the endothelium. In such a situation, the nutrition of the lens is altered and it becomes opaque. Glaucoma may develop due to hindrance in the filtration of the aqueous humor (consequent on the abnormal position).

Inflammatory changes may be produced by the unnatural position of the lens and so adhesions between the cornea, lens and iris may be brought about.

When the capsule of the lens is ruptured, the lens may be liquefied and resorted

Causes: 1. The condition may be a congenital anomaly.

2. Trauma is the most important cause in the acquired variety and found mostly in dogs, especially in Sealyheam and terrier breeds. Excessive barking may traumatise the ligaments.

3. Glaucoma which may be the result of cyclitis, may cause secondary luxation.

4. Since this condition occurs in older dogs (after the age of 3 years) degenerative changes(causes unknown)in the suspensory ligaments may be a cause.

5 A predisposing genetical factor may play a part.

Cataract : Opacity of the lens is cataract. Obviously, therefore, blindness or impairment of vision develops. Cataract may be partial or complete depending on its situation. It may be congenital or acquired, the former being the most common

Incidence : Cataract is common in dogs and rare in other animals. Among dogs, there is hereditary and breed susceptibility.

Pathogenesis : The lenticular tissue is capable of the following changes :— (1) proliferation of the capsular epithelium, (2) necrosis of the fibres of the lens and (3) increased sclerosis of the fibres forming the nucleus. So depending on the nature of the changes, cataract is classified as follows:

1 Subcapsular cataract : This is seen in horses, dogs and birds. In this condition there is abnormal proliferation of the lens epithelium Usually the proliferation occurs at the front surface of the lens, when it is known as *anterior polar cataract* Sometimes the proliferation of the cells may extend beyond to the posterior surface due to the degeneration of the lens. It is known as *posterior polar cataract* The cells, because of proliferation, become disorganised and form thicker layers producing opacity. This condition may be a result of posterior synechia or due to repeated attacks of periodic ophthalmia in horses and is usually associated with cortical cataract.

2 Cortical cataract: This is the most common form and involves the lens fibres. either at the front (anterior cortical cataract) or at the back (posterior cortical cataract). Due to accumulation of interstitial fluid consequent on the altered metabolism of the epithelial cells, the fibres become disintegrated and disorganised The cataract is stellate. spreading from the centre to the periphery. It is a progressive condition. Usually this type follows a corneal ulcer

3 Lamellar cataract : This occurs in young animals. It is non-progressive and is located between the nucleus and the cortex It may be congenital or acquired and results due to some injury during development and is seen in puppies following an attack of distemper or in those that have suffered from rickets (due fo vitamin D deficiency).

4. **Nuclear cataract** are probably the results of senile changes in which the fibres at the centre become more dense thereby making the nucleus dull or hazy.

Causes :

1. **Congenital** : (a) failure of the hyaloid artery (which is present in the embryo entering at the optic papilla and extending to the posterior surface of the lens and nourishing the vitreous humor) to regress and disappear completely, leaving remnants of its wall or its small branches.

(b) Impairment of translucence of the lens due to abnormal arrangement of the lens fibres or there may be fluid or droplets of fluid between the nucleus and cortex of the lens.

(c) A hereditary predisposition may precipitate the occurrence of cataract in later life.

(d) Deficiency of vitamin E : cataracts are found in chicks born of fowls fed vitamin E deficient diets.

2. **Acquired** : Degeneration of the lens due to :

(i) luxation.

(ii) impaired nutrition, as occurs in ophthalmitis and affections of the uveal tract.

(iii) degenerative ocular disease as in retinopathies and retinal detachment in dogs, as found in Pekingese.

(iv) Trauma.

(v) Senility—common in old stallions.

(vi) Diabetes mellitus : This is seen occasionally in dogs. Probably increased sugar content alters the osmosis in the aqueous humor.

(vii) Nutritional disease—deficiency of vitamin D; deficiency of vitamin C in the lens; deficiency of cystein, which plays great part in the oxidation-reduction processes

(viii) Toxins : (a) toxins circulating in diseases like influenza and periodic ophthalmia in horses and distemper in dogs may cause degeneration of the lens, b) toxins of uremia as occurs in chronic interstitial nephritis.

(ix) Poisons ı ergot in cattle and pigs.

(x) Absorbed radiation—(see page 131).

THE UVEAL TRACT

Anterior synechia is the condition in which there is adhesion of the iris to the posterior surface of the cornea. This results due to iritis. Anterior synechia causes glaucoma.

Posterior synechia is the adhesion of the posterior surface of the iris to the anterior surface of the lens capsule. This condition also results due to iritis. In this condition pupil cannot dilate.

Iridocyclitis. This is the inflammation of iris and ciliary body and is also known as **anterior uveitis**. This condition is best exemplified by specific condition in horses known as Periodic ophthalmia

Periodic ophthalmia or Equine recurrent iridocyclitis or
Moon Blindness

This is a disease of equidae, in which one or both eyes may be affected The idease may abate only to reappear (in the same or the other eye) in a mor severe manner causing greater damage to the eye, which finally becomes blind on repeated attacks.

Causes : The exact cause is obscure. The following are suggested.

1. Deficiency of riboflavin,
2. Infection by leptospira,
3. Hypersensitivity to leptospira (?)
4. Virus (?), transmitted by flies.

Symptoms and lesions: The disease starts with photophobia, belpharospasm, lachrymation and tightly contracted pupil, which does not dilate even in darkness The conjunctiva and sclera are congested and the cornea may show vascularization. The iris is dull and yellowish but not the normal brown. The pupil becomes clouded due to the presence of the exudate and leucocytes in the anterior chamber. Particulate matter settles at the lower border of the iris and posterior surface of the cornea forming the **keratic precipitates.**

Within a week or ten days, the disease subsides so that photophobia vanishes and the animal is able to see. Though the disease appears to subside, the eye is never completely normal since some amount of anterior synechia and posterior synechia are present.

Repeated attacks cause posterior synechia, subcapsular cataract (due to the alteration in the nutrition of the lens consequent on inflammatory changes of the uveal tract), gradual absorption of the lens after the capsule is ruptured, liquefaction of the vitreous humor, shrinkage and atrophy of the eye. The choroid and retina separate due to accumulation of the exudate. The eye is finally lost

THE RETINA

Progressive retinal atrophy in dogs : This condition is fouud in Irish setters due to a recessive inherited factor. The retina in both the eyes becomes atrophied, manifested in the early stages as night blindness, finally resulting in total blindness. The degeneration first starts in rods. Gradually the outer and inner nuclear layers become discharged and atrophied and finally the ganglion cell and nerve fibre layers become atrophied. It may take some years even for complete blindness to develop.

Detachment of retina : Retina lies just in contact with the inner surface of the choroid but not attached to it. So the retina can be lifted (detached) by accumulating edematous fluid or exudate or blood. This occurs in inflammatory conditions of the choroid or even of retina. Inflammatory exudate pours between the choroid and retina thus seperating them In injuries to the eye, especially thorn pricks in cattle, there may be hemorrhage resulting in the separation of the two structures.

Retina may be pulled away by the traction of shrinking fibrous band of the vitreous. In ophthalmitis, organising exudate may cause traction on the retina which may be datached from its normal position. Liquefaction or loss of vitreous humor may be another cause for detachment of the retina.

When the retina is detached, the rods and cones first become atrophied. and disappear, followed by the nuclear layer and the optic fibre layer Gliosis may also occur. Finally cataract, glaucoma and shrinkage of the eye result.

OPTIC NERVE

Atrophy :
  Causes :
  (a) Congenital.
  (b) Acquired : retinitis, papilledema, retinal degeneration, glaucoma; cho-
       roidoretinitis, trauma on occiput, hemorrhages; poisons: morphine,
       filix mas, arecanut; deficiency of vitamin A.

The optic papillae become thinned with disappearance of the interstitial capillaries. Retinal degeneration follows atrophy of the optic nerve. Total blindness results

GLAUCOMA

Glaucoma is a condition in which there is increased intraocular pressure
leading to secondary changes in the eyeball. Increased intraocular pressure may
result from (a) too excessive a secretion of the aqueous humor, or (b) hindrance
in its drainage. Among animals excessive production does not usually occur.
It is obstruction to drainage that is most commonly the cause of glaucoma.

If the causes that give rise to obstruction of the flow, leading to glaucoma
cannot be determined with certainty, the condition is known as **primary glaucoma**. Probably congenital glaucoma is of this category. If the causes for such
obstruction (see below) can be determined, the condition is known as **secondary
glaucoma**

Causes of secondary glaucoma : The pathological lesions causing obstruction may be due to (i) occlusion of the pupil, which may occur as a result of
posterior synechia. In this condition due to organisation of the inflammatory
products, the iris becomes adherent to the lens and so the fluid accumulates in
the posterior chamber, iris is pressed forward and the filtration triangle and the
spaces of Fontana are blocked by this protruding iris.

(ii) Occlusion of the filtration angle by the products of inflammation,
which may be acute or chronic. Majority of secondary glaucomas occur this
way. So obstruction may occur in : iridocyclitis in which anterior synechia
occurs; inflammation following trauma of the eye ball; luxation of the lens; intraocular hemorrhages: detachment of the retina and intraocular neoplasms.

Symptoms : Glaucoma may be unilateral or bilateral. The globe is enlarged
and exophthalmos may be noticed (buphthalmos). Cornea may be edematous
and opaque. Corneal vascularisation and pannus may result due to chronic
corneal edema. In man pain of the eye, headaches, toothache and earache
are felt.

Lesions and sequelae : The cornea is flattened and the iris is displaced anteriorly. There is opacity of the cornea, the lens and vitreous humor. Anterior
synechia may occur. Degenration of the lens may occur Blood vessels become
sclerosed. Atrophy of the choroid to a thin membrane may ensue. Due to
pressure a depression is excavated in the optic disc, which assumes the shape of a
cup (*Cupping of the disc*) The nerve fibres become atrophied Due to atrophy
of the nerve fibres, retina becomes degenerated and atrophied. Ganglion cell
layer disappears and blindness results

Ophthalmitis is seen as a symptom and lesion in various specific diseases :

Species of animal.	Disease.	Lesions seen in the eye.
Cattle	Mucosal Disease.	Conjunctivitis, keratitis and corneal opacity. In calves born of affected cows, optic neuritis, cataract, microphthalmia and retinal dysplasia seen.
	Infectious bovine rhinotracheitis (IBR)	Purulent conjunctivitis, keratitis and corneal ulceration.
	Malignant catarrhal fever.	Exudative non-suppurative retinitis, optic neuritis, iridocyclitis and acute conjunctivitis.

VIRAL DISEASES

Dog.	Infectious canine hepatitis.	Edema of cornea seen in convalescent stages.
	Rabies.	Negri bodies in ganglion cells.
	Canine distemper.	Retinitis and optic neuritis resulting in blindness when the optic tract is destroyed. Intranuclear and cytoplasmic inclusions seen in the ganglion cell layer; choroid loses its pigment and becomes thin. Pigment-laden cells invade the retina and the perivascular spaces.
Horse.	Borna disease.	Inclusion bodies in ganglion cells of retina. Retinitis and optic atrophy may occur.
Pigs.	African swine fever.	Blindness.
	Swine fever.	Inflammatory changes in retina and uveal
	Swine encephalo-myelitis.	tract; congestion, edema and hemorrhages in iris and infiltration by mononuclears of the ciliary body.
Fowl.	Marek's disease	Iridocyclitis

BACTERIAL DISEASES

Cattle.	Listeriosis.	Suppurative ophthalmitis and secondary corneal changes.
	Morexella bovis.	Infectious keratoconjunctivitis.
	Tuberculosis	Nodular uveitis affecting the iris or the choroid or both.
Calves.	Streptococcal meningitis; Polyarthritis; Coliform infection.	Purulent ophthalmitis
Sheep & goats	Mycoplasma agalactia	Opacity of cornea, keratitis with ulceration. conjunctivitis, turbidity of aqueous humor, glaucoma; exophthalmos, iridocyclitis; anterior syenchia; luxation of lens and loss of vitreous.

Species of animal.	Disease.	Lesions seen in the eye.
Cattle, pigs, cats, fowls and parrots.	Tuberculosis.	Hematogenous from lesions elsewhere.

RICKETTSIAL DISEASES

Sheep.	*Rickettsia ovis.*	Infectious keratoconjunctivitis.

MYCOTIC DISEASES

Dogs.	Coccidiodomycosis. (*C immitis*)	Uveal tract affected—detachment of retina occurs. Destruction of iris and ciliary body, infiltration of cornea by inflammatory cells; glaucoma.
	Blastomyces dermatidis	Diffuse uveitis—detachement of cornea
	Cryptococcus neoformans	Choroid affected; retina detached-
Horse.	*Cryptococcus neoformans.*	Keratitis and conjunctivitis.
Fowls.	*Aspergillus fumigatus.*	Panophthalmitis — fungus grows in the vitreous.

PROTOZOAL DISEASES

Dog.	Toxoplasmosis.	Inflammation of retina with infiltration of mononuclears in its layers; perivascular cuffing; hemorrhages and inflammatory exudate with mononuclears and plasma cells in vitreous and around the ciliary body.
	Leishmaniasis.	Conjunctivitis and keratitis.
Horse.	Trypanosomiasis.	Diffuse interstitial keratitis,

PARASITIC DISEASES

Dog.	Dirofilariasis.	These parasites, found in the aqueous humor cause keratitis.
Horses	Larvae of Habronema	Conjunctivitis. (The above are not natural inhabitants of the eye but occur only by chance).
in various. animals.	Thelaziasis.	This is a natural parasite of the eye and lives in the conjunctival sac and lachrymal ducts and cause mild conjunctivitis If the number of the parasites is more than 6, ulcerative conjunctivitis may be caused.
Fowl.	*Oxyspirura mansoni.*	Keratitis, conjunctivitis and ulceration of cornea.

Neoplasms of the eye :

Primary *:* Squamous cell carcinoma, especially in the bovines, is the most common neoplasm.

Adenomas and adenocarcinomas of the lachrymal gland and Harderian glands may be met with.

Adenoma and adenocarcinoma of the ciliary epithelium and iris may occur.

Secondary Matastases of carcinoma, sarcomas, melanoma, lymphosarcoma, meningioma and the venereal tumor may be met with.

DISEASES OF THE EAR

The external ear: This consists of the concha, the external auditory meatus and the ceruminous glands.

Otitis externa :

Causes : 1. Foreign bodies like awns may lodge in the ears of dogs and cause irritation and inflammation. This is a chronic condition in which there is hyperplasia of the epidermis, hyperkeratosis of the hair follicles and the infiltration of inflammatory cells. The skin becomes very much thickened and the sebaceous glands contain eosinophilic material. The foreign body may sometimes rupture the tympanum.

2. Ectoparasites

a) *Psoroptes communis*—causes profuse exudation into the meatus, which thus contains tenacious brown material. This is seen in sheep more often.

b) *Otodectes cynotis* causes otitis in dogs and cats. Due to irritation the dogs may shake their heads often and this leads, in the long-ear breeds (Dachshund) to hematomas. Secondary infection by bacteria may produce profuse exudate and tympanum may be ruptured.

c) *Otobius megnini* or the *spinose ear tick* causes otitis in cattle. Though only lymph is sucked by the larvae and the nymphs, secondary bacterial infection of the wounds caused results in otitis.

d) *Fungi* that produce dermatomycosis may also cause otitis.

e) *Stephanofilaria zaheeri* causes dermititis of the ears in buffaloes and may cause otitis.

3. Specific disease : In swine, *Actinomyces bovis* causes a typical actinomycotic granulomatous condition of the ears, which become thick and indurated.

The characteristic symptoms are the presence of thick pus in the external auditory meatus (Otorrhoea) and the thickening of the lining of the meatus. Shaking of the head is an important symptom of the presence of pus in the meatus. Stagnant pus may lead to rupture of the tympanum with subsequent occurrence of otitis media and even otitis interna.

Neoplasms of the ear are rare : Adenoma of the ceruminous glands may occur. Sarcoid in the equines and chondroma and chondrosarcoma may rarely be met with.

Middle ear consists of the tympanic cavity, the ossicles and the eustachian tubes. In horse, guttural pouches are diverticula of the eustachian tubes. The epithelium lining the tympanic cavity is continuous with the nasal mucosa through he eustachian tubes and so infection from the nose and pharynx can extend into the middle ear.

Otitis media :   Infection can occur through the external auditory meatus through the eustachian tubes. Normally, there is no communication between the external and middle chambers as, the tympanum seals the passage. But in condi- tions in which there is profuse exudate in the extrenal auditory meatus, ear drum can be ruptured by pressure and infection of the middle ear occurs.

The inflammatory exudate that accumulates in otitis media, unless drained becomes inspissated and organised, especially around the ossicles, immobilising them and so deafness may ensue.

As alresdy observed, infection can occur via the eustachian tubes from the nasal passages and the pharynx. The organisms found in such cases are *C: pyoge- nes* in calves, swine and sheep; *Pseudomonas aeruginosa* and *Streptococci* in swine; *Pasteurella* in cats and *Staphylococci* and *Mycobactertum   tuberculosis* in diffe- rent animals

Other sequelae of otitis media are : extension of infection   into the inner ear **(otitis interna)**, deafness, paralysis of the 7th cranial nerve, meningitis and ence- phalitis due to extension of infection into the cranial cavity through the 8th cranial nerve with resultant death.

CHAPTER 25

IMMUNOLOGICAL DISEASES

By Dr. B. B. Mallick,

Introduction

Whether pursued in Jenner's time or in our own, the immune process is clearly one of the body's defense mechanisms against infection. Ideally, therefore, the body should respond immunologically only to antigens (Ag) of pathogenic microorganisms, viruses, and parasites. However, it is not so since animals, including man will respond to injection of harmless Ag such as foreign serum proteins. Some persons also respond, often to their ultimate disadvantage, to otherwise harmless Ags of pollens, housedust and animal danders.

It used to be taken for granted that the body would at any rate never make the much graver mistake of responding immunologically to its own Ags. The mechanisms which prevent the formation of antibodies (Ab) to its own tissue components may on occasion break down. That is, body can, and in many cases does, produce an immunological response to its own Ags. This phenomenon is known as "Autoimmunity" and has rapidly assumed outstanding importance and is responsible for many of the immunological diseases. Thus autoimmunity is the general term used to describe an immune response, either Ab or cell mediated against normal body constituents. Earlier, the term immunological diseases was used to refer to this kind of conditions. But hypersensitive reactions caused by Exogenous allergens might well claim admission to the group.

ETIOLOGY

It may be postulated that the autoantibodies (AAb) may be manufactured under the following circumstances.

1. Alteration in antigenecity of tissue proteins :

This may be due to (a) degenerative lesions e.g., lens proteins in cataracts and skin in burns, (b) Attachment of hapten e.g., Allergic contact dermatitis.

Release of an Ag which has always been "isolated" from immunologically competent cells e.g., lens protein, thyroglobulin, proteolipid of myelin and carbohydrate of sperms   All these are separated from vascular bed and connective tissue by more or less impermeable cellular barriers, which prevent the Ags from reaching the blood stream so that the immune apparatus treats them as strangers and interlopers.

2. Loss of Tolerance :

The tolerance of self proteins might be expected to be lost in a variety of ways   (a) Forbidden clones might be formed as a result of some disease of lymphoreticular system. e.g., lymphoma   (b) or the defect might be a failure to eliminate a forbidden clone due to disorder of the regulating mechanism. (c) or it may be due to formation of an Ag which closely resembles self protein.   This can occur as a result of some degenerative process, the attachment of a hapten, or by infection with an organism which has an Ag in common with its host

3   Necrosis of tissue :

Tissue specific Ags have been identified in many organs and tissues and Abs against them have been reported in a wide range of diseases in which tissue destruction is occurring.   Thus Abs have been noted against the Ags of heart (in myocardial infarction, rheumatic carditis and post-cardiotomy syndrome), skin (in burns, exfoliative dermatitis) thyroid, lens (in cataract), pancreas (in pancreatitis) colon (in ulcerative colitis), liver (in hepatitis).   In most of these diseases it is unlikely that tissue specific AAbs are of any importance in the pathogenesis of the lesions.

4. Infections :

AAbs are also found as a result of some infections   These AAbs are formed as a result of tissue damage and though useful in diagnosis, do not contribute to the pathogenesis of the disease e g , in syphilis, complement fixing and precipitating Abs develop, which react with certain body constituents e.g., of heart muscle.

The sharing of Ag. between bacteria and host tissue has been proposed as a factor in acute rheumatic fever, acute nephritis and in ulcerative colitis. Some strains of E. coli share an Ag with human and rat colon.

The role of viruses in the etiology or pathogenesis of autoimmune diseases has long been suspected Theoretically, virus infections could precipitate autoimmune reactions in a variety of ways.

1. The virus itself can provide Ag cross-reacting with host Ags e g., viral nucleoprotein
2   The virus might unmask or release Ags from damaged cells---AAb directed against soluble nuclear components, for example, after infectious mononucleosis.
3. The virus might alter host cell Ags and act as "helper determinants".
4. The viruses might depress host cell Ags. e g , Embryonic antigens
5. The viruses might affect the proliferation or responses of immunocompetent cells or their precursors.

An immune complex is produced following interaction of Ag and Ab in the body. The formation of immune complex appears to be normal immune response, sometimes leading to pathological conditions and are then recognised as Immune complex Diseases. They involve untoward activation or inactivation of the effector system

Mechanism by which immune complexes can cause tissue injury :

Immune complexes with the greatest pathogenic potential are primarily those that can either inactivate plasma mediator systems notably complement or react with a number of cell types that have receptors or complex immunoglobulin (Fe receptors) or bound complement (C_3b & C_3d receptors). The two important mechanisms are :

i) Activation of plasma components

ii) Activation of cells by immune complex

The phlogistic effects of immune complexes

1. By virtue of their capacity to fix complement and to react with Fe receptor or platelets, neutrophils and other inflammatory cells, immune complexes can produce an acute inflammatory reaction.

2. Localization of immune complexes as seen in Glomerulonephritis (explained elsewhere).

3. Capacity of immune complex to block other immunological effector system.

    a) Immune complex can modulate the activation of T and B lymphocytes and may influence the presentation of Ag on the surface of macrophages.

    b) The effector function of T & B cells can be influenced by immune complex.

    c) Immune complex can induce production of anti-idiotypic antibody, which can function as antireceptor antibody and may either activate or inactivate T cells.

IMPORTANT AUTOIMMUNE DISEASES

It is not possible to discuss all the autoimmune diseases An attempt is made here to discuss, in brief, a few important such diseases which are of veterinary importance and have drawn the attention of clinicians and research workers.

CANINE SYSTEMIC LUPUS ERYTHEMATOSUS (SLE)

When canine idiopathic thrombocytopenic purpura (ITP), autoimmune hemolytic anemia (AHA) and membranous glomerulonephritis occur simultaneously or sequentially in the same animal, the syndrome is called canine S.L E. Polyarteritis, pleurisy and butterfly shaped facial eruptions and hepatic necrosis have also been observed in canine S. L.E. Histologically, in addition to changes accompanying I.T P. and A.H A., thickened glomerular basement membranes, so called "wire-loop" lesions occur. These are identical to those in human S L E. Glomeruli also contain deposits of PAS positive material and are often adherent to Bowman's capsule. Periglomerular lymphocytic and plasmacytic infiltration occurs, and fibrinoid change can sometimes be found in small renal arteries, but perivascular fibrosis i.e. 'onion-skin' lesions of splenic arteries, is not reported to occur in contradistinction to the situation in human S.L E.

The cause and pathogenesis of canine S.L.E. are unknown. In man experimental administration of hydralazine, an antihypertensive agent, produced a lupus-like syndrome But no sucn lesions could be produced with this drug in dogs

Anemia and thrombocytopenia may respond to corticosteroid treatment or splenectomy, but renal impairment is usually irreversible and eventually fatal.

Canine S L E. is associated with a severe Coomb's positive hemolytic anemia, Ab to thyroglobulin, rheumatoid factor and hyperglobulinemia.

A simle pattern of genetic inheritance is not noticed in canine S L.E. But it is shown that the offspring of the affected parent or the affected mother will exhibit multiple serological abnormalities and thymic lesions. Virus-like particles have been observed in renal glomerular endothelium, and so studies are now directed to find out the possibility of detecting a vertically transmitted infectious agent

AUTOIMMUNE HEMOLYTIC ANEMIA (AHA) :

The disease is marked by severe, recurring hemolytic anemia accompained by a positive reaction to the direct antiglobulin (Coomb's) test, which, in contrast to man, becomes negative during remission in dogs. The clinical signs include pallor, weakness, icterus, hemoglobinuria, anorexia, fever, and malaise. Splenomegaly, peripheral lymphadenopathy, and tachycardia may be detected during physical examination. Clinicopathologically, the anemia is macrocytic and normoblastic with polychromatophilia, anisocytosis, poikilocytosis. spherocytosis and hyperplasia of bone marrow. Most important feature is that eluates of erythrocytes from affected dogs can passively sensitize normal canine erythrocytes for indirect antiglobulin test, thus supporting an autoimmune pathogenesis for the condition. Nevertheless, the etiology and pathogenesis remain largely obscure.

Idiopathic Thrompocytopenic Purpura (I T P.) :

Canine I.T.P. is often, but not invariably associated with AHA. The combined disease has been linked to Evan's syndrome in man and may show the appearance of canine S L. E. About a third of affected dogs are reported to show spontaneous hemorrhage, usually as melena, epistaxis, hematuria or petechiae and ecchymoses on the skin and mucous membranes. Megakaryocytes may be plentiful in the bone marrow, but circulating platelet numbers frequently fall below 10,000/cu. mm Splenomegaly and lymphadenopathy may occur. Canine patients with I. Γ P. do respond to corticosteroid or splenectomy but the syndrome usually recurs. Thrombocytopenia has been produced experimentally in dogs by the administration of antiplatelet serum and by the isoimmunization of dogs with homologous platelets. As with AHA, the mechanisms that evoke spontaneous disease remain open for investigation. Isoimmune purpura thrombocytopenia in piglets has been reported. It is postulated that iso-Abs were transfered to piglets during early neonatal period via colostrum.

Glomerulonephritis :

Although spontaneous canine glomerulonephritis may be related to physical entrapment of immue complexes whose antigenic components are unrelated to kidney, Glomerulonephritis has been experimentally produced in dogs by injection of nephrotoxic antikidney Ab produced in homologous or heterologous species against canine renal Ags. This nephrotoxic serum nephritis appears to occur in two phases. Initially, nephrotoxic serum damages the renal giomerular base ment membrane; later deposition of electron dense material takes place beneath the glomerular capillary endothelium. These deposits are thought to be immune complexes comprised either of foreign serum protein Ag–Ab or of autologous glomerular Ag-Ab, Ag in the latter complex being released during initial exposure to nephrotoxic serum.

A rapidly progressing mesangio capillary glomerulonephritis in Finnish landrance lambs was described. The eitology may be immunological in origin.

Interstitial Nephritis :

Chronic interstitial nephritis (C I.N.) is the most frequently occurring, yet the least understood renal disease of dogs. It begins as an acute nonsuppurative interstitial nephritis but progresses inexorably to the subacute and chronic stages. Glomerular lesions usually are not present early in the syndrome, but membranous thickenings, fibrosis and adhesions are evident in many chronic cases. The natural history suggests that immunologic injury may contribnte to the pathogenesis of C. I. N?

Initial cause of interstitial nephritis is unknown although leptospirae are sometimes associated wilh the acute and subacute phases. It is proposed that these organisms invade the tubules from interestitium, since glomerular lesions do not accompany acute leptospiral nephritis If one assumes that leptospirae or some other infectious agents produce primary damage to tubular portion of the nephron, how can one account for the chronic progressive nature of C I.N.?

Firstly, the offending agent may remain in the tissues of carrier state individuals and become a source of constant irritation, thus provoking a chronic infla. mmatory response. Alternately, the causative agent may precipitate release of kidney specific Ag which can induce an autoimmune response against kidney by the lymphoreticular system.

The similarity of C.I.N. to pyelonephritis of man deserves mention. Both lesions are characterized by chronic non-suppurative interstitial inflammation, tubular atrophy and dilatation with intratubular casts and fibrosls. Human pyelonephritis often progresses in the absence of detectable Infectious agents So chronic pyelonephritis and C.I.N may develop through similar mechanisms.

Equine Infectious Anemia (E I. A.).

This is a disease produced by a virus transmitted mechanically by insects and characterized by vascular lesions anemia and glomerulonephritis. Splenomegaly and lymphadenopathy are accompanied by proliferation of atypical plasma cells ; serum IgM increases concomitantly with clinical episodes but high titer may presist in chronic cases The virus growing in cultured lymphocytes is very similar although not identical in ultrastructure to C-type particles of leukemia viruses The infected horses are C-3 deficient and complement coats the erythrocytes of some animals at various stages of the disease. It is possible that virus-Ab-complexes on the surface of erythrocytes account for the presence of complement. If this hypothesis is accepted, then the disease can be grouped under immune complex diseases

Autoimmune hemolytic anemia was encountered in seven horses in a study carried out over a period of nine years Six of these horses showed clinical anemia and peripheral evidence of remission was observed in each of these cases. The direct anti-equine globulin test was positive in six of the seven cases. Streptococcal infection was directly involved in one instance while three horses had purpura and other cases were suffering from neoplastic conditions involving cells of lympho-reticular series.

Thyroiditis :

Diseases of thyroid gland are manifested by enlargement (goitre) and by ncreased or decreased function Thyroiditis is an acute or chronic inflammation of the thyroid caused by infectious agents or by other unknown but apparently genetic factors.

Hashimoto's disease is the most common of the latter. It is a chronic disease, characterized by moderate, rubbery enlargement of the gland, with mild decrease in activity, extensive infiltration by lymphoid cells and accumulation of plasma cells and by the appearance of Abs against these thyroid specific Ags. The disease occurs about 30 times as frequently in females as in males, typically at the time of menopause. It was one of the first to be recognized as an autoimmune disease.

The three thyroid Ags that appear in course of Hashimoto's disease are microsomal Ag., thyroglobulin of the colloid and another colloid Ag designated CA_2. The microsomal Ag is localized within the cytoplasm of epithelial cells and is intimately associated with the lipoprotein membrane of microsomes. Its location can be ascertained by immunofluorescence with Hashimoto serum and fluorescent rabbit antihuman serum, using the sandwich technic. It is also detected by complement fixation. The Ab is rapidly cytotoxic for cultured thyroid cells in vitro, but its toxicity in vivo is considered doubtful.

The thyroglobulin comprises more than 75 percent of the protein in thyroid colloid and is an iodine containing molecule. CA_2 Ag constitutes less than one percent of proteins of thyroid colloid and contains no iodine. The antithyroglobulin Ab can be detceted by precipitation, agglutination of Ag-coated latex particles or tanned red blood cells, by immunofluorescence and by passive cutaneous anaphylaxis. CA_2 can be detected by immunofluoresence

Microscopic examination of thyroid tissue from a case of Hashimoto's disease or from experimental thyroiditis in animals reveals a mononuclear lymphatic infiltration. The infiltration is seen early and is often aggregated into well developed lymphoid follicles resembling those constituting normal lymph nodes. Large numbers of plasma cells later intermingle with the infiltrate. The cellular changes take place at the expense of normal thyroid structure and interfere with glandular function, so that a moderate hypothyroid condition results. The predominance of lymphocytes in early lesions suggest delayed type of hypersensitivity which is confirmed by the fact that thyroiditis can be transmitted via lymphoid cells from diseased animal to a normal animal.

Miscellaneous Autoimmune Diseases.

A case of **canine granulocytopathy** which closely resembled that observed in human neutrophyl dysfunction syndrome was described. The cause of the disease was attributed to the capacity to kill phagocytized bacteria. Ultrastructural studies did not reveal any abnormality in the affected neutrophils. It was concluded that this disease was similar to, but not homologus with the granulocytopathy syndrome described in man.

Myasthenia gravis pseudoparalytica in dogs was described.

Glomerulonephritis aud interstitial nephritis in cats were noticed. The disease in cats was similar to that seen in dogs in etiology, pathogenesis and lesions

"Cold hemagglutinin disease", a form of autoimmune hemolytic anemia caused by cold-acting erythrocytes autoantibodies was described. The AAb in this case was found to be IgM which acts at low temparatures only. The clinical symptoms noticed were cyanosis and gangrene of the extremities

High titres of **antispermatozoa Ab** were noticed in 35 cows of a swiss herd which had non-infectious abortion often, associated with fetal mummification. The Abs were present in the sera of these cows at the time of abortion.

LIST OF AUTOIMMUNE DISEASES OF ANIMALS

Dogs
1. Autoimmune hemolytic anemia
2. Systemic lupus erythematosus
3. Lymphocytic thyroiditis
4. Autoimmune thrombocytopenia
5. Demyelinating encephalitis following Canine Distemper
6. Canine interstitial nephritis
7. Idiopathic thrombocytopenic purpura
8. Immune complex glomerulonephritis
9. Rheumatoid arthritis.

Cats
1. Autoimmune hemolytic anemia
2. Diffuse membranous glomerulonephritis

Pigs
1. Isoimmune purpura thrombocytopenia
2. Rheumatoid arthritis
3. Autoantibodies against liver parenchymal cells.

Horses
1. Equine infectious anemia

Cattle
1. Autoimmune anemia due to Anaplasma
2. Isoconglutinin following C. B. P. P.

Sheep

1. Testicular degeneration and Epididymitis
2. Glomerulonephritis

Fowl

1. Autoimmune thyroiditis
2. Experimental allergic encephalitis

DISEASES APPARENTLY OF AUTOIMMUNE NATURE IN MAN

a) acquired hemolytic anemia
   1. 'Idiopathic'. acquired hemolytic anemia
   2. Hemolytic anemia following virus pneumonia or certain other infections.
   3. Paroxysmal cold hemoglobinuria.
b) Idiopathic thrombocytopenic purpura
c) Chronic leukopenia
d) Physiologic icterus
e) Periarteritis nodosa
f) Rheumatic fever
g) Rheumatoid arthritis
h) Lupus erythematosus
i) Glomerulonephritis
j) Certain diseases of eye
k) Multiple sclerosis
l) Certain diseases of the skin
m) Myasthenia gravis.

— + —

DISEASES CAUSED BY BACTERIA

Tuberculosis	Contagious bovine pleuropuemonia
Leprosy in buffaloes	Contagious caprine pleuropneumonia
Johne's disease	Actinobacillosis
Glanders	Actinomycosis
Ulcerative lymphangitis	Brucellosis
Epizootic lymphangitis	Vibrio affection
Anthrax	Winter dysentery of cattle
Black quarter	Vibrio dysentery
Malignant edema	Necrobacillosis
Braxy	Listeriosis
Black disease	Leptospirosis
Bacillary hemoglobinuria	Caseous lymphadenitis
Enterotoxemia	Contagious acne of horses
Tetanus	Botriomycosis
Botulism	Glasser's Disease
Swine erysipelas	Fowel cholera
Pasteurellosis	Bacillary White Diarrhoea in chicks
Bovine lymphangitis	Fowl spirochaetosis
Colibacillosis in new born animals	Chronic Respiratory Disease
Gut edema of swine	Infectious coryza
Paratyphoid in animals	
Strangles	

TUBERCULOSIS

Tuberculosis occurs universally in man and animals. By systemic tuberculin testing, examination of milk and effective culling, this disease has been eradicated from bovines in certain Scandinavian countries.

Tuberculosis is caused by an aerobic, acid-fast, rod-shaped organism, *Mycobacterium tuberculosis*. The organism can multiply only if plentiful oxygen is available. So in closed caseous lesions, the bacilli do not proliferate. But if the lesions were to open up into a bronchus, the organism proliferates rapidly.

Strains of tubercular organism : In nature several strains of *M. tuberculosis* are found that can be distinguished culturally, and by their pathogenicity to animals. These are :

The Human strain *(Myco. tuberculosis)*... Infective to humans and animals.

The Bovine strain *(Myco. bovis)* ... Infective to man and animals.

The Avian strain *(Myco. avium)* ... Infective to fowls and some animals (Pig and horse)

The cold-blooded strain *(Myco marinum)* ...Infective to cold blooded animals.

The Vole bacillus *(Myco. microti)* ... Infective to voles only.

The bacillus is composed chemically, of carbohydrates, proteins and lipids Of these lipid content is highest and forms 50 to 60% of the whole. These components have been isolated and to each is attributed certain properties.

The carbohydrate fraction, consisting mostly of polysaccharides is antigenic because it is able to provoke a neutrophilic reaction locally—an exudate rich in neutrophils is formed. Young neutrophils are mobilised from the bone marrow

The protein part, *tuberculoprotein*, on ingestion is also able to incite neutrophilic and monocytic infiltration. It is to this fraction of the organism that the development of tuberculin sensitivity is attributed. But it should be noted that tuberculin hypersensitivity is developed only if the tuberculoproteins are present in the bacilli or mixed with bacterial waxes but not if purified protein extracts are injected. But when once an animal is sensitised to the tuberculoproteins, the protein, even in a purified state, can incite an allergic reaction, which is of immense value in the diagnosis and control of the disease It is on this reaction that the *Tuberculin test* is based. The damage in tuberculosis is attributed to the acquired sensitivity of the tissues, which develops in 10 to 14 days.

The lipids get dispersed into the cytoplasm of the macrophages after ingestion by them and convert them into epithelioid cells. The name *epithelioid cells* is given to these cells because they resemble epithelial cells in having an oval, elongated, vesicular nucleus surrounded by abundant faintly pink granular cytoplasm, with a faintly outlined cell boundary. These cells are connected to the neighbouring ones by processes called epithelioid reticulum.

The lipids also are responsible for the formation of giant cells by the amtiotic division of the nuceli of the macrophages without attendant division of cytoplasm They may also be formed by the fusion of the epithelioid cells. To the *mycolic acid* fraction of the lipids is attributed the *acid-fastness* of the organism$_s$. The lipids are also responsible for the death of the neutrophils that accumulate around them. The higher the lipid content, the more acid-fast are the organisms and more virulent are they. Lipids render the bacillus resistant to therapeutic agents and they protect the organisms from being digested by the phagocytes.

The tubercular reaction : The reaction to infection of tuberculous organisms in an animal that has never been exposed to the bacilli differs from that evinced. in an animal that has become sensitised to the organisms. This is known as *Kochs' phenomenon.*

If a culture of the organisms is injected into a guinea-pig subcutaneously, within 10 to 14 days a hard nodule develops at the site of injection Subsequently this ulcerates and the ulcer persists till the death of the animal. From the site of injection, the organisms get disseminated first to the nearest lymph node and from there, generalised tuberculosis occurs, death occurring in 2 to 3 months.

On the other hand, if a culture were injected subcutaneously into a guinea-pig that is already tuberculous (i. e. when it is sensitised), at the site of injection, an acute fulminating inflammatory reaction develops with ulcer formation. There is no dissemination of the organisms even to the nearest lymph node and the ulcer subsequently heals.

Reaction of tissues to infection : By introduction of the tubercular organism into a tissue, a local reaction is set up consisting of edema, hyperemia and neutrophilic infiltration. The neutrophils are the earliest cells to come to the area. They are able to engulf the bacili but are not able to kill them. They are probably helpful in limiting the movement of the organisms into tissues. The neutrophils die and are ingested by macrophages which are attracted to the area. So within twenty four hours after infection, neutrophils disappear from the area.

The macrophages arrive at the site in response to the lipid content of the bacillus These cells are derived from the phagocytes of the reticuloendothelium and from the monocytes of the blood. In the lung, they are the phagocytes found in the alveolar walls. The macrophages ingest the neutrophills (containing the organism) as well as the bacilli. The bacilli continue to proliferate within the macrophages for some time (8 to 10 days). More and more macrophages collect around the earlier ones and so a nodule is formed. The bacilli are killed and the lipid substance gets dispersed into the cytoplasm of the macrophage converting it into an "epithelioid cell".

Under the influence of the tubercular lipid, giant cells are formed as detailed earlier. The giant cells contain numerous nuclei (sometimes up to 100), which are arranged as a horse shoe at the periphery of the cytoplasm This is known as *Langhans type* of giant cell. The nuclei may also be dispersed throught the cytoplasm as in a *foreign-body* giant cell Giant cells do not form until necrosis has occurred. The appearance of giant cells in this type of granulation tissue is almost pathognomonic of tubercuosis (*pathonomonic lesion* is one by which definite diagnosis can be made). In other words, if giant cells are seen in a section, tuberculosis must be suspected. Confirmation may be made by suitable staining (It is true that giant cells may be found in other granulomatous conditions like aspergillosis etc. But one should always suspect tuberculo.. sis unless otherwise, proved).

By the end of a week, a narrow zone of lymphocytes surrounds the lesion described above. Beyond these are found young fibroblasts which merge with the local tissues.

The lesion is microscopic and is known as a *tubercle*. It therefore consists now only of cells and is called a "hard tubercle" and is essentially a granuloma which is seen in other conditions also—Sarcoidosis, Syphilis, Glanders etc The visible tubercles are 1 mm to 2 mm. in size. These may become large by fusion of the neighbouring ones or a tubercle may grow by centrifugal expansion and become macroscopic. Such a tubercle is avascular. Because of the avascularity as well as due to the toxic action of the organism, the centre of the nodule undergoes necrosis followed by caseation and calcification and thus the tubercle becomes "soft". Caseation occurs in two weeks following infection Hence a well formed tubercle has the following microscopic appearance from within outwards:

1. A central calcified and caseated tissue. The calcified area stains blue with H & E while the caseated area is faintly bluish or colorless. 2. A zone of necrosis containing the nuclear remnants of the cells strewn over. 3. Surrounding the above is a zone of epithelioid cells. 4. Surrounding the epithelioid cells

and some times mingling with them are Langhans' giant cells. 5. Surrounding the above is a narrow zone of lymphocytes and 6. Finally, a zone of young fibroblasts which merge with the surrounding tissue.

Organisms can be seen, with suitable staining, within the epithelioid cells. One noteworthy feature is the virtual absence of polymorphs.

In animals which are resistant the fibroblastic activity is enormous and collogen is laid forming a capsule. But in those animals in which resistance is low as in the guinea-pig, fibrous tissue formation does not occur. Caseation occurs after about 2 weeks, when hypersensitivity is developed. The caseated material may be softened and liquefied and it is this process that facilitates formation of cavities and spread of disease by the pus-like material.

The lesion described above is known as "**productive**" or "**proliferative**" type and comes under the group "granulomas" (chronic type of inflammation). There is another type of lesion, which is much more acute, called **exudative type**. This is seen especially when the organisms reach the serous membranes like the peritoneum, pleura, pericardium and meninges with wide spaces. The exudate is voluminous and contains fibrin, neutrophils, lymphocytes and macrophages. This exudate may clot and may caseate. Occurring in the lungs, the exudative reaction produces *tuberculous pneumonia* The factors responsible for the exudative lesions are :

1. Sensitisation of animal with the production of a state of allergy. 2. Large dose of bacilli 3. Organisms with high virulence. 4 Susceptible host 5. Tissues with loose structure.

Primary focus We have studied the reaction of tissue to infection by the tubercular organism. The focal reaction, viz exudative inflammation with a great tendency to caseation and calcification is known as the *primary focus* This primary focus, becomes visible, experimentally, within eight days while calcification occurs in three weeks. The histology of this focus is the same as that of the tubercle studied earlier. The bacilli that proliferate in this focus are carried to the related lymph gland by lymphatics and there lesion similar to the primary focus, both in structure and age, is formed. So this combina. tion of primary focus and lesions in the lymph node is called *primary complex of Ranke*. It may often happen that the primary focus is either absent or not demonstrable because of the rapid transport of the organisms to the lymph nodes Such cases are known as *incomplete primary complex.*

Fate of primary focus :

1. **Healing** : A fibrous tissue capsule may form around the primary focus. The necrotic tissue may be removed and the lesion organised, healing the focus.

2. **Inactive quiescent foci** : The primary focus as well as the lesion in the related lymph node may be encapsulated. The organisms may remain in such lesions for months or years. In such animals certain amount of immurity develops, which is relative and never absolute The quiescent foci may be exacerbated due to diminished vitality of the animal when the organisms enter the blood and lymph streams and form foci in other organs.

3. **From the primary focus** the organisms may enter the lymph or blood streams and thereafter may either be destroyed by phagocytes or may cause fresh tubercular lesions in other organs This is called *early generalisation*.

Post primary infection : If the organisms are completely got rid of from a primary complex, then the animal develops no sensitivity and so behaves towards another infection like an animal that has never been infected at all. Tuberculin test is negative in such animals. This second infection is known as *reinfection* and is followed by the primary complex again.

If, on the other hand, lesions in the primary complex lie quiescent for some time, animal develops allergy. Now in such an animal, there may be new foci of tuberculosis due either to endogenous infection (by exacerbation of the primary focus) or exogenous or superinfection (i.e. bacilli enter afresh from outside, usually aerogenously). But such a tuberculous process differs from the primary infection in the following ways:

1. The spread of infection in this post-primary infection is not by lymph or blood but only intracanalicular That means to say, infection is through bronchi and bronchioles in lungs and via urinary tubules in the kidney.

2 Calcification does not occur in this process. The lesion undergoes softening and ruptures into different passages.

3. There is no involvement of the related lymph nodes and so no lesions are noticed in them. This is because spread is not by lymph stream.

Because of these factors, lesions are usually limited to organs in which the disease progresses slowly This is known as *chronic organ tuberculosis* and is a feature therefore, in sensitised animals. Among animals, *tuberculous mastitis* is the best example of chronic organ tuberculosis.

The following factors are responsible for the variation noticed in the course. and lesions of tuberculosis in different animals and species:

1. **Individual or Native resistance:** This is a hereditary characteristic. It is possible to raise resistant or highly susceptible guinea-pigs. Among men, Negroes are more susceptible to tuberculosis.

2· **Breed :** This is not a factor of consequence. Because within a species all breeds of animals are equally susceptible.

3. **Species:** There is variation in the course of the disease among different species of animals though all of them are susceptible. For example, unlike in cattle, the disease in horses, pigs, carnivores and fowls, is more acute, generalisation occurring early and terminating in death.

While caseation is a feature in the lesions of cattle, sheep and pigs, it is no so in horses and carnivores in which it is more of a productive nature.

Tuberculosis of bones is not a regular feature in animals while in fowls bone lesions are found in 90% to 95% of tuberculous birds.

In cattle, pigs and horses, tuberculosis of serous membranes is characterised by 'grapes' without much serous effusion. But in carnivores, there is plentiful serous exudate.

4. **Strain of tubercular bacillus :** The bovine strain of organism produces lesions described above under species. The avian strain, on the other hand, does not produce caseation in pigs while in fowls it produces caseous lesions.

5. **Immunity :** Presence or otherwise of immunity or allergy, we have noticed, alters considerably the course and form of the disease.

6. **Organ or tissue :** Different organs in the same animal react differently. In post-primary infection of lungs in cattle, there is profound softening and liquefaction of the lesion On the other hand, the lesion in the mammary glands (in the chronic lobular tuberculosis) does not caseate, but consists of proliferating granulation tissue that is rich in endothelioid cells.

Though tuberculosis of female genital organs is frequent in the cow. the genital organs in other animals are rarely affected.

7. **Age :** Incidence increases with age, probably, because of increasing possibility of and opportunity for infection.

8 **Hormonal influences:** Hyperthyroidism has a favourable effect on the course of tuberculosis while hypothyroidism has the opposite effect.

Cortisone and ACTH have a deleterious effect. Active disease is made worse and inactive lesions are exacerbated.

9. **Other miscellaneous factors :** Because of inadequate and deficient diet as well as unhygienic housing conditions there is great increase of tuberculosis in man, during war and famine.

Intercurrent and debilitating diseases may aggravate tuberculosis In people with silicosis, incidence of tuberculosis is very high.

Routes of spread :

1. **Direct spread :** The surrounding tissue becomes progressively involved and so the lesions become bigger. Also, the non-motile organisms may be carried by the phagocytes to other places.

2. **Lymphatic spread :** The organisms are carried along the lymphatics by the macrophages to the nearest lymph node where a tubercle may form. From there the organisms enter the lymphatics and ultimately reach the blood stream via thoracic duct and get disseminated widely. This lymphatic spread occurs invariably in the primary type of infection

3. **Spread by blood stream:** We have seen above that ultimately organisms enter the blood stream from lymphatics, But when a vein is eroded by a growing tubercle, caseous material with large number of bacilli may be discharged into the blood with the formation of tubercles in many organs giving rise to *acute miliary tuberculosis*. Miliary refers to a millet seed and so means a small tubercle of the size of that seed. More often hematogenous dissemination is not so massive and so the course is longer and the lesions may be few. Some organs like the thyroid, muscle, heart and pancreas seldom show any tubercles.

4. **By natural passages :** Infection may spread along the bronchi, ureters, vas deferens and the intestines. From the kidney, infection may spread to the bladder via ureters. Through aspiration by bronchi, infection may spread from one part of lung to another. By swallowing the aspirated sputum, infection may spread from lungs to the intestines.

5 **Spread in cavities :** If tuberculous focus has invaded a serous cavity, rapid spread is possible there because movements during respiration and peristalsis distribute the organisms quickly. A focus in mesenteric gland may infect the peritoneum while a sub-pleural focus may infect the pleura.

Hypersensitivity and immunity: Certain fundamental facts should be borne in mind; (i) Tubercular organism does not produce either an exotoxin (like clostridia) or an endotoxin (like the typhoid bacillus); (2) The organism is non-motile

How then is the organism able to produce such lethal effects ?

We have already studied Koch's phenomenon viz. the reaction in primary and secondary infections. In the primary infection the reaction is progressive and the animals die after the infection becomes generalised. But in the secondary infection the local reaction is violent, severe and attendant with much destruction of the tissue Yet, the process is localised and heals subsequently Now, these two processes are brought about by two different mechanisms The destructive process is due to an allergic reaction while the localisation and prevention of spread is due to immune bodies like agglutinins and opsonins. We have noted that tuberculoproteins sensitise the animals and produce a state of allergy. We have also noted that in such sensitised animals, reinfection by tubercular organisms results only in a localised lesion which may heal So, can we consider the state of allergy as immunity ? To the extent that allergy prevents the spread of infection and offers resistance, it may be considered to confer immunity But it must be understood that allergy is not immunity because, when the animal is desensitised by injecting large doses of tuberculo-protein, the resistance of the individual still remains.

Due to allergy, macrophages are brought to the site But it is due to immune property that these macrophages develop the ability to ingest and destroythem.

Hence, hypersensitivity and immunity together contribute to the resistance of the animal to infection. So, the resistance towards tuberculous infection can be reckoned simply as (a) mobilisation of macrophages in greater numbers with greater rapidity, (b) possession of greater phagocytic power by the macrophages, and (c) greater ability of the macrophages to destroy the organism,

Moderate infection and vaccination produce hepersensitivity but high degree of sensitivity can be produced *only* by living organisms. Rapid generalisation of infection, on the other hand, produces anergy.

Lastly we have to consider if a high degree of sensitivity is useful to the animals. The answer to this is NO, since it is harmful by causing massive edema and extensive necrosis. Low degree of sensitivity on the other hand is beneficial since, in the tubercles large number of organisms are either destroyed or inhibited

BOVINE TUBERCULOSIS

Tuberculosis in cattle is mostly caused by the *bovine* strain.

The *human* strain though infective causes only non-progressive lesions in the pharyngeal, thoracic and mesenteric lymph nodes.

The *avian* strain may cause non-detectable lesions. The lesions are found in the lymph nodes and serous membranes. Abortion in cow (even without macroscopic lesions) and meningitis are attributed to this strain.

Routes of infection : Source of infection is an affected animal The commonest routes of infection are :—

    a) **Aerogenous or inhalation :**
      i) By droplet infection.
      ii) Dust inhalation—the dust may be contaminated by sputum, feces, urine.

b) **Enterogenous or ingestion** :

i) Buccal mucosa ii) Pharyngeal mucosa iii) Intestinal mucosa	The contaminated materials may be milk or fodder or pastures. The latter may be contaminated by sputum, urine or feces from infective animals. Stagnant water may be infective for 18 days after contamination. Soil contaminated by infected dung can be infective for 6 to 8 weeks.

It is not yet clear which of the above two is the more important routes of infection. In the aerogenous infection the primary lesion or complex is found in the lungs, while in the latter the primary complex is found in the intestines and the mesenteric lymph nodes. But it may so happen that sometimes the organisms entering through the intestines may reach the lungs rapidly without leaving any lesions in the bowel. In such an event, the primary complex may be found in the lungs.

Unusual routes of Infection :

Cutaneous : This is very rare. Infection, if it occurs, is local and spreads to the concerned lymph node only. This occurs in *contact tuberculosis*.

Congenital : This is omphalogenic. Infection through this route viz. from maternal uterus to the fetus via umbilicus is seen and is possible only in the bovine.

The primary complex in the fetus is found in the liver and portal lymph nodes. In this type, generalisation occurs very quickly with death resulting within a very short period of time (few weeks or months). Lesions are found in the lungs, regional lymph nodes and spleen. In adult tuberculosis lesions of spleen are seldom seen. If encountered, then one must conclude that infection was congenital.

Genital : If animals suffer from tuberculous endometritis or tuberculous epididymitis, infection can be transmitted by copulation.

Mammary : Tuberculous mastitis can occur by infusion of contaminated medicines through the teats into the udder.

Usually animals over 6 months to 1 year old are affected by tuberculosis.

Symptoms : In the early stages when the lesions are localised no definite sympoms are noticed. There may be general malaise and the animals may become emaciated progressively, inspite of good appetite. In cows the symptoms become more severe after parturition. There may then be intermittent fever.

When the lungs are affected, there is dyspnoea, chronic cough, which is harsh and dry in the early stages but becomes moist and low later, denoting bronchopneumonia. On auscultation rales may be heard. Percussion reveals dull areas and pain may be evinced by such a procedure.

Intestinal lesions are accompanied by diarrhoea in some cases. Due to the pressure of enlarged mediastinal glands on the esophagus, bloat may occur. Dysphagia results due to enlargement of retropharyngeal glands. Enlarged mesenteric lymph nodes can be felt by rectal examination.

Lesions : Though tuberculosis can affect any tissue or organ the following are more often affected in cattle : lungs, pleura, liver, peritoneum, kidney and the regional lymph glands. The primary complex is mostly found in the lungs.

PART III

Infectious Diseases

Lungs : The lesions in lungs are either tubercular bronchopneumonia or miliary tuberculosis

The primary complex always starts as a smal sub-pleural lesion, usually in the diaphragmatic lobes. This lesion, having typical tuberculous structure starts in a bronchiole and spreads to the alveolus causing a bronchopneumonia-*caseous acinar bronchopneumonia.* If larger areas or whole lobes are involved, it is called *caseous lobular bronchopneumonia.*

The lesion spreads through the intrapulmonary lymphatics or the spread may be bronchogenic. That is, the infected material may be aspirated and then it may infect other areas. The lesion may also spread by continuity and expansion.

The lesions are usually encapsulated, caseous and gritty and vary in size. They may be within the substance of the lung or may protrude from the surface. Areas of consolidation may also be met with, which soon undergo necrosis and caseation But unlike in man these do not breakdown completely and form cavities or vomicae The lesion in cattle never becomes healed Rather, the tendency is for the disease to follow a more chronic course. The affected areas are enclosed in a thick fibrous capsule

Miliary tuberculosis of the lungs is secondary to lesions elsewhere. Organisms gaining entry in the blood-stream are disseminated widely. Both the lungs may be affected, showing scattered throughout its parenchyma, small, discrete, translucent nodules. This condition is usually acute.

Trachea and bronchi may show ulceration. Organisms may be implanted by coughed up sputum or they may be conveyed via lymphatics. First nodules form. The overlying mucosa gets necrosed and desquamated, leaving an ulcer.

Bronchi may contain mucopurulent material. Larynx may show ulcers or fungoid nodules.

Pleura and peritoneum : Tuberculous pleurisy and peritonitis are very common in cattle. Infection is usually lymphogenous. Infection can also occur when a lesion under the serous membrane opens on to the surface. When once the organisms gain entry onto the serous surface, they are rapidly disseminated throughout the area by means of respiratory and peristaltic movements.

The lesion starts as a diffuse, granulation tissue on the serous membranes. In this granulation tissue are formed tubercles which appear as clusters, resembling a bunch of grapes, or 'pearls'. Both the parietal and visceral layers may be thickened by the granulation tissue. The nodules have typical structure of a tuberculous nodule viz a caseous and calcified centre with epithelioid cells enclosed by fibrous tissue. Adhesion of two layers of the serous membranes may be present.

Lymphatic glands : The mediastinal and bronchial lymphatic glands are always affected and show, in early stages, small tubercular nodules, which become larger gradually. Finally the whole glands may be converted into a caseous and calcified mass.

Intestines : The primary complex may be found here only in the calves. In the adults, intestinal tuberculosis is usually secondary to the primary lesions.

In calves ulcers are found in the region of Peyer's patches and follicles. In the adult, nodules form. The organisms after infection enter the mucosa and reach the submucosa. It is the macrophages that carry the organisms to this site. In the submucosa, a nodule is formed with central caseous tissue. Because of pressure, the overlying mucosa undergoes necrosis and is cast off, leaving on oval ulcer with a raised, ragged border and firm base covered by caseous material.

The mesenteric lymph glands become affected when lesions in the intestines are present and so these are enlarged and caseous.

Tuberculous lymphangitis is common in animals suffering from chronic pulmonary tuberculosis. The sputum is swallowed and the organisms infect the intestinal mucosa from where the infection spreads to the lymphatics that run from the bowel wall to the mesenteric lymph nodes.

The lumina of dilated lymph vessels are filled with caseating tuberculous granulation tissue.

Liver : Infection is usually hematogenous :
   a) through the umbilicus in congenital infection.
   b) through the hepatic artery in generalised infection and
   c) through the portal vein from the intestinal lesions.

The lesions may be "miliary" or nodular. The nodules may be projecting over the surface, be yellow on section and have a thick capsule. The contents of the nodules may be caseated or calcified or they may be liquefied forming abscesses In some cases, instead of nodule formation, a diffuse cirrhotic form may be noticed.

Kidneys : Mostly adults are affected. The lesions which are "miliary" are found in the cortex and start in the interstitial tissue.

When the lesion erodes into the pelvis, descending infection via ureters to the bladder and urethra is possible.

Spleen : Lesions are seldom seen in the adult. But if found, they are due to congenital infection

Bones : Bones may rarely be affected. Young animals suffer more often. Infection is by hematogenous route. Lesion starts as an osteomyelitis initially and consists of tubercular granuloma. Eroding into joints tuberculous arthritis may be caused.

Central nervous system : Tuberculous meningitis is met with in cattle The piamater is more often infected and cerebral pia more frequently then the spinal pia. Infection is hematogenous and so is secondary to lesions found elsewhere in the body. The disease may spread to the brain via arachnoid spaces, choroid plexus and the space of Virchow-Robin The lesions resemble those of other serous membranes.

Mammary glands : Infection is almost always hematogenous. Three forms are distinguished

1. **Disseminated miliary tuberculosis :** This occurs in early generalisation of the disease. Typical tubercles, some caseated and others calcified are seen. All have thick capsules. These are always seen within the lobules, foci starting in the inter-acinar tissue. The supramammary lymph glands are always affected containing miliary tubercles or larger caseous areas.

2. **Lobular infiltrating tuberculosis** is the most common. This develops as *chronic organ tuberculosis* as a result of superinfection in a sensitised animas. 80 to 90% of all cases of tubercular mastitis are of this variety.

In this type, whole lobules may be affected. These are enlarged, projec beyond the surface of the gland and on section have lardaceous greyish-red appearance The lesion starts as a focus of granulation tissue in the lobule. These foci expand and coalesce and finally the whole lobule is involved. In this type the typical tubercles do not form though all cell types seen in a typical tubercle are encountered. The interlobular septa are thickened and are very prominent. The walls of the galactiferous ducts (intra and inter lobular) and the cysterns are much thickened with granulation tissue, which often shows caseated areas Their lumens are filled with masses of caseous material. The supramammary lymph nodes are not usually affected.

3. **Mastitis tuberculosa contagiosa:** In this condition the udder is greatly enlarged but no nodularity is seen. It appears to be hard. On section, the affected region has a mottled appearance Throughout the udder are dispersed. large map-like areas with hyperemic. jagged borders These lesions are yellow and caseous. Lobularity of the gland is lost. Supramammary lymph gland is always swollen and caseated.

Macroscopically, the udder in tuberculous mastitis. is increased in size and firm in consistency.

Normal udder on section has a grayish white silky appearance. It is soft and loose in texture. But a tuberculous gland is yellowish or brown in color and hard in consistency.

Though there may not be any alteration in the milk in early stages, it gradually changes. To start with, a few flecks of clot may be noticed. Later, the quantity diminishes and only an yellowish whey-like fluid may be seen. In advanced condition even this secretion stops altogether.

Buffaloes

The lesions in these animals are similiar to those of cattle but calcification is not common.

Sheep and Goats

Tuberculosis in sheep and goats, though reported, is very rare. It may be because of the outdoor life that these animals lead or it may be due to the action of some medicinal principles contained in the plants that these animals may injest or it may be because they are slaughtered young.

Sheep and goats are resistant to the human strain. Animals may be affected by bovine and avian strains.

Infection is always, probably, by inhalation since lesions are found more commonly in the lungs.

Symptoms: The commonest symptom noticed in sheep is cough due to bronchopneumonia. Respiratory distress may be noticed in advanced stages.

In the goat diarrhoea may be seen when intestines are involved and reveal ulceration and enlargement of mesenteric lymph nodes.

Lesions are similar to those in cattle. But the number of foci are fewer. In these animals calcification and fibrous encapsulation occur early. Cavitation is prominent in pulmonary lesions

CHAPTER 26

Horses

Tuberculosis in the horse is rare because of innate high resistance of the animal.

Majority of infections are by bovine type. Human and avian strains may also infect. The primary complex is found in the retropharyngeal or mesenteric lymph nodes since the route of infection is alimentary. Secondary lesions are found in lungs, liver, spleen and serous membranes. Very rarely are they found in the mammary gland and skin

Symptoms : Stiffness of the neck rendering it difficult for the animal to feed off the ground may be observed in horses in which the cervical vertebrae may be affected with tuberculous osteomyelitis. When lungs are affected, cough, intermittent fever, enlargement of glands and nasal discharge may be the symptoms seen.

The disease process in the horse is not similar to that of cattle, the following differences being seen in lesions :

In the horse	In the Cattle
1. Caseation and calcification rare, lesions are chronic and *Proliferative*. Hence gray, lardaceous appearance.	1. Extensive caseation and calcification are present.
2. Appearance is like "sarcoma"	2. Not so.
3. Early lesion does not have a zone of lymphocytes peripheral to the epithelioid cells and giant cells.	3. Zone of lymphocytes always present
4. Difficult to find tubercular organisms in the lesion which has abundant fibrous tissue.	4. Tuberculous organisms can always be demonstrated in the lesions.
5. Spleen very often affected.	5. Spleen seldom affected except in congenital tuberculosis.

Lungs : Usually secondary lesions are present, due to the hematogenous spread. The lesions are miliary, which may later coalesce so that whole lung becomes solidified and has a sarcoma-like appearance.

Spleen is often affected. Infection is hematogenous. The lesions consist of isolated nodules, upto the size of a fist. On sectiion a lardaceous appearance is found like a sarcoma but no caseation is present.

Liver : Nodules are found and the organ is enlarged. The nodules have similar structure like those of spleen

Vertebrae : Cervical and dorsal vertebrae may sometimes be affected.

Intestines : Small tuberculous ulcers may be present.

Lesions caused by avian strain in the intestines of the horse, resemble those of Johne's disease in cattle.

Serosae : Lesions in the peritoneum, pericardium, and pleura are common and are nodular, forming "pearls".

PIGS

Pigs are susceptible to all the three strains of tubercular organism. But majority of cases are of bovine strain. The characteristic features of porcine tuberculosis are :

1. Generalisation is not common. 2. Usually lesions are spotted at meat inspection and are mostly confined to the lymph nodes of head and neck 3. Serous membranes are not usually affected.

Infection is by ingestion mostly. Wound infection after castration and aerogenous infection may also occur. Infection by the human strain does not spread beyond the local lesion.

Symptoms similar to those of cattle may be noticed generally. Joints and meninges are more often affected

Lesions are mostly found in the retropharyngeal, portal and mesenteric lymph nodes as well as on the pharyngeal and intestinal mucosae.

Differences in the lesions produced by the bovine and avian strains in the pig are :

Bovine	Avian
1. Nodules show caseation and calcification and are surrounded by a fibrous capsule.	1. Lesions are proliferative and show tuberculous granulation tissue only; caseation is not a feature.
2. The caseous material in hepatic lesions liquefies.	2. Hepatic lesions do not caseate or liquefy or calcify.
3. Lymph nodes enlarged and show caseation and calcification.	3. Lymph nodes slightly enlarged and appear lardaceous.

Lungs Lesions are usually miliary, since infection is hematogenous. Infection with bovine strain causes extensive caseous bronchopneumonia o anterior lobes In infection by the avian strain caseation is not seen. In the interlobular septa are seen the characteristic string of pearls arrangement of the "amyloid-like" tubercles.

Liver : The bovine strain causes miliary tubercles. The avian strain causes infiltrating tuberculosis. The lesions consist of lardaceous, sarcoma-like areas, miliary to a pea size. From these tongue-like processes extend into the liver tissue between the lobules. Typical tubercular structure is lacking.

Spleen : This is frequeutly affected in generalised tuberculosis. Avian strain causes metastatic nodules, characterised by great tendency of the granulation tissue to undergo cicatrization, especially in the centre of the nodules. So, a depression appears in the nodules.

Meninges : These are frequently affected in generalised tuberculosis due to bovine strain.

DOGS AND CATS

Cats are highly resistant to human strain but highly susceptible to bovine. Dogs are susceptible to both equally.

The singular features of tuberculosis in carnivores are :

1. Typical tubercles are not formed. 2 Caseation is not a feature. 3. Granulation tissue is not tubercular granulation tissue but a non-specific granulation tissue in which are scattered at random histiocytes 4 Giant cels;

are rare. 5. Lesions are sarcomatous in appearance, 6. The lesions liquefy, forming milky pus. 7. When serous membranes are affected formation of exudate is plentiful. Route of infection is always alimentary in cats, but in dogs it may be alimentary or respiratory.

Lungs : The primary foci are found in the dorsal part of the diaphragmatic lobes. These are sarcoma-like nodules, pea to a walnut size. in which the central area is softened and a cavity occurs containing necrotic material. This cavity communicates with the pleural cavity through a fine passage. The pulmonary lymph nodes are swollen and on section the contrast between cortex and medulla is not present.

Generalised forms are of three kinds.

1. Acute miliary tuberculosis : One feature of this type in dogs and cats is that the tubercles are never found as dense clusters The grayish tubercles are scattered singly in the lung. Compensatory emphysema seen in other animals is also not present.

2. Large nodular tuberculosis in which the lesions are like those in the primary focus.

3. Multiple tuberculous bronchitis and peribronchial pneumonia.

The lesion starts as amyloid-like nodules with central cavities. These cavities are made up of only tuberculous bronchi, the walls of which are now composed of the granulation tissue.

Pleura and peritoneum : The lesions are unlike those of cattle and horses. Grapes are not found, but only plaque-like thickening of the serous membranes with copious serous or sero-fibrinous (sometimes even blood tinged) exudate. Lesions in the peritoneum are found when abdominal organs are affected.

Abdominal tuberculosis : When infection is by ingestion, lesions are found in the mesenteric lymph nodes. Lymph node near the ileo-cecal valve is most often affected The gland has a sarcoma-like appearance in the early stages later becoming purulent. Caseation does not occur

Infection may spread to spleen, kidney, liver, bones, joints, brain and reproductive system.

Avian Tuberculosis

This is not a problem in the tropics. Adult birds, those over one year, are more often affected. Infection occurs through infected feces, via digestive tract

Parrots may be infected with human strain.

Incubation period is serveral months.

Symptoms are vague; emaciation; comb pale and shrivelled; anemia; lameness; diarrhoea and decreased egg production. . Breast muscles are very emaciated so that keel is very prominent.

In parrots, nodules and ulcers may be found on the limbs and head (but not common in fowls).

Calcification is not a feature of avian T. B.

Lesions : The following organs are affected. Liver in 98%. spleen 90%; intestines 50% and bones 95% of cases.

Ovaries. kidneys, peritoneum and thymus are less frequently affected.

Death is due to cachexia or internal hemorrhage due to rupture of a disintegrating organ such as the liver or spleen.

Liver is enlarged and shows fatty degeneration. Caseous nodules of varying sizes are found and these project out of the organ. Ascites may be present. As the organ is friable hemorrhage is common under Glisson's capsule.

Spleen is enlarged and contains nodules varying in size. The whole organ may become caseated. Hemorrhages may occur under the capsule, which may rupture

Stomach : Nodules and ulcers are found at the junction of the glandular and muscular regions.

Intestines : Nodules may occur, which may later ulcerate.

Lungs : *Pulmonary tuberculosis* is uncommon. It is always secondary and so is miliary in type.

Bones : These are affected frequently as the liver. Long bones are affected. There may be small tubercles in the bone marrow or it may be completely caseated, causing atrophy of the cortex

In pigeons, joints of limbs may be affected. The cartilage is eroded and the joints contain dry caseous material.

Causes of symptoms in Tuberculosis.

1. Destruction of tissue, 2. Absorption of products of necrosis (a tissue toxemia). 3. A part becomes dysfunctioning 4. Secondary complications cause death 5. Hemorrhages from pulmonary tuberculosis causes death.

Diagnosis :

1. Tuberculin test (double intradermal or D.I.D. test) is very reliable and all animals suspected must be tested. The sites for DID test are :

Cattle—skin of neck region. Pigs—skin of ear, Fowl—wattle.

False negative reaction may be seen in (1) animals with advanced stages of tuberculosis, (2) animals in which the disease is very early, i. e. within 6 weeks after infection, (3) cows which have recently calved, (4) animals that have been desensitised. i. e. in animals injected by tuberculin during the preceding 8 to 60 days and (5) old animals.

Milk from a suspected udder may be centrifuged and cream or sediment examined for tuberculous bacilli.

3. Tuberculosis must be differentiated from aspiration pneumonia (with lung abscesses), pleurisy and pericarditis following traumatic reticulitis and contagious bovine pleuropneumonia.

4. Characteristic lesions may be observed at autopsy. Histopathological examination reveals the specific structure. Organisms can be seen in appropriately stained sections.

5. Biological test : Suspected sputum and discharges may be injected subcutaneously into the thigh of a guinea-pig. Necropsy after 6 to 7 weeks reveals tuberculous lesions.

LEPROSY IN BUFFALOES.

Leprosy in buffaloes caused by *Mycobacterium lepraebubalorum*, occurs in East Indies. It is characterised by chronic but non-progressive nodules in the dermis and subcutis. It is of particular interest because of the similarity of its histological features to those of human leprosy

Myco. lepraebubalorum, an acid fast bacillus, although demonstrable in tissue, has not been cultured sucessfully.

Lesions: The nodules present in the dermis and subcutis are firm, round and measure upto 5 cms. in diameter They occur as isolated or multiple nodules on the limbs and under surface of the abdomen The cut surface of the nodule is uniform and histologically it is an almost pure collection of large epithelioid cells without necrosis or calcification. The epithelial cells are large with voluminous, foamy, often vacuolated cytoplasm in which numerous acid-fast bacilli are demonstrable. The large vacuoles are believed to be the result of lipid production by the bacilli and are identical to the large "Lepra cells" of human leprosy. Giant cells of Langhan's type may be present.

Diagnosis : Demonstration of the collections of these "Lepra cells" laden with acid-fast bacilli in the lesions

JOHNE'S DISEASE
Syn : Paratuberculosis

This is a chronic fatal disease of ruminants caused by an acid-fast organism, *Mycobacterium paratuberculosis* Though only rumiuants are usually affected, horses and pigs may be affected very rarely.

Mice, hamsters, rats and rabbits can also be infected. Unweaned animals are better suited for experimental work Whatever may be the route of inoculation in the laboratory animals —intravenous, intraperitoneal or oral drenching— lesions are found only in the intestines. More successful infection can be obtained by oral method if macerated bowel mucosa is used instead of a pure culture. Laboratory animals do not develop diarrhoea and emaciation.

Three strains of *Mycobacterium paratuberculosis*, are known ‹

1. Bovine : This has no natural pathogenicity for sheep.
2. Ovine : This infects cattle also.
3   Scottish strain. encountered only in sheep of Scotland. Cattle may be affected This is a chromogenic strain producing intense orange pigment in tissues and cultures

All the strains are infective to cattle, sheep and goats and produce similar lesions.

In an animal infection is generalised since the organism can be cultured from various organs and lymph glands, even from gonads of both sexes.

Calves may be infected *in utero*. Organisms are excreted through milk.

Incubation period is long and protracted, even upto two years Animals are usually affected in their calfhood. The critical age appears to be six months. Adult cattle can be infected, but then they do not develop the disease but may recover.

Predisposing causes : Parturition. low nutrition, heavy milk yield and intercurrent diseases appear to be predisposing causes.

It is paradoxical to find the disease to be most severe in the best managed herds.

Clinical symptoms of diarrhoea or emaciation have no relation to the severity of lesions. Because animals with severe lesions may be clinically normal and vice versa.

Change in environment brings on clinical disease. Decrease in lime content of the soil appears to precipitate the onset of disease. Pastures may remain infected for one year.

Infection is by ingestion.

Pathogenesis : The bacteria enter the body through the tonsils and the intestinal mucosa and a state of bacteremia is set up during which period the organisms settle in the mucosa of the intestines, in the mesenteric lymph glands, the tonsils and other organs.

Wherever it settles, the organism incites a chronic granulomatous inflammation. Early in infection, lymphocytes, plasma cells and eosinophils infiltrate into the mucosa, soon followed by the macrophages, which engulf the organisms and become converted into epithelioid cells and occupy the mucosa, the lamina propria and breaking through the muscularis mucosae enter the submucosa.

Due to the pressure of these accumulating epithelioid cells the crypts become atrophied and disappear. Some surviving glands may become cystic. The villi become club shaped. The epithelium in some places may become atrophied. Because of the great infiltration of the epithelioid cells in the mucosa and submucosa the mucosa is thrown into folds, which resemble the corrugations of the brain.

In the submucosa, the epithelioid cells may fuse to form syncytial mass. Though in certain parts this may undergo necrosis, the process usually never progresses further but stops short at this stage called *symplasma stage*. Caseation is never seen in cattle though it may be noticed in sheep and goats. Calcification may be noticed in 25% of affected sheep and goats. A few Langhans type of giant cells may be seen.

Lymphangitis of the lymphatics of the area is seen and through these lymphatics the organisms may gain entry into the mesenteric lymph glands, where granulomatous lesions similar to those of the intestines are found.

Symptoms : Symptoms seen are brought out by fatigue and parturition. Emaciation is the most inportant clinical feature noticed. Animal may show diarrhoea, which is neither offensive nor blood stained. The animal, despite good appetite and feeding, becomes progressively weaker and emaciated with increasing thirst. The coat is staring and hide-bound. Though periods of improvement may be noticed, the animal succumbs ultimately.

In sheep and goats though diarrhoea is not a constant symptom, emaciation is observed.

Lesions : The carcass is very much emaciated. In such animals fat depots manifest gelatinous atrophy. Serous effusions in the peritoneal and pleural cavities and intermandibular edema may be noticed. The characteristic lesions are found in the intestines, especially the ileum, the wall of which is greatly thickened, two to twenty times its normal size. The mucosa is folded and shows corrugations which are transversely arranged and resemble the cerebral convolutions. These corrugations cannot be flattened out or stretched. The mucosa of the bowel is smooth and covered by a slimy material. The crests of the corrugations may show congestion. The earliest lesion may be observed at the ileocoecal valve. Intestinal corrugation is not a feature of ovine and caprine disease

Microscopically, the mucosa and submucosa are infiltrated by epithelioid cel's and eosinophils The villi and intestinal glands may disappear. Giant cells may be noticed. Advanced arteriosclerosis of the intestinal vessels is present In the sheep caseation may be seen in the submucosal lesions.

The mesenteric lymph nodes are swollen and juicy. They show the characteristic epithelioid granulation tissue with symplasma stage and giant cells. The lesions become caseous in sheep while in goats calcified nodules may be seen. There may be lymphangitis with thickening of the intestinal lymphatics which in sheep and goats may be knotty. In sheep and goats, tubercle formation (as in T. B.) is found in intestines and mesenteric lymph nodes.

Lesions have been described in the liver, tonsils and lymph nodes. In the liver, they consist of minute foci of epithelioid cells and lymphocytes in the triads and in the parenchyma. Bacilli can be demonstrated in these areas.

Diagnosis: Symptoms similar to those of Johne's disease may be encountered in gastro-intestinal helminthiasis, pyelonephritis, molybdenum poisoning, cirrhosis of the liver, pneumonia, metritis, salmonellosis, traumatic reticulitis, liver abscesses, and winter dysentery.

1. **Symptoms**
2. **Exmination of bowel washings :** After back racking, wash the bowel with saline and collect the bowel washings; centrifuge this and stain the sediment after making a smear of it. Acid-fast organisms may be found.
3. **Examination of rectal pinch :** Insert the hand into the rectum as far forward as possible and then with nail remove a small pinch of the mucosa. Make smears of this pinch, stain and examine for acid-fast organisms.

Note : *Inability to see the acid-fast organisms should not be construed as negative.*

4. Double intradermal test by Johnin in the neck.
5. Intravenous Johnin test.
6. Intravenous avian tuberculin test.
7. Complement fixation test and micro complement fixation test.
8. Lesions of the bowel at postmortem and microscopic demonstration of the organisms in the epithelioid cells in section.
9. Gel diffusion test using internal antigens of the organisms.
10 Lymphocyte activation test using peripheral blood lymphocytes. The lymphocytes from infected animals when exposed to antigen get stimulated but not the normal lymphocytes.

GLANDERS

Glanders is a disease of equines mostly. Man is susceptible. Though sheep and goats may be infected, cattle and pigs are immune. Dogs and zoo carnivores may be infected by feeding infected meat.

Cause : The causative organism is a gram-negative, rod-shaped, aerobic-non-motile, non-sporulating rod, known as *Actinobacillus mallei* or *Malleomyces mallei.*

Glanders organism is pathogenic to guinea-pigs, in which the disease runs an acute course. Following subcutaneous inoculation, generalised infection ensues and lymph glands, spleen, lung and liver reveal nodules.

If the organism is injected intraperitoneally into a male guinea pig suppurative orchitis results in 4 to 5 days (Strau'ss test).

Glanders is mostly chronic in horses, lesions being confined to lungs, the respiratory tract or the skin (farcy) and so the organism is voided in the discharges from these areas. In the donkey and mule, on the other hand, th's disease runs an acute course and the organism is found in all the tissues and so it may be voided through feces, urine, tears and saliva. Soil may be infective for 6 weeks after contamination.

Route of infection is mainly by ingestion. Aerogenous infection may occur but is rare.

Pathogenesis : Infection is by ingestion of contaminated feed or water. The organism may enter the lymphatics, penetrating the pharyngeal or intestinal mucosa. Ultimately they reach the general circulation from which they enter the pulmonary capillaries. During the septicemic phase, high temperature is caused.

Forming emboli in the pulmonary capillaries, hemorrhages are caused. The organisms incite an inflammatory reaction and a serous exudate forms in the perialveolar and peribronchial connective tissue. Penetrating the bronchial wall, the organisms may enter the lumen from where they may be voided with the secretion. The bacilli may be aspirated into the alveoli where a catarrhal inflammatory reaction is set up resulting in bronchopneumonia. Such lesions may soften terminating in suppurative pneumonia.

At the place where the bacilli are lodged in the alveoli, polymorphs accumulate, soon followed by alveolar histiocytes. The neutrophils become degenerated and their nuclei undergo karyorrhexis and the nuclear fragments become scattered in the lesion and are visible as stained particles in sections. This is a characteristic appearance in the glanders lesion. The macrophages become epithelioid cells, some of which may fuse to become giant cells. Peripheral to these may be hyperemia and fibrinous exudate walled off by fibrous granulation tissue. So the histology of glanders nodule may differ with age. In the young nodules are found a central degenerated area with particles of nuclei from poly. morphs surrounded by histiocytes, zone of red blood cells and fibrinous exudate and lymphocytes. In older nodules, the central degenerated area is surrounded by epithelioid cells and a few giant cells. Very old nodules, may even show scattered foci of calcification in the central necrotic area. Calcification is never complete. The above is the exudative form of miliary lesion, in which the central necrotic area is liquefied.

There may also be found a proliferative type of miliary lesion in which the hemorrhagic central area consists of epithelioid cells and giant cells and lymphocytes. In these areas suppuration does not occur.

Symptoms : In the acute form (as is seen in donkeys and mules and rarely in horses) the disease is a septicemia with high temperature.

In the chronic form, symptoms depend upon lesions.

In the pulmonary form, chronic cough and symptoms of pneumonia may be seen. If the nasal mucosa is affected, a purulent and oily nasal discharge may be present.

In the cutaneous variety, ulcers may be found on the affected parts with coiled and tortuous lymphatics. Oily discharge may occur from these ulcers.

Lesions :

Lungs In the lungs the lesions are usually small miliary nodules mostly of the exudative type described above. They are found in the parenchyma and under the pleura There may also be found pleurisy with fibrinous deposits.

On the Schneiderian membrane, trachea and larynx, are found ulcers. Organisms reach these areas via blood from a primary focus in the lung and form initially nodules in the submucosa. The histology of this nodule is similar to that found in the lung, consisting of central neutrophils surrounded by histiocytes. Due to suppuration in the center, the mucosa above dies and sloughs resulting in ulcers, which have sharp edges—"Punched out ulcers". Several of such ulcers may coalesce. Penetrating the cartilage the ulcers may perforate the septum. These ulcers may give rise to an oily, sticky, often blood-stained discharge and may heal by granulation tissue, the epithelium covering the scar finally. Large ulcers leave stellate scars.

Lymphatic glands : The submaxillary lymphatic gland is most commonly affected. Bronchial and thoracic glands may also show the lesions.

The submaxillary lymphatic glands become swollen and edematous with one or two yellowish grey centers. These never suppurate and open up as in strangles but become hard and indurated The changes may be exudative in which karyorrhexis is characteristic or proliferative which does not break down as in the former variety

Farcy : This is usually, secondary to internal lesions. Very rarely cutaneous infection may occur.

Farcy is essentially a lymphangitis and lymphadentitis of the limbs, more commonly of the hind limbs. Small nodules, called *Farcy buds*, are formed along the courses of lymph vessels which become thickened and corded—*farcy cords* The vessels are dilated and tightly packed with leucocytes. Leucocytes infiltrate focally into the walls also giving them a beaded appearance. The nodules may open and discharge an oily pus. The whole leg may become swollen (elephantiasis) The regional lymph nodes may be swollen and discharge the oily pus.

In *acute* glanders, lesions may be present in the lungs, liver and spleen In the lungs foci of catarrhal and croupous pneumonia as well as hemorrhagic infarcts are found. The lymph glands may be swollen and hyperemic.

The enlarged spleen may show nodules. Hemorrhages may be present in the serous membranes and the respiratory mucosa.

Diagnosis

1. I. D. P. test.
2 Serological tests—complement fixation test.
3. Isolation of organisms from lesions and conducting Straus's test.
4, Histopathological examination of lesions.

ULCERATIVE LYMPHANGITIS

This is a chronic contagious disease of horses mostly. Cattle also may be affected.

Etiology : *Corynebacterium pseudotuberculosis* (ovis), a gram positive, pleomorphic, non-sporulating organism is the cause.

Route of infection : Infection is always by wound infection of the limbs. Usually the hind limbs are more often affected. Only sporadic cases occur and that too in animals kept in unhygienic places.

Pathogenesis and Lesions : Infection usually commences at the fetlocks. From the wound, the organisms enter the lymphatics and cause lymphangitis, because of which the whole leg may be swollen. The lymphatics are swollen and become corded Along the course of these lymphatics abscesses form, which may open up and discharge a thick creamy pus which is sometimes blood tinged. The ulcer has a ragged edge. These ulcers may heal in one to two weeks leaving a small scar devoid of hair and pigment. As the old ulcers heal, new abscesses form along the course of the lymphatics. These may also heal, new ones forming high above. Slowly, the whole limb may be involved. Sometimes other places in the body and neck may be affected and death may occur because of widespread generalised affection. The regional lymph nodes are not affected and do not show any lesions. In some animals, the organism may enter the blood and produce metastatic abscesses in the internal organs. Usually kidneys are affected with abscesses in the cortex.

Diagnosis : See table on page 564.

Cattle may also be affected but in these animals the lymph nodes almost always show suppurative lymphadenitis. The abscesses discharge a gelatinous clear exudate. Usually, the skin of the neck and trunk is affected and not the limbs as in the horse. Clinically, it is difficult to find the organisms, *C ovis*, in the pus smears.

EPIZOOTIC LYMPHANGITIS

This is a chronic contagious disease of horses, caused by a dimorphic fungus, *Histoplasma* (Cryptococcus, Saccharomyces) *farciminosum*, which is a gram positive, coccoid organism with a double contoured wall Though this is a disease caused by a fungus, it is described here because the lesions caused are so similar to those in Glanders and Farcy and Ulcerative Lymphangitis that it is thought more advantageous for the students to study these diseases together.

Only horses and mules are affected. Man and cattle are said to be susceptible.

Route of infection Infection is by way of wound infection and it may be carried from an affected animal by fomites such as utensils. bedding, blankets, harness etc. Biting flies are also incriminated in the transmission of the disease. Transmission by copulation is suggested as one of the routes of infection.

A saprophytic stage of the organisms in the soil is said to exist.

Pathogenesis and lesions The cryptococcus does not appear to produce a toxin *in vitro*. It may be possible *in vivo*.

The fungus entering the subcutaneous tissue, incites an inflammation, and so ultimately an ulcer is formed. The organism spreads to other parts and deeper tissue by way of lymphatics A granulomatous reaction is elicited at the site Large number of macrophages infiltrate which engulf the fungus. Small nodules may form in the course of the lymphatics and they may rupture discharging thick exudate which later becomes purulent. The lesions have little tendency to heal. The skin in the area becomes thickened because of the granulation tissue in which yellow foci of purulent centers and macrophage infiltration are seen, The lymphatics are thickened and corded and contain pus. The regional lymph nodes are swollen and contain encapsulated softened foci. Cutaneous lesions though mostly found on the legs, may also be found on the back, trunk, neck, vulva and scrotum.

From nibbling at the lesion on the trunk and neck infection may extend to the nasal mucosa and lips. In the nose, small nodules may form which on breaking down give rise to a blood tinged discharge. The lesions may also extend to the pharynx, larynx and nasal sinuses. The nasal septum is not involved (compare with glanders). Suppurative lymphadenitis of the regional lymph nodes is present.

Lesions may be found in the eye, the cornea, the conjunctiva and the nictitating membrane being affected. Keratitis and conjunctivitis are thus caused.

Lungs may sometimes reveal lesions which may be solid granulomatous areas containing liquefied material.

When joints are affected the joint capsule contains large amounts of seropurulent exudate. The synovial membrane is covered with villous proliferation. The adjacent bone reveals osteoperiostitis or suppurative osteomyelitis The course of the disease is long and the animal becomes emaciated. Sometimes spontaneous recovery may occur.

Diagnosis : I Examination of the unstained pus smear reveals the double contoured globules of the fungus, which are gram positive on staining. Many of these will be found within the macrophages.

2. D. I. D. test with Histoplasmin can be done.

3. Epizootic lymphangitis must be differentiated from cutaneous glanders and ulcerative lymphangitis. In the table below are listed differences between these diseases.

TABLE SHOWING DIFFERENCES BETWEEN

Glanders	Ulcerative lymphangitis	Epizootic lymphangitis
1 Pus is oily	1. Pus is thick and creamy	1. Pus is thick
2. Pus contains gram negative bacilli	2. Gram positive rods in pus.	2 Gram positive coccoid bodies in pus
3. Ulcers do not heal	3. Ulcers heal	3. Ulcers may sometimes heal, but usually do not
4. Not amenable to treatment	4. Can be treated easily	4. Cannot be treated easily
5 Lesions are found on the nasal septum	5 Not found	5. Not found
6. Positive for mallein test	6. Negative	6. Negative
7. Negative for histoplasmin	7. Negative	7. Positive
8. No *C. ovis* antitoxin in the sera of animals	8. Sera of animals contain *C. ovis* antitoxin, which neutralise known *C. ovis* toxin	8. No antitoxin in sera

| 9. Lesions in lungs frequent | 9. Not present | 9. Lesions in lungs rare |
| 10 Margins of ulcers undermined. | 10. Margins thickened | 10. Margins not undermined. |

ANTHRAX

Syn. : Splenic fever; Charbon

Anthrax is an acute septicemic disease caused by *Bacillus anthracis*, a gram positive, capsulated. spore forming, rod-shaped aerobic organism.

Animals affected: Though goats, sheep and cattle are probably the naturally affected animals, all animals including fowls are susceptible. Disease occuring in poultry is rare. Infection by contact from animal to animal does not usually occur.

Routes of infection :

(a) **Ingestion** of contaminated feed and water. Bone meal and offals may also be sources of infection.

(b) **Wound infection:** this is the route in man causing *Malignant carbuncle or pustule*.

(c) **By biting flies :** biting flies — *Stomoxys calcitrans* and *Tabanus* may mechanically transmit the vegetative organisms.

(d) **Inhalation:** This is of importance in man — *'wool sorter's disease'*. The spores are inhaled while handling contaminated wool. Cattle may also suffer by inhaling dust containing spores.

(e) **Vaccination :** In some animals, the spores may vegetate and cause disease if the vaccine is not sufficiently attenuated.

Spread : This is brought about by streams, biting flies. carnivores and wild birds, movement of infected hides, bone meal, hair and wool.

Incubation period : One to 14 days. But when transmitted by biting flies, it may be 24 to 48 hours. Spores can remain viable for as long as 15 years in the soil. Drought and alkalinity of soil favour out breaks.

Pathogenesis : The bacilli that may be ingested reach the tonsils where they proliferate, reaching subsequently the lymphatic glands via lymphatics. Such of those bacilli that may reach the stomach are killed by the gastric juice.

The spores, unharmed by the gastric juice, develop into bacilli in the intestines. They may penetrate the intestinal glands and lymphoid follicles, reaching the mucosal and submucosal lymph spaces where they multiply. Bacilli or spores entering the skin, through wounds proliferate in the connective tissue.

The cells of the reticuloendothelial system as well as the bactericidal substances produced by the body kill and inhibit the organisms. The organism, therefore, develops a capsule with which it is able to resist the phagocytosis and iysis by antibodies. The capsule contains a glutamyl polypeptide. This substance inhibits the phagocytic activity of neutrophils, prevents clotting of blood (since it has fiobrinolytic property) and acts as a spreading factor facilitating invasion and so has a local effect. A diffusible toxin which is not formed *in vitro*, is supposed to be formed *in vivo* having local and general effects. The toxin consists of three components : Factor I, the edema factor, Factor II, the protective antigen and factor III, the lethal factor. It is only a few hours before death that the organism becomes septicemic in cattle and sheep. In horses, pig and dogs, the organisms are localised and do not become septicemic (so examination of blood in these animals will be negative).

The capsule of the bacteria swells by absorbing tissue fluid and so becomes gelatinous. This gelatinous fluid cannot be absorbed and so at those places where bacteria accumulate edematous swellings form. The bacilli that proliferate may clog the capillaries and with the toxin may injure the endothelium causing hemorrhages This is the cause of effusion of blood in organs in which the circulation is slow (spleen, liver and brain). The anthrax toxin causes nerve cell degeneration and this may be the cause for apoplectic form of death. The toxin acts on the respiratory centre in the brain and death is due to asphyxia.

Death occurs in 12 to 36 hours after onset of symptoms. Spores do not form in living animals. It is only when the carcass is opened that spores form. So carcass should not be cut open.

Symptoms : The animals may die without exhibiting any symptoms. They are found dead with oozing of dark colored, tarry blood from the natural orifices. This is the peracute type

In animals with acute type of the disease, elevation of temperature, depression, diarrhoea, dyspnoea and edematous swelling in the brisket, flank and throat regions are observed. In the horse, colic is a frequent symptom In swine and dogs pharyngitis is seen with swelling of the throat in the former, In birds are found swelling of throat, comb, mouth, wattles and head, fever, tremors, convulsions and dysentery The course is acute lasting 24 hours. Recovery may take place in some affected ostriches

Lesions :

Cattle : In animals that die of anthrax, putrefaction sets in early since the blood is fluid and so the carcass is bloated. Rigor mortis is never complete As hemolysis occurs (due to the presence of hemolysins) signs of asphyxia are noticed. (It is due to asphyxia that blood is dark in color). In the subcutis of various parts may be found yellow gelatinous fluid. The spleen is very much enlarged and soft. The capsule is tense. The pulp is dark and soft or even fluid. Rupture may occur sometimes. The enlargement is due to trapping of blood Liver and kidneys show congestion and degenerative changes. Hemorrhagic gastro-enteritis with thickened walls due to the presence of gelatinous fluid is seen. The mesenteric lymphatic glands are swollen and hemorrhagic. The lungs are dark, edematous and reveal hemorrhages. Bronchi contain bloody froth. Congestion of brain is noticed.

Sheep : The disease is peracute as sheep are more susceptible Splenomegaly is not very constant Subcutaneous edema is found

Horses : Edema may be noticed extensively over the abdomen, thorax and limbs The external genitalia may also reveal the edema Lesions may be localised in the intestines and pharyngeal region. In the former gastro-enteritis is found. In the latter pharyngitis and swelling of the neck and throat regions are noticed. Septicemia does not occur. Splenomegaly is not a lesion of equine anthrax

Swine : Intestinal lesions are not common Lesions may be localised iit the pharynx Edema of the throat is seen. Spleen is not enlarged Rarely enterius with diarrhoea may be seen. There is hemorrhagic necrotic, lymphadenitis of local glands.

Dogs : It is peracute. Pharyngitis and edema of face, head and neck may be noticed.

Man : Cutaneous anthrax occurs resulting in a malignant carbuncle. The organisms proliferate locally But when spores are inhaled–"wool sorter's disease"—the condition may become septicemic and generalised. Infection by alimentary tract is rare in man.

Birds : There may be hemorrhagic enteritis. The blood is tarry. Spleen is enlarged. There may be petechiae of peritoneum and heart; liver and kidneys may be congested. Pericardial sac is filled with serosanguineous fluid.

Bacterial examination of blood of birds may be negative since the bacilli may have become lysed.

Cause of death : The exact cause of death is not known. Toxins, asphyxia, and hypoglycemia are thought to bring about death. Degeneration of nerve cells is caused by anthrax toxin and this may be the cause of apoplectic form of death

Diagnosis : Anthrax must be differentiated from lightning stroke, black quarter, hemorrhagic septicaemia, babesiosis, leptospirosis, acute lead poisoning, acute bloat and snake bite. Animals that die suddenly with blood stained discharges through the mouth, nose and anus must be suspected to have died of Anthrax. Such animals should not be opened. First of all a blood smear must be made from the bloody discharge or from the tip of ear that is on the lower side (because by gravity all blood would have drained from the ear on the upper side). It may be stained by Leishman's stain or by methylene blue. Rod-shaped, capsulated organisms may be seen.

If putrefaction has commenced, anthrax bacilli must be distinguished from post-mortem invaders. The following differences may be noticed

Anthrax bacilli	P. M invaders
1. Never in chains of more than 2 or 3 organisms. If long chains are seen, it is not anthrax.	1. Long chains of bacilli noticed.
2. Spores never noticed.	2. Spores are present.
3. The ends are truncated (by methylene blue staining).	3. Ends of bacilli round.
4. Bacteria stain less intensely	4 Bacteria stain deeply.
5 Capsules present.	5 Capsules not present.
6. Segments of chains (bacilli) regular.	6 Bacteria segments not regular Some longer and others shorter.

If the carcass has become putrefied, the anthrax bacilli might have become disintegrated and so may be difficult to identify. The capsular material that is dispersed lines along the borders of the erythrocytes. In a smear stained with Leishman, if one sees the borders of the erythrocytes stained pink, then one may be reasonably correct to give a positive diagnosis.

In swine, dogs and horses, smears from the edematous fluid may be made and examined for the bacilli.

Fluorescent antibody technique may be applied on blood smears and tissue sections.

Biological test For this a swab of blood from the ear or swab of the blood-stained discharge may be utilised. A suspension is made of the material in a small quantity of normal saline. This is heated for ½ hour at 60°C, when all vegetative bacteria are killed. A piece of the muzzle may also be used. This is cut into small pieces and triturated with sterile saline and filtered. The filtrate is heated for 1/2 hour at 60° C. Then ½ml. of the suspension or filtrate is injected subcutaneously into the thigh of a guinea-pig. The animal will die in 36 to 48 hours, if the material is from an Anthrax animal. The blood of the guinea-pig will be teeming with anthrax bacilli. Gelatinous exudate will be found where the guinea-pig was inoculated.

Precipitation test : This is the Ascoli test, that is used in some countries to detect anthrax in hides and skins. For this test an immune serum with a high titre (manufactured from donkeys) is needed. The hide or skin is triturated with saline. This contains the anthrax protein. When on this is layered the serum in a narrow tube, a white precipitate must form *at once* to give a positive test

Materials needed to be sent to the laboratory : 1. Smears from blood obtained from ear. 2. Smears from the edematous swelling throat) in the case of dogs, pigs and horses. 3. A sterile swab impregnated with the blood from ear (or if the animal is opened by mistake, from the heart) 4. A dried piece of the muzzle.

BLACK QUARTER
Syn : **Black leg, quarter ill, symptomatic anthrax.**

Black quarter is an acute disease of cattle and sheep caused by *Clostridium chauvoei*, a rod shaped, gram positive, motile, sporulating, non-capsule forming, saccharolytic anaerobe

The disease occurs sporadically and is enzootic in certain localities. The organisms thrive in swampy grounds and persist indefinitely. In India, this disease is enzootic in Andhra Pradesh, Madras, Mysore and Bomday states. Hot and humid climate is favourable for the occurrence after the onset of rains.

Though sheep are susceptible, in India ovine black quarter caused by *Clostridium chauvoei* is rare. It is only cattle that are mostly affected. Swine may very rarely suffer. Dogs are seldom affected.

Among cattte, animals between the ages of six months and 2 years are most affected; younger animals do not have access to the swampy infected areas, while older ones have acquired immunity. It is the well nourished animals that are affected. Mortality rate is high, being almost 100%.

Routes of Infection

1. The infection is only, probably, by ingestion.
2. Wound infection may occur in sheep (in castration, shearing and docking) This is of greater importance in malignant edema

Incubation period is 1 to 5 days.

Pathogenesis : The spores that are ingested may be carried by macropha-ges of the intestines across the bowel, from where they may be distributed to the muscles (especially those of thigh and shoulder) where they le dormant for undetermined period. The factors that are responsible for the spores to vegetate and cause the disease are not fully known. Probably when the museles become devitalised by trauma or fatigue (due to forced walking for long distances) and when hemorrhages are present, the spores become activated and vegetate.

The growing organisms liberate toxins which prevent the phagocytic activity of neutrophils, injure the capillary endothelium (and so cause edema) and may directly cause necrosis of muscles. The edematous fluid which is low in protein, separates the muscle fibres and so generally facilitates spread of the organisms. Muscle necrosis may also be due to toxic thrombosis of the veins and capillaries.

The organisms may invade the blood circulation and set up metastatic lesions of gas edema and necrosis in liver, lungs, heart, kidneys and spleen.

The lesion in the muscle is a necrotising myositis with the formation of gas (the organism is able to ferment sugars) and this has the smell of rancid butter. Gangrene may snpervene

The toxins may diffuse into blood via the edema fluid and reach other organs causing degeneration. Affecting the thermoregulatory centre the toxin causes elevation of temperature. Death may occur within 24 to 60 hours.

Symptoms ; The first symptom noticed is lameness. The temperature may be high and depression and anorexia develop. Swellings, which are hot and painful in the early stages becoming cold and painless later, are found on the thigh, shoulder, neck and lumbar regions On pressure, the swelling crepitates due to gas. Laboured breathing may be present. The skin over the swelling is greenish or blue and is like a parchment (gangrene). Foamy discharge oozes from nostrils· Carcass putrefies and bloats rapidly. There is prolapse of the rectum.

Lesions : The carcass which is bloated due to putrefaction will be lying on its side. Where the swelling is the skin is dark When such a lesion is cut open, a sero-sanguineous, foul smelling fluid may exude. The subcutaneous tissue is edematous. The lesion is gas gangrene

The affected muscle has a porous appearance due to the presence of gas. It is dry in the centre while edematous at the periphery. This black color is due to the formation of iron sulphide a product of H_2S and Fe from blood and muscle. The regional lymph glands are swollen and edematous and may show hemorrhages. These lesions are found in the large muscles as well as in the diaphragm and tongue.

Fibrino-hemorrhagic pericarditis, pleurisy and endocarditis may be seen. In cattle the right auricle shows ulcerative endocarditis which may also be seen over the valves. Thrombi may be found in these places The heart is dark and friable and shows degenerative changes.

Lungs may reveal interlobular edema. Blood-colored froth oozes from nostrils. Kidney, liver and spleen may show degenerative changes and yellow necrotic foci containing gas. Intestinal mucosa may show inflammatory hyperemia and hemorrhages sometimes. Post-mortem decomposition of liver occurs early.

Microscopically, the affected muscle fibres reveal waxy degeneration and coagulation necrosis. Cross striations are retained however. Fibres of neighbouring muscle show cloudy swelling and fatty degeneration. Streaks of hemorrhage are evident between the fibres. Interstitial tissue is widened by edema Numerous bacteria can be seen. Peripheral to the necrotic lesion is found leucocytic infiltration.

Necrosis of muscle fibres is due to nutritional disturbances consequent on damage to the walls by the toxins. Formation of gas occurs only after the death of muscle fibres

Death is due to Toxemia.

Diagnosis : The disease must be differentiated from Anthrax, H S , bacillary hemoglobinuria, malignant edema, sweet-clover poisoning and acute lead poisoning.

1. By examination of smear made from the fluid of the swelling, gram positive rods with subterminal spores can be seen.

2 **Biological test :** Dried muscle piece from the lesion is cut up into small pieces, triturated with sterile saline and filtered. The filtrate containing spores is heated at 60°C. for 1/2 hour. One ml. of this is injected into the gluteal muscles of a guinea-pig. 10% sterile calcium chloride is injected along with the material. If the material contains spores of *Cl. chauvoei* the animal dies within 48 hours. At the site of inoculation the muscle is hemorrhagic, edematous and dark No gas is formed Liver impression smears reveal numerous organisms which are gram positive having sub-terminal spores.

3. **Fluorescent antibody technique :** This is useful for correct diagnosis and ,also to identify mixed infections. Fluorescent labelled specific antibodies are added to infected tissue, exudates or cultures that are fixed on slides. Then this slide is exposed to fluorescent antibody for 30 minutes. The slide is then washed in a phosphate buffer for 10 minutes, and mounted in 90% buffered glycerol. The slide is then examined under a microscope adapted for fluorescent microscopy. Specific organism is indicated by marked fluorescence.

MALIGNANT EDEMA
Syn. Gas gangrene.

Malignant edema is an acute febrile toxemia and is usually a wound infection occurring in horses, mules, sheep and swine. Dogs and cats may be rarely affected but poultry seldom.

Etiology : Malignant edema may be caused by several species of anaerobes : *Clostridium septicum, Cl. chauvoei, Cl, novyi, Cl. perfringens, Cl. gigas, Cl. sporogenes* and *Cl. sordellii.*

Routes of infection :
1 **Wound infection :**

a) This condition is usually a wound infection. Wounds may occur while hearing, docking, castration and during parturition. In rams infection may occurs through the wounds of their horns broken while fighting. b) Infection may occur through vaccination or venepuncture wounds. c) Umbilicus may be infected in the new-born. d) After shearing, dipping in a contaminated dip is another source of infection.

2. **Through digestive tract:** The anaerobes may normally be present in the intestines from where they may be transported to the muscles and infection set up

Pathogenesis : The organisms gain access to the heavy muscles (seats of predilection) and proliferate. The ideal conditions that appear to favour the proliferation of the anaerobes are the presence of exudations and blood clots. Anemia and congestion that reduce oxidation are favourable factors for the growth of the organisms.

The anaerobes produce powerful exotoxins which inhibit the activity of the neutrophils, injure the capillary endothelium and so cause edema and produce necrosis of the muscle fibres. Fermenting glycogen, gas is produced. The toxin diffuses into the blood and causes fatal toxemia. Organisms that may gain access into the blood reach various organs where gas edema is produced

Symptoms may be noticed within 12 to 48 hours after infection.

At the site of infection a soft red swelling is found. In infection by all organisms (except *Cl novyi)* gas may be found in the lesion giving a crackling or crepitating sound. So in these animals, a frothy blood-tinged exudate may be discharged from the wounds. In *Cl. novyi* infection there is no gas production. Due to toxemia, high fever, depression and anorexia are produced. Affected limbs are stiff and the animal goes lame. If infection occurs after parturition the vulva and perineal region may by swollen. In rams the edema is initially-restricted to the head ("swollen head") This is mostly caused by *Cl. novyi*.

Death occurs, due to toxemia, within 24 to 48 hours after the onset of symptoms

Lesions : The lesion in gas gangrene is essentially a cellulitis and the underlying muscle may be affected only to some extent. (Compare with black quarter in which the muscle is mainly affected).

The skin over the lesion is gangrenous—cold, bluish and like parchment. Sero-sanguineous fluid is present in the subcutis and inter-muscular tissue. Gas bubbles are present. Rancid odor occurs in infections with *Cl. welchii* and *Cl. sardellii.* Affected muscles are dark-red in color and edematous.

If the infection occurs after parturition, the walls of the uterus are edematous. Mucosa is swollen. There may be edema of the subserous tissues.

In animals in which alimentary tract is affected, the wall of the stomach is thickened with edema. The subserous and submucous tissues are edematous. The mucosa is red.

Effusions which are serosanguineous may be present in all serous cavities parenchymatous organs show degenerative changes; hyperemia may be generalised   Cyanosis may be present as lungs are edematous.

Diagnosis : From symptoms and lesions it is not difficult to diagnose this disease

BRAXY
Syn : Bradsot

This is an acute, toxemic and fatal disease of sheep caused by *Clostridium septicum*. Both O and H antigens have been described. Four subgroups are dentified on the basis of O agglutinins.

Incidence : Though rare in India, a disease similar to Braxy was reported from Madras. Eating frostly grass seems to predispose the animal to infection. Lambs and sheep under one year are mostly affected. The disease occurs in the winter.

Routes of infection : Ingestion : The organisms that are present in the soil are ingested with feed.

Pathogenesis : The organisms enter into the wall of the abomasum when its resistance is lowered by the presence of chilled grass. Wounds caused by nematodes may serve as portals of infection.

The organisms after entering the wall liberate toxins which are responsible for the symptoms and death.

Symptoms : Symptoms may not be noticed. If ailing animals could be studied, they show depression, signs of abdominal pain and diarrhoea. Affected animals stand aloof from others. Dyspnoea is also seen. Abdomen may be bloated. Death occurs within a few hours.

Lesions : The lesions may be confined to the abomasum. Focal areas of edema congestion and necrosis terminating in ulcers may be found. Gas may be evident.

The mucosa of the intestines may be congested and occasionally necrotic patches may be noticed.

Post-mortem changes supervene very quickly.

Diagnosis The causative organism can be isolated from the lesions, especially from animals destroyed in moribund condition.

BLACK DISEASE
Syn. Infectious Necrotic Hepatitis

This is an acute infectious toxemic disease of sheep mostly and occasionally cattle, caused by *Clostridium novyi* Type B. *(or Cl. oedematiens)*

Incidence : Sheep between the ages of 2 to 4 years, especially those in good condition, are affected.

Mode of transmission and pathogenesis : Black disease is always associated with liver fluke infection. The spores are ingested and are then transferred across the mucosa, probably by macrophages. The spores ultimately reach the liver, spleen and bone marrow where they exist in the R. E. Cells.

The cercaria after entering the host, have a migratory life. During the migration, they traverse through the liver before they reach the bile ducts, their final seats of habitat. During this migration in hepatic parenchyma, the cercaria cause tracks consisting of necrotic hepatic cells, leucocytes and blood. Peripheral to these tracks there may be a narrow zone of coagulative necrosis of the hepatic parenchyma surrounded by eosinophils and neutrophils. This tissue is ideal for the growth of *Cl. novyi* as anaerobic conditions are produced. The spores vegetate and liberate exotoxins which by diffusion cause further necrosis of the hepatic tissue and also cause widespread damage to the blood vessels. The damage to blood vessels is shown by congestion and edema of various organs and tissues. Nervous symptoms are also seen.

Symptoms : Since animals die within a few hours, no symptoms are seen. Ailing animals, if seen, are found to keep themselves aloof, are depressed and show respiratory distress. Cattle may show abdominal pain, absence of ruminal sounds and depression.

Lesions The disease is called "black disease" because the underside of the pelts is black due to severe congestion of subcutaneous vessels.

Liver : Liver is dark red due to engorgement and shows necrotic areas especially on the diaphragmatic lobe. These are circular areas, yellow in colour, ⅛ to ¾ inch in diameter surrounded by a zone of hyperemia Numerous liver flukes may be seen. In cattle the necrotic foci are linear. *Microscopically* these, areas show necrotic hepatic tissue together with dead leucocytes and numerous bacteria. A zone of leucocytes surrounds the focus.

Serous cavities—peritoneum, pericardium and pleura—contain large amounts of sero-sanguineous exudate. Pericardial fluid contains large amount of clots.

Hemorrhages are present under the endocardium and epicardium. Hyperemia of the abomasum and deuodenum may be noticed.

Diagnosis : This must be differentiated from Anthrax, Malignant edema Braxy, Black quarter (leg), and acute fascioliasis.

Organisms can be isolated from infected livers.

BACILLARY HEMOGLOBINURIA

Cattle are mostly affected but sheep and swine very rarely

It is an acute intoxication terminating fatally The causal organism is *Clostridium hemolyticum.*

This organism is found in swampy, low-lying areas where the disease may be endemic. Infection is by ingestion of contaminated feed and water. The spores, may be viable in the bones of carcases for as long as two years. It is possible that *Cl. hemolyticum* may exist in a dormant state in the livers of some cattle in endemic areas and may become active when the liver is damaged by liver flukes or some other stress.

The organisms produce necrotising and hemolysing toxins

The organisms enter the liver, cause anemic infarct by thrombosis of branches of portal vein and in the anaerobic environment thus produced, thrives, liberating its toxins. The hemolysin produces intravascular hemolysis and anemia. Due to resultant hypoxemia and action of toxins, capillary endothelium is damaged and so hemorrhages and edema are found in the tissues and serous cavities.

Symptoms : Abdominal pain, shallow respirations and hemoglobinuria are seen in those that could be observed. Death may occur quickly. **Pregnant** animals abort.

Lesions :

One or more anemic infarcts in liver with a hyperemic border is the constant lesion. Liver is enlarged, soft and friable. Rigor mortis sets in quickly. Low grade icterus may be present. Subcutaneous edema and hemorrhages are seen. Serous cavities contain blood-stained fluid. Hemorrhages under the serous membranes and endocardium are constant. Kidneys are swollen and show petechiae in the cortex.

Blood picture is not characteritic. Erythrocytes may diminish to as low as one million per cmm., while hemoglobin is reduced to 3 grams per 100 ml.

The disease has to be differentiated from : Cystic hematuria. anaplasmosis, babesiosis, anthrax, post-parturient hemoglobinuria, braken fern poisoning-leptospirosis, chronic copper poisoning, pyelonephritis and black disease.

Cause of Death : Toxemia and acute anemia.

Affections by Clostridium perfringens

Of the several types of *Clostridium perfringens*, types A, B, C and D cause enterotoxemia in animals   This is an intoxication.   The toxins liberated in the intestines produce the diseases.   Types B, C and D have been isolated in the various diseases affecting calves and sheep in India.

Cl. perfringens type D causes " Pulpy Kidney Disease ".

            type B causes " Lamb dysentery ".

            type C causes " Struck "

ENTEROTOXEMIA   Caused by *Cl. perfringens* type **D,**

Syn :   Pulpy Kidney disease; overating disease.

This is an acute toxemia of sheep and cattte.

Incidence *:*   Entertoxemia occurs in lambs and sheep.   Calves may also be affected.   Lambs which are well nourished and receiving ample amounts of milk are more often affected.   Similarly. enterotoxemia is found in fattening sheep maintained on a highly nutritious diet   Though morbidity is low, mortality is high (almost 100%)

Pathogenesis : *Clostridium perfringens* is a normal inhabitant of the alimentary tract. Infection is, therefore, by ingestion.   Normally, many of the organisms are destroyed in the rumen and abomasum and very few reach the intestines. The toxins that may be produced by these are removed by the normal movement of ingesta.   But under certain conditions, the clostridia multiply.   These conditions are (1) when animals are fed excessive amounts of starchy foods and (2) when there is certain amount of stasis and impaction in the alimetary tract.

   n the first condition, partly digested starch escapes into the intestines and there, because of their saccharolytic properties, the bacteria proliferate and grow luxuriously and liberate large quantities of toxin.   So the disease is called ' over eating disease ".   Heavy feeding on milk or sudden change of food from roughage to grain may be other causes.

In the second condition, because of atony of the intestines and stasis of food, there may be reverse passage of food from colon into ileum and this condition with impacted food favours the growth of the organisms and accumulation of large amounts of toxin.   Heavy tape worm infestation may be a cause for intestinal atony.

The toxins produced consist mostly of *epsilon* factor with smaller quantities of *alpha* and *theta* fractions.   The *epsilon* factor is the important and pathogenic factor. This acting on the intestinal mucosa increases its permeability and absorbing power and so *epsilon* toxin with other toxins are absorbed in greater quantities at a faster rate, causing a speedy, fatal toxemia. The toxins in the intestines cause diarrhoea.   Reaching the brain they may cause symmetrical encephalomalacia.

Symptoms : Since the illness is of very short duration in the lambs, symptoms may not be noticed.   But in those which are observed, the following symptoms may be noticed; diarrhoea. staggering, convulsions and orthotonus. In lambs that survive for a few hours clonic convulsions are found. Sheep that live for a day keep themselves aloof, are depressed with drooping heads, stagger and

have champing of jaws with salivation. Respiration is shallow and irregular.
Bloat and convulsions are seen. Hyperglycemia and glycosuria are observed·

Lesions may be indistinct. Congestion of the abomasum and intestines may
be present. So also, hemorrhages may be present on the intestinal mucosa.
The pericardium is distended with straw-colored fluid. Petechiae are presen.
on epicardium and endocardium. Congestion and edema of the lungs giving
rise to frothy discharge through the nostrils may be noticed.

The kidneys are enlarged and appear soft and pulpy. The condition resembles the postmortem change one may notice normally So. the kidneys should
be examined immediately after death. *Microscopically*, the change is
cloudy swelling and necrosis of the epithelium of proximal convoluted tubules.
The cortex and medulla may be congested. The kidneys in adult sheep and
cattle may not be "pulpy".

The liver and spleen may be congested. Hyperglycemia appears to be
due to increased glycogenolysis of liver glycogen. The brain shows lysis and
liquefaction of the white matter, while the grey matter is edematous.

Diagnosis :

1. The death in lambs with diarrhoea must always be suspected to be ente.
rotoxemic.

2. So also glycosuria is an important supporting symptom.

3 **Biological test:**

(a) The intestinal contents may be filtered through a bacterial filter and
.2 to .5 cc. of the filtrate may be injected into Swiss mice (weighing less than 25
grams) or young rabbits, intravenously. If the toxin is present, the mice will
die in a few minutes, to 4 hours.

(b) **Neutralisation test :**

2 cc. of the intestinal filtrate is mixed with 1 cc. of known *Cl. perfringens*
Type D antitoxin (Burrows Welcome & Co.,) and incubated for one hour at
37°C. Then 3 cc of this mixture is injected intravenously into Swiss mice. If
the filtrate contains the corresponding toxin, the mice do not die since the
toxin is inactivated by the known antitoxin.

This disease has to be differentiated from rabies, grass tetany, pregnancy
toxemia, acute lead poisoning or louping ill

ENTROTOXEMIA CAUSED BY CLOSTRIDIUM PERFRINGENS TYPE B.

This condition is known as *lamb dysentery* and occurs in lambs upto 3 weeks
of age   Mostly animals under 16 days of age are affected. Older animals may
rarely be affected but in them the course is chronic and is invariably fatal.

Pathogenesis : Infection is by ingestion. The beta toxin produced by the
organisms causes hemorrhagic enteritis and ulceration of the intestinal mucosa.
Beta toxin is inactivated by the proteolytic enzymes

Symptoms : In the *peracute* form sudden deaths may be met with without
any symptoms being manifested. In the *acute* form animals manifest severe abdominal pain, depression and reluctance to suckle   There may be dark colored
semifluid feces, which is often mixed with blood   Death usually occurs after
the animal goes into coma, within 24 hours after the onset of illness.

In the chronic form with which the older lambs are affected, the following symptoms are noticed : unthriftiness, depression and refusal to suckle

Lesions *Macroscopically*, in the *peracute* cases, a few small necrotic patches only may be seen in the intestines. In *acute* cases, there is extensive hemorrhagic enteritis. Ulcers may be seen later, if the animals live long enough. The peritoneal cavity contains serous or blood-stained fluid. When there is ulceration, adhesions of the intestines may be noticed. The mesenteric lymph nodes may be edematous or congested

The liver is pale and friable. Spleen may be slightly swollen. The kidneys which are slightly swollen are soft Subendocardial and subpericardial hemorrhages may be seen

Microscopically, lesions of acute hemorrhagic enteritis are seen. Infiltration by inflammatory cells is scanty

Clostridium perfringens type B also produces severe dysentery in calves 7 to 10 days of age, but the infection is supposed to be not so fatal as in lambs The course of the disease is 2 to 4 days. The affected animats are depressed and have dysentery Recovery, which is slow, requires 10 to 14 days.

The lesions in calves are similar to those described in lambs.

Diagnosis : from symptoms and lesions

ENTEROTOXEMIA CAUSED BY CLOSTRIDIUM PERFRINGENS TYPE C

Clostridium perfringens type C affects adult sheep, especially those fed abundantly and is known as "Struck" The mortality rate is 5 to 15%.

Symptoms : Usually animals suffering from struck die suddenly without showing any symptoms Rarely a few animals show abdominal pain

Lesions : *Macroscopically*, a large quantity of pale yellow fluid which clots on exposure to air, is seen in the peritoneal cavity. The abdominal blood vessels are congested and there may be large number of subperitoneal hemorrhages. The mucosa of the small intestines is intensely hyperemic and there may be ulcers with hyperemic borders in these areas. The pleura and pericardium contain large quantity of transudate. Subepicardial hemorrhages may be seen.

Type C Clostridium perfringens may also produce hemorrhagic enterotoxemia in young suckling lambs and calves, mostly within a few days after their birth Almost all the affected animals may succumb. In a few animals pain and dysentery may be noticed

Cl. perfringens type C may also produce hemorrhagic enteritis in suckling piglets. Most of the affected pigs become dull and depressed immediately after birth and will show diarrhoea. They may die within 24 hours In outbreaks whole litters may be affected Lesions inculde hyperemia of the intestinal mucosa and bloody coloration of the intestinal contents.

Diagnosis : from symptoms and lesions.

TETANUS
Syn : Lockjaw

Tetanus is a highly fatal infectious disease caused by the exotoxin of *Clostridium tetani*, a gram positive, sporulating anaerobe

The spores are terminal giving the organism a drum-stick appearance

Incidence: Tetanus is of world-wide distribution. Horses and mules are most susceptible. Sheep, goats, dogs and swine may also be infected. Cattle are least susceptible. Birds are resistant. Mice, rabbits, guinea-pigs and rats are susceptible.

Routes of Infection: Infection is almost always by way of wound infection. Deep punctured wounds, particularly, are more conducive for the growth of the organism affording anaerobic environment. In the horses gathered nail is therefore a common avenue. Wounds that occur during shearing, docking and castration may serve as portals of entrance. Similarly infection may occur after parturition in females and through the umbilical vein in the young. Surgical wounds may also be infected if sufficient care is not taken.

Infection through wounds in the alimentary tract is supposed to occur since the organism is a normal inhabitant of the intestines. The organism is voided with the dung and so it is found normally in the soil and stables.
Period of incubation is one to three weeks.

Pathogenesis: The spores that enter a wound require certain favourable conditions to vegetate and liberate the toxin. Toxin-free-spores are not pathogenic. They are engulfed by the leucocytes. When necrosis or hemorrhage or other aerobic and pyogenic bacteria are present, producing anaerobiasis, the spores vegetate. The spores may be in the tissues for a long time till favourable conditions are obtained. The organisms liberate the toxin locally. This toxin consists of three components.

1) a hemolysin—tetanolysin—which is not of much importance;

2) a neurotoxin—tetanospasmin, which is responsible for the nervous symptoms and

3) A fibrinolysin, which is not very potent.

The toxin is absorbed by the axons of the peripheral nerves. It passes with the intercellular fluid along the interneural spaces centripetally to the neurones of the spinal cord and the brain. This passage centripetally is brought about by the pressure exerted during muscular contraction. The toxin may reach the C N.S by way of blood also.

The exact action of the toxin on the nervous system is not known. The tetanus toxin gets fixed to a substance called "Protagon" in the nervous tissue. The protagon is a complex made of cerebroside plus oligosaccharides like N-acetyl galactosamine and galactose. It acts on the inhibitoty synapses interfering with the action of the inhibitory transmitter thus producing spastic action The toxin causes hyperirritability responsible for the tetanic spasms.

Symptoms: Since the muscles in general go into contraction, the symptoms exhibited are a result of this action. The third eye-lid droops due to paralysis. The muscles of jaw are firmly contracted (so lock jaw) and hence the animal may not be able to eat. The muscles of deglutition are also paralysed. So the saliva drools Any water or food ingested may come out through the nostrils. The back becomes arched, the tail is cocked-up and the neck may show opisthotonous or orthotonus. The fore legs are extended stiffly forwards and the hind legs backwards

The animal has an anxious look If sudden noise is made, it may go into convulsions. When backed, the animal falls and it cannot rise again.

Constipation and bloat may be present. The mucous membranes may be cyanotic.   Breathing may be shallow.

Temperature is high after death   Death is due to asphyxia since paralysis of respiratory muscles occurs.   Duration of sickness is 5 to 10 days in horses and cattle while it may be 3 to 4 days in sheep. The animals are conscious till the end.

Lesions : No characteristic lesions are found, Blood may be black and tarry Rigor mortis sets in immediately after death. *Microscopically*, degeneration of the neurones in the brain and spinal cord may be noticed (due to anoxemia).

Diagnosis : There are no tests to confirm a diagnosis. The organism is local and does not become septicemic. Symptoms alone are helpful. In its early stages tetanus must be differentiated from the following ; strychnine poisoning; grass tetany; milk fever; rabies; enterotoxemia in lambs; muscular rheumatism.

BOTULISM

Botulism is a highly fatal intoxication by the toxins of *Clostridium botulinum.* In man poisoning can occur by ingestion of canned and preserved meats, fruits and vegetables   Several antigenically distinct types of *Cl. botulinum* classified as A, B,C, D and E are described· Type D causes Lamsiekte, type A limber neck in fowls and type C forage poisoning in horses   Type E is seen in fish and fish products.

Incidence : Among animals, botulism has been noticed in cattle, horses, and fowls.   Sheep, dogs and swine, though susceptible, are not usually affected.

Pathogenesis : The organism is a normal inhabitant of the digestive tract of ruminants and horses.   It can thrive on rotting vegetation which gets contaminated by the dung of the animals.   It can also propagate on the carrion and bones. in warm and moist atmosphere   Under these conditions the anaerobe produces its exotoxin

Intoxication occurs by the ingestion of this preformed toxin. Infection therefore, occurs by the ingestion of the vegetation. In cattle, infection occurs when animals develop pica due to aphosphorosis, when they chew and eat bones of carcasses   The botulinum toxin present in the bones produces intoxication Fowls and ducks may succumb by eating contaminated larvae of flies.

The toxin is a neurotoxin.   The toxicity groups are Tryptophan, cystine and other amino acids which structurally simulate serotonin.   The toxin prevents the formation of acetylcholine from the parasympathetic and skeletal motor nerve endings   The toxin interferes with serotinin which is responsible for maintaining Ca-Mg ratio at the nerve endings of cells and the interference with Ca transport prevents the release of acetylcholine leading to muscular paralysis.   So paralysis of muscles supervenes and the symptoms noticed are therefore the result of this paralysis

Symptoms : The incubation period may be upto 5 days depending upon the concentration of the toxin.   Muscles of various parts of the body are paralysed especially those of pharynx. tongue, limbs and neck.   So the animal may not be able to eat and swallow. The head droops and there is incoordination of movement.   In birds it is called " limber neck " because the head and neck droop.

Respiration may be shallow or abnormal.   Death is due to asphyxia, Consciousness is normal till death.

Lesions : No characteristic lesions are noticed. Lesions in neurones may be due to hypoxia.   Degeneration of gnaglion cells, neuronophagia, necrosis, hemorrhages and focal glial proliferation and proliferation of vascular endothelium may be noticed. Hemorrhages may be present under the epicardium and endocardium Hyperemia of intestinal mucosa may be evident.

Diagnosis : Symptoms help in diagnosis Botulism must be differentiated from milk fever in cattle, equine encephalomyelitis and senecio poisoning in horses and louping ill in sheep.

Diagnosis may be confirmed by establishing the presence of the toxin in the feed and dead animals Filtrates of suspected feed may be injected into mice or guinea-pigs which may die of paralysis.

From the tissues of the dead animals filtrates may be prepared and these if injected into susceptible animals cause death in them.

Fluorescent antibody technique can be applied.

Neutralisation test using known antitoxins and the filtrates of suspected material in mice or guinea-pigs may also be applied. The unknown antigen and the specific antitoxin are injected simultaneously. The antitoxin protects the animal.

ERYSIPEALS IN ANIMALS

Erysipelas infection in animals is due to *Ersipelothrix (rhusiopathiae) insidiosa*, a gram positive rod. Swine are mostly affected but cattle and sheep rarely. Man is susceptible.

SWINE ERYSIPELAS

Incidence : Swine erysipelas is found in most places of the world and in some it may be enzootic since the organisms thrive in the soil. Young pigs' 3 to 6 months old, are most susceptible. But pigs of all ages may be affected• Recently farrowed sows are also more susceptible. Recovered animals though appearing healthy may pe carriers.

Routes of infection :

1. Through skin abrasions.

2. Alimentary tract Carrier animals harbour the organisms in their tonsils Infection of soil occurs through dung of affected and carrier animals.

Flies may transmit the disease.

Pathogenesis : It is the smooth strains that are pathogenic. The rough forms are avirulent and non-pathogenic. After the resistance of the animals is lowered, the organisms gain entry into the circulation through the tonsils or intestines and proliferate giving rise to a septicemia. Bacilli entering a skin wound ultimately reach the blood through lymph.

In the septic form the liberated toxins injure the walls of blood vessels giving rise to erythematous lesions of skin and hemorrhages in various parts of the body. This is the cause for enlargement of the spleen and lymph glands.

In some cases, the septicemia may subside and the organisms may localise in the skin, causing thrombosis of small vessels. This results in "diamonds" which are raised areas of erythema finally culminating in necrosis and gangrene.

The organisms may also settle in the joints and heart valves giving rise to chronic lesions viz sero-fibrinous arthritis and verrucose endocarditis respectively.

Symptoms : The incubation period is one to seven days Morbidity may be low but mortality is high upto 75%).

In very acute cases death may occur before any lesions and symptoms are seen. In less severe cases, there is high temperature initially, accompanied by anorexia, dullness, lethargy and unwillingness to move, congested conjunctiva with profuse lachrymation, constipation or diarrhoea and wobbly gait.

On the skin may be seen bright-red urticarial swellings having the form of a diamond. These may be found on the abdomen, neck, ears and thighs. In some animals the skin on the diamonds becomes necrotic, black in color and a crust may form. Gangrene may supervene.

In the chronic form, when the joints are affected, lameness is noticed. The affected joints may be swollen, hot to touch and painful.

Animals suffering from cardiac lesions show dyspnoea and cyanosis. Recovered animals are immune for life.

Lesions : The lesions in the acute stage are attributable to the damage of the blood vessels. So edema under the belly as well as serofibrinous exudate of pericardium and pleura may be noticed. Hemorrhages, ecchymotic or petechial are found on the stomach, intestines, epicardium and urinary bladder. Hemorrhagic enteritis may be noticed also. Mesenteric lymph nodes are enlarged and may show infarcts Hemorrhages are seen under the kidney capsule. Liver is enlarged and congested. Lungs are edematous. Spleen is enlarged due to hyperemia and hyperplasia of reticular cells.

In the chronic form of the disease, the joints reveal a nonsuppurative sero-fibrinous arthritis with increased synovial fluid. The articular cartilage may become ulcerated. The villi become thickened due to proliferation and the joint capsule also is thickened. Ankylosis may supervene and so the limb becomes stiff.

The cardiac lesions consist of either ulcers or more commonly, of large cauliflower-like vegetations on the valves. These vegetations cause stenosis of the A. V. opening or incompetence of the valves. Infarcts of kidney and spleen may be noticed. It is the mitral valve that is more often affected. Chronic venous congestion of the lungs with pulmonary edema may result

E. insidiosa causes a non-suppurative arthritis in lambs, after docking or due to umbilical infection after birth. The hock, stifle or carpal joints are swollen and the animals become lame. There is excessive turbid synovia but no suppuration occurs. Joint capsule is thickened and the articular cartilages are eroded.

Diagnosis : History, symptoms and lesions help in diagnosis. In the acute stage, the organism can be isolated from blood. When culture is inoculated into white mice, they die in 12 hours and the organism can be isolated from its spleen. The acute form of this disease must be differentiated from Hog cholera, salmonellosis and streptococcal infections.

PASTEURELLOSIS—HEMORRAGIC SEPTICEMIA
Syn. Stockyards disease.

Infection by *Pasteurella multocida* and *Pasteurella hemolytica* is very common among cattle and buffaloes. Sheep and goats are rarely affected Swine may also be affected. Fowls suffer frequently. Dogs are immune. Tularemia in rabbits and bubonic plague in man are caused by different strains of pasteurella

Since the organisms are septicemic and cause widespread hemorrhages, the condition is known as **hemorrhagic septicemia** The causative bacteria are short, gram negative, non-motile organisms which with Leishman's stain take a characteristic **bipolar** staining.

The organisms do not thrive long in soil. They are natural inhabitants of the tonsils and nasopharynx, from where they may invade the tissues when the vitality of the animals is reduced

It is also believed that the organism alone may not be able to cause the disease but that it is a secondary infection after the animal becomes weakened by a virus Other factors that may predispose the animal for infection are:- fatigue, transportation for long distances (shipping fever), confinement to damp, humid and close barns, worm infestation and starvation.

Routes of infection : Infection is mainly by ingestion of contaminated fodder, water etc. Carrier animals and clinical cases are sources of infection The organism is found in the salvia of affected animals. Passage through an animal increases the virulence of the organism.

Droplet infection may occur, especially in animals kept in crowded barns.

Infection may be introduced by new arrivals into the herd.

Pathogenesis : Incubation period is *two* to *five* days.

Depending on the virulence of the organism the nature of the disease caused varies.

If the organism is very virulent and the animal's resistance low, then *peracute* type of the disease results. The organisms entering the blood rapidly proliferate and spread throughout the body, death occurring in 10 to 24 hours. Symptoms shown are high temperature, prostration, diarrhoea and weaknes of heart. Lesions seen may be only petechiae on serous membranes and other organs. Subcutaneous edema containing gelatinous fluid may be present in the throat and brisket regions.

When the virulence of the organism is lower, the *acute form* of the disease is manifested. In this besides hemorrhages, fibrinous pneumonia and serofibrinous inflammation of the mucosa of gastro-intestinal tract are seen. Lymph glands may be swollen and hemorrhagic.

With organisms of low virulence, a more *chronic* type of the disease is met with. The organisms localise in certain organs (liver) where necrotic foci are seen.

The organisms are believed to liberate unidentified toxins which are responsibln for the pathology.

Lesions : Petechiae are found on all serous and mucous membranes especially on the epi and endocardium, peritoneum, pleura and gastrointestinal tract. Hemorrhagic gastro-enteritis may be found In the regions of head, neck and throat, there is swelling due to subcutaneous edema consisting of a gelatinous material. All lymph glands are swollen and hemorrhagic. Muscles may be edematous Peritoneum contains a large quantity of sero-sanguineous fluid. Myocarditis is present.

In acute and subacute cases, the predominant. lesion is fibrinous bronchopneumonia, in which various stages of hepatisation are found The interlobular septa are thickened and widened by infiltration with serous fluid produ. cing "marbling". *Corynebacterium pyogenes* may complicate by causing abscesses. The pleura shows fibrinous pleuritis The pleural cavity may contain a large quantity of sero-fibrinous exudate with fibrin flakes strewn on the visceral pleura. The mediastinal lymphatic glands as well as the bronchial nodes are swollenedematous and hemorrhagic Pericarditis and fibrinous tendinitis may be present. Spleen is normal Pregnant cows may abort

In buffaloes the subcutaneous edema in the region of throat is constant.

Pasteurella in swine. **Pasteurella multocida** is common casuing septicemic disease, while *pasteurella hemolytica* may be very rarely seen. The lesions found are fibrinous pneumonia with numerous organisms, acute pharyngitis (necrotising and ulcerative), edema of the throat, fibrinohemorrhagic polyarthritis and severe congestion of the mucosa of the stomach and intestines. The acute disease never subsides but becomes chronic in which the following lesions are seen : polyarthritis, adhesive pericarditis, pleurisy, widespread and extensive induration of the lung. The animals never recover.

Pasteurella hemolytica (of which two strains. A&T are described) is mostly associated with the pneumonic pasteurellosis encountered with shipping fever (see below)

In sheep *Pasteurella hemolytica* produces hemorrhagic septicemia. The lesions found in the disease, which is very acute without manifesting any specific symptoms, are : hemorrhages in the subcutaneous tissues, intermuscular tissues, under the serous membranes ; swollen and edematous lymph glands ; necrotizing pharyngitis ; intensely congested and cyanotic lungs, froth in the trachea, widespread bacterial embolism Lungs reveal fibrinous bronch-pneumonia similar to the bovine condition but without effusion into the pleural cavity.

Diagnosis :

1. Examination of blood smear, stained by Leishman's stain reveals bipolars.

2. A swab of heart-blood may be rubbed over the scarified abdomen of a rabbit. It will die in 24 to 40 hours showing hemorrhages, especially hemorrhagic tracheitis.

Hemorrhagic septicemia must be differentiated from anthrax, black quarter. contagious bovine pleuropneumonia, viral diseases and bronchopneumonia caused by several organisms.

Shipping fever : The condition is found in calves and young animals that are transported to great distances by rail, truck or ship In countries where there is great movement of animals, heavy losses are sustained.

Among calves, this is an acute respiratory disease

Etiology : The following are supposed to be the causal factors of the condition : stress plus viral infections plus bacterial infections. It is conjectured that stress factors predispose the animals to viral and or bacterial infections

Stress can be brought about by fright, heat, cold, fatigue, trauma, insufficient feed and water (giving rise to hunger and dehydration) and anxiety with probably endocrine exhaustion

The most common *virus* encountered and isolated and the presence of which is determined by serological tests is the myxovirus parainfluenza-3. Other viruses encountered are   The virus of infectious bovine rhino tracheitis, enterovirus bovine adenoviruses, viruses of Psittacosis-lymphogranuloma trachoma group and reovirus. A mycoplasma has also been isolated

Bacteria : The most commonly found is *Pasteurella multocida*. Other bacteria sometimes found are : *Streptococcus* sp, *Pseudomonas* sp, *Hemophilus* sp. and others.

Inoculation with *P. multocida* alone did not produce the disease But the efficacy of treatment of the condition with sulphonamides makes one suspect that the condition is a bacterial infection, which is a superimposed infection by bacteria on an animal made susceptible by the stress factors and viral infections.

Symptoms : The affected calves are depressed and stand aloof.   Anorexia, dry nose, copious mucopurulent nasal discharge, increased respiratory rate and sometimes presence of raeles are the other symptoms noticed.

Lesions: There may be edema and hemorrhages on the mucosa of the upper nasal passages, larynx and trachea The regional lymph nodes are edematous and hemorrhagic.

Pasteurella multocida has been found to cause fibrinopurulent meningitis in calves, 2 to 4 months of age. In these animals a fibrinopurulent polyarthritis may also be found. The symptoms seen are   muscle tremors, opisthotonus, rotation of eye balls, collapse, coma and death within a few hours. Both *Pasteu rella multocida* and *Pasteurella hemolytica* produce acute mastitis in cattle and sheep sporadically.

BOVINE LYMPHANGITIS

This is a chronic infectious disease of bovines in which there is abscess formation of the superficial lymph glands.

Cause : This disease is caused by *Yersinia pseudotuberculosis rodentium,* type III, which causes pseudotuberculosis in guinea-pigs and rabbits. This organism differs from other pasteurella in being motile at $22°C$. It it is related serologically to *P. pestis* (the organism causing bubonic plague in man) but differs from it in producing alkalinity in milk and H_2S in cultures.

Injected intraperitoneally into male guinea-pigs, the organism of Bovine lymphangitis causes suppurative orchitis—Straus's test.

Routes of infection : Wound infection is the most common method. Blood sucking insects, viz, ticks, mites and flies may transmit the disease.

Symptoms : The prescapular and or precrural lymph nodes are enlarged These become indurated later when they become painless. In the neighbouring lymphatics, may be found nodules which may soften later No systemic disturbances are ever noticed. There may be lameness of the limb affected. In untreated cases the animal loses condition The whole gland may become softened and converted into an abscess containing white, thick granular pus.

Extension of infection to lung through emboli may cause pneumonia when espiratory distress may be noticed.

Lesions : In the guinea-pigs yellow necrotic purulent foci may be found in the liver Spleen and mesenteric lymph nodes may not usually be affected.

In cattle, the condition is one of lymphangitis and lymphadenitis. The affected node is hot and painful in the early stages. Later it becomes softened and abscess is formed

Embolic suppurative pneumonia may be a complication.

Diagnosis : 1 Pus smears reveal bipolar organisms.

2. Isolate the organism and conduct Straus's Test,

COLIBACILLOSIS IN THE NEW BORN ANIMALS

The most important and severe condition affecting the new born animals is *Colibacillosis* caused by *E. coli* The livestock owner sustains considerable loss not only due to death of many young animals (mortality may be as high as 100% in some outbreaks) but also on the expensive treatment of the affected and the loss in condition of the surviving This condition is seen in calves, piglets, lambs and foals. The organism, *E. coli*, is ubiquitous and becomes pathogenic under certain conditions that predispose the animals to infection.

Routes of infection : The most common method of infection is through the alimentary canal by ingestion of contaminated feed or licking contaminated utensils or from a contaminated udder. Feces largely and to a lesser extent vaginal discharges are the infecting media The litter, udder and the environment are thus contaminated. Infection through umbilical stump may occur In foals intrauterine infection is common

Predisposing causes are errors in feeding and management, inadequate housing (over crowding is the most potent cause since such ovecrowded premises can never be rid of infection) exposure to extremes of weather, insufficiency of colostrum, congenital weakness and deficiency of vitamin A

Colostrum contains antibodies against this organism. It is found that absorption of colostrum stops by 48 hours after birth. Its absorption is at its peak by the 6th or 8th hour. However, colostrum should be made avilable to the new born within 24 hours after birth It has also been found that immunoglobulins continue to be secreted into the milk for the first week after parturition.

Pathogenesis : The organism is supposed to evolve 3 types of toxins : (i) that causes acute hypotension (characteristic of enteric-toxemic form) ; (ii) that damages the vascular endothelium and so there is great transudation from vessels into the serous cavities (as seen in the septicemic form and (iii) that causes the enteric form of the disease.

Three types of the disease are manifested : (i) the peracute enteric-toxemic colibacillosis; (ii) the septicemic colibacillosis and (iii) the enteric colibacillosis These different forms are caused by different sero types of the organism.

The peracute *enteric-toxemic colibacillos s* results from proliferation of certain strains of *E. coli* in the intestines producing a potent endotoxin, which causes hypotension, vascular collapse and hypothermia. The affected calves die within 2 to 6 hours after showing coma, subnormal temperature, pallid mucosa, cold and clammy skin, bradycardia and convulsive movements Diarrhoea is not present.

The *septicemic* collibacillosis occurs in animals which are deprived of colostrum or in those that become agammaglobulinemic inspite of ingestion of colostrum This form of the disease is found in animals in their first 4 days of li

The condition is acute and the course varies from 24 to 96 hours. The *symptoms* which are not very pathognomonic are depression, weakness, complete anorexia, tachycardia and elevated temperature that falls after diarrhoea sets in Diarrhoea and dysentery are noticed

In less severe cases septicemia subsides and the animal recovers but the organisms may be localised causing arthritis with swelling and pain of the joints, (Joint ill). Meningitis may also be seen, and such animals manifest recumbency and eye defects. Localising in other places abscesses are found in the liver, lung, kidney and tendons. In the lung pneumonia may be encountered.

In the majority of the calves infected, acute *enteric* colibacillosis may develop, when specific sero types of the organism infect, even in those that have received colostrum. In such animals the bacteria proliferate in the anterior part of the intestines The neonatal epithelium (which is permeable to the macromolecular globulins) is shed during the first 3 to 5 days of the life of the new born. So the mucosa with the loss of epithelium becomes vulnerable for inflammation if during this period the animal acquires infection The bacteria proliferate in the bowel and cause enteritis. Occurring usually during the first week of the life of the calf, the disease is manifested by watery or pasty feces, which is usually chalk-white *(White scours)* or yellow in color. Sometimes the feces is streaked with blood Defecation may be frequent and the tail and buttocks may be soiled with offensive and rancid feces.

Temperature may be high and pulse rate increased. Animal refuses to drink, is dull and restless. Rapid dehydration sets in Arched back and pain in the abdomen are noticed. If treatment is not given death may occur within 3 to 5 days.

In *lambs* the condition is not so frequent since they are usually reared naturally, (not crowded). But in those that suffer, the septicemic and peracute forms are met with. Though morbidity is lower, mortality is almost 100%. Some may show symptoms of meningitis.

Colibacillosis of the *piglets* is complicated by the viral transmissible gastroenteritis and iron deficiency anemia. The condition is an enteritis manifested by diarrhoea and rapid death Some animals may die, without showing any symptoms within 12 hours after birth, Others are dull and show profuse diarrhoea, soiling the hind quarters Dehydration is rapid. The animals are recumbent and have subnormal temperature. Such animals may die within 24 hours after birth, There is hypotonicity and hypomotility of the stomach, which becomes flatulent before diarrhoea starts.

Note : When infection occurs through the umblilical vein, the organisms first multiply in the clot of blood and liquefy it. Intima gets inflammed resulting in omphalophlebitis. The infection may spread to the adventitia. The clot disintegrates and emboli containing bacteria are thus carried by the portal veins or posterior vena cava to the liver and general circulation respectively. Ultimately septicemia is set up.

Lesions - Animals are dehydrated and show sunken eyes and loss of elasticity of skin In the acute form there are petechiae on all serous membrane swelling of the Peyer's patches, solitary follicles and mesenteric lymph nodes; degeneration of parenchymatous organs, catarrhal pneumonia and hemorrhages on the epicrodium and endocardium.

The umbilicus is swollen and contains serosanguineous fluid Adhesion of the peritoneum to abdominal organs is seen.

In the chronic metastatic form there may be purulent arthritis of the knee and hock joints Abscesses may be found in the periarticular connective tissue. Similarly, tendon sheaths may be affected. Purulent pneumonia may be found

Abscesses are found in the liver, kidney, mesenteric and bronchial lymph glands and brain Cerebrospinal fluid may be cloudy if brain is affected.

Diagnosis is not difficult from the symptoms. The organisms can be isolated from the mesenteric lymph glands in the acute cases and from the lesions in the joints or liver in the chronic.

Other organisms that may cause septicemia in new-born animals are: *Proteus morgani, Pseudomonas aeruginosa, Streptococci, Pasteurella multocida, Corynebacterium equi* (in foals), *Brucella abotrus, Erysipelothrix insidiosa* (lambs) and *Salmonella* (about this a detailed description is given later)

GUT EDEMA OF SWINE

Gut edema in swine is caused by hemolytic *E. Coli* among feeder pigs. In the affected herds morbidity may be 15% while mortality may vary from 20 to 100%.

Gut edema is usually found in well fed animals between the ages of 10 and 12 weeks. Probably feeding on heavy concentrates is the predisposing factor.

Pathogenesis : Excessive proliferation of the organism and the production of large amounts of the toxin cause an enterotoxemia Probably the lesions and symptoms are caused due to anaphylactic shock that has developed in the animals that have become sensitized to the toxin.

Symptoms : Clinically only nervous symptoms are evident, viz incoordination in the hind limbs, increased mobility, blindness, convulsions, edema of the eye lids and conjunctiva and recumbency. Protrusion of the eye ball may be seen The course of the disease is very short and death may occur in coma in 6 to 36 hours. In animals that may survive, incoordination may remain.

Lesions : Sometimes edema may not be encountered in the carcass because probably the course has been too short for the edema to develop. In others, edema is found in the eye lids ; subcutaneous tissue of forehead, throat, under surface of the belly, stomach wall (especially greater curvature) and the coiled protion of the colon

Hemolytic *E coli* can be cultured from the colon, rectum, and the mesenteric lymph nodes

Microscopically, necrotic arteritis of the vessels of the stomach wall and encephalomalacia, especially of the brain stem, may be noticed.

Enteric Colibacillosis of the feeder pigs

This condtion is found in pigs after weaning, upto 16 weeks of age and is caused by hemolytic *E. coli*. Usually pigs in very good condition and those on heavy grain feed are affected.

Symptoms seen are depression, pyrexia, anorexia and diarrhoea. Dehydration sets in rapidly.

Lesions include severe hemorrhagic gastro-enteritis This condition also is believed to be a manifestation of anaphylactic state.

PARATYPHOID IN ANIMALS

Infection by Salmonella is very common among young animals of all species and this is one of the causes for high mortality in dairy farms and piggeries in India.

The organisms responsible are : *Salmonella typhimurium* and *Salmonella dublin* in calvas and sheep ; *Salmonella cholarae-suis* and *S. typhimurium* in pigs and *S. typhimurium* in horses.

Among young animals these organisms cause a peracute septicemia or an acute enteritis. In older animals the infection may be less severe and among cattle, especially, it is often unnoticed. Adult cattle and pigs act as carriers though no symptoms are manifested.

Routes of infection : Infection is by ingestion of contaminated feed, water and milk. Mortality may be high in some outbreaks, reaching 100%. Recovered animals, especially adult cattle, are carriers

Predisposing causes : Comprise of transportation for long distances, faulty nutrition, parturition (in cows) and intercurrent diseases.

Pathogenesis : The organisms entering the wall of pharynx and intestines reach the lymph nodes and multiply. Within a week, entering the blood stream via the lymphatics, they produce septicemia and bacteremia and subsequently settle in other organs, especially the intestines, producing enteritis.

In animals that survive, especially calves and foals, the organisms may be localised in various joints, and also in the liver, spleen, mesenteric lymph nodes and gall bladder. From these areas. especially gall bladder, the organisms may be excreted in the feces and so such animals are sources of infection. The organisms may invade the lungs and meninges causing pneumonia and meningitis respectively, especially in pigs.

Symptoms : In the *peracute* cases, due to septicemia, death occurs in 24 to 48 hours after manifesting dullness, depression and high temperature. In *acute* cases symptoms of enteritis are seen, viz. high temperature, diarrhoea, loss of condition and progressive weakness and dehydration. Animals may die in 2 to 5 days. Those that survive (calves and foals) suffer from painful polyarthritis In pigs (and in cattle sometimes) chronic enteritis may be noticed with persisten diarrhoea, intermittent fever and severe emaciation.

Lesions : Hemorrhages and edema are seen on various serous surfaces. The intestines show enteritis, which is hemorrhagic sometimes. The spleen and mesenteric lymph nodes are swollen. In the liver may be seen the so called "typhoid nodules". These are foci of necrosis or reactive granuloma of Kupffer's cell hyperplasia. Lungs may reveal foci of pneumonia. Petechiae are found on the kidneys giving a "turkey egg" appearance

In swine, intestines may show "button ulcers" (that are also found in hog cholera) besides changes noticed in acute enteritis viz. necrosis and desquamation of the epithelium of the mucosa Skin shows erythematous patches, hemorrhages and raised plaques. Pneumonia and encephalitis may be present sometimes.

Diagnosis may be difficult and it is necessary to differentiate from other enteric diseases and poisoning by bracken fern, arsenic and chronic molybdenum poisoning In swine differentiate from hog cholera, pasteurellosis and swine erysipelas. A laboratory examination is needed for confirmation.

STRANGLES

Strangles is an acute disease of horses characterised by inflammation of the upper respiratory tract with abscess formation in the adjoining lymph nodes.

Cause is *Streptococcus equi* belonging to group C of Lancefield's classification. It forms a hemolysin and a leucocydin but does not produce erythrogenic toxin Young animals are more susceptible

Routes of infection : *Streptococcus equi* is a very resistant microbe and is able to withstand environmental influences.

Infected animals contaminate, through their nasal discharges the pasture, feeding and watering utensils, and persons handling the animals.

(a) Infection is mostly by ingestion. (b) Inhalation of droplets may be another route in some cases (c) Vulva may be infected by an infected nosing stallion or infection may be transmitted to it through penis from some other infected vulva (d) Scabs of cutaneous erruptions may infect other animals. (e) Udder may be infected by suckling colts (f) Wounds may also be portals of infection. (g) Lastly intrauterine infection may sometimes occur.

Incubation period : 4 to 8 days.

Pathogenesis: The streptococci enter the *mucous* glands of the nasopharyngeal mucosa and from there reach the local lymph node via the lymph spaces.   In the lymph node they grow and produce toxins, which exert positive chemotaxis and hence large number of neutrophils infiltrate. Lymphoid tissue undergoes necrosis and liquefaction, thereby forming an abscess.

If the abscess in evacuated, recovery may result. Sometimes the organism gains entry into lymph and blood vessels by penetration through their walls and so infects distant lymph glands and visceral organs, setting up suppurative inflammation.   Death may be due to septicemia and pyemia.

Symptoms : The disease develops suddenly with high temperature and anorexia attendant with a nasal discharge, which is serous at first becoming purulent subsequently. Pharyngitis and laryngitis develop with cough and difficulty in swallowing After the temperature subsides in 2 to 3 days, abscesses develop in the lymph nodes of the throat. These become hot, swollen and painful. The submaxillary lymph node may open discharging a thick creamy pus.   If no other complications occur, recovery ensues.

If the infection is severe, other lymph nodes in the body may be affected and metastatic abscesses may be found in the lungs liver, spleen and brain Infection may spread to the guttural pouches and into the thoracic cavity   Affection of the mesenteric lymph nodes is followed by colic. Involvement of lungs results in pneumonia

Lesions : Abscesses are found in the pharyngeal and submaxillary lymph nodes primarily. Oher lymph node , iz, mediastinal, bronchial and mesenteric, may also reveal the abscesses if infection is severe. Pericarditis, pleurisy and suppurative pneumonia may be noticed. If pyemia develops, abscesses may be noticed in the liver, spleen, brain, kidneys. testes, thymus, muscles of the neck, axillary and inguinal regions. Degeneration of parenchymatous organs is noticed as in septicemic diseases

Sequelae : Rarely, roaring and purpura hemorrhagica may develop in animals that survive.

Diagnosis is easy. Ordinarily it is not difficult to differentiate it with glanders, which is a more chronic infection and in which the pus is oily and yellow.

CONTAGIOUS BOVINE PLEUROPNEUMONIA

This is a highly infectious, intractable disease of cattle and encountered in Assam in India. Contagious bovine pleuropneumonia is caused by an organism belonging to the genus *Mycoplasma* and is known as *Mycoplasma mycoides*. This organism is polymorphic and is filtrable at one stage of its cycle of development. The following cyclic stages are noted in its development—elementary bodies, filaments, branching forms, chains and finally disintegration into elementary bodies. Bison, buffaloe, yak, reindeer and antelope may be affected rarely.

Routes of infection : Droplet infection occurs from an affected animal' So spread is facilitated by housing in closed byres and during transit by rall steamer or trucks

Animals which develop "sequestra" are carriers and so are potential sources of infection. Such sequestra may break down during violent coughing and so the animal becomes an "open" case.

Pathogenesis : Calves under one year of age are less susceptible than adults. Incubation period is 3 to 6 weeks, (In some cases it may be as long as six months).

The organisms enter the bronchioles via the respiratory tract. Inflammation of the bronchiolar walls is set up. Passing through these walls, the organisms enter the interlobular septa where again inflammation is set up followed by copious edema which causes dilatation and subsequent thrombosis of the lymph vessels

The inflammatory process, may subsequently spread to the lung alveoli setting up croupous pneumonia, which is manifested by red hepatisation followed by grey hepatisation Spread of inflammation to the branches of pulmonary arteries results' in thrombosis and subsequent anemia of the part, which hence becomes necrosed. Such necrosed areas become clearly demarcated and circumscribed by fibrous tissue. This isolated and enclosed lesion is called a "sequestrum".

In these sequestra, the mycoplasma may remain viable for years. During violent coughing the fibrous capsule may rupture. liberating the organism into the surrounding lymph spaces from where the surrounding tissue may be infected and thus disease process may be set up in other parts of the lung. Since the lesions are of different ages in different parts, the affected areas reveal different stages of the process—red hepatisation in some parts and gray in others.

In some animals a purely septicemic form may occur without any lung lesions and the organism may be excreted in the urine, milk, nasal discharge and amniotic fluid A toxin is supposed to be formed by the organism.

Inoculation of the pleural lymph containing the organism subcutaneously results in local inflammation and edema Death follows soon. Pulmonary lesions may not be noticed.

Symptoms :

Acute form :   In this form the disease runs a rapid course—about a week· Symptoms af acute pneumonia are evinced : high fever, anorexia, disinclination to move and loss in milk yield. Animals die of asphyxia.

Chronic form :   The acute condition may gradually assume a subacute or chronic form manifested by a painful cough, nasal discharge. drooping head and characteristic raeles on auscultation. Animals may die in 2 months or may recover incompletely, with the formation of sequestra. Such animals are potential danger as carriers.

In a herd about a quarter of the affected animals become ''Carriers''.

Lesions :   Lesions are found only in the thorax. Lung shows a marbled appearance on section due to the marked infiltration of the interlobular septa by a clear straw colored exudate—the lymph—which thereby makes the septa thickened. The ''lymph'' in time becomes gelatinous and organised. The thickened septa surround various colored lobules, which are in different stages of hepatisation—red and gray.

The lymph vessels may be prominent due to dilatation. Red infarcts due to thrombosis of pulmonary vessels may be seen.

Serofibrinous pleurisy may be seen with adhesion of the pleural surfaces in the chronic cases

Encapsulated necrotic tissue, sequestra, may be seen and these contain viable organisms and so act as carriers These may break down during violent coughing. The carriers are also known as ''lungers''. Rarely caseation and calcification may be noticed Other rare extra-pulmonary lesions seen are: serous or fibrinous pericarditis, peritonitis, serofibrinous arthritis and periarthritis.

Diagnosis :   The disease must be differentiated from chronic pasteurellosis, tuberculosis and parasitic pneumonia. Symptoms and lesions are sufficient to enable a correct diagnosis to be arrived at. Complement fixation test can be conducted for confirmation. Organisms can be cultivated from the affected lesions A plate complement-fixation test has been introduced which promises to be rapid and accurate For screening animals in herds an intradermal allergic test is made use of, the positive reactors showing edematous thickening.

Subcutaneous inoculation of ''lymph'' from an affected animal into a calf results in a large edematous swelling locally

CONTAGIOUS CAPRINE PLEUROPNEUMONIA

This is a very serious disease of goats prevalent in many states of India. Heavy mortality is caused among affected flocks. In some areas, the disease appears to be endemic.

The causative organism is specific, only goats being susceptible. It belongs to the genus *Mycoplasma*, and is known as *Borrelomyces peripneumoniae capri*.

As in the bovine disease, infection is by droplet infection.

Incubation period is shorter, being about 4 days.

Pathogenesis is similar to bovine pleuropneumonia.

Symptoms shown are those of pneumonia, anorexia, dullness, dry painful cough (especially when made to walk), laboured breathing and nasal discharge, which is watery in the early stages, but turns to be thick, mucopurulent and white later. Diarrhoea may be present.

The disease may last for three days to a week.

Lesions: Rhinitis and catarrhal inflammation of the upper respiratory tract are found. There is unilateral or sometimes bilateral fibrinous pleuropneumonia. Pleurisy is found with copious exudate which on exposure to air clots and becomes gelatinous. Marbling of the lungs though seen is not so prominent a in the bovine disease. Pericarditis is frequently seen and the lungs nearby are adherent to it. Sequestra, which are so characteristic in bovine pleuropneumonia are not a feature of the caprine variety, probably because the course is short and the disease is acute. The bronchial and mediastinal lymph nodes are congested and edematous.

Diagnosis: From the symptoms and lesions it is easy to diagnose the disease among goats since this malady occurs only in these animals. Subcutaneous inoculation of lymph into healthy goats results in an edematous swelling locally, death following in a week. Organisms can be isolated from the edema fluid.

Differences between C. B. P. P. and C. C. P. P.

C. B. P. P.	C. C. P. P
1 Subacute to chronic disease.	1. More acute.
2 Affects cattle, baffaloe and bison.	2 Affects goats only.
3. Involvement of upper respiratory tract less pronounced.	3. Upper respiratory tract involved more pronouncedly with rhinitis and nasal discharge.
4. Pneumonic lesions are diffuse.	4 Pneumonic leisons are focal and project out prominently from the surrounding areas.
5. Marbling of lungs prominent.	5. Marbling less common.
6. Sequestration is a characteristic feature.	6. This is not a feature.
7. Pericarditis is not a common finding.	7. Pericarditis is a common finding.
8 The exudate in the chest cavity is less and does not tend to clot.	8. The exudate in the chest cavity is more and tends to clot and becomes gelatinous within a few minues after exposure to the atmosphere.

ACTINOBACILLOSIS
Syn: Wooden tongue

This is caused by *Actinobacillus lignieresi*, which is a small, non-motile, rod shaped, gram negative, aerobic organism.

Cattle are mostly affected. Sheep are affected rarely.

Habitat : *A. lignieresi* is an obligatory parasite of the upper respiratory and alimentary mucosa. Infection to the soft tissues occurs through wounds and abrasions caused by foreign bodies. Sharp objects like awns etc. may pierc the mucosa through which infection may occur.

Lesions : In cattle, the tongue is mostly affected   Gums, pharynx, palate and the neighbouring lymph glands may also be affected   Other organs that are rarely affected are lungs, stomach, intestines, liver. peritoneum and pleural cavities

The organism affects only the soft structures   (while actinomycosis affects hard bone) and spreads by way of lymphatics and so the lymph glands are frequently affected.

Wherever the organism is lodged, it incites a granulomatous inflammation. Around the central colony of the organism are found a pallisade of 'Indian Club'-like structures, thought to be a product of the reaction of host to the invading organisms : probably arising from macrophages. Peripheral to these clubs neutrophils surround   Beyond these is a sheath of histiocytes which may eventually become epithelioid cells   In this zone may be found histiocytic giant cells Around this layer may be found lymphocytes, plasma cells and eosinophils and beyond a non-specific vascular fibrous tissue, This granulation tissue may radiate into the tongue along connective tissue

If the organisms die, neutrophils disappear and the giant cells invade and clean up the debris.

The neutrophils may migrate with bacteria *in situ*, thereby setting up secon. dary centres of infection Such neighbouring lesions may coalesce to form larger nodules in which "Sulphur granules", being the colonies of the organism can be seen   These nodules are surrounded by a dense fibrous tissue capsule. In some lesions, the centre may liquefy and suppurate. Coalescence of many such suppurating lesions may result in an abscess.

Spread is usually via lymphatics, which themselves may show lymphangitis. They are thickened like cords and seen under the mucosa along the lateral borders of tongue.

Nodules situated under the mucosa may erode the epithelium above forming ulcers. Coalescence of these may result in quite large ulcers

A third and less common variety of the lesion in the tongue is the diffuse sclerosing actinobacillosis (wooden tongue). In this condition there is diffuse overgrowth of connective tissue in which may be noticed, scattered, single nodules. The connective tissue replaces the parenchyma (muscle, gland etc.) and so the tongue becomes indurated, rigid and firm. It is the fixed portion of the tongue that is more often affected. Since the tongue is the organ of prehension, animals suffer from starvation as the lesion on the tongue is very painful. In the gums and palate, ulceration of the mucosa occurs with diffuse thickening of the sub - mucosa by granulation tissue. The lesion in the pharynx is polypoid in. character.

Lesions may be found in the walls of forestomachs(with digestive symptoms) in the skin, liver and lungs. The lesions in these places are not unlike those of tuberculosis But here necrosis and softening are not common as in tuberculosis. The affected organs become cirrhotic and the colonies are found in the lesions.

In the lungs the lesions extend along the inter - lobular septa. There may be cavities containing yellowish - green pus. Chronic pleuritis may occur.

Extending via the lymphatics, neighbouring lymph glands may be affected These may show discrete nodules standing out from their surface (in which sulphur granules may be seen) or they may become considerably enlarged and edematous In old standing cases, the glands become sclerosed, shrunken and become adherent to the overlying skin or the mucous membrane. The lymph glands more commonly affected are the retropharyngeal, submaxillary and the parotid Lesions in these glands cause dyspnoea and difficulty in swallowing.

In pigs the lesions are similar to those of cattle. In sheep which may be affected sometimes, the lingual lesions are not common. The subcutaneous tissue of the head, nose, lips, throat, cheeks and submaxillary region are affected, revealing the granulomata.

Diagnosis :   1. Symptoms   Differentiate from rabies, tuberculosis, lymphocytoma.
2. Examination of stained pus smear.
3. Histological examination of lesions.
4 Agglutination test.

ACTINOMYCOSIS
Syn : Lumpy Jaw

Cause is *Actinomyce* bovis—a gram positive, rod-shaped anaerobe It is pleomorphic, non-sporulating and non-motile. In the lesions branching forms are seen (ray fungus).

A. bovis is a strict, obligatory parasite of man and animals being present in the mucosa of the mouth and pharynx.

Route of Infection : Infection is usually through wounds in the oral cavity. The wounds may be caused by awns etc. or while cutting teeth or through lesions of Foot and Mouth Disease It is thought that necrosis which results by the penetrating objects, sets up necessary anaerobic conditions for the organism to grow and thrive.

Pathogenesis ; The basic lesion is a granuloma similar to that of actinobacillosis with a few minor differences In actinomycosis the sulphur granules are far bigger and the central part containing the organisms stains gram positive. Cattle are mostly affected. In swine, besides other tissues, mammary gland also is affected, infection occurring through wounds inflicted by the teeth of sucking piglets. In horses, *A. bovis* is supposed to cause poll evil and fistulous withers Among cats and dogs, actinomycosis runs a more acute form.

Lesions : In cattle, the most common lesion is the lumpy jaw, a suppurating osteomyelitis of the mandible The maxilla and other organs may rarely be affected. Infection occurs by direct extension from the gums and periodontium. If infection extends to periosteum directly, an actinomycotic periostitis may form. But it is more common for the medulla to be infected and the cavity to become filled with the specific actinocomycotic granulation tissue containing the isolated nodules. Softening and liquefaction occur and the pus gradually affects the cortex rarefying it and in which many fistulae form The periosteum whic is irritated forms a new subperiosteal bone (involucrum), thereby causing enlargement and thickening of the bone. But the actinomycotic granulation tissue invades this new bone also destroying it and forming cavities and fistulae. The structure and architecture of the whole bone is destroyed.

The process may extend to the adjacent structures; muscles, subcutaneous tissue and skin externally or mucous membrane internally and when the maxilla is affected, to the sinuses and ultimately to the duramater.

In the udder one may notice miliary nodules as in tuberculosis) scattered in the parenchyma. Infection may be hematogenous or by way of wound infection. In the sow, the organ may be indurated and fibrosed accompanied by suppuration.

Lungs : Infection may be hematogenous One finds miliary nodules as in tuberculosis.

Lymph glands : The neighbouring lymph glands may be affected. Infection is not via lymphatics but by direct extension.

The lesion in the bone is very painful preventing normal prehension and mastication. The animal may ultimately die of starvation. Infection of the esophagus and forestomachs gives rise to impaired digestion and diarrhoea.

Diagnosis : 1. Symptoms. 2 Examination of stained pus smea
              3. Histopathological examination of the lesions.

BRUCELLOSIS
Syn. Bang's disease; Contagious abortion

Brucellosis is a disease that is caused by different strains of Brucella, among cattle, sheep, pigs, horses, goats and man.

Strains of brucella : The different strains met with are : *Brucella abortus* (Bang), *Brucella suis, Brucella ovis, Brucella melitensis* and *Brucella canis*.

These strains can be differentiated by cultural methods.

The brucella organism is a short, non-motile, non-sporulating, rod-shaped gram negative organism Primary cultures require special atmospheric condition, for growth.

Growth on potatoe is characteristic, being yellowish-brown in color. This does not turn, on further incubation, to brown or chocolate color like *P. mallei. Brucella abortus* gives a positive Straus's test in male guinea-pigs. The Brucella grow well wherever erythritol (a carbohydrate) is available as a source of energy. Since erythritol is present in the fetal and placental tissues of cow, sheep, goat and pigs, Brucella infection of these tissues occurs in these animals only but not in others where erythritol is not found. Similarly erythritol is found in the testes and seminal vesicles of bull, ram, goat and boar and hence is the fact that Brucella infection occurs locally in these parts in these animals

Routes and sources of infection : Sources of infection are an aborted fetus, fetal membranes and discharges from uterus. Milk from affected udder, seminal fluids and urine in an affected male, meconium of a new born animal and discharges from infected bursae and joints are other sources

The important route of infection is by way of ingestion of infected feed and water. Infection may also be transmitted by flies, dogs, ticks, rats and fomites. But these should be considered as rare. Infection through coitus may occur if the bull is suffering from orchitis or epididymitis. Similarly artificial insemination of infected semen may be another source of infection Infection may also occur through intact or injured skin

Infection can also occur through conjunctiva. Tails of cows soiled with infected discharge may be responsible for spread of the disease to other animals through skin and conjunctiva

Udder may be infected through contaminated hands of milkers.

Though cows may be infected with *Br. suis* and *Br. melitensis*, abortions do not occur. The organisms are found in the mammary gland and supramammary lymph node and are excreted through milk. Once infected, cows are permanent carriers whether abortion occurs or not. Organisms continue to be excreted through milk for years.

Incubation period is 33 to 230 days.

Pathogenesis : The organisms after entry into the mucosa, reach the regional related lymph gland soon, where they multiply and cause acute lymphadenitis, hyperplastic enlargement, medullary hemorrhages, infiltration with neutrophils and eosinophils, accumulation of plasma cells in medullary sinusoids but without any fibrosis or necrosis. Subsequently they enter the blood stream in which they stay for sometime. During this bacteremic stage, the organisms reach all parts of the body but localise only in a few tissues; viz. spleen, mammary gland, lymph nodes of mammary gland, pregnant uterus, testes and accessory glands in the male. Wherever the organisms are localised, a granuloma develops The phagocytes that are attracted by the bacilli engulf them. The bacteria multiply within the cytoplasm of the phagocytes, which are transformed into epithelioid cells. Around these cells lymphocytes and plasma cells accumulate. In immature animals, the organisms may be found in the udder for a long time, even for years. The organisms may also reach and settle in joints, tendons, sheaths and bursae causing tendovaginitis, arthritis and bursitis.

The organism has special predilection for the embryonic tissues of the maternal and fetal placenta as well as the fetus When the animal becomes pregnant, the organisms invade the uterus from the mammary glands during one of bacteremic phases and multiply in the epithelium of the embryonic chorionic villi A severe ulcerative endometritis of the intercotyledonary spaces occurs. The villi undergo fatty degeneration and later a fibrino purulent exudate gradually loosens the connection of the villi with the maternal placental cells.

Entering into the connective tissue found between the chorion and allantois as well as into the umbilical cord a serous inflammation is produced in these places. The bacilli may reach the fetus either by way of blood stream or through swallowing of the infected amniotic fluid by the fetus. Lesions are produced in the fetus thereby.

The loosening of the fetal membranes from the maternal, results in their gradual separation resulting in stoppage of blood (so nourishment and oxygen supply are also stopped) to the fetus, which therefore dies. A dead fetus is a foreign body and so is expelled. Sometimes, the dead fetus instead of being expelled is retained, the amniotic fluid is absorbed and the fetus therefore becomes mummified. It is covered with a thick, tenacious exudate.

Abortion usually occurs during 6th or 7th month.

It is sometimes likely that a calf is born alive, But should it live, it does not develop any immunity since it is susceptible on attaining maturity.

In cases where the disease is slowly progressing or in animals which have had an earlier-abortion, the placenta is not shed, because, the connective tissue of the placenta proliferates and adhesion between the fetal and maternal placenta occurs. In animals which have had a prior infection some degree of resistance develops and so the reaction to a second infection is in the nature of productive type of inflammation instead of an acute necrotic inflammation seen in primary infections. So a chronic type of inflammation results in which there is necrosis and caseation in the center of the granuloma which consists of epithelioid cells, lymphocytes and plasma cells with a capsule of fibrous tissue. Though neutrophils are attracted by the necrotic tissue, abscess formation seldom occurs. In these cases, because of firm attachment, the membranes are not shed and so they should be carefully removed manually as otherwise suppurative organisms may invade and produce suppurative metritis.

Lesions : The placenta appears edematous, either focally or diffusely. It may be covered with fibrin flakes. The cotyledons are dull in appearance and some may be necrotic. In some cases the chorion may be thickened and leathery with a thick tenacious brown exudate.

Mammary gland and testes may reveal mastitis and orchitis respectively with granulomatous lesions described above. Hygroma of the knee is supposed to be caused by Brucella.

The fetus shows edema in all parts and cavities of the body. The walls of the serous cavities may have fibrinous flakes coated on. There is hemorrhagic gastroenteritis. Spleen and lymph nodes are swollen and hyperplastic and show necrotic foci.

No lesions are seen in the liver. Due to toxic reaction tubular nephrosis occurs.

Brucellosis in swine :

Brucellosis in swine in caused by *Brucella suis* mostly (though *Br. abortus* and *melitensis* can also infect) and is a chronic affection. Abortions are caused.

Routes of infection : Ingestion —uterine discharges and urine contaminate feed and water.

Copulation—from boars.

Pathogenesis : The pathogenesis of *Br. suis* is similar to *Br. abortus*. But *Br. suis* does not have any predilection to the pregnant uterus but is localised in all organs. More frequently localisation in the uterus and testes causes abortion and orchitis. All lymph nodes may be swollen but cervical lymph nodes are more commonly affected. Arthritis and osteomyelitis are caused by localisation in the joints and bones and so may cause lameness and paralysis.

The lesion is a typical granuloma. The phagocytes engulf the organism wherever it is localised. The organisms develop inside the phagocytes. Subsequently histiocytes and epithelioid cells accumulate, surrounded by lymphocytes and plasma cells. As hypersensitivity develops, the central area becomes necrosed and caseated. A few giant cells may be present. Calcification may occur subsequently.

Lesions : Uterus :- The uterus reveals whitish nodules which are firm or soft having the structure already described. The epithelium of the uterine mucosa may reveal coagulative necrosis and desquamation. The uterine glands may be swollen due to infiltration by the inflammatory cells. The endometrium is thickened, in some places as plaques, due to infiltration by macrophages, lymphocytes and plasma cells. A few neutrophils may be present.

The placenta may show areas of liquefactive necrosis and desquamation. There may be infiltration by leucocytes and hyperplasia of the fibrous tissue. Abortion occurs between 2nd and 3rd month.

Granulomata (with calcification sometimes), may be found in spleen, ovary, kidney, liver, brain, bone-marrow and joints. In the joints arthritis is caused resulting in lameness. Affecting the vertebrae, a suppurative spondylitis is caused resulting in paralysis.

Brucellosis in sheep : *Br. ovis* is found to cause abortion in sheep in New Zealand, Australia, Europe, U. S A. and South Africa. It is a comparatively mild organism and does not get generalised. Infection is by copulation. It produces only abortion. No other lesions are described. In the ram orchitis and epididymitis are produced. The semen is of poor quality and contains leucocytes and brucellae.

Brucella melitensis : This produces abortion in goats and 'Malta fever' in man.

Br. melitensis can affect sheep, cattle and pigs.

Infection is by ingestion.

In goats, there may be systemic disturbances with high fever, anorexia, depression, emaciation and even death during the bacteremic stage. The organism may localise in the mammary gland producing nodules and acute mastitis. The milk is watery and contains the organisms. Abortion occurs late in pregnancy. Lameness may be present when bones and joints are affected.

Brucellosis in Horses : Suppurative inflammation of ligamentum nuchae at its attachment to the occipital bone is known as "poll evil", and at its thoracic attachment it is known as "fistulous withers". These conditions are believed to be caused by *Brucella abortus*, though some authorities dispute it.

In man, **Malta fever** is characterised by undulant fever, weakness, articular rheumatism, night sweats and muscular stiffness.

Diagnosis :
1. Plate agglutination test by colored antigen.
2. Tube agglutination test. Positive titre is 1 in 40 and above. (To be conducted 3 weeks after abortion).
3. Skin sensitivity test by Brucellin.
4. Milk ring test, capillary milk ring test and rapid milk agglutination test.
5. Whey plate test.
6. Examination of semen in rams ; complement-fixation test and hemagglutination inhibition test in sheep
7. Isolation of organisms from abortus fetuses.
8. Card test. This is claimed to be superior to the agglutination test since (a) only one antigen-serum concentration is required, (b) lessesr time is consumed for the conduct of the test and (c) the disease can be detected earlier.

9   Fluorescent antibody technique.   This is a very rapid and reliable test.
     wherever facilities for such a test are available. Fetal stomach contends'
     uterine exudate and placenta can be used.

10   Rose Bengal plate test using a colored antigen of acidic pH detects bru-
     cella specific antibodies and helps to rule out cross reactions with other
     gram negative bacreria

11   A modified Ziehl-Nielson technique can be used for direct staining of
     Brucella in the stomach contents

12   Acid plate test, Rivanol test and mercapto ethanol test, help in differen-
     tiating antibodies due to vaccination and infection

VIBRION AFFECTION

Campylobacter fetus is a gram negative. motile, comma shaped organism
Motility is dart-like and is due to the flagella present at one end (lophotrichous)
Sheep and cattle are affected in which abortion is produced

Sheep : Infection in sheep is by way of ingestion   No coital infection
occurs. Spontaneous recovery in sheep may be noticed Abortion occurs late in
pregnancy (4th month)   Placenta is not retained.

The lesions seen are similar to brucella lesions. The placenta of a pregnant
animal is invaded by the organisms, becomes necrotic and separated. The organism causes endometritis, cervicitis and vaginitis in which there is degeneration
of the epithelium of the mucosa with infiltration of lymphocytes and neutrophils
into the submucosa   Arteriolitis of the vessels of the placenta with thrombosis
occurs very frequently. In the fetus edema of serous cavities and perirenal
tissue is found   There may be fibrin flakes in the edematous fluid. In the liver
may be seen necrotic foci

The fetus dies because of the bacteremia and toxemia as well as separation of
the fetal placenta from the maternal resulting in arrest of oxygen supply.

Cattle : Infection is by coitus or artificial insemination with contaminated
semen. Affected bull does not show any lesions though harbouring the organism
and being able to transmit it to cows. Cow to cow infection by contact does
not occur. In artificial insemination studs, infection from bull to bull can occur
through contaminated equipment, teasing animals and persons working there.

Infected cows are repeat breeders. Probably there is early embryonic death
and so animal comes into heat again.

Abortions occur between 4th and 6th months of pregnancy.

The lesions seen are similar to those described for sheep. Genital tract may
not show much change. There may be small nodules and cystic glands distributed in the uterine mucosa which also reveals lymphocytic infiltration and edema.
Cotyledons and placenta reveal necrotic areas and infiltration by inflammatory
cells. The membranes are opaque and leathery. Purulent exudate is found
between the endometrium and chorion.

Infected cows may throw off infection and may become immune   The main
difficulty is sterility and repeat breeding rather than abortion.

In the bulls the organism is located in the prepuce and does not cause any
pathological changes. They grow in the epithelial crypts of the prepuce and
glans penis. Bulls mostly throw off infection, but some may be infective for
3 to 5 years.

Diagnosis :
1.   By examination of smears from uterus and identification of organisms
2.   Culture and isolation of the organism. Cervical mucus may be used for cultural examination. But it must be remembered that the best time for culturing is between the 2nd week and 2nd month after exposure. Beyond the 2nd month many animals recover from infection and so may be negative.
3.   Mucus agglutination test using cervical mucus.

Differences between *Br. abortus* and *C. fetus*.

Br abortus	C. fetus
1. No flagellum.	1 Flagellum present.
2. Non - motile.	2. Motile.
3. Requires 10% CO_2 for primary culture	3. Requires 10% CO_2 and Nitrogen.
4 In shake cultures zone below surface broader and less dense.	4 Zone narrower and denser.
5. Surface growth comparatively easy.	5. Surface growth only after prolonged subculturing.
6. Younger animals affected	6. Older animals affected.
7. One or two abortions only.	7. Three or four successive abortions
8 Diagnosis by Serum agglutination tests.	8. Serum Agglutination tests not useful.
9. Growth on potato chromogenic	9. No growth on potato.
10. Abortion at 6th or 7th month.	10. Abortion at 4th to 6th month.
11 Gives positive Straus's test.	11. Straus's test negative.

WINTER DYSENTERY OF CATTLE

In some parts of the world a mild type of diarrhoea and dysentery is believed to be caused by a vibrio, known as *Vibrio jejuni*. There are conflicting reports as to whether *V. jejuni* can produce the disease *per se*. It is thought that a precipitating factor may be a primary viral infection.

The loss in a herd is by way of decrease in milk production. Though the disease is mild. it is explosive in character and most of the animals in the herd may be affected within a few days. So the drop in milk production will be enormous.

Route of infection is through ingestion of contaminated feed and drinking water. The source of infection is either a clinical case or a carrier. Incubation period is short, being 3 to 7 days.

Symptoms : An initial pyrexia is followed by diarrhoea and anorexia The temperature becomes normal as diarrhoea starts. The feces are watery, foul smelling and profuse. Though the condition is termed dysentery, blood in the feces is rarely seen. Milk yield is reduced. The feces become normal within 2 to 4 days Usually no deaths occur. Occasionally the disease may be severe with dysentery and dehydration.

Lesions reveal catarrhal enteritis with enlargement of the Peyer's patches and mesenteric lymph nodes The wall of the intestines is thickened due to edema and there may be petechiae on the mucosa. Deaths that occur may be due to complication by coccidiosis or necrobacillosis.

Diagnosis : This disease must be differentiated from coccidiosis and muco-sal disease.

VIBRIO DYSENTERY

Vibrio dysentery is a contagious disease of swine caused by *Vibrio coli* and is found in many parts of the world.   Morbidity rate is 30 to 40% while mortality is 60 to 70 percent.

Animals affected :   Pigs between 8 and 12 weeks of age are affected. Older animals may also be affected but in them the disease is mild   Recovered animas do not develop any immunity since the same animal may be affected again and again.

Route of Infection : Infection is by ingestion of feed contaminated by feces of Infected pigs.   Infected pig is always the source of infection.   When once introduced into a farm, it is infected indefinitely.

Pathogenesis : After ingestion, the organism causes enteritis, acting locally on the mucosa of the bowel.   Systemic invasion does not seem to occur.   Death of the affected animals is due to dehydration and toxemia.

Symptoms : Incubation period is 4 to 12 days   The disease starts with acute diarrhoea, rise in temperature and anorexia, Temperature falls when diarrhoea starts.   The feces are thin, watery and passed continuously without any straining.   Yellow at first the feces later turn to b ack or blood – colored due to hemorrhages.   Blood clots may be present.   Animals become depressed and dehydrated.

In peracute cases, death may occur in 24 hours.   In acute cases death or recovery may occur 2 to 4 days after the onset of symptoms.   Sometimes a chronic form may follow with diarrhoea and failure in growth.

Lesions :   There is enteritis of the colon and lower parts of the small intestines characterised by edema and swelling of the wall of the bowel due to hemorrhages, congestion and infiltration with inflammatory cells.   The goblet cells of the colon are hypertrophied and prominent.   Increased mucus secretion occurs   The mesenteric lymph nodes are swollen and edematous.   Later in the course a diphtheritic membrane may be seen adherent to the mucosa of the lower reaches of the bowel.   Liver is congested Hepatic cells show hydropic degeneration.   Cardiac lesions are similar to those found in fatal syncope.

Diagnosis : Smear made of feces reveals the organisms   Vibrio dysentery must be differentiated from salmonellosis.

NECROBACILLOSIS

Under this term are included various conditions supposed to be caused by a non-motile, non-sporulating, gram negative anaerobic organism which occurs in long filamentous form, called *Spherophorus (Fusiformis) necrophorus.*
The following conditions are produced :

In cattle and sheep — necrosis of the liver, foot rot.

In calves — calf diphtheria

In horses — quittor, poll evil, fistulous withers with metastatic lesions in
            internal organs

In foals — ulceration of intestines

In swine — necrosis of snout, tongue, foot rot and other parts of the body.

Calf diphtheria :

Usually calves under 6 months are affected. The organism which is ubiquitous gains entry through wounds in the mouth and during cutting of teeth. Foul and unhygienic surroundings are, therefore, a predisposing factor. Many, calves in the byre may be affected as an enzootic.

Pathogenesis and lesions : The organisms cause, at the point of entry, coagulation necrosis. Secondary bacterial infection by pyogenic bacteria aggravates the condition, Around the necrotic area is an intense hyperemic zone with neutrophilic infiltration. - Such necrotic areas may be found on the tongue, inner aspects of cheeks, gums, palate and pharynx. In the mouth and fauces, the necrotic lesions are covered by firmly adherent membrane which is dirty-white in color. When removed, a raw red ulcer is left. Much of the tongue may be necrosed. Infection may spread to the bone from the palate.

Infection may spread to the nasal cavities and to the larynx and trachea in which the characteristic dry necrotic areas are seen. Following aspiration of the material from the laryngeal lesions, infection may be set up in lungs and suppurative or gangrenous pneumonia results.

Necrotic lesions may be seen on the esophagus, rumen, omasum, the skin of the hoof and mucosa of the colon.

Generalisation by emboli of organisms results in dry necrotic foci being found in the liver, spleen, brain and heart muscle. Fibrinous or fibrinopurulent pleurisy may be caused.

Death occurs due to pneumonia, toxemia or asphyxia due to obstruction of respiratory passages.

Necrobacillosis affecting different organs :

Heart : Necrobacillary myocarditis is seen in calves and cattle in association with virulent necrobacillosis of liver, uterus and vagina. Dry, greyish necrotic foci are seen in the myocardium. These are demarcated from the healthy tissue by a red zone of inflammation.

Nasal cavity : Diphtheritic rhinitis characterised by dry yellowish exudate is caused by *S. necrophorus.*

Lungs : Metastatic necrobacillosis can occur from lesions of calf diphtheria, or necrobacillary omphalophlebitis. liver necrosis, necrotic endometritis and vaginitis. and foot rot.

In the lungs are found numerous yellow necrotic foci which on rupture into the pleural cavity cause fatal pleurisy. The foci are minute areas of necrotising pneumonia.

Necrotising bronchopneumonia can arise when the necrophorus bacillus is conveyed by the larvae of the lung worms.

Liver : Mostly cattle are affected The bacilli are transported from the intestines by the portal vein. Liver may also be infected during generalisation of the disease in calf diphtheria. At the point where the organisms settle circumscribed but progressing coagulation necrosis is produced by the toxins fo the organ. sm.

Macroscopically, the liver in enlarged and icteric In its parenchyma are studded dry, sharply circumscribed, yellowish grey, round or oval, areas of necrosis which are bordered by a zone of hyperemia The foci may undergo softenin and liquefaction It is not an abscess since pus cells are not present Organisms are present at the junction of necrotic and healthy areas.

Feet and tail : Wounds of the feet and tail may be infected with the necrotising organism producing dry gangrene. In the foot, the condition caused is "foot rot" and renders the animal lame. Surgical intervention is called for.

Occasionally mammary glands are affected by *S, necrophorus* as a metastatic process The lesions are dry and greyish-white. *Microscopically.* the characteristic necrotic appearance is seen without fibrosis No systemic symptoms are manifested.

LISTERIOSIS
Syn : Circling disease

This is an acute infectious disease of man and animals caused by *Listeria monocytogenes,* a gram positive, rod-shaped, non-sporulating, motile, betahemolytic organism, which is very resitant. It resists pasteurisation and can thrive in the soil for over one year

Animals affected : Sheep, cattle, goat, swine, rabbit, fowl and man may be affected. Rats, skunks guinea-pig, fox and dog may be infected. Rats are suspected to be reservoirs. Birds may be affected and may transmit the disease to other animals. Morbidity is about 10% of the herd and mortality in untreated animals is very high—nearly 100%

Routes of infection and disease manifested : Animals manifest three distinct syndromes— i) the meningo-encephalitis or nervous form; (ii) the infection of pregnant uterus and resulting abortion and (iii) septicemic or visceral form. These never overlap in the same animal. And when an outbreak occurs, all the animals affected manifest only one of these forms.

The nervous form is supposed to arise when infection is through nasal mucosa or through conjunctiva. Infection via trigeminal nerve from mouth is also suggested. Experimentally this form can be produced by intracarotid injection of a culture. But injection or infection by other routes fail to produce the nervous form. Ingestion may cause the visceral form of the disease and in pregnant animals abortion. Transmission through coitus may also result in abortion. Outbreaks of the disease have been encountered when silage was fed. It is thought that the bacteria stay viable in the silage made from contaminated forage. Organism is excreted through milk. So, the pathogenesis of the disease depends on the method of infection.

Pathogenesis ; The organism is voided in the feces, urine, milk, uterine discharges and aborted fetuses When ingested, the organism penetrates the intestinal mucosa, enters the blood and a state of bacteremia is produced and then the organism localizes in different organs A fatal septicemia may develop in some animals The pregnant uterus is highly susceptible to infection. So metritis occurs and abortion takes place late in pregnancy. In this visceral form brain is not affected.

Intranasal or intraconjunctival infection results in meningoencephalitis. The organisms are present only in the brain and nowhere else. The brain stem (pons medulla-oblongata and the spinal cord) is particularly affected where encephalitis occurs. From this region, infection extends to the meninges, ependyma and the eye. Infection via optic nerve produces endophthalmitis (especially in sheep and goats).

L. monocytogenes produces a potent hemolysin with lipse or phospholipase activity, leading to red cell destruction. This results in the release of ferritin which stimulates further production of hemolysins This interferes with RE system The hemolysin also acts on pace maker and contractile cardiac muscle fibres leading to electrical arrest of heart in septicemic cases.

Symptoms :

The nervous form : This is found in all animals but more commonly in ruminants. In sheep it is more acute, death occurring in 3 to 4 days while in cattle death occurs in 1 to 2 weeks

With rise in temperature, the animal is dull, lethargic, has a stiff gait, an arched back, rough hair - coat, constipation and weakness. Animals may go to feed trough normally but back out, fall down and go into convulsions Anorexia is common. The neck is pulled to one side and so it moves round and round in circles — hence the name ' circling disease '. There is dyspnoea and the animals may frequently show panophthalmitis. There may be one sided paralysis of the face with drooping of the ear, eyelids and lips. Finally animals become recumbent and die of respiratory failure.

In the visceral form : No lesions are found in the brain.

i) Abortion may occur sporadically when infection is hematogenous, Retention of placenta is frequent Organisms can be recovered from the stomach of the fetus. Unsolved abortions must be attributed to listeriosis.

ii) Septicemic form This does not occur in adult ruminants. Only foals, lambs, calves and piglets are affected, The brain is not affected and so no nervous symptoms are seen There will be dullness, emaciation, pyemia, diffuse gastroenteritis and hepatic necrosis

Lesions ; In the brain, gross lesions may not be present. A few hemorahages may be observed in the meninges and foci of softening are seen in the medulla

Microscopically, micro abscesses related to vessels are found in the brain stem. These start as collections of mononuclear cells (microglia) with a few neutrophils. The centres of the foci liquefy. Usually there is little necrosis of the nervous tissue. The white matter is edematous and shows foci of softening. Perivascular cuffing occurs with infiltrating lymphocytes, histiocytes, eosinophils and neutrophils. Infection may spread to meninges through the space of Virchow-Robin. The meninges are heavily infiltrated by lymphoid cells — lymphocytic leptomeningitis.

In the viscera where the organisms settle, small granulomas with necrotic or **purulent centres** form. Organisms can be demonstrated in these places. The cerebrospinal fluid which is turbid contains increased quantities of sugar and greater number of lymphocytes.

In fowls, sporadic cases occur. Few clinical symptoms are seen. Birds may die suddenly or there may be slow wasting. Nervous symptoms as seen in animals are not noticed. The following lesions are seen. There may be massive necrosis of cardiac muscle. Pericarditis with increased pericordial fluid may be noticed. Liver is enlarged and friable and shows necrotic foci. Fibrinous peritonitis and enteritis are observed. Spleen may be enlarged. Organisms can be cultured from abdominal organs and blood.

In man there may be acute purulent meningitis, encephalitis, septicemia and in the fetus may be found miliary glanulomas in all the organs and tissues. Fetus is infected in *utero*.

Diagnosis :

1. Symptoms — circling movements.
2. Histopathological examination of brain — observe microabscesses, perivascular cuffing.
3. Culture of organisms .
   (a) from blood in acute stage.
   (b) from brain stem—keep brain in ice box for 3 weeks and then culture on blood agar plates.
4. Biological : Drop infected material on the eye of a rabbit—conjunctivitis is produced. Intracerebral injection of suspected brain material into white mice kills them in 2 to 3 days if *L. monocytogenes* is present in the brain.
5. Surface fixation test is used for detection of antibodies.
6. Fluorescent antibody test can be done and with this a positive and rapid diagnosis can be made
7. Differential diagnosis : Differentiate from the following diseases :

   Cattle : (i) acetonemia — nervous symptoms (ii) brain abscesses:
   (iii) Brucella abortion (iv. Rabies and (v) lead poisoning.
   Sheep : Pregnancy toxemia : enterotoxemia.
   Pigs : (i) encephalitis of Hog cholera, (ii) Pasteurellosis, (iii) Aujeszky's. disease and (iv) middle ear infection.

LEPTOSPIROSIS
Syn: Weil's disease : Stuttgart disease :

This is a disease of dogs and large domesticated animals, caused by several species of Leptospira and is of zoonotic importance.

Incidence : Leptospirossis is now world - wide in distribution and has been seen in cattle, pig, sheep, dog and man. The rat is probably a reservoir of the organisms from which dogs and man may be infected, The condition caused by *L. icterohemorrhagiae* in man is know as *Weil's disease*.

Routes of infection :

1. Ingestion : Source of infection is usually an infected animal which infects the pastures, feed and drinking water by its urine. feces, fetuses that may be aborted and uterine discharges Licking of animals, among bovines, is a method of transmission During acute phase, the organism may be voided in the milk and so calves may get infected by drinking such milk.

2 Through coitus : The semen of an infected bull may contain the organisms which may be transmitted by coitus or by artificial insemination;

3. Through abrasions of skin or mucous membrane of nose and mouth

4. Thorugh intact skin, conjunctiva and nasal mucosa.

5 Infection in utero may occur

6 Man can be infected while swimming in infected water. Wild animals may act as carriers.

Outside the body the organisms may thrive in water for abont three months. Recovered animals may void the organism for as long as one year in the urine.

Leptospirosis of large animals : Though morta!ity is low, morbidity is high. Among calves mortality rate may be higher.

The economic loss among cattle. is mainly due to abortion that results. Among pigs abortions and deaths do occur while in sheep and goats loss is by way of deaths and loss of condition.

Organisms encountered :

Cattle : *L. icterohemorrhagiae* ; *L pomona* ; *L canicola* ; *L grippotyphosa*

Pigs : *L. pomona* ; *L canicolo* ; *L. icterohemorrhagiae.*

Sheep ; *L. pomona.*

Goats : *L grippotyphosa*

Horses : *L. pomona*; *L. canicola: L. iterohemorrhagiae*; *L. grippotyphosa.*

These organisms cannot be differentiated or identified by morphological examination. Only serological tests are useful.

The spirochaetes can be seen by dark field examination of fluid media. In sections they are stained black by silver stains of Levaditi and Warthin - Starry.

Pathogenesis : From the point of entry the organisms invade the blood stream and multiply rapidly producing septicemia. During this period the temperature rises This phase lasts for several days. If the animal does not die during the septicemic phase, the organisms settle down in the liver, kidney and the pregnant uterus, while this phase subsides. They localise nowhere else. The acute form is common in calves, piglets and lambs. In sheep and goats, the organism may localise in the nervous tissue producing encephalitis. During the acute phase jaundice is seen in all animals due to intravascular hemolysis and hepatic necrosis. Anemia,t icterus and hemoglobinuria are noticed

-In the animals that survive, the disease runs a subacute course in which the liver and kidney are affected. Leptospirae may be found in the urine. Albuminuria is present due to interstitial nephritis — focal or diffuse. Uremia may supervene and death follows in some cases

If the organisms localise in the gravid uterus abortion results due to degenerative lesions in the epithelium of the placenta.

Symptoms:

Cattle: Incubation Period: 2—10 days. Though mortality is low (5%),morbidity may be as high as 100%

Acute form; this is essentially a septicemia with initial high temperature, anorexia. petechiae on all visible mucous membranes, hemoglobinuria, jaundice and anemia Because of anemia, there is tachycardia and dyspnoea. If lactating, milk which is bloody, is reduced. Mammary gland is soft and milk is curdled. Abortion may occur during this period, or even later if the animal survives. Death may occur in 2 to 7 days.

Subacute form : The symptoms are similar to those in acute form, but milder. Icterus may not be noticed.   Fever, slight hemoglobinuria and dyspnoea may be present.   Milk secretion stops.   Abortions may occur.

Chronic form : No clinical symptoms are seen. Abortions occur, Chronic form can be detected only by serological tests.   Recovered animals require a prolonged period for convalescence. They void the organisms in urine for nearly two months after recovery.

Lesions :

Acute form : *L. pomona produces a hemolysin which causes* icterus and severe anemia   Severe hypoplasia of the bone marrow is observed, Petechiae may be observed on the serous membranes.

Liver : **Macroscopically**, the lesions may not be visible except some hemorrhages near the central veins

Microscopically, a centrilobular necrosis, due to hypoxia is evident.   The liver cells become dissociated and separated from the sinusoidal epithelium due to accumulation of edema fluid. The Kupffer's cells, which become hyperplastic contain hemosiderin crystals.   Lymphocytic infiltration occurs in the portal tracts.

Organisms are seen intracellularly or free in the sinusoids.

Kidneys :   *Macroscopically* the kidneys are swollen and red.

Microscopically, the tubular epithelium shows degenerative changes varying from hydropic degeneration to necrosis   The necrosed epithelium may become desquamated and thus forms casts – granular, cellular and hyaline.   Hemoglobin and bile pigment may be seen in the lumen.   In the interstitial tissue can be seen edematous fluid, lymphocytes and plasma cells Organisms can be seen within the epithelial cells or in the lumen of tubules.   There is an attempt at regeneration of the tubular epithelial cells, which in some places form giant cells similar to foreign body type.

Spleen in some cases may show hemosiderosis.   Acute hemorrhagic meningitis may be observed.   There is severe necrosis of the placenta due to hypoxia, which is a result of anemia.

Subacute form : In animals that survive, the organisms localise in the Kidneys where they grow in the tubules and are excreted in the urine.   Interstitial nephritis is produced with infiltration of lymphocytes and plasma cells. The inflammatory raction may subside and the focal lesion becomes scarred These are seen as grayish - white foci on the surface.   The aborted fetus does not show any specific lesions but for edema in the umbilical cord, pericardium, subcutaneous tissue and perirenal tissue   There may be interstitial focal nephritis   The fetus may be in a state of advanced putrefaction suggesting thereby that it was dead for sometime prior to abortion.   Placenta shows lesions of placentitis. Leptospira can be found sometimes in the fetus After abortion placenta is retained.

Sheep and goat : In sheep the disease may be peracute. Some may be found dead. Others may show jaundice besides dullness and may die in 12 hours. Abortion may occur in the acute form. In goats abortions may be noticed besides icterus.

Pigs ; *L. pomona* is the chief organism producing a comparatively mild chronic disease. Abortions may occur as well as birth of weak piglets, sometimes in outbreak proportions and so the rearing rate in an infected herd is low. Some animals may manifest nervous symptoms. Urine of sows and aborted fetus contain the organisms. Recovered animals are carriers for a long time and excrete the organisms in the urine. The urenal lesion is a focal interstitial nephritis.

Horses : Some workers believe that *periodic ophthalmia* of horses is caused bp *L. pomona*. This is characterised by conjunctivitis, keratitis, photophobia, intense lachrymation and iridocyclitis. (See page 529)

Leptospira may also produce mild subacute disease with fever, abortion, icterus and hemoglobinuria. The course of the disease is brief and the animals do not remain as carriers.

Dog : *L. canicola* is the common organism that infects most of the canine cases while *L. icterohemorrhagiae* may affect a few. A large proportion of the dogs have antibodies to leptospira in their serum.

L. canicola infection occurs from dog to dog while *L. icterohemorrhagiae* infection occurs from rats to dogs. Animals in the age group of 1 to 4 years are most affected. Recovered animals excrete the organisms in the urine for as long as three years.

Symptoms : From the symptoms manifested, it *is* not possible to determine if the causative agent is *L. icterohemorrhagiae* or *L canicola*. But, broadly it can be said that the former causes lesions of the liver and so jaundice, while the latter affects the kidneys and causes renal lesions and attendant symptoms.

Peracute form : This occurs in puppies affected by *L. icterohemorrhagiae*, death occurring of septicemia in a few hours to 2 or 3 days. The symptoms seen are fever and hemorrhages giving rise to epistaxis, hematemesis and melena. Mucous membranes reveal petechiae. Icterus may not develop in this stage.

Acute form : This is manifested by fever, anemia, icterus, diarrhoea vomition, dehydration, emaciation, acceleration of erythrocyte sedimentation rate, albuminuria, leucocytosis and foul odour of the animal.

Subacute or mild form : When the organisms localise in the kidneys, the symptoms manifested are those of progressive renal failure. Uremia may develop with death supervening.

Lesions :

Liver : *Macroscopically*, no lesions may be seen.

Microscopically, the normal columns of cells become disrupted as the hepatic cells become shrunken, rounded, dissociated from one another and discrete. The nuclei become pyknotic and hyperchromatic while the cytoplasm becomes granular and eosinophilic. Though the changes of hepatic cells are not specific for leptospira infection only, yet they are characteristic. Regeneration of hepatic parenchyma occurs and this is evidenced by increase in the size of cells (cytomegaly), binucleate cells, hyperchromatic nuclei and mitotic figures   Bile canaliculi are plugged with bile pigment and Kupffer's cells are loaded with hemosiderin. The spirochaetes can be seen either in the sinusoids or hepatic cells by silver impregnation method.

Kidney :

Acute form :   the glomeruli may not reveal any changes. Only the tubular epithelium reveals degenerative changes. These cells swell, becoming granular and eosinophilic. The cytoplasm may become vacuolated. Some may become necrotic and desquamated into the lumen. In some tubules, no epithelium may be present. The lumen may contain eosinophilic debris, some nuclei and a few erythrocytes. In some tubules regeneration of epithelium is evident by the presence of large cells with hyperchromatic nuclei, mitosis and large syncytial giant cells   The interstitial tissue is infiltrated by edematous fluid, lymphocytes and plasma cells. The organisms can be demonstrated in the tubular epithelial cells as well as in the lumens of the tubules

Spleen and lymph nodes are swollen, edematous and show hemorrhages. Mature lymphocytes are depleted while there is increase in the reticular cells of sinusoids

Gastro - intestinal tract : The gastric mucosa is swollen, dark - red, loose and lies in folds   The submucosa is very much thickened, hemorrhagic and edematous.

Microscopically, the glandular epithelium as well as the propria are necrosed, due probably to extensive thrombosis of vessels. Hemorrhages are found even in the muscular layer of the stomach wall.

The intestinal mucosa shows diffuse congestion and petechiae. Hemorrhages may be present on the serous surface

Hemorrhages are seen in other organs — lung, adrenal, pancreas, urinary bladder, myocardium and pleura.

Purulent necrotic laryngitis is almost a pathognomonic lesion of canine leptospirosis.

Subacute and inapparent form : The lesions are present only in the kidney.

Macroscopically, the kidneys are swollen. The capsule which peels off easily is tense. Petechiae are present on the cortex and are visible through the capsule.

Microscopically, the picture is one of chronic focal interstitial nephritis. The convoluted tubules become heavily surrounded by lymphocytes, plasma cells macrophages, a few neutrophils and erythrocytes.   Due to pressure, the tubular epithelium undergoes degeneration   The glomeruli may largely be unaffected. The spirochaetes may be seen in the tubular epithelial cells or in the lumens of tubules, singly in the former and as clusters in the latter

In animals in which uremia develops, extrarenal lesions are seen and these are : gastric hemorrhages and ulcers; calcium deposition on the gastric mucosa and in walls of the aorta, large arteries, laryngeal mucosa, endocardium of the left atrium and the costal pleura: rubber jaw syndrome.

L. canicola may cause nervous symptoms associated with hyperemia and petechiae of brain and spinal cord, purulent lympyocytic meningo - encephalitis and gliosis.

Diagnosis

1. Symptoms — icterus, anemia, hemoglobinuria, abortion, petechiae on mucous membranes
2. Demonstrate organism in sections by Levaditi's stain.
3. Agglutination test, microscopic agglutination test.
4. Agglutination lysis test
5. Inoculation into guinea-pigs intrperitoneally, of blood or milk or urine collected from ailing animals at the height of disease and recovering from or demonstration of the organisms in the blood of the guinea-pigs.
6. Dark field examination of urine.
7. Differential diagnosis : Must differentiate from other diseases

    Cattle : Babesiosis, Anaplasmosis, Brucellosis, post-parturient hemoglobi-
          nuria.

    Pigs Brucellosis; Eperythrozoonosis.

    Sheep : Chronic copper poisoning, Anaplasmosis

    Dog : Black tongue. Infectious viral hepatitis.

CASEOUS LYMPHADENITIS
Syn : Pseudotuberculosis of sheep

This is a chronic condition of sheep caused by *Corynebacterium ovis (C. pseudotuberculosis)* a gram positive, coccobacillary, pleomorphic, nonsporulating, nonmotile, noncapsule forming organism

This organism also casues ulcerative lymphangitis in horses and cattle.

Routes of Infection : Infection is usually by wounds. Wounds sustained while shearing, docking and castration are the usual portals of entry. Umbilicus in the new-born may be an avenue Source of infection is an infected animal, the organism not being able to thrive in the soil for long During shearing a node may rupture, infecting the shears, which may infect other animals if not sterilised before using on them.

In a few cases, ingestion may also be a route of infection. But then infection may be confined to the lymph nodes of the head.

Though young lambs may be affected incidence is greater in adults because of greater incidence of wounds in them during shearing. In some herds, morbidity may be as much as 70% but mortality is low. It is only those with extensive involvement or those with pulmonary lesions that may die.

Goats may, sometimes, be affected.

Pathogenesis : Gaining entry into subcutaneous tissue the organism is transported to the regional lymph node by macrophages.

By the action of the exotoxin necrosis of the parenchyma occurs. The macrophages may also be converted into epithelioid cells. (No giant cells are formed) The necrotic foci may coalesce, become liquefied and encapsulated and so an abscess may be formed The greenish purulent material becomes inspissated, the color changes to white and so looks like a caseated mass. Macrophages may transport the organisms beyond the capsule and so the lesion enlarges. Fresh foci of necrosis occur and around these another fibrous capsule is formed. In this way consentric layers of caseated tissue and capsule develop giving the lesion an 'onion skin' appearance.

The organisms may enter the lung, where micro abscesses or even broncho-pneumonia (especially in lambs) may occur.

If the organisms gain entry into the blood, metastatic abscesses may be set up in the kidney, liver and spleen.

Infection from supramammary lymph node may extend into the mammary gland with decrease or cessation of lactation and so lambs may be deprived of their nutrition and so die.

Symptoms In many instances affected sheep may not manifest any symptoms. Others show swelling and suppuration of the superficial lymph nodes, viz, subm-axillary, prescapular, precrural, prefemoral, and supramammary lymph glands. The abscesses may open discharging thick green pus.

When lungs are involved symptoms of chronic pneumonia may be seen.

When kidneys are affected symptoms of pyelonephritis may be manifested.

Lesions : Though an abscess may form subcutaneously at the point of entry of the organism, this usually resolves and heals. It is in the nearest lymph node that lesions are seen and which persist for a long time. The affected node becomes swollen and is completely converted into an abscess containing greenish pus in the early stages and later into a caseated mass showing the lami-nated structure described above.

Microscopically, the lesions consist of an abscess containing caseous or caseopurulent material surrounded by a thick fibrous tissue capsule. Sometimes, calcium salts may be deposited in the caseous material. In the periphery of the lesion may be seen inflammatory cells in which eosinophils may be found in large numbers. Epithelioid cells, neutrophils, lymphocytes and plasma cells are seen. Pus reveals many bacteria.

In the lungs may be seen tiny abscesses or large ones containing greenish pus in which numerous organisms can be seen. Bronchopneumonic lesions may be seen in lambs and debilitated animals. Overlying pleura may reveal pleuritis and adhesions. Bronchial lymph nodes reveal the characteristic caseous lesions.

Among goats the submaxillary lymph nodes are more often affected.

Diagnosis :

   1) Examine the pus for the organism.
   2) Isolate the organism and study the cultural characters.
   3) Perform Straus's test.
   4) Inject suspected material into guinea-pigs. Death occurs in 4 to 10 days showing caseous nodules in liver, spleen, lungs and associated lymph nodes.
   5) Agglutination test, using the serum, can be done.
   6) Differentiate from Tuberculosis

CONTAGIOUS ACNE OF HORSES

Pustules arise on the skin of horses, especially in such places where harness comes into contact. The cause is *C. ovis*.

Infection is from an ailing animal   Infection can be transmitted through harness or grooming kit.   Pressure of the harness blocking sebaceous glands resulting in folliculitis or seborrhoea may be a predisposing factor.

Symptoms : On the skin papules form which develop into pustules in 2 or 3 days   A crust may form when the pustules rupture and the greenish pus dries. Healing is spontaneoss and may take place within a week.   Rarely wounds may persist for 3 or 4 weeks.   The essential lesion is folliculitis with the formation of pustules.

Diagnosis   Pus reveals the *C, ovis.*

BOTRIOMYCOSIS

This is a granulomatous lesion caused by *Staphylococcus aureus* and is more commonly seen in horses, especially at the shoulder and sternal regions (following cutaneous lesions produced by harness) and tail after docking; in the spermatic cord following castration   In cows and pigs the mammary gland is usually affected with chronic mastitis

Macroscopically, the lesion is a granuloma made of dense granulation tissue in the centre of which are found abscesses which communicate with the exterior by means of sinuses.   Yellow pus is found to be discharged.   The pus contains botriomycotic granules, which are only conglomerations of the colonies of the organism

Microscopically, the lesion starts as a tiny abscess surrounded by granulation tissue.   Neutrophils are numerous in the centre, surrounding the organisms· A zone of macrophages collects around the granulocytes.   Spread is by means of lymphatics and fresh foci may occur in the neighbourhood.   These may coalesce and then bigger nodules develop by the formation of large amount of collagenous connective tissue and these ultimately form large abscesses.   In the mammary lesions of cows and pigs the picture is not unlike that of actinomycosis viz; a central bacterial mass surrounded by a zone of clubs.   In fact many of the cases diagnosed as actinomycosis of udder may very well be only botriomycosis.

Metastases from the primary lesion may occur in the regional   lymph node and also in internal organs.   Matastases have been   seen in bone marrow, brain, duramater, spinal cord and rarely in kidneys, liver, lungs, heart, peritoneum and uterus.

Diagnosis :   Stain pus and see gram positive cocci.

GLASSER'S DISEASE

Syn. Infectious   polyarthritis ; porkine polyserositis,

Glasser's Disease is an acute affection of young pigs, 5 to 12 weeks of age, caused by *Hemophilus suis.*   The characteristic manifestations are fibrinous meningitis, acute polarthritis and polyserositis

Etiology :   Though *Hemophilus suis* is identified as the causal organism, in some outbreaks a gram negative coccoid or coccobacillary organism, a pleuro pneumonia like organism (isolated from pigs and goats) and the organism causing enzootic pneumonia in pigs have been isolated individually or in association with *Hemophilus suis.*

Predisposing factor is exposure to stress during transportation.

Symptoms : This condition is peracute, manifested by initial high fever accompanied by complete anorexia, paresis stupor and hypersensitivity. All the joints are swollen and painful to the touch. Animals, therefore, stand on their toes and have short strides. The skin may be bluish in color. Most of the animals unless treated in time, die in 2 to 5 days Mortality rate is very high Those that survive, develop chronic arthritis and may suffer from intestina' obstruction as a result of peritoneal adhesions

Macroscopically, there is meningitis and the meninges, joints, the peritoneum the pleura and the p ricardium contain serofibrinous or in some cases, fibrino- purulent exudate.   The cerebro - spinal - fluid is increased in volume and is turbid due to the presence of leucocytes). The synovial fluid in the joints is increased in quantity and may be turbid. Sometimes there may be fibrin flakes in the fluid. The serous membranes contain increased quantity of sero–fibrinous fluid. The fibrin deposit gives a grey or yellowish color to the serosa. Conges- tion of the gastric mucosa may be present and in some cases the renal cortex may have a few petechiae.

Dignaosis : is based upon the symptoms, the macro and microscopical lesions, and isolation of the organism. This disease must be differentiated from streptococcal arthritis (in which the exudate is fibrinopurulent) erysipelas and mycoplasmal arth ritis (which is a very mild affection comparatively).

FOWL CHOLERA
Syn : Avian pasteurellosis

This is usually an acute septicemic disease of fowls. Other poultry including pigeons, turkeys and ducks are affected. The disease may also run a mild chronic course.

Etiology : Fowl cholera is caused by *Pasteurella multocida* (P. aviseptica).

Routes of infection : Infection is usually by ingestion of contaminated food and water Feces and saliva of affected birds contaminate the feed, water crates and poultry houses. Infection by contact and of wounds is also known to occur.

Incubation period : 4 to 9 days

Pathogenesis ; The organism, on being ingested, enters the circulation through the mucous membrane of the pharynx and upper respiratory passages. Rarely infection may pass through the intestinal wall. Entering the blood, the bacilli proliferate The pathogenic activity of the organism is attributed to some unknown toxin. This acting on the vascular endothelium is the cause of hemorr- hages in various organs and effusions into the serous cavities. Body temperature is elevated by the action of the toxin on the thermoregulatory centre.

In cases where the virulence of the organism is low or where the resistance of the bird is high the septicemic phase subsides and the organisms settle in various organs : liver, wattles, joints, lungs, ear or base of brain.

Symptoms : In the *peracute* form the bird may die within a few hours with- out showing any symptoms.

In the *acute* form in which the birds may live for a few days the following symptoms are observed; diarrhoea (often blood stained), cyanotic comb and wattles, pyrexia, depression, drowsiness and dyspnoea. Discharge form beak and nostrils may be noticed. Bird may be drowsy and becomes comatosed terminating in death

In the *chronic* form, the wattles become swollen and edematous. There may be discharge from the nose due to rhinitis or sinusitis. Joints may be enlarged. Middle ear infection results in torticollis.

Lesions : In the *peracute* form, apart from ecchymoses no other lesions may be found.

In the *acute* type, one always finds hemorrhages in the intestines and lungs together with petechiae on the serous membranes, which may sometimes be covered by fibrinous exudate. Hemorrhages are also frequent in the epicardium, endocardium and gizzard. The intestinal mucosa is bright red. Croupous pneumonia may be present.

In the less acute cases the liver is enlarged and shows small yellowish necrotic foci. In the *chronic* form, the hepatic foci are larger and are caseous. Similar lesions may be noticed in the lungs. The wattles and comb are enlarged and edematous. The connective tissue becomes necrotic later becoming caseous. The joints are swollen and contain caseous or serofibrinous exudate in the articular cavity. Similar caseous material may be found in the abdominal cavity.

Fowls become emaciated and debilitated.

Diagnosis : Fowl cholera must be differentiated from Ranikhet disease, fowl typhoid and fowl plague

Blood smears reveal bipolar organisms. Blood of a suspected bird may be inoculated into pigeons by scarification and the pigeon dies within 18 hours revealing hemorrhagic lesions. Bipolar organisms are seen and can be cultivated from its blood and tissues. Injection of 0.3ml. of ground up tissue from an infected bird into rabbits or mice introperitoneally or subcutaneously, will kill the anmals within 24 hours to 48 hours and the organisms can be recovered from such animals in pure culture.

BACILLARY WHITE DIARRHOEA OF CHICKS
Syn: Pullorum Disease

This disease though not widespread in India, is met with in some poultry farms. Baby chicks are mostly affected, mortality being high in them. Adults also may suffer in which it runs a milder course affecting the ovaries, resulting in decreased fertility and reduced hatchability of eggs laid by such fowls. These fowls act as carriers.

Cause : The causative organism is *Salmonella pullorum*, a non - motile member of the coli - typhoid group. The antigenic components of the organism are 9, 12_1, 12_2, 12_3

Routes of infection: The organisms are voided through the excreta. Infection may therefore occur by ingestion of contaminated food and water.

The most important route of infection is by way of eggs. The eggs laid by carriers are infected and the chicks that hatch out are therefore infected. These chicks may transmit the disease to other healthy chicks. The shells of such infected eggs are also sources of infection. Feeding infected eggs to chicks and fowls is yet another method of infection

Period of incubation : 4 to 10 days.

Pathogenesis : The organisms entering blood, multiply giving rise to bacteremia. In young chicks this may be fatal.

Or, the bacteremia may subside with the organisms becoming localised in various organs in which necrotic foci are set up From these areas. exacerbation may occur and the disease may flare up when the animals are exposed to debilitating influences such as changes in weather, prolonged transport, feeding excessive proteins and sudden changes in feed.

Symptoms : In acute cases chicks may be found dead without showing any symptoms. This is the fatal septicemic form. In the subacute cases chicks show malaise, drowsiness, anorexia and disinclination to move. Diarrhoea may be present, soiling the vent. The feces are white in color and hence the name "Bacillary white diarrhoea" to the disease. Mortality may be heavy, ranging from 30% to 90%. The duration of the disease may be 2 to 3 days

Adult birds may not manifest any symptoms but they may become carriers, the infective bacteria being present in the ovaries from which, eggs that may be laid get infected.

Lesions : Chicks dying of the acute type, may not show any lesions except congestion of the lungs. In the subacute variety may be noticed the following : yellowish necrotic nodules in the lungs and heart, enlargement of the liver with hemorrhagic streaks, cattarhal enteritis, presence of semisolid yellow mass in the ceca.

In the adult birds suffering from acute disease are found necrotic lesions in the heart, lung, liver, spleen and pancreas. Enteritis and pericarditis are also encountered. In the carrier hens the ova become discolored (many colored), angular, faceted and develop long stalks. In the cock necrotic foci and abscesses may be seen in the testes which are often atrophied

Diagnosis ;

In the chicks :

1. Symptoms of white diarrhoea.

2. Isolation of *S. pullorum* from viscera.

In the adult : Rapid whole blood agglutination test using a stained antigen.

Tube aggultination test using serum can also be done. Titer 1/40 and above indicates a carrier.

Blood is collected from the wing vein for these tests.

FOWL TYPHOID

This disease is caused by *S. gallinarum*, a non-motile organism. It has somatic antigens 1,9 and 12. Adults mostly suffer though chicks may also be affected.

Infection is by ingestion. Incubation peroid is 4 to 5 days.

Symptoms : In the adult the disease may be acute or chronic. In the acute form, the birds show symptoms of general malaise, viz. drowsiness, anorexia, fever, thirst and ruffled feathers. They may have greenish–yellow diarrhoea which pastes the fethers around the vent. Comb may be congested.

The duration of the disease may be 5 days. Mortality may be 4 to 30%

Egg transmission may occur. Recovered birds are carriers

In the chronic cases no symptoms may be seen. In the young chicks symptoms seen are similar to those of pullorum disease.

Lesions : The following are observed postmortem : cattarrhal enteritis, enlargement and congestion of the liver, which may assume a characteristic bronze discoloration. Other organs may be congested. Necrotic foci may be noticed in spleen and lungs. In chronic cases. spleen is markedly enlarged and there is severe hyperplasia of the reticulo-endothelial system.

Diagnosis : 1. Agglutination test. 2. By isolation and cultural examination of the organism.

It may be difficult to differentiate this condition from B.W.D. (Bacillary white diarrhoea). The following differences are seen.

Bacillary white diarrhoea	Fowl typhoid
1. Chicks mostly affected.	1 Adults mostly affected.
2. Diarrhoea, whitish, is a constant symptom.	2. Diarrhoea is not constant
3. Liver is of normal color.	3. Liver is bronze colored.
4. Ova abnormal.	4. Ova not usually affected.
5. Organism dees not ferment maltose or dulcite.	5. Organism ferments maltose and dulcite.
6 Organism turns litmus milk acid.	6. Organism turns litmus milk alkaline.

Differentiate from fowl cholera, avian monocytosis, ALC, Raniket Disease and fowl plague,

Isolate the organism and conduct a biological test in rabbits, which are susceptible to *Pasteurella multocida* but resistant to *Salmonella gullinarum*.

FOWL SPIROCHAETOSIS

This is an acute disease of fowls caused by *Borrelia anserina*, and now known as *spirochaeta gallinarum*, in India is important next to Raniket disease. Other species of birds affected are guineafowl, geese, turkey, sparrows and ducks.

Etiology : *S. gallinarum* is a spirochaete which is 8 to 20 μ in length, *has six spirals* and *is motile* (unlike Leptopira and Treponema)

Mode of infection : Infection is usually transmitted by the bites of the fowl tick, **Argas persicus** Infection may pass through the eggs of ticks to larvae and nymphs, which are also infective. Infection also occurs through ingestion of the infected ticks and their eggs or even contaminated food.

Infection may be spread by other biting flies and mosquitoes. An infected tick can be infective upto three years if kept at appropriate environmental temperature.

Incubation period 3 to 8 days

Symptoms : Intense thirst, high temperature, cyanosis of the head, diarrhoea and depression are the symptoms seen. The birds lose appetite, and their feathers are ruffled. They have drooping head and closed eyes. Blood examination may reveal the organism Anemia develops. Paralysis is seen before death. The duration of the acute illness is about 5 days.

Sometimes, the disease may take a chronic form when the symptoms seen are emaciation, anemia and depression. The course of this disease is about 21 days.

Lesions : The carcass is very weak and emaciated.

The spleen in enormously enlarged, as much as six times the normal size and shows hemorrhages. The liver also is enlarged, friable and contains white necrotic foci, Catarrhal enteritis and fibrinous pericarditis are also observed.

Microscopically, in the kidney there is coagulation necrosis of the tubular epithelium and interstitial infiltration by lymphoid cells. There may be hemorrhrges and infarcts, Lymphocytic infiltration of the cardiac muscle may also be seen. In some cases pneumonic areas with congestion of the lung may be met with.

Diagnosis :

1. Demonstrate organism in blood—Leishman or Giemsa stain.
2. Enlargement of spleen characteristic
3. Histological examination of spleen and liver stained with Levaditi's stain reveals the spirochaetes.
4. Agglutination and complement-fixation tests can be performed.

CHRONIC RESPIRATORY DISEASE (CRD)

This is chronic infectious disease not only involving respiratory system as obivious from the title but also reproductive system in the layers. Broilers and layer chicks in the age group of 4-8 weeks are affected most. Infected bird have poor feed intake, weight gains and show retarded growth. Layer birds show significant decrease in egg production. Turkeys, pigeons and pheasants are also affected.

Etiology: The main cause is cocco-bacillary organisms of PPLO group, known as *Mycoplasma gallisepticum.* Very often concurrent infection with Escherichia Coli, Infectious bronchitis, or New castle disease viruses aggrevate the disease. Stress, over rowding, exposure to inclement weather, imbalanced rations, post-vaccinal stress predispose and precipitate this disease.

Route of Infection : Infection occurs both vertically and horizontally. Adult birds acts as carrier and infection is transmitted through eggs. Infection occurs through contact or aerogenous routes dust and droplet infection. Incubation period is between 11-13 days.

Symptoms : Most important symptoms are respiratory rales, nasal discharges, and poor weight gains. The disease spreads slowly and duration is as long as 3 weeks to 2 months. Mortality in pure infections may be neglible. However, the condition in the field are seldom free from secondary infections, when the mortality due to complications can be as high as 25-30%. Fertile eggs from affected flocks with subclinical infections may show poor hatachibility rate.

Lesions : The organism entering through the respiratory system settle on the mucous membrane of nasal sinuses, trachea, lungs and air sacs and cause severe catarrhal inflammation. Thick creamy white membranous layer or cheesy white exudate may be found in trachea, bronchi and sinuses. Lungs may show focci of pneumonia. Air sacs will be thickened with cheesy material. In uncomplicated cases, air sacs may show beaded appearance. Fibrinous perihepatitis, pericarditis and pleuritis may also be observed. In older cases these membranes may be thickened due to chronic changes. In layers birds, Fallopian tubes may show varying degrees of inflammatory changes as evinced by oedematic thickening of walls, hyperemia and catarrh. In some, caseous exudate may be observed. Occasionally skeletal muscles, particularly breast muscles may show greyish white necrotic streaks.

Histopathology : In trachea, deceliation of mucosal epithelium, thickening of submucosa due to oedema and infiltration with mononuclear cells, hyperplasia of mucous glands is observed. Focal lymphoid hyperplasia is observed in the sub-mucosa of bronchi and air sacs which is considered to be pathognomic. Lungs reveal pneumonic changes with foci of granulomatous reaction containing giant cells.

Diagnosis (1) Isolate and identify *Mycoplasma galiseptiam* from the exudate in *trachea*, *air sacs* or *lungs* using biochemical and serological methods.

(2) Tube agglutination test / Slide agglutination test.

(3) *Heamagglutination test* : Pathogenic Mycoplasma agglutinates Fowl RBC's, Non-pathogenic donol and Heamagglutination inhibition test.

(4) Whole blood plate agglutination test.

(5) Staining the organisms in the section by F.A. technique.

(6) Immuno-peroxide test using impression smears and sections.

INFECTIOUS CORYZA

This is a disease of fowls caused by *Hemophilus gallinarum*, a gram-negative, pleomorphic, non-motile organism, exhibiting bipolar staining characteristics.

Though birds of all ages are susceptible, older birds are more commonly affected.

Route of infection : Infection is usualy aerogenous—droplet infection and dust. It may also occur through ingestion of contaminated water and feed. Carrier birds are usually the sources of infection. It may be associated with avitaminosis A, CRD and fowl pox.

Incubation period is 1 to 3 days.

Morbidity may be 100% and mortality may also be high.

Symptoms : The face is swollen and edematous. A foul smelling nasal discharge is present. Lachrymation, sneezing, coughing and dyspnoea are other symptoms seen. Feed consumption is decreased and egg production falls.

The duration of the disease in acute cases is one week while in chronic cases it may be one to two months. Secondary bacterial infection, undernourishment, parasitism and poor housing enhance the severity of infection.

Lesions : Acute catarrhal rhinitis and sinusitis are found. Airsacculitis is also observed. Subcutaneous edema of face and wattles together with conjunctivitis is often seen. In chronic cases there is cheesy exudate in the sinuses, nasal passages and conjunctival sacs.

Diagnosis :-

1. Demonstrate the bipolar, gram-negative organism in the nasal discharges.
2. Isolate *H. gallinarum* from suspected material.
3. Inoculate susceptible chicks with the suspected material. In the case of infectious coryza, symptoms will be seen in a few days.
4. Differentiate from CRD and vitamin A deficiency.

The following conditions were contributed by Dr. S. J. SESHADRI
SALMONELLOSIS (Paratyphoid)

The term Salmonellosis indicates infection of poultry due to various members of the genus Salmonella, other than infections caused by *S. pullorum* and *S. gallinarum*.

The disease is usually seen in chicks below one month old and usually runs an acute course. Adults are more resistant and remain carriers quite often, harbouring the organism in their reproductive organs.

Among various species of Salmonella the most important ones often encountered in this disease are *S. typhimurium*, *S. thompson* and *S. enteritidis*.

The organisms are shed in the feces and the eggs get contaminated with these. They are capable of penetration of the egg shell. Incubator temperature and high humidity are considered to be responsible in enhancing their penetrating properties. Man and most of the animals are susceptible to infection and may therefore act as carriers.

Symptoms : The symptoms are vague and in many ways are similar to the pullorum disease. No symptoms are seen in chicks over 3 weeks of age.

Lesions : In peracute cases lesions may be entirely absent. But in less acute cases, lesions similar to those found in pullorum disease are seen.

S. typhimurium causes widespread infections in most countries and produces focal necrotic lesions in liver, lungs and spleen. Liver is enlarged and may show hemorrhages.

Diagnosis : A definite diagnosis can be made only by bacteriolgical examination and in identification of the salmonella which are motile. Cultures are made from materials obtained from liver, bone marrow or yolk sac.

COLI BACILLOSIS

Coli bacillosis is caused by *Escherichia Coli* and manifests in chicken in several forms. It includes coli bacillosis, coli septocemia, omphalitis; air sacs disease, Hjarres disease, peritonitis, salpingitis synovitis, early chick and embryo mortality.

Incidence and Distribution : E. Coli is a common inhabitant of Intestinal tract and is present in water as feacal contamination. The most common sero types responsible for poultry diseases are 01, 02, 035 and 078. The organism contain O (somatic), K (capsular) and H (flagellar) antigens. The somatic antigen is endotoxin liberated on autolysis of smooth cell and consists of polysaccharides-phospho lipid complex with protein fraction resistant to boiling. It agglutinates with corresponding sera. The capsular antigens located on surface of the cell and are associated with virulence and interfere with 'O' agglutination. These antigens are polymeric acids containing 2% of reducing sugars. They can be destroyed by heating at 100°C for one hour.

Embryo and early chick Mortality : Infection by E.Coli of yolk sac by the way of feacal contamination of egg is an important manner. Ovarian infection or salpingitis is another source of infection. Chicks may also pick up infection after hatching in first few days. Low brooding temperature or fasting increases incidence of infection and mortality. Embryo death is characterised by watery, yellowish, brown or caseous yolk material. Such of the chicks which hatch from infected eggs may die shortly afterwards or continue to die upto 3 weeks; and show variety of lesions. (a) Musky chick is characterised by oedema and infection of yolk. (b) Inflammation of Navel (c) Chicks living more than four days have pericarditis due to systemic spread of infection from the infected yolk. (d) Chicks hatching with low grade yolk infection may show yolk retention and reduced weight gain.

Respiratory Infection : It manifests as a secondary infection following damage to the Respiratory tract by infectious bronchitis virus. New castle disease virus including vaccine strains and Mycoplasma. Lesions consists of Air sacculitis, pericarditis and perihepatitis.

Pericarditis : Septicemic infection results in acute pericarditis characterised by oedema and fibrinous exudate. Infection may also result in acute Myocarditis.

Chronic Salpingitis : Infection of fallopian tube occurs when greater abdominal air sacs are infected. Less frequently ascending infection via cloaca in laying hens may occur. High estrogenic activity appears to predispose fallopian tubes for coliform infection. Gross lesions consists of large caseous mass in thin walled dilated oviducts. Microscopically mild tissue reaction consisting of Heterophil infiltration under epithelium is observed.

Peritonitis : It is characterised by acute mortality. Infection is ascending type, organisms growing rapidly in the yolk material. Lesions consists of free yolk and fibrin in the peritoneal cavity.

Acute Septicemia : Lesions are similar to fowl typhoid and fowl cholera. They consists of congested pectoral muscles and greenish liver with foci of necrosis.

Hjarre's disease (coli granuloma) : Chicks and Turke's are characterised by granulomatons lesions in liver, ceaca, duodenum and mesentery. No lesons are found in the spleen.

MUSHY CHICK DISEASE
(Septic omphalitis)

This condition is usually due to bad management. *E. coli* may be associated with the disease, causing retention of caseous yolk sacs and reduced weight gains. Most yolk sac infections result in death from septicemia.

Symptoms : The disease is seen in chicks less than 10 days old. The affected chicks are weak and retarded in growth. Mortality may sometimes be upto 50 percent though usually it is 5 to 10 percent. Death is attributed to toxemia.

Lesions : At post - mortem yolk sacs appear distendend often containing four smelling, yellow and curdled or brown and watery yolk. There may be pericarditis and perihepatitis in adition. Adhesions between the skin, abdominal wall and underlying yolk sac may be seen in omphalitis.

NECROTIC ENTERITIS

This condition is caused by A and B toxin, produced by *Clostridium perfringens type 'A'* and *Clostridium perfringens type 'C'* respectively. More common in broilers of 2 to 4 weeks storadic outbreaks may occur with less than 6% mortality.

Predisposing factors : High levels of fish meal, wheat in diet predispose or exacerbate the outbreak of this disease. Damage to the intestinal mucosa by other factors like high fiber in litter, coccidial infection, high population clostridium perfringens organism in the litter or water also dispose chicks to outbreak of this disease.

Symptoms : The natural outbreaks are characterised by depression, inappetence, diarrhea, and ruffled feathers. Clinically illness is acute and of very short duration of 7 days.

Gross lesions : Lesions are confined to small intestines, usually Jejunum and ileum. Intestines are friable, and distended with gas. The mucosa is lined by a loose or tightly adherent yellow or greenish pseudo membrane (Fibrinous enteritis). Fleeks of heamorrhage may be observed but these are not extensive and frank.

GANGRENOUS DERMATITIS

(Syn. Necrotic Dermatitis, Gangrenous Cellulitis, Gangrenous Dermatomyositis, Wing Rot, Avian Malignant Oedema)

Etiology : It is caused singly or as combined infection of *Clostridium septicum*, *Clostridium perfringens type A*, and *Staphylococcus aureus*. In many instances the disease is caused a sequelea to cross infections like infectious bursal disease, inclusion body hepatitis, chicken infectious anaemia avian adeno virus and Reovirus, which seriously undermines the efficacy of immune system.

Broiler chicken from 2-20 weeks are affected. However most susceptible age group is between 4-8 weeks. Clostridial organisms are picked up from soil, feaces, dust, contaminated litter and feed. Staphylococcal organisms are ubiquitous on the skin and gain entry under favourable conditions or injury.

Symptoms : Cases of natural outbreak exhibit varying degrees of depression, in coordination, leg weakness, ataxia, and inappetance. The duration of illness is very short and acute. Mortality may be as high as 60%.

Gross Lesions : The lesions consist of dark moist skin in areas devoid of feathers overlying the wings, breast, abdomen and legs. Skin is significantly intact. Extensive blood tinged exudate with or without gas is observed. Subcutaneous tissue and underlying musculature is discoloured, grey or tan with oedema and gas. Lesions in the internal organs consist of focal necrotic areas in liver. Bursa fabricius is flaccid, and atrophied with extensive follicular necrosis.

Histopathology : The changes consist of oedema, emphysema in subcutaneous tissue with Gram positive cocci or bacilli. Skeletal muscles shows necrosis.

Diagnosis : It is based upon the lesions and finding of concurrent viral infectious disease mentioned earlier.

Chapter 27

DISEASES CAUSED BY VIRUSES

Foot and Mouth disease	Infectious canine hepatitis
Vesicular stomatitis	Blue tongue
Vesicular exanthema	Scrapie
Pox	Equine encephalomyelitis
Pseudo-cowpox	African horse sickness
Bovine papular stomatitis	Equine infectious anemia
Lumphy skin disease	Infectious bovine rhinotracheitis
Contagious ecthyma	Ephemeral fever
Hog cholera	Trasmissible gastroenteritis in pigs
African swine fever	Ranikhet disease
Rinderpest	Fowl plague
Bovine viral diarrhoea mucosal disease	Infectious laryngotracheitis
Canine distemper	Infectious bronchitis
Rabies	Avian encephalomyelitis
Aujeszky's disease	Avian monocyosis
	Infectious bursal disease

The viruses by definition are submicroscopic organisms containing nucleic acid—either RNA or DNA—enclosed in a protein coat; the viruses contrary to the earlier belief, do contain enzymes but lack Lipmann enzyme system for energy production; thus they are obligate intracellular parasites.

The replication of viruses involves the following sequential steps :

(a) Adsorption to the surface receptors of the cells.
(b) Penetration into the cells.
(c) Uncoating inside the cell with the release of nucleic acid.
(d) Biosynthesis of early "messengers" and proteins.
(e) Biosynthesis of late "messengers" and proteins.
(f) Assembly of viruses.
(g) Release of the virus

The above events inside the host cells leads to cell destruction and hence the pathologic consquences However, all the viruses do not lead the same type of pathogenesis. The sequelae to virus infection can be classified as follows :

(a) Cytocidal—cell destruction Eg. F & M; Polio
(b) Cytocidal and cell proliferation—Pox viruses.
(c) Latent infection—Hepes simplex.
(d) Persistant infection—slow viruses—Equine infectious anemia.
(e) Cell proliferation—Tumor viruses.

Thus it can be seen that there is complete cell destruction on one hand and cell stimulation on the other.

Recently the viruses have been classified into families and genera, though not species, based on their morphology, chemistry and mode of replication

DNA Viruses   Families.

Pox viridae; Herepto viridae; Papova viridae; Adeno viridae; Parvo viridae

620

RNA Viruses.   Families.

Myxo viridae; Paramyxo viridae; Rhabdo viridae: Reo viridae; Retro viridae; Toga viridae; Bunya viridae.

FOOT AND MOUTH DISEASE
Syn : Aphthous fever

Foot and mouth disease is a very contagious disease, caused by an epitheliotropic enterovirus of the 'Picorna' virus group, occurring in all cloven footed animals. Cattle and pigs are mostly affected while sheep and goats are affected less so. Wild ruminants (reindeer, antelope, deer etc.) may be affected. So the camels, laboratory animals and man are susceptible· Horses cannot be infected.

Sero types : Seven immunologically distinct sero types of the virus are known—O, A, C, SAT-1, SAT-2, SAT-3, and ASIA-1. There are more than 61 subtypes or variants of the main types.

The virus has a single stranded RNA core with a protein capsid It does not have an envelope and is 23 mμ.

Routes of infection : Ingestion of contaminated materials. Virus is voided in the saliva, semen, urine, feces, milk and discharge from the wounds of teat and can be transmitted mechanically by man and other animals. This is one of the most highly contagious diseases known. The virus is present in the saliva even before vesicles form and in the secretions and excretions before the animal is clinically ill. Wild animals probably act as reservoirs of the virus.

Aerogenous or droplet infection is possible when the mucous membrane of the upper respiratory passages is affected.

It is estimated that India suffers a monetary loss of over Rs. 4 crores due to FMD for the following reasons : (1) Reduced working capacity, (2 Reduced cattle trade due to interference with movement of animals, (3) decreased breeding capacity, (4) loss to hides and skin industry, (5) loss due to deaths, (6) diminished milk yield, (7) reduced meat production.

Incubation period : A few hours to a few days.

Pathogenesis : At the place of entry (gastric or intestinal mucosa) the virus invades the epithelial cells and multiplies, producing focal areas of degeneration and inflammation. The cells undergo vacuolation. Pyknosis of nuceli and loosening of cell connections together with leucocytic infiltration are noticed in these areas. From here, the virus invades the lymph and blood and reaches the epithelium of other mucous membranes and skin, producing vesicles. If infection is via skin wounds, virus multiplies in the skin epithelium and then invades the blood, coming to the skin again

Histogenesis of vesicles : The most favourable cells for the reproduction of the virus appear to be those in the middle layer of *Stratum spinosum*. The cells become swollen, rounded and the cytoplasm becomes acidophilic. The nuceli are pyknotic and intercellular prickles are retracted The cells become loosened· Inflammatory exudate derived from the hyperemic vessels of the papillae of the corium collects between the loosened cells which, therefore, float on the micro_ pools. Theses cells may undergo liquefactive necrosis. The cells over these foci undergo hydropic degeneration forming thus microvesicles. Necrotic cells attract neutrophils to the area, and these cause liquefaction. Coalescence of the neighbouring foci results in vesicles, some of which may be as large as 6cm across.

The roof of the vesicles is made of the compressed stratum corneum, stratum lucidum and stratum granulosum. The base is composed of the basal cells of stratum germinativum. In some places these cells may also be denuded. The dermis underneath is highly hyperemic. The vesicular fluid contains degenerated epithelial cells, leucocytes and rarely erythrocytes. Healing occurs by the growth of epithelium from the stratum germinativum of the base as well as from cells at the periphery.

This lesion in the epithelium occurs where squamous epithelium is present, viz. tongue, buccal mucosa, rumen, reticulum, omasum. skin of the udder teat, cleft of the foot, coronary band and conjunctiva.

Symptoms : The disease starts with high fever (104°F to 106°F) accompanied by anorexia, depression and fall in milk yield. On the formation of vesicles of the mouth temperature falls. These vesicles rupture due to movement of jaws and tongue leaving angry ulcers with ragged and irregular edges. There is plentiful salivation and so the animal makes smacking noises. The saliva dribbles in strings

Along with lesions in the mouth, vesicles also appear at the cleft of the foot and these on rupture leave red sores, which being painful render the animal lame. Animals do not like to go about and so tend to have a recumbent position. Infection by secondary bacterial invaders or infection by maggots makes the wound deep and complicated, requiring a long time to heal.

Appearance of vesicles on the teats, especially at their orifices is attendant with risk of infection entering into the mammary gland and causing severe mastitis. Abortion of pregnant animals may occur and infertility may result.

Animals lose condition rapidly and become emaciated. Period of convalescence is long.

Because of affection of endocrine glands, recovered animals develop a dry and rough coat with long hairs Such animals become 'panters' and cannot be put to work in the sun. Such complications are met with more commonly in exotic and crossbred animals.

Lesions : In the mouth, the vesicles are found on the dorsum of the tongue, especially near its anterior part and lateral aspects; on the mucosa of lips, cheeks, gums, dental pad and hard palate Sometimes vesicles may be found on the muzzle and external nares. The roof of vesicles in mouth becomes softened by saliva and may rupture due to movement of the jaws and tongue. The base is red and may contain remnants of epithelium Because of erosions, profuse stringy saliva dribbles from the mouth. Sometimes the base of the erosion may contain a mucopurulent exudate. Healing occurs in two weeks. Secondary infection may complicate the course of the disease

In suckling calves and lambs death may occur even before vesicles form due to acute gastroenteritis and myocarditis.

Apart from vesicles in the rumen. reticulum and omasum, there may be catarrhal gastroenteritis In animals that die there may be petechiae on the abomasum and intestines. The abomasal hemorrhages turn into ulcers and the contents of gastro-intestinal tract is blood-stained. In the respiratory system may be seen catarrhal inflammation, edema of luugs and sub-pleural hemorrhages.

In animals, in which the disease assumes a severe and fatal form, the heart is affected which is dilated and flabby. The ventricular musculature, the septal wall of the left ventricle and the papillary muscle are affected in which are found grayish streaks giving the heart a 'tigroid' appearance. There is hyaline degeneration and necrosis of the muscle accompanied by intense infiltration by lymphocytes and a few neutrophils Should the animals survive, the necrotic muscle fibres are removed by dissolution and the lesion heals by scar tissue. Calcium salts may be deposited in such areas. In some skeletal muscles focal, demarcated lesions like those in the heart are noticed. In these areas the muscle fibres are necrosed and infiltrated with leucocytes. Spleen may be enlarged and soft. The cerebral ventricles may contain increased amount of fluid, which may be turbid, sometimes. Because of cardiac lesions, circulatary disturbances develop with subcutaneous edema over the region of thorax and abdomen.

Because of pain, the animal suffering from foot lesions is lame Secondary bacterial infection of these lesions may cause osteomyelitis and arthritis, often suppurative in character. In these conditions metastatic suppurative lesions may be seen in different parts of the body.

Secondary effects :

i) Mastitis due to secondary infection of vesicles on the udder and teat. ii) Cardiac cicatrices, if heart is affected iii) Anemia. iv) Disturbances in sexual function. v) Diabetes. vi) Over growth of hair vii) Panting (lack of heat tolerance) viii) Dyspnoea ix) Delayed growth of young animals. (vi to ix due probably to endocrine damage.)

Foot and Mouth disease in sheep, goats and pigs resembles the disease of cattle, but is mild and the lesions are smaller. In sheep incubation period is 3 to 8 days and oral lesions appear on the dental pad. In swine incubation period is a week and lesions are found on the snout also. The feet are more often affected, and very severely, leading to acute lameness. Suckling pigs and lambs die in large numbers due to gastroenteritis and myocardial lesions. However, the disease may be very mild in sheep, goats and especially in pigs. The clinical symptoms in these animals may be very transitory and are likely to escape notice in such cases.

Carriers : Recovered sheep harbour the virus upto 5 months while recovered cattle shed the virus upto 6 months. The virus is found in the pharynx.

Diagnosis :

1. Clinical symptoms and lesions of the disease may be primary guiding factors

2. Compliment fixation test (C.F.T.) can be performed with vesicular fluid against known diagnostic antisera This does not only confirm the disease but also gives the serotype of the virus involved in the outbreak If "variant strains' are involved C.F.T. will indicate such probability.

3. The virus can be isolated in tissue culture and suckling mice. Vesicular fluid can be injected intradermally into the plantar foot pads of guinea pigs in which vesicles develop in 1 to 7 days depending on various factors.

4. The disease should be differentiated from vesicular stomatitis, vesicular exanthema, Rinderpest and Mucosal Disease.

VESICULAR STOMATITIS

Syn. Sore mouth of cattle and horses; Male de Yerbe (Mexico)

This is an infectious disease occurring in horses, cattle and pigs. Sheep can be infected Man is susceptible in whom an influenza-like disease is caused. The virus belongs to Rhabdo virus and is similar to Rabies virus. It has a bullet-shape with single stranded RNA and envelope and is $65 \times 185m\mu$.

Vesicular stomatitis has a seasonal incidence and so it is conjectured that it is transmitted through some vectors—flies, mosquitoes etc. which transmit the disease Multiplication of the virus in the mosquito, *Aedes aegypti*, has been established. Three antigenically different strains are known: the more virulent New Jersey strain and the milder Indiana and Cocal (Trinidad) strains.

Incubation period is 24 to 48 hours.

Pathogenesis of the disease is similar to the Foot and Mouth disease.

Lesions : Lesions are found in the oral mucosa. Only in swine are foot lesions encountered. In cattle, vesicles may sometimes be seen on the teats. Myocardial lesions are never seen. The disease is not fatal.

Diagnosis :
Vesicular stomatitis cannot be produced in cattle by intramuscular inoculation of the virus. It can be produced only by the lingual route. Foot and Mouth disease on the other hand can be produced by intramuscular inoculation.

VESICULAR EXANTHEMA

This is an acute febrile, but non-fatal viral disease of pigs. Horses may be infected. Laboratory animals are not susceptible.

The virus is typed as Calici virus and is $35-40m\mu$

Infection is by ingestion of infected meat

Incubation period is 16 to 28 hours.

Lesions : The pathogenesis and morphology of the lesions are similar to those of Foot and Mouth disease.

The vesicles are found on the snout, nose, tongue, lips, gums, interdigital spaces, dew claws, coronary band and teats and udder of nursing pigs. Though mortality is less than 5%, affected pigs lose considerable body weight. Pregnant sows may abort and lactating sows may become dry.

Diagnosis : Compliment fixation test, viral neutralisation in cell culture and gel diffusion precipitin test.

Differential diagnosis of Foot and Mouth disease, Vesicular stomatitis and Ves. exanthema.

Inoculation and susceptibility.

	Cattle	Horse	Pig	Sheep	G. Pig	Man
F.M.D.	+	—	+	+	+	+
V.S.	+	+	+	+	+	+
	Cattle can be infected only if injected lingually					
V.E.	—	can be infected.	+	—	—	—

POX
Syn : Variola

Pox is an acute disease of cattle, horse, sheep, goats, bufalloes, swine, fowls and man, caused by *an epitheliotropic* virus. Though the condition is relatively mild in cattle, horses, swine and fowls, it is malignant in sheep and man.

The virus of human, bovine and equine pox are related and might have originated from a single variety. Pox viruses are double stranded DNA viruses having an envelope and complex symmetry and measure 220 to 260 mu × 150 to 180mμ.

Cow pox: (Vaccinia)

This is a rare disease affecting mostly cows The lesions are found on the skin of teats and udder. Infection is usually carried from animal to animal through contaminated milkers' hands and teat cups. Flies may also transmit the disease. Cattle may be infected by recently vaccinated milkers.

Incubation period is 2 to 3 days.

Pathogenesis : The virus appears to have special affinity for cells of the prickle cell layer of the skin First there is congestion in spots—the roseola stage. Due to the action of the virus, there is initial proliferation of the cells, giving rise to small papules—the papular stage. The virus subsequently produces balooning and reticulating degeneration of the epithelium of this layer. Simultaneously, there is inflammatory hyperemia followed by serous exudation of the corium. The serous exudate infiltrates into the epidermis, detaches the cells and so vesicles are formed in the degenerated foci of prickle cell layer—the vesicular stage. Because of pressure, some of the non-degenerated cells are compressed and drawn out as fiine threads forming a loose mesh work in the vesicle. These strands connect the floor and the roof of the vesicle, forming a pillar, as it were. So when the vesicle enlarges and bulges due to proliferation of the epithelial cells, this strand causes a dimple in the centre (at the place of its attachment to the roof of the vesicle) and so the vesicle is said to be umbilicated. Subsequently leucocytes enter into the vesicle and so the pustular stage results. Due to the action and digestion of the epithelial strands by the enzymes, the pustule is no longer umbilicated but is flat or rounded The pustule may become dry forming a crust, which may be cas. off—the desquamative stage. Fresh epithelium covers the lesion from the edges.   Cytoplasmic inclusions are found in the epithelial cells.   The whole course comprising of the above 5 stages takes place in 2 to 3 weeks.

Being mild cow pox is uneventful.   But sometimes, the vesicles may rupture due to calves suckling or while milking and then become infected, mastitis resulting.   One attack confers solid immunity.

Buffalo Pox :   The condition is even milder than cow pox and the disease is caused by both vaccinia virus and the buffalo pox virus itself.

Horse Pox :   This is also, a mild condition caused by the virus of cow pox, lesions occurring on the pasterns (the leg form), on the buccal mucosa (the mouth form) and nostrils   A slight elevation of temperature may be noticed in the beginning.   Healing occurs in 3 to 4 weeks,

Sheep Pox :   The virus of sheep pox is host - specific and affects mostly sheep, though goats may be affected in a milder form.   Goat pox immunises sheep against sheep pox but sheep pox does not immunise goats against goat pox,

The disease is highly contagious and infection may occur through skin by contact or by aerogenous route.

Sheep pox is very severe with many fatalities, Incubation period is one week

Differing from cow pox, sheep pox is a generalised infection.   The virus becomes viremic, whatever may be the portal of entry and so gets distributed throughout the body. Initial symptoms seen are general malaise, fever, salivation nasal discharge and lachrymation.   Later nodules may be seen on the skin not covered by wool, viz lips, nostrils, eyelids, underneath the tail, udder, vulva scrotum and prepuce.

Pathogenesis ;   In general, the lesion is similar to that of cow pox.   But there is great infiltration of the corium by inflammatory cells, which undergo necrosis subsequently   The cells of the prickle - cell layer proliferate and these may even extend into the corium.   Along with these changes are found hyperkeratosis, parakeratosis and acanthosis of the surface   Balooning degeneration of the proliferating cells occurs   But unlike in cow pox, the vesicles that form are never large but microscopic and are multilocular   The sheep pox vesicle differs from the cow - pox variety in that the floor of the former is formed by the corium while in the latter by the stratum germinativum   They are also umbilicated. Subsequently necrosis of the entire epidermis may take place with the formation of scab   Healing occurs by cicatrisation and may take 5 to 6 weeks

In the papular stage large cells, resembling histiocytes make their appearance and these subsequently become " sheep pox cells ",   These are large, basophilic and stellate, with oval or irregular nuclei and large nucleoli.   In these cells also are found the acidophilic granular inclusions,

Acidophilic cytoplasmic inclusion bodies are found in the epithelial cells

Besides the skin, pock lesions are found on the mucous membrane of the pharynx, trachea and abomasum.   In these places either pock vesicles or ulcers may be found. There may be hemorrhagic inflammation of the mucosa of the respiratory passage and gastrointestinal canal. Near the pleural surface of the lungs may be found inflammatory catarrhal areas and gray caseous nodules. Other septicemic lesions, viz. acute lymphadenitis, fatty degeneration of parenchymatous organs, petechiae on serous membranes and interstitial nephritis may be met with.

Clinically, sheep pox is malignant, mortility rate being upto 50%. Because of necrosis and ulcer formation of skin, secondary bacterial infection may occur resulting in septicemia or pyemia. Death may be due to the hemorrhages that may occur in the pocks or due to secondary septicemia or pyemia.

Goat Pox : This is a benign disease to which man is susceptible. Incubation period is 15 days. The lesions may be found on the udder and have the same pathogenesis as other pock lesions.

Swine Pox : Two distinct viruses are found to cause swine pox. (i) the cow pox virus which causes a more severe condition and (ii) a virus that is specific and affects the swine only. Piglets are more often affected by the latter variety while adults are affected by the former.

Pig louse, *Haematopinus suis* and other biting insectes may transmit the disease from pig to pig.

The lesions which are typical of pox are found on the eyelids, snout, back and under the belly. Swine pox is benign the course running for 3 weeks. Elementary bodies are found in the epithelial cells.

Fowl Pox : This is a mild condition caused by an epitheliotropic virus. Four distinct viruses are known affecting fowls, turkeys, pigeons and canaries.

In fowls incubation period is 6 to 14 days and in a flock the outbreak may persist for 2 are 3 months.

Routes of infection : By wound infection of wattles, comb and skin due to fighting or pecking or through abrasions of the mucosa of the mouth. (Infection of the intact skin or mucous membrane does not occur). Blood sucking insects like mosqutoes, flies, ticks and lice may transmit the disease.

The virus remains viable on the dried scabs for long periods. In birds of 5 to 12 months of age. the disease in more often seen.

Bad sanitation and over crowding are conducive for spread of the disease in flocks.

The disease is manifested in three forms:

(1) **The cutaneous or comb form :** In this are found wart - like nodules on the comb, wattles and eyelids, Other parts of the body are affected less frequently.

(2) **The diphtheritic form :** In this form a diphtheritic membrane is formed on the mucosa of the mouth and pharynx

(3) **The oculonasal form :** in which catarrhal inflammation of the eyes and nostrils is present, Sinuses may contain cheesy deposits.

Mortality may occur in the second and third varieties. The causes of mortality are :

(a)   asphyxia   when the mouth, larynx and nostrils are affected,

(b)   starvation, when the eyes are affected. The eyelids become glued with
the sticky inflammatory exudate, rendering the birds blind, as a result of which
they are not able to pick up their feed and so starve to death.

Pathogenesis : The ep dermal cells proliferate and become enlarged. Though
the endoplasm of these swollen cells becomes dissolved no vesicles are formed.
since the ectoplasm becoming keratinized resists the rupture of the cells   More.
over the degenerated cells do not become detached as in mammalian pox, since
in the absence of inflammatary process (unlike the mammalian pox) exudation
does not occur   Hence only warty nodules are formed on the skin of wattles.
comb and eyelids, This is mild form, scabs forming in the nodules, Scabs may
fall off after 2 to 3 weeks leaving white scars, which disappear subsequently
i e. in one to two weeks

In the mucosa of the mouth and larynx, on the other hand, exudation
process does occur in addition to the swelling and proliferation of the epithelial
cells, leading to the formation of diphtheritic membrane which causes asphyxia
by occlusion of respiratory passages.

Inclusion bodies are found in the epithelial cells at the base of the warty
nodules or in the diphtheritic membranes   They are intracytoplasmic and are
known as **Bollinger bodies**. Within these bodies which measure 2 to 3u in diameter are found granules 0 25μ in diameter. These are known as **Borrel granules**
Each Borrel granule is capable of reproducing the disease when inoculated into a
fowl and so is considered to be a colony of the virus

Diagnosis of pox is easy from the symptoms and lesions. However. to confirm the diagnosis, isolation and identification of the virus is necessary by gel
precipitation and serum neutralization tests with the isolates.   The diphtheritic
form may confuse diagnosis between fowl pox and infectious larynogotracheitis
of fowls.

PSEUDO-COWPOX

A virus belonging to the contagious ecthyma pox groups causes lesions on
the teats of cows and on the hands of man ('milker's nodule')   This virus is
similar to that of contagious ecthyma and papular stomatitis and is grown on
tissue cultures. Transmission is from an affected animal to another usually through
milker's hands and teat cups.   This is a benign condition

Symptoms : Erythema progressing to pustules occur on the teats. The pustule
ruptures within 48 hours and a thick scab forms. The scab falls off in 7 to 10
days. Sometimes a small granuloma persists. In some animals chronic lesions
persist for months.

In man vesicles form on the hands.

Diagnosis : Pseudo cowpox has to be differentiated from cow pox. The lesions of the former are larger, more numerous, less painful and more prolonged
than the later.

BOVINE PAPULAR STOMATITIS

This is caused by a virus of pox group in young animals up to 2 years of age. The mucosa of the lower and upper lips, palate, papillae of the buccal mucosa and the muzzle show erythematous patches at first. These appear as rounded papules which are hard and of millet size. There is a surrounding red zone. The center becomes necrotic and covered with a scab. Loss of the epithelium leaves an erosion. Healing takes place in 8 days.

Microscopically, the lesion shows hyperemia, inflammatory edema and infiltration of inflammatory cells into the mucosa. The prickle-cell layer shows balooning degeneration Acidophilic inclusion bodies can be seen in the cytoplasm of the epithelial cells. The keratinised layer of the skin becomes disrupted and later the whole may be shed.

Diagnosis : This condition has to be differentiated from Foot and Mouth disease. In the former the lesions are confined to the mouth only.

LUMPY SKIN DISEASE

Syn : **Knopvelsiekte ; Bovine nodular exanthema.**

Lumpy skin disease is an acute infectious disease of cattle characterised by rapid eruption of large number of cutaneous nodules and development of superficial lymphangitis and lymphadenitis.

Cause : A virus, probably belonging to the Pox virus group, is the cause. Two strains of the virus are recognised In tissue cultures, the virus produces changes typical of pox viruses. In natural lesions, cytoplasmic acidophilic inclusions are seen.

So far, the natural spread of the disease has not been determined but biting insects, especially mosquitoes, are incriminated. Experimentally, the disease can be transmitted by injecting blood collected prior to the eruption of the nodules and also by injecting emulsified tissue of the nodules. Prior to the development of the cutaneous lesions, a viremic phase is seen for a week and the virus is eliminated in high concentrations from the salivary glands.

Incubation period : 2 to 4 weeks in natural cases. In experimental animals this is only 1 to 2 weeks.

Symptoms : Biphasic fever is observed and the cutaneous eruptions accompany the second phase in about 7 to 10 days after the first. Profuse salivation and oculonasal discharge are noticed at the same time. The animal does not like to move but often seeks shade. The nodules, characteristic of the disease, appear on the skin of various parts of the body, but these are more conspicuous in places where the hair is short and skin smooth. The nodules are intracutaneous, firm, circumscribed and flat and vary from 0. 5 to 5 cms. in diameter. Rarely some may extend to or are confined to the subcutis. In mild affections the nodules may be few, while in more severe cases these are numerous. The nodules at the vulva, perineum, udder and oral cavity may be less elevated than the cutaneous ones. Those that are found in non-pigmented areas are surrounded by a zone of intense hyperemia. The nodules may be found in the scrotum, testes, glans penis, inner surface of the prepuce, on the eye lids and conjunctiva.

The fate of the nodule varies. It may either resolve rapidly and disappear or it may become indurated and persist as a hard lump for 12 months or longer. In some cases they may become necrotic and sequestered. When the sequester falls out with the necrotic tissue of the subcutis, a deep ulcer is formed which may be partially filled with granulation tissue.

The lymph nodes are enlarged invariably and those situated subcutaneously stand out prominently.

Secondary bacterial infection of the necrotic core of the nodules may occur and this complicates the disease, producing edema of the limbs.

Macroscopically, on incision, the nodules are found to affect the full width of the cutis as a nodular mass of tough, creamy gray tissue with caseous necrotic core. Such nodules of about 1 cm diameter, circumscribed by a zone of intense hyperemia, are found in the superficial muscles, in the outer part of renal cortex, lungs and testes. Similar nodules are also seen in the respiratory passages and in the alimentary canal upto the abomasum.

Microscopically, changes are found to develop parallely in all the layers of the skin and subcutis. The cells of the epidermis and sebacious glands swell and develop increased acidophilia. Large cytoplasmic vacuoles may be found in them. Microvesicles form by the confluence of two or three such vacuoles. Except in the early lesions, acidophilic inclusions are found in the affected cells. Majority of the inclusions are seen within the vacuoles and these inclusions are surrounded by a halo of faintly eosinophilic material.

Edema is found in the superficial areas of the corium and in the dermal papillae particularly. Reticulo-endothelial elements undergo proliferation and degeneration. The edema is of such magnitude that the overlying epidermis is detached. Perivascular cuffing of smaller vessels by macrophages lymphocytes and plasma cells is seen in the deeper parts of the corium. Macrophages contain intracytoplasmic inclusions. Edema may be due to the thrombosis of the deeper venules.

Being a mild affection, the course of the lumpy skin disease is generally favourable and spontaneous recovery, though slow, therefore occurs. Recovery is followed by solid immunity of but 3 months duration. Some inapparent infections may also be encountered.

Diagnosis: is based on the symptoms and lesions.

CONTAGIOUS ECTHYMA

Syn. Contagious Pustular dermatitis; Sore mouth; Orb.

This is a disease of sheep and goats caused by dermatotropic virus belonging to pox viridae. Lesions are usually found on the lips and udder.

The virus of contagious ecthyma is found in the scabs in which it may be viable for as long as 15 years. Man can be infected by handling affected animals.

Routes of infection: Infection is by contact with affected animals. Wounds and abrasions may be portals of entry and hence its prevalence in summer dry seasons and feeding on dry pasture, when wounds are likely to occur in the mouth. Udders may be infected from affected lambs.

Incubation period : 5 to 8 days. Morbidity is 90% while mortality in uncomplicated cases is low.

Pathogenesis : The pathogenesis of the lesions is similar to that of pox viz formation of papules, and vesicles followed by necrosis, formation and shedding of the scab in about 3 weeks. Cytoplasmic acidophilic inclusions are present in the epithelial cells

Secondary infection complicates the disease by forming deep necrotising lesions and even septicemia. Extension of lesions to respiratory tract may cause mortality. which is otherwise negligible.

Recovery is followed by immunity, which lasts for 2 to 3 years.

Symptoms : Initially reddened areas are found on the skin, especially at the corners of the mouth and lips. These later grow as warty papules which subsequently become vesicles and pustules. Drying of the pustules results in yellowish scabs which may be shed, healing occurring in 3 weeks. The scabs are infective for a long time.

Affected lambs are not able to suckle and so become debilitated, **L** oss therefore, in this disease is due to slow gain in weight.

Lesions : As described above the lesions consisting of papules, vesicles, pustules and scabs, depending on their age, are found on the lips and corners of the mouth.

If infected by bacteria, deep ulcers may be found in the oral mucosa

—— **Microscopically**, the changes are similar to those in pox viz proliferation of the basal layer of epidermis with, spongiosis, balooning degeneration, vesicle and pustule formation in the upper layers. Eosinophilic intracytoplasmic inclusions are seen in these cells.

Lesions similar to the above may also be seen in other hairless areas as the head, on the genitalia, feet and udders

Animals become debilitated as they are not able to feed and drink. Affected ewes may suffer from pregnancy toxemia. Secondary infection by *Spherophorus necrophorus* causes severe necrosis of affected parts.

Diagnosis : Differentiate from sheep pox (contagious ecthyma is immuno logically different from sheep pox).

HOG CHOLERA
Syn. Swine fever.

This is a febrile infectious viral disease of pigs characterised by an acute course with symptoms of hemorrhagic septicemia. During the course of the disease croupous pneumonia and diptheritic inflammation of the gastrointestinal tract may develop. Only swine of all ages are susceptible. The virus can be cultivated on tissue cultures. But the virus does not grow on chick embryo.

The virus is an RNA virus and is 38—46 mμ

The virus of hog cholera and mucosal disease are antigenically similar.

Incubation period : 1 to 4 days by artificial infection. 7 days in natural outbreak.

Routes of infection :

1.   Through digestive tract.   Urine, blood, discharges from eye and bronch are infective.
2   Through respiratory tract by inhalation.
3.   Through conjunctiva.
4.   Through nasal mucosa.

Source of infection is an ailing animal.   Infection may occur directly from animal to animal or contaminated pastures and pig houses may be sources of infection.

Attendants and other animals may carry infection from animal to animal,

Uncooked garbage is an important source of infection.   Recovered animals may act as carriers for at least one month.

Pathogenesis : The virus gains entry into blood through the tonsillar tissue, probably, since deposition of the virus directly in stomach does not produce the disease.   The virus multiplies rapidly in the blood and gives rise to the various clinical and pathological symptoms of septicemia, viz. fever, hyperemia, inflammatory swelling of various mucous membranes.   The virus has special affinity for the endothelial cells of blood vessels and reticulo-endothelial system.   The vascular endothelial cells swell, proliferate and occlude the lumen.   The walls of the blood vessels undergo hyaline degeneration with infiltration by lymphocytes, macrophages and plasma cells.   These changes of the blood vessels are the cause of hemorrhages, necrosis and infarction found in various organs.   The virus is also capable of causing inflammation in lungs, thereby producing croupous pneumonia

In the intestines, especially in the cecum and colon, are found the characteristic "button ulcers".   These are formed from the Peyer's patches and solitary follicles.   To begin with there is hemorrhage in the follicles as a result of which the follicles become necrosed and form small grayish - yellow nodules which are surrounded by a ring of sero - hemorrhagic mucosa,   Subsequently this mucosa also becomes necrosed and with the fibrin and coagulated exudate forms a diph. theritic, yellow, raised patch over the follicle.   These changes are primarily due to occlusion of the blood vessels causing infarction and so hypoxia.   The rim of the nodules is slightly raised.   Secondary bacterial invasion intensifies the process and the lesion   expands concentrically, producing concentric lamellae of thickening.   Later, the central necrotic area becomes softened, cast off and an ulcer results covered with thick purulent exudate and raised borders —the typical "button ulcer".   Granulation tissue may fill the ulcer and the scar may be covered by epithelium growing from the sides.

Symptoms :   The viremic stage is manifested by high temperature and loss of appetite followed by intense leucopenia.   Conjunctivitis with thick sticky discharge may be found.   Nasal catarrh may be present.

Skin shows erythematous patches which become cyanotic.   Later vesicles may form on the lips, vulva and edges of ears.   The animal becomes rapidly dehydrated and weak· Constipation followed by diarrhoea and vomition are also observed. Cerebral symptoms including convulsions, fits, wobbling gait and howling followed by coma may be seen.

Death may occur in 14 days.

Lesions : The lesions vary depending upon the acute or chronic course of the disease and also by the presence or absence of secondary bacterial infection,

Pure swine fever : When not complicated by secondary infection, pure swine fever is manifested by changes of septicemia only.

Peracute cases : In animals that die suddenly early in the disease no post-mortem lesions may be encountered, The mucous membranes may be congested. Petechiae may be found on the kidneys, mucous and serous membranes. Lymphatic glands may be hyperemic and swollen.

Acute cases : The carcass is dehydrated, emaciated and soiled with diarrhoeal feces. The condition is hemorrhagic septicemic in character. So hemorrhages are found on the serous membranes, subcutaneous fat, the skin, the pericardium, the pleura and pulmonary tissue, larynx and bladder as well as in the bone marrow.

In the kidney, when the capsule is removed, may be found petechiae on the cortex extending deeply into the parenchyma. These give a characteristic '' turkey egg '' appearance.

The lymph nodes are usually affected and are swollen and red in color· Hemorrhages are found in the periphery and so the zone just under the cortex is red.

The skin of the ventral surface of abdomen and thorax, of the ears and internal aspects of thighs, the perineum and snout, shows erythematous patches resulting from changes in the blood vessels. Sometimes in these areas may be found edema, necrosis and sloughing.

In the gastrointestinal tract may be found diffuse cartarh. The characteristic lesion is the button ulcer already described.

In the spleen. due to the proliferation of the endothelium and hyalinisation and necrosis of the wall of follicular arteries, thrombosis and infarction result. The splenic infarcts are wedge shaped, found on the edges of the organ and are brownish in color.

The changes in the brain may be observed even before the actual onset of symptoms. The pathological process encountered is disseminated non – purulent meningo - encephalomyelitis. Both gray and white matter are affected· The lesions are confined to the blood vessels and their supporting mesenchymal tissue. The most conspicuous lesion is perivascular cuffing — i e., accumulation of lymphocytes, monocytes, plasma cells and local histiocytes in the perivascular space of Virchow - Robin. Neutrophils are seldom seen. The lymphocytes, monocytes and plasma cells are derived from the blood while the histiocytes of the locality proliferate by mitotic division There may be degeneration of the vascular endothelium and fibrinoid changes of the wall of arteries The changes noticed in the brain tissue are due to the vascular lesions and consist of small nodules of proliferating microglia seen around blood vessels. Chromatolysis, satellitosis and neuronophagia may be seen sometimes but there is no demyelination.

In the lungs may be noticed croupous pneumonia with darkened [hepatized] areas. The pleura over the affected lungs is covered by fibrinous deposits Fibrinous pericarditis may also be seen.

The chronic form : Some acute cases may survive and become chronic. Such animals are thin and emaciated. In these animals acute inflammatory changes and hemorrhages are absent. The lesions are confined to the intestines and lungs.

In the intestines are found necrotic ulcers, especially in the colon. The lung may be adherent to the thoracic wall and the parenchyma may contain numerous necrotic foci. Sometimes whole lobes may be necrosed, caseated and enclosed in a fibrous capsule. Such animals suffer from constant cough.

In mixed infections, *Salmonella cholerae suis* often complicates the picture. The original name of the disease 'Hog Cholera' was derived from the discovery of the *Salmonella cholerae suis* as its etiology. The virus etiology was confirmed later. There is splenomegaly and serofibrinous peritonitis besides other lesions described in the pure infection. The intestinal lesions are more severe than in the pure form

Diagnosis

1. Intense leucopenia and thrombocytopenia are characteristic laboratory aids.

2. Symptoms and lesions.

3. Inoculation into a healthy pig and study of disease produced. No definite diagnosis is possible without animal inoculation.

4. Agar - gel precipitation test.

5. Fluorescent antibody tissue section test.

6. Fluorescent antibody tissue culture test.

7. Exaltation of the cytopathogenic effect (CPE) of Ranikhet Disease Virus Hog cholera virus stimulates the CPE of RD virus in cell culture.

8. Swine fever must be differentiated from swine erysipeals, salmonellosis, anthrax, acute pasteurellosis, salt poisoning, meningitis and African Swine Fever.

AFRICAN SWINE FEVER

This disease, though confined to Africa now, can be transported to other countries and hence the great danger attendant on movement of pigs.

The virus of A. S. F. is antigenically different from that of Hog cholera and it belongs to irido viruses. A. S. F. is a more acute and fulminating disease than Hog cholera. The warthog appears to be a reservoir of the virus. No clinical symptoms are observed in this animal. Incubation period is 5 to 15 days.

In A. S. F. the affected animals invariably die, death occurring within 7 days after onset of symptoms Therefore, though the mode of action of the virus of A. S. F. is similar to that of Hog cholera, viz damage to vascular endothelium, yet infarction is seldom observed as the duration of the disease is short.   The following are the differences ;

A . S. F.	Hog Cholera
1. Very acute, no chronic cases seen.	1. Chronic cases are seen
2. Mortality 100%	2. Mortality not so high.
3 Death occurs within 7 days after onset of symptoms.	3. Death occurs only after 14 days.
4. Hemorrahages and edema more severe	4. Hemorrahage and edema less severe.
5. Wasting not seen.	5. Carcass dehydrated and emaciated.
6. Infarction not seen in spleen.	6. Spleen contains infarcts,
7. Spleen enlarged.	7. Usually spleen not enlarged.
8. Pulmonary edema present.	8. Pulmonary edema not present.
9. "Button ulcers" not seen.	9 Button ulcers common.
10. In lymphocytes necrosis and karyorrhexis seen—lymphopenia.	10. Necrosis of lymphocytes not seen.
11. Necrosis of periportal hepatic cells seen.	11. No changes seen in the liver.

RINDERPEST
Syn : Cattle plague

Rinderpest is also known as cattle plague and is an acute, contagious viral disease of cattle and buffaloes. The disease is mild in sheep and goats and is reported in pigs also.

This disease has been known to cause great financial loss and so is the most important disease of cattle in India. Rinderpest Eradication Campaign that was instituted and successfully completed in many states has considerably reduced the incidence and loss.

Rinderpest virus has been adapted to grow on embryonated eggs. On monolayer cultures of bovine kidney cells cytopathic changes have been produced.   These are rounding of the cells, syncitia and formation of giant cells. Intranuclear and intracytoplasmic incluisons have been seen in these cells. Serial passage of the virus through tissue cultures sufficiently attenuates it so as to be useful as a safe vaccine.   The virus is related to the virus of measles and canine distemper. and belongs to paramyxo viridae.

The virus is present in the body attached to leucocytes. It is excreted in the saliva, urine, feces, nasal discharge and tears. During the height of fever, it is present in the blood. Later it becomes localised in a concentrated form in the the spleen, lymph glands and liver.

Animals affected : Rinderpes affects naturally cattle and buffaloes, but outbreaks of the disease among sheep, goats and pigs have been reported in some states of India.   The virus can be adapted to rabbits. in which the characteristic chalkwhite necrotic lesions stand out in the Peyer's patches and in the sacculus rotundus and the appendix   The virus has been grown on chorio-allantoic membrane of chick embryo.   But the passage through goats, rabbits and chick embryos attenuates the virulence.

In India, the exotic breeds are affected more severely, mortality being almost 100 percent, while the mortality in indigenous breeds is only 20 to 50%

since they are more resistant.   However, occurrence of mild outbreaks have been noticed in this country.   The virus seems to show modified virulence due to its survival either through an immune or partially immune population of livestoek.

Routes of infection :   Infection is by ingestion of contaminated feed and water.   Infecti n into a herd is usually introduced by an affected animal.   Infeetion may also be conveyed by attendants and fomites.   Inhalation may also be an important route

Incubation period :   By experimental inoculation incubation period is 2 to 3 days while that in contact infection is 6 to 9 days

Outside the body, the virus cannot thrive for more than twenty four hours. It is rapidly destroyed by sunlight and ordinary disinfectants.

Symptoms   Fever is not ced usually on the 4th to 6th day after infection,

Along with the rise in temperature may be noticed anorexia, lachrymation, dryness of muzzle, congestion of conjunctiva and leucopenia.   There may be severe abdominal pain, arched back, staring coat and constipation. Subsequently diarrhoea develops and the temperature falls

By about the 7th day the lesions in the mouth develop with dribbling of saliva   The diarrhoea, which is very offensive is very severe so that the animal becomes very much emaciated, almost reduced to a skeleton.   The feces may contain mucus and blood.   Animal dies within 6 to 12 days from the onset of symptoms.

Pathogenesis and lesions :   The virus of Rinderpest has great affinity for lymphoid cells and epithelial cells of alimentary system.   It causes pyknosis and fragmenatation of the nuclei in lymphocytes, their necrosis and subsequent disappearance.   The germinal centres of lymph nodes are bare and do not contain lymphocytes.   Only the reticulum meshwork remains.   Other cells—plasma cells, and macrophages—may infiltrate such an area.   These changes are seen in all lymph nodes, the spleen and Peyer's patches.

The virus produces lesions in the oral mucosa after settling in the cells following a viremic state.   The mucosa on the inside of lower lips, commissures of the mouth and on the underside of the free portion of tongue is often affected   Essophagus may reveal lesions in severe affections.   Fore-stomachs are free.   To begin with, in a few cells of the Stratum Malp ghi, the nucleus becomes pyknotic and fragmented.   The cells become shrivelled and necrotic with the cytoplasm becoming acidophilic and they become separated from the adjacent cells   No Vesicles form.   These foci may enlarge by coalescence and extend superficially and so the cornified layer becomes lifted up to be seen as pin-point grayish - white foci.   Due to movement the necrotic tissue is cast off revealing shallow erosions with a raw, red base (due to congested capillaries) and sharply demarcated edges, since normal epithelium is found on the bord rs   The small foci may enlarge and be covered by bran - like deposits. (fibrin deposits)   Ulcers do not form unless infected by secondary bacteria

Abomasum is always affected and the lesions are severe at the pyloric region. There are necrotic foci accompanied by hemorrhagic streaks.   The folds of the abomasum are thick and edematous and the contents chocolate in color due to the presence of blood. The folds reveal hemorrhagic streaks on their free borders The capillary congestion and hemorrhage give a bright red coloration to the whole mucosa. Erosions may form by sloughing of the necrotic epithelium. Sometimes frank ulcers may be present.

Lesions in the small intestines are not so severe as in the abomasum or large intestines But streaks of hemorrhage and erosions may be seen on the crests of folds of mucosa, especially in the duodenum and ileum. The Peyer's patches are affected becoming necrotic and leaving deep ulcers with a bright hemorrhagic border.

The large intestines are more severely affected. The lesions are particularly prominent at the ileo-cecal valve, at the cecocolic junction and in the rectum. In these areas are found diffuse hemorrhages and also ulceration with the pseudo-membranous bran-like deposits. The ulceration and diphtheritic deposits are not due to the direct action of the virus but to secondary bacterial infection as in hog cholera. The streaks of hemorrhage on the folds of mucosa of rectum are responsible for the so called "zebra markings". The mucosa, therefore, is bright red and the lumen contains dark clotted blood. The lymphoid follicles become necrotic and the sloughing of the necrotic materials leaves deep ulcers. *Microscopically*, it is seen that epithelium is shed leaving the lamina propria, revealing hemorrhages, edema and leucocytic infiltration that are responsible for the thickness of the wall. Multinucleate cells are seen in the mucosa, lymph and hemolymph nodes.

Lesions in other organs : Since the respiratory epithelium is suscepetible to the virus, lesions are found in the mucosa of upper respiratory tract (petechiae), in the larynx (petechiae and erosions) and in the trachea (hemorrhages). Because of laboured breathing due to diarrhoea, weakness and recumbent position, alveolar and interstitial emphysema of lungs may be noticed.

In a few long-standing cases, subepicardial and subendocardial hemorrhages are seen especially of the left ventricle.

Petechiae and erosions may be seen in the bladder and vagina. Skin lesions have been described but are not constant. Acute congestion and edema of conjunctiva followed by purulent conjunctivitis and ulceration of cornea may be noticed

Rinderpest in sheep and goats : The symptoms in sheep and goats are fever, dullness, pneumonia and diarrhoea. Heavy mortality in these animals is met with, mostly due to secondary pneumonia. The difference from bovine lesions is that in sheep and goats no mouth lesions are usually seen. Nor is lachrymation observed in the ovine disease.

Diagnosis :
1. Symptoms and lesions.
2. Rinderpest must be differentiated from coccidiosis, Mucosal disease and Foot and Mouth disease.
3. Complement-fixation test.
4. Agar-gel diffusion technique with needle biopsy material of lymph nodes as antigen, using hyperimmune serum produced in rabbits only.
5. Virus isolation and diagnosis on tissue cultures.

6   Virus neutralisation on tissue cultures.

7   Intranuclear and intracytoplasmic inclusion bodies in the epithelium and lymphoid tissue,

8   A direct immuno-peroxidase test to detect Rinderpest antigens in the infected tissues has been developed recently.

BOVINE VIRAL DIARRHOEA—MUCOSAL DISEASE

Formerly it was thought that mucosal disease (MD) and bovine virus diarrhoea (BVD) were caused by different viruses. But now it is considered that they are different manifestations of the same disease.

B. V. D —Mucosal disease is found in cattle and buffaloes Usually calves between 6 and 14 months of age are affected. The causative virus is immunologically different from that of Rinderpest.

In the epidemic form (virus diarrhoea) morbidity is almost 100% but the mortality rate is only 4 to 8%. On the other hand, in the sporadic form (mucosal disease) the morbidity may be only 2 to 5% (rarely 20%) but the mortality may be 90%

Routes of infection : Ingestion is the common and important route of infection. An infected animal is the source, the virus being voided in its dung

Wild antelope and deer spread the disease. Some cattle may be viremic for long periods without developing any antibodies and so may be source of infecton

Incubation period : 9 to 10 days.

Symptoms : The disease can be be divided clinically into three forms : the mild, acute and chronic forms.

In the mild form which is usually seen after experimental inoculation, are found slight feverish condition, leucopenia, diarrhoea that persists for a few days, seromucoid nasal discharge, variable appetite, unproductive cough and reduction in milk yield. This form passes unnoticed in the field.

The acute form is characterised by high fever, sudden fall in milk yield, severe depression and loss of appetite. Rapid pulse occurs. The dung is watery, profuse and foul smelling. Cough, nasal discharge and increased respirations may be observed. In some animals there may be stringy saliva, profuse lachrymation and corneal opacity. There may be necrotic erosions on the mucosa of the mouth, pharynx and on the skin of the nose. In severe cases, feces may be mixed with blood Advanced dehydration may ensue. Lameness develops in some animals due to laminitis or due to ulcerative dermatitis of the interdigital space. This acute form lasts for 2 to 4 weeks. Pregnant animals may abort or a mummified fetus may form.

In the chronic form, which may extend upto 5 months, the animals manifest failure to gain weight, lameness due to laminitis with elongated distorted hooves, arched back, rough hair coat, intermittent diarrhoea and gradual emaciation

Lesions : Erosions are found in the buccal mucosa over the muzzle, tongue, lips. hard and soft palate, cheeks and pharynx. Severe erosions may be seen in the esophagus covered with fibrin or necrotic tissue. Erosions are found on the leaves of omasum or pillars of rumen and these may show hemorrhages and turn into ulcers also. *Microscopically*, the cells of the Malpighian layer undergo vacuolar degeneration with the formation of small vesicles which turn soon into erosions. Neutrophils may infiltrate this area. Secondary bacterial infection may transform the erosions into ulcers.

In the abomasum the erosions are present at the fundus and measure 1 to 1 5 mm These have raised walls and are circumscribed by a ring of petechiae. The gastric glands atrophy and may become cystic. The walls are thickened due to edema, congestion and hemorrhages in the submucosa and mucosa.

Erosions are also present in the small intestines, particularly over the Peyer's patches and these have thick tenacious, mucohemorrhagic exudate covering them. The lymphoid tissue and mucosa reveal necrosis and sloughing.

Colon reveals catarrhal inflammation with retention cysts. There may be necrosis and ulceration or diphtheritic inflammation in some parts.

The mesenteric lymph nodes are grossly normal or may be slightly edematous. On the other hand the cervical and retropharyngeal lymph nodes are enlarged and swollen. Histologically, no evidence of lymphocytic destruction is noticed as in Rinderpest.

Sometimes, nares may show erosions covered by profuse mucopurulent exudate,

Other lesions noticed are :— Conjunctivitis, keratitis, corneal opacity, skin lesions around vulva and preputial orifice, fatty degeneration and focal necrosis of the liver.

An intense leucopenia is present. Death may occur in 4 to 15 days but in some the disease may be chronic extending to several months.

Calves born of affected cows show cleft palate, arthrogryposis, cerebellar hypoplasia, retinal atrophy, optic neuritis, cataract, microphthalmia and retinal dysplasia.

Diagnosis :

I. Mucosal disease must be differentiated from Rinderpest, Foot and Mouth disease, parasitic and bacterial enteritis.

Differences between :

	Rinderpest and	**Mucosal Disease**
1. Animals affected	Bovines, sheep, goats and wild animals (deer etc,)	Only bovines
2. Age group	All age groups	Younger groups of 6—14 months more susceptible
3. Morbidity rate	High, 100%	Low, 2 to 16%
4. Mortality rate	High	High, 80·/
5. Pathogenesis	Known	Not worked out
6. Lesions in general	Erosions usually become ulcers and covered by bran-like deposits or flakes	Seldom become ulcers. Not covered by flakes.
7. Ileum - Peyer's patch	Necrosis and deep ulcers.	Slight erosion and occasional ulcers
8. Large intestines	Zebra markings	Congestion and catarrhal inflammation

9. Mucosal cysts -	Not seen	Present :— 10 to 20 cms. distal to ileocecal valves: cysts of 1 to 4 mm dia. seen
10. Lymphoid tissue.	Severely destroyed and necrotic changes seen	Only mild changes
11. Respiratory system	Affected. Petechial and hemorrhagic streaks present, Lungs - congested and emphysematous.	Not affected
12. Vesicles in the Malpighian layer	No vesicles form. The epithelial cells become necrotic.	Vesicles form due to hydropic degeneration of the cells above the basal layer.
13 Transmissibility	Can be transmitted	Transmission not so easy May be present.
14 Corneal opacity.	Not present.	

II.   Confirmation may be made by agar - gel diffusion technique and viral neutralisation tests in cell culture.

CANINE DISTEMPER
(Syn : Carre's disease : Hard - pad disease)

This is an acute febrile infectious and contagious disease of dogs, caused by a pantropic virus (belonging to the family of paramyxo viridae). The characteristic symptoms are associated with fever, acute catarrhal inflammation of various mucous membranes, pneumonia and in some cases skin lesions and involvement of central nervous system. The virus is related antigenically to measle and Rinderpest virus.

Animals affected : Dogs, foxes, jackals, ferrets, mink, racoon and other wild canidae are affected. Young dogs, under one year of age, suffer more often.

Routes of infection : Most common route is by ingestion. Inhalation of air contaminated with the virus may also occur. All the excretions and secretions contain the virus

The disease is never seen as a purely viral infection but is always complicated by secondary bacterial invaders, the most important being *Bordetella bronchisepticus*, causing pneumonia and *Salmonella* species producing gastrointestinal lesions. These organisms are normal residents of the respiratory and digestive tracts respectively and invade the tissues when the resistance is lowered by the viral infection.

Incubation period : Five days

Pathogenesis ; The virus being pantropic is able to infect tissues of all the three germinal layers. So it is found in the epithelial cells, the riticulo-endothelial cells and the brain. From the portal of entry the virus is conveyed to the regional lymph node where it multiplies and from there it enters the blood and becomes viremic constituting the stage of septicemia. Very susceptible animals may die even at this stage Then virus settles and proliferates in the epithelium of the skin, alimentary, respiratory, urinary and biliary tracts. Infecting the vascular endothelium of the brain, the virus spreads to the brain tissue and proliferates in the nuclei and cytoplasm of the neurones and glial cells

On the skin of abdomen and thigh, vesicles and pustules form due to secondary invasion by pyogenic organisms.

Symptoms : The first symptom is fever, which persists for 3 to 4 days and then drops. The temperature is normal till 11th or 12th day when it again shows a rise due to secondary bacterial infection (Diphasic fever curve). Nasal discharge, conjuctivitis and bronchitis may be present to a slight degree. Respiratory symptoms (dyspnoea) are seen when lungs are affected with pneumonia or edema. Diarrhoea, sometimes bloody, may occur leading to dehydration and emaciation. Pustules may be seen in a great majority of animals, especially young puppies, on the ventral side of the abdomen, groins and inner side of the thighs. Hyperkeratosis of the digital pads may be seen—(Hard-pad disease). Eye infection results in keratitis, retinitis and blindness In half of the affected animals are found the following nervous symptoms : epileptic seizures and excessive chewing movements accompanied by salivation. (chorea)

Lesions : In most of the cases, lesions are found in the respiratory tract.

A catarrhal exudate, which turns purulent subsequently, is found over mucosa of nose and pharynx. The laryngeal and bronchial mucosa is congested and the bronchi may contain either serous fluid or purulent exudate depending on whether there is edema or pneumonia of lungs The bronchioles contain small purulent plugs—*Bronchiolitis capillaris*. In pure Canine Distemper, the lesions in lungs are that of interstitial pneumonia. But when infected by secondary bacterial invaders, there may be purulent bronchopneumonia with the alveoli infiltrated with neutrophils, alveolar macrophages and mucin Bordering on the pneumonic areas may be noticed emphysemtous areas In some areas may be found collection of cells resembling epithelioid cells, some of which may be fused, so giving the area an appearance of giant cell pneumonia In these pneumonic areas may be seen triangular, depressed, dark areas of atelectasis. The pleura over the pneumonic lesions may be dull and covered with fibrinous deposits.

Acute catarrh of the gastrointestinal mucosa is noticed Peyer's patches and solitary follicles may be swollen. The lymph nodes in general are swollen and juicy. Slight enlargement of the spleen may be noticed.

The pericardium may contain large quantity of fluid. Myocardium may show yellow foci of fatty degeneration and small hemorrhages. Parenchymatous degeneration may be found in the liver (which may show a 'nutmeg' appearance) and renal cortex. Miliary necrotic foci may sometimes be seen in the liver.

Catarrh of the conjunctiva ulceration of the cornea and panophthalmitis are noticed in the eyes.

Hyperemia of the meninges and hemorrhages on the pia or brain and spinal cord may be present, There is non-suppurative encephalomyelitis. This differs from other viral diseases affecting the brain in that the neurones are not primarily affected. It is the myelinated areas of the brain that are affected, producing demyelination.

In the brain the capillaries dilate and become congested The endothelial and adventitial cells proliferate and mingle with lymphocytes while forming perivascular cuffs Degenerative changes in the endothelial cells may also be seen

Hemorrhage by diapedesis is noticed in the spinal cord, the medulla oblangata, round the central canal and brain stem   Edema of the space of Virchow-Robin choroid plexus and meninges may be noticed. Similar edema may be seen round ganglion cells.

Following the circulatory changes may be seen degenerative changes in ganglion cells - swelling, shrivelling, pyknosis of nuclei and necrosis—followed by satellitosis and neuronophagia. Around areas of colliquative. necrosis and myelinoclasis are found the microglia which phagocytise the material and form "gitter cells". These leave small vacuoles after removing the debris. Demyelination of white substance occurs in the cerebellum, anterior medullary velum, the myelinated tracts of the cerebellum and the white columns of the spinal cord Foci may also be found in the ganglia of the brain-stem and optic tract.

There is swelling of the capillary endothelium and the nerve fibres are separated widely by fluid. Following this edema, the myeline sheath and axis cylinders swell, disintegrate and disappear. Microglia proliferate and accumulate to remove the necrotic tissue. *Gemistocytes* appear in these areas of softening. These are astrocytes which have become plump and rounded with a vesicular nucleus after the processes are withdrawn. Perivascular cuffing and vasculitis are seen in these foci of demyelination.

Degenerative and inflammatory processes may be seen in the cerebrospinal and sympathetic ganglia and these include infiltration by lymphocytes and plasma cells, degeneration and loss of ganglion cells, neuronophagia and satellitosis. The proliferating satellite cells form small nodules and replace the lost ganglion cells. In summary the lesions produced are nasopharyngeal catarrh, pneumonia, gastrointestinal catarrh (manifested by diarrhoea), pustules on the abdomen and thighs, perivascular cuffing, vasculitis, satellitosis, gliosis (with formation of gemistocytes). capillary proliferation in brain, demyelination and neuronophagia.

Inclusion bodies : Intracytoplasmic and intrauuclear, acidophilic inclusion bodies are noticed in various cells described below. They may be rounded or ovoid, are 5 to 20μ in diameter. From one to ten inclusions may be seen in each cell.

The inclusions are found in the epithelial cells of the mucosa of the nose and pharynx; the mucosa of bronchi and bronchioles; in the alveolar epithelium and the alveolar mononuclears: in the epithelium lining the urinary and genital passages, the ducts of liver, pancreas and salivary glands; in the neurones, microglia and gemistocytes of the brain; in the retinal ganglion cells, and glia; in the reticulo - endothelial cells of the lymph nodes and spleen: epithelium of the adrenal medulla and in the circulating neutrophils.

Diagnosis : (a) Symptoms. (b) Finding inclusion bodies in smears of cells from the respiratory system, urinary bladder or bile ducts, (c) Animal inoculation (ferrets or dogs).

RABIES
Syn : Lyssa, Hydrophobia

Rabies is a highly fatal, infectious, viral disease to which all warm blooded animals including man are susceptible. It is primarily a disease of carnivores that is transmitted to other animals by bites. Since the virus is present in the saliva, infection by bites is facilitated. Rabies has been eradicated from England, and some Scandinavian countries by strict quarantine of imported dogs and muzzling others. The wild carnivores —wolves, jackals etc—appear to be reservoirs for the virus

Nature of the virus : The rabies Rhabdovirus is neurotropic, but has an affinity for salivary glands. It reaches the brain from the site of bite centripetally through the nerves. A hematogenic spread to brain is also suggested when virus enters blood. The virus is 75×180 mμ.

After reaching the brain, it parasitises the ganglion cells, grows and then spreads through various nerves that emanate from the brain in a centrifugal fashion. So virus spreading along the nerves that supply the salivary glands, the facial (vii) and glossopharyngeal (ix) reaches them and from there it is excreted through the saliva, It may be present in the milk also. Infection may spread to the fetus *in utero* from an affected bitch. Infection is almost always by the bite of an affected animal. Usually it is spread in this way by carnivores. Licking of a wound by a rabid dog may also result in infection.

In South America, the blood sucking vampire bats are known to be reservoirs and transmit the disease. In other countries, wild canidae (skunk, fox, racoon) are sources of infection

Types of virus : There are two strains of the virus—the "street virus" and the "fixed virus". "Street virus" is that which occurs normally in nature. If this virus were to be passed through rabbit brain fifty times. or so the "fixed virus" may be obtained. It is called fixed because, its properties of reproducing the disease, incubation period etc., are all fixed without substantial variation. The differences are ;

Street virus	Fixed virus
1. Incubation period variable and long.	1. Incubation period fixed and short,
2 Following intracerebral inoculation death of rabbit occurs in 14 to 20 days	2 Death of rabbit occurs in 7 days.
3, Negri bodies present	3, Negri bodies not usually present.
4. Salivary glands affected	4 Salivary glands not affected.

The virus can be cultivated on tissue cultures and chicken embryo.

The insectivorous bats harbour the virus without manifesting any symptoms. In the vampire bats though recovery occurs the virus is still present in the salivary glands·

The virus is present in the salvia of dogs up to 5 days before clinical symptoms are manifested.

Incubation period depends on the site of bite. The nearer the bite to the head, the shorter the incubation period. So it may vary from 2 weeks to 9 months So quarantine period in U. K. is now for 12 months.

Pathogenesis ; Reaching the brain, the virus damages the nerve cells and he vascular endothelium. Irritation of the nerve cells causes increased excitability manifested by the furious form. Due to affection of medulla, fever, polyuria and glycosuria are observed. Subsequently, the nerve cells undergo degeneration and so paralysis of various muscles sets in. Due to paralysis of the muscles of deglutition. the animal is not able to swallow and so saliva dribbles. Because of paralysis of jaw muscles, jaws cannot close and so hang down, Paralysis of respiratory muscles results in asphyxia and death.

Due to affection of vascular endothelium perivascular cuffing is noticed.

Symptoms : Rabies is manifested in the dogs in two forms : the "**furious form**" and the "**dumb form**". Both the forms terminate in death following paralysis

In the furious form the animal is very excitable, bitting everybody indiscriminately, even its master. It snaps and barks at imaginary objects. It chews all sorts of objects - sticks, stones etc and may drink its own urine, The eyes are red and the animal may have a vacant look, There is champing of jaws and dribbling of salvia. The timber of bark changes.

In the dumb form the animal is in a morose condition and has a bland vacant look. It does not obey orders and does not recognise its master The lower jaw hangs and saliva dribbles.

Death occurs in 3 to 4 days after onset of symptoms. At any rate death occurs within ten days. In horses symptoms of colic are characteristic besides mania. Cattle and sheep bellow and bleat incessantly. In cats furious form only is seen while in rabbits only dumb form is observed Bulls and rams may show excessive sexual urge.

Lesions : The lesions in rabies are limited to the central nervous system. The only macroscopical changes noticed may be hyperemia and edema of the meninges with a few petechiae

Microscopically, non - suppurative encephalitis is noticed. Perivascular-cuffing by lymphocytes mostly but rarely by plasma cells is noticed. Around ganglion cells, the satellite (Microglia) cells proliferate and form small nodules, the *Babes nodules*. After the ganglion cells get degenerated, neuronophagia occurs. These changes are seen more particularly in hippocampus, the brain stem and the gasserian ganglion. The Babes nodules are found in the gasserian ganglion earlier than in other parts.

Similar changes as described above are also seen in the cerebrospinal and sympathetic ganglia. Ganglioneuritis in paravertebral ganglia is constant and significant in rabies These are swollen and red in color and reveal infiltration by lymphocytes, degeneration, neuronophagia of the ganglion cells and setellitosis (forming the Babes nodules).

Other lesions include, acute catarrh of the mucosa of the respiratory and digestive tracts: fullness of gall bladder; hyperemia of kidneys, spleen, liver and salivary glands.

Negri bodies : Inclusion bodies are found in the cytoplasm of the neurones These are round and have a clear halo round them They are 1—27 μ and one to twenty bodies may be found in each neuron. They are also found in the dendrites. These inclusions which are called Negri bodies are numerous in hippocampus. In cattle, on the other hand, they are more numerous in the Purkinje's cells of the cerebellum. The inclusions which are acidophilic contain basophilic granules. The consensus of opinion is that the Negri bodies are aggregates of the virus

Diagnosis

1. Symptoms.

2. A suspected dog must be isolated for at least 10 days. If rabid, it will die within that time.

3 Make impression smears of hippocampus major, stain by Seller's or modified Mann's method and look for Negri bodies.

Note :— Inability to find Negri bodies is not conclusive evidence that the animal is not rabid, for in some instances Negri bodies may not be seen in positive cases.

4 Habel's mouse inoculation test :—

Swiss mice, 2—3 weeks old, are needed. Make an emulsion of a piece of Ammon's horn and dilute 20 times with sterile water. This may be incubated with an antibiotic for ½ hour. 0.03 ce. of this emulsion is injected through the skull into 6 mice. On the 5th, 6th and 7th days one mouse is sacrificed and the brain examined for Negri bodies. Negri bodies are seen on the 5th and 6th days.

5. A biological test, injecting the brain tissue into rabbits intracranially, can also be done for comfirmation

6 A fluorescent antibody test is done. This test has now replaced others in many countries.

7. A complement fixation test can also be carried out.

AUJESZKY'S DISEASE

Syn; Pseudorabies; Infectious bulbar paralysis; Mad itch

This is a disease affecting many species of animals, caused by a pantropic virus belonging to the group of Herpes viruses. Man is susceptible

Probably, pigs and rodents are the natural hosts Some adult pigs may harbour the virus without showing any symptoms. Among animals. the disease has been seen in cattle, sheep and pigs. Brown rats are probably reservoirs of the virus.

Routes of infection : Infection occurs by rubbing of snout of pig on to an abraded skin. Intranasal infections can also take place. Infection through intestines also occurs

Period of Incubation : 7 days in natural infection.

Pathogenesis : The virus being pantropic affects tissues derived from all the embryonic layers. From the site of infection, virus travels via peripheral nerves to the spinal cord after causing a local reaction, at the initial site, where the muscles and facia undergo necrosis. In the spinal cord, the ganglia are affected in which the neurones show degeneration and proliferation of glia cells This change is responsible for the intense pruritus (itching) at the site of infection. The virus may travel to the brain from the spinal cord. In both the places, a non-purulent encephalomyelitis is produced.

Symptoms : In pigs paralysis and incoordination of movement are seen-Convulsions, muscle tremors, vomition and complete paralysis are noticed preceding death. Pruritus (itching)is not seen in pigs. Deaths follow within 12 hours after the onset of symptoms. Usually young pigs are affected.

In cattle there is violent itching as shown by biting and kicking at the site. Animals bellow continuously and show maniacal behaviour Paralysis sets in following pyrexia and death occurs in 6 to 48 hours. In some cases sudden death may occur without showing any symptoms.

Lesions : At the place of itching considerable inflammation and edema are evident. Pulmonary edema, sometimes pneumonia, laryngitis subendocardial hemorrhages, gastroenteritis, myocarditis with myocardial necrosis are seen. Necrotic foci in the spleen and liver may be evident.

Microscopically, in the pigs non-purulent encephalitis with perivascular cuffing in the gray and white matter are seen. Meningitis may also be observed. In cattle only degenerative changes of ganglion cells are noticed.

Intranuclear inclusions may be found in the degenerating neurones of cerebral cortex in pigs.

Diagnosis :

1   Symptoms.

2.   Subcutaneous inoculation of edematous fluid of the lesion, parts of spinal cord and portions of brain into rabbits causes pruritus in 48 hours. Later the animals become frenzied and bite furiously at the site of infection. Still later they become paretic and die following clonic spasms and exhustion.

3.   Fluorescent antibody technique.

4.   This must be differentiated from rabies. Differences between Rabies and Pseudorabies are :

Rabies	Pseudorabies.
1. Incubation period long.	1. Incubation period very short, 24-48 hours
2. Course upto 10 days.	2. Course very rapid, death in 2 days.
3. No itching present.	3. Considerable itching present.
4. Paralysis of larynx not often	4 Paralysis of larynx often.
5. Consciousness lost.	5. Consciousness never lost.
6. Virus found in saliva.	6 Virus not found in saliva.
7 Inclusion bodies intracyto-plasmic.	7. Inclusion bodies intranuclear.
8. Cross immunity tests in rabbits can be performed	8. Not performed.
9 Serological tests can be performed.	9. Not. performed
10. Infection cannot be obtained by subcutaneous inoculation of rabbits.	10. Infection can be obtained by subcutaneous inoculation of rabbits.

INFECTIOUS CANINE HEPATITIS
Syn : Rubarth's Disease: Hepatitis contagiosa canis

This is an acute viral disease of dogs. This virus is identical with that causing fox encephalitis and belongs to the group of adeno viruses. The disease was fully described by Rubarth in 1947.

Incubation period : When ingested incubation period is a few hours. But when dogs are kept together, it is 6 to 9 days.

75 to 90% of affected dogs recover within two weeks.

Symptoms : Peracute : The dog dies within 12 to 24 hours. The animals in this stage may not reveal any weight loss. Very slight icterus may be present.

Acute : If the animal survives 24 hours, it will recover after an illness lasting 4 to 10 days. The symptoms seen are high temparature and malaise with anorexia and intense thirst. There may be subcutaneous edema of the head, neck and abdomen. Vomiting and bloody diarrhoea are commonly seen. In some animals nervous symptoms manifested by hysterical seizures with barking, paralysis of hind limbs and clonic spasms of extremities and neck are noticed. The mucous membranes may be slightly icteric but pale. Petechiae are seen on the gums. There may be keratoconjunctivitis of one or both eyes with copious lachrymation as well as cloudiness of the cornea. Corneal opacity disappears as the animal recovers.

The tonsils are swollen and red. Albuminuria may be present.

Lesions : The virus has special affinity for the endothelial, mesothelial and hepatic parenchymal cells, in which large, basophilic, intranuclear inclusions are found. The virus causes necrosis of these cells and hence produces various changes seen in the organs,

In the peracute fatal cases, the tonsils and the lymph nodes throughout the body are enlarged and hyperemic. Hemorrhages may be noticed on the meninges. spinal cord. brain, thymus, heart, stomach and intestines There may be serosanguineous fluid in abdominal cavity with deposits of fibrin on the surface of the liver, especially between the lobes and other abdominal viscera.

Liver is enlarged, congested and friable. The sinusoids are widely dilated causing pressure on the adjacent hepatic parenchyma. There are focal areas of necrosis. Wall of gall bladder is edematous

Microscopically, necrosis is centrilobular. The hepatic cells lose their staining affinity and the nuclei are not visible. In the cells peripheral to the necrotic area can be seen enlarged nuclei containing the thick basophilic inclusion bodies without any definite structure. The sinusoidal endothelium is swollen and the nuclei of the endothelial cells are enlarged and pyknotic. Infiltration by macrophages, lymphocytes and neutrophils is found in the space of Disse.

Inclusions described in the hepatic cells can be demonstrated in the endothelial cells and Kupffer's cells.

The wall of the gall bladder is very much thickened due to edema and is dark due to hemorrhages.

Spleen is swollen and shows changes of acute splenitis. Involution of follicles is apparent. Inclusion bodies are seen in the reticular and endothelial cells

There may be serous pancreatitis. Slight glomerulonephritis with swelling of endothelial cells of glomerular tufts is seen with inclusion bodies in them. Acute catarrhal gastroenteritis with edematous swelling of the mucosa and hemorrhages are also observed

The lymphnodes which are swollen show serous lymphadenitis with swelling of germinal centers. Hemosiderosis is present.

In the central nervous system is found edema of piamater, perivascular edema and small hemorrhages. The epithelial cells of blood vessels contain inclusions. In some cases there may be mild nonpurulent encephalitis. Due to hepatic injury hypoglycemia develops.

Diagnosis :

1. Only way to confirm is to demonstrate microscopically the intranuclear inclusions and hepatic lesions.

2. Severe hypoglycemia—Blood glucose level falls to 20mg/100cc.

BLUE TONGUE

Syn; Fievre catarrhale du mount: catarrhal fever of sheep: sore muzzle of sheep

This is an acute viral disease of sheep. Cattle may be infected but the disease in them is not as severe as in sheep Cattle may act as reservoirs. The virus is a double stranded RNA virus of orbivirus group.

Mode of infection :   Biting insects of the genus *Culicoides* transmit the disease It is not transmitted by direct contact or feed or water,

Incubation Period is usually less than a week.

The sheep ked, *Melophagus ovinus* may transmit the disease mechanically.

Pathogenesis : Affecting the blood vessels, hyperemia, edema and hemorrhage are produced in various tissues. Affecting hemopoietic tissue, anemia and leucopenia are caused.

Symptoms : First there is a rise in temperature, which is only for a short duration, 24 to 72 hours. Usually in visibly sick animals, no temperature rise is seen. There may be redness of nasal and oral mucosa with profuse salivation. There may be a serous nasal discharge which later turns mucous and even bloody. This may dry up and form crusts in the nose, blocking partially the passage. There may be edematous swelling of lips, nose, ears, tongue, intermandibular space and the face. There is edema and intense cyanosis of the tongue (hence blue tongue). On the nasal and oral mucosa may be found petechiae and thickening. By shedding of the epithelium at these places, ulcers are produced which may be subsequently infected and become gangrenous

As the fever subsides, flushing of the skin and feet takes place, so that the coronets are warm and the periople which should be pink turns deep red. The animals therefore go lame as the condition is painful. There may be a streaky zone parallel to the periople like the lines seen in laminitis (pododermatitis of horses). The mucosa of the tongue may become necrotic and show hemorrhages. There may be edema under the skin on the neck and abdomen,

There may be pneumonic lesions and gastroenteritis. (diarrhoea) which are the offshoots of spread of infection down the respiratory and digestive tracts from the oral lesions Wool may be shed.

The disease finally terminates with extreme emaciation, cachexia and muscular weakness, There may be torticollis. Death may be due to pneumonia and occurs 6 days after commencement of symptoms,

The morbidity may range from 5 to 50% while the mortality may be less than 10% Mortality is higher in lambs.

Pregnant cattle may abort. Infection during pragnancy may cause congenital deformities in the fetus.

Lesions : The lesions at postmortem are seen around the mouth and these include hyperemia, edema, cyanosis, multiple hemorrhages and swelling of the epithelium and these result in erosions and ulcers   Ulcers on the tongue may be found on the lateral surface, adjacent to the teeth. Due to infection by saprophytes, gangrene may occur.

In the foot there is congestion of the skin papillae together with edema and neutrophilic infiltration producing lesions similar to laminitis. It is the pressure on the sensitive laminae by the infiltrating fluid and cells that causes pain and lameness.

In the skeletal muscles may be noticed hemorrhages and necrotic foci, In the muscles affected may be found hyaline degeneration with loss of striation and pyknosis of sarcolemmal nuclei, If the sarcolemma is intact, regeneration is possible. Otherwise the degenerated sarcoplasm is removed by macrophages and scar tissue forms. Sometimes dystrophic calcification of the affected muscles may occur

Hemorrhages may be found on the mucosa of abomasum and duodenum as well as in the myocardium. Pericardial space is filled with serosanguineous fiuid, Liver may show fatty changes while in the spleen may be noticed congestion and swelling. Edema of the lungs developing into pneumonia may be found.

Diagnosis :

1. Symptoms and lesions.

2. Isolation of virus by inoculation on embryonating eggs and identification by inoculation into young lambs in which   temperature rises to 106° F by 4th or 5th day. Then temperature drops followed by swelling, hemorrhages and vesicles of the lips.   Virus is present in the blood during fever but not when temperature drops.

3. The virus can be grown in lamb - kidney cell culture. Identification of the virus can be done on serum virus – neutralisation in this culture. Antigenically different viruses occur in nature with variable virulence. As many as 18 distinct antigenic types of blue tongue virus have been reported.

4   Differentiate from Foot and Mouth disease, contagious ecthyma, photosensitisation and pasteurellosis.

SCRAPIE

This is a disease of sheep found mostly in Scotland, Europe and America characterised by pruritus and other nervous symptoms

Etiology : The etiological factor is still not determined. Though many suspect that it is a virus, the fact that the materials used for inoculation do not become sterile after boiling for 30 minutes or exposure to 20% formalin or to ribonuclease or to acetylethyeneimine (which render any virus non-infective) compells one to suggest that the infective agent is something other than a virus Again unlike other viruses, the agent of scrapie is not composed of nucleoprotein, It is now considered to be a "viriod"—a naked circular RNA

There is strong conviction among some scientists that scrapie may be a hereditary disease.

Routes of infection: Since the disease can be infected through the oral route. the disease is suspected to be spread by **Contact**. Experimentally, intracerebral inoculation is most successful in transmission while other routes are less so

Coitus is also mentioned as a route of infection and so **Congenital** infection is also possible.   There is vertical transmission of the disease from ewe to lamb

Incubation period is very long   In natural infections it may be upto 3 years while in experimental animals it may be 4 to 6 months.

Animals affected : The disease occurs naturally in sheep. Goats can be infected experimentally.   Morbidity is upto 40% while mortality may be 100%.

Symptoms : Usually sheep over two years of age are affected.   The disease is manifested by **alertness** and restlessness.   They have great desire to **rub** against hard objects because of intense pruritus and so all wool is lost,   When back is scratched, animals have a startled look and grind their teeth.   The muscles of thigh and shoulders begin to twitch.   The gait of the animal is unsteady (wobbling), stiff and trotting. When startled, the animals may fall in an epileptic fit.   They become emaciated and paralysis of the hind limbs sets in. Death may occur within 10 days to several months after onset of symptoms.   No recovery occurs in those that manifest symptoms.

Lesions :   No gross lesions characteristic of the disease are seen.

Microscopically, lesions may be seen in the medulla oblongata, pons, brain and spinal cord.   The most notable change is the vacuolation of neurones.   Low degree of perivascular cuffing, neuronophagy, satellitosis and gliosis may be present.   Demyelination around smaller vessels may be noticed.   The cerebrospinal fluid may be increased in quantity.

Other lesions seen are degenrative changes in the thyroid, choroiditis and detachment of the retina.

Diagnosis

1.   Clinical symptoms.

2.   Histological examination of brain and spinal cord reveals vacuolation of neurones

3.   Differentiate from pseudorabies, louping ill, pregnancy toxemia, photosensitive dermatitis and parasitic dermatitis.

EQUINE ENCEPHALOMYELITIS
Syn : Blind staggers

Encephalomyelitis caused by virus is encountered in horses and mules of India.   The causative role of virus in the outbreaks of encephalomyelitis in Indian horses and mules has been fully established.

In America two viruses which are antigenically different, the 'Eastern' and 'Western' strains are known to affect horses.   In Venezuela, a third strain, the 'Venezuelan' strain is encountered.   Though the clinical symptoms are similar in all the strains, yet the Eastern strain is more virulent with 90% mortality than the Western strain with only 27% mortality.   Apart from these, Japanese B encephalitis virus, St. Louis encephalitis virus. virus of Borna disease, and virus of Russian Spring–Summer encephalitis are known to cause encephalomyelitis in different parts of the world.

Animals susceptible :   Horses, mules and man are affected.   Brids can be infected, but they may not show any symptoms. It is thought that wild birds and insects may act as reservoirs for the virus   Actually, it is thought that birds are the natural hosts but horses and man are accidentally infected.   Young horses are more susceptible. After recovery from natural affection, immunity lasts for 2 years.

Experimenally, the virus can be transmitted to calves, guinea pigs, white mice and dogs by intracerebral inoculation. The virus can be cultivated on embryonated eggs and in tissue cultures.

The disease is more prevalent in summer when insects abound in greater numbers Younger horses are more susceptible.

Mode of transmission : From the wild-bird-reservoir, horses and man become infected by bites of mosquitoes, blood sucking bugs, lice and mites of fowls. *Aedes* and *Culex* genera of mosquitoes are the vectors in which the virus may multiply. Infection may occur from contact in horses.

Incubation period :   3 days to 3 weeks.

Pathogenesis : From the bite of the insect, virus enters the blood giving rise to viremia. The viremia is transient in the American disease but it is always persistent in the Venezuelan disease. The virus is found in the saliva and nasal discharge. From the blood virus may enter the brain and affect the neurones producing the symptoms.

Symptoms ; The initial pyrexia and malaise may go unnoticed.

The symptoms noticed are usually due to derangement of the central nervous system. Animals become excitable and restless and do not seem to be aware of their surroundings. They are blind and move in circles They do not obey commands of the master. They may walk blindly, butting into obstacles Muscular tremors may be noticed. Subsequently paralysis sets in, the animal appears to be in a stupor, with head hanging. The animal may drop down and is unable to get up. Feeding and swallowing as well as micturi ion and defecation become difficult Finally temperature may become subnormal and the animal dies. The course in fatal cases is 2 to 4 days.

Lesions : No gross lesions are noticed except that brain and spinal cord may be hyperemic and edematous. Lymph nodes may be swollen and juicy. Liver and kidneys may show parenchymatous degeneration. The hepatic lesions may be responsible for icterus which may be noticed in some cases *Microscopical* changes are restricted to the gray matter where there is disseminated non-purulent encephalitis. The neurones manifest tigrolysis, chromatolysis, fragmentation and neuronophagia There may be infiltration by lymphocytes and neutrophils and a few red blood cells around the affected neurones. The neutrophilic infiltration is suggestive of malacia. Perivascular cuffing is prominent. Vasculitis may be observed Diffuse microglial proliferation is seen.

Most severe lesions are seen in the cortex, thalamus, hypothalamus, and in the dorsal and the ventral columns of the spinal cord.

The lesions produced by different strains of the virus differ only in intensity. The Eastern strain causes more severe lesions and greater neutrophilic infiltration.

No specific inclusions are found.

Diagnosis :

1. Symptoms and lesions in the brain.

2. Transmit to guinea pigs intracerebrally or subcutaneously or intraperitoneally; brain from an affected animal is the source of virus. The brain should

be collected within an hour after death. Guinea pigs show symptoms of paralysis in 3—5 days,

3. Inoculate into chick embroys which die in 18 to 24 hours.

4. Neutralisation tests : infect the virus and known strain-specific antisera intra-cerebrally into guinea pigs.   If virus and antisera are homologus, no disease occurs.

5. Complement-fixation and H I tests can be done with sera of affected animals.

6. Differentiate from Rabies, Mouldy-corn poisoning, Botulism and tumors of brain.

Borna disease affecting horses, cattle and sheep; *Louping ill* in sheep (and man) *bovine encephalomyelitis* are all caused by viruses charecterised by non-suppurative encephalomyelitis.   Of these specific inclusions are found intranu-clearly in the neurones only in Borna disease (Joest bodies).

AFRICAN HORSE SICKNESS
Syn ;   Equine plague; La peste du cheval
This is a highly fatal viral disease of horses, mules and donkeys.

Biology and mode of infection : Though the disease is found to affect nat-urally horses, mules and donkeys, other animals— guinea-pigs, mice ferrets, rats, dogs and goats—can be experimentally infected.   A severe out break occurred in India in 1960 and over 16000 equines died in this country alone due to the outbreak.

Biology and mode of infection: The causative agent is a viscerotropic virus and is found in all tissues and fluids in the body.   A large number of antigenic strains are recognised, which are related to each other.   It is a double stranded RNA virus belonging to orbi virus of Reoviridae family.

The disease is most severe in horses, in which 90% mortality may occur. But in mules and donkeys, it is milder.

The virus is fairly resistant to environmental conditions.

Transmission from animal to animal is through the gnat, *Culicoides*.   Mere contact does not spread the disease.   Experimentally, ready transmission can be effected by intravenous injection of blood.   Periodical outbreaks are supposed to occur by infection from a reservoir, in which the virus does not produce any symptoms (silent host).   Dog is suspected to be such a reservoir.

Incubation period :   5 –7 days in natural infection

Symptoms :

Acute pulmonary form DUNKOP form.

In this form which is most common in acute outbreaks, there is sudden rise of temperature followed by intense dyspnoea and coughing.   Because of pulmonary edema, there is frothy nasal discharge.   Developing profuse sweating and profuse nasal discharge, the animal may die within a few hours after the onset of symptoms.

Subacute cardiac form : DIKKOP form.

In this form the incubation period is longer and may be upto 3 weeks. Fever develops more slowly in this form and lasts longer. The most characteristic feature is the development of edema in the region of the head, particularly in the temporal fossa, lips, eyelids and on the neck and chest The tongue is swollen, cyanotic and may have petechiae on its ventral aspect. Paralysis of esophagus renders the animal unable to swallow. Cardiac failure produces pulmonary edema, hydropericardium and endocarditis. The course in fatal cases may be about 2 weeks but mortality is not high as in the pulmonary form. Death is due to cardiac failure and hypoxia

A **Mixed form** can occur in which a subacute form may suddenly develop into the acute pulmonary form and in such cases both pulmonary and cardiac lesions are seen

A **mild form** known as horse sickness fever may be noticed in partially immune animals especially donkeys. No symptoms may be seen and no fatalities are noticed. There may be initial rise of temperature which returns to norma in 1 to 3 days. Anorexia, dyspnoea and mild conjunctivitis may be the only symptoms observed.

Depending on the severity of outbreaks mortality varies from 25% to 95%.

Lesions :

Acute form : Hydrothorax and severe pulmonary edema is noticed with frothy exudate filling the bronchi, trachea, pharynx and the nasal passages.

In **the cardiac form**, there is marked hydropericardium, hemorrhages on the endocardium and ascites. Myocardium may reveal foci of necrosis scattered throughout along with hemorrhages. The gastrointestinal mucosa may be congested. Liver is enlarged and congested. Lymph nodes may be hemorrhagic. Spleen and lymph nodes reveal depletion of lymphocytes but increase in the R.E. and plasma cells. Paralysis of the esophagus noticed is due to edema around pharynx.

Diagnosis :

1. Clinical symptoms and lesions.

2. Recovery and identification of virus by intracerebral inoculation into mice and then conducting neutralisation tests using a known antiserum.

3. Differentiate from equine viral arteritis, equine infectious anemia, babesiosis and purpura hemorrhagica.

EQUINE INFECTIOUS ANEMIA
Syn : Swamp fever, Equine malarial fever

This is a disease of horses caused by a mesenchymotropic virus. The virus is now considered to be a RNA tumour virus.

Animals affected : The disease is found only in horses, mules and donkeys. Pigs are susceptible to experimental inoculation. Fowls may carry the virus. It is thought that man may be infected

Routes of infection : Since the virus is present in the blood and all secretions and excretions the disease is transmitted by biting insects — Tabanus,

Stomoxys and mosquitoes, There is no evidence that the virus multiplies in the vectors. Transplacental infection to the fetus occurs. Foals may become infected by drinking milk. Infection can occur by using the same hypodermic needle for injection of infected and healthy animals, Contaminated bedding and utensils may be sources of infection since the virus can invade through the intact skin and mucosa of the nose and mouth. The source of infection is usually an infected or a carrier animal. The virus can remain in an animal for as long as 18 years.

Since flies and mosquitoes breed in swampy ground, the disease is more prevalent in such localities and hence the name *Swamp fever*. Predisposing factors are : malnutrition, hot and humid climate, fatigue, parasitization and faulty husbandry practices.

Incubation period 2 to 4 weeks

Pathogenesis : The virus affects the reticulo - endothelial system primarily and so changes are found in the blood, the blood forming and blood destroying organs. The destruction of erythrocytes, stoppage of erythropoiesis and other regressive changes are attributed mainly to the toxic products of metabolism.

Hypoxia that results due to destruction of erythrocytes is responsible for the damage to the parenchymatous organs.

The action of the virus on R. E. system results in hypertrophy and hyperplasia of histiocytes and lymphoid cells.

Hormonal disturbances and the effect of the virus on the nervous system cause further circulatory collapse, In chronic stages, the spleen loses its erythrocyte destroying property. The metaplasia of the splenic R. E. cells into lymphoid cells confers on this organ the function of antibody production, The erythrocyte destroying function is taken up by R. E. cells of the liver. Morbidity is 100% Though mortality is said to be 50%. actually there appears to be no recovery in affected animals since all of them eventually die due to exacerbation of chronic and inapparent cases.

Symptoms : Clinically, the disease is arbitrarily divided into acute, subacute and chronic forms, though it must be understood, that it is due to the exacerbation that death in subacute and chronic cases occurs.

Acute form: In acute cases the symptoms seen are only those of septicemia, viz. fever, inappetance, excessive thirst, depression, edema of the abdomen and petechial hemorrhages under the tongue and in the anterior nares. The hemorrhages mentioned are considered to be pathognomonic of the disease. Later intermittent fever and icterus are noticed with edema of limbs and prepuce. There may be lachrymation and serosanguineous nasal discharge. Mucosa does not become pallid nor are immature forms of erythrocytes found in the blood. The animal gradually becomes weaker and may die in a month. Those that survive pass on to the subacute stage. Pregnant animals may abort. Slight albuminuria may be present,

Hematology : There is a great reduction of erythrocytes. from 8 000.000 to 1,000,000 in 2 weeks.

The erythrocyte sedimentation rate is increased and the coagulability of the blood is reduced. Life span of erythrocytes is reduced to 28 to 87 days (normal 119 to 153 days.)

Hemoglobin content of the blood is diminished while the bilirubin content is increased. There may be neutrophilia but in some neutropenia may be observed. Lymphopenia is noticed but there may be monocytosis No abnormal or immature erythrocytes are usually seen. Anemia is normocytic and normochromic.

The sub-acute form : In this form there is relapsing fever and other symptoms recur at intervals. The animals may become chronically affected or due to exacerbation they may die mnifesting the acute form. Packed-cell-volume diminishes with each relapse. Albuminuria may be present Conjunctiva is icteric.

The chronic form : This may be continuance of the sub-acute form or the chronic form may arise even without passing through an acute attack Animals may become progressively weaker and show depression, edema, pale mucosa and debility. Erythrocyte counts may be low. Or, the animals may appear to be in good health, with infrequent intermittent fever.

Hematology ; There is relative lymphocytosis and siderocytes are seen in the blood. Clotting time in increased

Lesions : In animals that die or are destroyed during the acute stage, the lesions seen are typical of septicemia, viz sub-serous petechiae in the epicardium. peritoneum, pleura and endocardium Petechiae are also seen on the mucous membranes and in the parenchyma of organs. Edema is found subcutaneously. Cloudy swelling of parenchymatous organs is present The spleen and lymph nodes may be enlarged.

In the sub-acute stage, anemia is more conspicuous than hemorrhages and edema, The liver, kidneys and spleen are enlarged and the bone marrow is hyperplastic.

In the chronic form no changes except swelling of the spleen and bone marrow hyperplasia are seen.

Spleen :

Acute stage ; The spleen is very much enlarged, almost to twice its normal size, with a tense capsule on which hemorrhages may be seen. It is soft to the touch and deep red in color. On section, the pulp bulges. The malpighian bodies are not conspicuous. Infarcts may be seen sometimes.

Microscopically, there is increased amount of blood in the organ There may be increased number of mononuclears which are supposed to have been derived from the R. E. cells by metaplasia and are considered to be lymphoid cells. These are more conspicuous and characteristic in the sub-acute and chronic stages. Hemosiderin content is still considerable in the earlier stages but diminishes later.

Subacute and chronic stages : The spleen is enlarged with a tense capsule. The organ is firmer in consistency and grayish-red in color. On section pulp does not ooze out. Small nodules may project from surface of the spleen and on section these appear to be infected areas filled with blood.

Microscopically, the characteristic appearance is the replacement of the pulp by the hyperplastic lymphoid cell tissue which is derived from the reticulum cells. Consequently, hemosiderin storing cells are considerably reduced or may even disappear completely. Erythrophagocytosis and hemosiderosis are absent.

Lymph nodes :- The lymph nodes, especially splenic, renal, mesenteric and portal are all enlarged in the acute stage. On section hemorrhages are noticed.

Microscopically, in the acute stage there is hyperplasia of the reticular cells from sinus endothelium and there is erythrophagocytosis and hemosiderosis. These are more conspicuous in the sinus

In the chronic stage, there is a gradual replacement of the pulp by lymphoid cells and the normal structure is lost. Hemosiderosis is absent.

Liver :

Acute stage ; The liver may be enlarged The capsule may be stretched due to edema. The organ is soft to touch.

Microscopically, there is edema and lymphocytic infiltration of the capsule. Sinusoids may be dilated. Kupffer's cells are swollen and contain hemosiderin Liver capillaries may have hemosiderin-containing macrophages, the siderocytes. Because of pressure by dilated sinusoids the cells in the centre of lobules may disappear.

Subacute and chronic stages : The organ is enlarged and slightly firm. On section lobular outlines are very distinct. A nutmeg appearance is obtained due to the presence of central dilated capillaries surrounded by lighter peripheral parts of the lobules.

Microscopically, the central veins are congested while the sinusoids are dilated with macrophages (containing hemosiderin and red cells), lymphocytes, plasma cells and R. E. cells. The hemosiderin containing cells are called *sidero-cytes*. The Kupffer's cells are highly swollen and these may be converted into macrophages These macrophages collect in the dilated capillaries of the central part of the lobule and become siderocytes. The R. E. cells may accumulate as small nodules of cells in the sinusoids. Hemosiderin can be demonstrated by Perl's stain Portal triads may contain lymphocytes and monocytes.

Chronic stage : In the chronic stage, the macrophages containing hemosiderin are gradually replaced by lymphoid cells. The capillaries in the centre of the lobule may contain siderocytes and a few lymphoid cells. Sometimes myelocytes and plasma cells may also be present-

Due to hypoxia and pressure, the hepatic cells, especially in the centre, undergo degenerative changes, viz. depletion of glycogen followed by vacuolar degeneration, dissociation and disappearance of the cells The hepatic cells may also contain hemosiderin and bilirubin.

Heart :

Acute stage : The myocardium is pale and shows cloudy swelling or fatty degeneration leading to dilatation of the ventricles. Petechiae are found on pericardium and epicardium Pericardial space contains increased fluid.

Microscopically, there is lymphoid cell infiltration around blood vessels, Extension of this infiltration into the muscle bundles causes their atrophy.

Subacute and chronic stages ; Heart muscle shows under the epicardium and endocardium white fatty or cicatricial foci.

Microscopically, proliferation of histiocytes and siderocytes is seen, Reticular fibre sclerosis occurs, finally culminating in cicatrization of foci. The myocardial fibres undergo granular and hyaline degeneration.

Kidneys; Acute stage ; The kidneys are much enlarged and petechiae on the cortex are seen.

Microscopically, infiltration by lymphocytes and plasma cells is observed in the interstitial space and around glomeruli. Siderocytes may be found in the glomeruli. No changes are found in the tubular epithelium. Glomerulitis with swollen tufts and increased cellularity due to proliferation of the endothelial cells is seen

In the subacute and chronic stages, changes similar to those in the acute type are seen but hemorrhages are not conspicuous. There may be interstitial fibrosis leading to atrophy of glomeruli and tubules.

Bone marrow : Acute stage : The yellow fatty marrow may contain reddish areas

Microscopically, hyperemia is noticed. Though erythropoiesis is reduced in the acute stage, increased erythropoietic activity is noticed just before and during febrile attack. Later, this activity is reduced.

Subacute and chronic stage The marrow appears red

Microscopically, there is proliferation of lymphoid cells and siderocytes. Erythropoiesis is not depressed but is maintained as in a normal state. Myeloid cells may be more obvious with a shift to the left.

Nervous system ; Non-purulent encephalomyelitis may observed.

Microscopically, perivascular cuffing and formation of granulomas are noticed. The granuloma consists of histiocytes, a few giant cells and mononuclears. Vasculitis is common.

Other organs : The same lesions noticed in liver and spleen are also found in adrenals, gonads, pituitary and pancreas, viz. in the acute stage, venous congestion and stasis, parenchymal necrosis due to hypoxia and histiocytic and lymphoid cell proliferation.

In the subacute and chronic stages, there is is greater formation of histiocytes and lymphoid cells. hemosiderosis and fibrosis replacing parenchyma.

Diagnosis :

1. Symptoms and lesions: hematology: more than 4 to 7 siderocytes per 100000 leucocytes is evidence of this disease.

2. Inoculate healthy horse with blood from a suspected animal and observe rise in temparature and other symptoms.

3. Complement - fixation test

4. Immunodiffusion test.

5. No characteristic inclusion bodies are noticed.

6. Differentiate from: Influenza, anthrax, glanders, purpura hemorrhagica. babesiosis, leptospirosis and strongylosis.

INFECTIOUS BOVINE RHINOTRACHEITIS

This is a viral disease of cattle. affecting mostly young animals in feed lots. This is a herpes virus and also causes infectious pustular vulvovaginitis (which has already been described in chapter 20, page 448) and Infectious Kerato Conjunctivitis (page 525). The virus may be cultivated on tissue cultures using bovine kidney cells. Inclusions are found in the nuclei of these cells.

Infection is usually introduced by an ailing animal and is probably by nasal route. Nasal discharges contain the virus. Morbidity is high and all animals in the herd may be affected while mortatity is low (being about 3%) Animals usually become infected soon after a journey. Recovered animals may shed the virus for a long time. **Incubation period** under natural conditions is 4 to 6 days but experimentally the disease can be manifested in 18 hours.

Symptoms : In most of the animals, symptoms are mild. The disease starts with fever and nasal discharge, which in the early stages is serous but may become mucopurulent and hemorrhagic. There may be coughing and lachrymation. Abortion of pregnant cows may occur. Usually the course is short, being only 3 to 7 days. If it is prolonged, respiratory distress may develop leading to pulmonary emphysema Milk yield drops or stops.

Deaths occur due to secondary bacterial infection of the lungs.

In calves under 6 months of age. an encephalitic form may be encountered characterised by ataxia and depression followed by frenzied movements, convulsions, frothing at the mouth. opisthotonus and recumbency. There may be threshing of the legs and grinding of teeth. This is a rapid and fatal form.

Lesions : The muzzle is covered with crusts. The nasal mucosa is highly congested and edematous and covered by thick mucopurulent exudate. Similar congestion and exudation are present in the mucosa of the trachea and paranasal sinuses.

In some cases, the inflammation may extend to the pharynx, larynx and to the bronchi.

In more severe affections, there may be hemorrhages as well as suppurative and diphtheritic inflammation of the trachea.

An untoward complication is bronchopneumonia. Dyspnoea is due to this condition as well as to stenosis of trachea (because of exudate and edema of its walls), Death is due to bronchopneumonia.

Microscopically, mucosa reveals edema, infiltration by leucocytes and in some cases hemorrhages· In severe cases there may be pseudomembranous deposits and the mucosa may be destroyed. Acidophilic intranuclear inclusions are found in the epithelial cells of nasal and tracheal mucous membrane.

Diagnosis : I. Isolate virus from nasal swabs and identify by cross protection tests in cattle. 2. A serum neutralisation test can be conducted for detection of antibodies. (3) F. A. test is also done This is quick and reliable.

EPHEMERAL FEVER
Syn : Three day sickness

This is an infectious disease of cattle caused by a virus. It is called "three day sickness" because the duration of this mild affection is only three days However, loss in a herd is great since milk production is depressed.

Routes of infection : The virus is intimately associated with the leucocytes-platelet fraction of blood. Artificially animals can be infected by inoculation of whole blood or the leucocytes but not serum.

There is some evidence that cyclic development of the virus occurs in vectors, which are insects, most important being sand flies (*Ceratopogonidae species*). Recovery is followed by solid immunity lasting for two years.

Incubation period : 2 to 10 days.

Pathogenesis: After an initial viremic state, the virus localises in the joints, lymph nodes and muscles.

Symptoms : The first symptoms is high fever, which becomes normal in 3 days. During this period the animal is lethargic, anorexia develops, milk yield stops and there may be constipation or diarrhoea. Lachrymation and nasal discharge are present.

The animals feel stiff in their joints and there may be tremors. The standing posture resembles that of animals suffering from laminitis. Though animals start eating after 3 days, weakness and stiffness persist for 2 or 3 days more Animal likes to lie down. In 3 to 6 days it becomes well

Occasionally in some animals presistent recumbency occurs when they may develop pneumonia.

Lesions : The virus affects the mesenchymal tissue :— joints, muscles and lymph nodes. The lymph nodes are all swollen and juicy Serous cavities may contain increased amount of fluid together with congestion or petechiae on their surfaces

Diagnosis is not usually difficult. In individual cases it may have to be differentiated from milk fever, traumatic reticulitis and laminitis.

CF. test can be done.

TRANSMISSIBLE GASTROENTERITIS OF PIGS

This is an infectious disease of pigs caused by a virus belonging to corona virus group containing RNA. Though the virus can be cultivated on tissue cultures, growth on chick embroys does not occur.

Routes of infection : Infection is usually by ingestion and the source of infection is usually introduction of infected pig that may manifest no symptoms Infection may also be transmitted by visitors, vehicles and birds. Animals may excrete the virus upto 7 weeks after clinical recovery, But they do not become permanent carriers, Infection through nasal mucosa may occur.

Pathogenesis : After an initial viremic stage, the virus localises in the ki.neys and gastrointestinal mucosa.

The virus, withstanding the acid pH of the stomach, finally reaches the cols umnar epithelial cells of the small intestines. The virus entering these causes their destruction or may also alter function as, the cells become cuboidal or squamous.. This change in the epithelial cells alters the absorbing capacity of the intestines because of lack or loss of some enzymes. Lactose is not hydrolysed and so also other nutrients are not digested. This leads to deprivation of nutrition to the animal which thus becomes emaciated. The undigested lactose increases the osmotic pressure in the bowel and so fluid is not only retained but is absorbed into the intestines resulting in diarrhoea and dehydration.

Incubation period : is 12 to 48 hours.

Symptoms : Some piglets may die without showing any symptoms. In others there is sudden vomiting and diarrhoea. Pyrexia may be present. Diarrhoea may be profuse, watery and yellowish-green in color. Rough hair coat, dehydration and emaciation follow quickly, with weakness and death occurring in 2 to 5 days. In those that survive again in weight is slow.

Older animals usually do not show any symptoms Some may manifest mild symptoms but recover in 10 days.

Lesions. Kidneys Cortex is pale while the medulla is congested.

Gastrointestinal tract: Gastric vessels are congested and the deeper epithelium of crypts shows necrosis. Hemorrhages are found at the pylorus. The intestines show acute inflammation The gut loses its tonus and so is dilated. Necrosis of the mucosa may be noticed in some places. There may be nonpurulent encephalitis.

Petechiae may be seen on the kidneys, liver, spleen, larynx and lymph nodes. Protein casts are found in the distal convoluted tubules. Villi atrophy and this is evident within 24 hours after infection, but epithelial cell differentiation does not occur. So the normal villus to crypt ratio of 7 to 1 is reduced to 1 to 1. Regeneration occurs after 5 to 7 days

Death may be due to dehydration and metabolic acidosis coupled with hyperkalemia (which causes abnormal cardiac function.)

Diagnosis: This disease must be differentiated from swine fever. Confirmation is made by positive serological tests—hemagglutination and H.I. tests and virus neutralization test in tissue culture.

VIRAL DISEASES OF POULTRY
(Revised by Dr. S. J. Seshadri
RANIKHET DISEASE
Syn : **Newcastle disease: Doyle's disease: Pseudo-fowl plague:**
Pneumoencephalitis

Ranikhet disease is an acute disease of fowls with high mortality caused by a virus, belonging to the Paramyxo virdiae. It is called Ranikhet disease because the first case in India was recorded at a place called Ranikhet near Almora, U. P. It is known as Newcastle disease as it was first investigated by Doyle among fowls at Newcastle - on - Tyne in England.

The virus can be cultivated on embryonated eggs Ranikhet disease causes great loss in poultry farms if systemic vaccination is not done. Because of the availability of an excellent vaccine, the losses now are negligible.

Routes of infection; Infection is by ingestion of contaminated food and water. Source is usually an ailing bird or one in the incubation period; The disease can be spread by attendants, wild birds and fomites.

Incubation period : 4 to 11 days
The virus is believed to cause conjunctivitis and general malaise in man.

Pathogenesis : Three strains of the virus are recognised : (1) **Velogenic strain** which causes the acute septicemic form of the disease with hemorrhagic lesions in proventriculus and diphtheritic ulcerations in intestines.

2) **The Mesogenic strain** which causes the pneumo-encephalic form of the disease, characterised by respiratory symptoms and nervous symptoms and having a prolonged course compared to the acute septicemic form mentioned above. This form of the disease probably occurs in connection with *Mycoplasma gallisepticum infection* 3) **Lentogenic strain** which causes the inapparent form of the disease which is asymtomatic except for the complaints of the farmer that

the egg production in the farm has considerably gone down inspite of best management

The virus has great affinity for the vascular endothelium and lymphoid tissue. Entering the endothelial cells of the blood vessels, the following changes are brought about; damage and necrosis of endothelial cells, hydropic swelling of the media of the vessel wall, hyaline changes of cepillaries as well as hyaline thrombi in arterioles Lymphoid tissue becomes necrotic while the reticulum undergoes lipoid hyperplasia. These changes of the blood vessels are responsible for various lesions noticed viz. anemia of the skin and muscles, hyperemia, edema, hemorrhages and necrosis.

While morbidity is 100% mortality is 80 to 90%.

Symptoms vary according to the age of the birds and the strain of virus involved.

Infection by Velogenic strain :

In chicks under 4 weeks of age, there is difficulty in respiration. This is an acute form and the duration of the disease is 3 to 4 days. Some may die within a day, The typical symptoms seen are; dull and depressed condition, respiratory rales, gasping with outstretched neck, nasal discharge, greenish-watery diarrhoea and nervous symptoms such as paralysis, tremors and torticollis. In this variety mortality may be high,

In older and growing birds. the symptoms seen in chicks are also seen, viz, respiratory symptoms, ruffled feathers. Loss in egg laying and profuse watery greenish diarrhoea may be noticed. There is profuse salivation and the salvia may obstruct respiration producing sounds, Nervous symptoms may also be seen.

Infection by Mesogenic strain: This form of the disease is less severe. Mortality is variable and may range from 5 to 15%. Respiratory distress, greenish diarrhoea, marked loss of egg production and nervous symptoms characterised by paralysis of wings and legs and torticollis are the salient features observed.

Infection by Lentogenic strain : Mild form of disease occurs with mild respiratory symptoms and rapid drop in egg production, Mortality in adult birds may be negligible but in young chicks it may reach 50% The disease may be asymptomatic and its existence recognised only by serological tests.

Lesions : Hemorrhages are found in various parts of the body—in the proventriculus around the orifices of the glands (This is a pathognomonic lesion), in the cecal follicles, in the subserous tissue of the proventiculus and gizzard, in the cardiac fat and the serous membrane covering the xiphisternum. Edema is present subcutaneously and the serous cavities contain fibrinous exudate. Ovarian follicles show large hemorrhages, There may be hemorrhages, edema and lymphocytic infiltration of the mucosa of larynx and trachea (laryngo-tracheitis) and the lumen contains mucus. Catarrh of the nasal and conjunctival mucosa may also be seen. Interstitial pneumonia with exudate in the alveoli and bronchioles and proliferating inflammation of the air sacs may also be seen.

Gastrointestinal catarrh, with focal diphtheritic lesions in the mucosa of the pharynx, esophagus and intestines may be noticed. In the intestines, ulcers may be covered by bran-like deposits, comparable to 'button ulcers' of swine fever.

Microscopically. hyperemia, edema and hemorrhages are seen in various tissues. The media of vessels shows hydropic degeneration. Hyaline changes are noticed in capillaries and arterioles together with hyaline thrombi in the latter and necrosis of the endothelial cells   Regressive changes in the lymphopoietic system, evidenced by breakdown of lymphoid tissue and lipoid hyperplasia of the reticulum and dissolution of the cells of the spleen may occur.   In subacute cases, there may be proliferation of reticulo - histiocytic cells of liver and other organs   Retinal atrophy and detachment, retinal hemorrhages and inflamm- ation of the semicircular canals may be noticed. The semicirculitis may probably be responsible for the abnormal movements of the head.

Non-purulent lymphoid - cell neuritis and perineuritis may be met with (similar to that in Marek's Disease)-

Focal areas of necrosis may be observed in muscles. Inclusions may be seen in the reticulum cells of the spleen.

In the brain, lesions are found mostly in cerebellum and medulla where a non-purulent encephalomyelitis is seen with hyperemia, perivascular cuffing, neuronal degeneration, gliosis and neuronophagia.

Focal degeneration and necrosis of the parenchymal cells of liver, kidney and heart may be seen. presumably due to hypoxia.   Recovered birds may very rarely be carriers.

Diagnosis :

1. Symptoms.

2   Lesions—petechiae in the proventriculus and diphtheritic ulcers of small intestines.   Histopathological examination of lung and brain for identification of proliferative lesions, in the pneumoencephalic form of the disease.

3.   Hemagglutination and Hemagglutination-inhibition tests.

4.   Inoculation of a fowl with triturated liver, brain and spleen from a dead bird.   If the death is due to Ranikhet disease, the inoculated bird dies in 4 to 5 days.

5.   Inoculation on 9 to 11 day-old egg embryo by allantoic route and study of the characteristic lesions produced on embryo and chorioallantoic membrane. Depending upon the strain involved the embryos die as follows :

a)   in 24 to 48 hours—with the Velogenic strain.

b)   in 48 to 72 hours—with Mesogenic strain.

c)   100 hours and after—with Lentogenic strain.

The allantoic fluids can be tested for the presence of virus by HA and HI tests.

6.   Cross immunity test.

7.   Virus neutralisation test in chick embryos or tissue culture.

8.   Differentiate from fowl cholera sallmonellosis, spirochaetosis, laryngot- racheitis, hemorrhagic disease and vitamin K deficiency.

FOWL PLAGUE

This is an acute disease of fowls caused by a virus, belonging to the Myxovirus group   The disease resembles Ranikhet disease but the course is more acute.   Fowl plague has not been encountered in India. The following differences may be noticed

Fowl plague	Raniket disease
1, Incubation period is 36 hours to 3 days.	1. Incubation period longer—4 to 11 days.

2. No respiratory symptoms seen

2. Respiratory symptoms with laboured breathing coupled with expulsion of tenacious mucus seen.

3. No nervous symptoms seen

3. Nervous symptoms seen

4. Hemorrhages diffuse and widespread

4. Hemorrhages not so diffuse

5. Course very acute, death occurring in 24 to 48 hours

5. Course longer. Death in 2 to 4 days

6. Pigeons not susceptible

6. Pigeons highly susceptible

7. Comb and wattles black.

7. Comb and wattles not black.

Diagnosis can be confirmed by H. I. test and other serological tests.

INFECTIOUS LARYNGOTRACHEITIS (I. L. T.)

This is a highly infectious disease of fowls infecting the respiratory tract caused by a virus of the Herpes group.

Infection is by way of respiratory tract   Incubation period is 6 to 12 days. Comparatively older birds, above 10 weeks of age are affected.

Symptoms of respiratory affection and distress are noticed viz. gasping, rales and coughing. Bloody mucus may be expelled from trachea.

Birds may be dull and depressed with lachrymation. Drop in egg production occurs. The course of the disease may be about 2 weeks and morbidity is 90% while mortality is 10 to 50°/₀ in the classical from of the disease.

Survivors are carriers for two years and void the virus.

Cause of death is asphyxiation.

A mild form of the disease is also known to be caused by certain strains of the virus. Though the morbidity is very high, mortality may not occur unless complicated with mixed infection with Avian Respiratory Mycoplasmosis, Infectious Bronchitis, E. coli or with some fungi. In India the disease is of mild nature.

Lesions : The beak and mouth of dead bird may be soiled with blood and mucus. The tracheal and laryngeal mucosa is covered with bloody, tenacious exudate. In some birds the lumen of the trachea may be occluded by a thick cheesy material.

Microscopically, desqumation of the tracheal mucosa is noticed. Hemorrhage and purulent exudate may also be seen.

Intranuclear inclusions are found in the epithelial cells of tracheal mucosa.

Diagnosis :
  1. Symptoms and lesions.
  2. Notice the intranuclear inclusion bodies in tracheal mucosa.
  3. Differentiate from infectious bronchitis and Ranikhet disease.
  4 Inoculale on 9 to 12 day old embryonated eggs and observe lesions in chorioallantoic membrane—small, necrotic, pock-like areas visible on the 3rd-day. These become larger by 5th to 6th day when embryo dies. Intranuclear inclusion bodies can be demonstrated in these lesions. 5. Gel diffusion test with CAM.

INFECTIOUS BRONCHITIS

This is a highly infectious and contagious viral disease of fowl. In young hicks, especially under 4 weeks of age, the disease is very acute while in th adult only reduction in egg production is noticed. The virus belongs to 'corona' group of virus being similar to human common cold virus.

Infection is by inhalation and may also be from bird to bird by direct transmission. Incubation period is 18 to 36 hours.

Predisposing causes : Intercurrent infections; Ranikhet Disease; Infectious Laryngotracheitis; infection by *M. gallisepticum, E. coli,* feed with high protein; cold weather.

Pathogenesis : The virus replicates in the trachea and lungs after entry there and then becomes viremic, reaching other organs like kidneys and oviducts-where they may replicate. In the kidneys the tubules are damaged so that reab, sorption of water, glucose and electrolytes is reduced resulting in dehydration and acidosis.

Symptoms : In the young chicks, the symptoms seen are nasal discharge, gasping, rales, coughing and lachrymation. The chicks huddle under the hover. There may be swelling of the sinuses. Mortality may be as high as 25% but in chicks above 6 weeks of age mortality is negligible.

In older chicks and adults, rales, gasping and coughing may be noticed but there is no nasal discharge. Production of eggs declines rapidly. In some chicks. the oviduct may not be developed completely or partially and hence egg production may be interfered with or may not even comence at all resulting in great economic loss. The eggs laid are mishapen, thin or soft-shelled, rough and small. The quality is poor with watery white yolk. The course of ths disease is 1 to 2 weeks.

Lesions : Cattarrhal tracheitis is observed with exudate in the lumen. Air sacs may reveal fibrinous inflammation. Chicks that die show plugs of yellow caseous material obstructing the bronchi and lower parts of trachea. There may be catarrh of the nasal passage and sinuses. A few pneumonic foci may be seen.

Microscopically, the tracheal and bronchial mucosa is thickened due to edema and cellular infiltration. The mucosa is intact. No inclusion bodies are noticed.

When the oviduct is affected, its size is reduced and metaplasia of its epithelium may be noticed. The glands dilate and there is infiltration by monocytes. Proliferation of lymphoid follicles is noticed. The affected kidneys are pale and swollen. Deposits of urates are found not only in the kidney but throughout the body as in visceral gout. The epithelium of the tubules which may become necrotic is shed into the lumen. Lymphocytic infiltration of the interstitial tissue may be noticed.

Diagnosis :

1   Symptoms

2.   Isolation of virus and study. For this scraped tracheal and bronchial mucosa is used. 9--10 day embryonated eggs when inoculated, reveal dwarfing and curling of embryos and mortality in 5--7 days.

3.   Serum neutralisation tests.

4.   Differentiate from I.L.T. and R.D.

RD	ILT	IB
1. Brooder chicks affected.	1. Brooder chicks not affected.	1. Brooder chicks affected.
2. Incubation period 5-11 days.	2 Incubation period 5-12 days.	2. Incubation period 18-36 hours.
3. Egg production—most adverse effect, even complete stoppage.	3 Least effect on egg production.	3. Adverse effect next to R.D.
4. Mortality not as high as in I L T.	4. Mortality highest.	4 No mortality in chicks over 2 months of age.
5. In traehca[hemorrhagic nad inflammatory lesions without ulcerations.	5. Hemorrhagic lesions with ulcerations.	5 Catarrhal lesions with narrowing of lumən & without ulcerations.
6 Lesions on embryonating eggs-on chorioallantoic membrane small, necrotic foci; hemorrhages on the embryo. Gross congestion and hemorrhages on yolk sac. Death of embryo in 48 hrs.	6 Pock-like necrotic areas on the chorioallantoic membrane. No congestion or hemorrhages of yolk sac. Death of embryo in 3—5 days.	6. Dwarfing and curling of embryo which dies in 5-7days. No gross congestion or hemorrhages of yolk sac.

AVIAN ENCEPHALOMYELITIS
Syn ; Epitdemic Tremor

This is a disease of fowls caused by a virus of picorna group containing ribonucleic acid and affects mostly young chicks, characterised by ataxia and tremors of head and neck muscles.

Incubation period is 5 to 50 days.

Infection passes through the eggs.

Symptoms : Morbidity is 10 −20%. Mortality among affected chicks may be upto 60%. When eggs become infected embryo mortality is high.

The characteristic symptoms are dullness, disinclination to move and ataxia of the leg muscles. This condition is progressive so that the chick sits on its haunches and does not move. Then tremors of the head and neck start. Birds die because of stravation. Fowls that may survive develop a bluish coloration of the eye which makes the bird blind.

Lesions Lesions are microscopic and consist of perivascular cuffing, neuronal degeneration (consisting of tigrolysis, nuclear eccentricity and vacuolation) nerosis and gliosis. Purkinje cells are more often affected. Demyelination is not seen.

There may be hyperplasia of lymphoid follicles of the heart muscle, proventriculus and pancreas. No inclusion bodies are present.

Diagnosis :
1. History and symptoms.

2.   Histological appearance of brain tissue.

3.   Virus can be isolated from the brain and spinal cord; disease can be reproduced by intracerebral inoculation into susceptible chicks. The virus can also be isolated by inoculating by the yolksac route of 6-days old embryos. The embryos show paralysis. If no deaths are seen, the embryos are allowed to hatch and clinical symptoms are seen after 10 days. Serum-virus neutratisation test can be conducted with the virus isolated.

4.   Differentiate from Ranikhet disease, Vitamin E and Riboflavin deficiency.

AVIAN MONOCYTOSIS
Syn :   Pullet disease :   Blue comb :

This is an infectious disease of young adults caused by a virus which can be grown on 8 day chick embroys which die in 36 to 72 hours.

Symptoms :   Usually birds over 21 weeks of age are affected Disease start. with depression, anorexia and whitish watery diarrhoea. These are followed by dehydration and cyanosis (blue color) of the comb. Egg production drops. Course of the disease is 10 to 14 days.

Lesions :   The bird is dehydrated and the skeletal muscles become degenes rated resembling fish flesh. Necrosis of the liver and pancreas as well as hemorrhages on the serous membranes, heart and ovary are the features of the lesions and are populary described as "spottiness" of the liver, "chalkiness" of the pancreas and fish-flsh as in keel muscles. There may be severe enteritis. The crop is bulged and the contents are sour smelling.

Microscopically, the skeletal muscles lose their striations, and undergo fragmentation and Zenker's degeneration.

In the liver may be noticed foci of necrosis with infiltration by leukocytes. Regeneration may be evident.

Kidneys show changes as in gout—uric nephritis The epithelium of proximal convoluted tubules shows cloudy swelling, pyknosis and desquamation There may be hyaline casts in the dilated tubules. The characteristic radiating crystals are seen.

Ovary may reveal some broken egg follicles.

Blood :   There is relative and absolute increase in monocytes-monocytosis- the number being 8000 per cmm instead of a normal 1700. In blood smears, the monocytes show highly vacuolated or punched-out cytoplasm and this ie considered as a diagnostic feature. Hemoconcentration with increased hemoglobin is seen.

Blood N. P. N. and uric acid contents are increased.

Diagnosis :

1.   Symptoms and lesions.

2.   Differentiate from fowl cholera, pullorum disease and fowl typhoid.

INFECTIOUS BURSAL DISEASE (IBD)
(Gumboro disease)

It is a virus disease caused a virus belonging to Birna virus group. Virulent IBD virus is a pathotype variant of classical serotype-I strain. Isolates from recent outbreaks in 1993 in Namakkal area (Tamil Nadu) belonged to the above serotype.

Transmission, carriers and vectors : It is highly contagious and persistent virus in the environment of poultry houses. Virus survives upto 120 days in poultry sheds. Water, feed, droppings from infected birds are viable for 52 days in the poultry houses. Mealworm and *Aedes vexan* (Mosquito) appear to act as carriers. Egg trays, vehicles, used in the transport of birds eggs, personal handling birds in sheds and elsewhere are very important source of carriers of infection. Virus is very resistant to heat and disinfectants.

Morbidity and Mortality : Young chicks—Layers and broilers upto 8 weeks are succeptible. Morbidity is 100%. Earlier it was considered to be disease of low mortality. However, outbreaks of 1993 in India and elsewhere has proved it otherwise. Mortality as high as 80-90% has been observed. In some places the entire flock has been wiped off. Recent investigation has revealed that high mortality is associated with concurrent infection like Ranikhet disease, Inclusion body hepatitis, Coccidiosis and presence of Mycotoxins in the feed.

Incubation Period : Incubation period is short. Clinical signs are observed in 2-3 days following infection.

Pathogenesis : The target cell of infection is Lymphocyte. It undergoes degeneration and necrosis. There is also evidence of infection affecting the coagulation mechanism of blood in birds older than 6 weeks. The coagulation time is greatly increased.

Symptoms : One of the earliest sign of infection is self vent pecking. Other symptoms observed are anorexia, depression, trembling, watery and whitish diarrhea, and prostration followed by death.

Gross Lesions : At necropsy paint brush heamorrhages are observed on pectoral and thigh muscles. There is severe catarrh in intestines. Kidneys are blanched and pale. Lesions in bursa fabricius consist of gelatinous yellow transudate on serosal surface with colour changing from white to cream. Longitudinal striations on Bursa fabricius are prominent. Later Bursa fabricius is enlarged due to oedema and hyperemia to almost double its normal size. After 4-5 days Bursa starts progressively getting atrophied to 1/3 its normal size. Cut section reveal necrotic focci apart from fibrinous and heamorrhagic exudate on the mucosal surface. Spleen is slightly enlarged with greyish necrotic focci. Heamorrhages at the junction of proventriculus and gizzard is another important lesion.

Histopathology : All lymphoid organs are affected and show oedema, heamorrhage, and hyperemia. Degeneration and necrosis of lymphocytes particularly in the medullary region occurs in four days post infection. These are replaced by heterophils, pyknotic debris and hyperplasia of Reticulo-endothelial cells. As inflammation recedes cystic cavities develop in medullary areas. Proliferation of interfollicular connective tissue and bursal epithelium takes place giving a glandular appearance. Spleen shows R.E. cell hyperplasia around the arteries Thymus and ceacal glands shows lesser damage. Significant changes are observed in Harderian glands. Normally gland is infiltrated with plasma cells as the chicken grows. This is prevented when chicks are infected with IBD virus and fewer by 5-10 fold. In the chicks infected at 3 weeks age, necrosis of plasma cells occurs and reduced by 50%, 14 days post infection. This however is transient and returns to normal after another 14 days.

Lesions of kidney are non-specific.

Liver shows perivascular infiltration.

Diagnosis

(a) Gross and Histopathological lesion.

(b) Detection of virus in tissues by using Flouroscent antibody technique, Nucleic acid probes, or antigen capture enzyme immunoassays using monoclonal antibodies.

(c) Isolation of virus.

(d) Since the disease is more severe when there is concurrent infection simultaneous investigation for the presence of Raniketh, Inclusion body hepatitis, Mycoplasma, Coccidiosis and Mycotoxicosis should be undertaken.

Control and Prevention :

I. Strict biosecurity measures such as :

(a) Disposal of litter, dead birds, used gunny bags, curtains, and other disposables by incineration or deep burial with slaked lime.

(b) Restricting vehicular moments with crates, egg trays, and culled birds.

(c) Treating feeders, and waterers with 5% formalin.

(d) Fumigating new poultry sheds with Formalin fumes.

(e) Restricting personell to their sheds for work.

II. Vaccination

(a) Primary vaccination with mild or intermediate strain after 2 weeks.

(b) Booster vaccination with intermediate strain (live) after 5 weeks.

(c) Vaccination of breeder stock and seromonitoring by hatcheries to ensure adequate levels of maternal antibodies in the chicks.

EGG DROP SYNDROME (EDS-76)

It is due to group-III type of avian adeno virus. It was first reported by Dutch Workers Van Eck *et. al.* in 1976. Since then it has been reported in several countries. In India it was first reported by Mohanty in 1987. The outstanding features of this infection is severe drop in egg production at the peak production level in layer birds. Ducks, Geese, Owls, Stork and Swan are natural host from whom it spreads to chicken.

Pathogenesis and Trasmission : In experimental oral infection in adult layers, there is a limited viral replication in nasal mucous membrane, followed by viremia. After 3-4 days, virus replication takes place in Lymphoid tissue particularly in Spleen and Thymus. Massive replication takes place in Pouch shell gland 7-20 days post infection and to a lesser extent in other parts of oviduct.

Transmission of infection occurs in three ways :

(a) Vertically through embryonated eggs from infected layers.

(b) Horizontal spread occurs through contaminated egg trays, transport of infected and uninfected birds together, contaminated feeds inadequately sterilized vaccination needles and blades.

(c) Through use of drinking water contaminated by feaces, of domestic or wild ducks, Geese etc.

Symptoms : Pigmented eggs loose their colour. This is followed by production of thin shelled, soft shelled, or shell less eggs. Thin shelled eggs have a rough granular or sand paper texture. At times egg sizes is reduced. Albumin is thin and watery. Another very important symptom is fall in production upto 40-50%. The fall in production may be sudden or spread over 10 weeks.

Gross lesions : In natural outbreaks inactive flaccid ovaries and atrophied oviducts are observed. In experimental infection oedema of uterine folds, exudate in the ponch shell gland and mild splenomegaly are observed.

Histopathology : Pouch shell gland is the major site of change Virus multiplied in the surface epithelium produce intra-nuclear inclusion bodies after 7 days post infection. Many cells are sloughed off accompanied by severe inflammatory changes in the Lamina propria and epithelium with infiltration of macrophages, plasma cells, lymphocytes and heterophils.

Diagnosis : Symptoms, Lesions and HI test.

Control : An oil adjuvant inactivated vaccine is available and gives good protection. Birds are to be vaccinated between 14–16 weeks of age.

INCLUSION BODY HEPATITIS

It is caused by virus belonging to Group-I adeno-virus. Several sero type F_1 to F_{10} have been incriminated to cause this condition. Onset of this disease is associated with immuno suppressive diseases like Infectious Bursal disease and chicken anemia agent.

Broilers in the age group of 3–7 weeks are most susceptible Morbidity is low. Mortality rate is between 10–30%. The onset of the disease is characterised by sudden deaths. Mortality peaks 3–4 days after onset of this condition and stops on 5th day. However sporadic deaths continue for further 2–3 weeks.

Gross Lesion : The most important changes are seen in liver, which is swollen, pale and friable with petichea or echymotic heamorrhages. Similar heamorrhages are observed in skeletal muscles.

Histopathology : Hepatic cells show's degeneration, necrosis and cellular infiltration. Besides the hepatic cells reveal basophilic or eosinophilic inclusion bodies.

Prevention and control : Emphasis should be laid on control of concurrent immunosuppressive diseases like infectious bursal and chicken anemia agent; since these disease potentiate pathogenecity of adeno virus infection.

INFECTIOUS RUNTING SYNDROME (IRS)

The outstanding feature of this condition is pronounced retardation of growth and weight gain to the extent of 40–50% inspite of good apetite. It is also characterised by watery of mucoid droppings, indicating an enteric infection.

The etiological agent responsible for this syndrome appears to be multiple. Several viruses have been implicated and identified either with isolation studies or Electron Microscope studies which are as follows : (i) Calcivirus; (b) Entero like virus; (iii) FEW virus particle; (iv) Toga virus and (v) Parvo virus.

The infectious runting syndrome caused by Parvo virus has been studied and reported in detail. This virus causes stunting syndrome in broilers. Infection usually takes place in chicks younger than 7 days. In white leg horn chicken symptoms are mild. Infection spreads vertically and horizontally. Viral particles are 19.25 nm in diameter and are hexagonal.

In experimental studies, in antibody free 3 days old broiler chicks, infected orally, clinical signs appear in 3–5 days. They are characterised by onset of diarrhea with watery or mucoid droppings of mustard yellow colour. Apetite is normal or even voracious. Inspite of good apetite there is serious retardation of growth. Growth of juvenile feathers is delayed due to growth of abnormal new feathers. Chicks develop "Helicopter like appearance". Morbidity is 50–80% with 5–10% mortality.

There is pronounced anema. Heamatocrit and total RBC count values are greatly reduced.

Necropsy lesion consist of reduced body size, abnormal feathers. Small intestines are extremely pale. Ceace is distended with foul smelling gas. Pancreas are shrunken, white and firm. Bone marrow of Femur is yellowish or white. Metatarsal bones are soft and can be bent.

Control : No effective control measure are available at present.

MALABSORPTION SYNDROME

It is an enteric viral infection caused by corona virus like particles and Reo virus.

Broilers aged between 1–3 weeks are susceptible. Incubation period is 4 days by oral route, and 9–13 days by intra-trachial route. Infection spreads horizontally. Egg transmission rate is slow. A cent percent morbidity is observed in affected flocks. Mortality rate is 10%.

Symptoms : Uneven growth, poor pigmentation, abnormal feathering, lameness and skeletal abnormalities are seen.

Gross pathology : At necropsy enlarged proventriculus with heamorrhage and necrosis, catarrhal enteritis, atrophy of Bursa fabricius, necrotic foci in liver, Osteoporosis of bones, and inflammation of gastroenemins muscles are observed.

Histopathology : Intestines shows catarrh with atrophy of intestinal villi and lymphocytic infiltration which is a pathognomic change. Proventricultis with necrotic changes, non-suppurative myocarditis, pancreatitis are observed in visceral organs. Femur shows transverse and vertical cleft of its growth plate and necrosis of cartilage. The superficial portion of bony cortex shows fragmentation.

Diagnosis : It is made by gross and histopathological studies. Serologically it can be diagnosed by Agar gel precipitation test, indirect flouroscent antibody and ELISA test. Virus neutralisation test is based upon plague reduction in kidney or chicken embryo liver cell cultures.

Control : Organic Iodine solution (0.5%) and lye are effective inactivating agents. Attenuated Reo virus vaccine given subentaneously is effective. However, it interferes with efficacy of Marek's vaccine if used simultaneously.

CHICKEN ANEMIA AGENT (CAA)

It is caused by a virus and studies of isolates from Japan, U.K., Sweden, Germany indicate it to be a Parvovirus. Serological surveys indicate a world wide distribution of this condition.

2-3 weeks old chicks are highly susceptible. A mortality of 10% is seen in affected flocks. Disease is transmitted horizontally and vertically even though parents do not show any clinical signs of the disease.

Pathogenesis : Pathogenesis is not fully understood. However, it appears to be due to interference in heamopoiesis. It may be due to destruction of thymic cortical cells, which are supposed to be necessary for production of RBC's.

Severity and virulence of the disease is enhanced when birds are infected simultaneously with other immunosuppressive viruses like Marek's disease, Reticulo endotheliosis, and infectious bursal disease.

Maternal antibodies have a protective role in the young chicks. Although these chicks do not suffer clinically, they shed the virus. However, the protective effect of the maternal antibodies and the age resistance can be overwelhmed if there is dual infection by immuno suppressive viruses.

Pathology : Aplasia of bone marrow and replacement by adipose tissue is observed. Severe depletion of lymphoid cells in the cortex and medulla in Thymus, Spleen, Bursa fabricius and ceacal tonsils, with atrophy of these organs is seen. Liver is enlarged. Gangrenous dermatitis may be observed. Heamorrhages in proventriculus and generalised subcutaneous heamorrhages might be present.

Heamatological studies reveal a very low total RBC count and heamatocrit values. White blood cells and Thrombocytes are also decreased. Clotting time is prolonged.

Sequelea : The infection leads immunosuppression and birds become more susceptible to Marek's disease, Bacterial and Fungal infections.

Control : No effective and safe vaccines are available commercially at present.

ASCITES CAUSED BY RIGHT VENTRICULAR FAILURE
(Ascites due to RVF)

This is a disease of fast growing broiler. It is also observed in birds which are reared at high altitudes.

Pathogenesis : A broiler chick is ideally developed for rapid growth of muscle for meat, However, lungs of chicks are rigid and moulded in thoracic cavity with little possibility to expand unlike that of mammalian lung. Further the lungs of chicks grow much slower than the muscle. This results in limited capacity for capillaries to dilate for accommodating increased blood flow. Besides, the thin walled right ventricle is not strong enough to overcome the resistance in the pulmonary vessels. Keeping this in view, it will be interesting to trace as to how the Ascites develops.

The broiler chick fed on high energy rations with ideal environment and management is all set for fast growth. However, the high energy ration results in high BMR with resultant demand for increased level of oxygen requirement. This is further aggravated by stress factors like cold, dust and toxic fumes in overcrowded poultry houses. Hypoxic state leads to polycythemia, increased viscocity of blood. Consequently there is increased work load for Right ventricle with resultant pulmonary hypertension. This is over a period leads to decompensation of the Right ventricle, congestion of liver and abdominal vessel, ascites, and eventually right sided-failure and death.

HYDROPERICARDIUM (LEECHI DISEASE, ANGARA DISEASE)

This disease has been reported recently in Pakistan and Northern India from Jammu and Kashmir, Punjab and around Delhi. Broilers of 3–6 weeks are affected most. Adeno virus type has been incriminated as casual agent. Mortality is as high as 70%.

Lesions : Lesions observed are Hydropericardium with clear straw coloured fluid and enlarged mottled liver. Histological changes consists of Myocarditis, basophilic intra-nuclear inclusion bodies in the hepatic cells and distorted renal tubules. A formalised liver homogenates from affected birds has been tried with encouraging results in control of this condition.

CAGE FATIGUE IN COCKS

Cocks which have been reared on the floor for 30-40 weeks and than put in the cage, have been observed to suffer from this malady. Soon after being put in the cages, birds are dull, develop in appetance, stop drinking water and die within few days. Necropsy reveals birds to be extremely light, the gastro-intestinal tract is empty and proventriculus show congestion around proventricular glands. It may be mistaken for Ranikhet disease. Such birds when put back on the floor recover uneventfully.

MISCELLANEOUS CAUSES

Mortality in flock of birds may be due to non-infectious and non-specific reasons. At times it is alarming and poses a formidable challenge to the pathologist. It can be effectively met if one is acquinted with managemental procedures.

I. 1) **Early chick mortality** : Besides infectious causes mentioned earlier managemental factors also contribute to early chick mortality.

(a) *Poor hatchery management* : Chick from poorly managed hatchery show higher rate of early chick mortality. This could be due to lack of monitoring the temperature, humidity and ventilation system of the hatchery during incubation.

(b) *Commercial hatcheries transport* chicks to farmers to distant places involving long journeys. This results in exhaustion, dehydration and exposure to adverse conditions.

(c) *Brooder House Management* : Considerable mortality of young chicks may due to poorly managed brooder houses. (i) Too many chicks may be housed per brooder resulting in overcrowding. Ideal strength per brooder is between 150-175 chicks (ii) Lack of monitoring brooder temperature. A brooder may have too low or high temperature. This results in huddling and piling-up of chicks leading to suffocation and death. Such chicks may also show unabsorbed yolk.

2. Insufficient feeders and waterers.

3. Feeders and waterers of improper heights leading to inability of chicks to pick up feed and water with resultant death.

4. Improper ventilations.

II. **Cannabolism** : This is observed in overcrowded poultry houses particularly in adult cocks. The adult cocks have a peculiar habit of establishing superiority over each other called "peck order." In the process cocks resort to pecking each other. This result in severe injuries on the back leading to heamorrhage, infection and loss of feathers and death due to trauma. Remedy dies in housing few cocks preferably in individual cages or along with hens if fertile eggs are required

III. **Post-vaccinal Reactions** : This is observed in birds after vaccinating with R_2B or Komarov strain of vaccine around 8-10 weeks. This is characterised by progressive paralysis of the limbs, starvation and death. This can be remedied or minimised by giving second F_1 or any mild vaccine strain around 4th week.

DISEASES CAUSED BY FUNGI
By Dr. D S. Kalra

Mycoses
    Superficial mycoses
        Ringworm
        Favus
        Cutaneous streptothricosis
    Systemic mycoses
        Aspergillosis
        Coccidioidomycosis
        Rhinosporidiosis
        Sporotrichosis

Mucormycosis
Histoplasmosis
Candidiasis
Blastomycosis
Paracoccidioidomycosis
Mycotoxicosis
    Aflatoxicosis
    Degnalla Disease
    Other mycotoxicoses (Table)

A : MYCOSES

The diseases caused by fungi are termed as Mycoses (Singular, mycosis). Mycoses have been broadly classified as :

i) Superficial mycoses (Dermatomycoses or dermatophytoses).

ii) Systemic (deep) mycoses

SUPERFICIAL MYCOSES

Ring worm (dematomycosis) is a superficial infection of keratinized layers of the skin and its appendages, affecting all species of animals and man. It is of world wide distribution, with particular preference for hot and humid areas.

Ring worm is caused by a closely related group of fungi, commonly referred to as dermatophytes. The different species of fungi, causing ringwom in animals, belong to genera *Trichophyton* and *Microsporum*. (Genus *Epidermophton* is found only in man causing athelet's foot) *Trichophyton* organisms can grow within the shaft of hair (*endothrix*) as well as outside the hair fibre (*ectothrix*) and produce spores in long chains, whereas *Microsporum* fungi only grow around the root of hair and form mosaic pattem of spores, (ectothrix only)

Many species are described in different animals but there is no rigid host specificity. Man may be infected from animals.

Mode of infection : Contact with infected animals is a common method of infection. But infection can occur by fomites; grooming kit, bedding, harness and horse blankets. Since the spores may exist on animals without causing any lesions, carriers may be sources of infection. as the spores are very resistant. Stables and sheds may remain infected for years.

Animals susceptible : Ringworm is common in cattle, horses, goats, cats and dogs. Cattle are more susceptible than horses. It is less commn in swine, sheep, birds and rabbits. Young ones are more susceptible than adults.

Incubation period is one week to one month.

Pathogenesis ; The fungus grows on the keratinised tissue and in the hair follicles and fibres. The affected fibre becomes splintered and breaks off due to autolysis. The organism excretes an exotoxin which incites an inflammatory reaction in the epidermis and corium resulting in hyperemia, capillary dilatation and edema. Thus vesicles may form There is parakeratosis with increased desquamation. The lesions progress in a centrifugal manner and a ring of tiny vesicles is found at the boundary of the lesion and hence the name 'Ringworm'. The fungi are strict aerobes and so they may die in the centre of the lesion when covered by thick crusts. The absorption of the toxins leads to hypersensitive state and there will be allergic rashes on the skin. This is called " Derma tophytid reaction " or 'Id' lesions.

Secondary bacterial infection results in suppuration of the hair follicles

Lesions :

Cattle : The lesions are more commonly found on the skin of head, neck and the anal regions. Other parts of the body may be affected, especially in calves. Legs are seldom affected. The lesions appear as raised, round, crusty patches and have a diameter of 2-3 cm. Moist at first, the lesion becomes dry later. Alopecia may develop.

Horses: The skin of the head (especially around the eyes and nostrils) neck, chest, shoulders and back (especially the area coming into contact with harness and saddle) is more often affected. The lesions may be manifested in two forms, the superficial and deep. In the superficial form initial vesicles soon give place to thick crusts wich may become desquamated. Alopecia may develop. In the deeper form, the hair follicles are affected, vesicles and pustules develop.

Pigs : The lesions, which are superficial and mostly located on the trunk, are relatively mild and slow progressing These are in the form of dry, reddish-brown, crusty areas, usually not raised above the skin. Alopecia is not a common feature.

Sheep and Goats : The lesions, which are in the form of eruptions, appearing as scaly or crusty patches. are usually found on the head, back, shoulders, neck and chest in case of sheep and on the facial region and pinna in goats.

Dogs and Cats : In dogs, the lesions may appear on any part of the body. They are circular, upto 3 cm, in diameter, and look like crusty eruptions. In cats, the lesions. which are frequently confined to the face and extremities, are relatively inconspicuous. The lesions vary from mild non - inflammatory reaction, represented by bare patches with broken hairs and scaling, to severe inflammatory reaction resulting in thick crust formation and loss of hairs.

Camel : The lesions are usually in the form of dry, scaby, irregular, circumscribed patches and are mostly found on the legs and neck

The histopathological features may vary from little (or no) inflammatory reaction associated with mononuclear cell infiltration to a marked destructive reaction, along with infiltration of neutrophils, lymphocytes and plasma cells, with or without the presence of fungal elements in the stratum corneum and or on the hair shafts

The mild lesions, associated with little inflammatory reaction, reveal slight to moderate hyperkeratosis of the epidermis, including that of follicles, and later acanthosis. As the inflammatory [reaction becomes more marked, the epidermis shows exudation and ulcerative lesions, to be replaced by inflammatory crusts, besides marked infiltration of cellullar elements. In a severe reaction, hair follicles get destroyed and the underlying dermis shows inflammatory changes

Diagnosis : The diagnosis of ringworm depends upon characteristic lesions and demonstration of causative fungi, which can be done by direct microscopic examination of skin scrapings and by cultural methods,

Use of Wood's Lamp (ultraviolet light) can also help in the differentiation of *Trichophyton* and *Microsporum* infections, though it is of limited diagnostic value.

Histopathology and demonstration of fungal elements by special stains may also be helpful.

Favus

Favus, commonly known as 'honey comb ringworm, is also a chronic dermatophytosis caused by *Trichophyton schoenleinii (Achorion schoenleinii), Microsporum gypseum (A. gypseum)* and *M- gallinae (A. gallinae).* Favus is seen in man (due to *T. schoenleinii*), dogs and cats (*M. gypseum*) and fowls (*M gallinae*). In dogs the head and paws are mostly affected, while in cats the ears·

Lesions : The characteristic lesions are saucer-like favus scutulae; these are dense, felt-like aggregates of the fungus on the skin, having a central depression This appearance is due to greater growth at the periphery. When the crusts are removed, cupshaped depressions with raised borders are seen on the skin. Confluence of several such lesions gives a honey - combed appearance,

Microscopically, the fungs is found only in the horny layer. An inflammatory reaction consisting of intense hyperemia and cellular infiltration is found in the corium.

In the fowls, favus is found on the comb and wattles as white patches· Later feathered parts may also be affected. Crusts may eventually form by coalescence of the lesions giving them a white, thick appearance.

Diagnosis : Microscopic examination of skin scrapings, after treating with 20% sodium hydroxide, for the presence of spores and mycelia. Examination of affected hairs by Woods lamp (U. V. light with cobalt filter) gives fluorscence in positive cases

CUTANEOUS STREPTOTHRICOSIS

This is a dermatitis occurring in all species of animals caused by infection with fungus, *Dermatophilus congolensis* This disease is found prevalent in Africa, but has also been reported from Canada, U. S. A., and Great Britain. This has also been encountered in India The disease is refered as ' Strawberry foot ' in sheep in Scotland.

Animals of all ages are susceptible and moist, humid temperature is found to be conducive for infection. In the animals, areas of the skin which are always wet are more often affected. Injuries especially tick-bites, fly bites and raum t caused by scratches from thorns are portals of infection.

Usually, an infected animal is a source of infection. The fungus is viable in the soil for more than four months. Contact infection from animals is also frequent.

Pathogenesis : Entering cutaneous injuries, the fungus causes bacterial dermatitis. The exudate together with the epithelial debris and the mycelia of the organism produces crusts resembling those of the Ringworm. Infection by secondary bacterial invaders may produce suppuration and severe toxemia but in uncomplicated cases the lesion is self limiting and the scab separates from the lesion. To start with, the hair becomes erect with greasy amber colored exudate. This later turns into dirty yellow scales that may be greasy These finally become hard and horny and are shed causing alopecia. Mostly, the condition is a chronic one

Symptoms ;   In cattle the lesions are found on the neck, body, back of the udder and extend to the legs. In calves the muzzle may be affected, from where the lesions spread to the head and neck. In goats the lesions are found in the muzzle from where they spread to the feet and scrotum. Among sheep the dorsal parts of the body are mainly affected.

No pruritus is evident. The characteristic lesion is a thick horny crust, 2 to 5 cm. in diameter, dirty yellow in color. These lesions are close to each other giving a mosaic appearance. In the early stages pain may be evinced if crusts were raised. Beneath the crusts may be found pus and granulation tissue. Later, when the lesion heals, the crust can be removed.

Though heavy mortality can occur in young lambs, by and large, the health of the animals is unaffected. The lesions may be quiescent in the chronic stages in drier weather but may flare up during wet season. In places like groin which are always wet, necrotic or gangrenous dermatitis may develop, some with fatal results,

Microscopically, the organisms can be seen in large numbers in the hair follicles, above the stratum granulosum, as mycelia. The filaments of the mycelia may fragment longitudinally to form coccoid 'spores'.

Diagnosis ;
1. Isolate the causative organism from scrapings.
2. The branching mycelia can be demonstrated by suitable staining on impression smears made directly from the under surface of the scabs
3. Flourescent antibody technique can be utilised.

SYSTEMIC MYCOSES

Systemic mycoses usually occur as sporadic infections, though occasionally these may appear in the form of outbreaks, especially those infections which involve skin, tissues or mucous membranes of alimentary/respiratory tract.

ASPERGILLOSIS

Aspergillosis, caused by different species of *Aspergillus* (chiefly *A. fumigatus* and *A. flavus*), is a disease primarily of respiratory tract, producing characteristic granulomatous lesions. It is usually encountered in poultry birds, but cattle, buffaloes, horses, sheep, goats, dogs, and cats may also get this disease

Routes of Infection :

Infection is by inhalation of the spores, which are usually found in the litter, mouldy straw and grains. Exposure to inclement weather, faults in feeding, close confinement in damp houses, and prolonged antibiotic therapy are said to be predisposing factors. The fungus may invade and enter the egg shell.

Since aspergillosis is mostly found in chicks artificially hatched and raised in brooders, the condition is known as 'brooder pneumonia'.

Incubation period is 3 to 6 days.

Symptoms : In chickens, the symptoms are variable, depending upon the involvement of the respiratory, nervous or digestive systems. The acute form of the disease, which is usually seen in young chicks, is typically characterised by dullness, loss of appetite, rise in temperature, increased thirst, ruffled feathers, droopy wings, and rapid loss of condition. The birds may show darkening of comb and dyspnoea followed by sneezing, diarrhoea and emaciation. The fungus sometimes invades the brain causing paralysis and convulsions. Unilateral eye infection may also be seen which results in the accumulation of cheesy exudate in the conjunctival sac. The birds undergo coma before death. The duration of the disease is about a week. Of the affected birds, 50 to 90 percent may die.

The chronic form of the disease, usually seen in adult birds, occurs sporadically with low mortality. The signs are loss of appetite, anemia, gasping or coughing and rapid loss of body weight.

Clinical signs in mammals, in general, are rise in temperature, mucopurulent nasal discharge, slight cough, laboured breathing, sneezing, bloody foetid diarrhoea and conjunctivitis.

Lesions : In fowls, lesions may involve, besides respiratory tract and alimentary canal, liver, spleen, kidneys, heart, ovaries, skeletal muscles, eyes and brain.

The lungs often contain multiple nodules, small or large (usually 1 to 3 mm in diameter), having a dull grey color surrounded by a narrow zone of hyperemia. In the centre of the nodules may be seen bluish material containing the organisms. Sometimes diffuse pneumonic lesions may be present. The affected air sac is thickened and has white mouldy growth on the surface. The lesions in this case are usually saucer shaped with depressed centres.

Microscopically, the lesion is a granuloma with a central area of caseation containing septate branching hyphae of the organisms. Spores are not seen in the tissues. The hyphae are slender and long in the early lesions while they are plump and short in the older ones. In the early lesions, the colonies of the organisms are surrounded by neutrophils and macrophages. As the lesion ages, eosinophils, epithelioid cells and foreign body giant cells are seen and a fibrous capsule may be found. The bronchi and trachea may also show dense mycelial growth. The walls of bronchi are thickened with inflammatory exudate. If the fungus dies, granulation tissue replaces the inflammatory process.

Th organisms are poorly stained by hematoxylin and eosin method. Special stains like PAS, Bauer's, silver methanamine or Gridley's are required to bring them out clearly.

In cattle, placentitis occurs in pregnant animals resulting in abortion. Infection originates from the respiratory tract. The organism can be isolated from the stomach of the fetus. In lambs the pulmonary nodules resemble those caused by *Mullerius capillaris*.

In cattle and buffaloes, the lesions in lungs are usually nodular in character; sometimes diffuse pulmonary form is also seen. Generalized systemic aspergillosis involving heart, liver, lungs, kidneys, spleen, intestines and lymphnodes has also been seen.

Microscopically, in the lungs, there are multiple, discrete granulomas having centre of caseation necrosis with hemorrhagic infiltrating zone surrounding it. In the necrotic area, somtimes eosinophilic club-like structures, as observed in actinomycosis, are seen surrounding the fungi. In chronic cases, the necrotic tissue may be sloughed and get stagnated, giving rise to bronchiectasis.

The cutaneous lesions resemble those of tuberculous skin lesions.

In horses, systemic generalized infection involving brain and kidneys may be seen. Guttural pouch is at times affected. Epistaxis, abnormal respiratory noise, ocular defect, colic and facial paralysis may be noticed.

In pigs, lobar pneumonia or nodule formation may be seen. Spleen, kidneys, and mesenteric lymphnodes may also be affected owing to generalised infection.

Diagnosis: Diagnosis of the disease is difficult during life. At necropsy, it is made on the basis of characteristic lesions and demonstration of fungi by (i) direct microscopic examination of cotents of the nodule (ii) cultural isolation and (iii) histological examination of tissue sections on special staining.

COCCIDIOIDOMYCOSIS

Coccidioidomycosis is a granulomatous disease caused by *Coccidioides immitis*. It affects man and animals such as dogs, cattle, sheep, horses and pigs. It is however, comparatively benign in animals.

Routes of infection : Infection mostly occurs by inhalation of the spores Direct transmission from animal to animal is not believed to occur Infection by ingestion and cutaneous wounds is also suspected. Rodents are thought to be reservoirs of the fungus.

The fungus on cultures develops mycelia, but in tissues mycelia are not seen; only spherules called sporangia, 5 to 50μ in diameter, with double contoured wall are observed Reproduction in tissue is by endosporulation.

Symptoms : In cattle symptoms may not be observed. Lesions in lungs and lymph nodes are noticed postmortem

In horses, a chronic course is observed with emaciation, edema of legs, a mild fever, anemia and leucocytosis. Some animals may show colic

In dogs, an initial fever is followed by cough and dyspnoea. When the fungus becomes disseminated, diarrhoea and cachexia result If bones are affected lameness may occur

Lesions : The lesions resemble those of tuberculosis. Granulomatous nodules, some with caseous centres and others with calcification, may be noticed. The lymph nodes may show foci of suppuration in a mass of granulation tissue and the whole surrounded by a fibrous tissue capsule. On section thick yellow pus may be seen.

In dogs, a disseminated form may be noticed with greyish nodules in the lungs, lymphnodes, liver, spleen, bone marrow, meninges and other organs.

Microscopically : the structure of the nodules is a granuloma with the spherules surrounded by epithelioid cells. In cattle, the central part resembles the 'rosetti' of actinomycosis, with delimiting acidophil clubs.

The nature of cellular reaction depends on the phase of the organism. When the sporangium ruptures, liberating the young spherules, a neutrophilic exudative inflammation predominates. But when the spherules mature and grow bigger epithelioid cells predominate, with g ant cells and lymphocytes. But usually both exudative and proliferative reactions are seen since all stages of the organism may be present. The organism can be demonstrated within the Langhan's giant cells. Special stains (Gridley's) bring out double contoured wall.

Diagnosis : 1. Demonstration of the fungus in the lesions. 2. Intradermal sensitivity test using coccidioidin can be used. 3. Differentiate from tuberculosis, actinobacillosis, actinomycosis, infection with *C. pyogenes* and lesions caused by *Linguatula serrata.*

RHINOSPORIDIOSIS

Rhinosporidiosis, a chronic granulomatous disease of nasal mucosa, is caused by *Rhinosporidium seeberi* and affects cattle, buffaloes, goats, horses and man. In India, this disease is endemic to certain areas.

The organism invades the subepithelial stroma of the nasal mucosa through some trauma and incites a chronic inflammation, resulting in the formation of single or multiple polypoid granulomata, which may sometimes occlude the nasal passages.

Symptoms ; Affected horses have noisy breathing (snoring sound) and blood stained nasal discharge.

Lesions : The lesion is a nasal polyp, which may be single, multiple and unilateral. It may be sessile and cauliflower - like, soft to touch and pink in color. It bleeds easily.

Microscopically, the granulation tissue may consist of fibrous or fibromyxoid tissue covered by nasal mucosa. In this tissue are found numerous spherules with double contoured wall measuring from 200 μ to 300 μ in diameter. Each such sporangium may contain many endospores, which may be liberated by the rupture of the wall of the sporangium at its thinnest part. These endospores may infect nearby tissues and develop into new sporangia or they may be voided with the nasal discharge and infect other animals. The sporangium does not seem to incite much of leucocytic reaction and so very few lymphocytes and epithelioid cells are found to be attracted.

Diagnosis : Demonstration of spores / sporangia in nasal discharge and/ or in tissue sections of the polyp.

SPOROTRICHOSIS

Sporotrichosis, caused by *Sporotrichum schenkii,* is a suppurative granulomatous disease affecting man and animals, particularly equines. The causative fungus is known to exist as a saprophyte in the soil and plant debris in which it has septate mycelial form, though in lesions only cigar shaped, single walled, gram positive yeast - like forms occur.

The disease is slow spreading and only sporadic cases are found.

Routes of Infection : Infection is through wounds and so the lesions are usually found on the legs and trunk. Infection can occur directly from the discharges of an affected animal or from contaminated fomites, since the fungus can persist in organic matter.

Pathogenesis : At the portal of entry in the subcutaneous tissue, an inflammatory reaction is set up resulting in granulomatous nodules, abscesses and ulcers

Lesions : The lesions are in the form of nodules, found in the skin, measuring 1 to 4 cm in diameter. These are found on the lymphatics, which appear thickened and corded. The lymph nodes are not affected. The nodules may soften and open, discharging thick creamy pus.

Microscopically : the lesion is a granuloma consisting of a necrotic centre containing pus, surrounded by epithelioid granulation tissue with giant cells, plasma cells and lymphocytes. The whole lesion is enclosed by a dense fibrous tissue capsule The nodules may open and form ulcers which heal slowly. Usually, the general health of the animal is unaffected. Very rarely, suppurative granulomas may be found in the internal organs of horses, dogs and man.

Diagnosis : 1. Gram positive spores can be seen in the pus. 2. Injection of pus intraperitoneally into male guinea pigs produces purulent orchitis. 3 Differentiate from farcy, ulcerative and epizootic lymphangitis

MUCORMYCOSIS (PHYCOMYCOSIS)

Mucormycosis is the term used to describe the disease produced by fungi of the genera *Mucor, Rhizopus and Absidia*. It is known to cause granulomatous lesions in different parts of the body and may produce abortion in cattle. Usually lesions are noticed on post-mortem examination.

Animals affected : Pigs and cattle are usually affected. Man is also susceptible.

Lesions : In young pigs ulcers are seen in the gastric mucosa, especially at the ventricular diverticulam. They may also be found in the regions of pylorus and fundus. The wall of the stomach is thickened due to edema. The ulcerated mucosa consists of necrotic tissue surrounded by a thick zone of inflammatory cells including giant cells. The hyphae may invade the blood vessels producing thrombosis Healing of ulcer may take place by granulation tissue which contains giant cells. The intestinal wall may be thickened considerably and the mucosa ulcerated due to fibrous granulation tissue consisting of mycelia surrounded by histiocytes, foreign body giant cells and eosinophils. Mesentric lymphnodes reveal granulomatous lesions. Diarrhoea is seen in affected animals.

In cattle, the granulomatous lesions may be seen in the mesenteric and mediastinal lymphnodes and resemble tubercular lesions The rumen may sometimes be affected and the lesions may be noticed in the ventral sac. The ruminal wall is thickened, edematous and black. Metastatic lesions may also be seen in the liver. Affected animals may show symptoms of indigestion and ruminal atony.

In man lesions are seen mostly in the lungs and ears. Meninges, brain, eye and gastric mucosa may also be affected.

Diagnosis : Lesions reveal the presence of hyphae which are nonseptate, branching and coarse.

HISTOPLASMOSIS

Histoplasmosis is an infectious (but not contagious) granulomatous disease of man, dogs, cattle, horses, sheep, rats, mice and other wild animals. Among domesticated animals, dog is most often affected.

The causative fungus, *Histoplasma capsulatum*, is whitish to brown in color, having spherical spores of two sizes ; (i) the *microconidia*, 3 to 4 μ in diameter and (ii) the *macroconidia*, 8 to 12 μ in diameter. The fungus thrives in the soil which is, therefore, the source of infection for animals and man. In the soil and at room temperature it grows as a mycelium, while in cultures at 37° C and in the body it occurs as a yeast cell form.

Routes of infection : Infection is by inhalation of fungus. Histoplasma is shed with bronchial and tracheal secretion, feces, urine and bile.

Symptoms ; Two forms are seen–the benign inapparent and the fatal disseminating forms. In the benign form no symptoms are seen, inspite of extensive localised infiltration in the lungs. Radiograms may show pulmonary nodules.

In the disseminated form, which is chronic, emaciation, diarrhoea, vomiting, ascites, chronic cough, apathy, anorexia, weakness, irregular pyrexia and anemia are noticed. In some animals swelling of lymphnodes, splenomegaly and hepatomegaly may be seen.

Lesions :

Macroscopically, since the fundamental tissue change is the proliferation of the reticuloendothelial system and infiltration by lymphocytes and plasma cells, discrete nodules and enlargement of organs are seen. Lungs contain discrete nodules or in the malignant form diffuse white patches. There is enlargement of the bronchial, mediastinal and mesenteric lymph glands and other lymph nodes of the body. The wall of the intestines is thickened and the mucosa shows either rugae or nodules. Adrenals may be enlarged and may have nodules. Ulceration of the buccal mucosa may be seen. Liver may be enlarged, mottled in color and firm. Spleen is enlarged to many times its normal size. It is light in color, firm in consistency and may have projecting nodules. Nodules may also be seen on the skin, pancreas, heart, genitals and kidneys.

Microscopically, the lesions are characterised by marked proliferation of the reticulo-endothelial cells with formation of macrophages and epithelioid cells, which may contain the organisms. The cells swell and many undergo degeneration and karyolysis. There may be infiltration with varying number of rymphocytes and plasma cells No giant cells are formed, nor necrosis is a legular feature. In the benign variety there may be fibrous capsule but in the malignant variety capsulation is not a feature.

In the lungs, in the benign form there may be a few discrete encapsulated nodules or a group of epithelioid cells as islands. But in the disseminated form the alveoli and the interstitial stroma are flooded with epithelioid cells. Many of the epithelioid cells contain the organisms.

The enlargement of the lymph nodes is due to the proliferation of R. E. cells. The normal lymphatic tissue is replaced by the above mentioned cells and so only a few lymphocytes and plasma cells are left. A few necrotic foci may be seen.

R. E. cell proliferation is responsible for the enlargement of the spleen. The splenic architecture is obliterated.

In the liver, there is intralobular and interlobular proliferation of R. E cells resulting in enlargement of the organ. The hepatic cells get displaced and the function of the organ is interfered with. Lymphocytes and plasma cells are present in varying number.

The mucosa of the intestines has rugae or nodules and the wall is thickened because of proliferation of the R. E. cells in the lamina propria and the submucosa. The glands become atrophied due to pressure. Ulceration of the mucosa does not usually occur. The mesenteric lymph nodes are enlarged and reveal loss of architecture.

The parenchyma of the adrenals may be largely replaced by macrophages which contain the organisms in their cytoplasm

Diagnosis : 1. Histological examination of the lesions reveals the fungus in the R. E. Cells.

2. Cultural isolation of the organisms.

3. Skin test with histoplasmin.

4   Complement-fixation test.

5. Agar plate precipitation test.

6. Differentiate this from tuberculosis, blastomycosis, coccidioidomycosis and lymphocytoma.

CANDIDIASIS (Moniliasis)

Candidiasis, caused by different species of *Candida*, is a sporadic disease involving chiefly upper alimentary tract of man, swine and poultry (thrush) and mammary glands of dairy animals. Candida species may be normally present in the alimentary tract of pigs and other animals, but under certain conditions, they may assume the role of pathogens, as in hypo - vitaminosis A and a prolonged use of antibiotic supplements in feed.

This disease occurs most frequently in poultry—young chickens and turkeys, affecting usually esophagus, crop and proventriculus. The symptoms in poults are not characteristic. Affected birds become listless and stunted with poor feathering. In acute cases, no symptoms may be noticed. Mortality may be as high as 75%   The crop, which is most often involved, shows typical turkish towel appearance, with raised patches of ulceration and pseudo membranes. In tissue sections the organisms (Pseudohyphae) can be seen penetrating, more or less perpendicularly, the keratinized squamous epithelium of the crop, while lymphocytes and histiocytes infiltrate the submucosa

The symptoms and lesions in domestic animals vary depending upon the organs/tissues involved. The lesions are usually associated with keratinization, ulceration, lymphofollicular hyperplasia, neutrophilic and eosinophilic infiltration and edema   Symptoms in pigs are diarrhoea, intense thirst and encephalitic symptoms.   On post mortem diphtheritic membrane is found in the small intestines.

Characterist'c lesions of the disease and demonstration of causative fungus in the lesions, on microscopic examination or cultural isolation, help in the diagnosis.

BLASTOMYCOSIS

Blastomycosis (North American blas*omycosis) is a suppurative granulomatous disease of man and animals, especially dogs. It is caused by *Blastomyces dermatidis* and is endemic to North America. The causative agent, a mycelial yeast‑like fungus, is dimorphic. In tissues and in cultures at 37°C, it is yeast-li (large, round, budding cells.) and at 20-25°C, it is in mycelial form.

In man, the infection may disseminate following pulmonary involvement (it may remain confined to the skin (cutaneous form). In animals, howeve systemic form is often seen.

This disease in aogs usually has slow onset and chronic wasting course with respiratory symptoms. Later subcutaneous abscesses may develop, which on rupture appear as purulent ulcers   The lesions in the lungs may be in the form of grayish-white, circumscribed, nodular or diffuse consolidated areas, which on incision yield pus like exudate. In advanced cases such lesions may also be present in the liver, spleen, kidneys, adrenals. brain and lymphnodes. Micros copically the lesions are characteristically suppurative granulomas— intense pr, liferation of reticulo endothelial cells associated with necrosed areas, having foci of neutrophils and a few multinucleated giant cells, lymphocytes and yeast cells (causative fungi), with little tendency towards encapsulation.

Diagnosis is based on the demonstration of the causative fungi in the purulent exudate (from the skin lesions) or coughed up sputum during life and in the tissue sections of the lesions after death.

PARACOCCIDIOIDOMYCOSIS

Paracoocidioidomycosis (South American blastomycosis), caused by *Paracoccidioides brasiliensis*, is a disease more or less similar to North American blastomycosis. The causative fungi have many similarities. The tissue reaction to both these fungi is also similar. This condition can be differentiated by the demonstration of multiple budding forms (in contrast to single budding forms in case of *B. dermatitidis)* of the organisms in the lesions.

B: MYCOTOXICOSIS

The fungi have been known to produce toxic metabolites, i. e., mycotoxins, and the consumption of such toxins leads to development of disease syndromes known as mycotoxicoses.

The mycotoxins are capable of causing deleterious biological effects in man, animals and birds. The toxins may be contained within the fungi producing them or these may be elaborated in the substrate (feed stuffs) on which the fungi are growing. It is believed that contamination of food and feeds with mycoto-xins occurs following the invasion of crops, before or after harvesting, or feed and fodder, during their storage, by the toxigenic fungi. The fungi when placed under suitable environmental conditions produce toxic metabolites. This, in general, depends upon temperature and humidity in the environment, moisture content and quality of the substrate, micro and mycoflora present, length of storage etc.

The consumption of such toxins, even in small amounts, by man and animals may lead to the development of toxic effects, though the toxicity will depend upon the quantum of the toxin consumed and species, age, sex and nutritional status of the animals/man involved. The major mycotoxicoses so far recognised/suspected are given, along with their symptoms and lesions, in a tabulated form.

In India, aflatoxicosis is the condition most frequently encountered Ducks are more often affected than other species of birds and animals. Outbreaks of Degnalla disease have been reported to occur among buffaloes and cattle in the rice growing areas following the consumption of mouldy toxic rice straw. Cases of ergotism in animals have also been seen.

Diagnosis : Diagnosis of the mycotoxic conditions is rather difficult. The epidemiology of the disease outbreaks may suggest the possibility of mycotoxi-cosis, but the diagnosis is based on the clinical symptoms, pathological changes. gross and microscopic, and demonstration of toxins in the feed/fodder and, where possible, residue of toxins in the tissues/urine of the animals. It may be emphasized that the presence of suspected toxin in the feed/fodder is significant only when it is in appreciable concentration. Similarly, simply the presence of toxigenic fungi in the feed is no proof of involvement of mycotoxins.

AFLATOXICOSIS IN ANIMALS

Aflatoxicosis is a condition characterised by severe liver damage due to fee-ding animals and birds with ground – nut cake containing a toxin known as afla-toxin of a mould, *Aspergillus flavus* Other fungi incriminated are *A. parasiticus* and *Penicillum puberlum*.

Cattle, pigs and poultry are affected. Buffaloes appear to be more suecep-tible than white cattle. Ducks and dogs are also more suceptible.

Etiology : The condition results from feeding groundnut cake containing the aflatoxin. High humidity and storage for long periods appear to facilitate the growth of the fungus on the cake. Probably the toxin interferes with protein synthesis by the liver. It was found that the fungus to produce the toxin must be present in a pure culture in the feed, not contaminated with other fungi. For in the presence of other fungi, the toxin is not formed. This may be the cause for some feeds to be toxic while others not, though stored under similar condi-tions

There are 6 types of aflatoxins — B^1, B_2, G_1, G_2, M^1 and M_2. The first 4 are so named depending on the blue or green fluorescence produced on thin layer chromatograph plates and on their position on migration Excreted through the milk, B_1 and B_2 are known as M^1 and M_2. Of these B^1 is the most toxic.

Symptoms : Cattle The first symptom noticed is slight loss in appetite shown by the animal leaving away the concentrates in feed troughs. Subsequently it refuses the concentrates There may be periodical diarrhoea with dark colored fetid feces. During this period the animal is dull and grinds its teeth. Later ascites, emaciation, anemia and tenesmus are seen. Fluid accumulation in the abdomen increases gradually and the animal finally dies. No rise in body temperature is seen. There may be icterus and pneumonia occasionally. Pregnant animals may abort.

Poultry : In ducklings and turkey poults, loss of weight, oculo-nasal discharge, staggering gait and nervous symptoms may be noticed resulting in heavy mortality. In white skinned ducks, purplish discoloration of shank and web and hemorrhages on serous membranes and muscles (hemorrhagic syndrome) may be noticed. Though similar symptoms may be noticed in chicks, mortality is low. Reduced egg production occurs in layers.

Lesions :

Macroscopically, visible mucous membranes are pale or slightly icteric. Ascites is seen with variable quantity of straw colored fluid in the peritoneal cavity.

Liver is enlarged, pale and firm In some advanced cases it may be hard to cut. Section reveals necrotic greyish areas in some animals. In early cases liver may be yellow and fatty.

Microscopically, in the early stages the hepatic cells are loaded with fat. Subsequently centrilobular necrosis sets in. Later the whole lobule may be involved, The fibrous tissue proliferates and invading the parenchyma isolate the surviving hepatic cells into small groups of two or more cells or into cords having either single or double rows of cells This appearance is characteristic of pericellular cirrhosis. The hepatic cells assume different shapes but most of them are oval.

There is marked focal infiltration by lymphocytes and plasma cells in the interlobular connective tissue, occasionally, simulating chronic interstitial hepatitis.

There is marked bile duct proliferation, sometimes to such an extent as to suggest neoplasia.

Fibrosis causes occlusion of the central veins and so the name 'Veno occlusive disease' is given to it. The media of the hepatic arteries is hypertrophied which is a characteristic feature in cirrhosis.

In ducks hepatic carcinoma is very common,

Piglets born of sows that have ingested cantaminated cake may be affected. The piglets may either be poisoned *in utero* or they may have ingested the poison through the milk of the sows since aflatoxin is found to be excreted through milk The piglets become emaciated, icteric and show diarrhoea.

Macroscopically, the visible mucous membranes are icteric, petechiae are found in the gastro-intestinal tract, the liver is enlarged, soft, friable and very yellow.

Microscopically, the hepatic cells show intense fatty changes.

In a recent outbreak of acute aflatoxicosis in a piggery, piglets of 4 to 5 months of age suffered heavy mortality.   The animals were off feed first and so their weight gain was reduced from 3.2 Kg/per day to 1.2 Kg per day   The animals became very much emaciated and some had subnormal temperature before death.   Majority showed jaundice which was intense during the last stages of the disease.   Neutrophils increased. A few animals showed 90% neutrophils with shift to the left.

Lesions : Icterus of the mucous membranes and subcutaneous fat: wide spread hemorrhages in the stomach, intestines and large amount of free blood in the large intestines and abdominal and thoracic cavities. Petechiae and ecchymosis were found on the serous membranes, subcutaneous fat,   pericardium and endocardium. Liver was enlarged and was pale tan to yellow in color.   In some the liver was covered with yellow flakes.   The wall of the gall bladder was edematous

In some animals serous cavities contained half to two litres of yellow colored transudate. In some this was serosanguineous. Hemorrhagic gastroenteritis was prominent in the majority of animals.   Microscopical picture was similar to that described earlier.

In summary, the aflatoxin is hepatotoxic causing successively fatty infiltration, necrosis and postnecrotic cirrhosis.   So care should be taken not to feed groundnut cake too much.   It should be understood that aflatoxin is excreted through milk and there is a possibility of hazard in feeding such milk to children.

GANGRENOUS SYNDROME IN BUFFALOES & CATTLE
Syn : Degnalla Disease.
(By Dr. M. S. Kwatra)

Gangrenous syndrome is characterised by the necrosis of the tips of the er , tail, tongue and swelling of the extremities. Subsequently the skin from the swollen extremities is necrosed and peels off forming open wounds   It occurs in the paddy growing areas of Punjab, Haryana, possibly also in Kerala and Uttar Pradesh as also from the Degnalla area of District Sheikhpura in Parkistan. In severely affected cases the hooves and phalanges are cast off.   The buffaloes are more commonly and severely affected than cattle.   The exact etiology of the disease is still obscure. However it is generally prevalent in the winter months and can be produced in experimental buffalo calves by feeding them on the paddy straw collected from the premises of the owners having affected animals   The conditions closely resemble ergot fescue grass lameness in cattle which has   been reported from other countries,   The available evidence indicates that mycotoxins produced by *Fusarium tricinctum* infesting the grass having vasoconstrictive effect are probably responsible for the disease.   In India the paddy straw is also believed to be contaminated with mycotoxins and other vasoconstricting factors. The cold weather possibly exerts synergistic effect on the extremities. Another school of thought is that the symptoms are due to chronic selenium poisoning

The affected animals have normal appetite and there is generally no rise of temperature.   The gangrene of the extremities continue to progress till the necrosed extremities are shed off   Mortality rate may be 25% or more.   If paddy feeding is continued the animal dies a crippling death

Lesions :

Externally, necrosis of the tail and ears, swelling of limbs and necrosis of the skin of the extremities can be observed   The necrosed tip of the ear is hardened and difficult to cut .   The gangrenous extremities are clearly demarcated from the unaffected portion.   The swollen limbs on incision show straw coloured transudate collected in the subcutaneous tissues and between the tendons and muscle fibres.

Microscopically ;

The skin of the gangrenous extremities shows complete loss of architectural details.   At places saprophytic organisms embeded in the necrosed portion may be seen without any cellular reaction around them.   Edema is commonly seen in the skin sections taken from the swollen limbs.   The blood vessels do not show any significant microscopic change in the prenecrosed zone though some workers have noticed development of thrombi in a few cases.

The muscles and subcutaneous tissue from the affected area show edema and slight mononuclear infiltration.

MYCOTOXICOSES FOUND IN ANIMALS AND MAN

Mycotoxicosis	Toxins involved	Fungi incriminated	Man/animals affected	Symptoms/ Lesions
Alimentary toxic aleukia(ATA)	Sporofusarin poae - fusarin,(T2)	*Fusarium sporhtrichioides F, poae*	Man, Animals	Hyperemia, hemorrhages, stomatitis. gastroenteritis, leucopenia, agranulocytosis.
Ergotism	Ergot alkaloids	*Claviceps purpurea C. fusiformis*	Man, animals	Acute : nervous symptoms. chronic: vascular damage, gangrene of extremities, abortion,
Stachybotryotoxicosis	Stachybotry otoxin. (Roridin E)	*Stachybotrys alternans* (*S. atra*)	Man, horse, cattle, sheep, poultry	Hemorrhage, widespread necrosis, thrombocytopenia. Scrotal skin lesions, abortion.
Estrogenism	Zearalenone (F · 2)	*F. roseum* (*Gibberalla zea*)	swine	Tumefaction of vulva. enlargement of mammary glands, uterine edema and hyperplasia.
Facial eczema	Sporidesmin	*Pithomyces chartarum*	Sheep, cattle.	Photosensitization, liver damage, icterus.
Hemorrhagic Syndrome (Trichothecene toxicosis)	Trichothecenes	*F. tricinctum Aspergillus Penicillium Alternaria*	Cattle, pig, poultry,	Hemorrhages, epistaxis, blood tinged diarrhoea.
Trembling Syndrome	Penitrem A (Tremor-	*Penicillium cyclopium*	Cattle, sheep	Tremors, ataxia, convulsions.

Mycotoxi-cosis	Toxins involved	Fungi in-criminated	Man/animals affected	Symptoms/Lesions
(Penitrem intoxication)	tin A)	*P. palitans*		
Salivary syndrome (Slobber syndrome)	Slaframine	*Rhizoctonia leguminicola*	Cattle, sheep goat, horses	Excessive salivation, lacrimation and urination, diarrhoea, bloat and abortion.
Sweet clover disease	Coumarin	*Aspergillus Penicillum, Mucor*	Cattle, sheep	Prolonged clotting time of blood, hemorrhages, liver damage.
Fescue foot Syndrome	Butenolide(?) T-2 toxin (?)	*F tricinc-tum*	Cattle	Lameness, necrosis and gangrene of extremities,
Yellowed rice toxicity	Luteoskyrin, Cyclochloro-tine, Citreo-viridin	*P. islandi-cum; P. Ci-treoviride*	Man cats, non-human primates, chickens.	Cardiac beri beri (Man), vomiting, convulsion, ascending progressive paralysis, respiratory disorders
Mouldy corn toxicosis	T2 (?)	*F. culmo-rum; F. tri-cinctum*	Cattle & other animals.	Wide spread hemorrhages, edema.
Leucoence-phalomalacia (Mouldy corn poisoning)	?	*F. Monili-forme*	Horses, donkeys	Blindness, ataxia, circling, paralysis, recumbency, death, liquefactive changes in white matter of the brain.
Paspalum staggers	?	*Claviceps paspali*	Cattle, sheep horses.	Nervous symptoms, incoordination, shaking of legs with head nodding.
Nephropathy (Ochratoxicosis)	Ochrato-xin - A	*P. virdica-tum: A, och-raceus*	Cattle, sheep, pigs	Liver damage, abortion, progressive paralysis.
Rubratoxin poisoning (Rubratoxicosis)	Rubra-toxins A & B	*P. rubrum P. purpu-rogenum*	Farm animals pigs	Depression, severe damage of liver and kidneys and hemorrhages on serous membranes and viscera.

- - - -

CHAPTER 29

DISEASES CAUSED BY RICKETTSIA

Rickettsiosis in dogs	Heartwater
Rickettsial diseases of ruminants	Eperythrozoonosis
Q fever	Hemobartonellosis
Anaplasmosis	Ornithosis and Psittacosis

RICKETTSIOSIS IN DOGS

It is a disease caused by *Rickettsia canis* (Syn : *Ehrlichia canis*) in dogs and widely reported in Africa and India. It is transmitted to dogs by ticks of genus *Rhipicephalus Sanguineus* Hereditary transmission of the organism in the vector may be present. The incubation period is about 1 to 3 weeks, and the disease is manifested in three forms – cutaneous, septicemic and nervous forms.

The cutaneous form is manifested by ulcers which are round and shallow or superficial pustules on abdomen, resembling those of canine distemper, The septicemic form is characterised by diphasic fever which may last for 10 to 14 days.

The nervous form is characterised by convulsions, muscular weakness and partial paraplegia.

Diagnosis of the disease is by examining the blood smear. There is monocytosis and eosinopenia during febrile stage. Blood smears can be stained with Giemsa. Colonies of organisms are found in the cytoplasm of monocytes. About 1 to 40 colonies may be found in each cell and so the nucleus is indented. Each colony contains initial bodies with granules measuring 0.5 to 1.5 μ or elementary bobies 0.2 to 0.3 μ in diameter. The granules are pleomorphic.

RICKETTSIAL DISEASES IN RUMINANTS

These are caused by *Rickettsia bovina* and *Rickettsia ovina*, and have been reported in U K. and India in bovines and sheep respectively. The disease is commonly known as Tick Borne Disease because of its transmission by Ticks. *Rhipicephalus hemophysaloides* is incriminated in India and *Ixodus ricinus* in U. K. The disease itself is mild but it predisposes the animal to secondary infections by Staphylococci in lambs and to Louping ill in adult sheep. Adult cattle, sheep and goats are more susceptible than young animals. The incubation period is about 3 to 7 days.

In bovines the disease is characterised by polypnoea and high temperature (105°F) which persists for about 2 to 8 days. The temperature falls gradually and may be followed by a second attack. Milk yield falls during the febrile stage.

In sheep the symptoms are similar to those of cattle but respiratory distress is absent. In late pregnancy, animals abort In young lambs the symptoms are very mild and only moderate increase in temperature is seen.

The diagnosis of the disease in affected animals is made by examining blood. There is marked leucopenia, mostly due to depression of the neutrophils during the height of fever.

In the blood smears stained by Giemsa or polychrome methylene blue, bluish - purple organisms can be detected in the cytoplasm of neutrophils (to the extent of 95% of cells), monocytes, and large lymphocytes during febrile period for a few days in cattle and for several weeks in sheep. The organisms are round, rod shaped or ring-shaped or irregular masses measuring 0.75 to 3μ in diameter. These may be present singly or in clusters

Postmortem lesions are not characteristic except for splenomegaly in sheep. Histologically characteristic lesion is the destruction and disappearance of lymphoid tissues.

R. belgaumii causes, in sheep and goats of some parts of India, a mild feb-rile condition The parasite is found in the monocytes of the blood. *Hemophysalis bispinosa* is the vector.

The following diseases are caused by rickettsia in animals and found in other countries.

Contagious ophthalmia in sheep caused dy *R, conjunctiva* (See page 526).

Salmon disease in dogs and foxes is caused by *Neorickettsia helminthoeca*. The parasite is transmitted by encysted metacercaria (of the fluke *Troglotrema salmincola*), when dogs are fed fish of the family *Salmonidae*, which are the intermediary hosts of the fluke.

Q FEVER

This is a disease of importance to man, since it is transmitted from domestic animals which are reservoir hosts. It is manifested by pyrexia and acute pneumonia in man and is common in persons working in slaughter houses and handling carcases. The disease is usually not fatal. This is called Q-fever because it was first encountered and studied in the province of Queensland, Australia.

The etiological agent is *Rickettsia burnetii* (Syn. Coxiella burnetii). Man is susceptible to infection. Cows, seeep, dogs, goats, horses, rats, rabbits, mice and hedgehogs serve as reservoir hosts They do not show any symptoms of infection.

The vectors are ticks (*Dermacentor andersoni, D. occidentalis, hipicephalus sanguineus*) mites, lice and flies. There is evidence that the disease may be transmitted through contaminated milk and by inhalation of dust contaminated by *R. burnetii* in dairy barns and sheep and goat pens.

Animals show no symptoms or signs of infection Postmortem examination of those from which organisms have been recovered, did not reveal any specific lesions

In man, after an incubation period of 14 days. the disease is characterised by fever lasting for 3 to 6 days, headache, chill, generalised malaise, and respiratory symptoms due to acute pneumonia. The lesions consist of lobar pneumonia similar to those caused by pneumococci. Microscopically, lung shows alveolar exudate consisting principally of mononuclear cells as seen in psittacosal or viral pneumonia. Superimposed bacterial infection may. however, obliterate these specific lesions. Besides the alveolar exudate interstitial tissue may also show inflammatory reaction with infiltration by macrophages, lymphocytes and plasma cells and occasionally neutrophils. The septa may be necrosed. The bronchioles contain exudate with large numbers of neutrophils. Convalescence is long.

Apart from pneumonic lesions two rare manifestations of Q Fever have been reported. One is vegetative endocarditis involving mitral and aortic valves and the second is miliary granulomas in liver.

Diagnosis of this disease is done by inoculating saline suspension of suspected material into mice by intransal or intraperitoneal route. The lesions noticed are nodular or patchy granulomatous foci in liver, spleen, kidney and adrenals. Patchy or nodular areas of aplasia are seen in bone marrow. Proliferative changes in lung and exudation of mononuclear cells are observed in mice inoculated intransally. The organisms can be seen in tissues stained by Giemsa or Machiavello methods as plemorphic organisms in the form of lanceolate rods ($0.05 \mu \times 1.5\mu$). They are usually present within the cytoplasm of mononuclear cells and occasionally outside the cells.

In guinea-pigs which are experimentally infected with organisms the lesions seen are perivascular exudation of lymphoid cells, with a few monocytes and fibroblasts. Proliferation of vascular endothelium is seen in myocardium, lungs, adrenals, renal cortex and medulla and epididymis In lungs small foci of epithelioid cells are seen in alveoli. In later stages of the disease small nodules of epithelioid and giant cells are found in spleen, liver, vertebral marrow, and sometimes in heart, pancreas, kidney, mediastinal and mesenteric fat etc.

In man Weil-Felix test is used for diagnosis of other rickettsial diseases. This is an agglutination test in which to the patient's serum is added one of the strains of Proteus as an antigen. But in Q-fever this is not useful since antisera against *R. burnetii* do not agglutinate the Proteus organism. This test is also not useful in animal rickettsiosis.

ANAPLASMOSIS
(Syn : Gall Sickness)

Anaplasma is an organism included in the Rickettsial group based on its structure and behaviour. It affects cattle, sheep and goats, in which erythrocytes are parasitized.

A. marginale affects cattle, sheep and goats and is very virulent.

A. centrale, which is quite distinct from the former, affects cattle and causes a milder disease, but does not exist in India.

A. ovis affects only sheep and goats in USA, Africa and USSR.

The present consensus is that the organisam is a member of the Rickettsia. Electron microscopic study showed that the bodies in the erythrocytes though resembling inclusion bodies are in fact organisms only.

A. marginale infects cattle, deer and wild animals. The wild ruminants act as reservoirs, from which cattle become periodically infected.

Young calves upto one year, if infected, do not manifest any symptoms and the disease is mild. These are in a state of premunity and become carriers. In animals upto 2 years, the disease is acute but it is not fatal. In animals upto 3 years the disease is acute and occasional fatalities may occur. In animals above 3 years the disease is peracute and many fatalities occur. Recovered animals are carriers for life and they never become sterilised and become free of the disease.

Routes of infection :   Source of infection is an ailing animal or a carrier. Blood sucking insects-ticks, flies (Tabanus) and mosquitoes transmit the disease. Probably it is only the tick that is a -biological carrier while others transmit mechanically. Infection may also be spread mechanically by infected hypodermic needles and unsterilised instruments used in castration, ear notching. vaccination, spaying and amputation of the horn.   Infection can also occur by blood transfusion.   Intrauterine infection is also possible.

A. marginale ;   This parasite occurs usually at the pheriphery in the erythrocyte and as a coccoid form, measuring 0^-3 to 1.0μ.   The ticks that transmit the disease are of the genera *Boophilus, Hyalomma, Ixodes, Rhipicephalus* etc. It has a soluble antigen which is lipo-proteinaceous and a particulate antigen.

Pathogenesis :   The parasite entering the blood invades the erythrocyte It does not cause hemolysis but the infected red blood cells are phagocytised by the reticulo-endothelial cells, especially in the spleen.   Later the phagocytes of the spleen and bone marrow phagocytise the healthy red cells (those not containing the parasites) also due to autoimmune processes that develop during an infection. Babesia and plasmodium cause hemolysis, unlike the anaplasma, because probably due to their more vigorous metabolism.   During the disease there is increased erythropoietin titres of the blood plasma and this elevation continues during the period of convalescence also.   The convalescent period may be 2 to 3 months.   During recovery period there is increased hemopoiesis— reticulocytosis, macrocytemia and granulocytosis

In this disease there is significant decrease in the acetyl chlolinesterase activity and increased catalase activity of the erthrocytes than normal.

Symptoms :   Incubation period is longer than in babesiosis and is about a month in artificial infection but 1 to 3 months after bite of a tick.

In the young under 1 year. the disease being mild is inapparent.

In the adult. there is sudden rise in temperature, upto 105° F, soon followed by dullness, anorexia, rough hair coat, cessation of milk yield, dyspnoea, tachycardia, lachrymation, constipation alternating with diarrhoea.   Pregnant animals abort and then become srerile.   Jaundice appears, rumination stops. animal becomes week, trembles and dies in 2 days to 3 weeks. Mortality may be as high as 80% in imported cattle.   Hemoglobinuria is not observed.   Some animals may manifest aggressive symptoms attacking attendants.

Hematology : The red cell counts fall to about 1 to 2 millions Anisocytosis, poikilocytosis, punctate basophilia, Howell—Jolly bodies and nucleated erythrocytes are found in the blood smear.   The hemoglobin content and PCV are reduced. Parasites are found in the erythrocytes. In acute cases nearly 50% to 60% of the cells may be infected.

Calves may show anemia without showing any symptoms.

Lesions : The carcass is emaciated and the mucous membranes may be icteric.   Tissues are pale and the blood is thin and watery.   Serous cavities contain large amount of fluid.

The spleen is enlarged, reddish and the Malpighian corpuscles are prominent.   Liver is enlarged and brownish-yellow in color. Gall bladder is distended with granular bile   Petechiae may be noticed under the epicardium Myocardium is degenerated. Catarrhal gastroenteritis is encountered.   Kidneys are congested and show degenerative changes. Hemosiderosis may be noticed in spleen and liver, Extramedullary hemopoiesis is evidednt in the liver and spleen. The hepatic degeneration is not due to the parasities but to the anemia.

Diagnosis 1. Symptoms; 2. Observe parasites in a blood smear, 3. Transmit the disease in splenectomised animals. 4. C. F. test, 5. Direct and indirect Fluorescent antibody technique, 6 A capillary tube agglutination test can be done. This test is found to be very reliable and can be utilised for detecting the carriers also. The antigen for this test is obtainable commercially (Anatest of Diamond laboratories).

A ovis infects only the sheep and goats but not cattle, In sheep and goats the disease is mild and subclinical. Progressive anemia may occur.

HEARTWATER

Heartwater is a septicemic, non-contagious disease of sheep, goats and cattle characterized by high fever and nervous symptoms. The name 'heartwater' is derived from the characteristic pericardial effusion which is rather consistently observed in small ruminants but often absent in the disease of cattle.

The causative agent, *Cowdria* (formerly Rickettsia) *ruminantium* is an intracellular parasite transmitted by ticks—*Amblyomma* sp.

Though transmission does dot occur through the egg (differing from Rickettsia in which transasmission occurs through the eggs) there is transmission from the larval stage to the nymph and from the nymphal stage to the adult. Cowdria is a tiny rod-shaped, often diplococcal organism. It is gram negative, cannot be cultivated on artificial media and is not demonstrable in the circulating blood. It selectively parasitizes the endothelial cells of jugular vein, venacavae, renal glomerular capillaries and cerebral grey matter. In the blood it is present firmly attached to the red cells. The colonies of the microorganisms are found in the cytoplasm at the poles of the nucleus.

Exotic breeds are more susceptible than indigenous ones.

Calves and lambs below 3 weeks of age are highly resistant. Ruminants are more susceptible at about the age of early maturity. In cattle mortality may be 60% while in the European breeds of sheep it may be 100%.

Course and Symptoms : The incubation period in natural infection is 2 to 3 weeks but considerable variation is noticed. In the *peracute* form death occurs rapidly showing convulsions. In the *acute* cases the course of the disease is upto 6 days. High fever and other attendant signs are observed. As the course advances, nervous symptoms develop. These are especially prominent in cattle and consist of chewing movements, protrusion of the tongue, unsteady gait, twitching of eyelids and blinking, circling and terminal convulsive phase characterised by cutaneous hyperasthesia.

Though diarrhoea is commonly seen in cattle, it is not a regular feature in sheep.

Lesions : The lesions of cattle are less prominent than in sheep. The most prominent changes are ascites, hydrothrorax and hydropericardium. Hydropericardium may be absent in some animals and the fluid in bovines is proportionately lesser than in sheep. Retroperitoneal and mediastinal tissues are edematous. The lymph nodes of head neck and anterior mediastinum, are swollen and

edematous.   In sheep and goats the spleen is much enlarged but in cattle swelling is rather moderate.   Hyperemia and mucosal edema are prominent in the
abomasum and may also be found in the lower alimentary canal   Lungs are
edematous. small hemorrhages may be present in the endocardium, lymph nodes,
mucosa of the respiratory tract and the mucosa of the stomach and bowel.

Brain may be edematous with hemorrhages scattered throughout.   Necrotising changes in the small blood vessels of the brain may be the cause for these
hemorrhages.

Apart from the presence of the organisms in the vascular endothelium
no other specific histological changes are observed since no local inflammatory
reaction is encountered.

Diagnosis :
1.   Symptoms.
2,   Postmortem lesions.
3.   Demonstration of the organism in the endothelium of blood vessels.
4.   Heartwater should be differentiated from the following diseases, which
are not caused by rickettsia; Tetanus, Anthrax Babesiosis, Verminous
infestations.

EPERYTHROZOONOSIS

Eperythrozoon is found in sheep, pigs and cattle as a parasite of the erythrocytes, *E ovis* is found attached to the erythrocytes of the sheep and occurs as
a ring-shaped bacillary or coccobocillary form 0.3 μ to 1.0 μ in diameter.

After an incubation period of a week, parasites are found in the blood
Various biting insects have been incriminated to transmit the disease   Contaminated needles and instruments may also transmit infection. *In utero* infection
appears to exist.

As the organisms multiply in the erythrocytes, anemia develops and in
severe infections, icterus may be observed.

Hematological examination reveals fall in total erythrocyte counts (as low
as one million per c. mm), monocytosis and leucocytosis, the total white blood
cells reaching as high as 20,000 per c. mm.

Though mortality rate is very insignificant, repeated relapses may occur.

Lesions ; Anemia (hemolytic), regenerative bone marrow, depositoin of hemosiderin in the proximal convoluted tubules of the kidney, enlargement of the
spleen and lymphoid hyperplasia are the lesions observed.

Eperythrozoon suis (0.8 μ in diameter) is found in pigs and the affected animals
show mild fever, weakness of hind limbs, pallor of mucosae and emaciation. Jaundice may also be present. Extreme anemia, which is hemolytic, together with
hemorrhages may be found   Mortality is very low.   Recovered animals become
carriers for life.

Lesions ; Include jaundice, soft and enlarged spleen, yellow liver, flabby
and pale heart, serous effusions into the pericardium and peritoneum and thin
watery blood.

Microscopically, bone marrow is hyperplastic, liver shows hemosiderosis
and there may be some necrosis of liver lobules.

Eperythrozoon wenyoni is found in cattle but does not seem to cause any serious symptoms.

Diagnosis : 1 Examination of blood smear and identification of the organisms in the erythrocytes.

2. A complement-fixation test similar to the one used in anaplasmosis can be used, Sera from affected animals give positive test on the third day of clinical infection, remain positive for 2 to 3 weeks and later become negative.

HEMOBARTONELLOSIS

Hemobartonella, grouped under rickettsia, are found parasitizing the erythrocytes They occur as minute coccoid or bacillary forms. These parasites have been found In cattle, goats and dogs, in which hemolytic anemia may be produced.

Hemobartonella canis has been met with in India among dogs. Fleas are suspected to be the vectors. Anemia caused is of the hemolytic type and the number of erythrocytes may be reduced to as low as one million per c. mm. Blood picture reveals anisocytosis, basophilia and normoblasts

Lesions include pallor of the internal organs, sometimes icterus, enlargement of spleen and lymph nodes.

Diagnonis : Examine the blood smear and observe the small coccoid or bacillary organisms attached to the erthrocytes.

ORNITHOSIS AND PSITTACOSIS

This is a disease of man and birds caused by an organism known as *Chlamydia psittaci*, which is ordinarily grouped under psittocsis-lymphogranuloma group. These are midway between virues and rickettsia. Containing RNA, they subdivide by fission. They are intracellur obligate parasites.

The disease in man and psittacine birds (parrots, parakeets) is known as *psittacosis* while the disease manifested in pigeons, fowls, turkeys and ducks is known as *ornithosis*.

Man becomes infected from birds and infection may be by inhalation. In birds, mites and lice may introduce infection through skin. Egg transmission does not appear to occur, **Incubation period** is one to two weeks.

Pathogenesis : After inhalation, the organisms are found in the epithelial cells of the lung, air sacs and the pericardial membrane as elementary bodies which are known as Levinthal-Cole-Lillie (LCL) bodies. The elementary bodies on entry into a cell, become larger and larger and form plaques. Later, the large bodies divide repeatedly to become smaller and smaller to the size of original elementary bodies which again infest other cells. Entering the blood stream, the organisms reach the spleen, kidney and liver.

Birds affected : Ornithosis in pigeons. ducks and turkeys is severe, though mortality is low Though fowls can be infected, no symptoms are seen in them. Incubation period is 6 to 10 days.

Symptoms :

Man : The onset is sudden with fever, chills, headache, malaise, sore throat and cough. The cough is dry and non productive Though pnenmonia may be present, physical examination may be negative. There may be drowsiness and delirium. After 2 to 3 weeks recovery may be spontaneous. Mortality may be about 20%.

Among **pigeons**, though morbidity is high mortality is low. Young birds are more susceptible. There may be diarrhoea, nasal discharge and serous conjuctivitis.

In parrots and parakeets, symptoms seen are dullness and bloody diarrhoea.

Lesions :

Man : The main lesion is patchy pneumonia which is confined to the interstitial tissue and alveolar walls. The exudate does not contain any fibrin nor is there fibrinous pleurisy. The cells lining the alveoli swell and proliferate, revealing mitotic figures. There may be desquamation of the epithelial cells into the alveoli forming clumps. The elementary bodies can be seen in the alveolar epithelial cells. There is edema and mononuclear infiltration into the alveolar septa. There may be focal necrosis in liver and spleen. Heart and kidneys may reveal infiltration by mononuclears. There may be edema, congestion and hemorrhagic foci in the brain. The elementary bodies can be seen in the R. E. cells of liver and spleen

Birds :

Liver and spleen are very much enlarged. Spleen is dark-red and soft Proliferation of the reticulo-endothelial cells and infiltration of blood are the causes for the swelling. Sometimes, capsule may be ruptured due to large subcapsular effusions, resulting in hemorrhage into the body cavity. The enlarged liver is hyperemic and in some cases (especially chronic) it may be icteric. The lymphocytes, histiocytes and plasma cells in the periportal area proliferate. There may be serous or serofibrinous pleurisy, pericarditis and inflammation of air sacs (aerocystitis.) Serous conjunctivitis and rhinitis may be seen. There may be catarrhal gastroenteritis and bronchitis. Nephrosis may be seen occasionally. Pneumonic foci may be met with sometimes. In pigeons there may be enlargement of the pancreas, which may show small necrotic foci.

LCL bodies can be seen in the cytoplasm of mononuclears in spleen and tubular epithelium of kidney.

Diagnosis :

1. Examine exudates from pericardium and liver by impression smear and stain by Giemsa—see LCL bodies, each less than 0.5μ.

2. Inoculate white mice intraperitoneally with bacterial filtrates. Death in 5—10 days, On P. M. fibrinous exudate is seen which contains the organisms.

3. Inoculate 6—10 day old chick embryos. After 48 hrs. embryo dies. Examine yolk sac for organisms.

4. Complement-fixation test, cross-immunity and agglutination tests can be done.

— — — —

CHAPTER 30

DISEASES CAUSED BY PROTOZOA

Coccidiosis	Leishmaniasis
Toxoplasmosis	Amoebiasis
Babesiasis	Histomoniasis
Theileriasis	Globidiasis
East coast fever	Fowl Malaria
T. annulata	Leucocytozoon infection
T. mutans	*Hepatozoon canis*
Trypanosomiasis	
Surra	

COCCIDIOSIS

Coccidia are protozoa belonging to the class Sporozoa.

Two genera are pathogenic—Eimeria and Isospora.

Coccidia parasitize many animals and birds. The following species of coccidia are important.

Animal	Coccidia	Habitat
Cattle	*Eimeria zurnii*	Cecum, colon, terminal
,,	*E. bovis*	portion of ileum
,,	*E elipsoidalis*	small intestines
Sheep and Goats	*E. arloingi*	
,,	*E parva*	small intestines
,,	*E faurei*	
Swine	*E debleicki*	
,,	*Isospora suis*	small intestines
Dog and Cat	*Isospora bigemina*	
,,	*I. rivolta*	
,,	*I felis*	small intestines
,,	*E. canis*	
Fowl (See later)		
Rabbit	*E. stiedae*	Intrahepatic bile ducts
Geese	*E. truncata*	Renal tubules

The infective stages of the coccidium, the oocysts, are thick walled and very resistant. They may be viable, outside the body, under suitable conditions for nearly 2 years.

Routes of infection : Infection is by ingestion of the sporulated oocysts. So source of infection is an affected animal or a carrier. For understanding the pathogenesis of coccidiosis knowledge of the life cycle of the parasite is a requisite.

Oocysts containing sporocysts (4 with 2 sporozoites in each in the ease of Eimeria and 2 with four sporozoites in each in the case of Isospora) are ingested. By the action of intestinal juices sporozoites are released which enter the epithelial cells of the intestines—sporozoite (probably one per cell) grows and becomes bigger – the trophozoite. This becomes a schizont, occupying the whole of the cell—Crescent shaped merozoites similar to sporozoites are formed in the schizont by asexual division called schizogony—rupture of schizont and epithelial cell - merozoites liberated and these infect other epithelial cells forming schizonts which again give rise to fresh crop of merozoites. The asexual division occurs for a limited number of times (fixed for each species, for example in *E. bovis* there is only one and in *E. tenella* two or three). The asexual reproduction stops and the trophozoites enter into sexual reproduction. They become (instead of schizonts) gametocytes—male and female—the male gametocyte contains many microgametes which are motile. The female gametocyte becomes a single macrogamete The male microgamete leaves the gametocyte and entering the macrogamete, fertilizes it, forming a zygote. This with a wall formed is the oocyst. This oocyst escapes out of the host cell and is voided along with feces. Further development—sporogony—can occur only in the soil under suitable conditions. The zygote divides into 4 sporoblasts. These develop thick walls and become sporocysts. Each sporocyst divides into two sporozoites. At this stage the oocyst is infective and on ingestion the life cycle is repeated.

There is host specificity in coccidia and specific immunity may be provoked but no cross resistance is found. Infection confers certain amount of immunity,

Pathogenesis ;

The harm done to the host by the parasite is due to the destruction of the epithelial cells. In heavy infection this may be so great that there may be denudation of the mucosa and so severe hemorrhage and anemia may result. The organism may not invade into submucosal tissues but because of the infection. there may be intense inflammation of the lamina propria and even submucosa.

As the epithelial cells are destroyed the surviving ones are stimulated to proliferate and so hyperplasia of the intestinal epithelium occurs manifested by the enlargement of the villi which may show papillary projections and these cells are full of parasites.

Symptoms :

Cattle : Incubation period is 16 to 30 days in artificial infection, Infection by *E, zurnii* is more severe, Young animals under two years are more often affected. Initial fever may not be noticed. There may be sudden onset of bloody offensive diarrhoea with mucus. Animals evince considerable straining. Because of loss of blood. mucous membranes become pale. Weakness and dyspnoea follow because of anemia. Finally dehydration, emaciation and death occur. Animals that survive have a long period of convalescence.

In animals with milder infection. there may be diarrhoea but without blood.

Sheep and goats : Incubation period is 14 days. Young lambs under 6 months of age are affected and show diarrhoea and dysentery. Death may occur due to emaciation, anemia and dehydration.

Swine : Young piglets are affected. Dysentery is not a symptom. Necrotic patches may be seen.

Dog and Cat : The disease is not severe in these animals Rarely dysentery may be found with fatal results

FOWLS

The following table summarises the various species of Coccidia recorded from fowls in India.

Name of the species	Location	age	Endogenous stages	Oocyst characters	Percentage of incidence
E. tenella	Ceca	3-4 weeks	Mostly schizogenic generation	Broadly ovoid 22–23 × 17-20/µ Polar granule sporocytes measure 9—13 × 5—9µ. Sporulation time 22—30 hours.	Pathogenic 4.7%
E. necatrix	Small intestines proximal end. Sometimes throughout.	About 5 weeks	Schizogenic 2-3 generations & gametocytes in ceca and large intestines.	Ovoid : 20-21 × 16-18 µ Polar granule sprocysts 9-11 × 5-9µ steidae body sp' time: 24-30 h-s.	-do- 2 5%
E. maxima	Small intestines (middle)	-do-	-do-	Ovoid : 29-32 × 23-25µ polar granule sporocysts ; 13-20 × 7-12µ sp. time : 24-36 hours.	-do- 3 8%
E. acervulina	Small intestines (proximal)	-do-	-do-	Ovoid : 18-10 × 14 15µ Polar granule Sporocyst 7-10 × 4-7µ sp. time: 20-27 hrs.	1.2%
E mitis	Small intestines lower part	-do-	2 to 3 schizogenic generations	Subspherical 15-17 × 14-17µ Polar granule. Sporocysts measure 7-10 × 4.7 µ sp. time: 18-24 hrs.	3%
E brnnetti	Small intestines lower part	do	do	Ovoidal : 25 to 26 × 28 to 20 µ Steidae-body present Sp. time: 24 to 28 hrs:	2.3%
E praecox	Small intestines proximal	do	do	Ovoidal : 21 to 22 × 16 to 19 µ Steidae body present Sp. time: 24 to 40 hrs.	3.6%
E. hagani	Small intestines proximal portion	do	do	Ovoidal 17 to 23 × 16 to 17 µ Sp. time: 25 to 30 hrs.	1.6%

Eimeria tenella is the most pathogenic followed by E. *necatrix* and E. *brunetti* \ Birds aged 4 weeks are most susceptible Older birds are generally resistant. Cecal coccidiosis is produced only when heavy infections are acquired over a relatively short period of time (not exceeding 72 hours.) Number of oocysts required to produce disease depends on age—1 to 2 weeks—200,000 oocysts.

Pathological changes include petechial hemorrhages during the first 3 days. By 5th and 6th day the ceca are dilated and the contents consist of clotted or unclotted blood, schizonts and merozoites. From 7th day onwards, the gametogenous stages start appearing. By this time the cecal contents had become more consolidared. By 8th day the consolidated caseous plug completely fills the lumen of the cecum. The cecal core detaches from the mucous membrane by 8 to 10 days and may be shed in the feces.

E. *necatrix* causes a more chronic type of disease than E. *tenella*. Principal lesions are found in the middle third of the intestines. Sub-mucosal hemorrhage occurs on 5th to 6th day. The most characteristic feature is the balooning of the intestines. In contrast to E. *tenella*, the birds affected by E. *necatrix* may remain emaciated for several weeks or months afterwards. The chronic form of infection is a marked contrast to the acute type of cecal coccidiosis.

E. *burnetti* causes disease between 4 to 9 weeks of age. Lesions are confined to posterior part of the small intestines and the condition is typically rectal coccidiosis.

In E. *praecox*, the pathogenecity is low.

In E. *hagani* also the pathogenecity is low and the catarrhal inflammation of the duodenum on the 6th day is a marked feature.

E. *mitis* is also a mild pathogen.

E. *maxima* and E *acervulina* are common but moderately pathogenic.

Affected chicks show general symptoms of sickness viz. drooping head, hudding together. Feces contain blood. In layers reduction of egg production occurs Oocysts are seen in the feces from 7th day after infection. In some outbreaks mortality may be heavy.

Recovered birds may continue to pass oocysts upto 7 months.

Lesions : The lesions in some cases may not be obvious. The large intestines contain thin porridge-like feces mixed with blood and mucus. There may be catarrhal enteritis whih hyperemia and petechial hemorrhages on the mucosa, which may be edematous. In places the mucosa may be denuded. Secondary bacterial infection may produce necrotising, purulent or diphtberoid inflammation. Intestinal contents reveal oocysts.

Microscopically: the superficial epithelium of the intestine is desquamated The epithelial cells that are still left contain various stages of coccidia. There may be leucocytic infiltration of the submucosa. The villi appear elongated. Becau*e of congestion, edema and leucocytic infiltration the mucous membrane appears thickened. Healing can occur with regeneration of the epithelium

Rabbits : It is the epithelium of the intrahepatic bile ducts that are affected. The sporozoites reach the bile ducts through portal circulation or via lymphatics.

Macroscopically, the liver is studded with small or large nodules which may be round and grayish in color. On section these nodules are seen to have a connective tissue wall enclosing a soft material containing numerous oocysts.

Microscopically, the nodules are dilated bile ducts, the walls of which have become altered due to chronic inflammation Because of toxins produced as well as due to mechanical irritation, the epithelium proliferates and so papillary growths with fibrous tissue core are formed, which project into the lumen. Because of chronic inflammation there is connective tissue overgrowth of the bile ducts and this tissue destroys the elastic and muscle fibres and so dilatation of the duct occurs. The dilated bile ducts cause pressure atrophy of the adjacent hepatic parenchyma,

Horses : Coccidiosis is rare in Horses.

Diagnosis :
1. Dysentery coupled with straining should make one suspect coccidiosis.
2. Observe oocysts in the feces.
3. In very acute cases, many chicks die before oocysts are found. In such cases the crescent-shaped merozoites are found in plenty in the feces.
4. Differentiate from enteritis caused by *E. coli, Salmonella* and *Vibrio,*

TOXOPLASMOSIS

This is a disease caused by *Toxoplasma gondii*, a crescent shaped organism, 4μ to 7μ by 2μ to 4μ. It has a pointed and a blunt end. The single nucleus is nearer the pointed end.

Toxoplasma gondii is presently regarded as a coccidian parasite akin to *Isospora* It shows a double heteroxenous life cycle embracing both ideas of Toxoplasmosis and Coccidiosis. The intestinal phase (isosporan) has been found so far, in domestic cat only. whereas the extra - intestinal (toxoplasmic) phase is well documented in various domestic and wild animals, birds and man.

In the life cycle of *Toxoplasma*, the mature oocyst produced in the gut of domestic cat (homologous or primary host) is the key - stage When such oocysts are ingested by the homologous host (domestic cat) sporozoites released in the gut will initiate conventional coccidian cycle. However, if the oocysts are ingested by heterologous hosts (i e, hosts other than domestic cat) the sporozoites initiate the typical toxoplasmic phase. This is followed by the formation of pseudocysts (intra-cellular) and resistant cyst (extra-cellular), particularly in brain, muscles etc.

Various methods of reproduction have been observed. Endogeny, a special kind of 'internal budding' may operate within the pseudocysts (intracellular) or cysts (extra cellular) resulting in crescent shaped small uninucleate bodies called endozoites and cystozoites respectively. The endozoites with central nucleus are easily differentiated from the cystozoites with terminal nucleus, **(Note by : Dr. T. N Gosh)**

Toxoplasma affects man and all species of animals—domestic and wild. Probably the same species of the organism affects all the animals and man indicating no host specificity. *T gondii* is peculiar that it may affect any nucleiated cells in the body.

Routes of infection : The one route, that has been established with certainity, through which infection can occur is transplacental. This is probably the only and important route of infection in man, since babies are born with congenital infection.

Ingestion of infected materials is another important route of infection. Contamination may occur from feces, sputum, tears and milk

Droplet infection is suggested in some cases. Though the organism quickly dies outside the body, it is viable within the body in the pseudocysts for as long as 2 to 3 years.

Pathogenesis ; *T. gondii* is an intracellular parasite and affects the R. E. system especially.

The organism gaining entry into the body, invades a cell and multiplies there by binary fission and subsequently destroys it, The liberated parasites enter the blood stream and thus reach various organs. As the immunity of the host develops, a cyst is formed around the parasite and these cysts are viable for a long time. These pseudocysts are found in the brain, myocardium and the eye. In this encysted stage no symptoms are shown, prabably due to a balance of existence that has been reached between the host and the parasite

The clinical symptoms vary with the organs affected. Affecting the brain encephalitis is produced. Pneumonia and enteritis are caused, when lungs and intestines are involved.

The organism produces an exotoxin which induces a granulomatous reaction wherever it is located.

In pregnant animals, the organism may cause placentitis with resultant death of the fetus and abortion. This is especially the case in ewes.

In sections the crescent form may not be seen but rounded or ovoid forms only, Each pseudocyst may have upto 50 organisms

Symptoms :

Cattle ; In the acute stage the animals manifest fever, dyspnoea and nervous symptoms. Later animals become very weak and dull. There may be still-births or if calves are born alive, they die soon. Calves that are born with congenital infection show fever, cough, dyspnoea, grinding of teeth, tremors, shaking of head, cycling movements. nasal discharges, depression, prostration and death. Death may occur in 2 to 6 days.

Sheep ; Fever, dyspnoea, nasal discharge and nervous symptoms are seen, the animal becoming comatose. Death may occur in 2 weeks. Pregnant animals abort. Lambs may be born dead.

Pigs : Dyspnoea, cough, diarrhoea, incoordination of movement are seen. Pregnant animals abort If a living piglet is born, it may die soon.

Dogs ; The animals become emaciaied gradually and the lymyh nodes become enlarged. There may be dyspnoea and bloody diarrhoea. Female dogs may abort.

Cats : In the acute form there may be anorexia, pyrexia, lethargy, dyspnoea and death. Leukopenia, anemia. jaundice, abortion and cerebral symptoms may also be seen.

In the chronic form, there may be wasting and diarrhoea.

Lesions · Wherever the organism is present necrotic granulomas are formed. Ascites and hydrothorax are seen   The following changes may also be noticed.

Brain ; There may be congestion of meninges or even meningitis. Brain may be edematous Small areas of- softening are noticed with a whitish centre and a hyperemic periphery. There may be hemorrhages in these areas.

The ventricles are dilated with increased and turbid cerebrospinal fluid, the cellular and albumin contents of which are increased. Parasites can be found in it.

Microscopically, granulomatous leptomeningitis is seen with histiocytes, plasma cells, lymphocytes and eosinophils. Necrosis is found in the centre. The parasites, singly or as pseudocysts are found at the periphery of the lesions.

There may be perivascular cuffing. Foci of necrosis are found in the gray and white matter, walled off by glial cells. In some places may be found nodules comprising of proliferated glial cells. Vascular mineralisation was found in chronic cases.

Pseudocysts may be present without the animal manifesting any symptoms.

Heart : Pseudocysts may be present within the cardiac muscle fibres, without causing any disturbances But when the pseudocysts rupture, focal mycorditis is caused with necrosed centre surrounded by histiocytes, neutrophils and lymphocytes This lesion may be calcified.

Liver : In the liver may be noticed focal areas of coagulation necrosis consisting of acidophilic centre, surrounded by normal hepatic tissue. The parasite may be found within the hepatic cells or Kupffer's cells, as cysts, either singly or in pairs.

Spleen and lymph nodes : These are enlarged and show focal areas of necrosis surrounded by leucocytes. The parasite can be found within the endothelial cells of the veins and mononuclears of the pulp.

Lungs : Lungs reveal small discrete gray nodules or there may be areas of confluent pneumonia with hemorrhages.

Microscopically, the lesion reveals changes in the alveolar walls, the cells of which assume a cuboidal or columnar shape, resembling a fetal lung. The appearance is also similar to that found in pulmonary adenomatosis. The alveoli contain numerous mononuclear cells and neutrophils. The parasite is found in alveolar lining cells.

Gastrointestinal tract: Ulcerative inflammation is found in the gastrointestinal tract.

The lesion starts as a necrotising lesion of the submucous lymphoid tissue and ultimately becomes ulcerated. The ulcer is surrounded by granulation tissue in which the toxoplasma and its pseudocysts can be seen The parasites can be seen within the muscle cells of the intestinal wall.

Pancreas: When pancreas is affected, there may be necrotising lesions with intense leucocytic infiltration. There may be necrosis of peripancreatic fat. The parasitie can be seen in the ductal and acinar cells.

Eye : The eye lesions consisting of chorioretinitis and uveitis that are common in man are not seen in animals.

Diagnosis :

1.   Observe the organism in the lesion histologically or in smear preparations.

2.   Inoculate the suspected material into mice intraperitoneally.   The organism can be recovered from the animal.

3. The complement-fixation test, Sabin-Feldman dye test, neutralisation
test, hemagglutination and ELISA tests can be performed.

4.   An intradermal test using toxoplasmin has not been found to be useful
in animals.

Differentiate from sarcocysts, leptomonas, Hamondia.

BABESIASIS
Syn : Piroplasmosis : Tick fever

Babesiasis is a protozoan disease of animals in which the parasite inhabits
the erythrocytes producing hemolysis, hemoglobinemia, hemoglobinuria and
icterus.

Asexual life-cycle of the parasite takes place in the erythrocytes, consisting
of binary fission. The sexual life-cycle occurs in the ticks which are the intermediate hosts.

In animals the organism proliferates by budding, a reduced form of schizogony, forming pairs during which process the red cell breaks up.   The parasites
thus liberated enter fresh erythrocytes and repeat the asexual life cycle.   During
the asexual cycle great destruction of erythrocytes occurs resulting in anemia
(hemolytic anemia).   Erythrophagocytosis and auto-immune hemolysis may also
play a part, especially in chronic cases in the causation of anemia.

There are still no definite conclusions of sexual reproduction of Babesia in
ticks.   The disease is transmitted through heredity.   Transmission can also be
from stage to stage.   That means to say, if nymphal stage is infected, then when
it becomes an adult, it can be infective.   The infected ticks are capable of transmitting the disease even after raising a few generations on non—susceptible
animals.   In rare cases transplacental infection from mother to fetus can occur.

Premunity : In recovered animals, some organisms may be found lurking in
internal organs, especially the liver and spleen.   So long as the animal harbours
these organisms, it resists fresh infection and is immune.   This stage of resistance is known as premunity.   Such premune animals are carriers and can be
sources of infection to other animals.   This premunity may last for 1 or 2 years.
But it can be broken down by inter-current diseases or a too heavy infection.   If
an animal is treated effectively so as to rid it of all organisms the animal is no
longer immune and is suceptible for fresh infection.

Routes of infection :

1.   By bites of ticks.

2.   Animals can be infected by injection of blood from an affected one.

3.   Infection can occur through improperly sterilised needles etc.

Babesiasis in cattle
Syn : Red water fever : Texas fever.

Three species are known.

B. bigemina transmitted by *Boophilus microplus, a one host tick*

B bovis transmitted by *Ixodus ricinus,* a three host tick.

Of these *B, bigemina* is large. *B bovis* is not found in India.

B argentina is transmitted through *B microplus* and occurs in India.

B. bigemina is pear shaped and is $4 \mu - 5 \mu \times 2 \mu$. Rounded or irregular forms $2 \mu - 3 \mu$, may also be seen In each erythrocyte two parasites in contact with each other at their pointed ends are found.

Calves under 6 months of age are more resistant than elder animals, probably due to the ingestion of specific antibodies with colostrum.

Symptoms : Incubation period in natural infection is 1 to 2 weeks. In experimental animals it is 5 to 10 days. The disease starts with high temperature, 104° to 106°F. On the 2nd or 3rd day can be seen hemoglobinuria, the urine being red or black colored. Anorexia, depression, stoppage of rumination and decreased milk yield follow. Constipation may be observed. Conjunctiva are highly congested. The thumping heart beat can be heard from at a distance. The mucous membranes become pale. As the anemia progresses dyspnoea develops. Icterus may develop later. Pregnant animals abort. Ascites may develop in some. The febrile state lasts for about 12 days Death may occur in 10 to 12 days.

Rarely convulsions and delirium may be observed In the animals showing the above the brain is congested and the capillaries of the brain are packed with parasitized erythrocytes. Capillary hemorrhages together with edema of the white matter is present.

Recovered animals are very weak and the period of recuperation is long. Such animals are carriers. Exacerbations may occur without any possible reason.

Hematology In the acute phase, the organisms can be seen in the erythrocytes of peripheral blood. More than 50% of the erythrocytes may be infected. The blood is thin and watery. The infected erythrocytes are bigger than normal and also lighter and so they tend to concentrate at the margins of the blood smear.

Erythrocyte counts diminish from a normal 8 millions to one or two millions per c. mm. Hence the hematocrit and hemoglobin values diminish considerably. Anemic changes—anisocytosis, poikilocytosis, punctate basophilia— are found. Later normoblasts may be seen Bone marrow is hyperplastic.

In chronic or carrier animals parasites may be very few or may not be present at all in the peripheral blood. It is only by animal inoculation that the carrier state can be proved.

Lesions : Animals may be emaciated. There may be petechiae and ecchymoses on the serous membranes, under the cardiac serosa and mucosa of stomach and intestines There may be accumulation of fluid, often blood tinged, in the pericardium and peritoneum. Lungs may be edematous. There may be gastroenteritis. All organs are discolored by icterus.

Spleen is enlarged, soft and the pulp is dark red. The liver is enlarged and reveals fatty degeneration and centrilobular necrosis. Gall bladder is distended with yellow granular bile.   The kidneys are dark red.   Tubular epithelium may reveal degenerative changes and the lumens contain casts and hemoglobin. Bladder may contain blood stained urine.

In young animals, hemoglobinuria may not occur.   But anemia and jaundice may be seen with serous atrophy of fat.   Edema and splenomegaly are present.

Blood capillaries in certain vital organs may be clogged by the parsitized erythrocytes or by free parasites and may cause death.   Embolism of cerebral capillaries causes encephalitis.   The diseaase may then be mistaken for rabies.

Anemia is partially due to destruction of the erthrocytes by the parasites. Partly it is due to autoimmunisation with serum, autohemagglutinins acting as opsonins in erythrophagocytosis. Because the severity of anemia is not commensurate with the degree of parasitemia, the presence of autoimmunisation is suggested and later proved.

B. *argentina*, which are smaller than B *bigemina*, occur as round or oval bodies, 1μ to 1.5μ in diameter   Pear shaped forms may also be present.   These are found as pairs in each erythrocyte.   The two parasites are attached to each other at a more obtuse angle than in the case of B. *bigemina* and or located in the centre of the erythrocytes (while B. *bigemina* are nearer the rim).

The pathogenesis, symptoms and lesions caused by B. *argentina* are similar o those caused by B *bigemina*.

Babesiasis in sheep and goats

Two species are known.

1.   B. *motasi* transmitted by *Rhipicephalus bursa*, a two host tick.
2.   B *ovis*, transmitted by various types of ticks.

B. *motasi* is very virulent, is larger than B. *ovis* and the disease caused by it is typical of babesiasis described nnder B. *bigemina* of cattle which it resembles morphologically.

Symptoms caused by B. *ovis* are less severe though anemia and *icterus* are produced

BABESIASIS IN HORSES
Syn : Biliary fever.

Two species are found.

1.   B. *equi* (Nuttalia equi) smaller of the two, transmitted by *Rhipicephalus. evertsi*, *R. bursa*, *Dermacentor* and *Hyalomma*.
2.   B. *caballi*, larger, transmitted by *Dermacentor* and *Rhipicephalus*.

B. *equi* infection is more common in India though it is not rare to find both infections in the same animal

B *equi* are small and 4 parasites are found as Maltese Cross in each erythrocyte.

Symptoms :   Horses, mules and donkeys are affected.   Fever, acute anemia and jaundice accompanied by anorexia, constipation alternating with diarrhoea and high colored urine are seen. Hemoglobinuria is very rarely noticed   Nucleated red cells may be found in the peripheral blood   In untreated animals death may supervene in a week.   Recovered animals are carriers.   Mortality is not as high in horses as in cattle.   Death may be due to pneumonia, endocarditis or septicemia   The disease is milder in foals.

B. *caballi* infection is milder than that of B. *equi*.   Hemoglobinuria is not seen.

BABESIASIS IN DOGS
Syn: Tick fever, Malignant jaundice

Two species are known.

1. B. *canis* transmitted by *Rhipicephalus sanguineus*; a three host tick.
2. B. *gibsoni* tranmitted by *Hemophysalis bispinosa*, a three host tick. Pups are more susceptible than adults.

B. *canis* is the larger of the two, pear shaped and resembles B *bigemina* in morphology. Usually pairs may be found in the erythrocytes but it is not uncommon to see 12 to 16 in one red cell. Wild carnivores are also affected. About five days after infection, there is rise in temperature followed by anemia and icterus. Hemoglobinuria is inconstant. The blood smear reveals the parasites Acute cases die within a week manifesting emaciation and weakness.

In chronic cases. severe anemia may be noticed but icterus is not present. Animals may die after several weeks showing emaciation Lesions are similar to those described under bovine babesiasis Blood is thin and watery and hyperplasia of bone marrow is seen. Demyelination may be found Peripheral blood reveals immature erythrocytes. Parasites may be found free in blood. Recovered animals are carriers.

B. *gibsoni* is far smaller than B. *canis* and occurs as rings or ovoid forms in the erythrocytes and not as paired pears. Wild carnivora are also susceptible.

B *gibsoni* causes the chronic type of disease with progressive anemia and gradual wasting. Hemoglobinuria and jaundice are not seen.

BABESIASIS IN CATS

B *felis* infects cats and wild felidae, Vector is not known.

Diagnosis of babesiasis in Animals:

1. Examine a blood smear—search at the margins where the parasitized cells are, as these are lighter.
2. Centrifuge in a hematocrit as for estimation of packed cell volume and make smears from the layer just below the buffy coat, since the parasitized erythrocytes accumulate here being lighter due to lower specific gravity.
3. Complement-fixation test : for equines the antigen is stromata of erythrocytes of an infected animal. In cattle the antigen is water lysed erythrocytes.
4. Fluorescent antibody technique.
5. Indirect fluorescent antibody technique.
6. Gel precipitation test.
7. Passive hemagglutination test.
8. Carrier animals can be diagnosed by animal inoculation and by tests mentioned in 3, 4 and 5 above.

THEILERIASIS

Theileria are protozoa found in the erythrocytes. They differ from Babesia in that their schizogony takes place in the lymphocytes and monocytes of blood, lymph nodes and spleen.

On entry into the body through the bite of a tick, the sporozoites infect the lymphocytes and monocytes of the blood. lymph glands and spleen, where they become schizonts.

The schizont is about 3 μ to 10 μ in diameter and contains many red coccoid dots, which are infective organisms. These schizonts are called 'Koch's blue bodies' and are path⸱ gnomonic of this disease. These bodies may be found in the peripheral blood or in a biopsy material from a lymph gland pulp. Some ime after the onset of symptoms, the parasites enter the erythrocytes following the break-up of the Koch's blue bodies. Some of these may also enter the lymphocytes and monocytes to repeat the schizogony. Division can also occur within the erythrocytes, which, however, are not damaged and so hemloysis does not occur (hemoglobinuria, therefore, is not a symptom of theileriasis)

In the tick, though stage-to-stage transmission of the parasite is obtained in Theileriasis, hereditary transmission is thought to be not found. That means to say, the ovaries of the ticks are not infected and so infection is not passed on from an adult tick to its eggs

Three species are important :

1. *T. parva* found only in Africa : causing East Coast Fever, transmitted by *Rhipicephalus sps*
2. *T annulata*, transmitted by *Hyalomma sps*
3. *T. mutans*, transmitted by *Rhipicephalus sps*.

Of these 2 and 3 are also known as *Gonderia*. *T. parva* is not found in India. But the symptoms and lesions caused by T. *annulata* and *T. mutans*, are similar to those of *T. parva*

In erythrocytes the parasites may have bacillary (rod shaped) or a ring shape. The parasites divide at the time when the lymphocytes divide by mitosis and so daughter lymphocytes become parasitized

East coast fever

Symptoms : After an incubation period of about 2 weeks, there is rise in temperature, accompanied by dullness, salivation, lachrymation and swelling of the lymphnodes. Later diarrhoea may be present. Dyspnoea may develop due to pulmonary edema and this is the cause of death (asphyxia). In some cases there may be emaciation and coma followed by death. Icterus may be seen in terminal stages. The disease runs a course about a month and mortality is 90-95%.

Koch's blue bodies can be demonstrated in the peripheral blood or lymph nodes. The parasites can be seen in erythrocytes.

Artificial infection by injection of blood is possible only if it contains the Koch's blue bodies. Recovery is followed by solid immunity and no organisms are found in the host.

Lesions The characteristic appearance is swelling of lymph nodes, spleen and Peyer's patches. These are edematous and hemorrhagic. The serous sacs are filled with fluid and there may be pulmonary edema. Connective tissue also reveals edema, which may be hemorrhagic.

Microscopically, the germinal centres of the lymph nodes disappear. The lymphocytes are reduced in number due to their destruction and hence the reticulum is more conspicuous. The interstitial spaces are filled with remnants of the lymphocytes. Some of the lymphocytes may contain the schizonts. Similar changes are also found in the Malpighian corpuscles of the spleen, and in thymus, tonsils and Peyer's patches.

In the kidneys and liver may be found perivascular proliferation of lymphoid tissue from the adventitia. These foci appear as small graysh-white patches erroneously called infarcts. These are supposed to be very characteristic. The nodular accumulation of the cells causes atrophy of renal tubules.

Erosions and ulcers may be found in the abomasum. Hyperemia and acute catarrhal enteritis may be observed.

Hemorrhages may be present on the epicardium, endocardium and other serous membranes. Foci of hyaline degeneration and fibrosis may be found in the muscles.

Hematology : Oligocythemia may be noticed. Absolute monocytosis may be seen. Severe leucopenia occurs with total white cell counts as low as 2000/cmm.

T. annulata : This also causes a malignant affection. Acute and severe infection occurs with fatalites when Koch's blue bodies and parasites in erythrocytes can be found.

T. mutans ; This is an ubiquitous parasite in the blood of cattle, producing no symptoms. Sometimes acute conditions may occur. The parasites are very small and occur as rods and rings in the erythrocytes. Koch's blue bodies are not seen or are seen very rarely.

Inoculation of blood is always successful (not so with *T. parva* since infection does not occur unless Koch's blue bodies are present)

One attack by *T. annulata* and *T. mutans* confers life-long premunity since the animals retain the theileria for the rest of their lives.

Theileriasis in sheep and goats

T. ovis affects sheep and is benign. Lambs may sometimes become ill with pyrexia, weakness and death may supervene within two days. Koch's blue bodies are very few.

T. hirsi affects sheep and goats. Among goats the disease resembles East coast fever.

Diagnosis :

1. Blood smears reveal (i) Koch's blue bodies (ii) Parasites in erythrocytes.
2. Biopsy materials of superficial lymph nodes--make a smear of the pulp and observe Koch's blue bodies.
3. Differentiate from Babesiasis.

TRYPANOSOMIASIS

Trypanosome is a spindle shaped, flagellate affecting man and animals. But so far in India, infection in man has not been met with.

Trypanosomes may be pathogenic or non-pathogenic. Of the pathogenic only one, *T. evansi* causing Surra is found in India. Dourine caused by *T. equiperdum* has been stamped out. Other pothogenic Trypanosomes are :

T. congolense—causing Nagana in cattle, sheep, goats and dogs in South Africa.

T. vivax—causing Nagana in S. Africa.

T. brucei - dogs, horses, mules and donkeys are affected in that order of susceptibility in S. Africa. Cattle and pigs are resistant.

T. simiae—pigs and camels are highly susceptible. Present in S. Africa,

T. equinum—causing Mal de caderas in S. America.

T. cruzi—causing Chagas Disease in man and dog in South America

The non-pathogenic trypanosomes are ;

T. lewisi—Ubiquitous in rats

T. melophagium..Ubiquitous in sheep.

T. theileri—in cattle. Reported to be present throughout the world.

Mode of transmission : All trypnosomes except *T. equiperdum* are transmitted by biting insects—flies, bugs and fleas. *T. equiperdum* is transmitted directly from animal to animal by copulation.

Some vectors transmit the organisms mechanically as in *T. evansi* infection by *Tabanus Stomoxys* and *Lyperosia*. There is no cyclical development in the vector. With other trypanosomes a cyclical development occurs in the vectors. The sites of such development vary with different species of Trypanosomes. The following are the developmental sites in Glossina :

T. vivax—Proboscis only.

T. congolense—Proboscis and stomach. ⎫ Anterior station develo-

T. brucei—Proboscis, stomach and salivary ⎬ ping trypanosomes

glands, ⎭

T. cruzi and *T. lewisi* are examples of posterior station developing trypanosomes.

Therefore flies are not infective in these cases immediately after they suck blood though mechanical transmission may also occur with Glossina. Sufficient time must elapse for the development to be completed. During the cyclical development in the flies the trypanosomes pass through the leishmanial, leptomonas and crithidial stages.

In the vertebrate host the parasite multiplies by binary fission.

SURRA

This is caused by *T. evansi* which is 25 μ to 38 μ by 1.5 μ to 3.5 μ. Horse, mule, donkey cattle, camel, dog and elephant may be naturally affected. The vectors are *Tabanus* and *Hematopota*. Camels and cattle act as reservoirs. Incubation period is a few days to 2 or 3 weeks.

Symptoms ;

Horses : The first symptom is rise in temperature to 104°—106°F when the blood is teeming with trypanosomes. There may be urticarial eruptions on the skin. The fever becomes intermittent Conjunctiva shows petechiae and later may be icteric. Watery nasal discharge is present. Inspite of good appetite, the animal becomes progressively weaker. Edematous swellings develop on the lower parts of the limbs, extending later to throat, breast, belly and sheath. The superficial lymph nodes are swollen and paralysis of hind quarters may develop. In such cases death usually occurs in 6 to 8 weeks

Camel : In camels acute and chronic forms occur. In the acute form, the animal may die within a few weeks or months. Old camels are susceptible to this form The animal becomes quickly tired at work and the hump may be atrophied. Intermittent fever is observed, but it does not rise as high as in the horse. Trypanosomes are always present in the blood. The animal is dull, with lusterless eyes and staring coat. It becomes progressively weaker. Cerebral symptoms may sometimes be seen.

In the chronic form, which is more common there is intermittent fever, the intervals being short in the beginning (about 3 or 4 attacks a month) but becoming longer gradually. Trypnosomes can be found in the blood only during the febrile stage. The visible mucous membranes become pale and are studded with petechiae. Edema of the dependent parts and skin swellings develop. Affected animals become highly susceptible to and actually suffer from mange. They become weaker progressively and the hump disappears. Animal dies after a course exceeding three years if not treated, manifesting anemia, debility, pulmonary edema and bronchopneumonia.

Cattle : In cattle and buffaloes, the condition is comparatively mild and the animals may be harbouring the parasites without manifesting any symptoms (carriers). But when the animals become debilitated due to some other intercurrent disease, the parasites multiply and produce clinical symptoms.

An acute form may occur as an outbreak and it may be mistaken for anthrax or hemorrhagic septicemia. The symptoms shown are dullness, rise in temperature ($103° — 106°F$) with no inclination to work. Animals in milk do not yield any. The animal is dull, appears sleepy and may move in circles aimlessly and later falls down. There may be delirium, salivation, distressed breathing, frequent urination and defecation, grinding of teeth and coma followed by death in a day or two.

A subacute form may occur lasting for some days or weeks. The animals show dullness, intermittent fever, progressive anemia, gradual loss of condition, tenderness over the loins and back, edema of legs, diarrhoea, ulcerative keratitis, prostration and death.

Hematology : There is progressive anemia. Number of erythrocytes and hemoglobin content are reduced to 25%. Anemia may not only be due to a failure in production due to inhibition, but also to increased erythrophagocytosis and enlarged mononuclear phagocytic systems associated with autoimmune processes.

Hypoglycemia develops because the parasites consume large amount of glucose and the liver is unable to lay down glycogen reserve. Blood lactic acid is increased, blood potassium is decreased but the globulin fraction is increased.

Lesions : The animal is markedly emaciated. The spleen and lymph glands are enlarged and show hyperplasia of the follicles. There may be multiplication of the macrophages in the sinuses and infiltration by lymphocytes and plasma cells. Later organisation occurs with suppression of glandular function.

Congestion of bone marrow and of the gastro-intestinal mucosa may be present. Subcutis shows gelatinous infiltration. Petechiae are found on serous membranes. Serous exudate into pericardial cavity and peritoneum is seen. Ulceration of tongue and gastric mucosa is met with. Death in Trypanosomiasis may be attributable to

(1) Endotoxin that may be set free by the lysis of the parasites.
(2) Asphyxia due to increased blood lactic acid.
(3) Hypoglycemia
(4) Toxemia that develops due to dysfunction of liver and the cause for hepatic dysfunction is destruction of large quantities of glucose.

    (5) Erythrocyte production is inhibited by toxins.

    (6) Trypanosomes may liberate proteolytic ferments, which digesting proteins may liberate toxic products.

Diagnosis :

    1. Examination of the wet film—obtain a small drop of blood from the ear puncture, put the cover slip and examine under high dry. The motile trypanosomes can be seen easily. The method is useful in weeding out affected from a herd in an out break

    2. Examine a stained blood smear and see the trypanosomes.

    3 **Mercuric chloride test** : In carrier camels, this is very useful to diagnose. It is 99% accurate in these animals. Mix one drop of serum with **1** c.c. of 1 in 25.000 solution of mercuric perchloride in water. A white precipitate appears within a few minutes in positive cases.

    4. Formol-gel test : Mix 2 drops of formalin with 1 c.c. of serum and keep for 24 hours A complete gelation is positive. This is only 75% accurate

    5. A compliment-fixation test can be performed.

    6. **Biological test** : In animals in which the parasites are too few to be seen inoculate 2 to 5 cc. of blood from such animals into a rabbit or a mouse. A few days after inoculation animals show numerous parasites in blood.

LEISHMANIASIS

This disease has been recorded in dogs and in a bovine in India. Infection of horses, goats and sheep has been reported in other countries.

    In man 3 forms exist.

    1. Visceral form, in which there is splenomegaly and found in India, caused by *Leishmania donovani* (Kala azar).

    2 Cutaneous form seen in adults in the Mediterranean countries, caused by *L. tropica* – Oriental sore or Delhi Boil

    3 Cutaneous and visceral form seen in Brazil caused by *L. braziliensis*

These are morphologically indistinguishable but may be differentiated serologically.

In the vertebrate host, the lieshmania occur as non-flagellate, ovoid bodies about $4\,\mu \times 2\mu$ with a rod-shaped kinetoplast. In the invertebrate host and in cultures, the motile flagellate, leptomonad form, develops.

The leishmania reproduce by binary fission in the vertebrate. But in the invertebrate a life cycle is undergone which revitalises the parasite

Dogs can be infected with all the forms. Probably they act as reservoirs. It is the cutaneous form that is more frequent among dogs while the visceral form is rare. The vector for dogs is the sandfly (*Phlebotomus*) Infection is through the saliva of vector.

The organisms are ingested by the macrophages and these can be seen in peripheral blood. But they are more numerous in the spleen and lymph glands.

The infected macrophages may break up mechanically due to the proliferation of the parasite inside. The liberated organisms, the **Leishman-Donovan bodies**, may be found free in the lymph or plasma The reticulo-endothelial cells of dody, especially of spleen, liver, bone marrow and lymph nodes are parasitized.

Visceral type: This may be acute or chronic. In the acute type there is fever, loss of weight, anemia, enlargement of the spleen, liver and lymph glands. Ascites develops sometimes. Paralysis may sometimes be seen. Leucopenia is observed On necropsy, intestinal ulcers, cloudy swelling of kidneys and straw-bery-red color of bone marrow are seen.

Cutaneous form: The organisms that enter the skin from the bite of the vector are ingested by the histiocytes, within which active proliferation takes place. The macrophages are broken liberating the parasites which are ingested by other phagocytes and the cycle is repeated. Lymphocytes and plasma cells are found infiltrating around such a lesion. Neutrophils infiltrate the area and thus a papule is formed. In time the inflammation extends to the overlying epithelium and ulcer results.

A characteristic feature is the stimulation of the epithelium so that it becomes hyperkeratotic and acanthotic. The rete pegs become elongated deeply into the underlying tissue, resembling epidermoid carcinoma.

These lesions are found on the face, especially on the ears, mouth and nares. The organisms (Leishman-Donovan bodies) can be demonstrated in smears made from deep tissues at the edges of the lesions.

Diagnosis :

1. **From dead animal** make smears from the spleen or bone marrow.

2. **In living animals:**

   (a) Trephine bone for making smear—too painful (sternal puncture in man)

   (b) Make smears from a splenic puncture—dangerous in the dog.

   (c) In skin lesions—make smear from the edges, going deep into tissue.

In all these cases, stain by Giemsa's method and observe the Leishman-Donovan bodies.

AMOEBIASIS

Amoebae are unicellular organisms, most of which are non-pathogenic. *Entamoeba gingivalis* in the teeth of man, horses and dogs, E. *bovis* in the intes-tines of cattle and E. *coli* in the intestines of man are non-pathogenic. It is only E. *histolytica* that is pathogenic and is found in man and dogs. In man it causes dysentery and abscesses in the liver (and also in brain). In dog only dysentery is caused. Only one case of amoebic hepatic abscess in a dog has been reported.

The organisms secrete a proteolytic enzyme that can dissolve the epithelial ining of the gut and hence its name histolytica.

Infection is by ingestion of mature cysts of E *histolytica* through contamina-ted food and water. Many people can be carriers without showing any symptoms

The cysts which measure 20μ-30μ have a thick wall that protects them from environment.

The intestinal juices dissolve the cyst wall, liberating 4 trophozoites. They divide and become 8 amoebulae which penetrate the mucosa [and multiply by binary fission and cause destruction of the cells by their proteolytic ferments. They do not have cilia but move by pseudopodia The inflammatory reaction around the area of penetration by the parasite is minimal because it is only chemical in nature. Severe inflammatory reaction may occur when there is secondary bacterial infection.

The organisms easily enter crypts and other epithelial cells but the muscular coat acts as a barrier and so the trophozoites spread along the surface or border of the muscular coat, and so an ulcer with a broad base and narrow neck is formed—the flask shaped ulcers. Ulceration may be very severe and between the ulcers may be found healthy mucosa. The lesions, in man, are found in the cecum, ascending colon and rectum.

The trophzoites may enter the circulation and reach the liver producing abscesses. Embolic amoebic abscesses may also be found in the lungs and brain. The organisms must be encysted in the bowel before being voided Unencysted entamoebae die.

Diagnosis :

Examine the feces. The pathogenic *Enatomoeba histolytica* contains 1 to 4 nuclei and also engulfed erythrocytes. But the nonpathogenic cysts contain upto 8 nuclei but no erythrocytes.

HISTOMONIASIS
Syn : Enterohepatitis : Black head

Black head is an acute infectious disease of turkeys, especially young ones between two and 12 weeks of age, caused by a flagellate protozoan. *Histomonas meleagridis*, which has an amoeboid form in its life cycle.

Fowls may suffer from the disease but recover.

Route of infection : Natural infection is by ingestion of embryonated aggs of *Heterakis gallinae* containing the protozoan.

Pathogenesis : Incubation period is 15 to 21 days. After reaching cecum the parasite causes severe inflammation and there is intense cellular infiltration of the submucosa and the muscle coat resulting in necrosis. The overlying mucosa is destroyed.

Entering the portal vein, the organisms reach the liver, producing necrotic foci.

Symptoms : The symptoms shown are droopiness of head and wings and sulphur-colored diarrhoea. The bird becomes emaciated and dies after several days. Young turkeys die early. Mortality may be 50-100%.

Lesions : The lesions are found mostly in the ceca (often one cecum may be affected) and liver,

The affected cecum in enlarged with thickened wall. The surface is covered with dense necrotic and diphtheritic material, which may sometimes fill the lumen of the cecum. Ulcers may be seen and sometimes these pierce through the wall.

The lesions in the liver, which is enlarged, have a characteristic appearance Large, circular areas of coagulative necrosis are seen surrounded by red zones of hyperemia. Sometimes the lesions are green due to the presence of disintegrated blood pigment. These areas are slightly depressed below the surfrce Older lesions heal by scar tissue. Necrotic foci may sometimes be seen in the kidneys.

Because of cardiac weakness, the featherless parts of the head have a cyanotic blue color –hence *Black head*.

Diagnosis : Lesions are suggestive. Demonstrate the organism microscopically.

Differentiate from tuberculosis, tumors and fungal infections.

GLOBIDIASIS
Syn : Besnoitiosis

Primarily this is a disease of cattle, but may also affect horses, goats and sheep. It is a chronic debilitating and occasionally fatal disease, characterised by lesions in the skin and systemic manifestation. The organism responsible is *Besnoitia besnoiti*.

This disease is more common in the summer and biting flies like Tabanus etc. are supposed to act as mechanical carriers.

The mature parasites lie dormant encysted in the aereolar connective tissue of the dermis They may also be found encysted in the intima of the superficial veins of the head, neck and limbs. They may be seen encysted in the lamina propria of the upper alimentary and respiratory systems as well as of the male genitalia. Less commonly the parasites may be found in the connective tissue of other places. Within the macrophages the parasites multiply by binary fission. The host cell also may show nuclear divison and around this cell a homogeneous capsule develops.

Symptoms and Lesions : The disease in cattle is characterised by fever, catarrhal oculo-nasal discharge, which later becomes mucopurulent and may contain blood and hyperemia of the sclera Edema in the subcutis, muzzle and lips may be noticed in varying degrees Mortality rate is about 10% and period of convalescence is long. Pregnant animals may abort and bulls may become sterile for long periods.

Microscopically, there is epidermal hyperplasia, congestion and edema of the corium and infiltration by mononuclears around the blood vessels. Macrophages in these places usually engulf the parasites. Crescentic forms of the parasites are demonstrated in the vessels and the tissue spaces

As the parasites become encysted, edema of the connective tissue diminishes. On the the other hand the skin becomes thickened, scurfy (or formation of the crusts) and fissured Hair is lost and tissue fluids ooze out. In the chronic stage the skin becomes markedly thickened and wrinkled, especially on the perineum' eyelids, upper extremities, thighs and scrotum.

Histologically the cyst is smoothly spherical, 0.1 to 0.5 mm. in diameter and surrounded by a dense uniformly, esoinophilic wall which may be homogeneous or concentrically laminated. This is probably derived from the host. Inside, two giant nuclei, ovoid in shape are seen. The cysts contain tiny spores which are crescent in shape resembling toxoplasma.

Diagnosis : Lesions and demonstration of organisms Presence of cysts in the scleral conjunctiva is of diagnostic importance.

In certain countries, including India, alimentary globidiosis being causative of enteritis has been reported The cause is *Globidium fusiformis*

Diarrhoea and dysentery resembling coccidiosis infection is met with.

The spores resemble the schizonts of coccidia, and the two may be differentiated by floatation examination of the feces

FOWL MALARIA

Plasmodium gallinaceum is fouud affecting the domestic fowl in India. This protozoan is transmitttd by mosquitoes of the genera *Aedes* and *Armageres*.

Pathogenesis : The erythrocytes show schizonts which are round and irregular and may produce 8 to 30 signet-ring shaped merozoites. Later round gametocytes containing pigment granules may be found. Mortality among affected birds is high and may be upto 80%.

Symptoms : The clinical symptoms are vague. Birds become progressively emaciated and anemic. There may be fluctuation of temperature and paralysis may be seen in some cases, (due to massive numbers of exoerythrocytic forms in the endothelial cells of the blood vessels of the brain). There may be enlargement of the liver and spleen.

Diagnosis : Identify the parasites in the blood smears or in impression smears from the liver or spleen.

Note : This parasite is of interest because its life cycle resembles that of the malarial parasite of man and so is of great use in the research on human malaria.

LEUCOCYTOZOON INFECTION

Leucocytozoon is a protozoan parasitizing the red blood cells of the fowl. In India, *Leucocytozoon caulleryi* is met with. The vector is the black fly, of the genus *Simulium*.

The peripheral blood of the fowl contains the large gametocytes which render the erythrocytes very much enlarged and spindle shaped with two elongated processes at either end. The nucleus of the host cell may be compressed to one side of the cell forming a band which occupies about one third of its length.

Very young gametocytes appear as signet-rings. Formerly it was thought that the gametocytes developed in the leucocytes and so the name was erroneously given as leucocytozoon.

During the life cycle, schizogony may occur in the lymphoid cells or macrophages and may also be found in the brain, liver, kidneys, lungs and intestinal tissue in which megaloschizonts are observed.

Symptoms : The symptoms are vague. Anemia may be seen but deaths do not occur normally. In severe infections, hemorrhage may be found in the muscles, legs and internal organs.

Lesions : Include anemia, hemorrhages in the lungs. kidneys, liver, muscles and legs, enlargement of the liver and spleen.

Diagnosis : Identify the parasites in the blood.

HEPATOZOON CANIS

This protozoan is found to infect the dogs in India, and causes mild symptoms.

Rhipicephalus sanguineous is the vector in which sporogony occurs. Schizonts are found in the endothelial cells of the spleen, liver and bone marrow as round or ovoid bodies. Each schizont completely fills the host cell and contains 30 to 40 nuclei.

Gametocytes are found in the leucocytes as rectangular bodies, 8μ to 12μ by 3μ to 6μ. These have a delicate capsule, have dark purplish red nucleus and a number of pink granules in the cytoplasm.

Infection is by the ingestion of an infected tick. The sporozoites are liberated in the intestines of the dog from the tick and reach the spleen, liver and bone marrow via blood stream.

Symptoms : Slight elevated temperature, anemia with edema. Progressive emaciation and death may be seen in severe cases in 4 to 8 weeks after onset of symptoms.

Lesions include enlargement of the spleen and liver.

Diagnosis : Identify the gametocyte in the leucocytes or schizonts in the spleen or bone marrow.

DISEASES CAUSED BY HELMINTHS

Trematodes
  Fascioliasis
  Euretremiasis
  Paragonimiasis
  Opisthorchosis
  Amphistomiasis
  Schistosomiasis
  Flukes of poultry
Cestodes
  Echinococcosis
  Taeniasis of poultry
Nematodes
  Ascariasis
  Equine strogylosis
  Ancylostomiasis
  Hemonchosis
  Oesophagostomiasis
  Spirocercosis
  Dirofilariasis

Trichinellosis
Stephanuriasis
Dracunculosis
Hebronemiasis
Stephanofilariasis
Onchocercosis
Setariasis
Thelaziasis
Parasitic gastritis of pigs
Diactophymiasis
Thorny-headed worm of pigs
Strongyloidosis
Other nematodes in poultry
  Capillaria annulata
  Tetrameres fissipina
  Syngamus trachea

TREMATODES

LIVER FLUKES :

The flukes that inhabit the liver are :

Fasciola gigantica : *F. hepatica* : *Dicrocoelium dendriticum*.

Opisthorchis tenuicollis in cats, dogs and foxes. *O. sinensis* in man.

The disease caused by these flukes is known as *distomiasis*,

Fasciola gigantica (7.5 cms × 1.2 cms). This is found in cattle, sheep, goats and wild ruminants. Sometimes it may be found in horses, swine, rabbits and dogs while *F. hepatica* occurs in man also (outside India).

The metacercariae that encyst on the plants or grass blades are swallowed by the host in whose gut the cyst wall is dissolved The young flukes burrow through he intestinal mucosa and the wall and emerge on the peritoneum. Sometimes they crawl up the intestines and enter the bile duct at its opening in the intestine. During this migration some may enter the veins and thus get conveyed to the liver. Some enter the lymphatics and reach the mesenteric lymphatic glands.

During their migration the flukes feed on tissues and possibly on blood. On the peritoneum they may cause inflammation. In cases with heavy infection, peritonitis may occur which may be exudative when acute or proliferative when chronic. The flukes can be seen in the exudate. From the peritoneum they reach the liver burrowing through the capsule. After wandering about in the liver parenchyma, the flukes reach the bile ducts where they become mature and lay eggs. In this situation they may exist for a long time. upto 15 months normally, But some may live for longer periods and one was found to live for over 10 years

Some flukes may enter the hepatic veins and thus reach the heart and lungs where lesions caused by them may be found (Aberrant situation). Escaping into general circulation, the flukes may reach other organs where they become encysted. It is only flukes which reach the bile ducts that complete the life cycle.

The pathological changes produced by the flukes in the liver can be divided into two groups.

1 Changes caused by migrating flukes in the liver parenchyma and so resultant functional derangement

2. Changes produced in the bile ducts by the habitation of the adult flukes there and consequent results. These will be described one after the other. It should be understood that both types of changes can and do occur together.

1 **Changes produced by migratory flukes** : The flukes migrate within the the hepatic parenchyma, in search of their seat of predilection, viz the bile duct, This phase may last about 56 days. During this migration, tracks are formed by the burrowing of the fluke and these. in the initial stages, are filled with blood and liver tissue debris. On the cut surface these appear as red lines. The flukes which at this period measure 0.4 to 1.0 mm. feed on hepatic tissue and can be seen at the ends of these burrows. Older tracks are greenish-yellow due to infiltration by eosinophils. These are surrounded by a yellow-brown zone. The flukes may rupture the serous coat of the liver and if the perforations are many, hemorrhage can occur into the peritoneal cavity.

Microscopically, the early tracts contain blood and liver tissue debris. Later large number of eosinophils migrate to the area. Still later the hepatic tissue surrounding the tract undergoes cogulative necrosis and this is surrounded by macrophages and giant cells (in a radial manner) which remove the necrotic tissue The necrotic area becomes invaded by granulation tissue and this is heavily infiltrated by eosinophils and lymphocytes. Finally scar tissue is formed: If the infection is mild no evidence of damage may be found. But in heavier infection the tracks become confluent with the formation of large amount of granulation tissue. The liver in such a case appears enlarged, friable and greyish-green in color due to presence of large number of eosinophils. Sheep may die at this stage in heavy infections due to acute hepatitis resulting in failure of hepatic function. But should the animals survive, fibrous tissue is laid down and cirrhosis of the liver results.

Such of those flukes that are not able to reach the bile ducts become encapsulated and appear as small nodules, greyish-white in color. On section the nodules appear to have a dense connective tissue capsule enclosing a dirty-brown mass consisting of debris, blood, excretions of flukes and one or more flukes, which are 1 0 to 1.5 cm. long. The nodules which are present on the visceral surface project from under the capsule. The contents of these nodules may become caseated and calcified. Or sometimes they may be replaced by granulation tissue and so tiny scars are left. Such nodules are found in the lungs where flukes may be located as an aberration.

Changes produced by adult flukes residing in bile ducts ; The flukes that reach the bile ducts become sexually mature. They produce chronic cholangohepatitis and this is brought about by :

(1) The mechanical irritation of the suckers and cuticular scales/spines.

(2) Toxic irritation of the metabolic products.

(3) Irritation of secretions of special glands of the flukes.

(4) Secondary bacterial infection that may occur causes irritation by their metabolic products. These organisms *(F coli, Salmonella)* may either reach the bile ducts along with the flukes or they may reach the bile ducts later. These bacteria may cause decomposition of the contents of bile ducts and decomposition products may also be irritants

(5) Retention of bile that occurs may be an irritant.

Usually the left lobe is more severely affected.

Macroscopically, on the visceral surface can be found the thickened whitish corded bile ducts. Some of these may be an inch wide The affected organ may be shrunken if cirrhosis has occurred, with an uneven surface. Its color changes to grayish-brown or yellowish red. The liver is firmer and harder to the touch and cuts with difficulty.

The ducts in advanced cases, especially in cattle, are hard and calcified to merit the description **'pipe stem'** appearance, and when cut a grating sound is elicited

In sheep, swine and horses the dilatation of the bile ducts is due to ectasia and their lumens are dilated mechanically by the presence of flukes, bile and debris. There is little fibrosis and seldom any calcification. But on the other hand, in cattle there is diffuse connective tissue thickening of the bile ducts, which may be stenosed by the projection of mucosa in a papillary fashion. In certain places dilatation may also be due to distention by accumulated bile.

Microscopically, the epithelium of the bile ducts, undergoes degeneration and may be shed after becoming necrosed. The propria grows inwards in the form of villi and this contains granulation tissue consisting of young fibroblasts with infiltration by lymphocytes, neutrophils, plasma cells, mast cells and macrophages. The macrophages engulf blood and bile pigments There may be hyperplasia of the glands of the bile ducts and when the mouths of these glands are closed by the granulation tissue, retention cysts are formed

In the cattle the collagenous tissue undergoes hyaline degeneration and subsequently gets calcified Sometimes metaplastic bone formation may occur on the wall. The connective tissue in the walls proliferates and this may invade into the nearby surrounding parenchyma to a short extent

From the bile ducts the irritant may reach the periportal and interlobular connective tissue through the lymphatic vesseis. Hence the connective tissue together with that produced during the migration of flukes produces extensive cirrhosis. The connective tissue is infiltrated by eosinophils. The small bile ducts and capillaries may proliferate.

The portal lymph nodes in severe infection are enlarged. On section, the cortex appears narrowed while the medulla is expanded. This is due to sinus catarrh with hyperplasia of the reticular cells. The lymphoid tissue may be atrophied. Reticular cells and macrophages are filled with hemosiderin and bile pigments.

Since *Clostridium novyi* thrives well in the leisons caused by the wandering flukes, 'Black disease' is usually associated with and is a complication of liver fluke infection in sheep. (See Black disease, page 572).

Symptoms : In adult cattle no symptoms may be noticed, except weakness and anemia. But in sheep there may be weakness, dullness, anorexia, pale conjunctiva, edema and death in 2 to 3 weeks in acute cases. In some instances death may occur in 38 hours manifesting blood stained feces and nasal discharge.

In chronic cases there may be emaciation and edema in the submaxillary region (bottle jaw), anemia, ascites, hydrothorax, hydropericardium and generalised edema. Wool may be shed and there may be diarrhoea In those that die, course may be 2 to 3 months

Sudden death may occur in acute infections due to :
1. Hemorrhages into the abdominal cavity and liver.
2. Acute hepatic insufficiency as a result of massive hepatic destruction.

Cachexia is caused by :
1. Absorption of gland secretions and metabolic products of flukes.
2. Absorption of break-down products of disintegrated flukes.
3. Absorption of bacterial toxins and decomposed bile.
4, Digestive disturbances due to lack of bile supply into the intestines.

Diagnosis :
1. The operculated, yellow colored eggs can be identified in the feces.
2. Lesions and the appearance of parasites in the liver.

Dicrocoelium dendriticum, the lancet fluke is small, 4–12 mm. by 1 mm. and infects cattle, sheep, goats, horses, camels, deer, pigs and man. Intermediate hosts are land snails and ants. The effect of *Dicrocoelium* infection is similar but of lesser magnitude to that of *F. gigantica* Inhabiting the bile ducts they cause thickening of the ducts. Periportal fibrosis is also produced but large scale destruction of liver parenchyma does not occur.

Fluke of pancreas :

Eurytrema pancreaticum, is a fluke found in the pancreatic ducts and more rarely in the bile ducts and the duodenum of sheep, goats, cattle and buffaloes· It measures 8 to 16 by 5 to 8.5 mm It has a wide flat body and a peculiar conical projection to the tail. It has spines on the cuticle. The life history of the parasite has not completely been worked out.

Pathogenesis : Though a few flukes may not produce much damage, there may still be catarrhal inflammation with destruction of the epithelium of the duct. Granulomata are found wherever the eggs penetrate into the wall of the ducts. Plasma cells, lymphocytes and eosinophils predominate in these granulomata. In some animals chronic interstitial pancreatitis is produced in which the acinar tissue may be destroyed and replaced by fibrous tissue.

Paragonimus westermanii is a fluke which occurs in the lungs mostly and very rarely in the brain, spinal cord and other organs of pig, dog, cat, goat cattle, fox, mink, wild carnivorous animals and man. The parasite is reddish brown in color and measures 7.5 to 16 mm by 4 to 8 mm. The eggs are yellowish brown and are operculated. Two intermediate hosts, the snail and the crab or crayfish are required. In the final host, after ingestion of the crab, the metacercariae are liberated in the intestines and then they migrate across the peritoneal and pleural cavities to the lungs.

In the lungs the parasites live in pairs in inflammatory cysts which are surrounded by diffuse connective tissue and infiltrated with leucocytes and giant cells. The cysts are supposed to be communicating with the bronchioles for the eggs to pass out. Sometimes the eggs may also be encysted. In some cases frank pneumonia may be caused with thickening of the alveolar septa, hyperplasia of the bronchial epithelium, which may be denuded in some places and presence of exudate in the bronchi and bronchioles. The exudate contains the eggs. The parasites in the brain may cause nervous symptoms.

Diagnosis :

1 Observe the operculated eggs in the feces or sputum
2. A complement-fixation test has been developed
3, An intradermal test can also be used with the skin sensitising antibodies·

Opisthorchis tenuicollis (*O. felineus*) is a fluke occurring in the bile ducts mostly and rarely in the intestines and pancreatic ducts of dog, cat, pig, fox and man. It measures 7 to 12 by 1.5 to 2.5 mm. and is reddish when fresh. The cuticle is smooth. The eggs are operculated. Two intermediate hosts are present—a snail – *Bithynia leachi* and a fish

Opisthorchis sinensis (*Clonorchis sinensis*) :   This is otherwise known as Oriental or Chinese liver fluke and occurs in the bile ducts and sometimes in the pancreatic ducts and duodenum of dog, cat, pig, mink and man. Its size may be 25 by 5 mm. The eggs are thick and light brown in color, and are operculated. Two intermediate hosts are present · various species of operculated snails and fish of the family *Cyprinidae*

Infection in both the above is by ingestion of raw improperly cooked infected fishes.   Metacercariae that are liberated in the duodenum, reach the liver of the final host by way of the bile duct. Eggs begin to be passed out of the final host from the 16th day after infection

Pathogenesis : Living in the narrow bile ducts, the parasites incite a catarrhal cholecystitis. The epithelium of the bile ducts may be desquamated and so there may be bile stasis resulting in icterus. There may be papillomatous or adenomatous proliferation of the epithelium of the bile ducts. The surrounding hepatic tissue becomes cirrhotic. The eggs and worms become encysted.

Affected animals may not show symptoms unless heavily infected. Ascites. icterus and diarrhoea may be seen. In man *O sinensis* may be responsible for hepatic carcinoma.

Diagnosis ; Observe the operculated eggs in the feces.

AMPHISTOMIASIS

The amphistomes found are ;

Paramphistomum epiclitum
                 "   *orthocoelium*

Gigantocotyle explanatum
Gastrothylax crumenifer
Fischoederius elongatus
F. cobboldi
Cotylophoron cotylophorum
Ceylonocotyl thapari
Gastrodiscus secundus
Pseudodiscus collinsi

The adult flukes are found mostly in the fore stomachs. A few may be seen in the liver. These do not cause any symptoms being non-pathogenic.

The metacercariae are swallowed and the young flukes are found in the intestines, from where they migrate to rumen and reticulum via duodenum and abomasum, before becoming mature in 6 weeks to 4 months.

The immature flukes attach themselves to the mucosa and ingest it. Hence irritation is caused resulting in enteritis. Some flukes may even bore through to the peritoneum where hemorrhages may occur. In heavy infection there is persistent foetid diarrhoea and the animal becomes exhausted and emaciated and may die in a few days. The clinical condition caused by the immature parasites is known as " immature amphistomiasis".

Symptoms : Animal is weak, dull, off feed and becomes emaciated because of the offensive diarrhoea present. Mucous membranes are pale. Sub-maxillary edema is present Some animals may suffer from hypoglycemia. Mortality can be heavy, especially in sheep, in heavy infection

Lesions The duodenum and jejunum contain numerous flesh colored flukes. Subcutaneous edema is present together with ascites and hydropericardium. Fat in depots is gelatinous. The mucosa of the duodenum is thickened and shows fibro-cattarrhal inflammation.

Diagnosis :

1 Observe the operculated eggs with distinct yolk and germ cells in the dung. In heavy infections immature amphistomes are also seen in the feces.

2. Symptoms of emaciation, diarrhoea and bottle jaw in sheep must always make one suspect amphistomiasis, especially, if mortality is heavy in a flock.

Microscopically. the immature flukes may be seen in the deeper layers of the mucosa and in the wall of the gut.

SCHISTOSOMIASIS

Schistosomes are elongated flukes living in the veins of various animals. These are 1.5 to 2.0 cms. long.

Important blood flukes found in India are :

Schistosoma nasale in the veins of nasal mucosa of cattle and buffaloes; rarely in sheep, goats and horses.

S. spindale in the mesenteric and portal veins of cattle, buffaloes, sheep, goats and dogs,

S indicum in the mesenteric and portal veins of sheep, goats, cattle, elephants, horses and camels.

S incognitum in the portal veins of pigs and dogs.

Ornithobilharzia bomfordi and *O. dattai* in cattle and buffoloes.

Bivitellobilharzia nairi in the portal and hepatic veins of elephants

S. nasale : The adult parasites are found in the veins of the nasal mucosa and cause the condition 'Nasal granuloma' or Snoring disease' in cattle. Though the worms inhabit the nasal veins of buffaloes no symptoms are caused in them

Mode and route of infection : The cercariae pierce through the intact skin and visible mucosa of the host These enter the small cutaneous veins and are thus swept away by venous circulation. They ultimately reach the nasal mucosa via arterial system and finally reach the nasal veins. Here they become mature and eggs are laid.

The eggs become attached to and then enclosed by the capillary endothelium. The miracidium within the ovum secretes proteolytic enzymes, which passing out through the pores of egg-shell act on the basement membrane which becomes ruptured The eggs thus reach the surrounding tissue where a reaction is set up resulting in the formation of 'Pseudotubercles' This consists of the schistosome eggs in the centre surrounded by a pink stained area. This is known as 'actino body'. There are 4 stages in the formation and fate of pseudotubercles.(Rao 1933)

I Stage : A freshly laid egg attracts macrophages, lymphocytes and eosinophils which become arranged radially around the shell of the egg These cells fuse to form a sheath of the cytoplasm around the egg. The nuclei recede away from the shell to the periphery.

II Stage : Because of the proteolytic enzymes secreted by the miracidium, the cytoplasm of the cells immediately in contact of the shell becomes degenerated and this is arranged as rays emanating from the egg-mass and stains pink, giving the characteristic appearace of 'actino body'.

III Stage : Around this 'actino body' accumulate lymphocytes neutrophils, eosinophils and a few giant cells. These cells so crowd around the 'actino body' that it is reduced to a very narrow zone. So the infiltrating cells form a small abscess in the centre of which is the ovum. Ripening of the abscess eventually brings the egg to the surface of the nasal mucosa where it bursts and theryby the egg escapes with the pus.

IV Stage : Consists of the healing of the abscess by granulation tissue,

Symptoms : In the nasal cavity are found minute ulcers and the granulomatous growths occlude the lumen so much as to interfere with normal respiration A snoring sound is heard due to the brushing of air with the growths. Profuse nasal discharge, sometimes mixed with blood, is present Much economic loss occurs through the animals b coming unthrifty and not being able to do sustained work:

No symptoms are noticed in buffaloes because 'complete compatability' is established between the host and the parasite.

Diagnosis : Examine the nasal discharge for the characteristic boomrang-shaped ova which have a spine at one end.

Other schistosome infections :

The cercariae may reach the portal and mesenteric veins through the systemic orterial circulation

It is only those cercariae that reach the mesenteric or portal veins that develop into adults. Those going elsewhere die.

The pathology caused by the parasites is due to :

(1) injury produced by adult parasite in the veins.

(2) injury by the ova.

(3) injury to the skin by cercariae and

(4) injury caused by migrating cercariae.

Injury by the adult parasite in the veins

The adult worms do not appear to cause much damage. They may cause anemia by feeding on blood. There may be phlebitis and venous thrombosis occasionally. The blood pigments are regurgitated by the parasite and these are taken up by the cells of the R B system, especially those of liver and regional lymph nodes and spleen (bilharzial pigment). The excretions of the parasite may have some detrimental effect on the animal. After the parasite dies its disintegration products may be toxic

Injury by the eggs

Most of the damage to the host is caused by the eggs. After they are deposited in the venule they adhere to the endothelium which subsequently envelops them. By the action of the enzymes secreted by the miracidia as well as by means of the spines, the ova are able to rupture the wall of the venule and come into the tissues, where they act as irritants and incite the formation of a 'Pseudotubercle' consisting of infiltration by neutrophils, eosionophils, endothelioid cells and giant cells. Sometimes the ova may be found engulfed by giant cells. Those nodules that are formed in the submucosa of the intestines rupture into the lumen and so the ova are voided with the feces. During this process there may be hemorrhage in the dung. Animals may suffer from bloody diarrhoea Healing occurs by scar tissue.

But those ova that chance to enter the circulation may reach various organs where endarteritis, periarteritis, phlebitis and pseudotubercles may be formed. Much fibrous tissue proliferation occurs around the nodules. In time the miracidia may die in the eggs and may be calcified. Such pseudotubercles may be found in the mesenteric lymph glands or in the liver. Portal phlebitis and granulomatous hepatitis may be produced. Cirrhosis of the liver may ensue as a result of granulomatous lesions described above around the ova conveyed by portal vessels.

Symptoms : Animals with heavy infection suffer from anemia, debility, emaciation and diarrhoea. ending fatally. There may be small nodules and ulcers in the intestinal mucosa and the intestinal wall is much thickened due to inflammation and scar formation. There may be papillomatous growth of the intestinal mucosa. Serous cavities contain excess of fluid. Liver may be shrunken and hard in advanced cases with nodular (hobnail) surface due to cirrhosis Heavy infection of lungs by cercariae may cause fatal pneumonia.

Diagnosis : Feces contain the ova.

Cutaneous lesion in man is called 'Swimmers itch' or cercarial dermatitis and is caused by the penetration of the cercariae. In man and animals at the place of penetration a small nodule is formed consisting of infiltration of neutrophils, lymphocytes and eosinophils. Intense itching is produced at the site and an ulcer may form, healing taking place after a long time.

FLUKES OF POULTRY

Echinostoma revolutum:

This is a fluke upto 22 mm. long, occurring in the intestines, ceca and cloaca of fowls, pigeon, turkey and ducks. The intermediate hosts are snails, fishes and tadpoles.

Symptoms and Leisons: In mild infections no symptoms are seen. In heavy infections, especially in pigeons, there may be diarrhoea and dysentery due to enteritis, which may be hemorrhagic. Fatalities may occur.

Prosthogonimus macrorchis :

This is 5 to 7 m m. long and is found in the Bursa Fabricii and oviduct of duck and fowl.

Symptoms ; Affected birds are dull and lose their appetite. There is deep drop in egg-laying and the eggs may have thin shells.

Microscopically, the birds are anemic and emaciated. Peritonitis may be noticed. The oviduct is dilated and contains exudate and egg material. Rupture of oviduct may sometimes occur when the yolk material and flukes may be found in the abdominal cavity

CESTODES

All animals and man are infected with tape worms. Usually in mild infections, especially in adult animals, no great harm is caused. It is only in the young animals that heavy infestation may produce unthriftiness, constipation and diarrhoea

The following are the most common and important tape-worms found in domesticated animals.

Cattle ; *Moniezia expansa; Avitellina centripunctata;*

Sheep and goats: *Moniezia benedeni; Moniezia expansa; Thysanosoma actinioides; Avitellina centripunctata.*

Dogs: *Dipylidium caninum, Taenia pisiformis; T. hydatigena; Multiceps multiceps; M. serialis; Echinococcus granulosus; Diphyllobothrium latum.*

Cats: *Taenia taeniaformis; Dipylidium caninum; Diphyllobothrium latum*

The adult tape worms are all found in the intestines except *Thysanosoma actinioides* which is found in the biliary and pancreatic ducts. Tape worms may affect the host in the following ways.

1) The mucosa may be injured by the suckers and hooks causing enteritis. This is not of much consequence unless the number of parasites is great.

2) The parasites, because of their length and great body surface may deprive the host of its nutrition The parasite lacking an alimentary system may absorb by its body surface the digested nutritive substances from the gut of the host. This also will not be of much consequence unless the number of parasites is great.

3) The effete products of the parasites and the possible toxins formed by the tape worms may be toxic to the host.

4) A large number of tape worms may obstruct the intestines and cause sometimes volvulus.

5) Blocking bile ducts, obstructive jaundice may be caused.

6) Anemia of the sprue-type (macrocytic) is seen, especially in infection by *Diphyllobothrium latum* since it absorbs vit. B_{12}, Folic acid and iron causing deficiency of these in the host, It is also thought that some substances secreted by the parasites may destroy the activity of the gastric intrinsic factor and thus macrocytic anemia may be caused. In such cases there may be megaloblastic hyperactivity of the bone marrow, deposits of hemosiderin in the liver and spleen and fatty degeneration of the central nervous system.

Symptoms in dogs and cats : In heavy infestations animals may evince abdominal pain and enteritis. Abdominal irritation may be present indicated by rolling and biting at the belly by the animals. Epileptic seizures may be present, especially in cats. Appetite is diminished and the animal becomes unthrifty with a shaggy coat. It becomes progressively emaciated

The proglottides that may be evacuated irritate the anus and so the animals drag themselves on the ground sitting on their buttocks. This is a very characteristic posture of the animals with tape-worm infection.

Emaciation debility and diarrhoea are the symptoms seen in calves, lambs and kids in heavy infestations.

Bladder worms :

By far the greatest damage is done by the larval stages called the bladder worms of the parasites. Many animals serve as intermediate hosts. The larval stages may be of three varieties.

1) **Cysticercus**—monosomatic and monocephalic. This consists of only a bladder, containing one scolex. So each cysticercus can give rise to only one tapeworm eg, *Cysticercus cellulosae, C bovis* etc.

2) **Coenurus**—monosomatic and multicephalic. This consists of a bladder containing many scolices, each of which can give rise to a tape worm, eg *Coenurus cerebralis*

3) **Hydatid cyst**—multisomatic and multicephalic. In this the bladder has within it daughter and grand—daughter bladders, which in their turn contain brood capsules each containing many scolices, eg., Hydatid cyst of *Echinococcus*.

These bladder worms are found in various organs including the central nervous system. Cysticerci, which are small, usually do not cause much damage unless present in very large numbers in vital organs.

The following cysticerci are met with :

Cysticercus bovis ; The adult is *Taenia saginata* in the intestines of man The bladder worm is found in the muscles, liver, heart, lungs, diaphragm etc. of cattle. The affected meat is called 'measly beef', The cysticerci are seen as white spots. Usually no symptoms are seen unless infection is very heavy. In long standing cases the parasite may die followed by encapsulation and scar formation.

Cysticercus collulosae The adult is *Taenia solium* in the intestines of man. The bladder worms are found in the muscles, heart, lungs and brain of pigs The affected meat is known as 'measly pork'. Older bladder worms may be calcified.

Cysticercus ovis : The adult is *Taenia ovis* found in dogs and other carnivores. The bladder worm is found in the muscles of sheep and goats.

Cysticercus tenuicollis : The adult is *Taenia hydatigena* found in dogs. The
bladder worm is found in the liver, mesentery and omentum of sheep, goats and
pigs.

Coenurus cerebralis : Adult is *Multiceps multiceps* of dogs. The bladder
worm is found in the C. N. S. of sheep and causes 'Gid' (See page 414.)

ECHINOCOCCOSIS

The bladder worm of the tape worm *Echinococcus granulosus* is called
hydatid cyst. The adult is found in the dog, fox, wolf and other carnivora and
is characterised by having only 4 proglottides and is 1.0 cm. in length.

As described earlier the hydatid cyst has daughter and grand-daughter cysts
with numerous brood capsules and each of these brood capsules contain up to 40
scolices. So each hydatid can potentially give rise to hundreds of tape worms.

The hydatid cyst is found in all organs including the C.N S of cattle, sheep,
goats, swine and man. The onchosphere after being released in the intestines
burrows into the mucosa of the bowel and reaches the liver via portal vessels
and there develops into the hydatid. Entering general circulation it may reach
any organ. Liver, spleen and lungs are the most affected

Wherever found, the hydatid causes destruction of the tissues of the host
by pressure A dense fibrous capsule is formed by the host around the cyst.
Hepatic cysts cause hepatic insufficiency viz., digestive disturbances and ascites,
while the cysts in the lungs produce dyspnoea. Cysts in the brain produce cerebral symptoms, (viz paralysis, blindness etc.)

Diagnosis :

1. A complement-fixation test can be done.
2. Usually the hydatids are seen postmortem.
3. Intradermal allergic test in man is reliable (Casoni's test).

TAENIASIS IN POULTRY

The important tape worms found in poultry are *Davainea proglottina, Rail-
lietina echinobothrida R tetragona* and *R. cesticillus*. Taeniasis is a problem
only in the chicks kept under unhygienic conditions since ants and flies serve as
intermediate hosts.

Symptoms : Birds show anemia and emaciation preceded by anorexia, droopiness, increased thirst and diarrhoea Egg laying is decreased. In Davainea
infection paralysis may be noticed. In heavy infections, deaths may occur. *D
proglottina* is the smallest tape worm but most pathogenic since it burrows deep
into the wall of the intestines.

Lesions : Enteritis is seen in heavy infections. *R echinobothrida* produces
nodules on the wall of the bowel, resembling those of tuberculosis These
nodules can be seen from the external or serous surface of the gut The intestinal wall may be thickened due to congestion and infiltration by inflammatory
cells consisting of neutrophils, lymphocytes and eosinophils

Diagnosis is made by seeing the segments or eggs of the parasites in the
feces

NEMATODES (Round worms)
Ascariasis

Ascarids are very common in man, animals and birds. Younger animals are more susceptible. The following are the important worms encountered.

Cattle : *Neoascaris vitulorum* (especially in calves).

Horses : *Parascaris equorum.*

Pigs : *Ascaris suum* (Ascaris lumbricoides-var-suis).

Dogs, foxes, etc. *Toxocara canis, T. cati, Toxascaris leonina.*

Cats : *Toxocara cati, Toxascaris leonina.*

Fowls : *Ascardia galli, Heterakis gallinae.*

The adult worms are usually found in the intestines. Sometimes they may migrate to the stomach (and from there to esophagus), the bile duct and pancreatic duct. It is thought that deficiency of glucose may compel the worms to be erratic in their movements.

Infection is by ingestion of eggs in which the 2nd stage larvae are present. The larvae that are liberated in the stomach and intestines burrow through the gastric and intestinal wall and reach the liver via portal vein. Form there they reach the heart and from thence lungs. Some larvae may reach lungs directly without passing through the liver, if they enter lymphatics of the intestines. In the lungs the larvae leave the capillaries and enter the alveoli from where they travel up the bronchi into the trachea and from there to pharynx. They are coughed up and swallowed and so finally enter the intestines to commence their adult and sexual life. But *Toxocara canis* and *Neoascaris vitulorum* have somatic migration in contrast to the tracheal migration of other ascarids They are responsible for causing prenatal infection in puppies and calves. During the itinerary through the lung the larvae undergo two moults and so become the 4th stage larvae which grow into adults. Sometimes, the larvae may reach the left side of heart from the lung and then get dispersed to other organs where they die, or are encapsulated and calcified. The time that elapses between the ingestion of infective eggs and the appearance of eggs in the feces of the host is called **prepatent period** and this varies with different species of parasites. In A *lumbricoides* it may be 80 to 90 days.

The host is affected by

1. The adult in the intestine and
2 The migration of the larvae.

Damage caused by the adult : The damage done is due to mechanical and oxic effects. Mechanically, the parasite may gnaw the mucosa and cause injuries and so hemorrhages, erosions and ulcers may develop. Sometimes the parasite may perforate the intestinal wall and cause fatal peritonitis. Large number of parasites may become entangled with each other forming a mass and cause intestinal obstruction, intussusception and even rupture. In their erratic movements, some round worms may enter the bile duct and produce obstructive jaundice and cholangitis Some may enter the stomach and may be vomited. The toxic effects are due to loss of nutrient materials and the action of toxic substances excreted by the parasites. There may be, therefore, symptoms of anemia, emaciation, nervous symptoms. edema and central necrosis of hepatic lobules.   The results of these may be stunted or retarded growth in the younger animals and even death in some cases

Macroscopically, there may be catarrhal enteritis. Sometimes it may be hemorrhagic or even fibrinous

Injury by migrating larvae

In the intestines :   Passage through the intestinal wall is uneventful unless there is heavy infection when hemorrhage may be present.

In the liver :   As the larvae burrow through, the parenchyma is destroyed and the tract formed thus is filled by blood and debris. Neutrophils and eosinophils crowd around such a track. Healing by scar tissue occurs and the foci appear as 'milk spots' on the capsular surface. In heavy infections the lesions may become confluent and fibrosis may occur to such a degree as to obliterate whole lobules.   Due to development of allergy, in reinfections, the larvae may die and then incite a foreign body reaction So a small nodule with the central caseous material containing dead larvae surrounded by neutrophils, lymphocytes, eosinophils, epithelioid cells and giant cells may be seen. The larvae may convey bacteria to the liver producing secondary infection. But at the present day, due to better husbandry, lesions are very negligible.

In lungs When the larvae puncture through the capillaries, hemorrhage may be present. In heavy infections, together with edema this may cause death· Pneumonia may be caused by the irritation produced with loss of bronchiolar epithelium.   A granulomatous reaction is produced with infiltration of neutrophils, lymphocytes, eosinophils, histiocytes and giant cells. Some of the larvae may be killed. During this reaction some bronchioles may be destroyed. The inter and intralobular septa are thickened by infiltrating cells.

Pneumonic lesions cause dyspnoea and cough and this is known as 'Thumps' among swineherds.   The lesions produced by the larvae may increase the severity of viral pneumonia and influenza in pigs if concurrently present.

Symptoms in dogs and cats : It is the young pups and kittens under 5 weeks of age that show symptoms; viz. general malaise, inanition, lethargy, discinclination to play with fellows.   They may vomit and sometimes parasites may be expelled. The animals become progressively emaciated and the visible mucous membranes are anemic and pale.   There may be potbellied condition. Diarrhoea alternating with constipation may be noticed probably due to the intestinal irritation and to the irritation of nerves by the toxins excreted by the parasites. The condition progresses to severe anorexia, weakness and death, which may occur in 5 to 8 weeks

Older dogs are not usually much affected except that growth may be impaired.   The young pigs may vomit the fully grown worms.

The adult animals may get rid of the infection suddenly probably due to development of immunity.

Diagnosis :

1   Observe the eggs in feces.
2   Ascardies may be seen in dead animals.

Man and mice may act as transport hosts for *T. canis*. *Toxocara cati* has certain peculiarities: (1) In this infection there is no transplacental infection. (2) Earth worms, beetles, mice and cockroaches etc. act as transport hosts. (3) The larvae either develop within the wall of the gut without a migratory phase or migrate through the tissues before becoming an adult in the gut.

Visceral larval migrans is a condition caused mostly by the larvae of the dog and cat ascarides of the genus *Toxocara* in abnormal hosts like man and other animals.

The condition is characterised by eosinophilic granulomatous lesions in liver and lungs caused by the migrating nematode larvae. The larvae are commonly found in the liver, lungs, brain and eye etc. in children, who being in close association with dogs and cats, are liable to ingest the eggs of ascarides.

Intermittent fever, eosinophilia, persistent cough, pains of muscles and joints are the common symptoms seen.

Ascariasis in fowls
ASCARIDIA GALLI

The large round worm. This is found in the intestines of fowl, turkey and other birds.

The larvae from the eggs, after ingestion, penetrate into the duodenal mucosa from where they emerge after 7 days to become adults in 5 to 8 weeks No migratory phase is seen.

Symptoms : Young birds are more often affected, especially when their diet is deficient in vitamin A and Vit. B complex, Birds develop diarrhoea, become anemic and emaciated There may be drop in egg laying and paresis. In mild infection, birds recover and develop resistance to further infection.

Lesions : During the period when the larvae enter the duodenal mucosa, in heavy infections, enteritis may be produced

The worms may form, when in large numbers, a tangled mass literally occluding the intestinal lumen. Death may occur in such heavy infections.

HETERAKIS GALLINAE
Syn : Cecal worm

This is smaller than *A. galli* and is found in the ceca of fowl, duck, turkey and other birds. In this parasite the second stage larvae after coming out of ova go directly to the ceca without having any migration. Prepatent period is 1 month.

Comparatively *Heterakis gallinae* is a less pathogenic parasite. In heavy infections typhlitis may be caused showing diarrhoea. Nodules may be formed in the ceca.

Mention has already been made while discussing 'Black Head' that the protozoan, *Histomonas maleagridis* may be conveyed through the eggs of *H. gallinae* and so infection can occur in healthy birds by ingestion of such eggs.

Diagnosis : Fowl ascariasis can be detected by seeing eggs in feces Autopsy of dead birds reveals the worms.

EQUINE STRONGYLOSIS

Strongyles are very common in horses, mules and donkeys These are slender, elongated worms that are found mostly in cecum and large colon   They do
not have any intermediate hosts and so infection is direct by ingestion of infective larvae The strongyles are broadly divided into 'large' and 'small' ones.

The large strongyles consist of *Strongylus vulgaris*, *Strongylus edentatus* and
Strongylus equinus These are blood suckers and are 1-2 inches long The small
strongyles are less pathogenic and do not suck blood. They consist of many
species of *Trichonema* etc In large numbers, however, they may be harmful.

Strongylus vulgaris : (Has two teeth) The ingested larvae leave their sheath
while passing through the stomach and intestines   These then enter the mucosa
of the cecum and ventral colon where they develop into the fourth stage larvae
in 7 days by moulting in the mucosa In this location a focal inflammatory
reaction may be set up. The fourth stage larvae migrate into the arterioles of
the submucous tissue and then travel against the flow of blood, along the endothelium. Some may reach the superior mesenteric artery and from thence the
aorta. Along the track formed by the movement of the larvae, a linear deposit
of thrombus is noticed.

The parasite burrows into the intima of the artery, where hemorrhage and
necrosis pave the way for the formation of a thrombus by the deposition of
fibrin and accumulation of cells and debris. This may sometimes occlude the
lumen.

In chronic cases the parasite initiates the proliferation of the fibrous tissue
in the intima and adventitia so that the wall is very much thickened In this
tissue lymphocytes and edema fluid infiltrate and 'Verminous arteritis' results
Because of pressure of blood, loss of elastic tissue and degeneration and loss of
tone of muscle fibres aneurysm develops This is more common in the anterior
mesenteric artery. Thrombosis and aneurysm can also be met with in the aorta,
renal arteries, coeliac artery and its branches. Upto 50 parasites may be found
in each lesion. The larvae reside in these situations for several months (6 to 7)
and then drift along the blood flow, reach the intestinal wall, form nodules there
and then burrow through them and enter the lumen of cecum and colon where
they settle and become sexually mature.

Strongylus edentatus : (Has no teeth). The infective larvae burrow through
the wall of the large intestines and reach the sub-peritoneal tissue along the
vessels. Here they incite the formation of nodules into which bleeding may
occur sometimes. The parasite may develop in these nodules for 3 months.
These lesions may cause colic and anemia. Later the larvae reach the wall of
the large intestines, moving between the two layers of the peritoneum and there
nodules are again formed in which they develop for some more time. Later the
larvae leave the nodules and enter the lumen of the large intestines.

Strongylus equinus : (Has three teeth) The third stage larvae reach the
serous membrane by burrowing through the wall of the cecum and colon and
there incite the formation of nodules in which they develop and become fourth
stage larvae by moulting in about 11 days. They then migrate through the peritoneal cavity, causing subserous tunnels, into the right lobe of the liver and

wander about in that organ, causing tunnels, for about 6 to 7 weeks. Later they enter the peritoneal cavity and the pancreas where they undergo (118 days after infection of the host) the final ecdysis becoming sexually mature. After remaining in the abdominal cavity and the pancreas for a long time, they enter the wall of the cecum and colon and finally reach their lumen. (prepatent period 260 days).

In the 'small' strongyles, development of the larvae is in the mucosa of the wall of the large intestines where small nodules form. By the surface ulceration of the nodules the larvae enter the lumen of the bowel

Symptoms : Mild infection may pass off unnoticed In heavy infections the dung is loose and foul smelling. Diarrohea with loss of appetite, debility and weakness may develop. Animals become anemic showing edematous swelling on dependent parts.

Thrombosis of anterior mesenteric artery may cause colic while thrombosis of the aorta and iliac arteries may cause weakness of hind limbs and this becomes severe when the animal is exercised. The harm caused by stongyles can be divided into (a) harm done by the adult worms in the bowel, (b) damage done by the larvae during their migration.

The adults : 1 The adults draw into their mouths plugs of mucosa and so cause injuries and ulcers which may be subsequently infected by bacteria causing catarrhal and even purulent enteritis. 2. By sucking blood, anemia may be caused.

The larvae : 1 Larvae of *Strongylus vulgaris* may cause thrombosis of the anterior mesenteric artery and so colic may be produced. 2. (a) Nodules which are formed by the larvae may become infected by bacteria after the larvae come out leaving ulcers. (b) Nodules on the peritoneum may burst causing peritonitis and fatal hemorrhage. (c) Damage to the liver by the larvae of *S. equinus*.

Lesions : The lesions are seen usually in heavy infestation and these are associated with emaciation, cachexia and edema Mucosa of the cecum may reveal petechiae, erosions and catarrhal enteritis. There may be submucosal edema Subserous hemorrhages are caused by migrating larvae. There may be purulent foci in the liver and lymph nodes. The nodules formed in the mucosa and submucosa of the cecum mostly and colon less frequenlty, are small, pea-sized and firm. These known as 'worm nests', consist of a central debris containing the larvae and infiltration by eosinophils and surrounded by a connective tissue capsule.

Sometimes the nodules are larger and resemble lymphoid nodules They may contain blood elements and blood pigment together with neutrophils and macrophages. These may be bluish in color and are called 'haemomelasma'. The larvae may come out of the nodules leaving ulcers, which heal by scar tissue. Heavy scar formation, especially round the bowel, may cause stenosis. The larvae that die may be calcified

Diagnosis :

1. Examine the feces and identify the eggs,

2 Worm burden may be assessed before treating the animal since ligh infections do not harm the animals and every animal harbours a few parasites.

CHAPTER 31

ANCYLOSTOMIASIS

Syn : Hook worm disease

Hook worms are found in man and animals except the horse. These are all blood suckers and cause severe anemia. The following are the various species encountered.

Ancylostoma duodenale *Necator Americanus*	} Man.
Ancylostoma caninum *A. braziliense* *Uncinaria stenocephala*	} Dogs, foxes
Bunostomum phlebotomum	} Cattle
Bunostomum trigonocephalum *Gaigeria pachyscelis*	} Sheep and Goats
Necator suillus *Globocephalus urosubulatus* and other species of Globocephalus	} Swine.
Bunostomum sangeri	— Elephant.

Most of these are host specific, but the larval forms may penetrate the skin of aberrant hosts causing dermatitis known as 'Creeping eruption'.

Infection can occur in two ways :

(1) Either by the ingestion of infective larvae or

(2) The larvae can burrow through the intact skin.

These methods of infection occur in all the hook worms except *Uncinaria stenocephala* in which the first method, viz, ingestion alone occurs.

The infective larvae penetrate through the skin and enter the small veins. They reach the lungs and enter the alveoli burrowing through the capillaries. Ascending through the tracheobronchial tree, the larvae ascend the epiglottis and the pharynx and esophagus and then go to stomach and intestines The pre-patent period is about 10 weeks. Even the larvae that may be ingested probably go through the migratory phase before settling down in the small intestines. Prenatal infection by transplacental passage can occur.

How the parasite causes damage

1   **On the skin :** The larvae penetrate the stratum corneum and enter the hair follicles through which they enter the dermis via sebaceous glands. So, intense itching is produced due to local inflammation set up by the larvae. There is infiltration of lymphocytes, eosinophils and macrophages into the dermis and the lesion has a weeping discharge. The skin then becomes thickened and scaly : **"cooly's itch, water itch, ground itch".**

2   During the larval migration, verminous bronchopneumonia may be caused

3   The adult in the intestines takes a plug of the intestinal mucosa into its powerful buccal cavity and with the teeth damages the epithelium and sucks blood. An anticoagulant is secreted which prevents clotting. So, even if the worms were to leave that place, blood continues to flow. The worms attach

themselves to other places, shifting positions, and produce punctures afresh. It is thought that *Ancylostoma caninum* sucks blood from mucosal arterioles. Thus there is loss of a large quantity of blood since each parasite is estimated to suck 0.8 to 1.0 ml. of blood in 24 hours. Hypochromic and microcytic anemia develops (iron deficiency anemia). The bone marrow is hyperplastic but soon becomes aplastic either due to exhaustion or due to toxic inhibition by the toxins liberated.

The wounds of the mucosa may be infected by the secondary bacterial invaders.

Symptoms are those of anemia, viz weakness. shedding of wool, pallor of mucous membranes, hypoproteinemia, pot belly, diarrhoea, ascites and edema of the dependent parts—'bottle jaw'. Death may occur, in some cases, of cardiac weakness. Hook worm disease is seen more in young animals. Its manifestation depends upon the nutritious state of the animal. Well nourished animals last longer. But even in these, infection by large number of parasites is fatal.

Lesions : The carcase is cachectic. There may be enteritis and the intestinal contents may be blood tinged Other lesions seen are pneumonia, pleurisy, necrosis of the buccal mucosa, centrilobular hepatic necrosis, degeneration of the myocardium and serous atrophy of adipose tissue.

Diagnosis :
1. Demonstrate the eggs in feces.
2. Adult worms can be seen at autopsy.

STOMACH WORM DISEASE
Syn : Parasitic gastritis

Species of *Trichostrongylus, Hemonchus, Ostertagia, Cooperia, Hyostrongylus* and *Nematodirus* occur in the abomasum of ruminants. Some of these occur in the intestines also. The disease is more severe in sheep and goats than in cattle and in young animals than in the old. The following parasites are important.

Hemonchus contortus in cattle, sheep and goats

H. similis in cattle

H. longistipes in sheep and camel.

Mecistocirrus digitatus in cattle. sheep and goats.

Ostertagia ostertagi in cattle, sheep and goats.

Hyostrongylus rubidus in swine,

Various species of *Trichostrongylus* and *Cooperia* are found in cattle, sheep and goats.

In the same animals, more than one species may be found.

Infection is direct, by the ingestion of infective larvae. These become sexually mature in the abomasum and attach themselves to the mucosa. The prepatent period is 19—21 days. The parasites are not attached to the mucosa always. They are free in the stomach, puncturing the mucosa at times of feeding. These are blood suckers.

Larvae of *Ostertagia* enter deep into the mucosa and form tiny nodules from which the adult worms may be projecting.

Symptoms ; Since the parasites suck blood, anemia and attendant symptoms are seen. Animals are unthrifty, do not feed and do not gain weight. Wool becomes broken and falls off. Feces are soft and in the sheep, especially, they may be dark or black in color. Later diarrhoea develops. Animals become progressively weaker and emaciated. Edema in dependent parts develops and bottle jaw is conspicuous Mucous membranes are pale. Deaths occur especially in young animals lambs and calves). The phenomenon of 'self cure' occurs as the animals develop immunity.

    Leisons : Depending on the number of parasites present, there may be acute, subacute, or chronic catarrhal gastritis. There is hyperplasia of the cells of mucosa. In sheep and goats there may be hemorrhages in the mucosa as well as erosions. In cattle there is sub-mucosal edema. Abomasal contents may be reddish In fatal cases, the animals are emaciated and there is ascites, hydropericardium and hydrothorax. The blood is watery and bone marrow hyperplastic. The fat is gelatinous having undergone mucoid degeneration

    In infection by *Ostertagia*, small nodules can be seen in the mucosa The larvae get surronnded by infiltrating leucocytes and small micro abscesses may form. After the larvae leave, healing occurs by scar tissue. In heavy infections the mucosa appears pitted due to scarring and is very much thickened due to metaplasia and hyperplasia of glandular epithelium. In many places, the adult worms may be projecting from the nodules in which they are half embedded.

    In infection by *Trichostrongylus* and *Cooperia* there is catarrhal enteritis and thickening of the mucosa of the duodenum. Petechiae may be present. Mesenteric lymph nodes may be enlarged and juicy. Cooperial larvae may form granulomas on the serosa.

    In swine, diphtheritic gastritis is caused by *Hyostrongylus rubidus*. The worms are found attached to mucosa beneath the pseudomembranous deposits. Anemia and cachexia may lead to death. The liver may be friable and fatty.

Diagnosis :
1. Symptoms
2. See eggs in feces.
3. In sheep differentiate from anthrax, enterotoxemia, coccidiosis, deficiency of copper and cobalt.

OESOPHAGOSTOMIASIS
(Nodular worm disease)

    Oesophagostomes are called nodular worms because the larvae produce nodules in the walls of intestines.

The following species are important ;

Cattle : *Oesophagostomum radiatum.*
         Chabertia ovina.

Sheep : *Oesophagostomum columbianum.*
         O. venulosum
         Chabertia ovina.

Pigs : *Oesophagostomum dentatum.*
         O. quadrispinulum (Syn : O. longicaudum).
         O. brevicaudum.

Goats : *Those mentioned in sheep and also O. asperum.*

The adult Oesophagostomes live in the lumens of the large intestines. They do not attach themselves to the mucosa nor do they suck blood. The secretions of their esophageal glands irritate the mucosa causing inflammation The parasites live on the mucous exudate The adult Chabertia, on the other hand, draw plug of mucosa into their buccal capsule and digest it by means of digestive enzymes secreted by their esophageal glands.

The infective larvae of the parasites, except those of *O. venulosum* and *C ovina*, on ingestion by the host, penetrate the intestinal mucosa (any where between the pylorus and the rectum) and encyst. Usually they encyst on the lamina propria on the deep side of muscularis mucosae. It is usualy those that go deep into the submucosa that produce nodules described below. They moult and become 4th stage larvae and may return into the lumen in five days. The prepatent period is about 6 weeks.

In animals which are not sensitised by prior infection the entry of larvae into mucosa is uneventful They enter into the submucosa, moult once, leave their sheaths and return to the lumen of the bowel. Only very little inflammatory reaction occurs. The larvae are surrounded by histiocytes and lymphocytes. The cells also infiltrate into the edematous submucous tissue. Complete resolution occurs after the larvae leave the lesion Some deeper nodules may contain pus but even this lesion heals after the emergence of the larvae. So in these cases the intestines contain large number of adult worms but the gut wall shows no nodules.

But in animals which have had a prior experience of infection, the infective 3rd stage larvae penetrate the wall and a granulomatous reaction is incited Large number of eosinophils, lymphocytes, macrophages and foreign body giant cells infiltrate into the tissues and surround the larvae and thus a nodule or pimple is formed which is encapsulated Due to the presence of eosinophils the nodule is greenish in color. Central caseation may occur in this lesion, which now resembles a tubercle. In such a nodule, the larvae may die and the nodule may become calcified or the live larvae may escape into the bowel. In heavy infestations, the wall of the gut is studded with these nodules and such an intestine is known as 'pimply gut'. In such animals the lumen of bowel may contain very few adult worms. *O. columbianum* produces more severe lesions as its larvae burrow deep into the wall.

Symptoms : It is in the young animals that symptoms are noticed. Symptoms are particularly observed in sheep and consist of severe persistent diarrhoea. Animals become emaciated and dehydrated Normocytic and normochromic anemia is caused due to malnutrition and anorexia but not due to blood loss. Some animals may suffer from intussusception Death may supervene after animals become cachectic Sometimes the nodules may become infected when fatal peritonitis may occur The 'pimply gut' is not useful for making sausages and so economic loss is incurred.

Larvae of *O. venulosum* and *Chabertia* do not penetrate the wall but develop within the lumen of the bowel Since adult *Chabertia* attach themselves to the mucosa and digest it, injury is caused These injuries may be subsequently infected and septic ulcers may be formed. The parasite may, therefore, cause ulcerative and hemorrhagic colitis with edema of the bowel wall. Large amount of blood may be ingested by the parasites in heavy infections and so anemia may develop. The dung is black in color.

Diagnosis ; The condition can be confirmed by a postmortem examination.

SPIROCERCOSIS

Spirocerca lupi (Syn S sanguinolenta) is very common in dogs and is found to live in nodules formed in the esophagus. Fox, cat and wolf may also be affected.

Life cycle : The embryonated eggs are passed in feces. These hatch in certain coprophagus beetles when ingested by them. In these intermediate hosts, the larvae develop into 3rd stage larvae. These beetless may be eaten by frogs, snakes, lizards, rodents and fowls in which the larvae remain viable for some time. These are just **transport hosts.** The final host gets infected by eating either the beetles or the transport hosts.

In the final hosts, the larvae penetrate the gastric mucosa and travel along the gastric arteries in their adventitia and media and finally reach the aorta. There they colonise in the adventitia of the arch or thoracic part. In this situation the larvae develop and then pass on to the wall of the adjacent esophagus. They may reach this place through the connective tissue tissue or walls of veins. In the esophageal wall, nodules are formed around the worms and these nodules have tiny pores through which the worm may be projecting and through which eggs pass into the esophagus.

Lesions :

Aorta : Where the larvae reside, granulomatous inflammatory reaction is set up and the normal tissue of the adventitia and media is destroyed and so an aneurysm develops. Sometimes, this may rupture resulting in fatal hemorrhage. The worm may, sometimes, burrow into the intima and so the intimal surface of the aorta is, therefore, roughened and plaques may be seen.

Esophagus : The nodules are seen in the thoracic portion of the esophagus and occur as elevations on the luminal surface with a central pore through which an adult worm extrudes. The nodule comprises of a central mass consisting of the worms, neutrophils and debris surrounded by a thick connective tissue capsule in which are found lymphocytes, macrophages and plasma cells. Eosinophils are not present.

'Deforming ossifying spondylitis' is described to have been formed by the aberrant location of the larva in the thoracic vertebrae. Exostoses are seen on the ventral surface of the vertebrae.

There seems to be some connection between the spirocerca lesions and fibrosarcomas and osteosarcomas that have been found in the vicinity of the lesions. Pulmonary hypertrophic osteopathy has been suggested to be associated with the spirocerca granulomas of the esophagus and are supposed to be their metastases. Aberrant nodules may be found in the trachea and bronchi.

Diagnosis: 1. Observe the gelatin-capsule-like eggs containing the coiled larvae. 2. At autopsy examine the esophagus for nodules and the aorta for lesions.

DIROFILARIASIS

Two species, *Dirofilaria immitis* (the heart worm) and *D. repens* affect dogs, foxes, cats etc. The former is found mostly in the right ventricle and often in the pulmonary artery, while the latter is found in the subcutaneous connective tissue and is not pathogenic.

Dirofilaria immitis : These are slender filarid worms. 12 to 31 cms long. After copulation microfilaria are shed which circulate in blood. These are ingested by the mosquitoes when they bite the hosts and development takes place in their body cavities in about 10 days. The infective larvae migrate to the mouth parts of the mosquito. When this bites a fresh host the infective larvae enter the subcutis and muscles where they grow for 3 to 4 months. Later they enter the veins and reach the right ventricle where they may live for years.

In mild infections no great damage may be caused. But in large numbers the worms can cause mechanical obstruction to the flow of blood and so hypertrophy of the right ventricle and chronic venous congestion of liver, spleen and lungs may ensue. Ascites may result.

Dead worms may form emboli and cause pulmonary infarction. Worms in pulmonary arteries may cause fibrosis of the intima and hypertrophy of the arteries. Microfilaria that circulate in blood do not appear to cause any disturbances. If they die, they may incite small foci of granulomatous inflammation.

Diagnosis

Examine blood (especially blood taken at night since microfilaria are nocturnal) for microfilaria. Fresh blood can be examined, as in surra, for the microfilaria. Stained smears can also be examined in which case search at the borders.

It may sometimes happen that only males may be in the heart when no microfilaria can be seen in the peripheral blood. In such cases diagnosis is difficult. Symptomatic diagnosis must be resorted to.

TRICHINELLOSIS

Trichinellosis or Trichiniasis is a serious condition in man for whom the source of infection is animals. Animals themselves do not suffer much. The causative parasite is *Trichinella spiralis*. Adults live in intestinal wall, while the larvae are found encysted in the muscles. The disease is found, besides man, in swine, rats, dogs and cats. This infection does not occur in man and pigs in India. However, it has been recorded from cats and rodents in Calcutta and Bombay.

Life cyle : Infection occurs by ingestion of raw or undercooked meat containing the encysted larvae. These are released by the action of digestive juices and they quickly moult and become sexually mature. The males die immediately after copulation. The fertilised females burrow into the mucosa and lay eggs which hatch and liberate larvae into lymphatics. Entering the blood, the larvae reach muscles and invade them. It is only those that reach cross-striated muscles that develop. Usually the larvae reach those muscles which are most and continuously active viz masseter, tongue, diaphragm and intercostals. Those that reach other organs including heart die.

In the muscles the larvae come out of the capillaries and enter the muscle fibres in which serous myositis is caused with loss of cross striations. The place where the parasite resides has a spindle shaped bulge. The larva: may probably secrete an enzyme by the action of which the sarcoplasm becomes degenerated, homogeneous and then undergoes granular degeneration. The larvae in these fibres are coiled or take figure of eight shape. The portion of the fibre that has not been parasitized shows reparative change by a basophilic staining material with central migration of nuclei There may be local reaction manifested by edema and infiltration by eosinophils, plasma cells, lymphocytes and histiocytes

After 2 to 3 months, the larvae may be encapsulated by the proliferation of sarcolemma. Around this capsule may be formed a thin connective tissue sheath. Each capsule may contain one or more larvae. As the capsule is formed the local inflammatory reaction subsides and only a few eosinophils may be present. Later the capsule may be partially or completely calcified and in these the larvae may remain viable for a long time, as long as 30 years These calcified capsules may be visible to the naked eye as tiny flecks. Infection occurs by ingestion of muscles containing such cysts.

Symptoms ; In man and animals there may be severe gastro-intestinal catarrh which may sometimes cause death, if infection is heavy. In man after the larvae have entered the muscles, there may be severe muscular pain and swelling. There may be edema of eyelids and face together with fever, vomiting and exhaustion.

Diagnosis : Blood smear reveals eosinophilia which may persist for a long time.

Muscle may be examined for larvae.

STEPHANURIASIS

This is caused in pigs by the kidney worm *Stephanurus dentatus* which is found in the perirenal fat, adjacent tissue and in ureters.

Stephanurus dentatus is a thick round worm about 1 to 3 inches long and white in color. The adult worms inhabit cysts that communicate with the renal pelvis or the lumen of the ureter for the discharge of eggs which are voided through urine

Life cycle: Infection is usually by penetration of the infective larvae through skin. When they reach the abdominal mucles they lose their sheaths and become 4 th stage larvae. They reach liver via lungs and systemic circulation in 1 to 6 weeks.

Sometimes, infection: may occur by ingestion of the larvae which leave their sheaths in the wall of stomach. Through portal vessels they reach the liver in 3 or 4 days. The larvae wander about in the liver for 3 or 4 months, moult once and then piercing the capsule reach the perirenal fat, where copulation occurs, the female laying eggs in the cysts.

Effects on host

A By the migratory larvae :

(i) At the point of entry into the skin, small nodules and subcutaneous edema are found There may be infiltration by leucocytes and eosinophils. Similar nodules may be found in the lymphatic glands.

(ii) **Liver :** The liver becomes enlarged, firm in consistency and hard to touch : pale in color, the lobulation is exaggerated by the proliferation of perilo. bular fibrous tissue (The changes are similar to those produced by the larvae of ascaris). There is extensive portal fibrosis which by spread obliterates many of the lobules. Eosinophils and other leucocytes infiltrate into this fibrous tissue. The scars produced in the liver by the larvae of this worm are elevated (while those caused by ascaris larvae are depressed) Death may occur due to hepatic dysfunction The affected livers are unfit for human consumption.

(iii) **Other tissues :** The larvae and young adults may wander in various tissues. Invading the spinal canal, posterior paralysis is caused. It may be found in the lungs causing catarrhal pneumonia. In the kidney infarcts are caused. In the lumbar muscles and psoas muscles the larvae may cause pain and stiffness. Entering the portal veins, hepatic artery and posterior vena cava thrombi may be produced.

B. By the adult : The adult worms are found in cysts filled with foul smelling pus in the perirenal fat. The above lesions cause unthrifty condition in the animals which do not gain weight and become emaciated. Ascites develops due to hepatic lesions.

Diagnosis : Observe the embryonated eggs in the urine.

DRACUNCULOSIS

Syn : Guinea worm infection

This is found in man and dogs of many parts of India and is caused by *Dracunculus medinensis*, which is found subcutaneously The male is small and measures 1 to 3 cms. while the slender female is larger and may be upto a meter in length.

The female is found in the subcutaneous tissues. A small nodule is formed with an orifice at its apex through which the female deposits the larvae. The intermediate host is a crustacean, 'cyclops', in which after certain development the larvae become infective

Infection is by ingestion of the cyclops in water. The larvae which are liberated in the intestines of the final host enter the connective tissue of the abdominal wall, where copulation takes place. Males die probably at once. The female migrates through the tissues and ultimately arrives at the seat of predilection viz. subcutaneous tissue, where it may become mature in one year. Usually those parts which may come into contact with water may be affected viz. legs. A papule is formed on the skin which later becomes an ulcer. Through this ulcer the anterior end of the worm protrudes. When it comes in contact with water it wriggles, protruding through the pore at the same time and lays larvae. The worm causes pruritus and edema locally. Low grade fever and urticarial rashes may be present due to the metabolic products of the worms. Probably hypersensitivity develops, if the worm dies or if larvae enter tissues. Severe inflammatory reaction with abscess formation results.

The parasite of animals is designated *D. insignis* by some authors and is smaller than *D. medinensis*.

CHAPTER 31

HABRONEMIASIS
Syn : Summer sores—Bursatee

Parasites of *Habronema* species are found in the stomach, in the skin and in the conjunctiva causing various lesions

The parasites are :

1. *Habronema muscae*—intermediate host is common house fly, *Musca domestica*.
2. *Habronema microstoma*—intermediate host is *Stomoxys calcitrans*.
3. *Habronema megastoma*—intermediate host is *Musca domestica*.

H. megastoma is smaller than the other two but is more pathogenic since it is this parasite that causes nodules in the stomach and lesions on the skin.

Life cycle ; The larvae that are passed in the dung invade the maggots of the respective flies and develop. In the pupal stage of the fly they become infective and migrate to the proboscis. Infection can occur by swallowing infected flies with feed or water or the larvae may be deposited on the lips or on wounds. From the lips the larvae may be swallowed by the animal.

The larvae that reach the stomach become mature there.

Those larvae that are deposited on wounds cause cutaneous habronemiasis.

Effects on the host

Stomach :

H. muscae and H. microstoma may lie in the lumen of the stomach and may not do much harm except for a few ulcerations and mild gastritis with thickening of the mucosa.

H. megastoma enters the submucosa of the stomach. Many worms may lie entangled in the cavity of a nodule formed by chronic inflammatory tissues. These nodules which may be upto the size of a golf ball raise the muscularis mucosae and the mucous membrane. Each has a pore in the centre communicating with the cavity of the nodule by a fistulous tract. The cavity contains, besides the worms, soft purulent debris.

These nodules are common at the fundus. The fibrous tissue wall is infiltrated by eosinophils, lymphocytes and plasma cells.

No symptoms are seen. The nodules are seen only at autopsy. Perforation and suppurative gastritis can occur by secondary infection of pyogenic bacteria.

Skin : The larvae may be deposited on wounds on the skin. The usual sites are : below the medial canthus of the eye, prepuce, glans penis, middle line of abdomen, pectoral region, neck and limbs.

The larvae which penetrate deeply incite granulomatous inflammation on the skin. The lesions may be large about 12 inches in diameter.

The lesion develops as a papule covered by a scab. It rapidly becomes enlarged and ulcerated with a depressed centre and raised edges. There may not be much discharge but the red angry surface may be covered by a necrotic membrane. Small caseous foci may be scattered in the lesion and these may be calcified. Lesions do not have any tendency to heal.

Microscopically, the lesion is a highly vascular granulation tissue (and so bleeds easily) in which eosinophils infiltrate in large numbers. Larvae may be found in the deeper portions. Because of pruritus, the animal may rub against hard surfaces or if attainable. they may bite and nibble at the lesion, which thus becomes larger and larger.

Eye : Conjunctivitis which is serous first becoming purulent soon may be seen. The eye lids and nictitating membrane may be edematous. Small granulomas containing necrotic debris and caseated foci may be noticed in the nictitating membrane. The lesions do not heal with ordinary medication.

The condition is called bursatee because the infection is common in rainy season. Pulmonary granulomatous lesions have been encountered caused by the larvae and the characteristic picture is, besides other chronic changes, infiltration by a large number of eosinophils.

STEPHANOFILARIASIS

Dermatitis is caused in cattle by several species of Stephanofilaria. They are .

1. *Stephanofilaria dedosi*—in Indonesia— lesions are on the neck, dewlap, withers and around eyes.
2. *Stephanfilaria kaeli*—in Malaya—lesions on the legs.
3. *S. assamensis*—in India—'hump-sores'.
4. *S. zaheeri*—**ear sore** of buffaloes in India.
5. *S. stilesi* – in North America—in the mid-line of abdominal skin
6. *Unidentified*—causes 'Summer sore' in German and Danish cattle.

Life cycle : The intermediate hosts are various flies belonging to the genera Lyperosia, Hematobium and Musca.

Leisons : The Lesion caused by the parasite is dermatitis This appears as a circular dry area that is raised above the surrounding and may be upto 15 cms. in diameter. In the acute lesions there may be oozing of serum or even blood. These may dry forming crusts. In chronic stages there is hyperkeratosis and the lesion is covered by scabs. Because of itching the animals rub on hard surfaces and so the lesions may become hemorrhagic and enlarged and hair may be lost.

Microscopically, the parasite lies in cysts lined by squamous epithelium which is probably derived from the hair follicles There they incite inflammation and so eosinophils, lymphocytes, histiocytes and neutrophils may infiltrate. The dermal papillae may be elongated and the hair follicles and glands become atrophied. Epidermis undergoes hyperkeratosis.

Usually the reaction is not severe, unless the parasite dies or the animal develops hypersensitivity when there will be severe inflammatory reaction.

Scrapings from the lesions reveal the adult worms and larvae. Care must be taken that the material for examination is collected from the deep portion of the lesion.

ONGHOCERCOSIS

Various species of Onchocerca produce subcutaneous nodules and lesions eleswhere in the body. These are slender worms upto 75 mm. in length. The males are not easily visible being very thin.

Onchocerca gibsoni—causes subcutaneous nodules in the brisket region and other parts in cattle.

O. cervicalis—affects the ligamentum nuchae and causes poll evil and fistulous withers in horses.

O. reticulata—occurs in the flexor tendons and suspensory ligaments in horses.

O. faciata—occurs in the camels affecting ligamentum nuchae and subcutaneous tissues.

O. armillata—causes atheroma of aorta in cattle and buffaloes.

The microfilariae are found in the skin and around the affected tissues except in the case of *O. armillata* in which they are found in the blood.

The intermediate hosts are blood sucking insects belonging to the genera of *Culicoides* and *Simulium* (midgets and black flies)

Lesions : The nodules, 'Worm nests', caused by *O. gibsoni* are about 1—3 mm. in diameter and are freely moving in the subcutaneous tissue. The nodules have a thin fibrous capsule. The meat in which these nodules are seen is unwholesome and so the affected portions have to be discarded.

In the aorta, especially at the arch, *O. armillata* causes tracks and cysts, in which the worms inhabit. The males are shorter, being 7cm. long, while the females are nearly 70 cm. The vaginal opening of the females is near the pore of the cysts, through which the microfilariae may escape. The lesions produced do not appear to cause much damage, since all old cattle reveal the tortuous elevated streaks of the aorta at postmortem, without having shown any symptoms while alive. However, it is suspected that a large number of microfilariae may cause collapse and tetanic convulsions. Recurrent ophthalmitis occurred in animals exhibiting convulsions.

Elaeophoriasis : *Elaeophora polei* is a filarid worm, inhabiting the aorta of cattle in India. This worm differs from *Onchocerca armillata* in that the female of *E polei* is attached by its head to the wall of the aorta, while the rest of the worm is found freely swimming in the lumen of the vessel. (But with *O. armillata*, the whole of the worm is whithin the wall of the aorta). The affected portion of the aorta becomes diffusely thickened and less elastic due to replacement of the elastic tissue by connective tissue. In this infection also no clinical symptoms are observed.

SETARIASIS

Setaria cervi and *Setaria digitata* are found in the abdominal cavity of cattle while *Setaria equina* in the abdominal cavity of horses. No lesions or symptoms are seen. Intermediate hosts are various species of mosquitoes and *Stomoxys calcitrans*.

S digitata is important in that immature forms in aberrant hosts—sheep, goats and horses—may cause cerebrospinal nematodiasis (Kumri). See chapter 19 (Page 414) for details.

Parafilariasis

Parafilaria multipapillosa in horses and *P. bovicola* in cattle produce subcutaneous nodules, which open up and bleed. They may heal spontaneously. These blood nodules' are seen in summer when flies are abundant.

THELAZIASIS

The following parasites are found in the conjunctival sacs of animals.

Thelazia rhodesii—in cattle,
buffaloes, sheep and goats
T. lacrymalis in horses.
T. callipaeda in dogs,
T. leesi in camels.

Intermediate host is
Musca domestica

Oxyspirura mansoni in fowls (Intermediate host is cockroach).

The flies deposit the larvae in the conjunctival sac of the animals while feeding around the eyes. Usually no harm is done But in heavy infections conjunctivitis and ophthalmia can be caused.

Giant kidney worm

Dioctophyma renale is found usually in the renal pelvis but may occur encysted in the body cavity, uterus, mammary gland or bladder of dogs and minks It has been seen in cattle and horses also.

The female is 3 to 103 cms. long and 5 to 12 mm. in diameter (the longest pathogenic nematode) while the male is 35 mm. long and 3 to 4 mm. in diameter.

The female lays eggs which have thick, pitted, shells and are very resistant to environmental conditions, surviving for more than 5 years outside the host. The parasite requires two intermediate hosts, an annelid worm and a fish. Infection occurs by eating an infected fish. The prepatent period is about 2 years.

Lesions : Pyelitis is caused which may become suppurative. Due to mechanical obstruction, hydronephrosis may result. Compensatory hypertrophy of the other kidney occurs.

PARASITIC GASTRITIS IN PIGS is caused by spirurid worms, Ascarops strongylina; Physocephalus sexalatus and Symondsia paradoxa. The intermediate hosts are coprophagous beetles. The larvae complete their development in the mucosa into which they burrow.

The adult worms may also bore into the gastric mucosa and so are partly embedded in it. Usually unless found in large number, Ascarops and Pyhsocephalus do not cause much damage. Mild gastric catarrh may be present. But in larger numbers chronic catarrh and pseudomembranous inflammation may also be produced, especially near the pylorus, resulting in cachexia and death.

Symondsia is more pathogenic, especially the female worm which bores into the mucosa. Its posterior portion lies in connective tissue cysts while the anterior projects into the stomach. The male is free in the lumen. Severe gastritis may be caused with many fatalities.

Gnathostoma hispidum is another parasite that burrows into the gastric mucosa deeply and becomes embedded there, causing ulcers which have inflammatory borders. The cavity is filled by necrotic material. There is cellular

infiltration around the ulcer of which eosinophils predominate.   Migration by
young worms in the liver may cause hepatitis and necrotic tracts.

Gnathostoma spinigerum is found in the stomach of dog, cat and wild
carnivora Cyclops and fresh water fish (frog, snake etc. also) are the intermediate
hosts.   The larvae migrate through the body to reach the stomach wall. So they
may wander into liver and lung. In the stomach the parasites live in cavities or
cysts cantaining blood and purulent material closed by a thick wall.   These
cysts communicate with the lumen of the stomach by fistulae or canals.   Sometimes the cysts may open into the peritoneum causing fatal peritonitis.

As in G. hispidum necrotic lesions are also found in the liver.

This infection is very serious in cats which may subsequently die.   They
usually vomit while alive and sometimes the parasites may be found in it   In
man the larvae of G. spinigerum migrate to the skin forming nodules and infection is by eating uncooked infected fish.

THORNY-HEADED WORM OF PIGS

This is Macracanthorhynchus hirudinaceus found in the small intestines
of pigs and measures upto 15 inches.   It is thick being $\frac{1}{4}$ to $\frac{2}{3}$ inch in diameter.
The cuticle is transversely ridged.   The worm has retractile proboscis provided
with hooks.   Intermediate hosts are coprophagous beetles.   The parasite has
no digestive system and nourishment is obtained through cuticle.

Lesions :   The parasite attaches itself to the mucosa of the duodenum and
jejunum by means of the hooked probocis and so intensive mechanical irritation
may be caused.   Sometimes the wall of the bowel may be perforated resulting in
peritonitis and death.

Ulcers bordered by inflammed and thickened edges are formed.   These are
visible through the serosa as nodules.   The proboscis lies in necrotic tissue,
surrounded by granulation tissue with eosinophilic infiltration.   Necrotising
enteritis occurs.   Animals have poor growth and are emaciated. Affected gut
is not useful for sausage making

TRICHURIASIS

Various species of Trichuris (whip worms) are found in the cecum of animals
and usually are innocuous. But heavy infections may cause typhlitis, diarrhoea
and anemia

The different species seen are :

Trichuris ovis
T. globulosa   } In sheep and other ruminants
T. vulpis in dog and fox.
T. trichuira in pigs and man.
T. suis in pigs.
T. serrata
T. campanula   } In cats.

Life cycle is direct. The eggs are passed in the feces in an unsegmented
stage: Development of these to an infective stage takes 3 to 4 weeks. Infection
is by ingestion of the infective eggs. The larvae that emerge in the intestines
enter its glands in 2 to 8 days. Prepatent period is 4 to 5 weeks. The adult buries
its head deeply into the mucosa of the cecum.

Symptoms ɪ Only dogs may show symptoms if heavily infected. Affected dogs are unthrifty, lose weight and have diarrhoea. In heavy infections blood may be found in feces and there may be anemia, jaundice, microcytic, hypochromic anemia.

Diagnosis : Observe eggs with plug at both ends. Typhlectomy may have to be performed in dogs in heavy infections.

STRONGYLOIDOSIS

The species found in animals are :

Strongyloides papillous in ruminants.

S. westeri in horses.

S. suis
S. ransomi ·   }   in pigs

S. stercoralis in dog and man,

S. cati in cats

This parasite is peculiar in that it has a parasitic and a nonparasitic (free living) existence. These are small parasites about 2 to 3 mm. long.

Life cycle : The female lays eggs, which may hatch in the intestines.

But more often they hatch outside into larvae which may become either free living (non-parasitic) adults or parasitic females. The non-parasitic adults copulate and produce several generations and finally parasitic larvae are formed These penetrate the skin, like the ancylostome larvae. The larvae reach the right side of the heart and from there go to lungs, penetrate through the capillaries, enter the alveoli, crawl up the bronchioles and bronchi to trachea and then descend the esophagus to stomach and intestine, where they become adults. In the duodenum and jejunum the females burrow into the mucosa and lay eggs parthenogenetically. S. stercoralis is pathogenic to puppies The pathogenic effects are caused in three ways.

1. **Cutaneous lesions :** Repeated infection by larvae may sensitise the animal and may cause inflammatory reaction manifested by intense pruritus and erythema at the local site. In man 'creeping eruption' is the lesion seen There may be pustules at the site. Edema of the corium, dilatation of lymphatics and perivascular inflammatory reaction are observed histologically.

2. **Pulmonary lesions :** When the larvae break through the alveolar capilaries acute, localised inflammatory reaction may be set up around the alveoli. In heavy infection bronchopneumonia may be set up.

3. **Intestinal lesions ;** The worms burrow into the mucosa and cause chronic inflammation. There may be congestion, edema and erosion of the mucosa. There may be infiltration by eosinophils and in heavy infestation necrosis and sloughing of large areas occur.

Symptoms ɪ

Dogs : Symptoms are observed only in heavy infections. There may be diarrhoea which is hemorrhagic and later dehydration and death may occur. The symptoms in puppies resemble those of distemper.

Pigs : Abdominal pain, weight loss due to anorexia and diarrhoea may be seen.

Sheep : The infective larvae convey Fusiformis nodosus into the clefts of the foot and cause foot-rot and so lameness. In lambs enteritis may be caused. Source of infection in man is dog.

Diagnosis : Observe the embryonated eggs in dung.

OTHER ROUND WORMS IN POULTRY

Capillaria annulata : This parasite is found in the crop and esophagus of fowls, turkeys and other birds. These are long and slender, measuring 1 to 6 cms, in length. The earth worm is the intermediate host. Infection is by ingestion of the infected earthworm. *Prepatent period* is 1 to 2 months.

Lesions : In heavy infections the wall of the crop and esophagus becomes thickened. There is mucopurulent exudate and sloughing of mucosa. Since the worms bury themselves anteriorly into the mucosa, tracks may be seen histologically in it. Pseudomembranes may form and cover the walls. The contents of crop may be blood stained. Birds become emaciated and anemic. They manifest droopiness of head, weakness, and a peculiar penguine-like posture.

Diagnosis :

1. Symptoms.
2. Observe eggs with bilateral plugs in feces.
3. At autopsy observe the adult worms.

Gongylonema ingluvicola : This is found in the crop and beetles act as intermediate hosts. No pathology is caused usually

Acuria Spiralis : This parasite is found in the proventriculus of fowls, turkeys, pigeons and other birds. It may also be found in the esophagus and intestines. These are small worms measuring upto 10 mm in length. Intermediate hosts are grasshoppers, beetles and weevils.

The parasite lies deeply in the mucosa of the proventriculus and so in heavy infection may cause ulcers. The wall is hypertrophied The glands may be dilated. The mucosa in some infections may be completely destroyed In others there may be increased production of mucus.

Microscopically, there is inflammatory reaction with infiltration by eosinophils. The glandular epithelium reveals adenomatous proliferation. Desquamation of the epithelium and ulcer formation may be noticed.

Birds may show retarded growth, droopiness, weakness, anorexia and emaciation. Death may supervene in 2 to 3 weeks after the onset of symptoms.

Diagnosis : Examine feces for eggs. At autopsy adult worms may be recognised.

TETRAMERES (FISSIPINA) MOHETEDA

T. moheteda is the species found in India. These round worms are found in the proventriculus and are upto 6 m.m. in length. The male is free in the lumen while the globular female is found within the glands.

Intermediate hosts are grasshoppers. The parasite sucks blood and so is very pathogenic.

Symptoms seen are dullness, anemia and emaciation culminating in death.

lesions : Since the parasite inhabits the glands, the glandular epithelium undergoes degeneration and desquamation. There is periglandular edema and infiltration by inflammatory cells of which eosinophils may predominate. Sometimes there may be increased production of mucus. Along with these changes there is adenomatous proliferation of the glandular epithelium together with fibrosis. These changes are responsible for the thickening of the wall of the proventriculus.

Diagnosis : Postmortem examination is the reliable method of diagnosis. The red worms may be seen embedded in the glands of proventriculus.

Acuria hamulosa : This is a parasite of the gizzard of the fowl and other birds The worm is upto 25 mm. in length and requires grass hoppers and weevils as intermediate hosts.

Symptoms : Birds become dull, lose their appetite, become weak and emaciated and finally die.

Lesions : The parasite is found in the tunica propria of the gizzard and produces hemorrhagic inflammation of the mucosa. Where it lodges the parasite incites a granulomatous inflammation and a nodule is formed in which caseous and purulent material is found along with the red colored worms Due to pressure the mucosa and musculature of the gizzard become atrophied and thinned. The lumen may be narrowed. Sometimes rupture can occur due to loss of elasticity.

Microscopically, the parasitised glands become atrophied, while the nearby ones not parasitised show hypertrophy and increased secretion. The musculature is atrophied. Infiltration by lymphocytes and eosinophils is seen.

Diagnosis : Autopsy is helpful in arriving at a diagnosis.

SYNGAMUS TRACHEA
Syn : Gape worm

This worm inhabits the trachea of fowl, turkeys, pheasants and other birds. The male and female worms are permanently joined in copulation. The male is 5 mm long while the female is 5 to 20 mm. long. Sucking blood, the worms are bright red in color.

Life cycle : The female lays eggs in the trachea. These are coughed up and swallowed, to be voided with feces. The eggs become infective in 2 days. Infective larvae hatch out in 9 days. These larvae may be swallowed by snails slugs, earthworms, flies and other arthropods, which act as transport hosts only, in which the larvae may be encysted and be viable for years. Infection of the birds can occur by (1) ingesting eggs containing the infective larvae (2) ingesting the infective larvae that have hatched out of eggs. and (3) ingesting the transport hosts containing the encysted infective larvae. Passage through the earthworms appears to enhance the infectivity of the larvae.

The larvae reaching the bowel, penetrate it and enter the blood stream by which they ultimately reach the lung alveoli. After undergoing 2 ecdyses, the larvae migrate, after copulation in the bronchi. to their seat of predilection, the trachea. The prepatent period is 17 to 20 days.

The worms attach themselves to the mucosa of the trachea firmly. The anterior end may penetrate deeply as far as the submucosa.

Pathogenesis At the place where the parasites attach themselves ulcers are formed and a productive type of inflammation is caused. Small, raised warty growths occur, the centres of which contain caseous material. This is followed by tracheitis with blood-stained mucous exudate. Birds may become suffocated if the worms and the mass of mucus obstruct the lumen. Because the parasites are blood suckers, birds become anemic and emaciated. During the migratory phase, pneumonia may be caused.

Symptoms : Symptoms are noticed only in the young birds between 1 and 3 months. The older birds act as carriers. Presence of the worms and mucus in the trachea, obstructing the lumen, render the birds to make violent inspiratory efforts. The neck is extended and beak is partially or widely open. This syndrome is known as "gapes", which is characteristic of this infection. There may be cough. In an effort to get rid of the worms the birds may frequently shake the head or throw it upwards. Death may occur of asphyxia in some cases with heavy infection.

Lesions : Catarrhal tracheitis with blood-tinged exudate containing the red worms is noticed. In heavy infestation, the birds may be anemic and emaciated.

Diagnosis ;

1. Symptoms of 'gaping' are characteristic.
2. Observe eggs in the feces. These are smaller than those of *Capillaria* sps.
3. At autopsy red worms and mucous exudate can be seen in the trachea.

Chapter 32

DISEASES CAUSED BY INSECTS AND THEIR LARVAE

Bot flies   Scabies
Oestrus ovis   Burrowing mites—Sarcoptes,
Hypoderma sp   Cnemidocoptes
Blow flies, screw worm flies and Sucking mites—Psoroptes.
                flesh flies   Scale eating mites—Chorioptes
Sheep ked   Demodex
Lice   Marge in fowls
Ticks   Linguatula

BOT FLIES

Infection by Gasterophilus species :

The larvae of these flies are parasitic and live in the stomach and intestines of horses.

Several species are known :—

Gasterophilus intestinalis (equi)

Gasterophilus nasalis

Gasterophilus pecorum.

The adult flies do not feed. After laying eggs they die. Infection occurs during summer when fly population is high.

G. intestinals : This is the most common bot fly. Eggs which are yellowish-white and having an oblique operculum, are laid on the hairs at the fetlocks of fore legs and on the sternum. These are attached to the distal end of the hair by a parallel flange extending from its posterior extremity for about half its length. Due to slight irritation the animals lick the part. Moisture and friction provided during licking make the eggs hatch and the liberated larvae enter the mouth and wander about in the mucosa of the tongue for 3 to 4 weeks during which period they moult twice. Then they emerge beyond the pharynx and reach the stomach. The larvae, which are red attach themselves to the mucosa of the cardiac region by their two chitinous hooks. Feeding on blood, lymph and cell debris these larvae live in the stomach for 6 to 10 months during which period several ecdyses occur. Then the larvae leave the mucosa and pass out of the animal with feces and pupate in soil. The adult fly emerges 3 to 40 days later.

G nasalis : (The 'Chin' or 'Throat' fly).

The eggs which are yellow, brown or black are cylindrical and have terminal dome-like operculum. These are deposited on the hair in the intermandibular region, near the throat. The eggs lie along the hair and do not stand away from it. The larvae reach the mouth and from there finally they travel and settle near the pylorus and in the duodenum. These are pale-yellow in color. Some may be found in the pharynx

G. pecorum : The egg which is black is similar to that of *G intestinalis* but has a short stalk at the posterior extremity. The anterior end stands away from the hair. These are attached to the hairs around the hooves of the animals. The larvae on licking pierce the buccal mucosa, reach the pharynx, and migrate to the stomach. The red larvae attach themselves to the cardia. On their way out of the animal they are attached to the mucosa of the rectum for a few days.

Pathogenesis and lesions : Usually one species of larvae may be present. But 2 or more species may also be found.

The larvae bore through the mucosa with their chitinous hooks and live on blood, exudate and debris. At the place of attachment, therefore, erosions and ulcers are formed which appear as pits. These have raised margins. The erosions are red and covered by debris. In heavy infestations, chronic inflammation occurs and so the mucosa is thickened and there may also be hyperplasia of the epithelium and glands and so, sometimes, papillomatous and adenomatous growths are found.

After the larvae leave, healing occurs by scar tissue. But secondary bacterial infection may complicate the lesion and delay healing.

Microscopically, the lesions may be only superficial or may extend deep into the submucosa forming ulcers. When ulcers form the propria and submucosa are in a state of chronic inflammation with local eosinophilic infiltration. The fibrous tissue may undergo hyaline degeneration and the granulation tissue has plentiful blood capillaries. The thickening of the edges of the cardiac lesions is due to epithelial over growth. Rete pegs may form and extend deep into the tissue. Keratinisation may occur in the centre and pearls may be seen. This is supposed to be only a precancerous stage but not neoplastic. Glandular hyperplasia may be seen as papillomatous hyperplasia of the epithelium.

Sometimes the larvae may penetrate the whole wall when hemorrhage and gangrenous peritonitis may occur.

Lesions in the pylorus may lead to its stenosis with resultant dilatation of the stomach. Duodenal catarrh may extend into biliary and pancreatic ducts. Heavy infection at the pyloric end may cause obstruction with fatal results. *G. nasalis* may cause pharyngitis.

Symptoms : Usually no symptoms are manifested. But disturbances, if any, may be due to the toxic action of the metabolic products of the parasites. So the animals may appear unthrifty with a rough coat and have bad temper. Colic may be seen sometimes.

OESTRUS OVIS

This fly is responsible for nasal myiasis in sheep. The adult female lays larvae near the external nares and the larvae crawl towards the ethmoid region where they develop for some months and from there go to the frontal sinus or the maxillary sinus where they become mature in some more weeks. Altogether the larvae may be in the sheep for about 10 months. These leave the host during spring. They are sneezed out and drop on the ground and pupate. The adult fly emerges in about 6 weeks.

Pathology : The larvae attach themselves to the mucosa by means of their chitinous hooks and so cause irritation, inflammation and erosions. There is copious nasal discharge. Sheep therefore sneeze constantly and shake their heads. There is danger of spread of bacterial infection to the meninges by lymphatic sheaths of olfactory nerves.

Damage to the hosts is by way of annoyance since they become restless, hiding their noses and so do not find sufficient time to feed. So they become emaciated and even die.

Hypoderma species (Warble flies)

These flies cause considerable damage to the skin, and so great loss to leather industry ensues. The following species are met with.

Hypoderma bovis
H. lineatum   } In bovines.
H. crossi   In goats.

The adult fly lays eggs on the hairs of legs and rarely on other parts of the body. In 4 days the eggs hatch and the larvae burrow into the skin. The larvae of H bovis make their way along the nerves to the epidural tissue in the spinal canal. After attaining development to a certain stage here, the larvae leave the spinal canal through upper vertebral foramina and the spinal muscles and arrive at the subcutaneous tissue of the back. In this location the larvae mature after undergoing several ecdyses and then leave the host through holes they make in the skin, pupate and become adults.

In the case of H lineatum, the migration of the larvae is different. The larvae migrate through subcutaneous tissues, reach the pharyngeal and esophageal connective tissue and from there they reach the back. These never reach the epidural space of spinal canal.

Pathology : No harm seems to be done by the migrating larvae. As the larvae reach the back, an inflammatory reaction is set up and a small swelling occurs (warble) in the centre of which is a tiny pore through which the larvae breath and may finally emerge after enlarging it.

Because of irritation, there is edema at the place with infiltration by numerous eosinophils. As the larva matures, the pore becomes bigger and finally a big hole is left on the skin after the emergence of the larva. So multiple infestation spoils the skin and renders it useless. The most useful part of the skin is damaged.

Adult cattle try to avoid the flies and in so doing 'Gad' about (run pellmell) and are likely to injure themselves They lose condition, may abort and yield less milk. The larvae make tracks in the muscles and these tracks are greenish yellow and are gelatinous. This substance is known as 'butcher's jelly'. Such a material accumulates at the seat of predilection also and sometimes this is abundant. Meat is spoiled and so is unfit for human consumption.

BLOW FLIES, SCREW WORM FLIES AND FLESH FLIES

These are responsible for producing cutaneous myiasis as their larvae burrow into the subcutaneous tissues and produce myiasis, resulting in wounds, with foul smelling discharge.

The following flies cause myiasis.

Chrysomia bezziana—besides animals this attacks man also laying eggs on wounds and sores (screw worm fly).

Lucilia sericata (green bottle fly).

Calliphora erythrocephala   }
and *C. vomitoria*.   **Blue bottle flies**.

The screw-worm fly larvae feed on living tissues and produce serious effects on the hosts.   Large number of the maggots by feeding on tissues can produce cavities 4 to 5 inches in diameter.   All animals are affected.   The wounds infected are the naval of new born animals, wounds caused by castration, docking, dehorning; wounds produced by sharp grasses and objects and cuts while shearing.

Lucilia and Calliphora are important in sheep causing 'strike'.   In some herds heavy losses may be sustained.   The following wounds are usually infected: wounds made during castration and docking; wounds sustained by rams while fighting.   But mostly, the flies are attracted to the area that is soiled by urine and feces and where excoriation may occur.   This occurs on the breeches where the wool is soiled and emits offensive smell.   The larvae may burrow, feed on tissues and become mature in 2 days to 3 weeks depending on the environment. Then they drop on to the ground and pupate from which adult flies emerge in 3 days

Pathology :   The larvae destroy tissues and the products of this destruction may cause toxemia.   The excretory products of larvae may also be toxic   There may be loss of fluid, if there is extensive loss of skin.   Secondary bacterial infection may supervene.   Animal becomes very weak and in heavy infection may even die.

The affected animals are restless because of the irritation and pain caused.

They may rub or bite at the area from which a fowl smelling discharge may flow.   Wool in patches may be shed (The wound may be covered with scabs).

Hippobosca maculata attacks cattle and horses and sucks blood. Usually no harm is done but these can transmit anthrax and *Trypanosoma theileri* mechanically.

Melophagus ovinus, the sheep ked – a wingless fly that is universally found on sheep sucking blood.   Through blood sucking it causes irritation and discomfort to the host. The animals scratch continuously and so do not feed satisfactorily. Therefore loss of condition can occur. Wool may be lost by rubbing.   Infestation by keds predisposes the animals to be affected by blow flies.   The ked transmits *Trypanosoma melophagium* to sheep.

Pseudolynchia canariensis, is a dark brown fly resembling the sheep ked but having a pair of transparent wings. This is a blood sucker transmitting the protozoan parasite, *Hemoproteus columbae* among pigeons.

LICE

These are ectoparasites,   the life cycles of which occur entirely on the bodies of hosts.   They are host specific.

Lice are divisible broadly into (1) Biting lice and (2) sucking lice.

The biting lice feed on the epithelial debris of the skin while the sucking lice suck blood and tissue fluid.

Biting lice:	Sucking lice:

Cattle

    Bovicola (Demalina) *bovis*   *Hematopinus eurysternus*

                                                       Linognathus vituli

Buffaloes   *Solenoptes capillatus*

                                                       Hematopinus tuberculatus

Sheep

    Bovicola ovis   *Hemotopinus qurdripertusis*

                                                       Linognathus pedalis

Goats

    Bovicola caprae   *Linognathus steponsis*

Horse

    Bovicola equi   *Hemotopinus asini*

Dog

    Trichodectis canis   *Linognathus setosus*

    (intermediate host for

    Dipylidium caninum—the dog tape worm)

    Hetarodoxus spiniger

Cats

    Felicola subrostratus

Pig   *Hematopinus suis*; spreads swine pox.

Poultry

    Menopon gallinae (the shaft louse)

    Menacanthus stramineus (body louse)

    Lipeurus heterographus (head louse)

    L. caponis (the wing louse)

    Goniocotes gallinae—fluff louse.

Lice cause irritation and so animals scratch themselves because of itching, become restless and may lose condition. Wool and hides become damaged and milk-yield goes down. They develop rough coat. In heavy infestation anemia may be caused.

In heavy infections, there may be loss in egg production in poultry.

It is usually the young and debilitated animals that are heavily infested.

TICKS

Ticks are important because :

    (1) They transmit protozoal, viral, bacterial and rickettsial diseases.

    (2) They cause irritation, worry and uneasiness to the animals.

    (3) They may cause paralysis and even death in some cases.

    (4) They damage the skins and hides decreasing their market value.

1. TICKS TRANSMITTING DISEASES

Disease	Animal	Protozoan parasite	Vector
Piroplasmosis	Cattle	*Babesia bigemina*	*Boophilus microplus*—1 host tick.
,,	,,	*b. bovis*.	*Ixodus ricinus* – 3 host tick
,,	Sheep	*B. motasi*.	*Rhipicephalus bursa*—2 host tick.
,,	,,	*b. ovis*.	*Hemaphysalis bispinosa* 3 host tick.

Piroplasmosis	Horse	B. equi.	Rhipicephalns evertisi ⎫ 2 host R bursa ⎬ ticks Dermacentor Sps. Hyalomma Sps.
		B. cabbali	Darmacentor Sps. Rhipicephalus Sp.
	Dog	B. canis	Rhipicephalus sanguineus—3 host tick.
		B gibsoni	Hemaphysalis bispinosa
Theileriasis	Cattle	Theileria parva	Rhipicephalus Sp.
	,,	T. annulata	Hyalomma Sp.
	,,	T mutans.	Rhipicephalus Sp.
Anaplasmosis	Cattle	Anaplasma marginale	Various Sps of ticks and blood sucking insects.
	Sheep & Goats.	A. ovis	—do—
Encephalo myelitis	Horse	Virus	Dermacentor.
Tickborne fever.	Cattle	Rickettsia Sp.	Rhipicephulus hemophysaloids
Hepatozoon.	Dogs.	Hepatozoon canis	R sanguineus.
Spirochaetosis	Fowls.	Borrelia gallinarum	Argas persicus.
Tick fever.	,,	Aegyrianella pullorum	—do—

2. Ticks causing uneasiness

Otobius megnini—the spinose ear tick. This is found in the ears of horses, cattle, sheep, goats and dogs and causes otorrhoea. Ticks in large numbers may cause anemia because they are blood suckers. Causing restlessness and worry, they are responsible for decreased milk yield and loss in condition. Tick bites spoil the hides and skins.

3. Ticks causing paralysis

Some species of ticks of genus Dermacentor, Ixodus, Rhipicephalus, Hyalomma and Hemophysalis cause paralysis. The salivary glands of the female ticks secrete a toxin which appears to interfere with synthesis or release of acetylcholine at motor end plates of muscle fibres. There may be incoordination of movement. Death is due to respiratory failure. If the ticks are removed, recovery occurs. In heavy infection death may occur, especially in dogs The course of the disease may be 4-5 days.

SCABIES

Psrasites of families Sarcoptidae and Demodicidae cause scabies or mange. Sarcoptid mites can be divided into

(a) Burrowing mites (Sarcoptes and Cnemidocoptes).

(b) Sucking mites (Psoroptes).

(c) Scab-eating mites (Chorioptes).

Burrowing mites

Sarcoptes scabei in animals.

Cnemidocoptes gallinae and C. mutans in fowls.

Source of infection : Ailing animals. Infection may be by direct contact or through fomites; bedding, grooming kit, blankets, clothing. Animals in poor condition are more susceptible. Over crowding facilitates spread.

Pathogenesis: The adults feed on the succulent cells of stratum corneum. The female mites penetrate deep into the epidermis causing tunnels which run in various directions. The mites are found at the farther ends of these tunnels where they lay eggs. Larvae that hatch crawl out and may infect other animals or may become adults on the same animals burrowing into new places. Males live only superficially.

The epidermis is much atrophied because of the formation of the burrows in which the mites live, There is hyperkeratosis and sometimes parakeratosi, In the corium may be found inflammatory reaction with infiltration by leucocytes and histiocytes.

Symptoms: Since the mites produce itching, animals scratch themselves or even bite at the place .Red papules and vesicles form at the infected area. These are soon covered by crusts formed by drying of lymph.

Because of proliferation of fibrous tissue and increased keratinisation the skin becomes thickened and wrinkled. Due to lack of blood supply hair and wool fall off.

Complication may occur by secondary infections.

The following parts of the animals are affected first. Later infection may spread to other parts and the whole body may be affected

Sheep and goats: Face and ears

Cattle: Head,neck,brisket. inner side of thigh and scrotum.

Horses: Head and neck.

Pigs: Nose. ears,root of the tail, around the eyes, back and feet

Dogs: Muzzle, elbow, hock and root of the tail, lower surface of abdomen, inner surface of axilla.

So infection occurs first in areas not covered by hair or wool and where the skin is thin. If only small areas are affected with mild infection no effect on the general condition of the animals is felt. But large areas of scabies cause loss of condition in animals which may become progressively emaciated and die. Skin is spoiled and wool is lost. In sweating animals sweat glands are destroyed and so greater load is thrown on kidneys. which may therefore show degenerative changes.

Sucking Mites

These live on tissue fluid Through their long mouth parts. these mites bore through the epidermis as far as the corium and suck tissue fluid. *Psoroptes communis* is found on various animals and is host specific Different varieties are known as.

Psoroptes communis ovis (in sheep).

P. communis caprae (in goats)

P. communis equi (in horses)

P. communis bovis (in cattle)

The parasite is of greater importance in sheep in which it causes 'Sheep scab' Psoroptes does not burrow into the epidermis but only pierces through to suck the fluid During this process inflammation may be caused with exudation of lymph which dries into a crust under which are found the mites. The lesions may be found on the shoulders and sides in wooly sheep and on the bacdk and

tail of the hairy ones   The lesions start as papules oozing serum. Animals
scratch or bite at these places and so wool may be shed   Older lesions have
yellow crusts.   The mites are usually found at the periphery of these lesions,

Animals become weak in heavy infestations and may even die. The lesions
spread slowly and masses of wool are shed. Sometimes, there may be hyperkeratosis and thickening of the skin.

P. communis var caprae is found in the ears of goats producing lesions
similar to those in sheep

P. communis var equi is found on the mane, saddle region and base of tail
The skin is not thickened.

P. communis var bovis is found around the base of the tail, neck, inside of
thighs, and withers. The skin is thickened and wrinkled. The lesions may be
covered with scabs. Very severely affected animals may become weak and emaciated and die.

Scab eating mites   (Chorioptes)

These feed on epidermal scabs and so do not penetrate into the skin, but live
only superficially.   By their presence. they may cause irritation and so animals
scratch and become uneasy.   Dermatitis may thus result.

Various species are :

Chorioptes equi occurs in the horse and affects the feet at the region of fetlocks. Because of irritation the horses may stamp their feet and rub their legs on
posts or wires.   They kick often, especially at night.   The skin may be cracked
and the bones of the legs may be fractured even.

Chorioptes bovis is seen mostly at the base of tail, udder and on the escut-
cheon where scabs are found.   Only slight irritation may be caused

Otodectes cynotis is closely related to chorioptes and is found in the external
auditory meatus of dog, fox, cat and ferret causing ear-mange. Animal affected
becomes restless, scratching very frequently because the mites, sucking tissue
fluid cause inflammation and exudation.   Crusts are formed and later otorrhoea
and rupture of ear-drum may result. (See Page 533)

Comparatively sarcoptes mange is more widespread (Sometimes the whole
body being affected), than the psoroptic and chorioptic, in which the lesions are
localised and smaller.

Diagnosis : Scrapings from the affected area reveal the mites.

DEMODECTIC MANGE

This is a very intractable disease of animals especially of dogs caused by a
mite, Demodex folliculorum.

Demodex mange is found in all domestic animals including man.   Though
the mites appearing in different animals cannot be differentiated morphologically they are given different names according to the host on which they cause
lesions :- Demodex canis of dog: D bovis of cattle D equi of horse D ovis of
sheep. D caprae of goats D cati of cats, D pylloides of pigs. D cuniculi in
rabbits and D folliculorum in man. There does not seem to be much host
specificity since the disease can be communicated from one species of animal to
another, For example, goats can be infected by an ailing pig or dog.

Dogs : The mite lives in hair follicles and sebaceous glands and so in this respect differs from other mange mites. D modex can be seen in dogs in hair follicles as a latent infection, without causing any lesions. But the disease is precipitated in young animals which are undernourished or highly pedigreed, in animals with some other skin disease that has damaged the skin, in animals that have lost their teeth, in animals suffering from deficiency disease such as rickets and in animals that have become weakened by other diseases such as distemper.

Infection is by means of direct contact with the diseased. Usually short haired breeds of dogs are affected such as dachshunds, bull dogs, foxterriers. dalmatian etc. There is some evidence of congenital infection also.

Pathogenesis : The mites enter the hair follicle, reach the root and multiply. Each root may have about 20 parasites. Due to inflammation the papilla is destroyed and the hair is lost. The parasite then enters the sebaceous glands where conditions appear to be very congenial for the growth of the mites, since the altered sebum is a very favourable medium for them to grow. Sweat glands may become occluded and so may rupture giving rise to microscopic foci of inflammation since the sweat causes irritation locally The sebaceous glands may be dilated and become cystic and lined by squamous epithelium. These cells proliferate and a horny layer is formed. Thus the skin comes to be covered by scaly material which may be desquamated. The sebaceous cysts may rupture, the spilled sebum causing inflammation locally. The blood vessels are congested and dilated.

Secondary bacterial infection by staphylococci causes the formation of pustules.

Symptoms . In the dog the skin around the eyes, on the forehead and nose is frequently affected. The feet, axillary area and skin on neck and body are affected rarely.

Since the organism is deeply situated pruritus is not manifested. In animals this mange is observed in three forms.

Sqamous form : The lesion starts as a small patch of the size of a pea In this area the hair is lost and the skin is red in color and covered by bran-like scales. The area widens slowly hair being shed progressively The color of the skin turns to copper color and the skin is thickened, wrinkled and even cracked. This form may remain as such or may become squamo-papular or pustular

Squamo-papular form : This occurs due to development of small papules in the squamous form. In these two forms healing of the lesion is followed by regrowth of hair and increased pigmentation.

The pustular form : This is found in the same locations as the sqamous form. Small pustules are formed in the hair follicles and sebaceous glands and so acne develops. The skin becomes thickened, wrinkled and deep red in color Pustules develop in these areas discharging blood-stained pus containing the mites. This pus later dries to form crusts. Because of the redness this form is called 'Red mange'.

Regional lymph-nodes may be swollen and may contain the mites. In animals with extensive involvement emaciation and weakness may be noticed. Sometimes spontaneous healing may occur.

Microscopically, inflammatory and exudative reaction is found in the surrounding corium. In the pustular form leucocytic infiltration is extensive. Because of chronic inflammation there is hyperplasia of the fibrous tissue and epithelium and so the skin becomes thickened and wrinkled.

Cattle : The lesions are seen on the neck, brisket, dewlap, withers, shoulder, forearm and along the back. In some areas demodectic mange becomes generalised and even fatal. In some localities it may be enzootic. The lesions are pustular. Small nodules on the skin are seen in animals between the ages of 4 to 10 years. From these lesions a thick white waxy material can be expressed and mites can be found in this caseous material. This material may liquefy and an abscess formed discharging pus leaving small pores on the hide. Thus the value of the hide is lost. Sometimes squamous form may be seen with thick crusts on the neck, inner side of thighs and udder. In mild cases spontaneous recovery may occur.

Horses : Horses are very rarely affected. Alopecic lesions may be found on the face, nose and around the eyes. There may be exudative dermatitis. Papules may sometimes be seen but no clinical symptoms are shown.

Pigs Demodectic mange is ubiquitous in pigs and commences on the face where from it may spread down the neck and chest to the abdomen. Small cystic nodules or pustules may be formed. Sometimes pustules may coalesce to form bigger abscesses.

Goats: Demodectic mange may be particularly seen in certain parts of the world. A few fatalities may occur. Pustules or nodules may be found on the neck, chest and shoulder. The nodules contain caseous material in which the mites can be seen. In some localities the condition may occur as a latent form.

Shep: In this animal also demodectic mange can occur as a latent form. Lesions consisting of bald thick skin covered by scabs may be seen on the back and flanks. If pruritus is present, which is rare, itching is followed by rubbing and loss of hair or wool.

Cats: Rarely suffer. Lesions are found on the head as a desquamative form.

Diagnosis: Express the nodules or pustules and look for the long mites with four pairs of stumpy legs

MANGE IN FOWLS
Leg mange

This is caused by *Cnemidocoptes mutans* (scaly-leg mite) and affects the legs of older fowls and turkeys. The mites burrow under the epidermal scales from the tibio-tarsal joint downwards and produce inflammation. The exudation that occurs hardens into a powdery material which displaces and raises the scales. This with the marked keratinisation that occurs, is responsible for the thickened scaly nature of the skin. On removal of the crusts, raw, rough and thickened plaques may be noticed. Because of the itching nature of the lesions, the fowls scratch and peck at the spots. The accumulation of the crusts interferes with the flexion of the joints and the birds, therefore, become lame. Hence they feel difficulty in moving about and do not feed adequately, resulting in loss of condition and egg production. Sometimes, stupor, emaciation and death may occur. Scaly leg is usually found in overcrowded and unhygienic places.

Depluming itch

In places covered by feathers *Cnemidocoptes gallinae* invades and entering the epidermis causes irritation and so the birds pull out the feathers. This affection usually starts at the back and wings and spreads to head, neck and abdomen Later the affected parts become bare

This affection is found in spring and summer but disappears in winter, to reappear in the following summer.

The skin at the bare areas becomes thickened and wrinkled. Pustules may form there.

Dermanyssus gallinae

This common red mite of the fowl lives in the crevices of the pen, attacking the birds at night. Sucking blood the mite causes intense anemia which may result in death of young chicks This mite transmits the following diseases:—

Fowl spirochaetosis, Fowl cholera, Fowl pox and Equine encephalomyelitis.

LINGUATULA

Linguatula serrata, the tongue worm of dog is flat ventrally and slightly convex dorsally These measures upto 13 cms in lengh (females are longer). It is a parasite of dog but may be found in man, sheep, goat, cattle or horses.

Life history The adult is found in nasal passages attached to the mucosa by its two pairs of claws. It may also be found in the pharynx, larynx, esophagus or may pass into the maxillary sinus or inner ear (via Eustachian tube).

The female lays its egg which are sneezed or coughed out. These eggs are swallowed by intermediate hosts—herbivores (cattle, sheep, goat, pig, horse, rabbit and buffalo). Hatching out in the intestines, the first stage larvae bore through the bowel wall and get carried by blood and lymph to various organs—liver, lungs, kidneys and mesenteric lymphnodes etc,. In these organs the larvae are found in nodules in which they become second stage larvae. After undergoing several ecdyses, they become the infective nymphs. These may break out of the nodules and wander about in the tissues of intermediate host.

Infection of definitive host occurs by :

(1) Ingestion of tissues infected with nymphs. The nymphs migrate to the pharynx from stomach and from there to the nasal passages.

(2) Or the definitive host may sniff and inhale the nymphs.

Pathology :

(1) The larvae during their migration in the intermediate host may produce lesions similar to those produced by ascaris and strongyle larvae. Hemorrhagic peritonitis may occur.

(2) In the definitive host due to irritation caused by the parasite sneezing and coughing induced by rhinitis and pharyngitis occur. There may be violent rubbing. Mucus or blood tinged nasal discharge may be present. In some there may be epistaxis. Dyspnoea may develop in others. The adult parasites live only on mucus. But it is the irritation that makes the animals restless. Some may become enaciated. The parasite may live for 15 months.